CAMBRIDGE STUDIES IN ADVANCED MATHEMATICS 155

MARTINGALES IN BANACH SPACES

This book is focused on the major applications of martingales to the geometry of Banach spaces, but a substantial discussion of harmonic analysis in Banach space valued Hardy spaces is presented. Exciting links between super-reflexivity and some metric spaces related to computer science are covered, as is an outline of the recently developed theory of non-commutative martingales, which has natural connections with quantum physics and quantum information theory.

Requiring few prerequisites and providing fully detailed proofs for the main results, this self-contained study is accessible to graduate students with basic knowledge of real and complex analysis and functional analysis. Chapters can be read independently, each building from introductory notes, and the diversity of topics included also means this book can serve as the basis for a variety of graduate courses.

Gilles Pisier was a professor at the University of Paris VI from 1981 to 2010 and has been Emeritus Professor since then. He has been a distinguished professor and holder of the Owen Chair in Mathematics at Texas A&M University since 1985. His international prizes include the Salem Prize in harmonic analysis (1979), the Ostrowski Prize (1997), and the Stefan Banach Medal (2001). He is a member of the Paris Académie des sciences, a Foreign Member of the Polish and Indian Academies of Science, and a Fellow of both the IMS and the AMS. He is also the author of several books, notably *The Volume of Convex Bodies and Banach Space Geometry* (1989) and *Introduction to Operator Space Theory* (2002), both published by Cambridge University Press.

CAMBRIDGE STUDIES IN ADVANCED MATHEMATICS

All the titles listed below can be obtained from good booksellers or from Cambridge University Press. For a complete series listing, visit: www.cambridge.org/mathematics.

Already published

116 D. Applebaum *Lévy processes and stochastic calculus (2nd Edition)*
117 T. Szamuely *Galois groups and fundamental groups*
118 G. W. Anderson, A. Guionnet & O. Zeitouni *An introduction to random matrices*
119 C. Perez-Garcia & W. H. Schikhof *Locally convex spaces over non-Archimedean valued fields*
120 P. K. Friz & N. B. Victoir *Multidimensional stochastic processes as rough paths*
121 T. Ceccherini-Silberstein, F. Scarabotti & F. Tolli *Representation theory of the symmetric groups*
122 S. Kalikow & R. McCutcheon *An outline of ergodic theory*
123 G. F. Lawler & V. Limic *Random walk: A modern introduction*
124 K. Lux & H. Pahlings *Representations of groups*
125 K. S. Kedlaya *p-adic differential equations*
126 R. Beals & R. Wong *Special functions*
127 E. de Faria & W. de Melo *Mathematical aspects of quantum field theory*
128 A. Terras *Zeta functions of graphs*
129 D. Goldfeld & J. Hundley *Automorphic representations and L-functions for the general linear group, I*
130 D. Goldfeld & J. Hundley *Automorphic representations and L-functions for the general linear group, II*
131 D. A. Craven *The theory of fusion systems*
132 J. Väänänen *Models and games*
133 G. Malle & D. Testerman *Linear algebraic groups and finite groups of Lie type*
134 P. Li *Geometric analysis*
135 F. Maggi *Sets of finite perimeter and geometric variational problems*
136 M. Brodmann & R. Y. Sharp *Local cohomology (2nd Edition)*
137 C. Muscalu & W. Schlag *Classical and multilinear harmonic analysis, I*
138 C. Muscalu & W. Schlag *Classical and multilinear harmonic analysis, II*
139 B. Helffer *Spectral theory and its applications*
140 R. Pemantle & M. C. Wilson *Analytic combinatorics in several variables*
141 B. Branner & N. Fagella *Quasiconformal surgery in holomorphic dynamics*
142 R. M. Dudley *Uniform central limit theorems (2nd Edition)*
143 T. Leinster *Basic category theory*
144 I. Arzhantsev, U. Derenthal, J. Hausen & A. Laface *Cox rings*
145 M. Viana *Lectures on Lyapunov exponents*
146 J.-H. Evertse & K. Győry *Unit equations in Diophantine number theory*
147 A. Prasad *Representation theory*
148 S. R. Garcia, J. Mashreghi & W. T. Ross *Introduction to model spaces and their operators*
149 C. Godsil & K. Meagher *Erdős–Ko–Rado theorems: Algebraic approaches*
150 P. Mattila *Fourier analysis and Hausdorff dimension*
151 M. Viana & K. Oliveira *Foundations of ergodic theory*
152 V. I. Paulsen & M. Raghupathi *An introduction to the theory of reproducing kernel Hilbert spaces*
153 R. Beals & R. Wong *Special functions and orthogonal polynomials (2nd Edition)*
154 V. Jurdjevic *Optimal control and geometry: Integrable systems*
155 G. Pisier *Martingales in Banach Spaces*

Martingales in Banach Spaces

GILLES PISIER
Texas A&M University

Shaftesbury Road, Cambridge CB2 8EA, United Kingdom

One Liberty Plaza, 20th Floor, New York, NY 10006, USA

477 Williamstown Road, Port Melbourne, VIC 3207, Australia

314–321, 3rd Floor, Plot 3, Splendor Forum, Jasola District Centre, New Delhi – 110025, India

103 Penang Road, #05–06/07, Visioncrest Commercial, Singapore 238467

Cambridge University Press is part of Cambridge University Press & Assessment, a department of the University of Cambridge.

We share the University's mission to contribute to society through the pursuit of education, learning and research at the highest international levels of excellence.

www.cambridge.org
Information on this title: www.cambridge.org/9781107137240

First published 2016

A catalogue record for this publication is available from the British Library

ISBN 978-1-107-13724-0 Hardback

Contents

Introduction

Martingales (with discrete time) lie at the centre of this book. They are known to have major applications to virtually every corner of probability theory. Our central theme is their applications to the geometry of Banach spaces.

We should emphasize that we do *not* assume any knowledge about scalar valued martingales. Actually, the beginning of this book gives a self-contained introduction to the basic martingale convergence theorems for which the use of the norm of a vector valued random variable instead of the modulus of a scalar one makes little difference. Only when we consider the 'boundedness implies convergence' phenomenon does it start to matter. Indeed, this requires the Banach space B to have the Radon-Nikodým property (RNP). But even at this point, the reader who wishes to concentrate on the scalar case could simply assume that B is finite-dimensional and disregard all the infinite-dimensional technical points. The structure of the proofs remains pertinent if one does so. In fact, it may be good advice for a beginner to do a first reading in this way. One could argue similarly about the property of 'unconditionality of martingale differences' (UMD): although perhaps the presence of a Banach space norm is more disturbing there, our reader could assume at first reading that B is a Hilbert space, thus getting rid of a number of technicalities to which one can return later.

A major feature of the UMD property is its equivalence to the boundedness of the Hilbert transform (HT). Thus we include a substantial excursion in (Banach space valued) harmonic analysis to explain this.

Actually, connections with harmonic analysis abound in this book, as we include a rather detailed exposition of the boundary behaviour of B-valued harmonic (resp. analytic) functions in connections with the RNP (resp. analytic RNP) of the Banach space B. We introduce the corresponding B-valued Hardy spaces in analogy with their probabilistic counterparts. We are partly motivated by the important role they play in operator theory, when one takes for B the space of bounded operators (or the Schatten p-class) on a Hilbert space.

Hardy spaces are closely linked with martingales via Brownian motion: indeed, for any B-valued bounded harmonic (resp. analytic) function u on the unit disc D, the composition $(u(W_{t\wedge T}))_{t>0}$ of u with Brownian motion stopped before it exits D is an example of a continuous B-valued martingale, and its boundary behaviour depends in general on whether B has the RNP (resp. analytic RNP). We describe this connection with Brownian motion in detail, but we refrain from going too far on that road, remaining faithful to our discrete time emphasis. However, we include short sections summarizing just what is necessary to understand the connections with Brownian martingales in the Banach valued context, together with pointers to the relevant literature. In general, the sections that are a bit far off our main goals are marked by an asterisk. For instance, we describe in §7.1 the Banach space valued version of Fefferman's duality theorem between H^1 and BMO. While this is not really part of martingale theory, the interplay with martingales, both historically and heuristically, is so obvious that we felt we had to include it. The asterisked sections could be kept for a second reading.

In addition to the RN and UMD properties, our third main theme is super-reflexivity and its connections with uniform convexity and smoothness. Roughly, we relate the geometric properties of a Banach space B with the study of the p-variation

$$S_p(f) = \left(\sum\nolimits_1^\infty \|f_n - f_{n-1}\|_B^p\right)^{1/p}$$

of B-valued martingales (f_n). Depending on whether $S_p(f) \in L_p$ is necessary or sufficient for the convergence of (f_n) in $L_p(B)$, we can find an equivalent norm on B with modulus of uniform convexity (resp. smoothness) 'at least as good as' the function $t \to t^p$.

We also consider the strong p-variation

$$V_p(f) = \sup_{0=n(0)<n(1)<n(2)<\cdots} \left(\sum\nolimits_1^\infty \|f_{n(k)} - f_{n(k-1)}\|_B^p\right)^{1/p}$$

of a martingale. For that topic (exceptionally) we devote an entire chapter only to the scalar case. Our crucial tool here is the 'real interpolation method'. Real and complex interpolation in general play an important role in L_p-space theory, so we find it natural to devote a significant amount of space to these two 'methods'.

We allow ourselves several excursions aiming to illustrate the efficiency of martingales, for instance to the concentration of measure phenomenon. We also describe some exciting recent work on non-linear properties of metric spaces analogous to uniform convexity/smoothness and type for metric spaces.

We originally intended to include in this book a detailed presentation of 'non-commutative' martingale theory, but that part became so big that we decided to make it the subject of a (hopefully forthcoming) separate volume to be published, perhaps on the author's web page. We merely outline its contents in the last chapter, devoted to non-commutative L_p-spaces. There the complex interpolation method becomes a central tool.

The book should be accessible to graduate students, requiring only the basics of real and complex analysis (mainly Lebesgue integration) and basic functional analysis (mainly duality, the weak and strong topologies and reflexivity of Banach spaces). Our choice is to give fully detailed proofs for the main results and to indicate references to the refinements in the 'Notes and Remarks' or the asterisked sections. We strive to make the presentations self-contained, and when given a choice, we opt for simplicity over maximal generality. For instance, we restrict the Banach space valued harmonic analysis to functions with domains in the unit disc D or the upper half-plane U in $\mathbb{C}$ (or their boundary $\partial D = \mathbb{T}$ or $\partial U = \mathbb{R}$). We feel the main ideas are easier to grasp in the real or complex uni-dimensional case.

The topics (martingales, H^p-space theory, interpolation, Banach space geometry) are quite diverse and should appeal to several distinct audiences. The main novelty is the choice to bring all these topics together in the various parts of this single volume. We should emphasize that the different parts can be read independently, and each time their start is introductory.

There are natural groupings of chapters, such as 1-2-10-11 or 3-4-5-6 (possibly including parts of 1 and 2, but not necessarily), which could form the basis for a graduate course.

Depending on his or her background, a reader is likely to choose to concentrate on different parts. We hope probabilist graduate students will benefit from the detailed introductory presentation of basic H^p-space theory, its connections with martingales, the links with Banach space geometry and the detailed treatment of interpolation theory (which we illustrate by applications to the strong p-variation of martingales), while graduate students with interest in functional analysis and Banach spaces should benefit more from the initial detailed presentation of basic martingale theory. In addition, we hope to attract readers with interest in computer science wishing to see the sources of the various recent developments on finite metric spaces described in Chapter 13. A reader with an advanced knowledge of harmonic analysis and H^p-theory will probably choose to skip the introductory part on that direction, which is written with non-specialists in mind, and concentrate on the issues specific to Banach space valued functions related to the UMD property and the Hilbert transform.

The choice to include so much background on the real and complex interpolation methods in Chapter 8 is motivated by its crucial importance in Banach space valued L_p-space theory, which, in some sense, is the true subject of this book.

Acknowledgement. This book is based on lecture notes for various topics courses given during the last 10 years or so at Texas A&M University. Thanks are due to Robin Campbell, who typed most on them, for her excellent work. I am indebted to Hervé Chapelat, who took notes from my even earlier lectures on H^p-spaces there, for Chapters 3 and 4. The completion of this volume was stimulated by the Winter School on 'Type, cotype and martingales on Banach spaces and metric spaces' at IHP (Paris), 2–8 February 2011, for which I would like to thank the organizers. I am very grateful to all those who, at some stage, helped me to correct mistakes and misprints and who suggested improvements of all kinds, in particular Michael Cwikel, Sonia Fourati, Julien Giol, Rostyslav Kravchenko, Bernard Maurey, Adam Osękowski, Javier Parcet, Yanqi Qiu, Mikael de la Salle, Francisco Torres-Ayala, Mateusz Wasilewski, and Quanhua Xu; S. Petermichl for help on Chapter 6; and M. I. Ostrovskii for advice on Chapter 13. I am especially grateful to Mikael de la Salle for drawing all the pictures with TikZ.

Description of the contents

We will now review the contents of this book chapter by chapter.

Chapter 1 begins with preliminary background: we introduce Banach space valued L_p-spaces, conditional expectations and the central notion in this book, namely Banach space valued martingales associated to a filtration $(\mathcal{A}_n)_{n \geq 0}$ on a probability space $(\Omega, \mathcal{A}, \mathbb{P})$. We describe the classical examples of filtrations (the dyadic one and the Haar one) in §1.4. If B is an arbitrary Banach space and the martingale (f_n) is associated to some f in $L_p(B)$ by $f_n = \mathbb{E}^{\mathcal{A}_n}(f)$ $(1 \leq p < \infty)$, then, assuming $\mathcal{A} = \mathcal{A}_\infty$ for simplicity, the fundamental convergence theorems say that

$$f_n \to f$$

both in $L_p(B)$ and almost surely (a.s.).

The convergence in $L_p(B)$ is Theorem 1.14, while the a.s. convergence is Theorem 1.30. The latter is based on Doob's classical *maximal inequalities* (Theorem 1.25), which are proved using the crucial notion of *stopping time*. We also describe the dual form of Doob's inequality due to Burkholder-Davis-Gundy (see Theorem 1.26). Doob's maximal inequality shows that the convergence of f_n to f in $L_p(B)$ 'automatically' implies a.s. convergence. This, of course, is special to martingales, but in general it requires $p \geq 1$. However, for martingales that are sums of independent, symmetric random variables (Y_n) (i.e. we have $f_n = \sum_1^n Y_k$), this result holds for $0 < p < 1$ (see Theorem 1.40). It also holds, roughly, for $p = 0$. This is the content of the celebrated Ito-Nisio theorem (see Theorem 1.43), which asserts that even a weak form of convergence of the series $f_n = \sum_1^n Y_k$ implies its a.s. norm convergence.

In §1.8, we prove, again using martingales, a version of Phillips's theorem. The latter is usually stated as saying that, if B is separable, any countably additive measure on the Borel σ-algebra of B is 'Radon', i.e. the measure of a Borel subset can be approximated by that of its compact subsets. In §1.9, we prove the strong law of large numbers using the a.s. convergence of reverse B-valued martingales. In §1.10, we give a brief introduction to continuous-time martingales. We mainly explain the basic approximation technique by which one passes from discrete to continuous parameter.

To get to a.s. convergence, all the preceding results need to assume in the first place some form of convergence, e.g. in $L_p(B)$. In classical (i.e. real valued) martingale theory, it suffices to assume *boundedness* of the martingale $\{f_n\}$ in L_p $(p \geq 1)$ to obtain its a.s. convergence (as well as norm convergence, if $1 < p < \infty$). However, this 'boundedness $\Rightarrow$ convergence' phenomenon no longer holds in the B-valued case unless B has a specific property called the Radon-Nikodým property (RNP in short), which we introduce and study in Chapter 2. The RNP of a Banach space B expresses the validity of a certain form of the Radon-Nikodým theorem for B-valued measures, but it turns out to be equivalent to the assertion that all martingales bounded in $L_p(B)$ converge a.s. (and in $L_p(B)$ if $p > 1$) for some (or equivalently all) $1 \leq p < \infty$. Moreover, the RNP is equivalent to a certain 'geometric' property called 'dentability'. All this is included in Theorem 2.9. The basic examples of Banach spaces with the RNP are the reflexive ones and separable duals (see Corollary 2.15).

Moreover, a dual space B^* has the RNP iff the classical duality $L_p(B)^* = L_{p'}(B^*)$ is valid for some (or all) $1 < p < \infty$ with $\frac{1}{p} + \frac{1}{p'} = 1$ (see Theorem 2.22). Actually, for a general B, one can also describe $L_p(B)^*$ as a space of martingales bounded in $L_{p'}(B^*)$, but in general, the latter is larger than the (Bochner sense) space $L_{p'}(B^*)$ itself (see Proposition 2.20). In many situations, it is preferable to have a description of $L_p(B)^*$ as a space of B^*-valued measures or functions (rather than martingales). For that purpose, we give in §2.4 two alternate descriptions of the latter space, either as a space of B^*-valued vector measures, denoted by $\Lambda_{p'}(B^*)$, or, assuming B separable, as a space of weak* measurable B^*-valued functions denoted by $\underline{\Lambda}_{p'}(B^*)$.

In §2.5, we discuss the Krein-Milman property (KMP): this says that any bounded closed convex set $C \subset B$ is the closed convex hull of its extreme points. This is closely related to dentability, but although it is known that RNP $\Rightarrow$ KMP (see Theorem 2.34), the converse implication is still open.

In §2.6, we present a version of Choquet's representation theorem of the points of C as barycenters of measures supported by the extreme points. Choquet's classical result applies only to convex compact sets, while this version requires that the closed bounded separable convex set C lies in an RNP space. The proof is based on martingale convergence.

In §2.7, we prove that, if B is separable, a bounded B-valued vector measure admits a RN density iff the associated linear operator $L_1 \to B$ factorizes through the space ℓ_1 of absolutely summable scalar sequences. This fact (due to Lewis and Stegall) is remarkable because ℓ_1 is the prototypical example of a separable dual (a fortiori ℓ_1 has the RNP). In other words, if B has the RNP, the bounded B-valued vector measures 'come from' ℓ_1 valued vector measures, and the latter are differentiable since ℓ_1 itself has the RNP. There is also a version of this result when B is not separable (see Theorem 2.39).

In Chapter 3, we introduce the Hardy space $h^p(D; B)$ $(1 \le p \le \infty)$ formed of all the B-valued harmonic functions $u \colon D \to B$ (on the unit disc $D \subset \mathbb{C}$) such that

$$\|u\|_{h^p(D;B)} = \sup\nolimits_{0<r<1} \left(\int \|u(re^{it})\|^p dm(t) \right)^{1/p} < \infty.$$

When $p = \infty$, this is just the space of bounded harmonic functions $u \colon D \to B$.

In analogy with the martingale case treated in Chapter 1, to any f in $L_p(\mathbb{T}; B)$, we can associate a harmonic function $u \colon D \to B$ that extends f in the sense that $u(re^{it}) \to f(e^{it})$ for almost all e^{it} in $\partial D = \mathbb{T}$, and also (if $p < \infty$)

$$\int \|u(re^{it}) - f(e^{it})\|^p dm(t) \to 0$$

when $r \uparrow 1$ (see Theorem 3.1). The convergence at almost all boundary points requires a specific radial maximal inequality, which we derive from the classical Hardy-Littlewood maximal inequality. Actually, we present this in the framework of 'non-tangential' maximal inequalities. The term 'non-tangential' refers to the fact that we study the limit of $u(z)$ (in the norm of B) when z tends to e^{it} but staying inside a cone with vertex e^{it} and opening angle $\beta < \frac{\pi}{2}$.

This topic is closely linked with Brownian motion (see especially [176]). Indeed, the paths of a complex valued Brownian motion $(W_t)_{t>0}$ (starting at the origin) almost surely cross the boundary of the unit disc D in finite time, and if we define $T_r = \inf\{t > 0 \,|\, |W_t| = r\}$, then for any u in $h^p(D; B)$, the random variables $\{u(W_{T_r}) \mid 0 < r < 1\}$ form a martingale bounded in $L_p(B)$, for which the maximal function is closely related to the non-tangential one of u.

In general, the latter B-valued martingales do not converge. However, we show in Corollary 3.31 that they do so if B has the RNP. Actually, the RNP of B is equivalent to the a.e. existence of radial (or non-tangential) limits or of limits along almost all Brownian paths, for the functions in $h^p(D; B)$ $(1 \le p \le \infty)$ (see Theorem 3.25 and Corollary 3.31 for this). In the supplementary §4.7, we compare the various modes of convergence to the boundary, radial, non-tangential or Brownian for a general B-valued harmonic function. This brief §4.7 outlines some beautiful work by Burkholder, Gundy and Silverstein [176] and Brossard [148].

In Chapter 4 we turn to the subspace $H^p(D; B) \subset h^p(D; B)$ formed of all the *analytic* B-valued functions in $h^p(D; B)$. One major difference is that the (radial or non-tangential) maximal inequalities now extend to all values of p including $0 < p \leq 1$. To prove this, we make crucial use of *outer functions*; this gives us a convenient factorization of any f in $H^p(D; B)$ as a product $f = Fg$ with a *scalar* valued F in H^p and g in $H^\infty(D; B)$ (see Theorem 4.15). The Hardy spaces $H^p(D; B)$ naturally lead us to a more general form of RNP called the analytic RNP (ARNP in short). This is equivalent to the a.e. existence of radial (or non-tangential) limits for all functions in $H^p(D; B)$ for some (or all) $0 < p \leq \infty$. Thus we have RNP $\Rightarrow$ ARNP, but the converse fails, for instance (see Theorem 4.32) $B = L_1([0, 1])$ has the ARNP but not the RNP. On the martingale side, the strict inclusion

$$\{\text{analytic}\} \subsetneqq \{\text{harmonic}\}$$

admits an analogue involving the notion of 'analytic martingale' or 'Hardy martingale', and the convergence of the latter (with the usual bounds) is equivalent to the ARNP (see Theorem 4.30). In §4.6 we briefly review the analogous Banach space valued H^p-space theory for functions on the upper half-plane

$$U = \{z \in \mathbb{C} | \Im(z) > 0\}.$$

Chapter 5 is devoted to the UMD property. After a brief presentation of Burkholder's inequalities in the scalar case, we concentrate on their analogue for Banach space valued martingales (f_n). In the scalar case, when $1 < p < \infty$, we have

$$\sup_n \|f_n\|_p \simeq \|\sup |f_n|\|_p \simeq \|S(f)\|_p,$$

where $S(f) = (|f_0|^2 + \sum |f_n - f_{n-1}|^2)^{1/2}$, and where $A_p \simeq B_p$ means that there are positive constants C'_p and C''_p such that $C'_p A_p \leq B_p \leq C''_p A_p$. In the Banach space valued case, we replace $S(f)$ by

$$R(f)(\omega) = \sup_N \left(\int \left\| f_0(\omega) + \sum\nolimits_1^N \varepsilon_n (f_n - f_{n-1})(\omega) \right\|^2 d\nu \right)^{1/2}, \quad (1)$$

where ν is the uniform probability measure on the set Δ of all choices of signs $(\varepsilon_n)_n$ with $\varepsilon_n = \pm 1$.

In §5.2 we prove Kahane's inequality, i.e. the equivalence of all the L_p-norms for series of the form $\sum_1^\infty \varepsilon_n x_n$ with x_n in an arbitrary Banach space when $0 < p < \infty$ (see (5.16)); in particular, up to equivalence, we can substitute to the L_2-norm in (1) any other L_p-norm for $p < \infty$.

Let $\{x_n\}$ be a sequence in a Banach space such that the series $\sum \varepsilon_n x_n$ converges almost surely. We set

$$R(\{x_n\}) = \left(\int_D \left\| \sum \varepsilon_n x_n \right\|^2 d\nu \right)^{1/2}.$$

With this notation we have

$$R(f)(\omega) = R(\{f_0(\omega), f_1(\omega) - f_0(\omega), f_2(\omega) - f_1(\omega), \cdots\}). \tag{2}$$

The UMD$_p$ and UMD properties are introduced in §5.3. Consider the series

$$\tilde{f}_\varepsilon = f_0 + \sum\nolimits_1^\infty \varepsilon_n (f_n - f_{n-1}). \tag{3}$$

By definition, when B is UMD$_p$, (f_n) converges in $L_p(B)$ iff (3) converges in $L_p(B)$ for all choices of signs $\varepsilon_n = \pm 1$ or equivalently iff it converges for almost all (ε_n). Moreover, we have then for $1 < p < \infty$ and all choices of signs $\varepsilon = (\varepsilon_n)$

$$\|\tilde{f}_\varepsilon\|_{L_p(B)} \simeq \|f\|_{L_p(B)} \qquad (4)_p$$

$$\sup_{n \geq 0} \|f_n\|_{L_p(B)} \simeq \|R(f)\|_p. \qquad (5)_p$$

See Proposition 5.10. The case $p = 1$ (due to Burgess Davis) is treated in §5.6. The main result of §5.3 is the equivalence of UMD$_p$ and UMD$_q$ for any $1 < p, q < \infty$. We give two proofs of this; the first one is based on distributional (also called 'good λ') inequalities. This is an extrapolation principle, which allows us to show that, for a given Banach space B, $(4)_q \Rightarrow (4)_p$ for any $1 < p < q$. In the scalar case one starts from the case $q = 2$, which is obvious by orthogonality, and uses the preceding implication to deduce from it the case $1 < p < 2$ and then $2 < p < \infty$ by duality. We also give a more delicate variant of the extrapolation principle that avoids duality and deduces the desired inequality for any $1 < p < \infty$ from a certain form of weak-type estimate involving pairs of stopping times (see Lemma 5.26).

The second proof is based on Gundy's decomposition, which is a martingale version of the Calderón-Zygmund decomposition in classical harmonic analysis. There one proves a weak-type (1,1) estimate and then invokes the Marcinkiewicz theorem to obtain the case $1 < p < 2$. We describe the latter in an appendix to Chapter 5.

In §5.8 we show that to check that a space B is UMD$_p$, we may restrict ourselves to martingales adapted to the *dyadic* filtration, and the associated UMD-constant remains the same. The proof is based on a result of independent interest: if $p < \infty$, any finite martingale $(f_0, \ldots, f_N)$ in $L_p(B)$ (on a large enough probability space) can be approximated in $L_p(B)$ by a subsequence of a dyadic martingale.

In §5.9 we prove the Burkholder-Rosenthal inequalities. In the scalar case, this boils down to the equivalence

$$\sup_n \|f_n\|_p \simeq \|\sigma(f)\|_p + \|\sup_n |f_n - f_{n-1}|\|_p,$$

where $\sigma(f) = (|f_0|^2 + \sum \mathbb{E}_{n-1}|f_n - f_{n-1}|^2)^{1/2}$, valid for $2 \le p < \infty$.

Rosenthal originally proved this when f_n is a sum of independent variables, and Burkholder extended it to martingales. We describe a remarkable example of complemented subspace of L_p (the Rosenthal space X_p), which motivated Rosenthal's work.

In §5.10 we describe Stein's inequality and its B-valued analogue when B is a UMD Banach space. Let $(\mathcal{A}_n)_{n\ge 0}$ be a filtration as usual, and let $(x_n)_{n\ge 0}$ be now an *arbitrary* sequence in L_p. Let $y_n = \mathbb{E}^{\mathcal{A}_n} x_n$. Stein's inequality asserts that for any $1 < p < \infty$, there is a constant C_p such that

$$\left\| \left(\sum |y_n|^2 \right)^{1/2} \right\|_p \le C_p \left\| \left(\sum |x_n|^2 \right)^{1/2} \right\|_p \tag{6}$$

for any (x_n) in L_p.

For x_n in $L_p(B)$, with B UMD, the same result remains valid if we replace on both sides of (6) the expression $(\sum |x_n|^2)^{1/2}$ by

$$\left(\int \left\| \sum \varepsilon_n x_n \right\|_B^2 d\nu \right)^{1/2}.$$

See (5.89).

In §5.11 we describe Burkholder's geometric characterization of UMD spaces in terms of ζ-convexity (Theorem 5.64). We also include a more recent result (Theorem 5.69) in the same vein. The latter asserts that a real Banach space of the form $B = X \oplus X^*$ is UMD iff the function

$$x \oplus \xi \to \xi(x)$$

is the difference of two real valued convex continuous functions on B.

We end Chapter 5 by a series of appendices. In §5.12 we prove the hypercontractive inequalities on $\{-1, 1\}$. In §5.13 we discuss the Hölder-Minkowski inequality, which says that we have a norm 1 inclusion

$$L_q(m'; L_p(m)) \subset L_p(m; L_q(m'))$$

when $p \ge q$. In §5.14 we give some basic background on the space weak-L_p, usually denoted $L_{p,\infty}$, and §5.16 is devoted to a quick direct proof of the Marcinkiewicz theorem. In §5.15 we present a trick frequently used by Burkholder that we call the reverse Hölder principle. Hölder's inequality for a random variable Z on a probability space tells us that $Z \in L_q \Rightarrow Z \in L_r$ when $r < q$. The typical reverse Hölder principle shows that a suitable L_r bound involving independent copies of Z implies conversely that Z is in weak-L_p (or $L_{p,\infty}$) and a fortiori in L_q for all $r < q < p$.

In the final appendix to this chapter, §5.17, we explain why certain forms of exponential integrability of a function f can be equivalently reformulated in terms of the growth of the L_p-norms of f. This shows that the growth when $p \to \infty$ of the constants in many of the martingale inequalities we consider often implies an exponential inequality.

In Chapter 6 we start with some background on the Hilbert transform. We say that a Banach space B is an HT-space if the Hilbert transform on $\mathbb{T}$ defines a bounded operator on $L_p(\mathbb{T}; B)$ for some (or equivalently all) $1 < p < \infty$. We then show that UMD and HT are equivalent properties. To show HT $\Rightarrow$ UMD, we follow Bourgain's well-known argument (see §6.2). The converse implication UMD $\Rightarrow$ HT was originally proved using ideas derived from the beautiful observation that the Hilbert transform can be viewed as a sort of martingale transform relative to stochastic integrals over Brownian motion. We merely outline this proof in §6.4, and instead, we present first, in full details the more recent remarkable proof from Petermichl [378], which uses only martingale transforms relative to Haar systems, but with respect to a randomly chosen dyadic filtration (see §6.3).

In §6.5, following Bourgain, we prove that the Littlewood-Paley inequalities are valid in the B-valued case if B is UMD. More precisely, consider the formal Fourier series

$$f = \sum\nolimits_{n \in \mathbb{Z}} \hat{f}(n) e^{int}$$

of a function in $L_p(B)$, and let

$$V(f) = (\|\hat{f}(0)\|_B^2 + R(\{\Delta_n^+\})^2 + R(\{\Delta_n^-\})^2)^{1/2},$$

where

$$\Delta_n^+ = \sum\nolimits_{2^n \le k < 2^{n+1}} \hat{f}(k) e^{int} \quad \text{and} \quad \Delta_n^- = \sum\nolimits_{2^n \le -k < 2^{n+1}} \hat{f}(k) e^{ikt}.$$

The B-valued version of the Littlewood-Paley inequality is the equivalence of $\|f\|_{L_p(B)}$ and $\|V(f)\|_p$.

In Theorem 6.35 we show that the analogue of the Hilbert transform for the compact group $\{-1, 1\}^{\mathbb{N}}$ (in place of $\mathbb{T}$) is bounded on $L_p(B)$ for any $1 < p < \infty$ iff B is UMD. This is the analogue of the implication UMD $\Rightarrow$ HT, but for the Walsh system in place of the trigonometric one. The proof of this case is much more transparent.

In §6.7, we briefly review the analytic UMD property (AUMD). This is a weakening of UMD obtained by restricting to either analytic or Hardy martingales. The latter are discretizations of the B-valued martingales obtained by composing complex Brownian motion with a B-valued analytic function on D. The main novelty is that the space $B = L_1$ itself has the AUMD property. Moreover, the latter implies the ARNP.

In §7.1 we describe the B-valued version of Fefferman's famous duality theorem between H^1 and BMO on $\mathbb{T}$ (resp. $\mathbb{R}$). This requires that we carefully identify the various B-valued analogues of the Hardy spaces H^1 or BMO. If B is arbitrary, the duality holds provided that we use the atomic version of B-valued H^1, denoted by $h^1_{\rm at}(\mathbb{T}; B)$ (resp. $h^1_{\rm at}(\mathbb{R}; B)$); this is the so-called real variable variant of H^1, as opposed to the more classical Hardy space of analytic function, denoted by $H^1(D; B)$ (resp. $H^1(U; B)$) on the unit disc D (resp. the upper half-plane U). In general, for any $f \in H^1(D; B)$ (resp. $H^1(U; B)$), the boundary values of f and its Hilbert transform $\tilde{f}$ are both in $h^1_{\rm at}(\mathbb{T}; B)$ (resp. $h^1_{\rm at}(\mathbb{R}; B)$). When B is UMD (and only then), the converse also holds (see Corollary 7.20).

In §7.3 we discuss the space BMO and the B-valued version of H^1 in the martingale context. This leads naturally to the atomic version of B-valued H^1, denoted by $h^1_{at}(\{\mathcal{A}_n\}; B)$ with respect to a filtration $\{\mathcal{A}_n\}$. Its dual can be identified with a BMO-space for B^*-valued martingales, at least for a 'regular' filtration $(\mathcal{A}_n)$. Equivalently (see Theorem 7.32), the space $h^1_{at}(B)$ can be identified with $\widetilde{h}^1_{\max}(B)$, which is defined as the completion of $L_1(B)$ with respect to the norm $f \mapsto \mathbb{E}\sup_n \|f_n\|_B$ (here $f_n = \mathbb{E}^{\mathcal{A}_n} f$).

In §7.5 we show that the classical BMO space coincides with the intersection of two suitably chosen translates of the dyadic martingale-BMO space.

In Chapter 8 we describe successively the complex and the real method of interpolation for pairs of Banach spaces (B_0, B_1) assumed compatible for interpolation purposes. The complex interpolation space is denoted by $(B_0, B_1)_\theta$. It depends on the single parameter $0 < \theta < 1$ and requires B_0, B_1 to be both *complex* Banach spaces. Complex interpolation is a sort of 'abstract' generalization of the classical Riesz-Thorin theorem, asserting that if an operator T has norm 1 simultaneously on both spaces $B_0 = L_{p_0}$ and $B_1 = L_{p_1}$, with $1 \le p_0 < p_1 \le \infty$, then it also has norm 1 on the space L_p for any p such that $p_0 < p < p_1$. Since we make heavy use of this *complex method* in non-commutative L_p-theory, we review its basic properties somewhat extensively.

The real interpolation space is denoted by $(B_0, B_1)_{\theta,q}$. It depends on two parameters $0 < \theta < 1$, $1 \le q \le \infty$, and now (B_0, B_1) can be a pair of *real* Banach spaces. Real interpolation is a sort of abstract generalization of the

Marcinkiewicz classical theorem already proved in an appendix to Chapter 5. The real interpolation space is introduced using the 'K-functional' defined, for any $B_0 + B_1$, by

$$\forall t > 0 \;\; K_t(x) = \inf\{\|x_0\|_{B_0} + t\|x_1\|_{B_1} \mid x_0 \in B_0, x_1 \in B_1, x = x_0 + x_1\}.$$

When $B_0 = L_1(\Omega, m)$, $B_1 = L_\infty(\Omega, m)$, we find

$$K_t(x) = \int_0^t x^\dagger(s)ds, \tag{7}$$

where $x^\dagger$ is the non-increasing rearrangement of $|x|$ and (Ω, m) is an arbitrary measure space. We prove this in Theorem 8.50 together with the identification of $(L_1, L_\infty)_{\theta,q}$ with the Lorentz space $L_{p,q}$ for $p = (1-\theta)^{-1}$.

Real interpolation will be crucially used in the later Chapters 9 and 12 in connection with our study of the 'strong p-variation' of martingales. The two interpolation methods satisfy distinct properties but are somewhat parallel to each other. For instance, duality, reiteration and interpolation between vector valued L_p-spaces are given parallel treatments in Chapter 8.

The classical reference on interpolation is [6] (see also [51]). Given the existence of these excellent treatises, one could question the need for the compact but detailed presentation (essentially self-complete) we give in Chapter 8. We hope this will be appreciated at least by some readers (e.g. interpolation theory is not the usual background of probabilists). Moreover, several specific points which play an important role for Banach space valued L_p-spaces (commutative or not) are not completely proved in the existing references (which contain much more on other topics). For instance, we have in mind (8.15), Theorem 8.34, Theorem 8.39, §8.6, §8.7 and §8.11.

We devote a significant amount of space in §8.3 to describing the dual of the complex interpolation space $(B_0, B_1)_\theta$ in connection with the RNP and ARNP. The analogous question for the real method, involving the J-functional that is dual to $x \mapsto K_t(x)$, is treated in §8.8.

We devote much space to pairs of Banach space valued L_p-spaces; i.e. given a compatible pair (B_0, B_1), a measure space $(\Omega, \mathcal{A}, m)$ and $1 \le p_0, p_1 \le \infty$, we identify the interpolation spaces for the pair $(Lp_0(m; B_0), Lp_1(m; B_1))$. The complex case is treated in Theorem 8.21. For the real case, we give a formula for the K-functional (see §8.6), generalizing (7), from which the interpolation space can be derived easily (see §8.7) in connection with Lorentz spaces.

Lastly, we consider pairs with some special symmetries, e.g. pairs of the form (B^*, B) associated to a positive definite inclusion $B^* \subset B$ (or $\overline{B^*} \subset B$ in the complex case) for which we show that Hilbert space automatically appears at the centre of the interpolation scale.

In Chapter 9 we study the strong p-variation $W_p(f)$ of a scalar martingale (f_n). This is defined as the supremum of

$$\left(|f_{n(0)}|^p + \sum\nolimits_{k=1}^{\infty} |f_{n(k)} - f_{n(k-1)}|^p\right)^{1/p}$$

over all possible increasing sequences

$$0 = n(0) < n(1) < n(2) < \cdots .$$

The main results are Theorem 9.2 and Proposition 9.6. Roughly, this says that, if $1 \le p < 2$, $W_p(f)$ is essentially 'controlled' by $(\sum |f_n - f_{n-1}|^p)^{1/p}$, i.e. by the finest partition corresponding to consecutive $n(k)$s, while, in sharp contrast, if $2 < p < \infty$, it is 'controlled' by $|f_\infty| = \lim |f_n|$, or equivalently, by the coarsest partition corresponding to the choice $n(0) = 0$, $n(1) = \infty$.

The proofs combine a simple stopping time argument with the reiteration theorem of the real interpolation method.

In Chapter 10, we turn to uniform convexity and uniform smoothness of Banach spaces. We show that certain martingale inequalities characterize the Banach spaces B that admit an equivalent norm for which there is a constant C and $2 \le q < \infty$ or $1 < p \le 2$ such that, for any x, y in B,

$$\|x\|^q + C\|y\|^q \le \frac{\|x+y\|^q + \|x-y\|^q}{2} \tag{8}$$

or

$$\frac{\|x+y\|^p + \|x-y\|^p}{2} \le \|x\|^p + C\|y\|^p, \tag{9}$$

respectively. This is the content of Corollary 10.7 (resp. Corollary 10.23). We use this in Theorem 10.1 (resp. Theorem 10.25) to show that actually *any* uniformly convex (resp. smooth) Banach space admits for some $2 \le q < \infty$ (resp. $1 < p \le 2$) such an equivalent renorming. The inequality (8) (resp. (9)) holds iff the modulus of uniform convexity (resp. smoothness) $\delta(\varepsilon)$ (resp. $\rho(t)$) satisfies $\inf_{\varepsilon>0} \delta(\varepsilon)\varepsilon^{-q} > 0$ (resp. $\sup_{t>0} \rho(t)t^{-p} < \infty$). In that case we say that the space is q-uniformly convex (resp. p-uniformly smooth). The proof also uses inequalities going back to Gurarii, James and Lindenstrauss on monotone basic sequences. We apply the latter to martingale difference sequences viewed as monotone basic sequences in $L_p(B)$. Our treatment of uniform smoothness in §10.2 runs parallel to that of uniform convexity in §10.1.

In §10.3 we estimate the moduli of uniform convexity and smoothness of L_p for $1 < p < \infty$. In particular, L_p is p-uniformly convex if $2 \le p < \infty$ and p-uniformly smooth if $1 < p \le 2$.

In §10.5 we prove analogues of Burkholder's inequalities, but with the square function now replaced by

$$S_p(f) = \left(\|f_0\|_B^p + \sum\nolimits_1^\infty \|f_n - f_{n-1}\|_B^p\right)^{1/p}.$$

Unfortunately, the results are now only one sided: if B satisfies (8) (resp. (9)), then $\|S_q(f)\|_r$ is dominated by (resp. $\|S_p(f)\|_r$ dominates) $\|f\|_{L_r(B)}$ for all $1 < r < \infty$, but here $p \le 2 \le q$ and the case $p = q$ is reduced to the Hilbert space case.

In §10.6 we return to the strong p-variation and prove analogous results to the preceding ones, but this time with $W_q(f)$ and $W_p(f)$ in place of $S_q(f)$ and $S_p(f)$ and $1 < p < 2 < q < \infty$. The technique here is similar to that used for the scalar case in Chapter 9.

Chapter 11 is devoted to *super-reflexivity*. A Banach space B is super-reflexive if every space that is *finitely representable* in B is reflexive. In §11.1 we introduce finite representability and general super-properties in connection with ultraproducts. We include some background about the latter in an appendix to Chapter 11.

In §11.2 we concentrate on super-P when P is either 'reflexivity' or the RNP. We prove that super-reflexivity is equivalent to the super-RNP (see Theorem 11.11). We give (see Theorem 11.10) a fundamental characterization of reflexivity, from which one can also derive easily (see Theorem 11.22) one of super-reflexivity.

As in the preceding chapter, we replace B by $L_2(B)$ and view martingale difference sequences as monotone basic sequences in $L_2(B)$. Then we deduce the martingale inequalities from those satisfied by general basic sequences in super-reflexive spaces. For that purpose, we review a number of results about basic sequences that are not always directly related to our approach. For instance, we prove the classical fact that a Banach space with a basis is reflexive iff the basis is both boundedly complete and shrinking. While we do not directly use this, it should help the reader understand why super-reflexivity implies inequalities of the form either $(\sum \|x_n\|^q)^{1/q} \le C\|\sum x_n\|$ for $q < \infty$ or $\|\sum x_n\| \le (\sum \|x_n\|^p)^{1/p}$ for $p > 1$ (see (iv) and (v) in Theorem 11.22). Indeed, they can be interpreted as a strong form of 'boundedly complete' for the first one and of 'shrinking' for the second one.

In §11.3 we show that uniformly non-square Banach spaces are reflexive, and hence automatically super-reflexive (see Theorem 11.24 and Corollary 11.26). More generally, we go on to prove that B is super-reflexive iff it is J-convex, or equivalently iff it is J-(n, ε) convex for some $n \ge 2$ and some $\varepsilon > 0$. We say that B is J-(n, ε) convex if, for any n-tuple $(x_1, \ldots, x_n)$ in the unit ball of B, there is an integer $j = 1, \dots, n$ such that

$$\left\| \sum_{i<j} x_i - \sum_{i \ge j} x_i \right\| \le n(1 - \varepsilon).$$

When $n = 2$, we recover the notion of 'uniformly non-square'. The implication super-reflexive $\Rightarrow$ J-convex is rather easy to derive (as we do in Corollary 11.34) from the fundamental reflexivity criterion stated as Theorem 11.10. The converse implication (due to James) is much more delicate. We prove it following essentially the Brunel-Sucheston approach ([122]), which in our opinion is much easier to grasp. This construction shows that a non-super-reflexive (or merely non-reflexive) space B contains very *extreme finite-dimensional* structures that constitute obstructions to either reflexivity or the RNP. For instance, any such B admits a space $\widetilde{B}$ finitely representable in B for which there is a dyadic martingale (f_n) with values in the unit ball of $\widetilde{B}$ such that

$$\forall n \ge 1 \qquad \|f_n - f_{n-1}\|_B \equiv 1.$$

Thus the unit ball of $\widetilde{B}$ contains an extremely sparsely separated infinite dyadic tree. (See Remark 1.35 for concrete examples of such trees.)

In §11.4 we finally connect super-reflexivity and uniform convexity. We prove that B is super-reflexive iff it is isomorphic to either a uniformly convex space, or a uniformly smooth one, or a uniformly non-square one. By the preceding Chapter 10, we already know that the renormings can be achieved with moduli of convexity and smoothness of 'power type'. Using interpolation (see Proposition 11.44), we can even obtain a renorming that is both p-uniformly smooth and q-uniformly convex for some $1 < p, q < \infty$, but it is still open whether this holds with the optimal choice of $p > 1$ and $q < \infty$. To end Chapter 11, we give a characterization of super-reflexivity by the validity of a version of the strong law of large numbers for B-valued martingales.

In §11.6 we discuss the stability of super-reflexivity (as well as uniform convexity and smoothness) by interpolation. We also show that a Banach lattice is super-reflexive iff it is isomorphic, for some $\theta > 0$, to an interpolated space $(B_0, B_1)_\theta$ where B_1 is a Hilbert space (and B_0 is arbitrary). Such spaces are called θ-Hilbertian.

In §11.7 we discuss the various complex analogues of uniform convexity: When restricted to analytic (or Hardy, or PSH) martingales, the martingale inequalities characterizing p-uniform convexity lead to several variants of uniform convexity, where roughly convex functions are replaced by plurisubharmonic ones. Of course, this subject is connected to the analytic RNP and analytic UMD, but many questions remain unanswered.

In Chapter 12 we study the real interpolation spaces $(v_1, \ell_\infty)_{\theta,q}$ between the space v_1 of sequences with bounded variation and the space ℓ_∞ of bounded sequences. Explicitly, v_1 (resp. ℓ_∞) is the space of scalar sequences (x_n) such that $\sum_1^\infty |x_n - x_{n-1}| < \infty$ (resp. $\sup |x_n| < \infty$) equipped with its natural norm. The inclusion $v_1 \to \ell_\infty$ plays a major part (perhaps behind the scene) in our treatment of (super-) reflexivity in Chapter 11. Indeed, by the fundamental Theorem 11.10, B is non-reflexive iff the inclusion $\mathcal{J}\colon v_1 \to \ell_\infty$ factors through B, i.e. it admits a factorization

$$v_1 \xrightarrow{a} B \xrightarrow{b} \ell_\infty,$$

with bounded linear maps a, b such that $\mathcal{J} = ba$.

The remarkable work of James on J-convexity (described in Chapter 11) left open an important point: whether any Banach space B such that ℓ_1^n is not finitely representable in B (i.e. is not almost isometrically embeddable in B) must be reflexive. James proved that the answer is yes if $n = 2$, but for $n > 2$, this remained open until James himself settled it [281] by a counter-example for $n = 3$ (see also [283] for simplifications). In the theory of type (and cotype), it is the same to say that, for some $n \geq 2$, B does not contain ℓ_1^n almost isometrically or to say that B has type p for some $p > 1$ (see the survey [347]). Moreover, type p can be equivalently defined by an inequality analogous to that of p-uniform smoothness, but only for martingales with independent increments. Thus it is natural to wonder whether the strongest notion of 'type p', namely type 2, implies reflexivity. In another tour de force, James [282] proved that *it is not so*. His example is rather complicated. However, it turns out that the real interpolation spaces $\mathcal{W}_{p,q} = (v_1, \ell_\infty)_{\theta,q}$ $(1 < p, q < \infty, 1 - \theta = 1/p)$ provide very nice examples of the same kind. Thus, following [399], we prove in Corollary 12.19 that $\mathcal{W}_{p,q}$ has exactly the same type and cotype exponents as the Lorentz space $\ell_{p,q} = (\ell_1, \ell_\infty)_{\theta,q}$ as long as $p \neq 2$, although $\mathcal{W}_{p,q}$ is *not reflexive* since it lies between v_1 and ℓ_∞. The singularity at $p = 2$ is necessary since (unlike $\ell_2 = \ell_{2,2}$) the space $\mathcal{W}_{2,2}$, being non-reflexive, cannot have both type 2 and cotype 2 since that would force it to be isomorphic to Hilbert space.

A key idea is to consider similarly the B-valued spaces

$$\mathcal{W}_{p,q}(B) = (v_1(B), \ell_\infty(B))_{\theta,q},$$

where B can be an arbitrary Banach space. When $p = q$, we set

$$\mathcal{W}_p(B) = \mathcal{W}_{p,p}(B).$$

We can derive the type and cotype of $\mathcal{W}_{p,q}$ in two ways. The first one proves that the vector valued spaces $\mathcal{W}_{p,q}(L_r)$ satisfy the same kind of 'Hölder-Minkowski' inequality as the Lorentz spaces $\ell_{p,q}$, with the only exception of

$p = r$. This is the substance of Corollary 12.18 (and Corollary 12.27): we have the inclusion

$$\mathcal{W}_p(L_r) \subset L_r(\mathcal{W}_p) \quad \text{if} \quad r > p$$

and the reverse inclusion if $p < r$.

Another way to prove this (see Remark 12.28) goes through estimates of the K-functional for the pairs (v_1, ℓ_∞) and (v_r, ℓ_∞) for $1 < r < \infty$ (see Lemma 12.24). Indeed, by the reiteration theorem, we may identify $(v_1, \ell_\infty)_{\theta,q}$ and $(v_r, \ell_\infty)_{\theta',q}$ if $1 - \theta = (1 - \theta')/r$, and similarly in the Banach space valued case (see Theorem 12.25), we have

$$(v_1(B), \ell_\infty(B))_{\theta,q} = (v_r(B), \ell_\infty(B))_{\theta',q}.$$

We also use reiteration in Theorem 12.14 to describe the space $(v_r, \ell_\infty)_{\theta,q}$ for $0 < r < 1$. In the final Theorem 12.31, we give an alternate description of $\mathcal{W}_p = \mathcal{W}_{p,p}$, which should convince the reader that it is a very natural space (this is closely connected to 'splines' in approximation theory). The description is as follows: a sequence $x = (x_n)_n$ belongs to $\mathcal{W}_p$ iff $\sum_N S_N(x)^p < \infty$, where $S_N(x)$ is the distance in ℓ_∞ of x from the subspace of all sequences (y_n) such that $\operatorname{card}\{n \mid |y_n - y_{n-1}| \neq 0\} \leq N$.

In Chapter 12 we also include a discussion of the classical James space (usually denoted by J), which we denote by v_2^0. The spaces $\mathcal{W}_{p,q}$ are in many ways similar to the James space; in particular, if $1 < p, q < \infty$, they are of co-dimension 1 in their bidual (see Remark 12.8).

In §13.1 and §13.2 we present applications of a certain exponential inequality (due to Azuma) to concentration of measure for the symmetric groups and for the Hamming cube.

In Chapter 13 we give two characterizations of super-reflexive Banach spaces by properties of the underlying *metric spaces*. The relevant properties involve finite metric spaces. Given a sequence $\mathcal{T} = (T_n, d_n)$ of finite metric spaces, we say that the sequence $\mathcal{T}$ embeds Lipschitz uniformly in a metric space (T, d) if, for some constant C, there are subsets $\widetilde{T}_n \subset T$ and bijective mappings $f_n \colon T_n \to \widetilde{T}_n$ with Lipschitz norms satisfying

$$\sup_n \|f_n\|_{\mathrm{Lip}} \|f_n^{-1}\|_{\mathrm{Lip}} < \infty.$$

Consider for instance the case when T_n is a finite dyadic tree restricted to its first $1 + 2 + \cdots + 2^n = 2^{n+1} - 1$ points viewed as a graph and equipped with the usual geodesic distance. In Theorem 13.10 we prove following [138] that a Banach space B is super-reflexive iff it does not contain the sequence of these dyadic trees Lipschitz uniformly. More recently (cf. [289]), it was proved that the trees can be replaced in this result by the '*diamond graphs*'. We describe the analogous characterization with diamond graphs in §13.4.

In §13.5 we discuss several non-linear notions of 'type p' for metric spaces, notably the notion of Markov type p, and we prove the recent result from [357] that p-uniformly smooth implies Markov type p. The proof uses martingale inequalities for martingales naturally associated to Markov chains on finite state spaces.

In the final chapter, Chapter 14, we briefly outline the recent developments of 'non-commutative' martingale inequalities initiated in [401]. This is the subject of a second volume titled *Martingales in Non-Commutative L_p-Spaces* to follow the present one (to be published on the author's web page). We only give a glimpse of what this is about.

There the probability measure is replaced by a standard normalized trace τ on a von Neumann algebra A, the filtration becomes an increasing sequence (A_n) of von Neumann subalgebras of A, the space $L_p(\tau)$ is now defined as the completion of A equipped with the norm $x \mapsto (\tau(|x|^p))^{1/p}$ and the conditional expectation $\mathbb{E}_n$ with respect to A_n is the orthogonal projection from $L_2(\tau)$ onto the closure of A_n in $L_2(\tau)$. Although there is no analogue of the maximal function (in fact, there are no functions at all !), it turns out that there is a satisfactory non-commutative analogue (by duality) of Doob's inequality (see [292]). Moreover, Gundy's decomposition can be extended to this framework (see [373, 410, 411]). Thus one obtains a version of Burkholder's martingale transform inequalities. In other words, martingale difference sequences in non-commutative L_p-spaces ($1 < p < \infty$) are unconditional. This implies in particular that the latter spaces are UMD as Banach spaces. The Burkholder-Rosenthal and Stein inequalities all have natural generalizations to this setting.

1
Banach space valued martingales

We start by recalling the definition and basic properties of conditional expectations. Let $(\Omega, \mathcal{A}, \mathbb{P})$ be a probability space and let $\mathcal{B} \subset \mathcal{A}$ be a σ-subalgebra. On $L_2(\Omega, \mathcal{A}, \mathbb{P})$, the conditional expectation $\mathbb{E}^{\mathcal{B}}$ can be defined simply as the orthogonal projection onto the subspace $L_2(\Omega, \mathcal{B}, \mathbb{P})$. One then shows that it extends to a positive contraction on $L_p(\Omega, \mathcal{A}, \mathbb{P})$ for all $1 \leq p \leq \infty$, taking values in $L_p(\Omega, \mathcal{B}, \mathbb{P})$. The resulting operator $\mathbb{E}^{\mathcal{B}} : L_p(\Omega, \mathcal{A}, \mathbb{P}) \to L_p(\Omega, \mathcal{B}, \mathbb{P})$ is a linear projection and is characterized by the property

$$\forall f \in L_p(\Omega, \mathcal{A}, \mathbb{P}) \, \forall h \in L_\infty(\Omega, \mathcal{B}, \mathbb{P}) \qquad \mathbb{E}^{\mathcal{B}}(hf) = h\mathbb{E}^{\mathcal{B}}(f). \tag{1.1}$$

Here 'positive' really means positivity preserving, i.e. for any $f \in L_p$,

$$f \geq 0 \Rightarrow T(f) \geq 0.$$

As usual, we often abbreviate 'almost everywhere' by 'a.e.' and 'almost surely' by 'a.s.'

We will now consider conditional expectation operators on Banach space valued L_p-spaces.

1.1 Banach space valued L_p-spaces

Let $(\Omega, \mathcal{A}, m)$ be a measure space. Let B be a Banach space. We will denote by $F(B)$ the space of all measurable simple functions, i.e. the functions $f\colon \Omega \to B$ for which there is a partition of Ω, say, $\Omega = A_1 \cup \cdots \cup A_N$ with $A_k \in \mathcal{A}$, and elements $b_k \in B$ such that

$$\forall \omega \in \Omega \qquad f(\omega) = \sum\nolimits_1^N 1_{A_k}(\omega) b_k. \tag{1.2}$$

Equivalently, $F(B)$ is the space of all measurable functions $f\colon \Omega \to B$ taking only finitely many values.

Definition 1.1. We will say that a function $f\colon \Omega \to B$ is Bochner measurable if there is a sequence (f_n) in $F(B)$ tending to f pointwise.

Let $1 \le p \le \infty$. We denote by $L_p(\Omega, \mathcal{A}, m; B)$ the space of (equivalence classes of) Bochner measurable functions $f\colon \Omega \to B$ such that $\int \|f\|_B^p\, dm < \infty$ for $p < \infty$, and $\text{ess sup}\|f(\cdot)\|_B < \infty$ for $p = \infty$. As usual, two functions that are equal a.e. are identified. We equip this space with the norm

$$\|f\|_{L_p(B)} = \left(\int \|f\|_B^p\, dm\right)^{1/p} \quad \text{for} \quad p < \infty$$

$$\|f\|_{L_\infty(B)} = \text{ess sup}\|f(\cdot)\|_B \quad \text{for} \quad p = \infty,$$

with which it becomes a Banach space.

We will often use the following elementary consequence of Fubini's theorem:

$$\int \|f\|_B^p\, dm = \int_0^\infty p t^{p-1} m(\{\|f\|_B > t\})dt. \tag{1.3}$$

Of course, this definition of $L_p(B)$ coincides with the usual one in the scalar valued case, i.e. if $B = \mathbb{R}$ (or $\mathbb{C}$). In that case, we often denote simply by $L_p(\Omega, \mathcal{A}, m)$ (or sometimes $L_p(m)$, or even L_p) the resulting space of scalar valued functions.

For brevity, we will often write simply $L_p(m; B)$, or, if there is no risk of confusion, simply $L_p(B)$, instead of $L_p(\Omega, \mathcal{A}, m; B)$.

Given $\varphi_1, \ldots, \varphi_N \in L_p$ and $b_1, \ldots, b_N \in B$, we can define a function $f\colon \Omega \to B$ in $L_p(B)$ by setting $f(\omega) = \sum_1^N \varphi_k(\omega) b_k$. We will denote this function by $\sum_1^N \varphi_k \otimes b_k$ and by $L_p \otimes B$ the subspace of $L_p(B)$ formed of all such functions.

Proposition 1.2. *Let $1 \le p < \infty$. Each of the subspaces $F(B) \cap L_p(B)$ and $L_p \otimes B \subset L_p(B)$ is dense in $L_p(B)$. More generally, for any subspace $V \subset L_p$ dense in L_p, $V \otimes B$ is dense in $L_p(B)$.*

Proof. Consider $f \in L_p(B)$. Let $f_n \in F(B)$ be such that $f_n \to f$ pointwise. Then $\|f_n(\cdot)\|_B \to \|f(\cdot)\|_B$ pointwise, so that if we set $g_n(\omega) = f_n(\omega) 1_{\{\|f_n\| < 2\|f\|\}}$, we still have $g_n \to f$ pointwise and, in addition, $\sup_n \|g_n - f\| \le \sup_n \|g_n\| + \|f\| \le 3\|f\|$. Therefore, by dominated convergence, we must have $\int \|g_n - f\|_B^p\, dm \to 0$ and, of course, $g_n \in F(B) \cap L_p(B)$. This proves the first assertion. The second and third ones are then obvious since $F(B) \cap L_p(B) \subset L_p \otimes B$ (indeed, we can take $\varphi_k = 1_{A_k}$ with $m(A_k) < \infty$, as in (1.2)). □

Remark 1.3. If B is finite-dimensional, then $F(B)$ is dense in $L_\infty(B)$, but this is no longer true in the infinite-dimensional case, because the unit ball of B is not compact.

Remark 1.4. Let $(\Omega, \mathcal{A})$ be a compact space equipped with its Borel σ-algebra (e.g. $\mathbb{T}$) and a Radon measure m; then any continuous B-valued function on Ω is Bochner measurable and the space $C(\Omega; B)$ of all such functions is included in $L_\infty(\Omega, \mathcal{A}, m; B)$. Moreover, $C(\Omega) = C(\Omega; \mathbb{R})$ is dense in $L_p(m; \mathbb{R})$ for any $1 \leq p < \infty$. This remains true on a locally compact and σ-compact space (e.g. if $\Omega = \mathbb{R}$) provided $C(\Omega; B)$ is replaced by the space of compactly supported continuous functions.

We now turn to the definition of the integral of a function in $L_1(B)$. Consider a function f of the form (1.2) in $L_1(B) \cap F(B)$. We define

$$\int f\, dm = \sum\nolimits_1^N m(A_k) b_k.$$

This defines a continuous linear map from $L_1(B) \cap F(B)$ to B, since we have obviously, by the triangle inequality,

$$\left\| \int f\, dm \right\| \leq \sum m(A_k) \|b_k\| = \|f\|_{L_1(B)}.$$

By density, this linear map admits an extension defined on the whole of $L_1(B)$, which we still denote by $\int f\, dm$ when $f \in L_1(B)$. The extension clearly satisfies the following fundamental inequality, called Jensen's inequality:

$$\forall f \in L_1(B) \qquad \left\| \int f\, dm \right\|_B \leq \int \|f\|_B\, dm = \|f\|_{L_1(B)}. \qquad (1.4)$$

This extends the linear map $f \to \int f\, dm$ from the scalar valued case to the B-valued one. More generally, let $(\Omega', \mathcal{A}', m')$ be another measure space and let $T\colon\ L_1(\Omega, \mathcal{A}, m) \to L_1(\Omega', \mathcal{A}', m')$ be a bounded operator. We may clearly define unambiguously a linear operator $T_0\colon\ F(B) \cap L_1(m; B) \to L_1(m', B)$ by setting, for any f of the form (1.2),

$$T_0(f) = \sum\nolimits_1^N T(1_{A_k}) b_k.$$

We have clearly, by the triangle inequality,

$$\|T_0(f)\|_{L_1(m';B)} \leq \sum\nolimits_1^N \|T(1_{A_k})\| \|b_k\| \leq \|T\| \sum m(A_k) \|b_k\| = \|T\| \|f\|_{L_1(B)}.$$

Thus we can state the following:

Proposition 1.5. *Given a bounded operator $T\colon L_1(\Omega, \mathcal{A}, m) \to L_1(\Omega', \mathcal{A}', m')$, there is a unique bounded linear map $\widetilde{T}\colon L_1(\Omega, \mathcal{A}, m; B) \to L_1(\Omega', \mathcal{A}', m'; B)$ such that*

$$\forall \varphi \in L_1(\Omega, \mathcal{A}, m)\ \forall b \in B \qquad \widetilde{T}(\varphi \otimes b) = T(\varphi)b. \tag{1.5}$$

Moreover, we have $\|\widetilde{T}\| = \|T\|$.

Proof. By the density of $F(B) \cap L_1(B)$ in $L_1(B)$, the (continuous) map T_0 admits a unique continuous linear extension $\widetilde{T}$ from $L_1(m; B)$ to $L_1(m'; B)$, with $\|\widetilde{T}\| \le \|T_0\| \le \|T\|$. If φ is a simple function in L_1, then (1.5) is clear by definition of T_0. Approximating φ in L_1 by a simple function, we see that (1.5) is true in general. The unicity of $\widetilde{T}$ is clear since (1.5) implies that $\widetilde{T}$ coincides with T_0 on $F(B) \cap L_1(B)$. Finally, considering a fixed b with $\|b\| = 1$, we easily derive from (1.5) that $\|T\| \le \|\widetilde{T}\|$, so we obtain $\|T\| = \|\widetilde{T}\|$. □

It is not true in general that a bounded operator on L_p (or from L_p to L_q) extends boundedly to $L_p(B)$ as in the preceding proposition for $p = 1$. Nevertheless, it is true for *positive* operators, as follows:

Proposition 1.6. *Let $1 \le p, q \le \infty$. Let $(\Omega, \mathcal{A}, \mathbb{P})$ be an arbitrary measure space and let $T\colon\ L_p(\Omega) \to L_q(\Omega)$ be a bounded linear operator. Let*

$$T \otimes I_B\colon\ L_p(\Omega, \mathbb{P}) \otimes B \to L_q(\Omega, \mathbb{P}) \otimes B$$

be the unique linear operator such that

$$\forall \varphi \in L_p(\Omega, \mathbb{P}) \quad \forall x \in B \qquad (T \otimes I_B)(\varphi \otimes x) = T(\varphi) \otimes x.$$

If T is positive (i.e. if T preserves non-negative functions), then $T \otimes I_B$ extends to a bounded operator $\widetilde{T \otimes I_B}$ from $L_p(\Omega, \mathbb{P}; B)$ to $L_q(\Omega, \mathbb{P}; B)$, which has the same norm as T, i.e.

$$\|\widetilde{T \otimes I_B}\|_{L_p(B) \to L_q(B)} = \|T\|_{L_p \to L_q}.$$

Proof. It clearly suffices to show that

$$\forall f \in L_p(\Omega, \mathbb{P}) \otimes B \qquad \|(T \otimes I_B)f(\cdot)\|_B \overset{\text{a.s.}}{\le} T(\|f(\cdot)\|_B). \tag{1.6}$$

For that purpose, we can assume B separable (or even finite-dimensional) so that there is a countable subset $D \subset B^*$ verifying

$$\forall x \in B \qquad \|x\| = \sup_{\xi \in D} |\xi(x)|.$$

Clearly, for any ξ in B^*, we have

$$\langle \xi, (T \otimes I_B)f(\cdot) \rangle = T(\langle \xi, f(\cdot) \rangle),$$

and hence by the positivity of T for any finite subset $D' \subset D$,

$$\sup_{\xi \in D'} |\langle \xi, (T \otimes I_B) f(\cdot) \rangle| \overset{\text{a.s.}}{\leq} T(\sup_{\xi \in D} |\langle \xi, f(\cdot) \rangle|);$$

therefore we obtain (1.6), and the proposition follows. □

Remark. Let B_1 be another Banach space and let $u\colon\ B \to B_1$ be a bounded operator. Then, for any f in $L_p(\Omega, \mathcal{A}, \mathbb{P}; B)$, we have

$$\widetilde{T \otimes I_{B_1}}(u(f)) = u[\widetilde{T \otimes I_B}(f)].$$

In particular, if we set $g = \widetilde{T \otimes I_B}(f)$, then for any ξ in B^*, we have

$$T(\xi(f)) = \xi(g). \tag{1.7}$$

Indeed, this is immediately checked for f in $L_p(\Omega, \mathbb{P}) \otimes B$, and the general case is obtained after completion.

Note that now that $\widetilde{T \otimes I_B}$ makes sense, the preceding argument can be repeated to show that

$$\forall f \in L_p(\Omega, \mathbb{P}; B) \qquad \|(\widetilde{T \otimes I_B}) f\|_B \overset{\text{a.e.}}{\leq} T(\|f(\cdot)\|_B). \tag{1.8}$$

A priori, in the preceding (1.8), we implicitly assume that B is a real Banach space, but actually, if B is a complex space (and T is $\mathbb{C}$-linear on complex valued L_p), we may consider B a fortiori as a real space and then (1.8) remains valid.

Remark 1.7. Let $1 \leq p \leq \infty$. Let B_1, B_2 be Banach spaces. We will denote by

$$B_1 \oplus_p B_2$$

the direct sum equipped with the norm defined by

$$\forall (x_1, x_2) \in B_1 \oplus B_2 \quad \|(x_1, x_2)\| = (\|x_1\|^p + \|x_2\|^p)^{1/p}. \tag{1.9}$$

More generally, for any family $(B_i)_{i \in I}$ of Banach spaces, we denote by

$$\left(\oplus \sum\nolimits_{i \in I} B_i\right)_p$$

the direct sum, 'in the sense of ℓ_p', which means the subspace formed of those $x = (x(i)) \in \prod_{i \in I} B_i$ such that $\sum_{i \in I} \|x(i)\|^p < \infty$ equipped with the norm $x \mapsto (\sum_{i \in I} \|x(i)\|^p)^{1/p}$. When the spaces B_i are all identical to a space B, the latter space will be denoted by $\ell_p(I; B)$. When $I = \mathbb{N}$, we denote this simply by $\ell_p(B)$. Note that $\ell_p(B) = L_p(\mathbb{N}, \nu; B)$, where ν is the counting measure on $\mathbb{N}$.

Remark 1.8. Occasionally, mainly in Chapters 4 and 8, we will consider quasi-Banach spaces. Here is a brief discussion of this notion. When $0 < p < 1$, the space $E = L_p(m)$ over any measure m equipped with its usual distance

$d(f, g) = \|f - g\| = (\int |f - g|^p dm)^{1/p}$ is complete but fails the usual triangle inequality and is not even locally convex (except when finite-dimensional). However, for any $f, g \in E$, it is well known that

$$\|f + g\| \le (\|f\|^p + \|g\|^p)^{1/p}. \tag{1.10}$$

We call this the p-triangle inequality, and any space satisfying this is called p-normed. A space that is p-normed for some $0 < p < 1$ is often called quasi-normed. It suffices for this that there be a constant c such that $\|f + g\| \le c(\|f\| + \|g\|)$ for any $f, g \in E$. When it is complete, we refer to it as a quasi-Banach space. Clearly, for any normed (or merely p-normed) space B, the space $L_p(m; B)$, which can be defined as earlier, is again p-normed. Thus, for any quasi-Banach space B, the space $L_p(m; B)$ is a quasi-Banach space. We refer to [45] for more information on quasi-Banach spaces.

Remark 1.9. Let $(\Omega, \mathcal{A}, m)$ be a σ-finite measure space. We will denote by $L_0(\Omega, \mathcal{A}, m; B)$ (or simply $L_0(m; B)$) the set of equivalence classes (modulo equality almost everywhere) of Bochner measurable B-valued functions f. When m is finite, $L_0(\Omega, \mathcal{A}, m; B)$ is classically equipped with a metrizable topological vector space structure. One possible distance defining this topology is

$$\forall f, g \in L_0(m; B) \quad d(f, g) = \int \frac{\|f - g\|}{1 + \|f - g\|} dm.$$

Consider a sequence (f_n) in $L_0(m; B)$. Then we have

$$d(f_n, f) \to 0 \Leftrightarrow m(\{\|f_n - f\| > \varepsilon\}) \to 0 \quad \forall \varepsilon > 0.$$

When $m(\Omega) = \infty$, assuming m σ-finite, let $w \in L_1(m)$ be such that $w > 0$ a.e. so that the measure $m' = w \cdot m$ is finite and equivalent to m. We may obviously identify $L_0(m; B)$ with $L_0(m'; B)$. We then define the topological vector space structure of $L_0(m; B)$ by transplanting the one just defined on $L_0(m'; B)$. Then f_n tends to f in this topology iff, for any measurable subset $A \subset \Omega$ with $m(A) < \infty$, the restriction of f_n to A converges in measure to $f_{|A}$ in the preceding sense (this shows in particular that this is independent of the choice of w). We then say that f_n tends to f 'in measure' (or 'in probability' if $m(\Omega) = 1$). It is sufficient (resp. necessary) for this that f_n tends to f almost everywhere (resp. that (f_n) admits a subsequence converging a.e. to f).

In conclusion, for any σ-finite measure space, and any $0 < p \le \infty$, we have a continuous injective linear embedding $L_p(m; B) \subset L_0(m; B)$. All these facts are easily checked.

1.2 Banach space valued conditional expectation

In particular, Proposition 1.6 applied with $p = q$, is valid for $T = \mathbb{E}^{\mathcal{B}}$. For any f in $L_1(\Omega, \mathcal{A}, \mathbb{P}; B)$, we denote again simply by $\mathbb{E}^{\mathcal{B}}(f)$ the function $\widetilde{T \otimes I_B}(f)$ for $T = \mathbb{E}^{\mathcal{B}}$.

Proposition 1.10. *Let $(\Omega, \mathcal{A}, \mathbb{P})$ be a probability space. The conditional expectation $f \mapsto \mathbb{E}^{\mathcal{B}}(f)$ is an operator of norm 1 on $L_p(B)$ for all $1 \leq p \leq \infty$, satisfying*

$$\forall f \in L_p(\Omega, \mathcal{A}, \mathbb{P}; B)\ \forall h \in L_\infty(\Omega, \mathcal{B}, \mathbb{P}) \quad \mathbb{E}^{\mathcal{B}}(hf) = h\mathbb{E}^{\mathcal{B}}(f). \tag{1.11}$$

The operator $\mathbb{E}^{\mathcal{B}}$ is a norm 1 projection from $L_p(\Omega, \mathcal{A}, \mathbb{P}; B)$ to $L_p(\Omega, \mathcal{B}, \mathbb{P}; B)$, viewed as a subspace of $L_p(\Omega, \mathcal{A}, \mathbb{P}; B)$.

Proof. By (1.1), Proposition 1.6 applied to $T = \mathbb{E}^{\mathcal{B}}$ with $p = q$ produces an operator P of norm 1 on $L_p(\Omega, \mathcal{A}, \mathbb{P}; B)$, such that $P(hx) = hP(x)$ for any $x \in L_p(\mathcal{A}, \mathbb{P}) \otimes B$ and $h \in L_\infty(\Omega, \mathcal{B}, \mathbb{P})$ and also such that $P(x) = x$ for any $x \in L_p(\mathcal{B}, \mathbb{P}) \otimes B$. Therefore, by density, $P(hx) = hP(x)$ holds for any $x \in L_p(\Omega, \mathcal{A}, \mathbb{P}; B)$. Also $P(x) = x$ holds for any $x \in \overline{L_p(\mathcal{B}, \mathbb{P}) \otimes B} \subset L_p(\Omega, \mathcal{A}, \mathbb{P}; B)$. Since we may clearly identify the latter subspace with $L_p(\Omega, \mathcal{B}, \mathbb{P}; B)$, the statement follows. □

Remark 1.11. Let $1 \leq p, p' \leq \infty$ such that $p^{-1} + p'^{-1} = 1$. Let $f \in L_p(\Omega, \mathcal{A}, \mathbb{P}; B)$. Note that $g = \mathbb{E}^{\mathcal{B}}(f) \in L_p(\Omega, \mathcal{B}, \mathbb{P}; B)$ satisfies the following:

$$\forall E \in \mathcal{B} \quad \int_E g d\mathbb{P} = \int_E f d\mathbb{P}. \tag{1.12}$$

In particular, $\mathbb{E}(g) = \mathbb{E}(f)$. More generally,

$$\forall h \in L_{p'}(\Omega, \mathcal{B}, \mathbb{P}) \quad \int hg d\mathbb{P} = \int hf d\mathbb{P}, \tag{1.13}$$

and also

$$\forall h \in L_{p'}(\Omega, \mathcal{A}, \mathbb{P}) \quad \int hg d\mathbb{P} = \int (\mathbb{E}^{\mathcal{B}} h) g d\mathbb{P}. \tag{1.14}$$

Indeed (among many other ways to verify these equalities), they are easy to check by 'scalarization', by a simple deduction from the scalar case. More precisely, for each of these identities, we have to check the equality of two vectors in B, say, $x = y$, and this is the same as $\xi(x) = \xi(y)$ for all ξ in B^*. But we know by (1.7) that $\langle \xi, g(\cdot) \rangle = \mathbb{E}^{\mathcal{B}} \langle \xi, f \rangle$, so to check (1.12)–(1.14), it suffices to check them for $B = \mathbb{R}$ (or $B = \mathbb{C}$ in the case of complex scalars) and for all scalar valued f. But in that case, they are easy to deduce from the basic (1.1). Indeed, by the density of L_∞ in $L_{p'}$, (1.1) remains valid when $h \in L_{p'}$, so we derive (1.12) and (1.13) from it. To check (1.14), inverting the roles of g and h, we find by (1.12) with $E = \Omega$ (and with f replaced by hg) that

$$\int hg d\mathbb{P} = \int \mathbb{E}^{\mathcal{B}}(hg) d\mathbb{P} = \int (\mathbb{E}^{\mathcal{B}} h) g d\mathbb{P}.$$

In the sequel, we will refer to the preceding way to check a vector valued functional identity as the *scalarization principle.*

We will sometimes use the following generalization of (1.8).

Proposition 1.12. *Let $f \in L_1(\Omega, \mathcal{A}, \mathbb{P}; B)$. For any continuous convex function $\varphi : B \to \mathbb{R}$ such that $\varphi(f) \in L_1$, we have*

$$\varphi(\mathbb{E}^{\mathcal{B}}(f)) \le \mathbb{E}^{\mathcal{B}} \varphi(f). \tag{1.15}$$

In particular,

$$\|\mathbb{E}^{\mathcal{B}}(f)\| \le \mathbb{E}^{\mathcal{B}} \|f\|. \tag{1.16}$$

Proof. We may clearly assume B separable. Recall that a real valued convex function on B is continuous iff it is locally bounded above ([82, p. 93]). By Hahn-Banach, we have $\varphi(x) = \sup_{a \in C} a(x)$, where C is the collection of all continuous real valued affine functions a on B such that $a \le \varphi$. By the separability of B, we can find a countable subcollection C' such that $\varphi(x) = \sup_{a \in C'} f(x)$.

By (1.7), we have $a(\mathbb{E}^{\mathcal{B}}(f)) = \mathbb{E}^{\mathcal{B}}(a(f))$ for any $a \in B^*$, and hence, also for any affine continuous function a on B (since this is obvious for constant functions). Thus we have $a(\mathbb{E}^{\mathcal{B}} f) = \mathbb{E}^{\mathcal{B}} a(f) \le \mathbb{E}^{\mathcal{B}} \varphi(f)$ for any $a \in C'$, and hence, taking the sup over $a \in C'$, we find (1.15). □

Remark 1.13. To conform with the tradition, we assumed that $\mathbb{P}$ is a probability in Corollary 1.10. However, essentially the same results remain valid if we merely assume that the restriction of $\mathbb{P}$ to $\mathcal{B}$ is σ-finite (or that $L_\infty \cap L_1$ is weak*-dense in L_∞ both for $\mathcal{B}$ and $\mathcal{A}$). Indeed, we can still define $\mathbb{E}^{\mathcal{B}}$ on $L_2(\Omega, \mathcal{A}, \mathbb{P})$ as the orthogonal projection onto $L_2(\Omega, \mathcal{B}, \mathbb{P})$, and although it is rarely used, it is still true that the resulting operator on $L_p(\Omega, \mathcal{A}, \mathbb{P})$ preserves positivity and extends to an operator of norm 1 on $L_p(\Omega, \mathcal{A}, \mathbb{P}; B)$ for all $1 \le p \le \infty$, satisfying (1.11), (1.13) and (1.14). See Remark 6.20 for an illustration.

Remark. By classical results due to Ron Douglas [213] and T. Ando [102], conditional expectations can be characterized as the only norm 1 projections on L_p ($1 \le p \ne 2 < \infty$) that preserve the constant function 1. This is also true for $p = 2$ if one restricts to positivity-preserving operators.

1.3 Martingales: basic properties

Let B be a Banach space. Let $(\Omega, \mathcal{A}, \mathbb{P})$ be a probability space. A sequence $(M_n)_{n\geq 0}$ in $L_1(\Omega, \mathcal{A}, \mathbb{P}; B)$ is called a martingale if there exists an increasing sequence of σ-subalgebras $\mathcal{A}_0 \subset \mathcal{A}_1 \subset \cdots \subset \mathcal{A}_n \subset \cdots \subset \mathcal{A}$ (this is called 'a filtration') such that for each $n \geq 0$, M_n is $\mathcal{A}_n$-measurable and satisfies

$$M_n = \mathbb{E}^{\mathcal{A}_n}(M_{n+1}). \tag{1.17}$$

This implies, of course, that

$$\forall n < m \qquad M_n = \mathbb{E}^{\mathcal{A}_n} M_m.$$

In particular, if (M_n) is a B-valued martingale, (1.12) yields

$$\forall n \leq m \quad \forall A \in \mathcal{A}_n \quad \int_A M_n d\mathbb{P} = \int_A M_m d\mathbb{P}. \tag{1.18}$$

A sequence of random variables (M_n) is called 'adapted to the filtration' $(\mathcal{A}_n)_{n\geq 0}$ if M_n is $\mathcal{A}_n$-measurable for each $n \geq 0$. Note that the martingale property $M_n = \mathbb{E}^{\mathcal{A}_n}(M_{n+1})$ automatically implies that (M_n) is adapted to $(\mathcal{A}_n)$. Of course, the minimal choice of a filtration to which (M_n) is adapted is simply the filtration $\mathcal{M}_n = \sigma(M_0, M_1, \ldots, M_n)$. Moreover, if (M_n) is a martingale in the preceding sense with respect to some filtration $(\mathcal{A}_n)$, then it is a fortiori a martingale with respect to $(\mathcal{M}_n)$. Indeed, we have obviously $\mathcal{M}_n \subset \mathcal{A}_n$ for all n, and hence applying $\mathbb{E}^{\mathcal{M}_n}$ to both sides of (1.17) implies $M_n = \mathbb{E}^{\mathcal{M}_n}(\mathbb{E}^{\mathcal{A}_n} M_{n+1}) = \mathbb{E}^{\mathcal{M}_n} M_{n+1}$.

An adapted sequence of random variables (M_n) is called 'predictable' if M_n is $\mathcal{A}_{n-1}$-measurable for each $n \geq 1$. Of course, the predictable sequences of interest to us will not be martingales, since predictable martingales must form a constant sequence.

We will also need the definition of a submartingale. A sequence $(M_n)_{n\geq 0}$ of real valued random variables in L_1 is called a submartingale if there are σ-subalgebras $\mathcal{A}_n$, as previously, such that M_n is $\mathcal{A}_n$-measurable and

$$\forall n \geq 0 \qquad M_n \leq \mathbb{E}^{\mathcal{A}_n} M_{n+1}.$$

This implies, of course, that

$$\forall n < m \qquad M_n \leq \mathbb{E}^{\mathcal{A}_n} M_m.$$

For example, if (M_n) is a B-valued martingale in $L_1(B)$, then, for any continuous convex function $\varphi : B \to \mathbb{R}$ such that $\varphi(M_n) \in L_1$ for all n, the sequence $(\varphi(M_n))$ is a submartingale. Indeed, by (1.15), we have

$$\varphi(M_n) \leq \mathbb{E}^{\mathcal{A}_n} \varphi(M_{n+1}). \tag{1.19}$$

A fortiori, taking the expectation of both sides, we have (for future reference)

$$\mathbb{E}\varphi(M_n) \le \mathbb{E}\varphi(M_{n+1}). \tag{1.20}$$

More generally, if I is any partially ordered set, then a collection $(M_i)_{i\in I}$ in $L_1(\Omega, \mathbb{P}; B)$ is called a martingale (indexed by I) if there are σ-subalgebras $\mathcal{A}_i \subset \mathcal{A}$ such that $\mathcal{A}_i \subset \mathcal{A}_j$ whenever $i < j$ and $M_i = \mathbb{E}^{\mathcal{A}_i} M_j$.

In particular, when the index set is

$$I = \{0, -1, -2, \ldots\},$$

the corresponding sequence is usually called a 'reverse martingale'.

The following convergence theorem is fundamental.

Theorem 1.14. *Let $(\mathcal{A}_n)$ be a fixed increasing sequence of σ-subalgebras of $\mathcal{A}$. Let $\mathcal{A}_\infty$ be the σ-algebra generated by $\bigcup_{n\ge 0} \mathcal{A}_n$. Let $1 \le p < \infty$ and consider M in $L_p(\Omega, \mathbb{P}; B)$. Let us define $M_n = \mathbb{E}^{\mathcal{A}_n}(M)$. Then $(M_n)_{n\ge 0}$ is a martingale such that $M_n \to \mathbb{E}^{\mathcal{A}_\infty}(M)$ in $L_p(\Omega, \mathbb{P}; B)$ when $n \to \infty$.*

Proof. Note that since $\mathcal{A}_n \subset \mathcal{A}_{n+1}$, we have $\mathbb{E}^{\mathcal{A}_n}\mathbb{E}^{\mathcal{A}_{n+1}} = \mathbb{E}^{\mathcal{A}_n}$, and similarly, $\mathbb{E}^{\mathcal{A}_n}\mathbb{E}^{\mathcal{A}_\infty} = \mathbb{E}^{\mathcal{A}_n}$. Replacing M by $\mathbb{E}^{\mathcal{A}_\infty} M$, we can assume without loss of generality that M is $\mathcal{A}_\infty$-measurable. We will use the following fact: the union $\bigcup_n L_p(\Omega, \mathcal{A}_n, \mathbb{P}; B)$ is dense in $L_p(\Omega, \mathcal{A}_\infty, \mathbb{P}; B)$. Indeed, let $\mathcal{C}$ be the class of all sets A such that $1_A \in \overline{\bigcup_n L_\infty(\Omega, \mathcal{A}_n, \mathbb{P})}$, where the closure is meant in $L_p(\Omega, \mathbb{P})$ (recall $p < \infty$). Clearly $\mathcal{C} \supset \bigcup_{n\ge 0} \mathcal{A}_n$ and $\mathcal{C}$ is a σ-algebra, hence $\mathcal{C} \supset \mathcal{A}_\infty$. This gives the scalar case version of the preceding fact. Now, any f in $L_p(\Omega, \mathcal{A}_\infty, \mathbb{P}; B)$ can be approximated (by definition of the spaces $L_p(B)$) by functions of the form $\sum_1^n 1_{A_i} x_i$ with $x_i \in B$ and $A_i \in \mathcal{A}_\infty$. But since $1_{A_i} \in \overline{\bigcup_n L_\infty(\Omega, \mathcal{A}_n, \mathbb{P})}$, we clearly have $f \in \overline{\bigcup_n L_p(\Omega, \mathcal{A}_n, \mathbb{P}; B)}$, as announced.

We can now prove Theorem 1.14. Let $\varepsilon > 0$. By the preceding fact, there is an integer k and g in $L_p(\Omega, \mathcal{A}_k, \mathbb{P}; B)$ such that $\|M - g\|_p < \varepsilon$. We have then $g = \mathbb{E}^{\mathcal{A}_n} g$ for all $n \ge k$, hence

$$\forall n \ge k \qquad M_n - M = \mathbb{E}^{\mathcal{A}_n}(M - g) + g - M$$

and, finally,

$$\begin{aligned}\|M_n - M\|_p &\le \|\mathbb{E}^{\mathcal{A}_n}(M - g)\|_p + \|g - M\|_p \\ &\le 2\varepsilon.\end{aligned}$$

This completes the proof. □

Definition 1.15. Let B be a Banach space and let $(M_n)_{n\ge 0}$ be a sequence in $L_1(\Omega, \mathcal{A}, \mathbb{P}; B)$. We will say that (M_n) is uniformly integrable if the sequence of non-negative random variables $(\|M_n(\cdot)\|)_{n\ge 0}$ is uniformly integrable. More precisely, this means that $(\|M_n\|)$ is bounded in L_1 and that, for any $\varepsilon > 0$, there is a $\delta > 0$ such that

$$\forall A \in \mathcal{A} \qquad \mathbb{P}(A) < \delta \Rightarrow \sup_{n \geq 0} \int_A \|M_n\| < \varepsilon.$$

In the scalar valued case, it is well known that a subset of L_1 is weakly relatively compact iff it is uniformly integrable.

Corollary 1.16. *In the scalar case (or the finite-dimensional case), every martingale that is bounded in L_p for some $1 \leq p < \infty$ and is uniformly integrable if $p = 1$ is actually convergent in L_p to a limit M_∞ such that $M_n = \mathbb{E}^{\mathcal{A}_n} M_\infty$ for all $n \geq 0$.*

Proof. Let (M_{n_k}) be a subsequence converging weakly to a limit, which we denote by M_∞. Clearly $M_\infty \in L_p(\Omega, \mathcal{A}_\infty, \mathbb{P})$, and we have $\forall A \in \mathcal{A}_n$,

$$\int_A M_\infty d\mathbb{P} = \lim \int_A M_{n_k} d\mathbb{P},$$

but whenever $n_k \geq n$, we have $\int_A M_{n_k} d\mathbb{P} = \int_A M_n d\mathbb{P}$ by the martingale property. Hence

$$\forall A \in \mathcal{A}_n \qquad \int_A M_\infty d\mathbb{P} = \int_A M_n d\mathbb{P},$$

which forces $M_n = \mathbb{E}^{\mathcal{A}_n} M_\infty$. We then conclude by Theorem 1.14 that $M_n \to M_\infty$ in L_p-norm. □

Note that, conversely, any martingale that converges in L_1 is clearly uniformly integrable.

Remark 1.17. Fix $1 \leq p < \infty$. Let I be a directed set, with order denoted simply by $\leq$. This means that, for any pair i, j in I, there is $k \in I$ such that $i \leq k$ and $j \leq k$. Let $(\mathcal{A}_i)$ be a family of σ-algebras directed by inclusion (i.e. we have $\mathcal{A}_i \subset \mathcal{A}_j$ whenever $i \leq j$). The extension of the notion of martingale is obvious: a collection of random variables $(f_i)_{i \in I}$ in $L_p(B)$ will be called a martingale if $f_i = \mathbb{E}^{\mathcal{A}_i}(f_j)$ holds whenever $i \leq j$. The resulting net converges in $L_p(B)$ iff, for any increasing sequence $i_1 \leq \cdots \leq i_n \leq i_{n+1} \leq \cdots$, the (usual sense) martingale (f_{i_n}) converges in $L_p(B)$. Indeed, this merely follows from the metrizability of $L_p(B)$! More precisely, if we assume that $\sigma\left(\bigcup_{i \in I} \mathcal{A}_i\right) = \mathcal{A}$, then, for any f in $L_p(\Omega, \mathcal{A}, \mathbb{P}; B)$, the directed net $(\mathbb{E}^{\mathcal{A}_i} f)_{i \in I}$ converges to f in $L_p(B)$. Indeed, this net must satisfy the Cauchy criterion, because otherwise we would be able for some $\delta > 0$ to construct (by induction) an increasing sequence $i(1) \leq i(2) \leq \cdots$ in I such that $\|\mathbb{E}^{\mathcal{A}_{i(k)}} f - \mathbb{E}^{\mathcal{A}_{i(k-1)}} f\|_{L_p(B)} > \delta$ for all

$k > 1$, and this would then contradict Theorem 1.14. Thus $\mathbb{E}^{\mathcal{A}_i} f$ converges to a limit F in $L_p(B)$, and hence, for any set $A \subset \Omega$ in $\bigcup_{j \in I} \mathcal{A}_j$, we must have

$$\int_A f = \lim_i \int_A \mathbb{E}^{\mathcal{A}_i} f = \int_A F.$$

Since the equality $\int_A f = \int_A F$ must remain true on the σ-algebra generated by $\bigcup \mathcal{A}_j$, we conclude that $f = F$, thus completing the proof that $\mathbb{E}^{\mathcal{A}_i} f \to f$ in $L_p(B)$.

1.4 Examples of filtrations

The most classical example of filtration is the one associated to a sequence of independent (real valued) random variables $(Y_n)_{n \geq 1}$ on a probability space $(\Omega, \mathcal{A}, \mathbb{P})$. Let $\mathcal{A}_n = \sigma(Y_1, \ldots, Y_n)$ for all $n \geq 1$ and $\mathcal{A}_0 = \{\phi, \Omega\}$. In that case, a sequence of random variables $(f_n)_{n \geq 0}$ is adapted to the filtration $(\mathcal{A}_n)_{n \geq 0}$ iff f_0 is constant and, for each $n \geq 1$, f_n depends only on $Y_1, \ldots, Y_n$, i.e. there is a (Borel-measurable) function F_n on $\mathbb{R}^n$ such that

$$f_n = F_n(Y_1, \ldots, Y_n).$$

The martingale condition can then be written as

$$\forall n \geq 0 \qquad F_n(Y_1, \ldots, Y_n) = \int F_{n+1}(Y_1, \ldots, Y_n, y)\, d\mathbb{P}_{n+1}(y),$$

where $\mathbb{P}_{n+1}$ is the probability distribution (or 'the law') of Y_{n+1}.

An equivalent but more 'intrinsic' model arises when one considers $\Omega = \mathbb{R}^{\mathbb{N}_*}$ equipped with the product probability measure $\mathbb{P} = \bigotimes_{n \geq 1} \mathbb{P}_n$. If one denotes by $Y = (Y_n)_{n \geq 1}$ a generic point in Ω, the random variable $Y \to Y_n$ appears as the nth coordinate, and $Y \to F_n(Y)$ is $\mathcal{A}_n$-measurable iff $F_n(Y)$ depends only on the n first coordinates of Y.

The *dyadic filtration* $(\mathcal{D}_n)_{n \geq 0}$ on $\Omega = \{-1, 1\}^{\mathbb{N}_*}$ is the fundamental example of this kind: here we denote by

$$\varepsilon_n \colon\ \Omega \to \{-1, 1\} \qquad (n = 1, 2, \ldots)$$

the nth coordinate, we equip Ω with the probability measure $\mathbb{P} = \otimes(\delta_1 + \delta_{-1})/2$, and we set $\mathcal{D}_n = \sigma(\varepsilon_1, \ldots, \varepsilon_n)$, $\mathcal{D}_0 = \{\phi, \Omega\}$. We denote by $\mathcal{D}$ (and sometimes by $\mathcal{D}_\infty$) the σ-algebra generated by $\cup_n \mathcal{D}_n$. Clearly the variables (ε_n) are independent and take the values ± 1 with equal probability $1/2$.

Note that $\mathcal{D}_n$ admits exactly 2^n atoms and, moreover, $\dim L_2(\Omega, \mathcal{D}_n, \mathbb{P}) = 2^n$. For any finite subset $A \subset [1, 2, \ldots]$, let $w_A = \prod_{n \in A} \varepsilon_n$ (up to indexation, these are called Walsh functions) with the convention $w_\phi \equiv 1$. It is easy to check that $\{w_A \mid A \subset [1, \ldots, n]\}$ (resp. $\{w_A \mid |A| < \infty\}$) is an orthonormal basis of $L_2(\Omega, \mathcal{D}_n, \mathbb{P})$ (resp. $L_2(\Omega, \mathcal{D}, \mathbb{P})$).

Given a Banach space B, a B-valued martingale $f_n\colon\ \Omega \to B$ adapted to the dyadic filtration $(\mathcal{D}_n)$ is characterized by the property that

$$\forall n \geq 1 \qquad (f_n - f_{n-1})(\varepsilon_1, \ldots, \varepsilon_n) = \varepsilon_n \varphi_{n-1}(\varepsilon_1, \ldots, \varepsilon_{n-1}),$$

where φ_{n-1} depends only on $\varepsilon_1, \ldots, \varepsilon_{n-1}$. We leave the easy verification of this to the reader.

Of course, the preceding remarks remain valid if one works with any sequence of ± 1 valued independent random variables (ε_n) such that $\mathbb{P}(\varepsilon_n = \pm 1) = 1/2$ on an 'abstract' probability space $(\Omega, \mathbb{P})$.

In classical analysis it is customary to use the Rademacher functions $(r_n)_{n \geq 1}$ on the Lebesgue interval $([0, 1], dt)$ instead of (ε_n). We need some notation to introduce these together with the Haar system.

Given an interval $I \subset \mathbb{R}$, we divide I into parts of equal length, and we denote by I^+ and I^-, respectively, the left and right half of I. Note that we do not specify whether the endpoints belong to I since the latter are negligible for the Lebesgue measure on $[0, 1]$ (or $[0, 1[$ or $\mathbb{R}$). Actually, in the Lebesgue measure context, whenever convenient, we will identify $[0, 1]$ with $[0, 1)$ or $(0, 1)$. Let

$$h_I = 1_{I^+} - 1_{I^-}.$$

We denote $I_1(1) = [0, 1)$, $I_2(1) = [0, \frac{1}{2})$, $I_2(2) = [\frac{1}{2}, 1)$ and, more generally,

$$I_n(k) = \left[\frac{k-1}{2^{n-1}}, \frac{k}{2^{n-1}}\right)$$

for $k = 1, 2, \ldots, 2^{n-1}$ $(n \geq 1)$. We then set $h_1 \equiv 1$, $h_2 = h_{I_1(1)}$, $h_3 = \sqrt{2}\, h_{I_2(1)}$, $h_4 = \sqrt{2}\, h_{I_2(2)}$ and, more generally,

$$\forall n \geq 1\ \forall k = 1, \ldots, 2^{n-1} \qquad h_{2^{n-1}+k} = |I_n(k)|^{-1/2} h_{I_n(k)}.$$

Note that $\|h_n\|_2 = 1$ for all $n \geq 1$.

The Rademacher function r_n can be defined, for each $n \geq 1$, by $r_n(t) = \operatorname{sign}(\sin(2^n \pi t))$ (in some books this same function is denoted r_{n-1}). Equivalently,

$$r_n = \sum\nolimits_{k=1}^{2^{n-1}} h_{I_n(k)}.$$

Then the sequence $(r_n)_{n \geq 1}$ has the same distribution on $([0, 1], dt)$ as the sequence $(\varepsilon_n)_{n \geq 1}$ on $(\Omega, \mathbb{P})$. Let $\mathcal{A}_n = \sigma(r_1, \ldots, r_n)$. Then $\mathcal{A}_n$ is generated by the 2^n-atoms $\{I_{n+1}(k) \mid 1 \leq k \leq 2^n\}$, each having length 2^{-n}. The dimension of

$L_2([0,1], \mathcal{A}_n)$ is 2^n and the functions $\{h_1, \ldots, h_{2^n}\}$ (resp. $\{h_n \mid n \geq 1\}$) form an orthonormal basis of $L_2([0,1], \mathcal{A}_n)$ (resp. $L_2([0,1])$).

The *Haar filtration* $(\mathcal{B}_n)_{n\geq 1}$ on $[0,1]$ is defined by

$$\mathcal{B}_n = \sigma(h_1, \ldots, h_n)$$

so that we have $\sigma(h_1, \ldots, h_{2^n}) = \sigma(r_1, \ldots, r_n)$ or, equivalently, $\mathcal{B}_{2^n} = \mathcal{A}_n$ for all $n \geq 1$ (note that here $\mathcal{B}_1$ is trivial). It is easy to check that $\mathcal{B}_n$ is an atomic σ-algebra with exactly n atoms. Since the conditional expectation $\mathbb{E}^{\mathcal{B}_n}$ is the orthogonal projection from L_2 to $L_2(\mathcal{B}_n)$, we have, for any f in $L_2([0,1])$,

$$\forall n \geq 1 \qquad \mathbb{E}^{\mathcal{B}_n} f = \sum\nolimits_1^n \langle f, h_k \rangle h_k,$$

and hence for all $n \geq 2$,

$$\mathbb{E}^{\mathcal{B}_n} f - \mathbb{E}^{\mathcal{B}_{n-1}} f = \langle f, h_n \rangle h_n. \tag{1.21}$$

More generally, for any B-valued martingale $(f_n)_{n\geq\emptyset}$ adapted to $(\mathcal{B}_n)_{n\geq 1}$, we have

$$\forall n \geq 2 \qquad f_n - f_{n-1} = h_n x_n$$

for some sequence (x_n) in B.

The Haar functions are in some sense the first example of *wavelets* (see e.g. [75, 98]). Indeed, if we set

$$h = 1_{[0,\frac{1}{2})} - 1_{[\frac{1}{2},1)}$$

(this is the same as the function previously denoted by h_2), then the system of functions

$$\left\{2^{\frac{m}{2}} h((t+k)2^m) \mid k, m \in \mathbb{Z}\right\} \tag{1.22}$$

is an orthonormal basis of $L_2(\mathbb{R})$. Note that the constant function 1 can be omitted since it is not in $L_2(\mathbb{R})$. This is sometimes called *the Haar wavelet*. See §6.3 for an illustration of this connection.

In the system (1.22), the sequence $\{h_n \mid n \geq 2\}$ coincides with the subsystem formed of all functions in (1.22) with support included in $[0,1]$.

Remark 1.18. In the dyadic filtration, each atom of $\mathcal{D}_{n-1}$ is split into half to give rise to two atoms of $\mathcal{D}_n$ with equal probability. Consider now on $(\Omega, \mathbb{P})$ a filtration $(\mathcal{T}_n)$ where each atom of $\mathcal{T}_{n-1}$ is partitioned into at most $m(n)$ disjoint sets in $\mathcal{T}_n$ of unequal probability. We assume that the partition obtained by partitioning all the atoms of $\mathcal{T}_{n-1}$ generates $\mathcal{T}_n$. Assuming that $\mathcal{T}_0 = \{\phi, \Omega\}$, we find that the number of atoms of $\mathcal{T}_n$ is at most $m(1)\cdots m(n)$. If one wishes to, one can realize this filtration on $[0,1]$ simply by partitioning it into at most

$m(1)$ intervals in one-to-one correspondence with the partition of Ω that generates $\mathcal{T}_1$. Then one continues in the same way for each interval in the resulting partition.

We would like to point out to the reader that the resulting picture is a very close approximation of the general case. In fact, let B be any Banach space and let (f_n) be a B-valued martingale with respect to an arbitrary filtration $(\mathcal{A}_n)$. If (f_n) is formed of simple functions, then (f_n) is actually adapted to a filtration $(\mathcal{T}_n)$ of the kind just described. Indeed, we may replace $\mathcal{A}_n$ by $\mathcal{T}_n = \sigma(f_0, \ldots, f_n)$, since this obviously preserves the martingale property, and then for each n, let $T_{n-1} \subset B^n$ be the finite subset that is the range of $(f_0, \ldots, f_{n-1})$. For each $x \in T_{n-1}$ we set

$$A_x = \{\omega \mid (f_0, \ldots, f_{n-1})(\omega) = x\},$$

with the convention that negligible sets are eliminated, so that $\mathbb{P}(A_x) > 0$. Then $(A_x)_{x \in T_{n-1}}$ forms a finite partition of Ω, generating $\mathcal{T}_{n-1}$, and each A_x is partitioned into the sets

$$A_x \cap \{f_n = y\} \quad y \in f_n(\Omega).$$

If we denote by $m(n)$ a common upper bound, valid for all x with $\mathbb{P}(A_x) > 0$ for the number of $y \in f_n(\Omega)$ with $\mathbb{P}(A_x \cap \{f_n = y\}) > 0$, then we find that $(\mathcal{T}_n)$ is of the kind just described. Note, however, that the number of atoms of $\mathcal{T}_n$ of which A_x is the union (i.e. the number of such ys) depends on x.

The dyadic case is the special case of $(\mathcal{T}_n)$ when $m(n) = 2$ for all n and the partitions are into sets of equal probability.

When f_0 is constant equal to, say, x_ϕ, one can picture the values of the martingale as the vertices of a *rooted tree* with root at x_ϕ, in such a way that the vertices in the n th level of the tree are the values of f_n and each vertex is in the convex hull of its 'children'. Indeed, the martingale property reduces to

$$\forall x \in T_{n-1} \quad x = \sum\nolimits_{(x,y) \in T_n} \frac{\mathbb{P}(A_x \cap \{f_n = y\})}{\mathbb{P}(A_x)} y.$$

In the dyadic case, the tree is regular of degree 2, and in the case of $(\mathcal{T}_n)$ just described, the degree is at most $m(n)$ for the vertices of the $(n-1)$th generation.

As we will show in Lemma 5.41, the martingales formed of simple functions just described are sort of 'dense' in the set of all martingales.

Remark 1.19. We can refine $(\mathcal{T}_n)$ just like we did earlier for $(\mathcal{A}_n)$ with the Haar filtration $(\mathcal{B}_k)$. There is a filtration $(\mathcal{C}_k)$ and $0 = K(0) < K(1) < K(2) < \cdots$ such that $\mathcal{C}_{K(n)} = \mathcal{T}_n$ for any $n \geq 0$ and such that, for each $k \geq 0$, $\mathcal{C}_{k+1}$ has at most one more atom than $\mathcal{C}_k$.

More precisely, assuming that $\mathcal{T}_0$ is trivial, i.e. it has a single atom, and that $\mathcal{T}_1$ has $k(1)+1$ atoms ($k(1) \geq 1$), enumerated as $A_1, \ldots, A_{k(1)}, A_{k(1)+1}$, we define the σ-algebras $\mathcal{C}_1, \ldots, \mathcal{C}_{k(1)}$ such that $\mathcal{T}_0 = \mathcal{C}_0 \subset \mathcal{C}_1 \subset \cdots \subset \mathcal{C}_{k(1)} = \mathcal{T}_1$ as follows. For $1 \leq j \leq k(1)$, we define the σ-algebra $\mathcal{C}_j$ as the one generated by $A_1, \ldots, A_j$. Thus the list of its $(j+1)$ atoms is $A_1, \ldots, A_j, \Omega \setminus (A_1 \cup \cdots \cup A_j)$.

Now assume that, for each $n > 0$, $\mathcal{T}_n$ has $k(n)+1$ atoms ($k(n) \geq 1$) that are not atoms of $\mathcal{T}_{n-1}$. Let

$$K(n) = k(1) + \cdots + k(n).$$

By an inductive construction, this procedure leads to a filtration $(\mathcal{C}_k)_{k\geq 0}$ such that $\mathcal{C}_{K(n)} = \mathcal{T}_n$ for any $n > 0$. Indeed, assuming we already have constructed $\mathcal{C}_{K(n)}$, we enumerate the atoms of $\mathcal{T}_{n+1}$ that are not atoms of $\mathcal{T}_n$ as $\bar{A}_1, \ldots, \bar{A}_{k(n+1)}, \bar{A}_{k(n+1)+1}$ and for each j such that $K(n) < j \leq K(n) + k(n+1)$, we define $\mathcal{C}_j$ as generated by $\mathcal{C}_{K(n)}$ and $\bar{A}_1, \ldots, \bar{A}_j$. The resulting filtration $(\mathcal{C}_k)_{k\geq 0}$ has the property that $\mathcal{C}_{k+1}$ has at most one more atom than $\mathcal{C}_k$ for any $k \geq 0$.

In the dyadic case, for the Haar filtration, we set $\mathcal{B}_1$ trivial, so the correspondence with the present construction should be shifted by 1. Moreover, we did not make sure (although this is easy to do) that the filtration $(\mathcal{C}_k)_{k\geq 0}$ is strictly increasing, as in the Haar case for $(\mathcal{B}_k)_{k\geq 1}$.

Remark 1.20. We will see in Chapters 3 and 4 more examples of discrete or continuous filtrations related to the boundary behaviour of harmonic and analytic functions. In Chapter 6, we will consider a filtration $(\mathcal{A}_n)_{n\in\mathbb{Z}}$ on $\mathbb{R}$ equipped with Lebesgue's measure. Each $\mathcal{A}_n$ is generated by a partition of $\mathbb{R}$ into intervals of length 2^n ($n \in \mathbb{Z}$).

Remark 1.21. In [258], the following beautiful example appears. Let $T = \mathbb{R}/2\pi\mathbb{Z}$, equipped with its normalized Haar measure m. Let $q > 1$ be a prime number. For each integer $n \geq 0$, consider on $\mathbb{T}$ the σ-algebra $\mathcal{F}_{-n}$ generated by all the measurable functions on $\mathbb{T}$ with period q^{-n}. Note that $\mathcal{F}_0$ is simply the whole Borel σ-algebra on $\mathbb{T}$. The conditional expectation with respect to $\mathcal{F}_{-n}$ on $L_1(\mathbb{T})$ can be described simply like this: for any $f \in L_1(\mathbb{T})$ with formal Fourier series $\sum_{k\in\mathbb{Z}} \hat{f}(k)e^{ikt}$ the function $\mathbb{E}^{\mathcal{F}_{-n}}(f)$ has a formal Fourier series supported by the set of ks that are divisible by q^n, namely $\sum_{k\in\mathbb{Z}, q^n|k} \hat{f}(k)e^{ikt}$. Thus, if $\hat{f}$ is finitely supported, we have

$$\mathbb{E}^{\mathcal{F}_{-n}}(f) = \sum\nolimits_{k\in\mathbb{Z},\ q^n|k} \hat{f}(k)e^{ikt}.$$

We can also write

$$\mathbb{E}^{\mathcal{F}_{-n}}(f)(t) = q^{-n} \sum_{j=1}^{q^n} f(t + q^{-n} j).$$

Note that $\mathcal{F}_{-\infty} \cdots \subset \mathcal{F}_{-n-1} \subset \mathcal{F}_{-n} \subset \cdots \subset \mathcal{F}_0$, where $\mathcal{F}_{-\infty} = \cap_{n \geq 0} \mathcal{F}_{-n}$ is the trivial σ-algebra on $\mathbb{T}$. This example illustrates well the notion of reverse martingale described in §1.9. Let (x_n) be a finite sequence in a Banach space B. Let $f_0 = \sum_j x_j e^{iq^j t}$. Then the sequence formed by the reverse partial sums

$$f_{-n} = \sum_{j \geq n} x_j e^{iq^j t} \tag{1.23}$$

forms a martingale with respect to the filtration $(\mathcal{F}_{-n})$ with martingale differences $f_{-n} - f_{-n-1} = x_n e^{iq^n t}$.

1.5 Stopping times

To handle the a.s. convergence of martingales, we will need (as usual) the appropriate maximal inequalities. In the martingale case, these are Doob's inequalities. Their proof uses *stopping times*, which are a basic tool in martingale theory. Given an increasing sequence $(\mathcal{A}_n)_{n \geq 0}$ of σ-subalgebras on Ω, a random variable $T\colon \ \Omega \to \mathbb{N} \cup \{\infty\}$ is called a stopping time if

$$\forall n \geq 0 \qquad \{T \leq n\} \in \mathcal{A}_n,$$

or equivalently if

$$\forall n \geq 0 \qquad \{T = n\} \in \mathcal{A}_n.$$

If $T < \infty$ a.s., then T is called a finite stopping time.

For any martingale (M_n) and any stopping time T we will denote by M_T the random variable $\omega \mapsto M_{T(\omega)}(\omega)$.

Proposition 1.22. *For any martingale $(M_n)_{n \geq 0}$ in $L_1(B)$ relative to $(\mathcal{A}_n)_{n \geq 0}$ and for every stopping time T, $(M_{n \wedge T})_{n \geq 0}$ is a martingale on the same filtration.*

Proof. Observe that $M_{n \wedge T} = \sum_{k<n} M_k 1_{\{T=k\}} + M_n 1_{\{T \geq n\}}$ clearly is in $L_1(B)$. Moreover, we have $M_{n \wedge T} - M_{(n-1) \wedge T} = 1_{\{n \leq T\}}(M_n - M_{n-1})$, but $\{n \leq T\}^c = \{T < n\} \in \mathcal{A}_{n-1}$ and hence

$$\mathbb{E}^{\mathcal{A}_{n-1}}(M_{n \wedge T} - M_{(n-1) \wedge T}) = 1_{\{n \leq T\}} \mathbb{E}^{\mathcal{A}_{n-1}}(M_n - M_{n-1}) = 0. \qquad \square$$

Given a stopping time T, we can define the associated σ-algebra $\mathcal{A}_T$ as follows: we say that a set A in $\mathcal{A}$ belongs to $\mathcal{A}_T$ if $A \cap \{T \leq n\}$ belongs to $\mathcal{A}_n$ for each $n \geq 0$. Then one can easily check that $\mathcal{A}_T$ is a σ-algebra.

The following lemma is very easy. We leave the proof as an exercise.

Lemma 1.23. *A (scalar or Banach space valued) function defined on Ω is $\mathcal{A}_T$-measurable iff its restriction to each set $\{T = k\}$ $(0 \le k \le \infty)$ is $\mathcal{A}_k$-measurable. If S, T are stopping times, $T \wedge S$ and $T \vee S$ are stopping times. Moreover, if $S \le T$ then $\mathcal{A}_S \subset \mathcal{A}_T$.*

In the next statement, we collect a number of basic facts about stopping times. Although we include the proof, we encourage the reader to prove this as an exercise.

Proposition 1.24. *(i) Consider M_∞ in $L_1(\Omega, \mathcal{A}, \mathbb{P}; B)$ and let $M_n = \mathbb{E}^{\mathcal{A}_n} M_\infty$ be the associated martingale. Then if T is a stopping time, $M_T \in L_1(\Omega, \mathcal{A}, \mathbb{P}; B)$ and we have*

$$M_T = \mathbb{E}^{\mathcal{A}_T}(M_\infty). \tag{1.24}$$

Moreover,

$$\mathbb{E}^{\mathcal{A}_n}(M_T) = M_{T\wedge n} = \mathbb{E}^{\mathcal{A}_T}(M_n). \tag{1.25}$$

More generally, if S is any other stopping time, we have

$$\mathbb{E}^{\mathcal{A}_S}(M_T) = M_{T\wedge S} = \mathbb{E}^{\mathcal{A}_T}(M_S). \tag{1.26}$$

(ii) If $(M_n)_{n\ge 0}$ is a martingale in $L_1(\Omega, \mathcal{A}, \mathbb{P}; B)$ and if $T_0 \le T_1 \le \cdots$ is a sequence of bounded stopping times then $(M_{T_k})_{k\ge 0}$ is a martingale relative to the sequence of σ-algebras $\mathcal{A}_{T_0} \subset \mathcal{A}_{T_1} \subset \cdots$. This also holds for unbounded times if we assume as in (i) that $(M_n)_{n\ge 0}$ converges in $L_1(\Omega, \mathcal{A}, \mathbb{P}; B)$.

Proof. Let $A \in \mathcal{A}_T$. Let $A_k = A \cap \{T = k\}$. Then (here we incorporate the value $k = \infty$ in the sum)

$$\int_A M_\infty d\mathbb{P} = \sum\nolimits_k \int_{A_k} M_\infty d\mathbb{P} = \sum\nolimits_k \int_{A_k} M_k d\mathbb{P} = \sum\nolimits_k \int_{A_k} M_T d\mathbb{P} = \int_A M_T d\mathbb{P}.$$

This proves (1.24). Note that this equivalently means

$$\mathbb{E}^{\mathcal{A}_T}(M_\infty) = \sum\nolimits_{0\le k\le\infty} 1_{\{T=k\}} \mathbb{E}^{\mathcal{A}_k}(M_\infty), \tag{1.27}$$

from which it is clear that $M_T \in L_1(\Omega, \mathcal{A}, \mathbb{P}; B)$.

To prove (1.25), we first assume that T is bounded. Let $d_0 = M_0$ and $d_k = M_k - M_{k-1}$ $(k > 1)$). Then we can write $M_T = \sum d_k 1_{\{k\le T\}}$ and since $\{k \le T\} \in \mathcal{A}_{k-1}$ the martingale property implies $\mathbb{E}^{\mathcal{A}_n}(d_k 1_{\{k\le T\}}) = 0$ for all $k \ge n$. It follows that $\mathbb{E}^{\mathcal{A}_n} M_T = \sum_{k<n} d_k 1_{\{k\le T\}} = M_{T\wedge n}$.

Applying this to the bounded stopping time $T \wedge N$ with $n \le N$ we find $\mathbb{E}^{\mathcal{A}_n} M_{T\wedge N} = M_{T\wedge n}$ and letting $N \to \infty$ we obtain (1.25), provided we check the following:

Claim. $M_{T\wedge N} \to M_T$ in $L_1(\Omega, \mathcal{A}, \mathbb{P}; B)$ when $N \to \infty$.
Indeed, we have $M_{T\wedge N} - M_T = 1_{\{T>N\}}(M_N - M_T)$ and hence

$$\|M_{T\wedge N} - M_T\| \le 1_{\{N<T<\infty\}}\|M_N - M_T\| + 1_{\{T=\infty\}}\|M_N - M_\infty\|,$$

and a fortiori

$$\|M_{T\wedge N} - M_T\| \le 1_{\{N<T<\infty\}}(\|M_N\| + \|M_T\|) + 1_{\{T=\infty\}}\|M_N - M_\infty\|. \quad (1.28)$$

By Theorem 1.14, we know that $\|M_N - M_\infty\| \to 0$ in L_1, and a fortiori that $\|M_N\|$ converges in L_1, and hence forms a uniformly integrable sequence. Moreover, we have obviously $\mathbb{P}\{N < T < \infty\} \to 0$. Therefore

$$\mathbb{E}\left(1_{\{N<T<\infty\}}(\|M_N\| + \|M_T\|)\right) \to 0.$$

Thus the preceding claim follows from (1.28).

The other equality in (1.25) is easier: it follows from (1.24) applied to the 'stopped' martingale (f_k) defined by $f_k = M_{k\wedge n}$ for which $f_\infty = M_n$. We will deduce (1.26) from (1.25) using (1.27): we have

$$\mathbb{E}^{\mathcal{A}_T}(M_S) = \sum\nolimits_{0\le n\le\infty} 1_{\{T=n\}}\mathbb{E}^{\mathcal{A}_n}(M_S) = \sum\nolimits_{0\le n\le\infty} 1_{\{T=n\}}M_{n\wedge S} = M_{T\wedge S}.$$

The remaining assertion (ii) is now obvious. □

1.6 Almost sure convergence: Maximal inequalities

The following is arguably the most fundamental fact about martingales.

Theorem 1.25 (Doob's maximal inequalities). *Let $(M_0, M_1, \ldots, M_n)$ be a submartingale in L_1, and let $M_n^* = \sup_{k\le n} M_k$. Then*

$$\forall t > 0 \qquad t\mathbb{P}(\{M_n^* > t\}) \le \int_{\{M_n^* > t\}} M_n d\mathbb{P}, \quad (1.29)$$

and if $M_n^* \ge 0$, then for all $1 < p < \infty$, we have

$$\|M_n^*\|_p \le p'\|M_n\|_p, \quad (1.30)$$

where $\frac{1}{p} + \frac{1}{p'} = 1$.

Proof. We can rewrite the submartingale property as saying that for any A in $\mathcal{A}_k$ with $k \le n$ we have

$$\int_A M_k d\mathbb{P} \le \int_A (E^{\mathcal{A}_k} M_n) d\mathbb{P} = \int E^{\mathcal{A}_k}(1_A M_n) d\mathbb{P} = \int_A M_n d\mathbb{P}. \quad (1.31)$$

Fix $t > 0$. Let

$$T = \begin{cases} \inf\{k \le n \mid M_k > t\} & \text{if } M_n^* > t \\ \infty & \text{otherwise.} \end{cases}$$

Then T is a stopping time relative to the sequence of σ-algebras $(\mathcal{A}'_k)$ defined by $\mathcal{A}'_k = \mathcal{A}_{k\wedge n}$. Since $M_k > t$ on the set $\{T = k\}$, we have

$$t\mathbb{P}\{M_n^* > t\} = t\mathbb{P}\{T \le n\} = t\sum\nolimits_{k\le n} \mathbb{P}\{T = k\} \le \sum\nolimits_{k\le n} \int_{\{T=k\}} M_k,$$

whence by (1.31)

$$\le \sum\nolimits_{k\le n} \int_{\{T=k\}} M_n = \int_{\{T\le n\}} M_n.$$

This proves (1.29). To prove (1.30), we use an extrapolation trick. Recalling (1.3), if $M_n^* \ge 0$, we have

$$\begin{aligned} \mathbb{E}M_n^{*p} &= \int_0^\infty pt^{p-1}\mathbb{P}\{M_n^* > t\}dt \\ &\le \int_0^\infty pt^{p-2} \int_{\{M_n^*>t\}} M_n d\mathbb{P}\, dt \\ &= \int M_n \left(\int_0^{M_n^*} pt^{p-2}dt\right) d\mathbb{P} = \int \frac{p}{p-1} M_n (M_n^*)^{p-1} d\mathbb{P}, \end{aligned}$$

hence by Hölder's inequality,

$$\begin{aligned} &\le p' \|M_n\|_p \|(M_n^*)^{p-1}\|_{p'} \\ &= p' \|M_n\|_p (\mathbb{E}M_n^{*p})^{\frac{p-1}{p}}, \end{aligned}$$

so that after division by $(\mathbb{E}M_n^{*p})^{\frac{p-1}{p}}$, we obtain (1.30). □

The following inequality is known as the Burkholder-Davis-Gundy inequality. It is dual to Doob's maximal inequality. Indeed, by (1.30), we have for any x in L_p

$$\|(\mathbb{E}^{\mathcal{A}_n}x)\|_{L_p(\ell_\infty)} = \|\sup_n |\mathbb{E}^{\mathcal{A}_n}x|\|_p \le p'\|x\|_p. \tag{1.32}$$

Therefore it is natural to expect a dual inequality involving an 'adjoint mapping' from $L_{p'}(\ell_1)$ to $L_{p'}$, as follows:

Theorem 1.26. *Let $(\theta_n)_{n\ge 0}$ be an arbitrary family of random variables. Then, for any $1 \le p < \infty$,*

$$\left\|\sum |\mathbb{E}^{\mathcal{A}_n}\theta_n|\right\|_p \le p \left\|\sum |\theta_n|\right\|_p. \tag{1.33}$$

In particular, if $\theta_n \geq 0$,

$$\left\|\sum \mathbb{E}^{\mathcal{A}_n}\theta_n\right\|_p \leq p\left\|\sum \theta_n\right\|_p.$$

Proof. Since $|\mathbb{E}^{\mathcal{A}_n}\theta_n| \leq \mathbb{E}^{\mathcal{A}_n}|\theta_n|$, it suffices to prove this assuming $\theta_n \geq 0$. In that case, consider $f \geq 0$ in $L_{p'}$ with $\|f\|_{p'} = 1$ such that $\left\|\sum \mathbb{E}^{\mathcal{A}_n}\theta_n\right\|_p = \left\langle \sum \mathbb{E}^{\mathcal{A}_n}\theta_n, f\right\rangle$. Then

$$\begin{aligned}\left\langle \sum \mathbb{E}^{\mathcal{A}_n}\theta_n, f\right\rangle &= \sum \langle \theta_n, \mathbb{E}^{\mathcal{A}_n} f\rangle \\ &\leq \left\|\sum \theta_n\right\|_p \|\sup_n \mathbb{E}^{\mathcal{A}_n} f\|_{p'};\end{aligned}$$

hence, by Doob's inequality,

$$\leq p\left\|\sum \theta_n\right\|_p.$$

□

Remark 1.27. Note that it is crucial for the validity of Theorems 1.25 and 1.26 that the conditional expectations be *totally ordered*, as in a filtration. However, as we will now see, in some cases we can go beyond that. Let $(\mathcal{A}_n^1)_{n\geq 0}$, $(\mathcal{A}_n^2)_{n\geq 0}, \ldots, (\mathcal{A}_n^d)_{n\geq 0}$ be a d-tuple of (a priori mutually unrelated) filtrations on a probability space $(\Omega, \mathcal{A}, \mathbb{P})$. Let $I_d = \mathbb{N}^d$ and for all $i = (n(1), \ldots, n(d))$ let

$$\mathbb{E}_i = \mathbb{E}^{\mathcal{A}^1_{n(1)}}\mathbb{E}^{\mathcal{A}^2_{n(2)}} \ldots \mathbb{E}^{\mathcal{A}^d_{n(d)}}. \tag{1.34}$$

Then, by a simple iteration argument, we find that for any $1 < p \leq \infty$ and any x in L_p we have

$$\|\sup_{i\in I_d} |\mathbb{E}_i x|\,\|_p \leq (p')^d \|x\|_p.$$

A similar iteration holds for the dual to Doob's inequality: for any family $(x_i)_{i\in I_d}$ in $L_{p'}$ we have

$$\left\|\sum |\mathbb{E}_i x_i|\right\|_{p'} \leq (p')^d \left\|\sum |x_i|\right\|_{p'}.$$

To illustrate this (following [145]), consider a dyadic rooted tree T, i.e. the points of T are finite sequences $\xi = (\xi_1, \ldots, \xi_k)$ with $\xi_j \in \{0, 1\}$ and there is also a root (or origin) denoted by ξ_ϕ. We introduce a partial order on T in the natural way, i.e. ξ_ϕ is $\leq$ any element and then we set $(\xi_1, \ldots, \xi_k) \leq (\xi'_1, \ldots, \xi'_j)$ if $k \leq j$ and $(\xi_1, \ldots, \xi_k) = (\xi'_1, \ldots, \xi'_k)$. In other words, $\xi \leq \xi'$ if ξ' is on the same 'branch' as ξ, but 'after' ξ.

This is clearly *not* totally ordered since two points situated on disjoint branches are incomparable. Nevertheless, as observed in [145], we have the following: Consider a family $\{\varepsilon_\xi \mid \xi \in T\}$ of independent random variables and for any ξ in T let $\mathcal{A}_\xi = \sigma(\{\varepsilon_\eta \mid \eta \le \xi\})$, and let

$$\mathbb{E}_\xi = \mathbb{E}^{\mathcal{A}_\xi}.$$

We have then for any $1 < p \le \infty$ and any x in L_p

$$\| \sup_{\xi \in T} |\mathbb{E}_\xi x| \, \|_p \le (p')^3 \|x\|_p.$$

The idea is that $\mathbb{E}_\xi$ is actually of the form (1.34) with $d = 3$, see [145] for full details.

Remark 1.28. Let B be a Banach space and let $(M_n)_{n\ge 0}$ be a B-valued martingale. Then the random variables Z_n defined by $Z_n(\omega) = \|M_n(\omega)\|_B$ form a submartingale. Indeed, by (1.8) we have for every k and every f in $L_1(\Omega, \mathbb{P}; B)$

$$\|\mathbb{E}^{\mathcal{A}_k}(f)\| \overset{\text{a.s.}}{\le} \mathbb{E}^{\mathcal{A}_k}(\|f\|_B) \tag{1.35}$$

hence taking $f = M_n$ with $k \le n$ we obtain

$$\|M_k\| \le \mathbb{E}^{\mathcal{A}_k}(\|M_n\|),$$

which shows that (Z_n) is a submartingale. In particular, by (1.31), we have for any A in $\mathcal{A}_k$

$$\mathbb{E}(1_A\|M_k\|) \le \mathbb{E}(1_A\|M_n\|). \tag{1.36}$$

As a consequence, we can apply Doob's inequality to the submartingale (Z_n), and we obtain the following.

Corollary 1.29. *Let (M_n) be a martingale with values in an arbitrary Banach space B. Then*

$$\sup_{t>0} t\mathbb{P}\{\sup_{n\ge 0} \|M_n\| > t\} \le \sup_{n\ge 0} \|M_n\|_{L_1(B)} \tag{1.37}$$

and for all $1 < p < \infty$

$$\| \sup_{n\ge 0} \|M_n\| \, \|_p \le p' \sup_{n\ge 0} \|M_n\|_{L_p(B)}. \tag{1.38}$$

We can now prove the martingale convergence theorem.

Theorem 1.30. *Let $1 \le p < \infty$. Let B be an arbitrary Banach space. Consider f in $L_p(\Omega, \mathcal{A}, \mathbb{P}; B)$, and let $M_n = \mathbb{E}^{\mathcal{A}_n}(f)$ be the associated martingale. Then M_n converges a.s. to $\mathbb{E}^{\mathcal{A}_\infty}(f)$. Therefore, if a martingale (M_n) converges in $L_p(\Omega, \mathbb{P}; B)$ to a limit M_∞, it necessarily converges a.s. to this limit, and we have $M_n = \mathbb{E}^{\mathcal{A}_n} M_\infty$ for all $n \ge 0$.*

Proof. The proof is based on a general principle, going back to Banach, that allows us to deduce almost sure convergence results from suitable maximal inequalities. By Theorem 1.14, we know that $\mathbb{E}^{\mathcal{A}_n}(f)$ converges in $L_p(B)$ to $M_\infty = \mathbb{E}^{\mathcal{A}_\infty}(f)$. Fix $\varepsilon > 0$ and choose k so that $\sup_{n\geq k} \|M_n - M_k\|_{L_p(B)} < \varepsilon$. We will apply (1.37) and (1.38) to the martingale $(M'_n)_{n\geq 0}$ defined by

$$M'_n = M_n - M_k \text{ if } n \geq k \quad \text{and} \quad M'_n = 0 \text{ if } n \leq k.$$

We have in the case $1 < p < \infty$

$$\| \sup_{n\geq k} \|M_n - M_k\| \|_p \leq p'\varepsilon$$

and in the case $p = 1$

$$\sup_{t>0} t\mathbb{P}\{\sup_{n\geq k} \|M_n - M_k\| > t\} \leq \varepsilon.$$

Therefore, if we define pointwise $\ell = \lim_{k\to\infty} \sup_{n,m\geq k} \|M_n - M_m\|$, we have

$$\ell = \inf_{k\geq 0} \sup_{n,m\geq k} \|M_n - M_m\| \leq 2 \sup_{n\geq k} \|M_n - M_k\|.$$

Thus we find $\|\ell\|_p \leq 2p'\varepsilon$ and, in the case $p = 1$, $\sup_{t>0} t\mathbb{P}\{\ell > 2t\} \leq \varepsilon$, which implies (since $\varepsilon > 0$ is arbitrary) that $\ell = 0$ a.s., and hence by the Cauchy criterion that (M_n) converges a.s. Since $M_n \to M_\infty$ in $L_p(B)$ we have necessarily $M_n \to M_\infty$ a.s. Note that if a martingale M_n tends to a limit M_∞ in $L_p(B)$ then necessarily $M_n = \mathbb{E}^{\mathcal{A}_n}(M_\infty)$. Indeed, $M_n = \mathbb{E}^{\mathcal{A}_n} M_m$ for all $m \geq n$ and by continuity of $\mathbb{E}^{\mathcal{A}_n}$ we have $\mathbb{E}^{\mathcal{A}_n} M_m \to \mathbb{E}^{\mathcal{A}_n} M_\infty$ in $L_p(\mathcal{B})$ so that $M_n = \mathbb{E}^{\mathcal{A}_n} M_\infty$ as announced. This settles the last assertion. □

Corollary 1.31. *Every scalar valued martingale $(M_n)_{n\geq 0}$ that is bounded in L_p for some $p > 1$ (resp. uniformly integrable) must converge a.s. and in L_p (resp. L_1).*

Proof. By Corollary 1.16, if $(M_n)_{n\geq 0}$ is bounded in L_p for some $p > 1$ (resp. uniformly integrable), then M_n converges in L_p (resp. L_1), and by Theorem 1.30, the a.s. convergence is then automatic. □

The following useful lemma illustrates the use of stopping times as a way to properly 'truncate' a martingale.

Lemma 1.32. *Let $(M_n)_{n\geq 0}$ be a martingale bounded in $L_1(\Omega, \mathcal{A}, \mathbb{P}; B)$ where B is an arbitrary Banach space. Fix $t > 0$ and let*

$$T = \begin{cases} \inf\{n \geq 0 \mid \|M_n\| > t\} & \text{if } \sup \|M_n\| > t \\ \infty & \text{otherwise.} \end{cases}$$

Then

$$\mathbb{E}(\|M_T\| 1_{\{T<\infty\}}) \leq \sup_{n\geq 0} \mathbb{E}\|M_n\|, \tag{1.39}$$

and moreover, $(M_{n\wedge T})_{n\geq 0}$ is a uniformly integrable martingale.

Proof. By Proposition 1.22, we already know that that $(M_{n\wedge T})_{n\geq 0}$ is a martingale in $L_1(B)$. First we claim that for any $0 \leq k \leq n$, we have

$$\mathbb{E}(1_{\{T=k\}}\|M_k\|) \leq \mathbb{E}(1_{\{T=k\}}\|M_n\|).$$

Indeed, $\{T = k\} \in \mathcal{A}_k$, so this is a particular case of (1.36). Summing this with respect to $k \leq n$, we obtain

$$\mathbb{E}(1_{\{T\leq n\}}\|M_T\|) \leq \mathbb{E}(1_{\{T\leq n\}}\|M_n\|),$$

and taking the supremum over $n \geq 0$, we obtain (1.39).

Now recall that by definition $\sup \|M_n\| \leq t$ on $\{T = \infty\}$. More generally, we have $\sup_{n<T} \|M_n\| \leq t$, so that

$$\sup_n \|M_{n\wedge T}\| \leq \max\{1_{\{T<\infty\}}\|M_T\|, t\} \leq 1_{\{T<\infty\}}\|M_T\| + t. \quad (1.40)$$

Then $\sup_n \mathbb{E}(1_A\|M_{n\wedge T}\|) \leq \mathbb{E}(1_A Z)$ for any $A \in \mathcal{A}$ where $Z = 1_{\{T<\infty\}}\|M_T\| + t$. Thus $(\|M_{n\wedge T}\|)_{n\geq 0}$ is uniformly integrable (since the single variable Z is so). □

We will now show that the a.s. convergence of L_1-bounded scalar martingales in Corollary 1.31 holds even without the assumption of uniform integrability. For that purpose, we will use the following simple fact. For future reference, we state it in the Banach space valued case, so (i) and (ii) in what follows are equivalent properties of B and $(\mathcal{A}_n)_{n\geq 0}$, but, of course, when $B = \mathbb{R}$, both assertions hold.

Proposition 1.33. *Let $(\Omega, \mathcal{A}, \mathbb{P})$ be a probability space. Let B be a Banach space, and let $(\mathcal{A}_n)_{n\geq 0}$ be an increasing sequence of σ-subalgebras of $\mathcal{A}$. The following assertions are equivalent:*

(i) Every B-valued martingale adapted to $(\mathcal{A}_n)_{n\geq 0}$ and bounded in $L_1(\Omega, \mathbb{P}; B)$ is a.s. convergent.
(ii) Every B-valued uniformly integrable martingale adapted to $(\mathcal{A}_n)_{n\geq 0}$ is a.s. convergent.

Proof. Assume (ii). Let (M_n) be a martingale bounded in $L_1(B)$. Fix $t > 0$ and consider $(M_{n\wedge T})$ as in Lemma 1.32. Since $(M_{n\wedge T})$ is uniformly integrable, it converges a.s. by (ii). This implies that if $\{T(\omega) = \infty\}$ then $(M_n(\omega))_{n\geq 0}$ is a.s. convergent. But by Doob's inequalities,

$$\mathbb{P}\{T < \infty\} = \mathbb{P}\{\sup \|M_n\| > t\} \leq \frac{C}{t},$$

where $C = \sup \mathbb{E}\|M_n\|$. Therefore this probability can be made arbitrarily small by choosing t large, so that we conclude that the martingale $(M_n)_{n\geq 0}$ itself converges a.s. This shows that (ii) $\Rightarrow$ (i). The converse is trivial. □

Finally, we can state what is usually referred to as the 'martingale convergence theorem'.

Theorem 1.34. *Every L_1-bounded scalar valued martingale converges a.s.*

Proof. By Corollary 1.16, every scalar valued uniformly integrable martingale converges in L_1, and hence by Theorem 1.30 it converges a.s. Thus the present statement follows from the implication (ii) $\Rightarrow$ (i) from Proposition 1.33. □

Remark 1.35. There are well-known counter-examples showing that Theorem 1.34 does not extend to the Banach space valued case. For instance, let $\Omega = \{-1, 1\}^{\mathbb{N}}$ equipped with the usual probability measure $\mathbb{P}$ and let $\mathcal{A}_n$ be the σ-algebra generated by the $(n+1)$ first coordinates denoted by $\varepsilon_0, \varepsilon_1, \ldots, \varepsilon_n$. A classical example of a real valued martingale is $M_n = \prod_{k\leq n}(1+\varepsilon_k)$, which is positive and of integral 1. Note however that it does not converge in L_1. Another example is $M_n = \sum_{k\leq n} \alpha_k\varepsilon_k$, where (α_k) are real coefficients. This particular martingale is bounded in L_1 iff $\Sigma|\alpha_n|^2$ is finite. By the martingale convergence theorem, these two martingales must converge a.s. However, we can give very similar Banach space valued examples that do not converge. Take for instance $B = c_0$, and let (e_n) be the canonical basis of c_0. Let $M_n^1 = \sum_{k\leq n} \varepsilon_k e_k$. Then $\|M_n^1(\omega)\|_{c_0} = \sup_{k\leq n} |\varepsilon_k(\omega)| \equiv 1$ but clearly there is no point ω in $\{-1, 1\}^{\mathbb{N}}$ such that the sequence $(M_n^1(\omega))_{n\geq 0}$ is convergent in c_0, since we have

$$\forall \omega \in \Omega \quad \forall k < n \qquad \|M_n^1(\omega) - M_k^1(\omega)\|_B = 1.$$

We can give a similar example in L_1. Let $B = L^1(\Omega, \mathbb{P})$ itself and let

$$M_n^2(\omega) = \prod\nolimits_{k\leq n}(1 + \varepsilon_k(\omega)\varepsilon_k). \tag{1.41}$$

Then again $\|M_n^2(\omega)\|_B = 1$ for all ω, but also it is easy to check that

$$\forall \omega \in \Omega \quad \forall k < n \quad \|M_n^2(\omega) - M_k^2(\omega)\|_B \geq 1, \text{ and } \|M_n^2(\omega) - M_{n-1}^2(\omega)\|_B = 1,$$

so that $(M_n^2)_{n\geq 0}$ is nowhere convergent.

In the next chapter, we will show that the preceding examples cannot occur in a Banach space with the RNP.

We will also need the following.

Theorem 1.36. *Every submartingale (M_n) bounded in L_1 (resp. uniformly integrable) converges a.s. (resp. a.s. and in L_1).*

Proof. We use the so-called Doob decomposition: we will write our submartingale as the sum of a martingale $(\widetilde{M}_n)_{n\geq 0}$ and a predictable increasing sequence (A_n) (recall that this means that A_n is $\mathcal{A}_{n-1}$ measurable for each $n \geq 1$). Let us write $\Delta_0 = M_0$ and $\Delta_n = M_n - M_{n-1}$ if $n \geq 1$. Let $d_n = \Delta_n - \mathbb{E}^{\mathcal{A}_{n-1}}(\Delta_n)$ if $n \geq 1$ and $d_0 = \Delta_0$, and let $\widetilde{M}_n = \sum_{k\leq n} d_k$. Then $(\widetilde{M}_n)_{n\geq 0}$ is a martingale. Indeed, by construction, we have $\mathbb{E}^{\mathcal{A}_{n-1}}(d_n) = 0$ or equivalently $\mathbb{E}^{\mathcal{A}_{n-1}}\widetilde{M}_n = \widetilde{M}_{n-1}$. To relate $(\widetilde{M}_n)$ to (M_n), we note that

$$M_n = \sum\nolimits_{0\leq k\leq n} \Delta_k = \sum\nolimits_{0\leq k\leq n} d_k + \sum\nolimits_{1\leq k\leq n} \mathbb{E}^{\mathcal{A}_{k-1}}(\Delta_k);$$

hence

$$M_n = \widetilde{M}_n + A_n,$$

where

$$A_n = \sum\nolimits_{1\leq k\leq n} \mathbb{E}^{\mathcal{A}_{k-1}}(\Delta_k).$$

Moreover, by the submartingale property, $\mathbb{E}^{\mathcal{A}_{n-1}}(\Delta_n) \geq 0$ for all $n \geq 1$ so that

$$0 \leq A_1 \leq A_2 \leq \cdots \leq A_{n-1} \leq A_n \leq \cdots .$$

On one hand, $\mathbb{E}A_n = \sum_{1\leq k\leq n} \mathbb{E}\Delta_k = \mathbb{E}M_n - \mathbb{E}M_0$, and since (M_n) is assumed bounded in L_1, we have $\sup_{n\geq 1} \mathbb{E}A_n < \infty$. Therefore by monotonicity A_n converges a.s. and in L_1 when $n \to \infty$ (in particular it is a uniformly integrable sequence). On the other hand, we have

$$\mathbb{E}|\widetilde{M}_n| = \mathbb{E}|M_n - A_n| \leq \mathbb{E}|M_n| + \mathbb{E}A_n;$$

therefore $(\widetilde{M}_n)$ also is bounded in L_1 and is uniformly integrable if (M_n) is. By the martingale convergence theorem (Theorem 1.34), $(\widetilde{M}_n)$ converges a.s., hence $M_n = \widetilde{M}_n + A_n$ also converges a.s., and, in the uniformly integrable case, it also converges in L_1.

If we impose the initial condition $A_0 = 0$, the preceding proof also shows uniqueness: Indeed, $M_n = \widetilde{M}_n + A_n$ implies $A_n - A_{n-1} = \Delta_n - d\widetilde{M}_n$ and (assuming A_n $n-1$-measurable) this imposes $A_n - A_{n-1} = \mathbb{E}^{\mathcal{A}_{n-1}}(\Delta_n - d\widetilde{M}_n) = \mathbb{E}^{\mathcal{A}_{n-1}}(\Delta_n)$, which uniquely determines A_n if set $A_0 = 0$. □

Corollary 1.37. *Let B be an arbitrary Banach space, and let $(M_n)_{n\geq 0}$ be a B-valued martingale bounded in $L_1(B)$. Then $\|M_n\|_B$ converges a.s. Moreover, $(M_n)_{n\geq 0}$ converges a.s. in norm iff $\{M_n(\omega) \mid n \geq 0\}$ is relatively compact for almost all ω.*

Proof. The first assertion follows from Theorem 1.36 and Remark 1.28. Since any element (or any sequence) in $L_1(B)$ takes its values in a separable subspace of B, it suffices to prove the second one for a separable B. Assume that

$\{M_n(\omega) \mid n \geq 0\}$ is ω-a.s. relatively compact. Let $f(\omega)$ be a cluster point in B of $\{M_n(\omega) \mid n \geq 0\}$. Note that by Theorem 1.34, for any ξ in B^*, $\xi(M_n(\omega))$ converges ω-a.s., and hence it must converge to $\xi(f(\omega))$. (Incidentally: this shows that f is scalarly measurable, and hence by §1.8 is Bochner measurable.) Let $D \subset B^*$ be a countable weak-$*$ dense subset. Clearly, $M_n(\omega)$ tends ω-a.s. to $f(\omega)$ in the $\sigma(B, D)$-topology, but if $\{M_n(\omega) \mid n \geq 0\}$ is relatively compact, the latter topology coincides on it with the norm topology, and hence $M_n(\omega) \to f(\omega)$ in norm. Conversely, if $\{M_n(\omega) \mid n \geq 0\}$ is convergent, it is obviously relatively compact. □

Remark 1.38. The maximal inequalities for B-valued martingales can be considerably strengthened when $B = \ell_r$ for some $1 < r < \infty$: Consider a filtration $(\mathcal{A}_n)$ as usual, $f \in L_p(\ell_r)$ and let (f_n) be the martingale associated to f. Let (e_k) be the canonical basis of ℓ_r. We may develop f and f_n as $f = \sum_k f(k)e_k$ and $f_n = \sum_k f_n(k)e_k$. In accordance with previous notation, we set $f(k)^* = \sup_n |f_n(k)|$. Let then

$$f^{**} = \left\|\sum f(k)^* e_k\right\|_{\ell_r} = \left(\sum f(k)^{*r}\right)^{1/r}.$$

Then, for any $1 < p < \infty$, there is a constant $c(p, r)$ such that

$$\|f^{**}\|_p \leq c(p, r)\|f\|_{L_p(\ell_r)} = c(p, r)\left\|\left(\sum |f(k)|^r\right)^{1/r}\right\|_p.$$

Note that $p = r$ is an easy consequence of Doob's inequality. See [85] for $p \leq r$ and [326] for the general case and for a weak type-$(1, 1)$ inequality that can be proved using the Gundy decomposition described in §5.4. Finally, the extension to the case $B = L_r$ requires only minor modifications. See [136] for the case when B is a UMD space with unconditional basis, as defined in the sense of Chapter 5 (or a UMD Banach lattice).

Remark 1.39. Fix $1 \leq p < \infty$. Let I be a totally ordered set, with order denoted simply by $\leq$, and let $(f_i)_{i \in I}$ in $L_p(B)$ be a martingale as in Remark 1.17. In general when I is uncountable $\sup_{i \in I} \|f_i(.)\|_B$ is not measurable. However, assuming $p > 1$ for simplicity, if $(f_i)_{i \in I}$ is bounded in $L_p(B)$, by monotone convergence the family (f_J^*), indexed by finite subsets $J \subset I$ (ordered by inclusion) and defined by $f_J^* = \sup_{i \in J} \|f_i\|_B$, converges in L_p to an element $f^* \in L_p$, and the analogue of Doob's maximal inequalities (1.38) hold. Indeed, by (1.38), for any finite J we have $\|f_J^*\|_p \leq p' \sup_{i \in I} \|f_i\|_{L_p(B)}$ so it suffices to pass to the limit when $J \to I$, and we obtain $\|f^*\|_p \leq p' \sup_{i \in I} \|f_i\|_{L_p(B)}$. In fact, the element f^* is the supremum of the lattice bounded family $(\|f_i\|_B)_{i \in I}$ in L_p viewed as a Banach lattice. It is the smallest element of L_p such that for any $i \in I$ we have $\mathbb{P}\{\|f_i\|_B \leq f^*\} = 1$ (but in general we cannot control the probability of the subset $\cap_{i \in I}\{\|f_i\|_B \leq f^*\}$, which might not be measurable).

Note however that if $g_i \in L_p$ are such that $0 \le g_i \le \|f_i\|_B$ for all $i \in I$, and also such that $\sup_{i\in I} g_i$ is measurable, then we have $\sup_{i\in I} g_i = g^* \le f^*$.

1.7 Independent increments

There are cases where the maximal inequalities can be extended to L_p with $0 < p < 1$. For instance, let $(Y_n)_{n\ge 0}$ be a sequence of independent B-valued random variables, let $f_n = \sum_0^n Y_n$. If $Y_n \in L_1(B)$ is symmetric for all n (this implies $\mathbb{E}Y_n = 0$), then $(f_n)_{n\ge 0}$ is a martingale satisfying $\mathbb{P}(\sup_n \|f_n\| > t) \le 2\sup_n \mathbb{P}(\|f_n\| > t)$. More generally, we have

Theorem 1.40. *Let (Y_n) be a sequence of B-valued Borel measurable functions on $(\Omega, \mathcal{A}, \mathbb{P})$, such that, for any choice of signs $\xi_n = \pm 1$, the sequence $(\xi_n Y_n)$ has the same distribution as (Y_n). Let $f_n = \sum_0^n Y_k$. We have then:*

$$\forall t > 0 \qquad \mathbb{P}(\sup \|f_n\| > t) \le 2 \limsup \mathbb{P}(\|f_n\| > t) \tag{1.42}$$

$$\forall p > 0 \qquad \mathbb{E}\sup \|f_n\|^p \le 2 \limsup \mathbb{E}\|f_n\|^p. \tag{1.43}$$

If f_n converges to a limit f_∞ in probability (i.e. $\mathbb{P}\{\|f_n - f_\infty\| > \varepsilon\} \to 0$ for any $\varepsilon > 0$), then it actually converges a.s. In particular, if f_n converges in L_p $(p > 0)$, then it automatically converges a.s. Finally, if f_n converges a.s. to a limit f_∞, we have

$$\forall t > 0 \qquad \mathbb{P}(\sup \|f_n\| > t) \le 2\mathbb{P}(\|f_\infty\| > t). \tag{1.44}$$

More generally for any Borel convex subset $K \subset B$, we have

$$\mathbb{P}(\cup_n \{f_n \notin K\}) \le 2\mathbb{P}(\{f_\infty \notin K\}).$$

Proof. Let us first assume that f_n converges a.s. to a limit f_∞. We start by the last assertion. Observe that for each n, $f_\infty = f_n + (f_\infty - f_n)$ has the same distribution as $f_n - (f_\infty - f_n)$. Let T be the stopping time $T = \inf\{n \ge 0 \mid f_n \notin K\}$. Then

$$\mathbb{P}(\cup_n \{f_n \notin K\}) = \mathbb{P}\{T < \infty\} = \sum\nolimits_0^\infty \mathbb{P}(T = n).$$

Note that by the convexity of K, $f_n + (f_\infty - f_n) \in K$ and $f_n - (f_\infty - f_n) \in K$ imply $f_n \in K$, and hence $\{T = n\}$ implies that either $f_\infty \notin K$ or $f_n - (f_\infty - f_n) \notin K$. Therefore

$$\mathbb{P}(T = n) \le \mathbb{P}(T = n, f_\infty \notin K) + \mathbb{P}(T = n, f_n - (f_\infty - f_n) \notin K).$$

Choosing the signs $\xi_k = 1\ \forall k \le n$ and $\xi_k = -1$ for $k > n$, we see that

$$\mathbb{P}(T = n, f_n - (f_\infty - f_n) \notin K) = \mathbb{P}(T = n, f_\infty \notin K)$$

hence we conclude

$$\mathbb{P}(\cup_n\{f_n \notin K\}) = \sum\nolimits_0^\infty \mathbb{P}(T = n) \le 2\sum\nolimits_0^\infty \mathbb{P}(T = n, f_\infty \notin K) \\ \le 2\mathbb{P}(f_\infty \notin K).$$

Replacing K by the ball of radius t, we obtain (1.44) assuming the a.s. convergence of $\{f_n\}$. In particular (replacing Y_n by zero for all $n > N$) this yields

$$\mathbb{P}(\sup_{n\le N} \|f_n\| > t) \le 2\mathbb{P}(\|f_N\| > t). \tag{1.45}$$

Taking the supremum over all N's we obtain the announced inequality (1.42). Using $\mathbb{E}\|f_n\|^p = \int_0^\infty pt^{p-1}\mathbb{P}(\|f_n\| > t)$, we deduce from (1.45) that

$$\mathbb{E}\sup_{n\le N}\|f_n\|^p \le 2\mathbb{E}\|f_N\|^p$$

from which (1.43) follows immediately.

If f_n converges in $L_p(B)$ to a limit, say f_∞, we have for any fixed k

$$\mathbb{E}\sup_{n>k}\|f_n - f_k\|^p \le 2\sup_{n>k}\mathbb{E}\|f_n - f_k\|^p.$$

The a.s. convergence is then proved exactly as earlier for Theorem 1.30. More generally, if f_n converges in probability, we have, for any fixed k and $t > 0$, by (1.42)

$$\mathbb{P}(\sup_{n>k}\|f_n - f_k\| > t) \le 2\sup_{n>k}\mathbb{P}(\|f_n - f_k\| > t)$$

from which we deduce similarly that f_n converges a.s. □

Corollary 1.41. *Let (Y_n) be independent variables in $L_1(B)$ with mean zero (i.e. $\mathbb{E}Y_n = 0$) and let $f_n = \sum_0^n Y_k$ as before. Then, for any $p \ge 1$, we have*

$$\|\sup\|f_n\|\|_p \le 2^{1+1/p}\sup\|f_n\|_{L_p(B)}.$$

Proof. Let $(Y'_n)_n$ be an independent copy of the sequence (Y_n), let $\widetilde{Y}_n = Y_n - Y'_n$ and $f'_n = \sum_0^n Y'_n$. Note that $(\widetilde{Y}_n)$ are independent and symmetric. By (1.43), we have

$$\mathbb{E}\sup\|\widetilde{f}_n\|^p \le 2\sup\mathbb{E}\|\widetilde{f}_n\|^p,$$

but now if $p \ge 1$, we have by convexity

$$\begin{aligned}\mathbb{E}\sup\|f_n\|^p &= \mathbb{E}\sup\|f_n - \mathbb{E}f'_n\|^p \le \mathbb{E}\sup\|f_n - f'_n\|^p \\ &\le 2\sup\mathbb{E}\|f_n - f'_n\|^p \\ &\le 2\sup\mathbb{E}(\|f_n\| + \|f'_n\|)^p \\ &\le 2^p(\mathbb{E}\sup\|f_n\|^p + \mathbb{E}\sup\|f'_n\|^p) = 2^{p+1}\mathbb{E}\sup\|f_n\|^p.\end{aligned}$$ □

Corollary 1.42. *For a series of independent B-valued random variables, convergence in probability implies almost sure convergence.*

Proof. Let $f_n = \sum_0^n Y_k$, with (Y_k) independent. Let (Y'_k) be an independent copy of the sequence (Y_k) and let $f'_n = \sum_0^n Y'_k$. Then the variables $(Y_k - Y'_k)$ are independent and symmetric. If f_n converges in probability (when $n \to \infty$), then obviously f'_n and hence $f_n - f'_n$ also does. By the preceding theorem, $f_n - f'_n$ converges a.s. therefore we can choose fixed values $x_n = f'_n(\omega_0)$ such that $f_n - x_n$ converges a.s. A fortiori, $f_n - x_n$ converges in probability, and since f_n also does, the difference $f_n - (f_n - x_n) = x_n$ also does, which means that (x_n) is convergent in B. Thus the a.s. convergence of $f_n - x_n$ implies that of f_n. □

Remark. Let B be any Banach space. Any B-valued martingale (M_n) bounded in $L_1(B)$ that converges in probability must converge a.s. Indeed, if it converges in probability there is a subsequence $(M_{n(k)})$ that converges a.s., and we will show that a.s. convergence of the whole sequence follows. Fix $t > 0$ and consider $(M_{n\wedge T})$ as in Lemma 1.32. The a.s. converging subsequence $(M_{n(k)\wedge T})$ must converge in $L_1(B)$ to a limit f since by (1.40) the convergence is dominated. By Proposition 1.22 we have $M_{n\wedge T} = \mathbb{E}^{\mathcal{A}_n}(M_{n(k)\wedge T})$ for any $n \leq n(k)$ and hence by the continuity of $\mathbb{E}^{\mathcal{A}_n}$ we must have $M_{n\wedge T} = \mathbb{E}^{\mathcal{A}_n} f$. By Theorem 1.30 $M_{n\wedge T} \to f$ a.s. This shows that (M_n) converges a.s. on the set $\{T = \infty\} = \{\sup \|M_n\| \leq t\}$. But by Doob's inequality, we can choose $t > 0$ large enough so that $\{\sup \|M_n\| > t\}$ is arbitrarily small, and hence we conclude that (M_n) converges a.s.

We now come to a very useful classical result due to Ito and Nisio [272].

Theorem 1.43. *Consider a separable Banach space B. Let $D \subset B^*$ be a countable subset that is norming, i.e. such that, for x in B, we have* $\|x\| = \sup_{\xi\in D} |\xi(x)|$. *Let (f_n) be as in Theorem 1.40. Assume that ω-a.s. there is $f(\omega)$ in B such that $\xi(f(\omega)) = \lim \xi(f_n(\omega))$ for any ξ in D.*

Then $\lim \|f_n(\omega) - f(\omega)\| = 0$ *a.s.*

Proof. Exceptionally, we will use here the results of the next section. We start by the observation that f is Borel measurable. Indeed, our assumptions on D and f imply that for any $x_0 \in B$ and any $R > 0$ the set $\{\|f - x_0\| < R\} = \{\omega \mid \sup_{\xi\in D} |\xi(f(\omega)) - \xi(x_0)| < R\}$ is measurable (because the f_n's are assumed Borel measurable). Thus, $f^{-1}(\beta)$ is measurable for any open ball β, and since B is separable, f must be Borel measurable. Hence by Corollary 1.46, for any $\varepsilon > 0$ there is $K \subset B$ convex compact such that

$$P(f \notin K) < \varepsilon.$$

By the proof of the preceding statement, since $f_n - (f - f_n)$ clearly has the same distribution as f for any n, we have

$$\mathbb{P}\left(\bigcup_n \{f_n \notin K\}\right) \leq 2\mathbb{P}(f \notin K) < 2\varepsilon.$$

Since $\varepsilon > 0$ is arbitrary, it follows that $\{f_n(\omega) | n \geq 0\}$ is relatively compact ω-a.s. But then the norm topology coincides with the $\sigma(B, D)$-topology on $\{f_n(\omega) | n \geq 0\}$, and since $f_n(\omega) \xrightarrow{\sigma(B,D)} f(\omega)$, we conclude as announced that $f_n(\omega) \to f(\omega)$ in norm ω-a.s. □

Remark. Here is a nice application of the preceding statement to random Fourier series on $\mathbb{T} = \mathbb{R}/2\pi\mathbb{Z}$. Consider complex numbers $\{a_n \mid n \in \mathbb{Z}\}$, and let (ε_n) be an independent sequence of symmetric ± 1 valued variables as usual. Let us consider the *formal* Fourier series

$$\sum_{n \in \mathbb{Z}} \varepsilon_n a_n e^{int}.$$

We wish to know when this series is almost surely the Fourier series of a (random) function in B, where B is either $L_p([0, 2\pi], dt)$ $(1 \leq p < \infty)$ or $C([0, 2\pi])$. So the question is, when is it true that for almost all choices of signs $(\varepsilon_n(\omega))$, there is a function $f_\omega \in B$ such that

$$\forall n \in \mathbb{Z} \qquad \widehat{f_\omega}(n) = \varepsilon_n(\omega) a_n?$$

The preceding result shows that this holds iff the sequence of partial sums $S_N = \sum_{|n| \leq N} \varepsilon_n a_n e^{int}$ converges a.s. in norm in B. (Just take for D the intersection of the unit ball of B^* with the set of all finite Fourier series, with, say, rational coefficients. To verify that it is norming, one may use the Fejer kernel.) Actually, the norm convergence holds in any reordering of the summation. Thus we find that norm convergence of the partial sums (in any order) is automatic, which is in sharp contrast with the case of ordinary Fourier series of (non-random) continuous functions.

1.8 Phillips's theorem

Let $(\Omega, \mathcal{A}, \mathbb{P})$ be a σ-finite measure space. A function $f\colon\ \Omega \to B$ is called scalarly measurable if for every ξ in B^* the scalar valued function $\xi(f)$ is measurable.

Now assume B separable. As is well known, the unit ball of B^* is weak* compact and metrizable, and hence separable. Therefore, there is a dense countable subset $D \subset B^*$ such that

$$\forall x \in B \qquad \|x\| = \sup_{\xi \in D} |\xi(x)|. \tag{1.46}$$

Moreover B^* is $\sigma(B^*, B)$-separable. By (1.46) if $f\colon\ \Omega \to B$ is scalarly measurable, then $\|f\|_B$ is measurable. Similarly, for every x in B, the function $\|f - x\|_B$ is measurable. The following application of Theorems 1.34 and 1.36 is a classical result (due to Phillips) in measure theory on Banach spaces.

Theorem 1.44. *Let $(\Omega, \mathcal{A}, \mathbb{P})$ be σ-finite. Assume that B is a separable Banach space. Then every scalarly measurable function $f\colon\ \Omega \to B$ is Bochner measurable.*

Proof. Let D be as in (1.46). Since $\|f(\cdot)\|_B$ is a finite (measurable) function, we have $\Omega = \bigcup_n \{\|f\|_B \le n\}$; therefore we can easily reduce the proof to the case when $\|f(\cdot)\|_B$ is bounded, and in particular, we can assume $\|f\|_B \in L_1(\Omega, \mathcal{A}, \mathbb{P})$. Moreover, we can easily reduce to the case of a probability space $(\Omega, \mathcal{A}, \mathbb{P})$ and also assume that $\mathcal{A}$ is the σ-algebra generated by $(\xi(f))_{\xi \in D}$. Hence we can assume $\mathcal{A}$ countably generated. It is then easy to check that there is an increasing sequence $(\mathcal{A}_n)$ of finite σ-subalgebras of $\mathcal{A}$ such that the union $\bigcup_{n \ge 0} \mathcal{A}_n$ generates $\mathcal{A}$. For each $n \ge 0$, we can then define an $\mathcal{A}_n$-measurable simple function $f_n\colon\ \Omega \to B^{**}$ by setting

$$\langle \xi, f_n(\omega) \rangle = \mathbb{E}^{\mathcal{A}_n}(\langle \xi, f \rangle). \tag{1.47}$$

Indeed, for any atom A of $\mathcal{A}_n$, and any ω in A we set $f_n(\omega) = x^{**}$ where $x^{**} \in B^{**}$ is defined by $x^{**}(\xi) = \mathbb{E}(1_A \xi(f(\omega))$. Clearly, (1.47) holds and (f_n) is a martingale, so that $(\|f_n\|)_{n \ge 0}$ is a submartingale. Similarly, for every x in B, $(f_n - x)$ is a martingale and hence $(\|f_n - x\|)$ is a submartingale. By Theorem 1.36, $\|f_n - x\|$ converges a.s. for every x in B.

We first claim that $\lim \|f_n - x\| \le \|f - x\|$ a.s. (see also the next remark). Indeed note that by (1.47) we have for any ξ in the unit ball of B^*

$$|\xi(f_n - x)| = |\mathbb{E}^{\mathcal{A}_n}(\xi(f - x))| \le \mathbb{E}^{\mathcal{A}_n}\|f - x\|,$$

hence, taking the sup over all such ξ (and recalling that $f_n - x$ is a simple function), we find

$$\|f_n - x\| \le \mathbb{E}^{\mathcal{A}_n}\|f - x\|,$$

but by Theorem 1.30 (scalar case),

$$\limsup_{n\to\infty} \mathbb{E}^{\mathcal{A}_n}\|f - x\| = \|f - x\|;$$

therefore a.s. as announced,

$$\forall x \in B \qquad \limsup_{n\to\infty} \|f_n - x\| \le \|f - x\|. \tag{1.48}$$

This is enough to conclude. Indeed, let D_0 be a dense countable subset of B and let A be the set of all ω in Ω such that $\limsup_{n\to\infty} \|f_n(\omega) - x\| \le \|f(\omega) - x\|$

for all x in D_0. Then (since D_0 is countable)

$$A \in \mathcal{A} \quad \text{and} \quad \mathbb{P}(A) = 1.$$

But when a sequence of points x_n in B^{**} satisfies for some b in B for all x in a dense subset of B,

$$\limsup_{n\to\infty} \|x_n - x\| \leq \|b - x\|, \tag{1.49}$$

it is very easy to check that x_n must tend to b when $n \to \infty$. (Indeed, this inequality (1.49) remains true by density for all x in B, hence we simply take $x = b$!) In particular, for any ω in A we must have $\lim_{n\to\infty} \|f_n(\omega) - f(\omega)\| = 0$, which shows that f is Bochner measurable as a B^{**} valued function. Finally, the next lemma shows that f is Bochner measurable as a B-valued function (and hence a posteriori, the variables f_n actually take their values in B). □

Lemma 1.45. *Let B be a closed subspace of a Banach space X. Let f be an X valued Bochner measurable function. If the values of f are actually (almost all) in B, then f is Bochner measurable as a B-valued function.*

Proof. For any point x in X, let us choose a point $\tilde{x} \in B$ such that

$$\|x - \tilde{x}\| \leq 2\,\mathrm{dist}(x, B).$$

Let (f_n) be a sequence of X valued (measurable) simple functions tending pointwise to f. Let $g_n(\omega) = \tilde{f}_n(\omega)$. Note that the 'steps' of g_n are the same as those of f_n. Then,

$$\|f_n(\omega) - g_n(\omega)\| \leq 2\,\mathrm{dist}(f_n(\omega), B) \leq 2\|f_n(\omega) - f(\omega)\| \to 0.$$

Therefore (g_n) is a sequence of B-valued (measurable) simple functions tending pointwise to f. □

Remark. Actually, there is always equality in (1.48). Indeed, let $\{\xi_1, \xi_2, \ldots\}$ be an enumeration of the set D. Note that, for each fixed k, the variables $\sup_{j\leq k} |\xi_j(f_n - f)|$ tend to 0 when $n \to \infty$ a.s. and in L_1, since this is true for each single martingale $(\xi(f_n))_{n\geq 0}$ by the martingale convergence theorem. Hence

$$\begin{aligned} \mathbb{E}\|f - x\| &= \sup_k \mathbb{E} \sup_{j\leq k} |\xi_j(f - x)| \qquad (1.50)\\ &\leq \sup_k \mathbb{E} \limsup_{n\to\infty} \sup_{j\leq k} |\xi_j(f_n - x)| \\ &\leq \mathbb{E} \limsup_{n\to\infty} \|f_n - x\|, \end{aligned}$$

but since $\|f_n - x\|$ converges a.s., we have by Fatou's lemma

$$\mathbb{E} \limsup \|f_n - x\| = \mathbb{E} \liminf \|f_n - x\| \leq \liminf \mathbb{E}\|f_n - x\|, \tag{1.51}$$

and by (1.48),

$$\leq \mathbb{E}\|f - x\|. \tag{1.52}$$

The chain of inequalities (1.50), (1.51), (1.52) and (1.48) shows that

$$\mathbb{E} \lim \|f_n - x\| = \mathbb{E}\|f - x\| \text{ but } \lim \|f_n - x\| \leq \|f - x\| .$$

This forces $\lim_{n\to\infty} \|f_n - x\| = \|f - x\|$ a.s.

Corollary 1.46. *Let $f\colon \Omega \to B$ be a scalarly measurable (in particular a Borel measurable) function with values in a separable Banach space, then for any $\varepsilon > 0$, there is a convex compact subset $K \subset B$ such that*

$$\mathbb{P}\{f \notin K\} < \varepsilon.$$

Proof. Consider $\varepsilon_k > 0$ tending to zero and such that $\sum_k \varepsilon_k < \varepsilon$. By the theorem, there is a sequence of simple functions (f_n) tending to f. Now since $\|f_n - f\| \to 0$ a.s. for any $k > 0$, there is $n(k)$ large enough so that

$$\mathbb{P}\{\|f_{n(k)} - f\| > \varepsilon_k\} < \varepsilon_k. \tag{1.53}$$

Let T_k be the (finite) set of values of $f_{n(k)}$, and let T be the set of points $b \in B$ such that $d(b, T_k) \leq \varepsilon_k$ for all k. Clearly T is relatively compact (since for any $\delta > 0$ it admits a finite δ-net, namely T_k for any k such that $\varepsilon_k < \delta$). By (1.53), we have $\mathbb{P}\{d(f, T_k) > \varepsilon_k\} < \varepsilon_k$, and hence $\mathbb{P}\{f \notin T\} \leq \sum_k \varepsilon_k < \varepsilon$. To conclude, we simply let K be the closure of the convex hull of T. □

Remark. When the conclusion of the preceding corollary holds, one usually says that the distribution of f (i.e. the image measure of $\mathbb{P}$ under f) is a Radon measure on B, or that it is 'tight'. What the corollary says is that this is automatic in the *separable* case.

1.9 Reverse martingales

We will prove here the following.

Theorem 1.47. *Let B be an arbitrary Banach space. Let $(\Omega, \mathcal{A}, \mathbb{P})$ be a probability space and let $\mathcal{A}_0 \supset \mathcal{A}_{-1} \supset \mathcal{A}_{-2} \supset \cdots$ be a (this time decreasing) sequence of σ-subalgebras of $\mathcal{A}$. Let $\mathcal{A}_{-\infty} = \bigcap_{n\geq 0} \mathcal{A}_{-n}$. Then for any f in $L_p(\Omega, \mathcal{A}, P; B)$, with $1 \leq p < \infty$, the reverse martingale $(\mathbb{E}^{\mathcal{A}_{-n}}(f))_{n\geq 0}$ converges to $\mathbb{E}^{\mathcal{A}_{-\infty}}(f)$ a.s. and in $L_p(B)$.*

We first check the convergence in $L_p(B)$. Since the operators $(\mathbb{E}^{\mathcal{A}_{-n}})_{n\geq 0}$ are equicontinuous on $L_p(B)$ it suffices to check this for f in a dense subset of $L_p(B)$. In particular, it suffices to consider f of the form $f = \sum_1^n \varphi_i x_i$ with φ_i

an indicator function and x_i in B. Since $\varphi_i \in L_2(\Omega, \mathbb{P})$, we have (by classical Hilbert space theory) $\mathbb{E}^{\mathcal{A}_{-n}}\varphi_i \to \mathbb{E}^{\mathcal{A}_{-\infty}}\varphi_i$ in $L_2(\Omega, \mathbb{P})$ when $n \to \infty$. (Note that $L_2(\Omega, \mathcal{A}_{-\infty}, \mathbb{P})$ is the intersection of the family $(L_2(\Omega, \mathcal{A}_{-n}, \mathbb{P}))_{n\geq 0}$.)

Observe that $\|f - g\|_p \leq \|f - g\|_2$ if $p \leq 2$ and $\|f - g\|_p^p \leq 2^{p-2}\|f - g\|_2^2$ if $\|f\|_\infty \leq 1$, $\|g\|_\infty \leq 1$ and $p > 2$. Using this, we obtain that, a fortiori, $\mathbb{E}^{\mathcal{A}_{-n}}f \to \mathbb{E}^{\mathcal{A}_{-\infty}}f$ in $L_p(B)$ for every f of the earlier form, and hence for every f in $L_p(B)$.

We now turn to a.s. convergence. We first replace f by $\tilde{f} = f - \mathbb{E}^{\mathcal{A}_{-\infty}}(f)$ so that we can assume $\mathbb{E}^{\mathcal{A}_{-n}}(f) \to 0$ in $L_p(B)$ and a fortiori in $L_1(B)$. Let $f_n = \mathbb{E}^{\mathcal{A}_{-n}}f$. Now fix $n > 0$ and $k > 0$ and consider the (ordinary sense) martingale

$$M_j = \begin{cases} f_{-n-k+j} & \text{for } j = 0, 1, \ldots, k, \\ f_{-n} & \text{if } \ j \geq k. \end{cases}$$

Then, by Doob's inequality (1.37) applied to (M_j), we have for all $t > 0$

$$t\mathbb{P}\{\sup_{n\leq m\leq n+k} \|f_{-m}\| > t\} \leq \mathbb{E}\|f_{-n}\|;$$

therefore

$$t\mathbb{P}\{\sup_{m\geq n} \|f_{-m}\| > t\} \leq \mathbb{E}\|f_{-n}\|,$$

and since $\mathbb{E}\|f_{-n}\| \to 0$ when $n \to \infty$, we have $\sup_{m\geq n} \|f_{-m}\| \to 0$ a.s., or equivalently $f_{-n} \to 0$ a.s. when $n \to \infty$. □

As a corollary, we have the following classical application to the strong law of large numbers.

Corollary 1.48. *Let $\varphi_1, \ldots, \varphi_n$ be a sequence of independent, identically distributed random variables in $L_1(\Omega, \mathcal{A}, \mathbb{P}; B)$. Let $S_n = \varphi_1 + \cdots + \varphi_n$. Then $S_n/n \to \mathbb{E}\varphi_1$ a.s. and in $L_1(B)$.*

Proof. Let $\mathcal{A}_{-n}$ be the σ-algebra generated by $(S_n, S_{n+1}, \ldots)$. We claim that $S_n/n = \mathbb{E}^{\mathcal{A}_{-n}}(\varphi_1)$. Indeed, for every $k \leq n$, since the exchange of φ_1 and φ_k preserves $S_n, S_{n+1}, \ldots$, we have

$$\mathbb{E}^{\mathcal{A}_{-n}}(\varphi_k) = \mathbb{E}^{\mathcal{A}_{-n}}(\varphi_1).$$

Therefore, averaging the preceding equality over $k \leq n$, we obtain

$$\mathbb{E}^{\mathcal{A}_{-n}}(\varphi_1) = \frac{1}{n}\sum\nolimits_{1\leq k\leq n} \mathbb{E}^{\mathcal{A}_{-n}}(\varphi_k) = \mathbb{E}^{\mathcal{A}_{-n}}(S_n/n) = S_n/n.$$

Hence $(S_n/n)_{n\geq 1}$ is a reverse martingale satisfying the assumptions of the preceding theorem (we may take say $\mathcal{A}_0 = \mathcal{A}_{-1}$); therefore $\frac{1}{n}S_n \to \mathbb{E}^{\mathcal{A}_{-\infty}}(\varphi_1)$ a.s. and in $L_1(B)$. Finally, let $\mathcal{T} = \bigcap_{n\geq 0} \sigma\{\varphi_n, \varphi_{n+1}, \ldots\}$ be the tail σ-algebra. By the zero-one law, $\mathcal{T}$ is trivial. The limit of S_n/n is clearly $\mathcal{T}$-measurable, hence it must be equal to a constant c, but then $\mathbb{E}(S_n/n) \to c$, so $c = \mathbb{E}(\varphi_1)$. □

1.10 Continuous time*

Although martingales with continuous time play a central role in probability theory (via Brownian motion and stochastic integrals), in the Banach space valued case they have not been so essential up to now. So, although we occasionally discuss the continuous case, we will mostly avoid developing that direction. Actually, in the context of martingale inequalities (our main interest in this book), the discrete time case is almost always the main point. Furthermore, in many situations, once the discrete case is understood, the continuous case follows by rather routine techniques, which are identical in the scalar and Banach valued case. The latter techniques are now very well described in several books, to which we refer the reader, e.g. [18, 22, 61–64, 81, 90].

In this section we briefly outline some of the basic points of martingales with continuous parameter analogous to what we saw for the basic convergence results proved in the discrete case in the preceding chapter.

Let $(\Omega, \mathcal{A}, \mathbb{P})$ be a probability space. Let $I \subset \mathbb{R}$ be a time interval. Given a filtration $(\mathcal{A}_t)$ indexed by I of σ-subalgebras of $\mathcal{A}$, we say that $(M_t)_{t \in I}$ is a martingale if (as usual) it is a martingale relative to the ordered set I. Of course if I contains its right endpoint denoted by b then $M_t = \mathbb{E}^{\mathcal{A}_t} M_b$ for all $t \in I$.

We will say that a collection of B-valued random variables $(M_t)_{t \in I}$ is *separable* if there is a countable subset $Q \subset I$ (in practice Q will be the set of rational numbers in I) and $\Omega' \subset \Omega$ in $\mathcal{A}$ with $\mathbb{P}(\Omega') = 1$ such that for any $\omega \in \Omega'$ and any $t \in I$ there is a sequence (t_n) in Q such that $t_n \to t$ and $M_t(\omega) = \lim_{n \to \infty} M_{t_n}(\omega)$. (The usual definition is slightly more general.) This holds in particular if the paths $t \mapsto M_t(\omega)$ are continuous on I (or merely either left or right continuous assuming, say, that I is open).

This implies for instance that for any open subinterval $J \subset I$, we have a.s.

$$\sup_{t \in J} \|M_t\| = \sup_{t \in J \cap Q} \|M_t\| \text{ and } \sup_{t,s \in J} \|M_t - M_s\| = \sup_{t,s \in J \cap Q} \|M_t - M_s\|,$$

thus proving that suprema of this kind are measurable, even though each is a priori defined as an uncountable supremum of measurable functions.

It is important to emphasize that for the definition of separability to make sense, we must be given (M_t) as a collection of measurable functions and *not just as equivalent classes* modulo a.s. equality. While the distinction is irrelevant in the discrete time case, it is important in the continuous case.

When $I = [0, \infty)$, we denote by $\mathcal{A}_\infty$ the smallest σ-algebra containing all the σ-algebras $\mathcal{A}_t$. Without loss of generality, we may assume that $\mathcal{A} = \mathcal{A}_\infty$.

The basic martingale convergence Theorem 1.14 is easy to extend.

Theorem 1.49. *Assume $I = [0, \infty)$. Let $1 \le p < \infty$, let B be any Banach space and let $M \in L_p(\Omega, \mathbb{P}, \mathcal{A}; B)$. We have*

$$\lim_{t\to\infty} \mathbb{E}^{\mathcal{A}_t} M = \mathbb{E}^{\mathcal{A}_\infty} M$$

in $L_p(B)$. Moreover, if (M_t) is a separable martingale such that $M_t \overset{a.s}{=} \mathbb{E}^{\mathcal{A}_t} M$ for any $t > 0$, then $\lim_{t\to\infty} M_t = \mathbb{E}^{\mathcal{A}_\infty} M$ almost surely.

Proof. The set I is a particular instance of directed set, as considered in Remark 1.17. Thus the convergence in $L_p(B)$ has already been established there. Let $M_\infty = \mathbb{E}^{\mathcal{A}_\infty} M$. Assume first $p > 1$. We claim that for any $0 \le s < \infty$ we have

$$\| \sup\nolimits_{t>s} \|M_t - M_s\|_B \|_p \le p' \sup\nolimits_{t>s} \|M_t - M_s\|_{L_p(B)}. \tag{1.54}$$

Indeed, if the sup on the left-hand side is restricted to a countable subset $\{t \in Q \mid t > s\}$, then this is easy to derive from Doob's inequality (1.38) (applied to arbitrary finite subsets of this set). But now since (M_t) is separable, there is $\Omega' \subset \Omega$ with $\mathbb{P}(\Omega') = 1$ and Q countable such that for any $\omega \in \Omega'$ we have $\sup_{t>s} \|M_t(\omega) - M_s(\omega)\|_B = \sup_{t>s, t\in Q} \|M_t(\omega) - M_s(\omega)\|_B$. This proves our claim. Similarly, if $p = 1$, we obtain

$$\sup\nolimits_{c>0} \mathbb{P}\{\sup\nolimits_{t>s} \|M_t - M_s\|_B > c\} \le \sup\nolimits_{t>s} \|M_t - M_s\|_{L_1(B)}.$$

Then the proof can be completed exactly as we did for Theorem 1.30 by showing that $\lim_{s\to\infty} \sup_{t>s} \|M_t - M_s\| = 0$ a.s. □

Remark 1.50. It is instructive to observe that Doob's maximal inequalities for directed martingales can be exploited effectively to justify (1.54), as follows. Let us denote by Z_s^* the lattice supremum of the family $\{\|M_t - M_s\|_B \mid t > s\}$ in the sense of Remark 1.39. By the latter remark, we have for any $p > 1$ $\|Z_s^*\|_p \le p' \sup_{t>s} \|M_t - M_s\|_{L_p(B)}$. By definition, Z_s^* is the smallest Z such that $Z \in L_p$ and $\|M_t - M_s\|_B \le Z$ a.s. for any $t > s$. But by our separability assumption, the variable $Z = \widetilde{Z}_s^*$ defined by

$$\widetilde{Z}_s^*(\omega) = \sup_{t>s, t\in Q} \|M_t(\omega) - M_s(\omega)\|_B = \sup_{t>s, t\in Q} \|M_t(\omega) - M_s(\omega)\|_B \quad \forall \omega \in \Omega'$$

satisfies this and is obviously minimal. Thus we must have $Z_s^* = \widetilde{Z}_s^*$ a.s., and we obtain (1.54). A similar argument can be used for $p = 1$.

The definitions and the basic properties of stopping times and of the algebra $\mathcal{A}_T$ can be extended without any problem: we say that a function : $\Omega \to [0, \infty]$ is a stopping time if $\{T \le t\} \in \mathcal{A}_t$ for any $t \in [0, \infty)$, and by definition $A \in \mathcal{A}_T$ if $A \cap \{T \le t\} \in \mathcal{A}_t$ for any $t \in [0, \infty)$.

Since we will be mainly interested in Brownian martingales in §3.5 and §4.5, far from seeking maximal generality (for that we refer e.g. to [18, 64] or [81]) we allow ourselves rather strong assumptions, which are satisfied in the Brownian case. We will often assume, as is the case for Brownian motion, that the paths $t \mapsto M_t(\omega)$ are continuous for almost all ω. If, in addition, we also assume $t \mapsto M_t$ locally bounded on I in $L_p(B)$ for some $p > 1$, then by dominated convergence (by Doob's inequality), $t \mapsto M_t$ is continuous from I to $L_p(B)$.

We will now describe how the continuous time can be discretized in order to apply the discrete time theory that we just saw. The following basic facts are easy to verify at this stage: If (M_t) is a B-valued martingale with respect to $(\mathcal{A}_t)$, then $(\|M_t\|_B)$ is a non-negative submartingale. Assuming that $t \mapsto M_t$ is separable and bounded in $L_p(\Omega, \mathcal{A}, \mathbb{P}; B)$, then Doob's inequality (1.38) clearly extends: For any $1 < p < \infty$ and any $0 < R < \infty$ we have

$$\| \sup_{0<t\leq R} \|M_t\|_B\|_p \leq p' \| \|M_R\|_B\|_p, \tag{1.55}$$

and similarly for (1.37).

Fix $R > 0$. It will be convenient to consider the following assumption:

$(A)_R$ *The paths* $t \mapsto M_t(\omega)$ *are (uniformly) continuous on* $[0, R]$ *and we have*

$$\mathbb{E} \sup_{0\leq t\leq R} \|M_t\| < \infty. \tag{1.56}$$

Remark 1.51. By dominated convergence, if the paths are a.s. (uniformly) continuous on $[0, R]$ (here a.s. means except for a negligible measurable subset of Ω), we have necessarily

$$\lim_{\delta\to 0} \mathbb{E} \sup_{0\leq s,t\leq R,\ |t-s|\leq\delta} \|M_t - M_s\| = 0.$$

We start with an elementary discretization of stopping times.

Lemma 1.52. *Fix* $R > 0$. *Then, for any stopping time* T *(with respect to* $(\mathcal{A}_t)$*), there is a sequence of stopping times* $S_n \geq T$ *taking only finitely many values in* $[0, R] \cup \{\infty\}$ *such that almost surely*

$$T \wedge R = \lim S_n \wedge R.$$

Proof. Let $0 = t_0 < t_1 < t_2 < \cdots t_{N-1} < t_N = R$ be such that $|t_j - t_{j-1}| \leq \delta$ for all j. Let $Q = \{t_0, \ldots, t_N\}$. We define

$$S(\omega) = \inf\{q \in Q \mid q \geq T(\omega)\},$$

and we set $S(\omega) = \infty$ if $T > R$. This means that if $0 < S(\omega) \leq R$ then $S(\omega)$ is equal to the unique t_k such that $t_{k-1} < T(\omega) \leq t_k$. Then S is a stopping time with respect to the filtration $(\mathcal{A}_t)$ with $S \geq T$, and S is a simple function (admitting ∞ as one of its values) with values in $Q \cup \{\infty\}$, such that $|S \wedge R - T \wedge R| \leq \delta$. Choosing, say, $\delta = 1/n$, the lemma follows. □

The next result is an extension of Proposition 1.22.

Lemma 1.53. *Let T be a stopping time with respect to $(\mathcal{A}_t)$. Fix $R > 0$. Let B be a Banach space and let (M_t) be a martingale (with respect to $(\mathcal{A}_t)$) satisfying $(A)_R$. Then, for any $0 \leq s < t \leq R$ we have a.s.*

$$\mathbb{E}^{\mathcal{A}_s} M_{T\wedge t} = M_{T\wedge s}. \tag{1.57}$$

Proof. Let us first assume that T takes only finitely many values included in the set $[0, R] \cup \{\infty\}$. Let $Q = \{t_0, \ldots, t_N\}$, be a finite set containing $\{s, t\}$ and also containing all the possible finite values of T. We may assume $0 = t_0 < t_1 < t_2 < \cdots < t_{N-1} < t_N = R$. The idea of the proof is very simple: We consider the martingale $(M_t)_{t\in Q}$ adapted to $(\mathcal{A}_t)_{t\in Q}$ and we view T as a stopping time relative to the latter filtration and taking values in $Q \cup \{\infty\}$. Then Proposition 1.22 yields (1.57). More precisely, for any $0 \leq j \leq N$, we set

$$\mathcal{B}_j = \mathcal{A}_{t_j} \quad \text{and} \quad f_j = M_{t_j}$$

and also $\mathcal{B}_j = \mathcal{A}_R$ and $f_j = M_R$ for all $j > N$. Then it is easy to check that (f_j) is a martingale adapted to the filtration $(\mathcal{B}_j)$. With this new filtration, the time T corresponds to the variable $\theta : \Omega \to \mathbb{N}$ defined as follows. We set $\theta = j$ on the set $\{T = t_j\}$ and $\theta = N$ on the set $\{T = \infty\}$. Since Q contains all the finite values of T, this defines θ on Ω, and it is easy to check that θ is now a stopping time relative to the filtration $(\mathcal{B}_j)$. By Proposition 1.22 applied to the martingale (f_j) we have for any $0 \leq j \leq k \leq N$

$$\mathbb{E}^{\mathcal{B}_j} f_{\theta\wedge k} = f_{\theta\wedge j}. \tag{1.58}$$

In particular, this holds for j, k such that $t_j = s$ and $t_k = t$. In that case $\mathcal{B}_j = \mathcal{A}_{t_j} = \mathcal{A}_s$, $f_{\theta\wedge j} = M_{T\wedge s}$ and $f_{\theta\wedge k} = M_{T\wedge t}$. Thus (1.58) coincides with (1.57) in this case. This proves (1.57) under our initial assumption on T.

To prove the general case, we use a sequence S_n as in Lemma 1.52. By the first part of the proof we have for any n

$$\mathbb{E}^{\mathcal{A}_s} M_{S_n\wedge t} = M_{S_n\wedge s}. \tag{1.59}$$

But now since the paths are continuous we know that $M_{S_n\wedge s} \to M_{T\wedge s}$ and $M_{S_n\wedge t} \to M_{T\wedge t}$ a.s. when $n \to \infty$, and moreover by dominated convergence (recall (1.56)) the convergence holds also in $L_1(B)$. Thus, passing to the limit in n, we see that (1.59) yields (1.57). □

Remark. In the 1970s M. Métivier and J. Pellaumail (see [61, 62] for details) strongly advocated the idea that stochastic integrals should be viewed as vector measures with values in the topological vector space L_0 (resp. $L_0(B)$) of measurable functions with values in $\mathbb{R}$ (resp. a Banach space B). Equipped with convergence in measure the latter space is metrizable but not locally convex. The relevant vector measures are defined on the σ-algebra $\mathcal{P}$ of 'predictable subsets' of $[0,\infty)\times\Omega$, associated to a filtration $(\mathcal{A}_t)_{t>0}$ on Ω. The σ-algebra $\mathcal{P}$ is defined as the smallest one containing all subsets of $[0,\infty)\times\Omega$ of the form

$$[S,T] = \{(t,\omega) \mid S(\omega) < t \le T(\omega)\},$$

where S, T are stopping times such that $S \le T$, as well as all subsets of the form $\{0\}\times A$ with $A \in \mathcal{A}_0$. This is the same as the σ-algebra generated by all real valued adapted processes with continuous paths.

Moreover, $\mathcal{P}$ contains all the elementary processes of the form

$$f(t,\omega) = 1_{]t_0,t_1]}(t)g(\omega)$$

with g bounded and $\mathcal{A}_{t_0}$-measurable $(0 < t_0 < t_1 < \infty)$.

For instance, given a martingale (M_t) in $L_1(\Omega,\mathbb{P})$, we can define a vector measure μ_M on $\mathcal{P}$ by setting first

$$\mu_M(]S,T]) = M_T - M_S\,, \quad \mu_M(\{0\}\times A) = 0$$

and then extending this to all of $\mathcal{P}$.

In particular, if f is an elementary process as previously, we have

$$\int f d\mu_M = (M_{t_1} - M_{t_0})g.$$

This allows to make sense of $\int F d\mu_M$ when F is a linear combination of elementary processes and one then denotes simply

$$\int F dM = \int F d\mu_M.$$

In this way one defines the stochastic integral with respect to M of any bounded predictable process supported on a bounded interval. Actually it suffices for μ_M to extend to such a vector measure on $\mathcal{P}$ that M be a so-called 'semi-martingale' (roughly this is the sum of a martingale and a process with bounded variation) and, quite remarkably, it was proved (by Dellacherie-Mokobodzki-Meyer and Bichteler) that conversely only semi-martingales give rise to an L_0 valued vector measure (among all processes with sufficiently regular paths, called 'càdlàg'). This result is presented in detail in [62], to which we refer the reader for the precise definition of a semi-martingale.

1.11 Notes and remarks

Among the many classical books on Probability that influenced us, we mention [11, 22], see also [35]. As for martingales, the references that considerably influenced us are [26, 34, 71] and the papers [165, 175].

Martingales were considered long before Doob (in particular by Paul Lévy) but he is the one who invented the name and proved their basic almost sure convergence properties using what is now called Doob's maximal inequality.

In Theorem 1.40, we slightly digress and concentrate on a particular sort of martingale, those that are partial sums of series of independent random vectors. In the symmetric case, it turns out that the maximal inequalities (and the associated almost sure convergence) hold for 'martingales' bounded in $L_p(B)$ for $p < 1$. Our presentation of this is inspired by Kahane's book [44]. See Hoffmann-Jørgensen's [265] for more on this theme. We also include an important useful result (Theorem 1.43) due to Ito and Nisio [272] showing that, for such sums, even a very mild looking kind of convergence automatically guarantees norm convergence.

Although this looks at first rather eccentric, conditional expectations and martingales may be considered with respect to certain complex measures, see [190] for more details.

This chapter ends with a brief presentation of continuous time martingales. As already mentioned in the text, many excellent books are now available such as e.g. [18, 61–64, 81, 90]. We should warn the reader that we use a stronger notion of 'separability' for a random process than in most textbooks, such as [63], but the present one suffices for our needs.

2

Radon-Nikodým property

2.1 Vector measures

To introduce the Radon-Nikodým property (in short RNP), we will need to briefly review the basic theory of vector measures. Let B be a Banach space. Let $(\Omega, \mathcal{A})$ be a measure space. Every σ-additive map $\mu\colon\ \mathcal{A} \to B$ will be called a (B-valued) vector measure. We will say that μ is *bounded* if there is a finite positive measure ν on $(\Omega, \mathcal{A})$ such that

$$\forall A \in \mathcal{A} \qquad \|\mu(A)\| \le \nu(A). \tag{2.1}$$

When this holds, it is easy to show that there is a minimal choice of the measure ν. Indeed, for all A in $\mathcal{A}$ let

$$|\mu|(A) = \sup\{\Sigma \|\mu(A_i)\|\},$$

where the supremum runs over all decompositions of A as a disjoint union $A = \cup A_i$ of finitely many sets in $\mathcal{A}$. Using the triangle inequality, one checks that $|\mu|$ is an additive set function. By (2.1) $|\mu|$ must be σ-additive and finite. Clearly, when (2.1) holds, we have

$$|\mu| \le \nu.$$

We define the 'total variation norm' of μ as follows:

$$\|\mu\| = \inf\{\nu(\Omega) \mid \nu \in M(\Omega, \mathcal{A}), \nu \ge |\mu|\},$$

or equivalently

$$\|\mu\| = |\mu|(\Omega).$$

We will denote by $M(\Omega, \mathcal{A})$ the Banach space of all bounded complex valued measures on $(\Omega, \mathcal{A})$, and by $M_+(\Omega, \mathcal{A})$ the subset of all positive bounded measures. We will denote by $M(\Omega, \mathcal{A}; B)$ the space of all bounded B-valued measures μ on $(\Omega, \mathcal{A})$. When equipped with the preceding norm, it is a Banach space. Let $\mu \in M(\Omega, \mathcal{A}; B)$ and $\nu \in M_+(\Omega, \mathcal{A})$. We will write

$$|\mu| \ll \nu$$

if $|\mu|$ is absolutely continuous (or equivalently admits a density) with respect to ν. This happens iff there is a positive function $w \in L_1(\Omega, \mathcal{A}, \nu)$ such that

$$|\mu| \le w.\nu$$

or equivalently such that

$$\forall A \in \mathcal{A} \qquad \|\mu(A)\| \le \int_A w d\nu.$$

Recapitulating, we may state:

Proposition 2.1. *A vector measure μ is bounded in the preceding sense iff its total variation is finite, the total variation being defined as*

$$V(\mu) = \sup \left(\sum\nolimits_1^n \|\mu(A_i)\| \right)$$

where the sup runs over all measurable partitions $\Omega = \bigcup_{i=1}^n A_i$ of Ω. Thus, if μ is bounded, we have $V(\mu) = |\mu|(\Omega)$.

Proof. Assuming $V(\mu) < \infty$, let $\forall A \in \mathcal{A} \quad \nu(A) = \sup \left(\sum_1^n \|\mu(A_i)\| \right)$, where the sup runs over all measurable partitions $A = \bigcup_{i=1}^n A_i$ of A. Then ν is a σ-additive finite positive measure on $\mathcal{A}$, and satisfies (2.1). Thus μ is bounded in the earlier sense (and of course ν is nothing but $|\mu|$). The converse is obvious. □

Remark. Given $\nu \in M(\Omega, \mathcal{A})$ and $f \in L_1(\Omega, \mathcal{A}, \nu; B)$, we can define a vector measure μ denoted by $f.\nu$ (or sometimes simply by $f\nu$) by

$$\forall A \in \mathcal{A} \quad \mu(A) = \int_A f d\nu.$$

It is easy to check that $\mu = f.\nu$ implies

$$|\mu| = F.\nu \text{ where } F(\omega) = \|f(\omega)\|_B, \tag{2.2}$$

and therefore

$$\|f.\nu\|_{M(\Omega,\mathcal{A};B)} = \|f\|_{L_1(\Omega,\mathcal{A},\nu;B)}. \tag{2.3}$$

Indeed, by Jensen's inequality, we clearly have

$$\forall A \in \mathcal{A} \quad \|\mu(A)\| \le \int_A \|f\| d\nu,$$

hence $|\mu| \le F.\nu$. To prove the converse, let $\epsilon > 0$ and let g be a B-valued simple function such that $\int_A \|f - g\| d\nu < \epsilon$. We can clearly assume that g is supported by A, so that we can write $g = \sum_1^n 1_{A_i} x_i$, with $x_i \in B$ and A_i is a disjoint partition of A. We have

$$\Sigma \|\mu(A_i) - \nu(A_i)x_i\| = \Sigma \left\| \int_{A_i} (f-g)d\nu \right\| \le \int_A \|f-g\|d\nu < \epsilon;$$

hence

$$\int_A \|g\|d\nu = \Sigma \nu(A_i)\|x_i\| \le \Sigma \|\mu(A_i)\| + \epsilon,$$

and finally,

$$\int_A \|f\|d\nu \le \int_A \|g\|d\nu + \epsilon \le \Sigma \|\mu(A_i)\| + 2\epsilon,$$

which implies

$$\int_A \|f\|d\nu \le |\mu|(A) + 2\epsilon.$$

This completes the proof of (2.2).

Proposition 2.2. *Consider a bounded B-valued vector measure μ on $(\Omega, \mathcal{A})$. There is a unique continuous linear map $[\mu]\colon\ L_1(\Omega, \mathcal{A}, |\mu|) \to B$ such that $[\mu](1_A) = \mu(A)$ for all A in $\mathcal{A}$. By convention, for any g in $L_1(\Omega, \mathcal{A}, |\mu|)$ we will denote $[\mu](g) = \int g d\mu$. With this notation, we have*

$$\left\| \int g d\mu \right\| \le \int |g| d|\mu|. \tag{2.4}$$

Proof. Let V be the set of all linear combination of characteristic functions, i.e. $g = \sum_1^n c_i 1_{A_i}$ with $c_i \in \mathbb{C}$ and with $A_i \in \mathcal{A}$ mutually disjoint then we can define

$$\forall g \in V \qquad [\mu](g) = \sum_1^n c_i \mu(A_i).$$

This definition is clearly unambiguous and satisfies by the triangle inequality

$$\|[\mu](g)\|_B \le \sum |c_i|\ \|\mu(A_i)\| \le \sum |c_i| |\mu|(A_i) = \int |g| d|\mu|.$$

Thus $[\mu]$ is continuous on the dense linear subspace $V \subset L_1(\Omega, \mathcal{A}, |\mu|)$, hence $[\mu]$ has a unique extension to the whole of $L_1(\Omega, \mathcal{A}, |\mu|)$. □

We note if $|\mu| \le w.\nu$

$$\forall g \in L_\infty(\Omega, \mathcal{A}, \nu) \qquad \left\| \int g d\mu \right\|_B \le \int |g| w d\nu. \tag{2.5}$$

For future reference, we include here an elementary 'regularization' lemma concerning the case when $|\mu|$ is a finite measure on $\mathbb{T}$ or $\mathbb{R}$ (or any locally compact Abelian group) absolutely continuous with respect to the Haar-Lebesgue measure m.

Lemma 2.3. *Let B be a Banach space. Let $(\Omega, \mathcal{A})$ be either $\mathbb{T}$ or $\mathbb{R}$ equipped with the Borel σ-algebra and let $\mu \in M(\Omega, \mathcal{A}; B)$ be such that $|\mu|$ is absolutely continuous with respect to m. Then for any $\varphi \in L_1(\Omega, \mathcal{A}, m) \cap L_\infty(\Omega, \mathcal{A}, m)$ we define $\mu * \varphi \in M(\Omega, \mathcal{A})$ by setting for any Borel set A*

$$\mu * \varphi(A) = \int (\int \varphi(y) 1_A(x+y) dm(y)) d\mu(x).$$

Then there is a bounded continuous function $f \in L_1(\Omega, \mathcal{A}, m; B)$ such that for any Borel set A

$$\mu * \varphi(A) = \int_A f dm.$$

Proof. Let $\check{\varphi}(x) = \varphi(-x)$ and $\check{\varphi}_y(x) = \varphi(y-x)$. Let $g = 1_A$. By the translation invariance of m and a straightforward generalization of Fubini's theorem to this setting, we have

$$\begin{aligned} \mu * \varphi(A) &= \int (\int \varphi(y) 1_A(x+y) dm(y)) d\mu(x) \\ &= \int (\int 1_A(y) \check{\varphi}_y(x) dm(y)) d\mu(x) \\ &= \int_A f(y) dm(y) \end{aligned}$$

where $f(y) = \int \check{\varphi}_y d\mu$. We claim that $y \mapsto \check{\varphi}_y$ is a continuous function from Ω to $L_1(|\mu|)$. Using this, by (2.4) f is a continuous and bounded B-valued function and moreover

$$\int \|f\| dm \leq \int |\check{\varphi}_y| d|\mu| dm(y) \leq |\mu|(\Omega) \|\varphi\|_{L_1(dm)} < \infty.$$

To verify the claim, let w be the density of $|\mu|$ relative to m, so that $|\mu| = w.m$. By a classical result, for any $\varphi \in L_1(\Omega, \mathcal{A}, m)$ the mapping $y \mapsto \check{\varphi}_y$ is a continuous function from Ω to $L_1(m)$. Let $\varepsilon > 0$, let c be such that $\int_{w>c} w dm < \varepsilon$. We have

$$\int |\check{\varphi}_y - \check{\varphi}_{y'}| d|\mu| = \int |\check{\varphi}_y - \check{\varphi}_{y'}| w dm < c \int |\check{\varphi}_y - \check{\varphi}_{y'}| dm + 2\|\check{\varphi}\|_\infty \varepsilon$$

and hence $\limsup_{|y-y'|\to 0} \int |\check{\varphi}_y - \check{\varphi}_{y'}| d|\mu| \leq 2\|\check{\varphi}\|_\infty \varepsilon$. Since ε is arbitrary, the latter $\limsup = 0$, proving the claim. □

We will use very little from the theory of vector measures, but for more information we refer the interested reader to [21].

2.2 Martingales, dentability and the Radon-Nikodým property

Definition. A Banach space B is said to have the Radon-Nikodým property (in short RNP) if for every measure space $(\Omega, \mathcal{A})$, for every σ-finite positive measure ν on $(\Omega, \mathcal{A})$ and for every B-valued measure μ in $M(\Omega, \mathcal{A}; B)$ such that $|\mu| \ll \nu$, there is a function f in $L_1(\Omega, \mathcal{A}, \nu; B)$ such that $\mu = f.\nu$ i.e. such that

$$\forall A \in \mathcal{A} \qquad \mu(A) = \int_A f d\nu.$$

Remark 2.4. For the preceding property to hold, it suffices to know the conclusion (i.e. the existence of a RN derivative f) whenever ν is a finite measure and we have $|\mu| \le \nu$. In fact it all boils down to the case when $\nu = |\mu|$. Indeed, if we assume that this holds in the latter case, let ν be σ-finite such that $|\mu| \ll \nu$. We have then (scalar RN theorem) $|\mu| = w.\nu$ for some $w \in L_1(\nu)$, so if we find f' in $L_1(|\mu|; B)$ such that $\mu = f'.|\mu|$, we have by (2.2) $|\mu| = \|f'\|.|\mu|$ and hence $\|f'\| = 1$ $|\mu|$-a.s. and therefore if $f = wf'$ we have $\mu = f.\nu$ and $f \in L_1(\nu; B)$.

We will need the concept of a δ-separated tree.

Definitions. Let $\delta > 0$. A martingale $(M_n)_{n \ge 0}$ in $L_1(\Omega, \mathcal{A}, \mathbb{P}; B)$ will be called δ-separated if

(i) M_0 is constant,
(ii) Each M_n takes only finitely many values,
(iii) $\forall n \ge 1,\ \forall \omega \in \Omega \quad \|M_n(\omega) - M_{n-1}(\omega)\| \ge \delta$.

Moreover, the set $S = \{M_n(\omega) \mid n \ge 0, \omega \in \Omega\}$ of all possible values of such a martingale will be called a δ-separated tree. If the martingale is *dyadic*, i.e. it is adapted to the standard dyadic filtration on $\{-1, 1\}^{\mathbb{N}}$ described in §1.4, then we will call it a δ-separated dyadic tree.

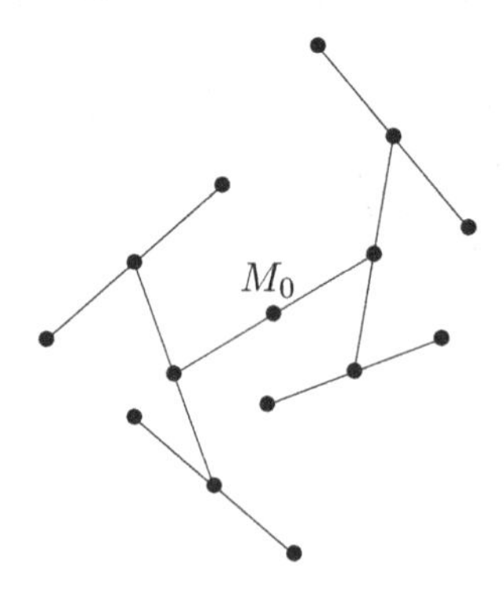

Figure 2.1. A δ-separated dyadic tree (M_0, M_1, M_2, M_3).

Remark (Warning). Some authors use the term δ-*bush* for what we call a δ-separated tree (e.g. [5, p. 111]). Some authors use the term δ-*separated tree* for what we call a δ-separated dyadic tree.

Another perhaps more intuitive description of a δ-separated tree is as a collection of points $\{x_i \mid i \in I\}$ indexed by the set of nodes of a tree-like structure that starts at some origin (0) then separates into N_1 branches, which we denote by $(0, 1), (0, 2), \dots, (0, N_1)$, then each branch itself splits into a finite number of branches, etc. in such a way that each point x_i is a convex combination of its immediate successors, and all these successors are at distance at least δ from x_i. We will also need another more geometric notion.

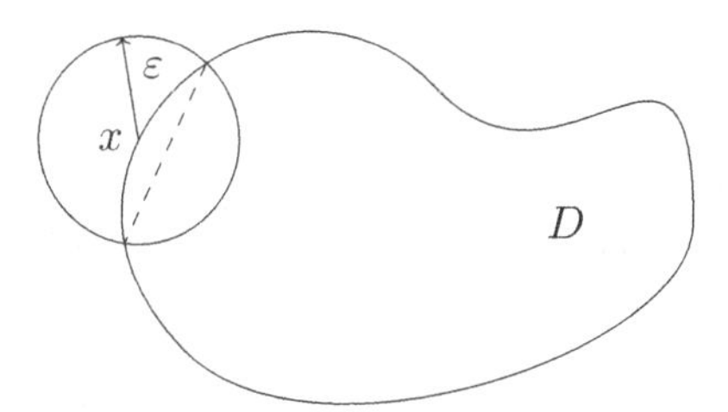

Figure 2.2. Dentable set.

Definition. Let B be a Banach space. A subset $D \subset B$ is called dentable if for any $\varepsilon > 0$ there is a point x in D such that

$$x \notin \overline{\text{conv}}(D \backslash B(x, \varepsilon))$$

where $\overline{\text{conv}}$ denotes the closure of the convex hull, and where

$$B(x, \varepsilon) = \{y \in B \mid \|y - x\| < \varepsilon\}.$$

By a slice of a convex set C (in a real Banach space B) we mean the non-empty intersection of C with an open half-space determined by a closed hyperplane. Equivalently, $S \subset C \subset B$ is a slice of C if there is $\xi \in B^*$ and $\alpha > 0$ such that

$$S = \{x \in C \mid \xi(x) > \alpha\} \neq \phi.$$

Remark 2.5. Let $D \subset B$ be a bounded subset and let C be the closed convex hull of D. If C is dentable, then D is dentable. Moreover, C is dentable iff C admits slices of arbitrarily small diameter. Note in particular that the dentability of all closed bounded convex sets implies that of all bounded sets. Indeed, the presence of slices of small diameter clearly implies dentability. Conversely, if C is dentable, then for any $\varepsilon > 0$ there is a point x in C that does not belong to the closed convex hull of $C \setminus B(x, \varepsilon)$, and hence by Hahn-Banach separation, there is a slice of C containing x and included in $B(x, \varepsilon)$, therefore with diameter less than 2ε. Now if $C = \overline{\text{conv}}(D)$, then this slice must contain a point in D, exhibiting that D itself is dentable.

The following beautiful theorem gives a geometric sufficient condition for the RNP. We will see shortly that it is also necessary.

Theorem 2.6. *If every bounded subset of a Banach space B is dentable, then B has the RNP.*

Proof. Let $(\Omega, \mathcal{A}, m)$ be a finite measure space and let $\mu\colon \mathcal{A} \to B$ be a bounded vector measure such that $|\mu| \ll m$. We will show that μ admits a Radon-Nikodým derivative in $L_1(\Omega, \mathcal{A}, m; B)$. As explained in Remark 2.4 we may as well assume that m is finite and $|\mu| \le m$. Now assume $|\mu| \le m$ and for every A in $\mathcal{A}$ let $x_A = \frac{\mu(A)}{m(A)}$ and let

$$C_A = \{x_\beta \mid \beta \in \mathcal{A},\ \beta \subset A,\ m(\beta) > 0\}.$$

Note that $\|x_A\| \le 1$ for all A in $\mathcal{A}$, so that the sets C_A are bounded. We will show that if every set C_A is dentable then the measure admits a Radon-Nikodým derivative f in $L_1(\Omega, \mathcal{A}, m; B)$.
Step 1: We first claim that if C_Ω is dentable then $\forall \varepsilon > 0\, \exists A \in \mathcal{A}$ with $m(A) > 0$ such that

$$\text{diam}(C_A) \le 2\varepsilon.$$

This (as well as the third) step is proved by an exhaustion argument. Suppose that this does not hold, then $\exists \varepsilon > 0$ such that every A with $m(A) > 0$ satisfies $\text{diam}(C_A) > 2\varepsilon$. In particular, for any x in B, A contains a subset β with $m(\beta) > 0$ such that $\|x - x_\beta\| > \varepsilon$. Then, consider a fixed measurable A with $m(A) > 0$ and let (β_n) be a maximal collection of disjoint measurable subsets of A with positive measure such that $\|x_A - x_{\beta_n}\| > \varepsilon$. (Note that since $m(\beta_n) > 0$ and the sets are disjoint, such a maximal collection must be at most countable.) By our assumption, we must have $A = \bigcup \beta_n$, otherwise we could take $A' = A \setminus \bigcup \beta_n$ and find a subset β of A' that would contradict the maximality of the family (β_n). But now if $A = \bigcup \beta_n$, we have

$$x_A = \Sigma(m(\beta_n)/m(A))x_{\beta_n} \quad \text{and} \quad \|x_A - x_{\beta_n}\| > \varepsilon.$$

Since we can do this for every $A \subset \Omega$ with $m(A) > 0$ this means that for some $\varepsilon > 0$, every point x of C_Ω lies in the closed convex hull of points in $C_\Omega \setminus B(x, \varepsilon)$, in other words this means that C_Ω is not dentable, which is the

announced contradiction. This proves the preceding claim and completes step 1. Working with C_A instead of C_Ω, we immediately obtain
Step 2: $\forall \varepsilon > 0\, \forall A \in \mathcal{A}$ with $m(A) > 0\ \exists A' \subset A$ with $m(A') > 0$ such that

$$\operatorname{diam}(C_{A'}) \leq 2\varepsilon.$$

Step 3: We use a second exhaustion argument. Let $\varepsilon > 0$ be arbitrary and let (A_n) be a maximal collection of disjoint measurable subsets of Ω with $m(A_n) > 0$ such that $\operatorname{diam}(C_{A_n}) \leq 2\varepsilon$. We claim that, up to a negligible set, we have necessarily $\Omega = \bigcup A_n$. Indeed if not, we could take $A = \Omega - (\bigcup A_n)$ in step 2 and find $A' \subset A$ contradicting the maximality of the family (A_n). Thus $\Omega = \bigcup A_n$. Now let $g_\varepsilon = \Sigma 1_{A_n} x_{A_n}$. Clearly, $g_\varepsilon \in L_1(\Omega, m; B)$ and we have

$$\|\mu - g_\varepsilon.m\|_{M(\Omega,\mathcal{A};B)} \leq 2\varepsilon m(\Omega). \tag{2.6}$$

Indeed, for every A in $\mathcal{A}$ with $m(A) > 0$

$$\mu(A) - \int_A g_\varepsilon dm = \Sigma m(A \cap A_n)[x_{A\cap A_n} - x_{A_n}]$$

hence

$$\left\| \mu(A) - \int_A g_\varepsilon dm \right\| \leq \Sigma m(A \cap A_n)\|x_{A\cap A_n} - x_{A_n}\|$$
$$\leq m(A)(2\varepsilon),$$

which implies (2.6).

This shows that μ belongs to the closure in $M(\Omega, \mathcal{A}, B)$ of the set of all measures of the form $f.m$ for some f in $L_1(\Omega, \mathcal{A}; B)$, and since this set is closed by (2.3) we conclude that μ itself is of this form. Perhaps, a more concrete way to say the same thing is to say that if $f_n = g_{2^{-n}}$ then $f = f_0 + \sum_{n\geq 1} f_n - f_{n-1}$ is in $L_1(\Omega, m; B)$ and we have $\mu = f.m$. (Indeed, note that (2.6) (with (2.3) and the triangle inequality) implies $\|f_n - f_{n-1}\|_{L_1(B)} \leq 6.2^{-n} m(\Omega)$.) □

To expand on Theorem 2.6, the following simple lemma will be useful.

Lemma 2.7. *Fix $\varepsilon > 0$. Let $D \subset B$ be a subset such that*

$$\forall x \in D \qquad x \in \overline{\operatorname{conv}}(D \setminus B(x, \varepsilon)) \tag{2.7}$$

then the enlarged subset $\widetilde{D} = D + B(0, \varepsilon/2)$ satisfies

$$\forall x \in \widetilde{D} \qquad x \in \operatorname{conv}(\widetilde{D} \setminus B(x, \varepsilon/2)). \tag{2.8}$$

Proof. Consider x in $\widetilde{D}$, $x = x' + y$ with $x' \in D$ and $\|y\| < \varepsilon/2$. Choose $\delta > 0$ small enough so that $\delta + \|y\| < \varepsilon/2$. By (2.7) there are positive numbers $\alpha_1, \ldots, \alpha_n$ with $\Sigma\alpha_i = 1$ and $x_1, \ldots, x_n \in D$ such that $\|x_i - x'\| \geq \varepsilon$ and

$\|x' - \Sigma\alpha_i x_i\| < \delta$. Hence $x' = \Sigma\alpha_i x_i + z$ with $\|z\| < \delta$. We can write $x = x' + y = \Sigma\alpha_i(x_i + z + y)$. Note that $x_i + z + y \in \widetilde{D}$ since $\|z + y\| \le \|z\| + \|y\| < \varepsilon/2$ and moreover

$$\|x - (x_i + z + y)\| = \|x' - x_i - z\| \ge \|x' - x_i\| - \|z\|$$
$$\ge \varepsilon - \delta \ge \varepsilon/2.$$

Hence we conclude that (2.8) holds. □

Remark 2.8. In the preceding proof, if we wish, we may obviously choose our coefficients (α_i) in $\mathbb{Q}$ or even of the form $\alpha_i = 2^{-N}k_i$ for some N large enough with $k_i \in \mathbb{N}$ such that $\sum k_i = 2^N$.

We now come to a very important result which incorporates the converse to Theorem 2.6.

Theorem 2.9. *Fix $1 < p < \infty$. The following properties of a Banach space B are equivalent:*

(i) B has the RNP.
(ii) Every uniformly integrable B-valued martingale converges a.s. and in $L_1(B)$.
(iii) Every B-valued martingale bounded in $L_1(B)$ converges a.s.
(iv) Every B-valued martingale bounded in $L_p(B)$ converges a.s. and in $L_p(B)$.
(v) For every $\delta > 0$, B does not contain a bounded δ-separated tree.
(vi) Every bounded subset of B is dentable.

Proof. (i) $\Rightarrow$ (ii). Assume (i). Let $(\Omega, \mathcal{A}, \mathbb{P})$ be a probability space and let $(\mathcal{A}_n)_{n\ge 0}$ be an increasing sequence of σ-subalgebras. Let us assume $\mathcal{A} = \mathcal{A}_\infty$ for simplicity. Let (M_n) be a B-valued uniformly integrable martingale adapted to $(\mathcal{A}_n)_{n\ge 0}$. We can associate to it a vector measure μ as follows. For any A in $\mathcal{A} = \mathcal{A}_\infty$, we define

$$\mu(A) = \lim_{n\to\infty} \int_A M_n d\mathbb{P}. \tag{2.9}$$

We will show that this indeed makes sense and defines a bounded vector measure. Note that if $A \in \mathcal{A}_k$ then by (1.18) for all $n \ge k$ $\int_A M_n d\mathbb{P} = \int_A M_k d\mathbb{P}$, so that the limit in (2.9) is actually stationary. Thus, (2.9) is well defined when $A \in \bigcup_{n\ge 0} \mathcal{A}_n$. Since (M_n) is uniformly integrable, $\forall \varepsilon > 0\, \exists \delta > 0$ such that $\mathbb{P}(A) < \delta \Rightarrow \|\mu(A)\| < \varepsilon$. Using this, it is easy to check that μ extends to a σ-additive vector measure on $\mathcal{A}_\infty$. Indeed, note that, by (1.14), we have

$\mathbb{E}(M_n 1_A) = \mathbb{E}(M_n \mathbb{E}^{\mathcal{A}_n}(1_A))$. Thus the limit in (2.9) is the same as

$$\lim_{n\to\infty} \mathbb{E}(M_n \mathbb{E}^{\mathcal{A}_n}(1_A)). \tag{2.10}$$

To check that this definition makes sense, note that if $\varphi_n = \mathbb{E}^{\mathcal{A}_n}(1_A)$, then

$$\forall n < m \qquad \mathbb{E}(M_n\varphi_n) - \mathbb{E}(M_m\varphi_m) = \mathbb{E}(M_m(\varphi_n - \varphi_m)), \tag{2.11}$$

and $\varphi_n \to 1_A$ in L_1 by Theorem 1.14. But by the uniform integrability (since $|\varphi_n - \varphi_m| \le 2$) we also must have $\|\mathbb{E}(M_m(\varphi_n - \varphi_m))\| \to 0$ when $n, m \to \infty$. Indeed, we can write for any $t > 0$

$$\|\mathbb{E}(M_m(\varphi_n - \varphi_m))\| \le 2 \sup_m \int_{\|M_m\|>t} \|M_m\| + t\mathbb{E}|\varphi_n - \varphi_m|,$$

so that $\limsup_{n,m\to\infty} \|\mathbb{E}(M_m(\varphi_n - \varphi_m))\| \le 2\sup_m \int_{\|M_m\|>t} \|M_m\|$ and hence must vanish by the uniform integrability. Thus by (2.11) we conclude that the limit in (2.9) exists by the Cauchy criterion.

By Theorem 1.36, the submartingale $\|M_n\|$ converges in L_1 to a limit w in L_1. By (2.9) and Jensen's inequality, we have

$$\|\mu(A)\| \le \lim_{n\to\infty} \mathbb{E}(\|M_n\| 1_A) = \int_A w d\mathbb{P},$$

from which the σ-additivity follows easily (the additivity is obvious by (2.9)).

Note that for all A in $\mathcal{A}$

$$|\mu|(A) \le \int_A w d\mathbb{P}. \tag{2.12}$$

Indeed, for all $A_1, \ldots, A_m$ in $\mathcal{A}$ disjoint with $A = \cup A_i$

$$\sum_1^m \|\mu(A_i)\| \le \sum_1^m \int_{A_i} w d\mathbb{P} = \int_A w d\mathbb{P}$$

and taking the supremum of the left-hand side, we obtain Claim 2.12. This shows $|\mu| \ll \mathbb{P}$. By our assumption (i), there is f in $L_1(\Omega, \mathcal{A}, \mathbb{P}; B)$ such that $\mu(A) = \int_A f d\mathbb{P}$ for all A in $\mathcal{A}$.

Recall that for any $k \ge 0$ and for any A in $\mathcal{A}_k$ we have by (1.18)

$$\forall n \ge k \qquad \mathbb{E}(M_n 1_A) = \mathbb{E}(M_k 1_A)$$

hence by (2.9) $\mu(A) = \mathbb{E}(M_k 1_A)$ for any A in $\mathcal{A}_k$. Therefore we must have

$$\forall k \ge 0 \quad \forall A \in \mathcal{A}_k \qquad \int_A f d\mathbb{P} = \int_A M_k d\mathbb{P}$$

or equivalently, since this property characterizes $\mathbb{E}^{\mathcal{A}_k}(f)$ (see the remarks after (1.7)) $M_k = \mathbb{E}^{\mathcal{A}_k}(f)$. Hence by Theorems 1.14 and 1.30, (M_n) converges to f a.s. and in $L_1(B)$. This completes the proof of (i) $\Rightarrow$ (ii).
(ii) $\Rightarrow$ (iii). This follows from Proposition 1.33.
(iii) $\Rightarrow$ (iv) is obvious. We give in what follows a direct proof that (iv) implies (i).
(iv) $\Rightarrow$ (v) is clear, indeed a bounded δ-separated tree is the range of a uniformly bounded martingale (M_n), which converges nowhere since $\|M_n - M_{n-1}\| \geq \delta$ everywhere.
(vi) $\Rightarrow$ (i) is Theorem 2.6, so it only remains to prove (v) $\Rightarrow$ (vi).

Assume that (vi) fails. We will show that (v) must also fail. Let $D \subset B$ be a bounded non-dentable subset. Replacing D by the set $\widetilde{D}$ in Lemma 2.7, we can assume that there is a number $\delta > 0$ such that

$$\forall x \in D \qquad x \in \text{conv}(D - B(x, \delta)).$$

We will then construct a δ-separated tree inside D. Let $(\Omega, \mathcal{A}, \mathbb{P})$ be the Lebesgue interval. We pick an arbitrary point x_0 in D and let $M_0 \equiv x_0$ on $\Omega = [0, 1]$. Then since $x_0 \in \text{conv}(D - B(x_0, \delta))$

$$\exists \alpha_1 > 0, \ldots, \alpha_n > 0 \quad \text{with} \quad \sum_1^n \alpha_i = 1 \quad \exists x_1, \ldots, x_n \in D$$

such that

$$x_0 = \sum_1^n \alpha_i x_i \quad \text{and} \quad \|x_i - x_0\| \geq \delta. \tag{2.13}$$

We can find in Ω disjoint subsets $A_1, \ldots, A_n$ such that $\mathbb{P}(A_i) = \alpha_i$ and $\cup A_i = \Omega$. We then let $\mathcal{A}_0$ be the trivial σ-algebra and let $\mathcal{A}_1$ be the σ-algebra generated by $A_1, \ldots, A_n$. Then we define $M_1(\omega) = x_i$ if $\omega \in A_i$. Clearly (2.13) implies $\mathbb{E}^{\mathcal{A}_0} M_1 = M_0$ and $\|M_1 - M_0\| \geq \delta$ everywhere. Since each point x_i is in D, we can continue in this way and represent each x_i as a convex combination analogous to (2.13). This will give M_2, M_3, etc.

We skip the details of the obvious induction argument. This yields a δ-separated martingale and hence a δ-separated tree. This completes the proof of (v) $\Rightarrow$ (vi) and hence of Theorem 2.9.

Finally, as promised, let us give a direct argument for (iv) $\Rightarrow$ (i). Assume (iv) and let μ be a B-valued vector measure such that $|\mu| \ll \nu$ where ν is as in the definition of the RNP. Then, by the classical RN theorem, there is a scalar density w such that $|\mu| = w.\nu$, thus it suffices to produce a RN density for μ with respect to $|\mu|$, so that, replacing ν by $|\mu|$ and normalizing, we may as well assume that we have a probability $\mathbb{P}$ such that

$$\forall A \in \mathcal{A} \quad \|\mu(A)\| \leq \mathbb{P}(A).$$

Then for any finite σ-subalgebra $\mathcal{B} \subset \mathcal{A}$, generated by a finite partition $A_1, \ldots, A_N$ of Ω, we consider the $\mathcal{B}$-measurable (simple) function $f_{\mathcal{B}} \colon \Omega \to B$ that is equal to $\mu(A_j)\mathbb{P}(A_j)^{-1}$ on each atom A_j of $\mathcal{B}$. It is then easy to check that $\{f_{\mathcal{B}} \mid \mathcal{B} \subset \mathcal{A}, |\mathcal{B}| < \infty\}$ is a martingale indexed by the directed set of all such $\mathcal{B}$'s. By Remark 1.17, if (iv) holds then the resulting net converges in $L_p(B)$, and a fortiori in $L_1(B)$ to a limit $f \in L_1(B)$. By the continuity of $\mathbb{E}^{\mathcal{C}}$, for each fixed finite $\mathcal{C}$, $\mathbb{E}^{\mathcal{C}}(f_{\mathcal{B}}) \to \mathbb{E}^{\mathcal{C}}(f)$ in $L_1(B)$, and $\mathbb{E}^{\mathcal{C}}(f_{\mathcal{B}}) = f_{\mathcal{C}}$ when $\mathcal{C} \subset \mathcal{B}$, therefore we must have $\mathbb{E}^{\mathcal{C}}(f) = f_{\mathcal{C}}$ for any finite $\mathcal{C}$. Applying this to an arbitrary $A \in \mathcal{A}$, taking for $\mathcal{C}$ the σ-subalgebra generated by A (and its complement), we obtain (recall that $f_{\mathcal{C}}$ is constant on A, equal to $\mu(A)\mathbb{P}(A)^{-1}$)

$$\mathbb{E}(1_A f) = \mathbb{E}(1_A f_{\mathcal{C}}) = \mathbb{P}(A) \times \mu(A)\mathbb{P}(A)^{-1} = \mu(A),$$

so that we conclude that $f.\mathbb{P} = \mu$, i.e. we obtain (i). □

Remark. If the preceding property (vi) is weakened by considering only dyadic trees (i.e. martingales relative to the standard dyadic filtration on say $[0, 1]$), or k-regular trees, then it does not imply the RNP: Indeed, by [143] there is a Banach space B (isometric to a subspace of L_1) that does not contain any bounded δ-separated *dyadic* tree, but that fails the RNP. Actually, that same paper shows that for any given sequence $(K(n))$ of integers, there is a Banach space B failing the RNP but not containing any δ-separated tree relative to a filtration such that $|\mathcal{A}_n| \leq K(n)$ for all n.

Corollary 2.10. *If for some $1 \leq p \leq \infty$ every B-valued martingale bounded in $L_p(B)$ converges a.s. then the same property holds for all $1 \leq p \leq \infty$.*

Remark 2.11. Note that for $1 < p < \infty$, if a B-valued martingale (M_n) is bounded in $L_p(B)$ and converges a.s. to a limit f, then it automatically also converges to f in $L_p(B)$. Indeed, by the maximal inequalities (1.38) the convergence of $\|M_n - f\|^p$ to zero is dominated, hence by Lebesgue's theorem $\int \|M_n - f\|^p d\mathbb{P} \to 0$.

Corollary 2.12. *The RNP is separably determined, that is to say: if every separable subspace of a Banach space B has the RNP, then B also has it.*

Proof. This follows from Theorem 2.9 by observing that a B-valued martingale in $L_1(B)$ must 'live' in a separable subspace of B. Alternately, note that any δ-separated tree is included in a separable subspace. □

Corollary 2.13. *If a Banach space B satisfies either one of the properties (ii)–(v) in Theorem 2.9 for martingales adapted to the standard dyadic filtration on [0,1], then B has the RNP.*

Proof. It is easy to see by a suitable approximation (see Remark 2.8) that if B contains a bounded δ-separated tree, then it contains one defined on a subsequence $\{\mathcal{A}_{n_k} \mid k \geq 1\}$, $(n_1 < n_2 < \cdots)$ of the dyadic filtration $(\mathcal{A}_n)$ in [0,1]. This yields the desired conclusion. □

Corollary 2.14. *If a Banach space B satisfies the property in Definition 2.2 when (Ω, ν) is the Lebesgue interval* $([0, 1], dt)$*, then B has the RNP.*

Corollary 2.15. *Any reflexive Banach space and any separable dual have the RNP.*

Proof. Since the RNP is separably determined by Corollary 2.12, it suffices to prove that separable duals have the RNP. So assume $B = X^*$, and that B is separable. Note that X is necessarily separable too and the closed unit ball of B is a metrizable compact set for $\sigma(X^*, X)$. Let $\{M_n\}$ be a martingale with values in the latter unit ball. For any ω, let $f(\omega)$ be a cluster point for $\sigma(X^*, X)$ of $\{M_n(\omega) \mid n \geq 0\}$. Let $D \subset X$ be a countable dense subset of the unit ball of X. For any d in D, the bounded scalar martingale $\langle M_n, d\rangle$ converges almost surely to a limit that has to be equal to $\langle f(\omega), d\rangle$. Hence since D is countable, there is $\Omega' \subset \Omega$ with $\mathbb{P}(\Omega') = 1$ such that

$$\forall\, \omega \in \Omega' \quad \forall\, d \in D \qquad \langle M_n(\omega), d\rangle \to \langle f(\omega), d\rangle.$$

In other words we have $M_n(\omega) \xrightarrow{\sigma(X^*,D)} f(\omega)$ or equivalently (since we are in the unit ball of B) $M_n(\omega) \xrightarrow{\sigma(X^*,X)} f(\omega)$ for any ω in Ω'. Notice that we did not discuss the measurability of f yet. But now we know that $\omega \to \langle f(\omega), x\rangle$ is measurable for any x in X, hence since X is separable for any $b \in B$, $\omega \mapsto \|b - f(\omega)\| = \sup_{x \in D} |\langle b - f(\omega), x\rangle|$ is measurable, so $f^{-1}(\beta) = \{\omega \mid f(\omega) \in \beta\}$ is measurable for any open (or closed) ball $\beta \subset X^*$, and finally since X^* is separable, for any open set $U \subset X^*$, the set $f^{-1}(U)$ must be measurable, so f is Borel measurable.

We claim that this implies that f is Bochner measurable. This (and the desired conclusion) follows from Phillips's theorem (see §1.8). Alternatively we can conclude the proof by the same trick as in §1.8, as follows.

For any b in B we have

$$\|b - M_n\| = \sup_{d \in D, \|d\| \leq 1} |\langle b - M_n, d\rangle| = \sup_{d \in D, \|d\| \leq 1} |\mathbb{E}_n\langle b - f, d\rangle| \leq \mathbb{E}_n\|b - f\|$$

(note that $\omega \to \|b - f(\omega)\|$ is bounded and measurable, so that $\mathbb{E}_n\|b - f\| \xrightarrow{\text{a.s.}} \|b - f\|$). Therefore, $\limsup_n \|b - M_n\| \le \|b - f\|$ a.s.We can assume that this holds on the same set of probability one for all b in a countable dense subset of B, hence actually for all b in B. But then taking $b = f(\omega)$ we have for almost all ω, $\limsup_{n\to\infty} \|f(\omega) - M_n(\omega)\| = 0$. Thus we conclude by Theorem 2.9 that B has the RNP. □

Remark. The examples of divergent martingales described in Remark 1.35 show that the separable Banach spaces $L_1([0, 1])$ and c_0 fail the RNP.

Remark. The RNP is clearly stable by passing to subspaces but obviously not to quotients Indeed, ℓ_1, being a separable dual, has the RNP but any separable space (e.g. c_0) is a quotient of it.

Remark 2.16. The notion of 'quasi-martingale' is useful to work with random sequences that are obtained by perturbation of a martingale. An adapted sequence $(F_n)_{n\ge 0}$ in $L_1(B)$ is said to be a quasi-martingale if

$$\sum_1^\infty \|\mathbb{E}_{n-1}(F_n - F_{n-1})\|_{L_1(B)} < \infty.$$

Given such a sequence, let

$$f_n = F_n - \sum\nolimits_1^n \mathbb{E}_{k-1}(F_k - F_{k-1}),$$

so that denoting $df_n = f_n - f_{n-1}$ we can write

$$df_n = dF_n - \mathbb{E}_{n-1}(dF_n).$$

Clearly (f_n) is then a martingale and for all $m < n$ we have pointwise

$$\|(f_n - f_m) - (F_n - F_m)\|_B \le \sum\nolimits_{m<k\le n} \|E_{k-1}(F_k - F_{k-1})\|_B$$

and hence

$$\|(f_n - f_m) - (F_n - F_m)\|_{L_1(B)} \le \sum\nolimits_{m<k\le n} \|E_{k-1}(F_k - F_{k-1})\|_{L_1(B)}.$$

Assume that (F_n) is bounded in $L_1(B)$ (resp. uniformly integrable). Note that this holds iff the same is true for (f_n). Therefore, if B has the RNP, (F_n) converges a.s. (resp. and in $L_1(B)$) by Theorem 2.9.

Remark 2.17. Let B be any Banach space. Let $F : [0, 1] \to B$ be a Lipschitz function. Assume for simplicity the Lipschitz norm is ≤ 1, by which we mean that

$$\forall s, t \in [0, 1] \quad \|F(s) - F(t)\| \le |s - t|.$$

Let λ denote the Lebesgue measure on $[0, 1]$. We claim that there is a bounded B-valued vector measure $\mu \in M([0, 1]; B)$ such that $|\mu| \leq \lambda$ and

$$\forall s > t \in [0, 1] \quad \mu([t, s)) = F(s) - F(t).$$

Indeed, we can use the preceding identity to define $\mu(I)$ for any interval $I = [t, s)$ (we also set $\mu(I) = 0$ whenever I is a singleton). Note that $\|\mu(I)\|/|I| \leq 1$. We will associate to F a martingale (f_n) adapted to the filtration $(\mathcal{A}_n)$ defined in §1.4. This is essentially the dyadic filtration, but realized on $[0, 1]$. We define a function f_n in the unit ball of $L_\infty([0, 1], \mathcal{A}_n; B)$ by setting f_n equal to $\mu(I)/|I|$ on any of the 2^n atoms $I \in \mathcal{A}_n$. By the argument given at the beginning of the proof of Theorem 2.9, μ extends to a σ-additive B-valued vector measure on the Borel σ-algebra of $[0, 1]$, such that $|\mu| \leq \lambda$. This proves our claim.

Conversely, let $\mu \in M([0, 1]; B)$ be such that $|\mu| \leq \lambda$. Then, let $F(t) = \mu([0, t])$. This clearly defines a B-valued Lipschitz function with Lipschitz norm ≤ 1.

The preceding correspondence $F \leftrightarrow \mu$ obviously extends to any bounded interval $[a, b] \subset \mathbb{R}$ in place of $[0, 1]$, and viewing $\mathbb{R}$ as the union of an increasing sequence of such intervals, we may extend this as well to $\mathbb{R}$ itself.

If (and only if) B has the RNP, this allows us to prove that all the B-valued Lipschitz functions F defined on any interval $(a, b) \subset \mathbb{R}$ (or on $\mathbb{R}$ itself) are differentiable almost everywhere. This is the B-valued version of the classical Rademacher Theorem, well known for $B = \mathbb{R}^n$. More precisely, if the measure μ associated to F admits a density $f \in L_1(\lambda; B)$ such that $\mu = f.\lambda$, we have

$$\forall s > t \quad F(s) - F(t) = \int_t^s f(x)dx,$$

and then the B-valued version of the classical Lebesgue differentiation theorem ensures that $F'(x)$ exists and $F'(x) = f(x)$ at a.e. x in (a, b). The B-valued version of Lebesgue's differentiation theorem can be deduced from the Hardy-Littlewood maximal inequality in much the same way that we deduced the martingale convergence theorem from Doob's inequality. See §3.3 and Remark 3.9 for more details.

Conversely, the a.e. differentiability of all B-valued Lipschitz functions F implies the RNP by Corollary 2.14.

The following complements the panorama of the interplay between martingale convergence and Radon-Nikodým theorems. This statement is valid for general Banach spaces, but we should emphasize for the reader that the ω-a.s. convergence of the variables $\omega \mapsto \|f_n(\omega)\|$ is considerably weaker than that of the sequence $(f_n(\omega))$ itself. The latter requires the RNP by Theorem 2.9.

Proposition. *Let B be an arbitrary Banach space. Consider $\mu \in M(\Omega, \mathcal{A}; B)$ such that $|\mu| = w.\mathbb{P}$ where $\mathbb{P}$ is a probability measure on $(\Omega, \mathcal{A})$ and $w \in L_1(\Omega, \mathcal{A}, \mathbb{P})$. Let $(\mathcal{A}_n)_{n\geq 0}$ be a filtration such that $\mathcal{A}_\infty = \mathcal{A}$, and such that, for each n, $\mu_{|\mathcal{A}_n}$ admits a RN density f_n in $L_1(\Omega, \mathcal{A}, \mathbb{P}; B)$ (for instance this is automatic if $\mathcal{A}_n$ is finite or atomic). Then $\|f_n\| \to w$ a.s.*

Proof. By Proposition 2.1, for each fixed $\varepsilon > 0$ we can find unit vectors $\xi_1, \ldots, \xi_N$ in B^* such that the vector measure

$$\mu_N\colon\ \mathcal{A} \to \ell_\infty^N$$

defined by $\mu_N(A) = (\xi_j(\mu(A)))_{j\leq N}$ satisfies $|\mu_N|(\Omega) > |\mu|(\Omega) - \varepsilon = 1 - \varepsilon$. Assume $|\mu|(\Omega) = \int w\ d\mathbb{P} = 1$ for simplicity. Note that $|\mu_{|\mathcal{A}_n}| \leq |\mu|_{|\mathcal{A}_n} = w_n.\mathbb{P}$ where $w_n = \mathbb{E}^{\mathcal{A}_n} w$. Therefore $\|f_n\| \leq w_n$. By the martingale convergence Theorem 1.14, $w_n \to w$ a.s. and in L_1, and hence

$$\limsup \|f_n\| \leq w \quad \text{a.e.}$$

and $\int \limsup \|f_n\| \leq \int w = 1$. We claim that

$$\int \liminf\nolimits_n \|f_n\| \geq \int \liminf\nolimits_n \sup\nolimits_{j\leq N} |\xi_j(f_n)| = |\mu_N|(\Omega) > 1 - \varepsilon.$$

Indeed, being finite-dimensional, ℓ_∞^N has the RNP and hence $\mu_N = \varphi_N.\mathbb{P}$ for some φ_N in $L_1(\Omega; \mathcal{A}, \mathbb{P}; \ell_\infty^N)$. This implies (by (2.2)) $|\mu_N| = \|\varphi_N\|.\mathbb{P}$. Clearly $\mathbb{E}^{\mathcal{A}_n}\varphi_N = (\xi_j(f_n))_{j\leq N}$ and hence when $n \to \infty$

$$\sup\nolimits_{j\leq N} |\xi_j(f_n)| \to \|\varphi_N\| \quad \text{a.s. and in } L_1.$$

Thus

$$\mathbb{E} \liminf\nolimits_n \sup\nolimits_{j\leq N} |\xi_j(f_n)| = \int \|\varphi_N\| d\mathbb{P} = |\mu_N|(\Omega) > 1 - \varepsilon,$$

proving the preceding claim.

Using this claim, we conclude easily: We have $\liminf \|f_n\| \leq \limsup \|f_n\| \leq w$ but $\int \liminf \|f_n\| d\mathbb{P} > \int w\ d\mathbb{P} - \varepsilon$, so we obtain $\liminf \|f_n\| = \limsup \|f_n\| = w$ a.e. □

2.3 The dual of $L_p(B)$

Notation. By analogy with the Hardy space case, let us denote by

$$h_p(\Omega, (\mathcal{A}_n)_{n\geq 0}, \mathbb{P}; B)$$

the (Banach) space of all B-valued martingales $M = \{M_n \mid n \geq 0\}$ that are bounded in $L_p(B)$, equipped with the norm

$$\|M\| = \sup_{n\geq 0} \|M_n\|_{L_p(B)}.$$

Remark 2.18. Note that, by Theorem 1.14, the mapping

$$f \to \{E_n(f) \mid n \geq 0\}$$

defines an isometric embedding of

$$L_p(\Omega, \mathcal{A}_\infty, \mathbb{P}; B) \text{ into } h_p(\Omega, (\mathcal{A}_n)_{n\geq 0}, \mathbb{P}; B).$$

Remark 2.19. Let $1 < p < \infty$. With this notation, Theorem 2.9 says that B has the RNP iff

$$h_p(\Omega, (\mathcal{A}_n)_{n\geq 0}, \mathbb{P}; B) = L_p(\Omega, \mathcal{A}_\infty, \mathbb{P}; B).$$

We now turn to the identification of the dual of $L_p(B)$.

Let p' be the conjugate exponent such that $p^{-1} + p'^{-1} = 1$.

Suppose that we are given a filtration $\mathcal{A}_0 \subset \cdots \mathcal{A}_n \subset \mathcal{A}_{n+1} \subset \cdots$ of finite σ-subalgebras and let us assume $\mathcal{A} = \mathcal{A}_\infty$. Let $L_p(B) = L_p(\Omega, \mathcal{A}, \mathbb{P}; B)$ with $1 \leq p < \infty$. Let φ be a bounded linear form on $L_p(B)$. By restriction to $L_p(\Omega, \mathcal{A}_n, \mathbb{P}; B)$, φ defines a linear form φ_n in $L_p(\Omega, \mathcal{A}_n, \mathbb{P}; B)^*$. But, since $\mathcal{A}_n$ is finite, we have $L_p(\Omega, \mathcal{A}_n, \mathbb{P}; B)^* = L_{p'}(\Omega, \mathcal{A}_n, \mathbb{P}; B^*)$ isometrically, hence φ_n corresponds to an element M_n in $L_p(\Omega, \mathcal{A}_n, \mathbb{P}; B^*)$. Moreover, since φ_n is the restriction of φ_{n+1} it is easy to see that $M_n = \mathbb{E}_n(M_{n+1})$, i.e. that $\{M_n\}$ is a B^* valued martingale. Moreover, we have

$$\sup_n \|M_n\|_{L_{p'}(B^*)} = \|\varphi\|_{L_p(B)^*}.$$

Proposition 2.20. *If $1 \leq p < \infty$, the preceding correspondence*

$$\varphi \to (M_n)_{n\geq 0}$$

is an isometric isomorphism from $L_p(\Omega, \mathcal{A}, \mathbb{P}; B)^*$ to the space

$$h_{p'}(\Omega, (\mathcal{A}_n)_{n\geq 0}, \mathbb{P}; B^*).$$

Proof. Indeed, it is easy to see conversely that given any martingale $\{M_n\}$ in the unit ball of $h_{p'}(\Omega, (\mathcal{A}_n)_{n\geq 0}, \mathbb{P}; B^*)$, M_n defines an element φ_n in $L_p(\Omega, \mathcal{A}_n, \mathbb{P}; B)^*$ so that φ_{n+1} extends φ_n, and $\|\varphi_n\| \leq 1$. Hence by density of the union of the spaces $L_p(\Omega, \mathcal{A}_n, \mathbb{P}; B)$ in $L_p(B)$, we can extend the φ_n's to a (unique) functional φ in $L_p(B)^*$ with $\|\varphi\| \leq 1$. Thus, it is easy to check that the correspondence is one-to-one and isometric. □

Remark 2.21. By Remark 2.18, we have an isometric embedding

$$L_{p'}(\Omega, \mathcal{A}_\infty, \mathbb{P}; B^*) \subset L_p(\Omega, \mathcal{A}_\infty, \mathbb{P}; B)^*.$$

Theorem 2.22. *A dual space B^* has the RNP iff for any countably generated probability space and any $1 \le p < \infty$ we have (isometrically)*

$$L_p(\Omega, \mathcal{A}, \mathbb{P}; B)^* = L_{p'}(\Omega, \mathcal{A}, \mathbb{P}; B^*).$$

Moreover for B^* to have the RNP it suffices that this holds for some $1 \le p < \infty$ and for the Lebesgue interval.

Proof. If $\mathcal{A}$ is countably generated we can assume $\mathcal{A} = \mathcal{A}_\infty$ with $\mathcal{A}_\infty$ associated to a filtration of finite σ-algebras $(\mathcal{A}_n)$, as earlier. Then Theorem 2.22 follows from Proposition 2.20 and Remark 2.19. The second assertion follows from Corollary 2.13. □

Remark 2.23. The extension of the duality theorem to the case when $\mathbb{P}$ is replaced by a σ-finite measure m is straightforward. Indeed, in that case, we can find a disjoint measurable partition of Ω into sets Ω_n of finite measure, so that

$$L_p(\Omega, m; B) = \ell_p(\{L_p(\Omega_n, m_n; B)\})$$

(where m_n denotes the restriction of m to Ω_n) and we are reduced to the finite case.

Incidentally, the same idea allows to extend the duality to the case when the space $L_p(\Omega, m; B)$ can be identified with a direct sum $\ell_p(\{L_p(\Omega_i, m_i; B)\}_{i \in I})$ over a set I of arbitrary cardinal, with each m_i finite. Using this one could also remove the σ-finiteness assumption but we will never need this level of generality.

By Corollary 2.15 we have

Corollary 2.24. *If B is reflexive, then $L_p(\Omega, \mathcal{A}, \mathbb{P}; B)$ is also reflexive for any $1 < p < \infty$.*

Remark. Of course the preceding isometric duality holds for any dual space B^* when the measure space is discrete (i.e. atomic).

Remark. The preceding theorem does remain valid for $p = 1$. Note however that, if $\dim(B) = \infty$, the B-valued simple functions are, in general, *not dense* in the space $L_\infty(\Omega, \mathcal{A}, \mathbb{P}; B)$ (see, however, Lemma 2.41). This is in sharp contrast with the finite-dimensional case, where they are dense because the unit ball, being compact, admits a *finite* ε-net for any $\varepsilon > 0$. We wish to emphasize that we defined the space $L_\infty(\Omega, \mathcal{A}, \mathbb{P}; B)$ (in Bochner's sense) as the space of

B-valued Bochner-measurable functions f (see §1.8) such that $\|f(.)\|_B$ is in L_∞, equipped with its natural norm. This definition makes sense for any measure space $(\Omega, \mathcal{A}, \mathbb{P})$, and, with it, the preceding theorem does hold for $p = 1$.

2.4 Generalizations of $L_p(B)$

We now turn to a slightly different description of $h_p(\Omega, (\mathcal{A}_n)_{n\geq 0}, \mathbb{P}; B)$ that is more convenient in applications. Let $(\Omega, \mathcal{A}, m)$ be a finite measure space. (The σ-finite case can be treated as in Remark 2.23.) We denote by $\Lambda_p(\Omega, \mathcal{A}, m; B)$, or simply by $\Lambda_p(m; B)$, or even simply $\Lambda_p(B)$ when there is no risk of confusion, the space of all vector measures $\mu\colon\ \mathcal{A} \to B$ such that $|\mu| \ll m$ and such that the density $\frac{d|\mu|}{dm}$ is in $L_p(\Omega, \mathcal{A}, m)$. We equip it with the norm:

$$\|\mu\|_{\Lambda_p(m;B)} = \left\| \frac{d|\mu|}{dm} \right\|_{L_p(m)}.$$

It is not hard to check that this is a Banach space.

Let $\mu \in \Lambda_p(B)$ so that $|\mu| = w.m$ with $\|w\|_p = \|\mu\|_{\Lambda_p(B)} < \infty$. Then for any scalar simple function $f = \Sigma 1_{A_k} f_k$ in $L_{p'}$ (A_k disjoint in $\mathcal{A}$, $f_k \in \mathbb{C}$), if we denote

$$\int f\, d\mu = \sum \mu(A_k) f_k,$$

we have obviously

$$\left\| \int f\, d\mu \right\| \leq \sum \|\mu(A_k)\| |f_k| \leq \sum \int_{A_k} w\, dm |f_k| = \int |f| w\, dm$$
$$\leq \|f\|_{p'} \|w\|_p.$$

Therefore, we can extend the integral $f \to \int f d\mu$ to a continuous linear map $I_\mu\colon\ L_{p'}(m) \to B$ such that

$$\|I_\mu\| \leq \|w\|_p. \tag{2.14}$$

For convenience, we still denote $I_\mu(f)$ by $\int f d\mu$. Note however that $\int f\, d\mu \in B$.

Remark 2.25. More generally, if $\mathcal{B} \subset \mathcal{A}$ is a σ-subalgebra, the analogue of the conditional expectation of μ is simply its restriction $\mu_{|\mathcal{B}}$ to $\mathcal{B}$. It is easy to check that the density of $|\mu_{|\mathcal{B}}|$ (with respect to $|m_{|\mathcal{B}}|$) is $\leq \mathbb{E}^{\mathcal{B}} w$. Thus conditional expectations define operators of norm 1 on $\Lambda_p(B)$ for any $1 \leq p \leq \infty$.

If $(\Omega, \mathcal{A}, m)$ is a countably generated probability space and if $(\mathcal{A}_n)_{n\geq 0}$ is a filtration of finite σ-subalgebras of $\mathcal{A}$ whose union generates $\mathcal{A}$, then if $p > 1$, $\Lambda_p(B)$ can be identified with the space $h_p(\Omega, (\mathcal{A}_n)_{n\geq 0}, m; B)$. Indeed, it is not hard to check that for any martingale (f_n) bounded in $L_p(B)$, the set function μ defined on $\bigcup_{n\geq 0} \mathcal{A}_n$ by the (stationary) limit

$$\mu(A) = \lim_{k\to\infty} \int_A f_k \, dm \tag{2.15}$$

extends to a vector measure in $\Lambda_p(B)$ with $\|\mu\|_{\Lambda_p(B)} = \sup_{n\geq 0} \|f_n\|_p$. Conversely, any vector measure μ defines such a martingale by simply restricting μ to $\mathcal{A}_n$ for each $n \geq 0$. See the proof of Theorem 2.9. Thus we have proved:

Proposition 2.26. *The correspondence $(f_n) \mapsto \mu$ defined by (2.15) is an isometric isomorphism from $h_p(\Omega, (\mathcal{A}_n)_{n\geq 0}, m; B)$ to $\Lambda_p(\Omega, \mathcal{A}_\infty, m; B)$ for any $p > 1$.*

We leave the proof as an exercise.

Remark 2.27. When $p = 1$, the space $\Lambda_1(\Omega, \mathcal{A}_\infty, m; B)$ can be identified with the subspace of $h_1(\Omega, (\mathcal{A}_n)_{n\geq 0}, m; B)$ formed of all martingales such that $(\|f_n\|_B)$ is uniformly integrable.

Consider now f in $L_p(\Omega, m; B)$ and μ in $\Lambda_{p'}(B^*)$. Assume first that f is a B-valued simple function i.e. $f = \sum 1_{A_k} f_k$ with (A_k) as before but now with f_k in B. We denote $\int f \cdot d\mu = \sum \langle \mu(A_k), f_k \rangle \in \mathbb{C}$. Again we find

$$\begin{aligned}\left|\int f \cdot d\mu\right| &\leq \sum \|\mu(A_k)\| \|f_k\| \leq \sum |\mu|(A_k) \|f_k\| \\ &\leq \int \|f\| d|\mu| = \int \|f\| w dm \leq \|w\|_{p'} \|f\|_{L_p(B)}.\end{aligned}$$

Therefore, we may extend this integral to the whole of $L_p(B)$, so that we may define a 'duality relation' by setting

$$\forall f \in L_p(B) \ \forall \mu \in \Lambda_{p'}(B^*) \qquad \langle \mu, f \rangle = \int f \cdot d\mu.$$

Proposition 2.28. *Let $1 \leq p < \infty$ and let p' be conjugate to p, i.e. $p^{-1} + p'^{-1} = 1$. Then, in the duality described earlier, we have an isometric identity $L_p(\Omega, \mathcal{A}, m; B)^* = \Lambda_{p'}(\Omega, \mathcal{A}, m; B^*)$.*

Proof. We may as well assume $m(\Omega) = 1$. If $\mathcal{A}$ is countably generated, then the proof follows from Propositions 2.20 and 2.26. The general case is similar using Remark 1.17. □

It is natural to wonder whether there is a alternate description of the dual of $L_p(B)$ as a space of B^* valued functions. As the next statement shows, this is possible, but at the cost of a weakening of the notion of measurability.

A function $f\colon \Omega \to B^*$ will be called weak* scalarly measurable if for every b in B the scalar valued function $\langle f(.), b\rangle$ is measurable. Assume B separable. Let us denote by $\underline{\Lambda}_p(\Omega, \mathcal{A}, m; B^*)$ the space of (equivalence classes of) weak* scalarly measurable functions $f\colon \Omega \to B^*$ such that the function $\omega \mapsto \|f(\omega)\|_{B^*}$ (which is measurable since B is assumed separable) is in L_p. We equip this space with the obvious norm

$$\|f\| = (\int \|f(\omega)\|_{B^*}^p)^{1/p}.$$

We have then

Theorem 2.29. *Assume B separable. Then for any $1 \le p' \le \infty$, we have (isometrically)*

$$\Lambda_{p'}(\Omega, \mathcal{A}, m; B^*) = \underline{\Lambda}_{p'}(\Omega, \mathcal{A}, m; B^*).$$

Proof. We assume as before that $\mathcal{A}$ is generated by a filtration $(\mathcal{A}_n)_{n\ge 0}$ of finite algebras. We may clearly reduce to the case when $m = \mathbb{P}$ for some probability $\mathbb{P}$. Assume first $p' > 1$. By Proposition 2.20, it suffices to show how to identify $h_{p'}(\Omega, (\mathcal{A}_n)_{n\ge 0}, \mathbb{P}; B^*)$ with $\underline{\Lambda}_{p'}(\Omega, \mathcal{A}, \mathbb{P}; B^*)$. Consider a martingale (f_n) in $h_{p'}(\Omega, (\mathcal{A}_n)_{n\ge 0}, \mathbb{P}; B^*)$. By the maximal inequality, (f_n) is bounded a.s. and hence a.s. weak* compact. Let $f(\omega)$ be a weak* cluster point of (f_n). Then for any fixed $b \in B$, the scalar martingale $\langle f_n(.), b\rangle$ converges a.s. Its limit must necessarily be equal to $\langle f(.), b\rangle$. This shows that f is weak* scalarly measurable. Let D be a countable dense subset of the unit ball of B. Since D is countable, and $\langle f(.), b\rangle = \lim_{n\to\infty}\langle f_n(.), b\rangle$ for any $b \in D$, we have a.s.

$$\|f\| = \sup_{b\in D} |\langle f, b\rangle| \le \lim_{n\to\infty} \|f_n\|$$

and hence by Fatou's lemma

$$\|f\|_{\underline{\Lambda}_{p'}} \le \|(f_n)\|_{h_{p'}}.$$

Note that for any $b \in D$, we have

$$\langle f_n(.), b\rangle = \mathbb{E}_n(\langle f(.), b\rangle). \tag{2.16}$$

Conversely, consider now $f \in \underline{\Lambda}_{p'}(\Omega, \mathcal{A}, \mathbb{P}; B^*)$. Fix n. Let A be an atom of $\mathcal{A}_n$. Then $b \mapsto \mathbb{P}(A)^{-1}\int_A \langle b, f\rangle$ is a continuous linear form on B with norm $\le \mathbb{P}(A)^{-1}\int_A \|f\|_{B^*}$. Let us denote it by f_A. Let f_n be the B^* valued function that is equal to f_A on each atom $A \in \mathcal{A}_n$. We have clearly $\mathbb{E}_n(\langle b, f\rangle) = \langle b, f_n\rangle$ and hence $\mathbb{E}_n(\langle b, f_{n+1}\rangle) = \langle b, f_n\rangle$ for any b in D. Since D separates points, this shows that (f_n) is a martingale, and moreover $\|f_n\| = \sup_{b\in D} |\langle b, f_n\rangle| \le \mathbb{E}_n\|f\|$. It follows that

$$\|(f_n)\|_{h_{p'}} \le (\int \|f\|^{p'})^{1/p'} = \|f\|_{\underline{\Lambda}_{p'}}.$$

By (2.16), the correspondences $(f_n) \mapsto f$ and $f \mapsto (f_n)$ are inverses of each other. This shows that $(f_n) \mapsto f$ is an isometric isomorphism from $h_{p'}$ to $\underline{\Lambda}_{p'}$ if $p' > 1$. When $p' = 1$, we use Remark 2.27. By the uniform integrability, (2.16) still holds so the same proof is valid. □

Corollary 2.30. *Assume B separable. Then B^* has the RNP iff any weak* scalarly measurable function f with values in B^* is Bochner measurable.*

Proof. If B^* has the RNP, we know by Theorem 2.22, that $\Lambda_\infty(B^*) = L_\infty(B^*)$. In other words any bounded weak* scalarly measurable function is Bochner measurable. By truncation, the unbounded case follows. Conversely, if the latter holds we have $\Lambda_\infty(B^*) = L_\infty(B^*)$ and B^* has the RNP by Theorems 2.22 and 2.29. □

2.5 The Krein-Milman property

Throughout this section, let B be a Banach space over $\mathbb{R}$.

Recall that a point x in a convex set $C \subset B$ is called extreme in C if whenever x lies inside a segment $S = \{\theta y + (1-\theta)z \mid 0 < \theta < 1\}$ with endpoints y, z in C, then we must have $y = z = x$. Equivalently $C\backslash\{x\}$ is convex. See [21] and [10] for more information.

Definition. We will say that a Banach space B has the Krein-Milman property (in short KMP) if every closed bounded convex set in B is the closed convex hull of its extreme points.

We will show that RNP $\Rightarrow$ KMP.

The converse remains a well-known important open problem (although it is known that RNP is equivalent to a stronger form of the KMP; see later). We will use the following beautiful fundamental result due to Bishop and Phelps.

Theorem 2.31 (Bishop-Phelps). *Let $C \subset B$ be a closed bounded convex subset of a (real) Banach space B. Then the set of functionals in B^* that attain their supremum on C is dense in B^*.*

We will need a preliminary simple lemma.

Lemma 2.32. *Let $f, g \in B^*$ be such that $\|f\| = \|g\| = 1$. Let $0 < \varepsilon < 1$ and $t > 0$. Let*

$$C(f,t) = \{x \in B \mid \|x\| \le t f(x)\}.$$

(i) Assume that $f(y) = 0$ and $\|y\| \le 1$ imply $|g(y)| \le \varepsilon/2$. Then either $\|f - g\| \le \varepsilon$ or $\|f + g\| \le \varepsilon$.
(ii) Assume that $t > 1 + 2/\varepsilon$ and that g is non-negative on $C(f,t)$. Then $\|f - g\| \le \varepsilon$.

Proof.

(i) Let $g_1 \in B^*$ be a Hahn Banach extension of the restriction of g to $\ker(f)$ such that $\|g_1\| \le \varepsilon/2$. Then $g - g_1 = af$ for some $a \in \mathbb{R}$. Since f, g have norm 1 we must have $1 - \varepsilon/2 \le |a| \le 1 + \varepsilon/2$, and hence either $|a - 1| \le \varepsilon/2$ or $|a + 1| \le \varepsilon/2$. Also $g \pm f = (a \pm 1)f + g_1$, and hence $\|g \pm f\| \le |a \pm 1| + \varepsilon/2$. This yields (i).
(ii) Assume $f(y) = 0$ and $\|y\| \le 1$. Let $z = (2/\varepsilon)y$. Since $\|f\| = 1$ there is x with $\|x\| = 1$ such that $f(x) \ge (1 + 2/\varepsilon)/t$ and $f(x) > \varepsilon$. Then $\|x \pm z\| \le 1 + 2/\varepsilon \le t f(x) = t f(x \pm z)$, and hence $x \pm z \in C(f,t)$. By our assumption on g, we have $g(x \pm z) \ge 0$, and hence $|g(y)| = (\varepsilon/2)|g(z)| \le (\varepsilon/2)g(x) \le \varepsilon/2$. By (i) we obtain $\|f - g\| \le \varepsilon$ since the other case is ruled out by our choice of x. Indeed, $\|f + g\| \le \varepsilon$ would imply $f(x) + g(x) \le \varepsilon$ and hence $g(x) \le \varepsilon - f(x) < 0$. □

Proof of Theorem 2.31. Let $t > 0$ and let $f \in B^*$ with $\|f\| = 1$. We claim that there is a point $x_0 \in C$ such that $C \cap (x_0 + C(f,t)) = \{x_0\}$. Taking this for granted, let us complete the proof. Let $0 < \varepsilon < 1$ and $t > 1 + 2/\varepsilon$. Note that $C(f,t)$ is a closed convex cone. Since $t > 1$, $C(f,t)$ has a non-empty interior denoted by U, which, by a well-known fact, is dense in $C(f,t)$. Note that $x_0 + U$ and C are disjoint (because $0 \notin U$). By Hahn-Banach separation (between the convex open set $x_0 + U$ and C) there is a $0 \ne g \in B^*$ and $\alpha \in \mathbb{R}$ such that $g \le \alpha$ on C and $g \ge \alpha$ on $x_0 + C(f,t)$. We must have $g(x_0) = \alpha = \sup\{g(x) \mid x \in C\}$, and after normalization, we may assume $\|g\| = 1$. Then $g \ge \alpha$ on $x_0 + C(f,t)$ implies $g \ge 0$ on $C(f,t)$. Therefore, by (ii) in Lemma 2.32, $\|f - g\| \le \varepsilon$ and g attains its supremum on C at x_0. This proves the Theorem.

We now prove the claim. We will use Zorn's lemma. We introduce the partial order on C defined by the convex cone $C(f,t)$: For $x, y \in C$ we declare that $x \ge y$ if $x - y \in C(f,t)$. Note that this means $\|x - y\| \le t(f(x) - f(y))$. We will show that there is a maximal element in C. Let $\{x_i\}$ be a totally ordered subset. Since $\{f(x_i)\}$ is bounded in $\mathbb{R}$ (because C is bounded), it admits a convergent subnet and since $\|x_i - x_j\| \le t(f(x_i) - f(x_j))$ if $x_i \ge x_j$, the subnet

(x_i) is Cauchy and hence converges to some limit $x \in C$ (because C is closed). Clearly we have in the limit $\|x - x_j\| \le t(f(x) - f(x_j))$, or equivalently $x \ge x_j$ for all j. Thus, the order is inductive and we may apply Zorn's lemma: There is a maximal element $x_0 \in C$, which means

$$\forall x \in C \quad \|x_0 - x\| \le t(f(x_0) - f(x)). \tag{2.17}$$

Then if $x \in C \cap (x_0 + C(f,t))$ we have $x - x_0 \in C(f,t)$, which means $\|x - x_0\| \le tf(x - x_0)$ and hence by (2.17) $\|x - x_0\| = 0$. This shows that x_0 is the only point in $C \cap (x_0 + C(f,t))$. □

Remark 2.33. Our goal in the next series of observations is to show that any slice of a closed bounded convex set C contains a (non-void) face.

(i) Let $x^* \in B^*$ be a functional attaining its supremum on C, so that if $\alpha = \sup\{x^*(b) \mid b \in C\}$, the set $F = \{b \in C \mid x^*(b) = \alpha\}$ is non-void. We will say that F is a *face* of C. We need to observe that a face enjoys the following property: If a point in F is inside the segment joining two points in C, then this segment must entirely lie in F.

(ii) In particular, any extreme point of F is an extreme point of C.

(iii) Now assume that we have been able to produce a decreasing sequence of sets $\cdots \subset F_n \subset F_{n-1} \subset \cdots F_0 = C$ such that F_n is a face of F_{n-1} for any $n \ge 1$ and the diameter of F_n tends to zero. Then, by the Cauchy criterion, the intersection of the F_n's contains exactly one point x_0 in C. We claim that x_0 is an extreme point of C. Indeed, if x_0 sits inside a segment S joining two points in C, then by (i) we have $S \subset F_1$, hence (since F_2 is a face in F_1 and $x_0 \in F_2$) $S \subset F_2$ and so on. Hence $S \subset \cap F_n = \{x_0\}$, which shows that x_0 is extreme in C.

(iv) Assume that every closed bounded convex subset $C \subset B$ has at least one extreme point. Then B has the KMP.

To see this, let $C_1 \subset C$ be the closed convex hull of the extreme points of C. We claim that we must have $C_1 = C$.

Indeed, otherwise there is x in $C \setminus C_1$ and by Hahn-Banach there is x^* in B^* such that $x^*_{|C_1} < \beta$ and $x^*(x) > \beta$. Assume first that this functional achieves its supremum $\alpha = \sup\{x^*(b) \mid b \in C\}$. This case is easier. Note $\alpha > \beta$. Then let $F = \{b \in C \mid x^*(b) = \alpha\}$, so that F is a face of C disjoint from C_1. But now F is another non-void closed bounded convex set that, according to our assumption, must have an extreme point. By (ii) this point is also extreme in C, but this contradicts the fact that F is disjoint from C_1.

In general, x^* may fail to achieve its norm, but we can use the Bishop-Phelps theorem (Theorem 2.31) to replace x^* by a small perturbation of itself that will play the same role in the preceding argument. Indeed, by Theorem 2.31, for any $\varepsilon > 0$ there is y^* in B^* with $\|x^* - y^*\| < \varepsilon$ that achieves its sup on C. We may assume $\|b\| \le r$ for any b in C. Let $\gamma = \sup\{y^*(b) \mid b \in C\}$ and note that $\gamma > \alpha - r\varepsilon$; and hence $y^*(b) = \gamma$ implies $x^*(b) > \alpha - 2r\varepsilon$. Hence if ε is chosen so that $\alpha - 2r\varepsilon > \beta$, we are sure that $\widetilde{F} = \{b \in C \mid y^*(b) = \gamma\}$ is included in $\{b \mid x^*(b) > \beta\}$ hence is disjoint from C_1. We now repeat the preceding argument: $\widetilde{F}$ must have an extreme point, by (ii) it is extreme in C hence must be in C_1, but this contradicts $\widetilde{F} \cap C_1 = \emptyset$. This proves the claim, and hence B has the KMP.

(v) The preceding argument establishes the following general fact: let $\mathcal{S}$ be a slice of C, i.e. we assume given x^* in B^* and a number β so that

$$\mathcal{S} = \{b \in C \mid x^*(b) > \beta\},$$

then if $\mathcal{S}$ is non-void it must contain a (non-void) face of C.

Theorem 2.34. *The RNP implies the KMP.*

Proof. Assume B has the RNP. Let $C \subset B$ be a bounded closed convex subset. Then by Theorem 2.9, C is dentable. So for any $\varepsilon > 0$, there is x in C such that $x \notin \overline{\text{conv}}(C \backslash B(x, \varepsilon))$. By Hahn-Banach separation, there is x^* in B^* and a number β such that the slice $\mathcal{S} = \{b \in C \mid x^*(b) > \beta\}$ contains x and is disjoint from $C \backslash B(x, \varepsilon)$. In particular, we have $\|b - x\| \le \varepsilon$ for any b in $\mathcal{S}$, so the diameter of $\mathcal{S}$ is $\le 2\varepsilon$. By Remark 2.33 (v), $\mathcal{S}$ must contain a face F_1 of C, a fortiori of diameter $\le 2\varepsilon$.

Now we can repeat this procedure on F_1 : we find that F_1 admits a face F_2 of arbitrary small diameter, then F_2 also admits a face of small diameter, and so on. Thus, adjusting $\varepsilon > 0$ at each step, we find a sequence of (non-void) sets $\cdots \subset F_{n+1} \subset F_n \subset \cdots \subset F_1 \subset F_0 = C$ such that F_{n+1} is a face of F_n and diam $(F_n) < 2^{-n}$. Then, by Remark 2.33 (iii), the intersection of $\{F_n\}$ contains an extreme point of C. By Remark 2.33 (iv), we conclude that B has the KMP. □

Let $C \subset B$ be a convex set. A point x in C is called 'exposed' if there is a functional x^* such that $x^*(x) = \sup\{x^*(b) \mid b \in C\}$ and x is the only point of C satisfying this. (Equivalently, if the singleton $\{x\}$ is a face of C.) The point x is called 'strongly exposed' if the functional x^* can be chosen such that, in addition, the diameter of the slice

$$\{b \in C \mid x^*(b) > \sup_C x^* - \varepsilon\}$$

tends to zero when $\varepsilon \to 0$. Clearly, the existence of such a point implies that C is dentable. More precisely, if C is the closed convex hull of a bounded set D, then D is dentable because every slice of C contains a point in D (see Remark 2.5).

We will say that B has the 'strong KMP' if every closed bounded convex subset $C \subset B$ is the closed convex hull of its strongly exposed points. It is clear (by (vi) $\Rightarrow$ (i) in Theorem 2.9) that the strong KMP implies the RNP. That the converse also holds is a beautiful and deep result due to Phelps, for the proof of which we refer to [381]:

Theorem 2.35. *The RNP is equivalent to the strong KMP.*

2.6 Edgar's Choquet theorem

Our goal is the following version due to Edgar [216] of Choquet's classical theorem for compact convex subsets of (metrizable) locally convex spaces for which we refer to [16]. We follow the very simple proof outlined in [247].

Theorem 2.36. *Let C be a bounded, separable, closed convex subset of a Banach space B with the RNP. For any $z \in C$, there is a Bochner-measurable function $\Phi_z : \Omega \to C$ on a probability space $(\Omega, \mathbb{P})$ taking values in the set $ex(C)$ of extreme points of C and such that*

$$z = \mathbb{E}(\Phi_z).$$

The proof is a beautiful application of martingale ideas, but there is a catch: it uses more sophisticated measure theory than used in most of this book, in particular universally measurable sets and von Neumann's lifting theorem, so we start by a review of these topics.

Recall that a subset $A \subset X$ of a topological space X is called universally measurable if it is measurable with respect to any probability measure m on the Borel σ-algebra of X, denoted by $\mathcal{B}_X$. This means that for any $\varepsilon > 0$, there are Borel subsets A_1, A_2 such that $A_1 \subset A \subset A_2$ and $m(A_2 \setminus A_1) < \varepsilon$. Clearly these sets form a σ-algebra, that we denote by $\mathcal{U}(\mathcal{B}_X)$. If we are given a probability $\mathbb{P}$ on $\mathcal{B}_X$ and $A \in \mathcal{U}(\mathcal{B}_X)$, taking suitable unions and intersections we find Borel subsets A_1, A_2 (depending on $\mathbb{P}$) such that $A_1 \subset A \subset A_2$ and $\mathbb{P}(A_2 \setminus A_1) = 0$. Therefore $\mathbb{P}$ obviously extends to a probability on $\mathcal{U}(\mathcal{B}_X)$ (for such a set A we set $\mathbb{P}(A) = \mathbb{P}(A_1) = \mathbb{P}(A_2)$).

A mapping $\rho : X \to Y$ between topological spaces will be called universally measurable if it is measurable when X, Y are equipped respectively with $\mathcal{U}(\mathcal{B}_X), \mathcal{U}(\mathcal{B}_Y)$.

A topological space is called Polish if it admits a metric (compatible with the topology) with which it becomes a complete separable metric space, or equivalently if it is homeomorphic to a complete separable metric space. We will use the basic fact that a continuous image of a Polish space is universally measurable to prove the following lemma.

Lemma 2.37. *For any C as in the preceding Theorem, the set of extreme points $ex(C)$ is universally measurable.*

Proof. Clearly C and hence $C \times C$ is Polish. Let $g: C \times C \to C$ be defined by $g(x, y) = (x + y)/2$. Let $\varepsilon_n > 0$ be any sequence tending to 0. Let $F_n = \{(x, y) \in C \times C \mid \|x - y\| \geq \varepsilon_n\}$. Clearly F_n is closed and hence Polish, and

$$C \setminus ex(C) = \cup_n g(F_n).$$

By the basic fact just recalled, $C \setminus ex(C)$ (or equivalently $ex(C)$) is universally measurable. □

In our context, von Neumann's classical selection theorem implies in particular the following fact that we admit without proof (see e.g. [337] for a detailed proof).

Lemma 2.38. *For C and g as earlier, let Δ denote the diagonal in $C \times C$. Then there is a universally measurable lifting $\rho : C \setminus ex(C) \to (C \times C) \setminus \Delta$, i.e. we have ('lifting property') $g(\rho(x)) = x$ for any $x \in C \setminus ex(C)$.*

Proof of Theorem 2.36. Let $I = \{1, 2, \dots\}$. Let $\Omega = \{-1, 1\}^I$ with $\mathbb{P}$ equal to the product of $(\delta_1 + \delta_{-1})/2$. We denote by $\mathcal{B}$ (resp. $\mathcal{B}_n$) the σ-algebra generated by the coordinates $\omega \mapsto \omega_k$ with $k \in I$ (resp. with $1 \leq k \leq n$), and let $\mathcal{B}_0$ denote the trivial σ-algebra. Then we 'complete $\mathcal{B}$' by setting $\mathcal{A} = \mathcal{U}(\mathcal{B})$.

Fix a point $z \in C$. We will define a dyadic martingale (f_n) indexed by the set I, such that $f_0 = z$ and formed of C valued Bochner measurable functions with $f_n \in L_1(\Omega, \mathcal{B}_n, \mathbb{P})$. Since C is bounded, this martingale will be bounded in $L_1(B)$, and by the RNP of B (and in some sense merely of C) it will converge in $L_1(B)$ and a.s. Moreover we will adjust f_n to make sure that $f(\omega) = \lim_{n\to\infty} f_n(\omega)$ takes its values in $ex(C)$ for a.a. ω.

The idea is very simple: if $z \in ex(C)$ we are done, so assume $z \notin ex(C)$, then we can write $z = (x + y)/2$ with $y - x \neq 0,\ x, y \in C$. This gives us f_1: we set $f_1(\omega) = x$ if $\omega_1 = 1$ and $f_1(\omega) = y$ if $\omega_1 = -1$. Then we repeat the same operation on each point x, y to define the next variable f_2 and so on. More precisely, assuming we have defined a dyadic C valued martingale $f_0, \dots, f_n$ relative to $\mathcal{B}_0, \dots, \mathcal{B}_n$ we define f_{n+1} like this: either $f_n(\omega) \in ex(C)$ in which case nothing needs to be done so we set $f_{n+1}(\omega) = f_n(\omega)$, or $f(\omega) \in C \setminus ex(C)$ and then we may write

$$f_{n+1}(\omega) = (x(\omega) + y(\omega))/2$$

with $x(\omega) \neq y(\omega)$ both $\mathcal{B}_n$-measurable. We then set $f_{n+1}(\omega) = x(\omega)$ if $\omega_{n+1} = 1$ and $f_{n+1}(\omega) = y(\omega)$ if $\omega_{n+1} = -1$. We can rewrite f_{n+1} as

$$f_{n+1}(\omega) = f_n(\omega) + \omega_{n+1}(x(\omega) - y(\omega))/2.$$

Note that at this point the measurable selection problem is irrelevant: since f_n takes at most 2^n values, we can lift them one by one.

Up to now, the construction is simple minded. There is however a difficulty. If no further constraint is imposed on the choices of $x(\omega)$, $y(\omega)$, there is no reason for the eventual limit $f(\omega)$ to lie in the extreme points. Edgar's original proof remedies this by repeating the construction 'trans-finitely' using martingales indexed by the set of countable ordinals instead of our set I and proving that one of the variables thus produced will be with values in $ex(C)$ before the index reaches the first uncountable ordinal.

The alternate proof in [247] uses a trick to accelerate the convergence. The trick is to define for any $z \in C$

$$\delta(z) = \sup\{\|x - y\|/2 \mid x, y \in C,\ z = (x + y)/2\}$$

and to select $x(\omega)$, $y(\omega)$ in the preceding construction such that

$$\|x(\omega) - y(\omega)\|/2 > \delta(f_n(\omega)) - \delta_n$$

where $\delta_n > 0$ is any sequence (fixed in advance) such that $\delta_n \to 0$ when $n \to \infty$. Since $f_{n+1}(\omega) - f_n(\omega) = \pm(x - y)/2$, this gives us

$$\|f_{n+1} - f_n\| > \delta(f_n) - \delta_n. \tag{2.18}$$

Then let $f = \lim f_n$. Let us now repeat the martingale step but now applied to f. The idea is to prove that we cannot continue further. We define (universally) measurable functions $y, z : \Omega \to C$ as follows. Either $f(\omega) \in ex(C)$ and then we do nothing: we set $y(\omega) = z(\omega) = f(\omega)$, or $f(\omega) \in C \setminus ex(C)$ and then we may write

$$f(\omega) = (x(\omega) + y(\omega))/2$$

with $x(\omega) \neq y(\omega)$ defined by $\rho(f(\omega)) = (x(\omega), y(\omega))$. By Lemma 2.38 (here a measurable selection is really needed), x and y are universally measurable. To show that $f(\omega) \in ex(C)$ a.s. it suffices to show that $x = y$ a.s. or equivalently that $\mathbb{E}(1_A(x - y)) = 0$ for any measurable $A \in \mathcal{A}$, with $\mathbb{P}(A) > 0$. Let us fix such a set A. The key lies in the following (we write $\mathbb{E}_n$ instead of $\mathbb{E}^{\mathcal{B}_n}$)

Claim: $\|\mathbb{E}_n((x - y)1_A)\| \leq \delta(f_n)$ a.s.
Let

$$\hat{x} = 1_{\Omega\setminus A} f + 1_A x \quad \text{and} \quad \hat{y} = 1_{\Omega\setminus A} f + 1_A y.$$

These $\mathcal{A}$-measurable functions take their values in C and we have $f = (\hat{x} + \hat{y})/2$. Therefore

$$f_n = \mathbb{E}_n f = (\mathbb{E}_n \hat{x} + \mathbb{E}_n \hat{y})/2,$$

and by the very definition of $\delta(.)$ we have $\|\mathbb{E}_n \hat{x} - \mathbb{E}_n \hat{y}\|/2 \le \delta(f_n)$ and our claim follows.

Now by (2.18) we have $\mathbb{E}\|f_{n+1} - f_n\| + \delta_n > \mathbb{E}\delta(f_n)$ and hence, since f_n converges in $L_1(B)$, we must have

$$\mathbb{E}\delta(f_n) \to 0.$$

Therefore, by the preceding claim $\mathbb{E}\|\mathbb{E}_n((x-y)1_A)\| \to 0$. But by Jensen (see (1.4))

$$\|\mathbb{E}(\mathbb{E}_n((x-y)1_A))\| \le \mathbb{E}\|\mathbb{E}_n((x-y)1_A)\|$$

and hence this tends to 0 but in fact $\mathbb{E}(\mathbb{E}_n((x-y)1_A)) = \mathbb{E}((x-y)1_A)$, so we conclude

$$\mathbb{E}((x-y)1_A) = 0.$$

Since this holds for any A, we have $x = y$ a.s. proving that $f \in ex(C)$ a.s. Of course this statement is relative to $\mathcal{A}$ equipped with the obvious extension of $\mathbb{P}$ to $\mathcal{A}$. More precisely, we just proved that there are subsets $A_1, A_2 \in \mathcal{B}$ with $A_1 \subset \{f \in ex(C)\} \subset A_2$ such that $\mathbb{P}(A_1) = \mathbb{P}(A_2) = 1$. Thus if we define Φ_z as equal to $f = \lim f_n$ on A_1 and, say, equal to 0 on $\Omega \setminus A_1$, then Φ_z satisfies the required properties. □

2.7 The Lewis-Stegall theorem

Let $\mu\colon\ \mathcal{A} \to B$ be a bounded vector measure and let m be a finite measure on $\mathcal{A}$ such that for some constant C

$$\forall A \in \mathcal{A} \qquad \|\mu(A)\| \le Cm(A). \tag{2.19}$$

Let $u_\mu\colon\ L_1(m) \to B$ be the bounded linear map defined by

$$\forall g \in L_1(m) \qquad u_\mu(g) = \int_\Omega g\, d\mu.$$

It is easy to verify that u_μ is bounded with

$$\|u_\mu\colon\ L_1(m) \to B\| = \sup\{m(A)^{-1}\|\mu(A)\| \mid A \in \mathcal{A} \quad m(A) > 0\}. \tag{2.20}$$

Moreover, if μ admits a RN density $\psi \in L_1(m; B)$ so that $\mu = \psi.m$, then (2.19) implies $\psi \in L_\infty(m; B)$ with

$$\|\psi\|_{L_\infty(m;B)} = \|u_\mu\colon\ L_1(m) \to B\|. \tag{2.21}$$

Indeed, (2.19) implies that for any closed *convex* subset β of $\{x \in B \mid \|x\| > C\}$ the set $A = \{\psi \in \beta\}$ must satisfy $m(A) = 0$ (because otherwise $m(A)^{-1} \int_A \psi dm \in \beta$, which contradicts (2.19)), and since we may assume B separable, this implies $m(\{\|\psi\| > C\}) = 0$ and hence we find $\|\psi\|_{L_\infty(m;B)} \le \|u_\mu\colon\ L_1(m) \to B\|$. The converse is obvious by (1.4).

We say that a linear operator $u\colon\ X \to Y$ between Banach spaces factors through ℓ_1 if there is a factorization of u of the form

$$u\colon\ X \xrightarrow{w} \ell_1 \xrightarrow{v} Y$$

with bounded operators v, w. We denote by $\Gamma_{\ell_1}(X, Y)$ the space of all such maps $u\colon\ X \to Y$. Moreover, we define $\gamma_{\ell_1}(u) = \inf\{\|v\|\,\|w\|\}$ where the infimum runs over all possible such factorizations. The space $\Gamma_{\ell_1}(X, Y)$ equipped with this norm is easily seen to be a Banach space (Hint: if $u = \sum u_n$ with $\sum \gamma_{\ell_1}(u_n) < \infty$, we may factorize u_n as $u_n = v_n w_n$ with $\|v_n\| = \|w_n\| \le (1 + \varepsilon_n)\gamma_{\ell_1}(u_n)^{1/2}$, then one can factor u through the ℓ_1-space that is the ℓ_1-direct sum of the ℓ_1-spaces through which the u_n's factor). The reason why we introduce this notion is the following result due to Lewis and Stegall.

Theorem 2.39. *Let μ be a vector measure satisfying* (2.19) *so that $u_\mu\colon\ L_1(m) \to B$ is bounded. Then μ admits a Radon Nikodým density in $L_1(m; B)$ (actually in $L_\infty(B)$) iff u_μ factors through ℓ_1.*

We will use two lemmas.

Lemma 2.40. *Consider $\psi \in L_\infty(\Omega, \mathcal{A}, m; B)$ taking only countably many values. Let $\mu = \psi.m$ be the associated vector measure. Then $u_\mu \in \Gamma_{\ell_1}(L_1(m), B)$ and*

$$\gamma_{\ell_1}(u_\mu) = \|\psi\|_{L_\infty(B)}.$$

Proof. By assumption, we have a countable partition of Ω into sets $(A_n)_{n\ge0}$ in $\mathcal{A}$ such that ψ is constant on each A_n, say $\psi(\omega) = x_n$ for all ω in A_n. We define $w\colon\ L_1(m) \to \ell_1$ and $v\colon\ \ell_1 \to B$ by $w(g) = \left(\int_{A_n} g\, dm\right)_{n\ge0}$ and $v((\alpha_n)) = \sum \alpha_n x_n$. Then obviously $\|w\| \le 1$ and $\|v\| \le \sup \|x_n\| \le \|\psi\|_{L_\infty(B)}$. Since $u_\mu = vw$ we obtain $\gamma_{\ell_1}(u_\mu) \le \|\psi\|_{L_\infty(B)}$. The converse follows from (2.21). □

Lemma 2.41. *Assume B separable. The subspace of $L_\infty(\Omega, \mathcal{A}, m; B)$ formed of functions with countable range (i.e. 'simple functions' but with countably many values) is dense in $L_\infty(\Omega, \mathcal{A}, m; B)$.*

Proof. Let $f \in L_\infty(m; B)$. Since B is separable, for any $\varepsilon > 0$ there is a countable partition of B into Borel sets (B_n^ε) of diameter at most ε. Let $A_n^\varepsilon = f^{-1}(B_n^\varepsilon)$, so that $\{A_n^\varepsilon \mid n \geq 0\}$ is a countable partition of Ω in $\mathcal{A}$. Let

$$x_n^\varepsilon = m(A_n^\varepsilon)^{-1} \int_{A_n^\varepsilon} f\, dm \quad \text{and} \quad f_\varepsilon = \sum\nolimits_n 1_{A_n^\varepsilon} x_n^\varepsilon.$$

Note that f_ε is the conditional expectation of f with respect to the σ-algebra generated by $\{A_n^\varepsilon \mid n \geq 0\}$. Since B_n^ε has diameter at most ε, we have $\|f(\omega) - f(\omega')\| \leq \varepsilon$ for a.a. $\omega, \omega' \in A_n^\varepsilon$ and hence $\|f(\omega) - x_n^\varepsilon\| \leq \varepsilon$ for a.a. $\omega \in A_n^\varepsilon$. Therefore we have $\|f - f_\varepsilon\|_{L_\infty(B)} \leq \varepsilon$. □

Proof of Theorem 2.39. If u_μ factors through ℓ_1, then $\mu = v\widetilde{\mu}$ for some $\ell_1(I)$ valued vector measure $\widetilde{\mu}$ (still satisfying (2.19)). Since $\ell_1(I)$ has the RNP, $\widetilde{\mu}$ and hence a fortiori $v\widetilde{\mu}$ admits a RN derivative in $L_1(m, B)$. Conversely, assume $\mu = f.m$ for some f in $L_1(m; B)$. Then we may as well assume B separable. By (2.21) our assumption (2.19) implies $\|f\|_{L_\infty(B)} \leq C$. By Lemma 2.41 for any $\varepsilon > 0$ there is $f_\varepsilon \in L_\infty(B)$ with countable range such that $\|f - f_\varepsilon\|_{L_\infty(B)} \leq \varepsilon$. Now taking $\varepsilon(n) = 2^{-n}$ (say) we find $f = \sum_0^\infty \psi_n$ with $\psi_0 = f_{\varepsilon(0)}$ and $\psi_n = f_{\varepsilon(n)} - f_{\varepsilon(n-1)}$ for any $n \geq 1$ so that

$$\sum\nolimits_0^\infty \|\psi_n\|_{L_\infty(B)} < \infty$$

and each ψ_n is a countably valued function in $L_\infty(B)$. Let $u_n\colon\ L_1(m) \to B$ be the linear operator associated to the vector measure $\mu_n = \psi_n.m$. Clearly $u = \sum u_n$. Therefore (since $\Gamma_{\ell_1}(L_1(m), B)$ is a Banach space) we conclude by Lemma 2.40 that u factors through ℓ_1. □

Remark. Note that for any f in $L_1(\Omega, \mathcal{A}, m; B)$ there is a countably generated σ-subalgebra $\mathcal{B} \subset \mathcal{A}$ and a separable subspace $B_1 \subset B$ such that $f \in L_1(\Omega, \mathcal{B}, m; B_1)$. Indeed, we already noticed that we can replace B by a separable subspace B_1. Then, let $\mathcal{B}$ be the σ-subalgebra generated by f (i.e. the one formed by the sets $f^{-1}(\beta)$ where β runs over all Borel subsets of B_1). Then, $\mathcal{B}$ is countably generated since f is Bochner-measurable and trivially $f \in L_1(\Omega, \mathcal{B}, m; B)$. This implies that, if $\mu = f.m$, we have $u_\mu = u_\mu \mathbb{E}^{\mathcal{B}}$.

2.8 Notes and remarks

For vector measures and Radon-Nikodým theorems, a basic reference is [21]. A more recent, much more advanced, but highly recommended reading is Bourgain's Lecture Notes on the RNP [8]. For more on the convexity issues related to the Bishop-Phelps theorem, see [20].

For martingales in the Banach space valued case, the first main reference is Chatterji's paper [178] where the equivalence of (i), (ii), (iii) and (iv) in Theorem 2.9 is proved. The statements numbered from 2.10 to 2.22 all follow from Chatterji's result but some of them were probably known before (see in particular [271]).

Rieffel introduced dentability and proved that it suffices for the RNP. The converse is (based on work by Maynard) due to Davis-Phelps and Huff independently. The Lewis-Stegall theorem in §2.7 comes from [329]. Theorem 2.34 is due to Joram Lindenstrauss and Theorem 2.35 to Phelps [381]. See [21] for a more detailed history of the RNP and more precise references.

In §2.6 we present Edgar's theorem (improving Theorem 2.34) that the RNP implies a Choquet representation theorem using Ghoussoub and Maurey's simpler proof from [247] (see also [416]). We refer to [314] for more illustrations of the use of Banach valued martingales.

Martingales have been used repeatedly as a tool to differentiate measures. See Naor and Tao's [358] for a recent illustration of this.

Our presentation of the RNP is limited to the basic facts. We will now briefly survey additional material.

Charles Stegall [428] proved the following beautiful characterization of duals with the RNP:

Stegall's Theorem ([428]). *Let B be a separable Banach space. Then B^* has the RNP iff it is separable. More generally, a dual space B^* has the RNP iff for any separable subspace $X \subset B$, the dual X^* is separable.*

In the 1980s, a lot of work was devoted (notably at the impulse of H.P. Rosenthal and Bourgain) to 'semi-embeddings'. A Banach space X is said to semi-embed in another one Y if there is an injective linear mapping $u\colon\ X \to Y$ such that the image of the closed unit ball of X is closed in B (and such a u is then called a semi-embedding). The relevance of this notion lies in

Proposition 2.42. *If X is separable and semi-embeds in a space Y with the RNP, then X has the RNP.*

Proof. One way to prove this is to consider a martingale (f_n) with values in the closed unit ball B_X of X. Let $u\colon\ X \to Y$ be a semi-embedding. If Y has

RNP then the martingale $g_n = u(f_n)$ converges in Y to a limit g_∞ such that $g_\infty(\cdot) \in \overline{u(B_X)} = u(B_X)$. Let now $f(\omega) = u^{-1}(g_\infty(\omega))$. We will show that f is Borel measurable. Let U be any open set in X. By separability, there is a sequence $\{\beta_n\}$ of closed balls in X such that $U = \cup \beta_n$. Then

$$\{\omega \mid f(\omega) \in U\} = \cup_n \{\omega \mid g_\infty(\omega) \in u(\beta_n)\}$$

but since $u(\beta_n)$ is closed and g_∞ measurable we find that $f^{-1}(U)$ is measurable. This shows that f is Borel measurable. By Phillips's theorem, f is Bochner measurable. Now, since $g_n = \mathbb{E}_n(g_\infty) = \mathbb{E}_n(u(f)) = u(\mathbb{E}_n(f))$ we have

$$f_n = u^{-1}(g_n) = \mathbb{E}_n(f),$$

and hence f_n converges to f a.s. This shows that X has the RNP (clearly one could use a vector measure instead of a martingale and obtain the RNP a bit more directly). □

We refer to [144] for work on semi-embeddings. More generally, an injective linear map $u\colon X \to Y$ is called a G_δ-embedding if the image of any closed bounded subset of X is a G_δ-subset of Y. We refer to [241, 244, 245, 247, 248] for Ghoussoub and Maurey's work on G_δ-embeddings. To give the flavor of this work, let us quote the main result of [245]: A separable Banach space X has the RNP iff there is a G_δ-embedding $u\colon X \to \ell_2$ such that $u(B_X)$ is a countable intersection of open sets with convex complements. The proof of Proposition 2.42 shows that the RNP is stable under G_δ-embedding.

As mentioned in the text, it is a famous open problem whether the KMP implies the RNP. It was proved for dual spaces by Huff and Morris using the theorem of Stegall [428], see [10, p. 91], and also for Banach lattices by Bourgain and Talagrand ([10, p. 423]). See also Chu's paper [181] for preduals of von Neumann algebras. Schachermayer [419] proved that it is true for Banach spaces isomorphic to their square. See also [420–423, 435] for related work in the same direction.

We should mention that one can define the RNP for *subsets* of Banach spaces. One can then show that weakly compact sets are RNP sets. See [10, 77] for more on RNP sets.

Many results from this chapter have immediate extensions to RNP operators: we say that an operator $u : B_1 \to B_2$ between two Banach spaces has the RNP if it transforms B_1 valued bounded martingales into a.s. convergent B_2 valued ones. With this terminology, B has the RNP iff the identity on B is an RNP operator. See Ghoussoub and Johnson's [242] for an example of an RNP operator that *does not* factor through an RNP space. This is in sharp contrast with the situation for weakly compact operators, which, by a classical result from [203], *do* factor through reflexive spaces.

A Banach space X is called an Asplund space if every continuous convex function defined on a (non-empty) convex open subset $D \subset E$ is Fréchet differentiable on a dense G_δ-subset of D. Stegall [429] proved that X is Asplund iff X^* has the RNP. We refer the reader to [77] for more information in this direction.

A metric characterization of the RNP based on Theorem 2.9 (see Chapter 13 for more on this theme) appears in [371].

In [179], Cheeger and Kleiner give a characterization of the RNP (for separable spaces) as those that can be represented as inverse limits of finite-dimensional spaces with what they call the 'determining property'. In [180], they establish differentiability almost everywhere for Lipschitz maps from a certain class of metric spaces (called PI spaces) to Banach spaces with the Radon-Nikodým property (RNP).

3

Harmonic functions and RNP

In contrast with the next chapter, we will mostly work in the present one with real Banach spaces (unless specified otherwise). Note however that the complex case is included since any complex space may be viewed a fortiori as a real one (but not conversely).

We will denote by D the open unit disc in $\mathbb{C}$. Its boundary ∂D, the unit circle, will be often identified with the compact group $\mathbb{T} = \mathbb{R}/2\pi\mathbb{Z}(\simeq \mathbb{R}/\mathbb{Z})$ by the usual map taking $t \in \mathbb{T}$ to e^{it}.

From now on in this book, we reserve the notation m for the normalized Haar (or Lebesgue) measure on $\mathbb{T}$ (or ∂D), correspondingly identified with the measure $dt/2\pi$. Possible exceptions to this rule will be clear from the context.

In the present chapter, we focus on functions on D, and postpone the discussion of harmonic functions on the upper half-plane to §4.6.

3.1 Harmonicity and the Poisson kernel

Let $V \subset \mathbb{C}$ be an open subset of the complex plane. A function $u\colon\ V \to \mathbb{R}$ is called harmonic if it is C^2 and if $\Delta u = 0$.

If V is simply connected (for instance if V is an open disc), every real valued harmonic function $u\colon\ V \to \mathbb{R}$ is the real part of an analytic function $F\colon\ V \to \mathbb{C}$. This implies in particular that every harmonic function is C^∞. Also, just like the analytic ones, harmonic functions satisfy the mean value theorem, as follows. We say that a continuous function $u\colon\ V \to \mathbb{R}$ satisfies the mean value property if for every z in V and every $r > 0$ such that $\overline{B(z, r)} \subset V$, we have (recall $dm(t) = dt/2\pi$)

$$u(z) = \int u(z + re^{it})\, dm(t). \tag{3.1}$$

Integrating in polar coordinates, it is easy to check that this implies

$$u(z) = \int_{B(z,r)} u(a+ib)\frac{dadb}{|B(z,r)|}. \tag{3.2}$$

The mean value property is closely connected to martingales, and in fact, as we will see in §3.5, fundamental examples of martingales are produced by composing a harmonic function with Brownian motion.

It is well known that the mean value property characterizes harmonicity. Actually, an even weaker formulation is enough, as follows. We say that u has the weak mean value property if for every z in V, there is a sequence of radii $r_n > 0$ with $r_n \to 0$ such that $\overline{B(z, r_n)} \subset V$ and

$$\forall n \qquad u(z) = \int u(z + r_n e^{it})\, dm(t). \tag{3.3}$$

It is known that this implies that u is harmonic (and hence has the ordinary mean value property).

Another classical property of harmonic functions is the maximum principle, which says that if $K \subset V$ is compact then $\sup_{z\in K} |u(z)|$ is attained on the boundary of K. (Of course, the case $K = \bar{D}$ follows immediately from the variant of the mean value formula appearing as (3.9).)

Now assume $\overline{D} \subset V$. The mean value property gives us

$$u(0) = \int u(e^{it})dm(t). \tag{3.4}$$

In fact, there is also a formula that gives $u(z)$ as an average of its boundary values on $\overline{D}$, this is given by the classical Poisson kernel. We have for all $z = re^{i\theta}$ in D

$$u(z) = \int u(e^{it})P_r(\theta - t)dm(t), \tag{3.5}$$

where

$$P_r(s) = \frac{1-r^2}{1-2r\cos s + r^2} = \sum\nolimits_{n\in\mathbf{Z}} r^{|n|}e^{ins}. \tag{3.6}$$

We will use the notation

$$P_z(t) = P_r(\theta - t) = \sum\nolimits_{n\geq 0} z^n e^{-int} + \sum\nolimits_{n<0} \bar{z}^{|n|} e^{-int} \tag{3.7}$$

when $z = re^{i\theta}$, $0 \leq r < 1$, $\theta \in \mathbb{R}$. Moreover, we will denote by P^z the corresponding measure on $\mathbb{T}$, i.e. we set (recall $dm(t) = dt/2\pi$)

$$P^z(dt) = P_z(t)dm(t), \tag{3.8}$$

so that (3.5) can be written

$$\forall z \in D \quad u(z) = \int u(e^{it})P^z(dt). \tag{3.9}$$

Two basic elementary properties of the Poisson kernel are

$$P_r(t) \geq 0 \quad \text{and} \quad \int P_r(t)dm(t) = 1,$$

so that P^z is (for every z in D) a probability measure on $\mathbb{T}$.

To check (3.5), one simple way is to apply (3.4) to $u \circ \varphi_z$ where φ_z is the Möbius transformation φ_z that maps 0 to z i.e.

$$\varphi_z(\zeta) = \frac{\zeta + z}{1 + \bar{z}\zeta}.$$

Another way to look at these formulae is to use Harmonic Analysis on the circle group $\mathbb{T}$, letting

$$f(t) = u(e^{it}),$$

and interpreting (3.5) as a convolution formula, as follows:

$$u(re^{i\theta}) = P_r * f(\theta). \tag{3.10}$$

If $F = \sum_{n \geq 0} a_n z^n$ is an analytic function such that $u = \Re(F)$ in a neighborhood of $\overline{D}$, we have for $0 \leq r \leq 1$

$$u(re^{i\theta}) = \sum\nolimits_{n \in \mathbb{Z}} b_n e^{int} r^{|n|} \tag{3.11}$$

where $b_0 = \Re(a_0)$, $b_n = \frac{1}{2}a_n$ for $n > 0$ and $b_n = \frac{1}{2}\bar{a}_{-n}$ for $n < 0$.

Let

$$u_r(\theta) = u(re^{i\theta}).$$

By (3.6) we have $\widehat{P_r}(n) = r^{|n|}$, by (3.11) $\widehat{u_r}(n) = b_n r^{|n|}$ and $\widehat{f}(n) = b_n$. Hence (3.5) boils down to the observation that

$$\forall n \in \mathbb{Z} \qquad \widehat{u_r}(n) = \widehat{P_r}(n)\widehat{f}(n).$$

Here (as often) we use the classical fact that two L_1-functions (or more generally two measures) coincide as soon as their Fourier transforms are equal.

Let us record in passing that P_r is a continuous function on $\mathbb{T}$ with

$$\|P_r\|_\infty = \frac{1+r}{1-r} \quad (0 \leq r < 1) \tag{3.12}$$

and also that if $0 \leq r, s < 1$

$$P_r * P_s = P_{rs}.$$

We will use repeatedly the basic inequalities valid for any $f \in L_p(\mathbb{T}, m)$ and any $\mu \in M(\mathbb{T})$, if $1 \le p \le \infty$ and $0 \le r < 1$

$$\|P_r * f\|_p \le \|f\|_p \tag{3.13}$$

$$\|P_r * \mu\|_M \le \|\mu\|_M. \tag{3.14}$$

These are easy to check using the fact that $\|P_r\|_1 = 1$. Indeed, $P_r * f$ (resp. $P_r * \mu$) appears as an average of translates of f (resp. μ), so this follows immediately from the convexity of the norm in L_p (resp. M).

The symmetry of the Poisson kernel, i.e. the fact that $P_r(t) = P_r(-t)$, also plays an important role in the sequel. For instance, using Fubini's theorem, this symmetry immediately implies that, if $1 \le p \le \infty$, $1/p + 1/q = 1$ and $0 \le r < 1$, we have

$$\forall f \in L_p(\mathbb{T}, m) \quad \forall g \in L_q(\mathbb{T}, m) \quad \int_{\mathbb{T}} g(P_r * f)dm = \int_{\mathbb{T}} (P_r * g)f dm, \tag{3.15}$$

$$\forall \mu \in M(\mathbb{T}) \quad \forall f \in C(\mathbb{T}) \quad \int_{\mathbb{T}} (P_r * f)d\mu = \int_{\mathbb{T}} (P_r * \mu)f dm. \tag{3.16}$$

Classically, the Poisson kernel is the fundamental tool to solve the Dirichlet problem in the disc, i.e. the problem of finding a harmonic extension inside the disc of a function defined on the boundary. The extension process is given by the 'Poisson integral' of a function, defined as follows. Consider f in $L_1(\mathbb{T}, m)$ real valued, we define, for all $z = re^{i\theta}$ in D

$$u(z) = \int f(t)P_r(\theta - t)dm(t) = P_r * f(\theta) \tag{3.17}$$

or equivalently

$$u(z) = \int f(t)P^z(dt). \tag{3.18}$$

More generally, given a real measure μ on $\mathbb{T}$, we can define

$$u(z) = \int P_r(\theta - t)\mu(dt) = P_r * \mu(\theta)$$

or

$$u(z) = \int P_z(t)\mu(dt). \tag{3.19}$$

Note that the series appearing in (3.7) converge uniformly (and even absolutely) on $\mathbb{T}$, hence

$$u(z) = \sum\nolimits_{n \ge 0} z^n \widehat{\mu}(n) + \sum\nolimits_{n<0} \bar{z}^n \widehat{\mu}(n). \tag{3.20}$$

In the sequel, we refer to this function u as the Poisson integral of f or of μ. It is classical that this process yields a harmonic function. In fact, since

$$P_z(t) = \Re[C_z(t)]$$

where $C_z(t) = \frac{1+ze^{-it}}{1-ze^{-it}}$ analytically depends on z, it is easy to deduce from (3.19) that u is the real part of an analytic function, hence is harmonic.

3.2 The h^p spaces of harmonic functions on D

Let $V \subset \mathbb{C}$ be an open set. Let B be a real Banach space. When dealing with a complex space, we will view it as a real one. A continuous function $u\colon V \to B$ is called harmonic if for all continuous linear forms $\xi\colon B \to \mathbb{R}$, the function $\xi(u)$ is a real valued harmonic function. In that case, it is well known that u is a C^∞ function from V into B for the norm topology on B. Assume $\overline{D} \subset V$. We again denote as before

$$u_r(z) = u(rz).$$

Then the formulae that are derived from the mean value property clearly extend to the Banach valued case, as long as they make sense. For example, we have

$$u(z) = \int u(e^{it})P^z(dt). \tag{3.21}$$

To check vector valued identities of this kind, we can use linear forms, for instance for all continuous linear forms $\xi\colon B \to \mathbb{R}$ we have by the scalar case

$$\xi(u(z)) = \int \xi(u(e^{it}))P^z(dt),$$

hence (3.21) follows. We will use this 'scalarization' principle repeatedly in the sequel, to check the identity of two vectors.

For all $0 < p \le \infty$, we denote by $h^p(D; B)$ the set of all harmonic functions $u\colon D \to B$ such that $\sup_{0\le r<1} \int \|u_r(e^{it})\|_B^p dm(t)$ is finite. We set

$$\|u\|_{h^p(D;B)} = \sup_{0\le r<1} \left(\int \|u_r(e^{it})\|^p dm(t)\right)^{1/p}.$$

When $1 \le p \le \infty$, the space $h^p(D; B)$ equipped with this norm clearly is a Banach space.

We use the notation

$$h^p(D) = h^p(D; \mathbb{R}).$$

We will see that the functions u in $h^p(D; B)$ do not necessarily have radial limits for the norm topology of B. However if we start from a 'good' function on the boundary, its Poisson integral does have radial or non-tangential limits.

Let $1 \leq p \leq \infty$. Given a function f in $L_p(\mathbb{T}, m; B)$, we define its Poisson integral $u(z)$ for each $z = re^{i\theta}$ in D as before

$$u(z) = \int P_r(\theta - t) f(t) dm(t). \tag{3.22}$$

Then u is clearly in $h^p(D; B)$. Indeed, harmonicity reduces to the scalar case (discussed after (3.20)) by 'scalarization'. Then by Jensen's inequality

$$\|u(z)\| \leq \int P_r(\theta - t) \|f(t)\| dm(t)$$

and hence

$$\|u(z)\| \leq v(z) \tag{3.23}$$

where v is the Poisson integral of the function $\|f(\cdot)\|_B$, and moreover by (3.13)

$$\|u\|_{h^p(D;B)} \leq \|v\|_{h^p} \leq \|f\|_{L_p(B)}. \tag{3.24}$$

The next result shows that in a certain sense f (resp. μ) is the boundary value of its Poisson integral u.

Theorem 3.1. *Let $1 \leq p \leq \infty$. Consider a function f in $L_p(\mathbb{T}, m; B)$, and let u be the Poisson integral of f defined in (3.22). For $0 \leq r < 1$, let*

$$u_r(\theta) = u(re^{i\theta}).$$

(i) If $p < \infty$, then $u_r \to f$ in $L_p(\mathbb{T}, m; B)$ when $r \to 1$.

(ii) If f is continuous on $\mathbb{T}$, then $u_r \to f$ uniformly on $\mathbb{T}$ (i.e. in the space $C(\mathbb{T}; B)$ of B-valued continuous functions on $\mathbb{T}$).

(iii) If $p = \infty$ and $B = \mathbb{R}$, then $u_r \to f$ in the weak topology $\sigma(L_\infty(m), L_1(m))$.*

Assuming $B = \mathbb{R}$, let u be the Poisson integral of a real measure μ in $M(\mathbb{T})$.

(iv) Then the measures $u_r.m$ tend to μ in the weak topology $\sigma(M(\mathbb{T}), C(\mathbb{T}))$ (recall $M(\mathbb{T}) = C(\mathbb{T})^*$), when $r \to 1$.*

(v) If $\{u_r \mid 0 < r < 1\}$ is uniformly integrable, then $\mu = f.m$ for some $f \in L_1(m)$ and u is the Poisson integral of f.

Proof. Recall that (by Proposition 1.2) the set of all trigonometric polynomials with coefficients in B is dense in $L_p(\mathbb{T}, m; B)$ if p is finite. If f is such a trigonometric polynomial, say $f(t) = a_0 + \sum_1^N a_n \cos(nt) + b_n \sin(nt)$, with $a_k, b_k \in B$, then $u_r(t) = f * P_r(t) = a_0 + \sum^N r^n(a_n \cos(nt) + b_n \sin(nt))$, and hence $u_r \to f$ for the $L_p(m; B)$-norm (or actually any norm since u_r remains in a fixed finite-dimensional space). Then the general case follows from the uniform boundedness of $f \to P_r * f$ together with the density of the set of trigonometric polynomials in the appropriate topology. More precisely, consider $f \in L_p(\mathbb{T}, m; B)$. If $p < \infty$, for any $\varepsilon > 0$ there is a B-valued trigonometric polynomial g, with Poisson integral v, such that $\|f - g\|_p < \varepsilon$. Let $v_r(\theta) = P_r * g(\theta)$. We have by the triangle inequality and (3.13)

$$\|f - u_r\|_p \le \|g - v_r\|_p + \|f - g\|_p + \|u_r - v_r\|_p \le \|g - v_r\|_p + 2\varepsilon$$

and hence $\overline{\lim}_{r\to 1} \|f - u_r\|_p \le 2\varepsilon$, so that finally $\overline{\lim}_{r\to 1} \|f - u_r\|_p = 0$, which proves (i). The same argument proves (ii). For (iii) and (iv), we use the selfduality of the Poisson integral, more precisely we use (3.15) and (3.16). Then (iv) can be easily checked, we have for all g in $C(\mathbb{T})$

$$\langle u_r.m, g\rangle = \int (P_r * \mu) g dm = \int P_r * g d\mu$$

and hence by part (ii) this tends to $\int g d\mu$, so that $u_r.m$ tends weakly to μ. A similar argument establishes (iii). Lastly, if $\{u_r \mid 0 < r < 1\}$ is uniformly integrable it is relatively weakly compact in $L_1(m)$, so there is a sequence $r(n) < 1$ increasing to 1 such that $u_{r(n)}$ converges weakly to some $f \in L_1(m)$. By (iv) we must have $\mu = f.m$. This proves (v). □

Remark. If $f\colon \partial D \to B$ is continuous, there is a function $u\colon \overline{D} \to B$ continuous on $\overline{D}$ and harmonic inside D such that $u|_{\partial D} = f$. Indeed, we simply extend the Poisson integral by letting $u(t) = f(t)$ for all t in ∂D. By (ii) u is continuous on $\overline{D}$.

Let B be a complex Banach space. Let μ be a B-valued bounded vector measure on $(\mathbb{T}, \mathcal{B})$, the torus equipped with its Borel σ-algebra. With the notation from Proposition 2.2, we can naturally define its Fourier transform as

$$\forall n \in \mathbb{Z} \qquad \widehat{\mu}(n) = \int e^{-int} d\mu,$$

and we find

$$\forall n \in \mathbb{Z} \qquad \|\widehat{\mu}(n)\| \le |\mu|(\mathbb{T}).$$

In particular, we may apply this to $\mu = fdm$: We set for any $f \in L_1(\mathbb{T}, m; B)$

$$\widehat{f}(n) = \int e^{-int} f(t) dm(t)$$

and we have

$$\forall n \in \mathbb{Z} \qquad \|\widehat{f}(n)\| \le \|f\|_{L_1(\mathbb{T},m;B)}. \tag{3.25}$$

Returning to the general case of a real Banach space B, we can also define the Poisson integral of μ by setting for all $z = re^{i\theta}$ in D

$$u(re^{i\theta}) = \int P_r(\theta - t)d\mu(t)$$

or equivalently, if $P_z(t) = P_r(\theta - t)$,

$$u(z) = \int P_z(t)d\mu(t).$$

Then u is a B-valued harmonic function in $h^1(D; B)$. Indeed, harmonicity follows as before by 'scalarization', and by (2.4) we have

$$\|u(z)\| \leq \int P_z(t)d|\mu|(t) \tag{3.26}$$

and hence $\|u(z)\| \leq v(z)$ where v is the Poisson integral of $|\mu|$. Since $|\mu|$ is a finite positive measure, v is in h^1, hence u must be in $h^1(D; B)$, and we have

$$\|u\|_{h^1(D;B)} \leq \|\mu\|.$$

Note in passing that for any z with $|z| \leq r < 1$ we have $|v(z)| \leq |\mu|(\mathbb{T})\|P_z\|_\infty = |\mu|(\mathbb{T})(1 + r)(1 - r)^{-1}$ and hence

$$\|u(z)\| \leq |\mu|(\mathbb{T})(1 + r)(1 - r)^{-1}. \tag{3.27}$$

Again, by 'scalarization', (3.20) remains valid for any complex Banach space B and any $\mu \in M(\mathbb{T}, \mathcal{B}; B)$, so that, when $|z| < 1$, we can rewrite $u(z)$ as an absolutely convergent series (now with coefficients in B)

$$u(z) = \sum\nolimits_{n\geq 0} z^n \widehat{\mu}(n) + \sum\nolimits_{n<0} \bar{z}^n \widehat{\mu}(n). \tag{3.28}$$

Moreover by (3.25) we have

$$\sup_{n\in\mathbb{Z}} \|\widehat{\mu}(n)\| \leq \|u\|_{h^1(D;B)}. \tag{3.29}$$

Indeed, (3.25) implies that for any $r < 1$ we have $\|\widehat{\mu}(n)\|r^{|n|} \leq \|u_r\|_{L_1(\mathbb{T},m;B)}$.

Theorem 3.2. *Let B be an arbitrary Banach space. Every u in $h^1(D; B)$ is the Poisson integral of a bounded vector measure μ on $(\mathbb{T}, \mathcal{B})$ such that $\|u\|_{h^1(D;B)} = \|\mu\|_{M(\mathbb{T},\mathcal{B};B)}$. The Poisson integral defines an isometric isomorphism between the spaces $M(\mathbb{T}, \mathcal{B}; B)$ and $h^1(D; B)$.*

Proof. Consider a function u in $h^1(D; B)$. Let $u_r(z) = u(rz)$ as usual. We can find $0 < r_n < 1$ with $r_n \uparrow 1$ such that the measures $\|u_{r_n}(t)\|dm(t)$ converge weakly to a positive measure ν on $\mathbb{T}$ with $\nu(\mathbb{T}) \leq \|u\|_{h^1(D,B)}$. We first claim that for every f in $C(\mathbb{T})$, $\int f(t)u_{r_n}(t)dm(t)$ converges in B when $n \to \infty$.

We will denote

$$L_n(f) = \int f u_{r_n} dm,$$

and $L(f) = \lim_{n\to\infty} L_n(f)$. To show that the limit $L(f)$ exists, we check the Cauchy criterion. If say $r_n < r_k$, then by the vector valued analogue of (3.15) (which can be checked by 'scalarization' as for (3.21)), we have

$$\int f u_{r_k} dm - \int f u_{r_n} dm = \int (f - P_{r_n/r_k} * f)\, u_{r_k} dm$$

so that $\|L_k(f) - L_n(f)\| \le \|f - P_{r_n/r_k} * f\|_\infty \|u\|_{h^1(D;B)}$ and this tends to zero by Theorem 3.1 (ii). Thus $L(f)$ does exist.

Since $\|L_n(f)\| \le \int |f| \|u_{r_n}\| dm$, we have in the limit necessarily

$$\|L(f)\| \le \int |f| d\nu.$$

By the density of $C(\mathbb{T})$ in $L_1(\mathbb{T}, \mathcal{B}; \nu)$ the linear map $f \to L(f)$ has a unique bounded linear extension defined on the whole of $L_1(\mathbb{T}, \mathcal{B}; \nu)$ and we still have

$$\forall f \in L_1(\mathbb{T}, \mathcal{B}; \nu) \qquad \|L(f)\| \le \int |f| d\nu.$$

We now set $\mu(A) = L(1_A)$, so that

$$\forall A \in \mathcal{A} \qquad \|\mu(A)\| \le \nu(A). \tag{3.30}$$

Clearly μ is σ-additive, and we have $|\mu| \le \nu$. Now for any f in $L_1(\mathbb{T}, \mathcal{B}; \nu)$ (a fortiori for any f in $C(\mathbb{T})$) we have

$$L(f) = \int f d\mu. \tag{3.31}$$

Indeed, both sides of (3.31) are continuous linear maps on $L_1(\mathbb{T}; \nu)$, which coincide on the linear span of indicators. In particular, we have for all z in D

$$L(P_z) = \int P_z d\mu.$$

But $L(P_z)$ is nothing but $u(z)$ since

$$L(P_z) = \lim_{n\to\infty} \int P_z(t) u_{r_n}(t) dm(t) = \lim_{n\to\infty} u(r_n z) = u(z),$$

therefore we conclude that u is the Poisson integral of μ. We have

$$\|\mu\|_{M(\mathbb{T},\mathcal{B};B)} \le \nu(\mathbb{T}) \le \|u\|_{h^1(D;B)}$$

and the converse inequality has already been checked before stating Theorem 3.2. This completes the proof, since the last assertion is merely a recapitulation.

In addition, note that we obtain $|\mu|(\mathbb{T}) = \|\mu\|_{M(\mathbb{T},\mathcal{B};B)} = \nu(\mathbb{T}) = \|u\|_{h^1(D;B)}$, and hence, since (a priori) $|\mu| \le \nu$, we find (a posteriori) $\nu = |\mu|$. □

By a simple variant, we have

Corollary 3.3. *In the same situation as in Theorem 3.2, let* $1 < p \le \infty$. *Every* u *in* $h^p(D;B)$ *is the Poisson integral of a* B*-valued bounded vector measure* μ *on* $(\mathbb{T},\mathcal{B})$ *such that* $|\mu| \ll m$ *and the Radon-Nikodým derivative* $w = d|\mu|/dm$ *is in* $L_p(\mathbb{T},m)$ *and satisfies*

$$\|w\|_p = \|u\|_{h^p(D;B)}.$$

Proof. We repeat the proof of Theorem 3.2, but this time we can assume that $\|u_{r_n}\|$ converges for the topology $\sigma(L_p, L_{p'})$ to a function $w \ge 0$ in L_p and satisfying

$$\|w\|_p \le \sup_n \Big(\int \|u_{r_n}\|^p dm\Big)^{1/p} \le \|u\|_{h^p(D;B)}. \tag{3.32}$$

Clearly this implies $|\mu| = \nu = w.m$ in the preceding argument. Finally, since $|\mu| = w.m$, then a simple convexity argument yields

$$\|P_r * \mu(\theta)\| \le P_r * w(\theta)$$

and hence $\|u\|_{h^p(D;B)} \le \|w\|_p$. □

Let $1 \le p \le \infty$, we will denote by $\widetilde{h}^p(D;B)$ the subspace of $h^p(D;B)$ formed by the functions u that are the Poisson integral of a function in $L_p(\mathbb{T},m;B)$.

Let us denote by $\mathcal{T}(D;B)$ the $\mathbb{R}$-linear subspace of $h^p(D;B)$ formed by all the functions of the form

$$u(z) = a_0 + \sum_1^n \Re(z^k)a_k + \Im(z^k)b_k \tag{3.33}$$

for some integer $n > 0$ and $a_0, \ldots, a_n,\ b_1, \ldots, b_n$ in B. If B is a complex Banach space then this is the same as the $\mathbb{C}$-linear subspace formed by all the functions of the form

$$u(z) = x_0 + \sum_1^n z^k x_k + \bar{z}^k y_k \tag{3.34}$$

for some integer $n > 0$ and $x_0, \ldots, x_n,\ y_1, \ldots, y_n$ in B. The harmonic functions in $\mathcal{T}(D;B)$ can be described as the Poisson integrals of the functions in $L_1(\mathbb{T},m;B)$ with finitely supported Fourier transform. We refer to the latter functions as B-valued 'trigonometric polynomials'. Recall that the set of these B-valued 'trigonometric polynomials' is dense in $L_p(\mathbb{T},m;B)$ for p finite. If $p = \infty$, its closure in $L_\infty(\mathbb{T},m;B)$ can be identified with the Banach space

$C(\mathbb{T}; B)$ of all the B-valued continuous functions on $\mathbb{T}$ equipped with the sup norm. Moreover, recall that $L_\infty(\mathbb{T}, m; B)$ is defined as the set of essentially bounded Bochner measurable functions, or equivalently, we have for any p finite

$$L_\infty(\mathbb{T}, m; B) = \{f \in L_p(\mathbb{T}, m; B) \text{ such that } \|f(\cdot)\|_B \in L_\infty(\mathbb{T}, m)\}.$$

In analogy with this, we obviously have for all finite p

$$\tilde{h}^\infty(D; B) = \tilde{h}^p(D; B) \cap h^\infty(D; B).$$

Then we have

Theorem 3.4. *Let B be an arbitrary Banach space and let $1 \le p \le \infty$. Then the Poisson integral defines an isometric isomorphism between $L_p(\mathbb{T}, m; B)$ and $\tilde{h}^p(D; B)$, and an isometric isomorphism between $C(\mathbb{T}; B)$ and the closure of $\mathcal{T}(D; B)$ in $h^\infty(D; B)$. Moreover, if $1 \le p < \infty$, $\mathcal{T}(D; B)$ is dense in $\tilde{h}^p(D; B)$.*

Proof. By the preceding theorem and the definition of $\tilde{h}^p(D; B)$ the map $f \to u$ is an isometry of $L_p(\mathbb{T}, m; B)$ onto $\tilde{h}^p(D; B)$ for $1 \le p \le \infty$. Moreover f is a B-valued 'trigonometric polynomial' iff u is in $\mathcal{T}(D; B)$. Therefore if p is finite (since the trigonometric polynomials are dense in $L_p(\mathbb{T}, m; B)$) the range of this isometry (i.e. $\tilde{h}^p(D; B)$) coincides with the closure of $\mathcal{T}(D; B)$ in $h^p(D; B)$. The same argument for $p = \infty$ yields an isometric isomorphism between $C(\mathbb{T}; B)$ and the closure of $\mathcal{T}(D; B)$ in $h^\infty(D; B)$. □

Actually, as the next theorem shows every real valued harmonic function u such that $\sup_{r<1} \|u_r\|_p < \infty$ for some $p > 1$ happens to be the Poisson integral of an L_p-function to which u_r converges in L_p. So, just like for martingales, we have a 'boundedness implies convergence principle'. Here we restrict to the real valued case. In the B-valued case, as we will soon see in §3.4, the corresponding fact requires that B has the RNP.

Theorem 3.5. *Let u be a real valued harmonic function on D.*

(i) *If $1 < p \le \infty$, u belongs to $h^p(D)$ iff u is the Poisson integral of a function f in $L_p(\mathbb{T}, m)$. Moreover, we have then $\|u\|_{h^p} = \|f\|_p$.*
(ii) *In the case $p = 1$, u belongs to $h^1(D)$ iff u is the Poisson integral of a real measure μ on $\mathbb{T}$. Moreover, we have $\|u\|_{h^1} = \|\mu\|_M = |\mu|(\mathbb{T})$.*
(iii) *We have $u \ge 0$ on D iff u is the Poisson integral of a non-negative measure μ in $M(\mathbb{T})$. Moreover, we have $\|u\|_{h^1} = u(0) = \mu(\mathbb{T})$.*

Proof. Assume $u \in h^p(D)$. If $1 < p \le \infty$, the unit ball of $L_p(\mathbb{T})$ is $\sigma(L_p, L_{p'})$-compact, and hence there is a sequence $0 < r_n < 1$ with $r_n \to 1$ such that the sequence (u_{r_n}) converges for $\sigma(L_p, L_{p'})$ to some function f in $L_p(\mathbb{T})$ and we have $\|f\|_p \le \|u\|_{h^p}$. We claim that $P_r * f = u_r$ for all $0 \le r < 1$. Indeed, for each fixed $0 \le r < 1$ the operator $g \to P_r * g$ is $\sigma(L_p, L_{p'})$-continuous (since by (3.15) it is self-adjoint) and hence $P_r * f = \lim_{n\to\infty} P_r * u_{r_n}$ for $\sigma(L_p, L_{p'})$, but by (3.5) (applied to the harmonic function $z \to u(r_n z)$) we have

$$P_r * u_{r_n}(\theta) = u_{rr_n}(\theta),$$

and by Theorem 3.2 since u_r is continuous on $\mathbb{T}$, $u_{rr_n} \to u_r$ uniformly on $\mathbb{T}$, hence we must have $P_r * f = u_r$, which shows our claim that u is the Poisson integral of f, with $\|f\|_{h^p} \le \|u\|_p$. By (3.13) equality holds. This proves (i).

The proof of (ii) is entirely similar, but using the topology $\sigma(M, C)$ for which the unit ball of $M(\mathbb{T}) = C(\mathbb{T})^*$ is compact, and $d\mu$ is obtained as the weak* limit of a sequence of measures $u_{r_n}.dm$. We skip the details. If $u \ge 0$, then $u_r.dm \ge 0$ for all $r < 1$, hence $\mu \ge 0$ being the weak limit of a sequence of positive measures. Conversely $\mu \ge 0$ implies $u \ge 0$ since $P_r \ge 0$. Moreover, we find $\|\mu\|_M \le \|u\|_{h^1}$ and (3.14) gives the converse inequality. If $u \ge 0$ we have by (3.4) $\int u_r dm = u(r.0) = u(0)$ and hence

$$\|u\|_{h^1} = \sup_{r<1} \int u_r dm = u(0) = \int d\mu = \mu(\mathbb{T}). \qquad \square$$

We set

$$\widetilde{h}^p(D) = \widetilde{h}^p(D; \mathbb{R}).$$

Equivalently this is the closed span in $h^p(D)$ of the functions $\Re(z^n)$ and $\Im(\bar{z}^n)$, $n \ge 0$. We conclude this section by a mere recapitulation:

Corollary 3.6.

(i) If $1 < p \le \infty$, the Poisson integral defines an isometric isomorphism from $L_p(\mathbb{T}, m; \mathbb{R})$ onto $h^p(D)$ and $\widetilde{h}^p(D) = h^p(D)$.

(ii) If $p = 1$, it defines an isometric isomorphism between $M(\mathbb{T}; \mathbb{R})$ and $h^1(D)$, and also an isometric isomorphism between $L_1(\mathbb{T}, m; \mathbb{R})$ and $\widetilde{h}^1(D)$.

In particular $\widetilde{h}^1(D) \ne h^1(D)$.

3.3 Non-tangential maximal inequalities: boundary behaviour

The behaviour of harmonic functions near boundary points is a classical subject. In the context of h^p (and H^p) spaces the maximal inequalities stated in the following are a fundamental tool. They control the non-tangential limiting behaviour at almost every boundary point. To define them we need some specific notation.

For any $t \in \mathbb{T}$, we introduce the so-called Stolz region

$$\Gamma_\alpha(t)$$

defined for $0 < \alpha < \infty$, with $0 < \beta < \pi/2$ such that $\tan\beta = \alpha$ as the interior of the smallest convex set containing e^{it} and the closed disc with centre zero and radius $\sin\beta$.

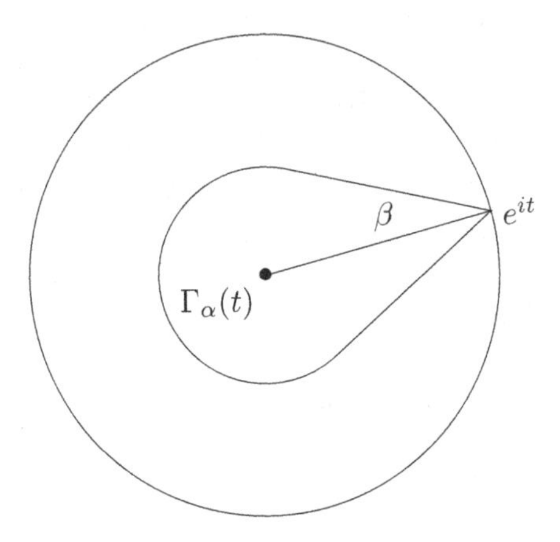

Figure 3.1. Stolz domain.

Equivalently, $\Gamma_\alpha(t)$ is the open domain bounded by the two tangents from the point e^{it} to the circle $\{|z| = \sin\beta\}$ and by the larger of the two arcs of that circle between the points of contact. Near the boundary point e^{it}, $\Gamma_\alpha(t)$ looks like a cone with vertex e^{it} and opening angle β.

Let B be any Banach space. Then, for $f \in L_p(\mathbb{T}, m; B)$, $1 \le p \le \infty$ with Poisson integral u harmonic on D, we define $M_\alpha f$ as follows:

$$M_\alpha f(t) = \sup_{z \in \Gamma_\alpha(t)} \|u(z)\|.$$

Note that we do need to 'truncate' the cone with vertex e^{it}, because otherwise we would be incorporating in the maximal function $M_\alpha f(t)$ some far away boundary points that are irrelevant for the behaviour of f near t.

Furthermore, we define the radial maximal function as

$$M_{\text{rad}} f(t) = \sup_{0 \le r < 1} \|u(re^{it})\|.$$

Note that obviously

$$M_{\text{rad}} f \le M_\alpha f.$$

The key observation about these is that they are dominated by the Hardy-Littlewood maximal function. This maximal function is related to Lebesgue's classical theorem that says that if f is a locally integrable scalar function on $\mathbb{R}$ (or on $\mathbb{T}$ in the periodic case) then,

$$\lim_{h\to 0}\frac{1}{2h}\int_{t-h}^{t+h} f(s)ds = f(t) \quad \text{for almost every} \quad t\in\mathbb{R}. \tag{3.35}$$

Moreover, for any finite measure μ on $\mathbb{R}$ admitting $f(t)dt$ as its absolutely continuous part we have

$$\lim_{h\to 0}\frac{1}{2h}\mu([t-h,t+h]) = f(t) \quad \text{for almost every} \quad t\in\mathbb{R}. \tag{3.36}$$

In particular if μ is singular with respect to Lebesgue measure, we have:

$$\lim_{h\to 0}\frac{1}{2h}\mu([t-h,t+h]) = 0 \quad \text{for almost every } t\in\mathbb{R}.$$

For any $f\in L_1(\mathbb{R},dx;B)$, its Hardy-Littlewood maximal function is classically defined as

$$Mf(t) = \sup_{h,h'>0}\frac{1}{h+h'}\int_{t-h}^{t+h'}\|f(x)\|dx. \tag{3.37}$$

The Hardy-Littlewood maximal function of $f\in L_1(\mathbb{T},m;B)$ is defined as

$$Mf(t) = \sup\frac{1}{m(I)}\int_I \|f(s)\|dm(s), \tag{3.38}$$

where the supremum runs over all possible open arcs $I\subset\mathbb{T}$ containing t. For a (vector) measure $\mu\in M(\mathbb{T};B)$ we set analogously

$$M\mu(t) = \sup\left\{\frac{|\mu|(I)}{m(I)} \mid I \text{ open arc, } t\in I\right\}.$$

Note that if $g(t) = \|f(t)\|_B$ we have clearly

$$Mf = Mg \quad \text{and also} \quad M\mu = M|\mu|. \tag{3.39}$$

We now state the Hardy-Littlewood Maximal Theorem. We should emphasize that, by (3.39), the B-valued case trivially reduces to the case $B=\mathbb{R}$, so the apparent generalization is straightforward. For the proof (for $B=\mathbb{R}$) we refer to any of the classical or modern treatises, such as [32, 36, 48, 85, 86, 89, 99].

Theorem 3.7. [(Hardy-Littlewood maximal inequalities)] *Let B be any Banach space. If $1\le p\le+\infty$ and $f\in L_p(\mathbb{T};B)$ then $Mf(t)$ is finite almost everywhere and there is a constant C_p depending only on p such that if $p>1$ we have*

$$\|Mf\|_p \le C_p\|f\|_p, \tag{3.40}$$

and when $p = 1$,

$$\forall c > 0 \quad cm\{Mf > c\} \leq 2 \int_{\{Mf > c\}} \|f\| dm \leq 2\|f\|_1. \tag{3.41}$$

Moreover, for any measure $\mu \in M(\mathbb{T}; B)$

$$\sup_{c>0} cm\{M\mu > c\} \leq 2\|\mu\|_{M(\mathbb{T};B)}. \tag{3.42}$$

This was surely a partial inspiration for Doob's inequalities. Note that, arguing as in the proof of (1.30), one deduces (3.40) from (3.41) with $C_p \leq 2p'$.

Remark 3.8. We skip the classical proof of this. Note however that the results of §7.5 (see Remark 7.36) allow to deduce the Hardy-Littlewood maximal inequalities on $\mathbb{R}$ from Doob's inequality.

Remark 3.9. The preceding result holds just as well for $f \in L_p(\mathbb{R}; B)$ and $\mu \in M(\mathbb{R}; B)$ with Mf as defined earlier, and $M\mu$ defined similarly using intervals in place of arcs. Note that for any $f \in L_1(\mathbb{R}; B)$ (resp. $f \in L_1(\mathbb{T}; B)$) we have for a.e. $t \in \mathbb{R}$ (resp. $t \in \mathbb{T}$)

$$f(t) = \lim_{h+h' \to 0} \frac{1}{h+h'} \int_{t-h}^{t+h'} f(x)dx$$

(resp.

$$f(e^{it}) = \lim_{h+h' \to 0} \frac{1}{h+h'} \int_{t-h}^{t+h'} f(e^{ix})dx \;),$$

where we restrict to $h, h' > 0$. This fact is an ancestor to the martingale convergence theorem. Indeed, this a.e. convergence can be deduced from the relevant maximal inequality (here the Hardy-Littlewood one) exactly as in the proof of Theorem 1.30. This is the so-called Banach principle. We apply the latter principle again in Corollary 3.18 and Corollary 4.21.

The next result reduces in many respects the study of non-tangential convergence to the Hardy-Littlewood maximal inequalities.

Theorem 3.10. *For any $\alpha > 0$, there is a constant A_α (depending only on α) such that for any $f \in L_p(\mathbb{T}; B)$ and $1 \leq p \leq +\infty$ we have*

$$M_\alpha f(t) \leq A_\alpha Mf(t) \qquad \forall t \in \mathbb{T}.$$

More generally, if u is the the Poisson integral of $\mu \in M(\mathbb{T}, \mathcal{B}; B)$, we have

$$M_\alpha \mu(t) = \sup_{z \in \Gamma_\alpha(t)} |u(z)| \leq A_\alpha M\mu(t).$$

Proof. By (3.39) this reduces to $f \geq 0$. By translation it suffices to prove this for $t = 0$. Let ψ be a positive measurable function on $\mathbb{T}$ that is even when viewed on $[-\pi, \pi]$ and non-increasing on $[0, \pi]$, so that the sets $\{\psi > c\}$ are all symmetric intervals. We will say that such a ψ is admissible. Let $z = re^{i\theta}$. Assume $P_z(t) \leq \psi(t)$ for all t. Let $\mu = f dm$. We have

$$|u(z)| = \left| \int P_z(t)\mu(dt) \right| \leq \int \psi(t) d\mu(t) = \int \left(\int_0^\infty 1_{\{\psi(t)>c\}} dc \right) d\mu(t)$$
$$= \int_0^\infty \frac{\mu(\{\psi > c\})}{m(\{\psi > c\})} m(\{\psi > c\}) dc \leq M\mu(0) \int \psi dm.$$

Thus if we take $z = r$ and $\psi = P_r$ we find $M_{rad} f(0) \leq Mf(0)$. But it is not too difficult to check (see later) that, for some $A_\alpha < \infty$ (depending only on α), for each $z \in \Gamma_\alpha(0)$ there is an admissible function ψ_z such that $P_z(s) \leq \psi_z(s)$ for all $s \in \mathbb{T}$ and moreover such that

$$\int \psi_z dm \leq A_\alpha. \tag{3.43}$$

By the preceding estimate, this implies $M_\alpha f(0) \leq A_\alpha M f(0)$ and the theorem follows.

To check (3.43) note that if $z = re^{i\theta}$

$$P_z(s) = P_r(s - \theta) = \frac{1 - r^2}{1 - 2r\cos(s - \theta) + r^2}$$

and $x \mapsto \cos x$ is decreasing on $[0, \pi]$. We may clearly restrict ourselves to the case $|\theta| \leq \pi/2$ (since otherwise a look at $\Gamma_\alpha(0)$ shows that $|z| \leq \sin\beta$ and the desired bound follows already from (3.27)). Then if $|\theta| \leq |s| \leq \pi/2$ we set $\psi_z(s) = P_r(|s| - |\theta|)$ and if $|s| \leq |\theta|$ we set $\psi_z(s) = P_r(0)$. The definition of $\psi_z(s)$ can be completed outside $[-\pi/2, \pi/2]$ by setting $\psi_z(s) = P_r(\pi/2 - |\theta|)$ there. Then ψ_z is admissible and dominates P_z. Moreover,

$$\int_{-\pi}^{\pi} \psi_z(s) ds \leq \int_{-\theta}^{\theta} \psi_z ds + \int_{|\theta|<|s|\leq\pi/2} \psi_z ds + 2\int_{\pi/2}^{\pi} \psi_z ds = I + II + III$$

where on the one hand $II \leq 2\int_\theta^\pi P_r(s) ds \leq \int P_r(s) ds = 2\pi$ and on the other hand $I \leq 2|\theta| \frac{1+r}{1-r} \leq 4|\theta|/(1-r)$. But it is easy to check that the constraint that $z \in \Gamma_\alpha(0)$ implies that for some $c_\alpha > 0$ depending only on α we have $|\theta|/(1-r) \leq c_\alpha$. Lastly, we have

$$III \leq \pi P_r(\pi/2 - |\theta|) = \pi \frac{1 - r^2}{1 - 2r\sin(|\theta|) + r^2}$$

and again since the constraint $z \in \Gamma_\alpha(0)$ implies that $\sin(|\theta|)$ is small when $r \to 1$, we have a bound $III \leq c'_\alpha$. Thus we obtain $\int \psi_z(s) ds \leq 2\pi + 4c_\alpha + c'_\alpha$. □

Corollary 3.11. *Let $f \in L_p(\mathbb{T}; B)$ and let $u \in h^p(D; B)$ be its Poisson integral. If $1 < p \le +\infty$, we have*

$$\|M_\alpha f\|_p \le A_\alpha C_p \|f\|_p = A_\alpha C_p \|u\|_{h_p(D;B)},$$

and if $p = 1$

$$\sup_{c>0} cm(\{M_\alpha f > c\}) \le 2A_\alpha \|f\|_1 = 2A_\alpha \|u\|_{h_1(D;B)}.$$

Moreover, if u is the Poisson integral of a measure $\mu \in M(\mathbb{T}, \mathcal{B}; B)$ we have

$$\sup_{c>0} cm(\{M_\alpha \mu > c\}) \le 2A_\alpha \|\mu\|_{M(\mathbb{T},\mathcal{B};B)} = 2A_\alpha \|u\|_{h^1(D;B)}. \quad (3.44)$$

Corollary 3.12. *Recall the notation $u_r(e^{it}) = u(re^{it})$. Let $A_0 = \inf_{\alpha>0} A_\alpha$. If $1 < p \le +\infty$ then for any $u \in h_p(D; B)$*

$$\| \sup_{0<r<1} \|u_r\|_B \|_p \le A_0 C_p \sup_{0<r<1} \| \|u_r\|_B \|_p, \quad (3.45)$$

and for any $u \in h_1(D; B)$

$$\sup_{c>0} cm(\{\sup_{0<r<1} \|u_r\|_B > c\}) \le 2A_0 \sup_{0<r<1} \| \|u_r\|_B \|_1. \quad (3.46)$$

Note the remarkable interchangeability between the sup and the integral expressed by the inequality (3.45).

Just like for martingale convergence, the maximal inequalities imply a.e. convergence statements: the next one describes the non-tangential behaviour of a real valued harmonic function u, i.e. it deals with the convergence of $u(z)$ as z goes to $t \in \mathbb{R}$ while remaining in the cone $\Gamma_\alpha(t)$, which forbids z to go 'tangentially' to t.

Theorem 3.13. *Consider u in $h_p(D)$ with $1 \le p \le +\infty$. If $p > 1$, let f be the L_p function on $\mathbb{T}$ of which u is the Poisson integral. If $p = 1$, let μ be the bounded measure of which u is the Poisson integral, and let f be the Radon-Nikodým derivative of the absolutely continuous part of μ with respect to the Lebesgue measure. Then u admits non-tangential limits a.e. and for almost every $t \in \mathbb{T}$,*

$$\lim_{z \in \Gamma_\alpha(t)\ z \to e^{it}} u(z) = f(t).$$

At this point the reader could prove this as an exercise, combining Theorem 3.1 with the maximal inequalities for $p > 1$ and using the next lemma for $p = 1$. But since this will be proved in greater generality as a consequence of Corollary 3.18 and Proposition 3.22, we postpone the complete details of the proof till the end of §3.4.

The ancestor of the preceding statement is of course the Lebesgue classical differentiation theorem (3.36). For emphasis, we reformulate the key underlying fact as a lemma:

Lemma 3.14. *For any singular measure $\nu \in M(\mathbb{T})$, with Poisson integral v we have a.e.*

$$\lim_{z \in \Gamma_\alpha(t)\ z \to e^{it}} v(z) = \lim_{re^{i\theta} \in \Gamma_\alpha(t)\ re^{i\theta} \to e^{it}} P_r * \nu(\theta) = 0. \tag{3.47}$$

Proof. By (3.26) we may clearly assume $\nu \geq 0$. Fix $t \in \mathbb{T}$ such that

$$\lim_{h \to 0} \frac{\nu([t-h, t+h])}{2h} = 0.$$

We will admit the classical fact that this holds for a.e. t (see (3.36)), and will sketch the proof that (3.47) holds for any such t.

We first tackle radial convergence. We claim that $\lim_{r \to 1} v(re^{it}) = 0$.

Let F be a positive, even, continuous function on $\mathbb{T}$, such that $s \mapsto F(e^{is})$ is decreasing on $[0, \pi]$. For example P_r is such a function for any $0 < r < 1$. Then clearly for any $b > 0$

$$\begin{aligned} \int F d\nu &= \int F \wedge F(b) d\nu + \int (F - F(b))^+ d\nu \\ &= \int F \wedge F(b) d\nu + \int_{F(b)}^{\infty} \nu(\{F \geq c\}) dc \\ &\leq F(b)\nu(\mathbb{T}) + \sup_{c > F(b)} \frac{\nu(\{F \geq c\})}{m(\{F \geq c\})} \int F dm. \end{aligned}$$

Note that if $c \geq F(b)$, we have $\{F \geq c\} = [-h, h]$ for some $0 \leq h \leq b$. Therefore

$$\int F d\nu \leq F(b)\nu(\mathbb{T}) + \sup_{0 < h \leq b} \frac{\nu([-h, h])}{m([-h, h])} \int F dm.$$

We now return to our fixed $t \in \mathbb{T}$ and apply this to $F = P_r$ with $b = b(r)$ such that $P_r(b(r)) \to 0$ when $r \to 1$, and with ν replaced by ν translated by t. We find

$$\lim_{r \to 1} v(re^{it}) = \lim_{r \to 1} \int P_r(t - s) d\nu(s) \leq \lim_{r \to 1} \sup_{h \leq b(r)} \frac{\nu([t-h, t+h])}{m([-h, h])}.$$

Thus the claim follows since we can find $b(r) > 0$ such that both $P_r(b(r)) \to 0$ and $b(r) \to 0$ when $r \to 1$. Indeed, since $P_r(t) = (1 - r^2)((1 - r)^2 + 4r \sin^2(t/2))^{-1}$, it suffices to choose $b = b(r)$ such that, say, $\sin^2(b/2) = \sqrt{1 - r}$, then $P_r(b(r)) \in O(\sqrt{1 - r})$, proving the claim and settling the radial case.

Now when $\alpha > 0$ is fixed and $z = re^{i\theta} \in \Gamma_\alpha(t) \to e^{it}$, a similar argument can be applied using the function ψ_z appearing in (3.43). Assuming $t = 0$ for simplicity, and $z = re^{i\theta} \in \Gamma(0)$, we use $F = \psi_z$. Then for any $b > 0$ we have

$$\|v(z)\| \leq \int \psi_z d\nu \leq \psi_z(b)\nu(\mathbb{T}) + \sup_{0 < h \leq b} \frac{\nu([-h, h])}{m([-h, h])} A_\alpha.$$

But now when $z = re^{i\theta} \in \Gamma(0) \to 0$ we can choose $b_z > 0$ tending to 0 but also such that $\psi_z(b_z) \to 0$ when $z = re^{i\theta} \in \Gamma(0) \to 0$. (For instance $b_z = |\theta| + |\theta|^{1/4}$ works.) We skip the details. □

We will see later on that the full force of boundary behaviour of Banach space valued functions requires the RNP. However, as for the maximal inequalities alone, the generalization to the Banach space valued case is immediate.

For any u in $h^p(D; B)$, we denote

$$M_\alpha u(t) = \sup_{z \in \Gamma_\alpha(e^{it})} \|u(z)\|_B.$$

We have then:

Theorem 3.15. *If $p > 1$,*

$$\|M_\alpha u\|_{L_p(\mathbb{T},m)} \leq C_{p,\alpha} \|u\|_{h_p(D;B)}, \tag{3.48}$$

and if $p = 1$

$$\sup_{c>0} cm(\{M_\alpha u > c\}) \leq C_{1,\alpha} \|u\|_{h_1(D;B)}.$$

Proof. Assume first that u is the Poisson integral of a function $f \in L_p(m; B)$ with $p > 1$. Let v denote the Poisson integral of $t \mapsto \|f(t)\|$. By (3.23), we have $\|u\| \leq v$ and hence by Corollary 3.11 $\|M_\alpha u\|_{L_p(\mathbb{T},m)} \leq \|M_\alpha v\|_{L_p(\mathbb{T},m)} \leq C_{p,\alpha} \|f\|_{L_p(m;B)}$. Applying this to the function $f = u_s$ for $0 \leq s < 1$ we obtain

$$\|M_\alpha u_s\|_{L_p(\mathbb{T},m)} \leq C_{p,\alpha} \|u_s\|_{L_p(\mathbb{T},m)} \leq \|u\|_{h_p(D;B)},$$

and taking the supremum over $s < 1$ we obtain (3.48) by monotone convergence. The same argument applies if $p = 1$. □

Corollary 3.16. *Let $1 \leq p \leq \infty$. Then, for any sequence (f_n) in $L_p(\mathbb{T}, m)$, with the corresponding Poisson integrals denoted by u_n, we have if $p > 1$*

$$\| \sup_n M_\alpha(u_n)\|_{L_p(m)} \leq C_{p,\alpha} \| \sup_n |f_n| \|_{L_p(m)}, \tag{3.49}$$

and if $p = 1$,

$$\forall c > 0 \quad cm(\{\sup_n M_\alpha(u_n) > c\}) \leq C_{\alpha,1} \| \sup_n |f_n| \|_{L_1(m)}. \tag{3.50}$$

Proof. It is enough to prove this for a finite family $(u_1, ..., u_N)$. But in the latter case we can view these inequalities as a particular case of Theorem 3.15 corresponding to the case $B = \ell_\infty^N$. (Recall that we denote by ℓ_∞^N the space $\mathbb{C}^N$ equipped with the norm defined by $\forall x \in \mathbb{C}^N$, $\|x\| = \sup_{n \leq N} |x_n|$.) Indeed, if we define $u : D \to \ell_\infty^N$ by setting $u(z) = (u_n(z))_{n \leq N}$, then we have

$$M_\alpha(u) = \sup_{n \leq N} M_\alpha(u_n) \text{ and } \|u\|_{h^p(D,\ell_\infty^N)} = \| \sup_{n \leq N} |f_n| \|_{L_p(m)}.$$

To obtain (3.49) and (3.50), we let $N \to \infty$ in the resulting inequality. □

It is somewhat instructive to observe that conversely Corollary 3.16 essentially implies Theorem 3.15 using the following classical fact.

Lemma 3.17. *Let B_0 be a separable closed subspace of a Banach space B. Then there is a sequence (ξ_n) in the unit ball of the dual B^*, such that*

$$\forall x \in B_0 \quad \|x\| = \sup |\xi_n(x)|.$$

Indeed, assume (3.49) and (3.50) and let us recover Theorem 3.15 as announced. Observe that any B-valued harmonic function must be separable valued, since it is continuous on D and D itself is separable. Let $B_0 \subset B$ be a separable closed subspace containing the range of u. Let (ξ_n) be as in Lemma 3.17. Let $u_n(z) = \xi_n(u(z))$, and let f_n be the radial limit of u_n. Clearly, $M_\alpha(u) = \sup_n M_\alpha(u_n)$. Moreover,

$$\begin{aligned} \| \sup |f_n| \|_{L_p(m)} &= \sup_N \| \sup_{n \le N} |f_n| \|_{L_p(m)} \\ &\le \sup_N \sup_{r<1} \| \sup_{n \le N} |(u_n)_r| \|_{L_p(m)} \le \|u\|_{h^p(B)}. \end{aligned}$$

Then we can deduce Theorem 3.15 from (3.49) and (3.50). We leave the remaining details to the reader.

The maximal inequalities imply the following important fact that considerably strengthens Theorem 3.1.

Corollary 3.18. *Consider u in $\widetilde{h}_p(D; B)$ with $1 \le p \le +\infty$. Let f be the $L_p(B)$ function on $\mathbb{T}$ of which u is the Poisson integral. Then u admits non-tangential limits a.e. and for almost every $t \in \mathbb{T}$,*

$$\lim_{z \to e^{it} \; z \in \Gamma_\alpha(t)} u(z) = f(t).$$

Proof. When u is in $\mathcal{T}(D, B)$ this is obvious. Fix $\varepsilon > 0$ and choose $v \in \mathcal{T}(D, B)$ with $\|u - v\|_{h_p(D;B)} < \varepsilon$. Assume v is the Poisson integral of g. We set

$$L(u, t) = \limsup_{z \to e^{it} \; z \in \Gamma_\alpha(t)} \|f(e^{it}) - u(z)\|_B.$$

Note that $L(v, t) = 0$ and $L(u - v, t) \le \|f(t) - g(t)\|_B + M_\alpha(u - v)(t)$. By the triangle inequality, we have

$$\begin{aligned} L(u, t) &\le L(v, t) + L(u - v, t) = L(u - v, t) \\ &\le \|f(t) - g(t)\|_B + M_\alpha(u - v)(t) \end{aligned}$$

and hence by (3.48)

$$\|L(u, t)\|_p \le \varepsilon + C_{p,\alpha}\varepsilon,$$

so we conclude that $L(u,t) = 0$ a.e. The proof for $p = 1$ is similar: using the notation $\|\varphi\|_{1,\infty} = \sup_{c>0} cm(\{|\varphi| > c\})$, we have

$$\|L(u,t)\|_{1,\infty} \leq 2(\|L(v,t)\|_{1,\infty} + \|L(u-v,t)\|_{1,\infty}),$$

and we conclude as for $p > 1$. □

We will also need to observe the following useful fact

Proposition 3.19. *Let $1 < p \leq \infty$. Consider u in $h^p(D; B)$. The following are equivalent.*

(i) $u \in \tilde{h}^p(D; B)$, that is to say u is the Poisson integral of a function f in $L_p(\mathbb{T}, m; B)$.
(ii) u admits radial limits a.e. on $\mathbb{T}$.
(iii) u admits non-tangential limits a.e. on $\mathbb{T}$.

Proof. (i) ⇒ (iii) follows from Corollary 3.18 and (iii) ⇒ (ii) is obvious. It remains to prove (ii) ⇒ (i). Assume (ii). Let $f = \lim_{r\to 1} u_r$. By the maximal inequalities (cf. Theorem 3.15) $\sup_{r<1} \|u_r\|_B$ is in $L_p(\mathbb{T}, m)$, hence, for p finite, by dominated convergence

$$\int \|f\|^p dm < \infty \quad \text{and} \quad \int \|f - u_r\|^p dm \to 0$$

when $r \to 1$, so that $f \in L_p(\mathbb{T}, m; B)$. Moreover, $P_r * f(\theta) = \lim_{s\to 1} P_r * u_s = \lim_{s\to 1} u(rse^{i\theta}) = u(re^{i\theta})$ so that u is the Poisson integral of f. Clearly, if $u \in h^\infty(D; B)$ then $f \in L_\infty(\mathbb{T}, m; B)$. This completes the proof even if $p = \infty$. □

The case $p = 1$, excluded from the preceding statement, will be covered by Proposition 3.21 (recall also Theorem 3.2).

Remark 3.20. Let $V \subset \mathbb{C}$ be a simply connected open set bounded by a Jordan curve. By a well-known theorem due to Carathéodory (refining the Riemann mapping theorem), there is a biholomorphic bijection $\psi : D \to V$ (in other words a conformal equivalence) that continuously extends to a homeomorphism from $\bar{D}$ to $\bar{V}$, see [15]. This fact is classically used to transfer to V the solution of the Dirichlet problem for D, which finds a harmonic extension inside V of any continuous function on ∂V.

More generally, we will need to consider unbounded simply connected open sets $V \subset \mathbb{C}$, for which there is a finite subset $F \subset \partial D$ such that the biholomorphic bijection $\psi : D \to V$ continuously extends to a homeomorphism from $\bar{D} \setminus F$ to $\bar{V}$.

If the boundary ∂V is assumed nice enough, so that, say, it admits tangents everywhere, then the notion of non-tangential convergence can be defined as before: Given $\xi \in \partial V$, where this tangent $T(\xi)$ exists, we say that a sequence (z_n) in V, which converges to ξ, does so *non-tangentially* if there is an integer N such that the size of the angle between $\xi - z_n$ and $T(\xi)$ is bounded below by a positive number $\gamma > 0$ for all $n \geq N$. (In the preceding case of D, we used $\beta = \pi/2 - \gamma$ and $\alpha = \tan\beta$.)

Let $\xi = \psi(e^{it})$ for some $e^{it} \in \partial D \setminus F$. Assume now that ψ extends to an analytic function in a neighbourhood of e^{it} in $\mathbb{C}$ with $\psi'(e^{it}) \neq 0$, and also that $\psi^{-1} : V \to D$ extends to an analytic function in a neighbourhood of ξ in $\mathbb{C}$ (necessarily such that $(\psi^{-1})'(\xi) \neq 0$). This assumption could be weakened but it suffices for the cases we use in this book. Then, near e^{it} (resp. near ξ), the map ψ (resp. ψ^{-1}) is 'conformal' in the original sense that it preserves angles (simply because infinitesimally $\psi(z) - \psi(e^{it})$ is the same as $\lambda(z - e^{it})$ with $\lambda = \psi'(e^{it})$, and similarly for ψ^{-1}). In particular, we conclude that z_n in D tends non-tangentially to e^{it} iff $\psi(z_n)$ tends non-tangentially in V to ξ.

3.4 Harmonic functions and RNP

In general, Banach space valued harmonic functions that are bounded on D do not necessarily admit radial limits. We give two simple and classical examples. The first one is the function $f_0 : D \to c_0(\mathbb{C})$ defined by

$$f_0(z) = (z^n)_{n\geq 0}.$$

Clearly f_0 is a bounded and analytic (hence harmonic) function on D with values in $c_0(\mathbb{C})$. However, it is clear that, when $r < 1$ tends to 1, the limit of $f_0(re^{it})$ exists for no point $e^{it} \in \mathbb{T}$. A second example is the function $f_1 : D \to L_1(m)$ defined by the Poisson kernels

$$f_1(z) = P_z(t).$$

Clearly f_1 is harmonic since $z \to \int \varphi(t)P_z(t)dm(t)$ is harmonic for any $\varphi \in L_\infty(m)$. Moreover, f_1 takes its values in the unit sphere of $L_1(m)$, and it is well known that for any $e^{it} \in \mathbb{T}$, we have

$$\limsup_{r,s\to 1} \|f_1(re^{it}) - f_1(se^{it})\|_{L_1} = \limsup_{r,s\to 1} \int |P_r(\theta) - P_s(\theta)| d\theta/2\pi = 2. \quad (3.51)$$

Therefore, the radial limits of the function f_1 exist for no point of $\mathbb{T}$. The idea behind (3.51) is that, when $r \to 1$, the probabilities $P_r(\theta)d\theta/2\pi$ tend to the Dirac mass at $\theta = 0$ and $P_r(\theta) \to 0$ for any $\theta \neq 0$. Indeed, when $s < r < 1$ with s close to 1 but with r much closer to 1 than s, almost the entire mass of the probability measure $P_r(\theta)d\theta/2\pi$ is concentrated on a subset that has small mass for $P_s(\theta)d\theta/2\pi$. More precisely, for any fixed small $\varepsilon > 0$ we have (dominated convergence) $\lim_{r\to 1} \int_{|\theta|>\varepsilon} P_r(\theta)d\theta = 0$ and hence $\lim_{r\to 1} \int_{|\theta|\le\varepsilon} P_r(\theta)d\theta/2\pi = 1$ but $\int_{|\theta|\le\varepsilon} P_s(\theta)d\theta \le 2\varepsilon(1+s)(1-s)^{-1}$. We leave the details to the reader.

However, we will show that such examples cannot occur in a reflexive Banach space or one with the Radon-Nikodým property (in short RNP). Nor can they occur if $u \in \tilde{h}^1(D; B)$, as follows:

Proposition 3.21. *Let u in $h^1(D; B)$ be the Poisson integral of a bounded vector measure μ such that $|\mu| \ll m$. Then u admits radial limits a.e. iff $u \in \tilde{h}^1(D; B)$. When this holds, u admits non-tangential limits a.e.*

Proof. Assume $|\mu| \le w.m$ with $w \in L_1(\mathbb{T}, m)$. Let v be the Poisson integral of w. By (3.26), we have $\|u(z)\| \le v(z)$ for all z in D.

Assume u admits radial limits a.e. on $\mathbb{T}$. Let $f(e^{it}) = \lim_{r\to 1} u(re^{it})$. Then $\|f(e^{it})\| \le v(e^{it})$ and hence $f \in L_1(\mathbb{T}, m; B)$. Since $\|u(re^{it})\| \le P_r * w(t)$ and $P_r * w$ converges in $L_1(m)$, the functions $t \to \|u(re^{it})\|$ form an equi-integrable family indexed by $0 \le r < 1$, and hence $\|f(t) - u(re^{it})\| \to 0$ not only a.e. but also in $L_1(\mathbb{T}, m)$. This ensures that u is the Poisson integral of f by the usual argument

$$u(z) = \lim_{r\to 1} u(rz) = \lim_{r\to 1} \int u_r dP^z = \int f dP^z,$$

and we conclude $u \in \tilde{h}^1(D; B)$. This proves the 'only if' part. The converse is clear by Theorems 3.4 and 3.19, and actually u admits non-tangential limits a.e. □

Proposition 3.22. *Let B be an arbitrary Banach space. Every function u in $h^1(D; B)$ can be written uniquely as a sum $u = u^a + u^s$ where u^a is the Poisson integral of a B-valued bounded vector measure μ^a such that $|\mu^a| \ll m$ and u^s is the Poisson integral of a vector measure μ^s satisfying $|\mu^s| \perp m$. Moreover, the function u^s tends non-tangentially to zero a.e. on $\mathbb{T}$. Furthermore, u admits radial limits a.e. iff u^a is in $\tilde{h}^1(D; B)$.*

Proof. By Theorem 3.2 there is a B-valued bounded vector measure μ and a measure $\nu \ge 0$ on $(\mathbb{T}, \mathcal{B})$ such that $|\mu| \le \nu$ and u is the Poisson integral of μ. Let $\nu = \nu^a + \nu^s$ be the decomposition of ν with $\nu^a \ll m$ and $\nu^s \perp m$. Let T^a, T^s be a partition of $\mathbb{T}$ into Borel subsets such that ν^s is supported by T^s and m by T^a, i.e.

$$m(T^a) = 1 \quad \text{and} \quad \nu^s(T^a) = 0.$$

We define

$$\forall A \in \mathcal{A} \qquad \mu^a(A) = \mu(A \cap T^a)$$
$$\mu^s(A) = \mu(A \cap T^s).$$

Then,

$$|\mu^a| \le \nu^a \quad \text{and} \quad |\mu^s| \le \nu^s.$$

Letting u^a and u^s be the corresponding Poisson integrals, this gives the announced decomposition. Unicity is clear (indeed, if $u = u'^a + u'^s$ is another decomposition then $u^a - u'^a = -(u^s - u'^s)$ is the Poisson integral of a measure μ' such that $|\mu'|$ is both singular and absolutely continuous hence is equal to zero). Let v be the Poisson integral of ν^s. By (2.5) we have for all $z = re^{i\theta}$

$$\|u^s(z)\| = \|P_r * \mu^s(\theta)\| \le P_r * \nu^s(\theta) = v(z),$$

and by Lemma 3.14 v has non-tangential limits equal to zero a.e., and hence the same is true for u^s. A fortiori, this shows that u^a has radial limits a.e. iff u also does, therefore the last assertion follows from Proposition 3.21. □

We will be mainly interested in the non-tangential behaviour of $z \mapsto u(z)$, that will require the RNP, but meanwhile here is one more property valid for an *arbitrary* Banach space B:

Proposition 3.23. *Let $1 \le p \le \infty$ and let B be arbitrary. Then for any function $u \in h^p(D, B)$, the function*

$$z \mapsto \|u(z)\|$$

tends non-tangentially (and radially) a.e. to the Radon-Nikodým derivative $w = d|\mu|/dm$, for the (bounded) vector measure μ admitting u as its Poisson integral.

Proof. By Proposition 3.22, we may assume that u is the Poisson integral of a bounded vector measure μ such that $|\mu| \ll m$. Note that the finite-dimensional case obviously reduces to the uni-dimensional one. Thus, by Theorem 3.13, if B is finite-dimensional we have non-tangential convergence of $u(z)$ to $f(t)$ and hence $\|u(z)\| \to \|f(t)\|$, $\mu = f \cdot dm$ and $w(t) = \|f(t)\|$. Moreover, we have in that case $\int \lim_{z \to t, z \in \Gamma_\alpha(t)} \|u(z)\| \, dm = |\mu|(\mathbb{T})$.

We will deduce the general case from this. Assume $|\mu|(\mathbb{T}) = 1$. Fix $\varepsilon > 0$. Recalling Proposition 2.1, it is easy to see that we can find unit vectors $\xi_1, \ldots, \xi_N$ in B^* such that the bounded vector measure

$$\mu_N \colon \mathcal{B} \to \ell_\infty^N$$

defined by $\mu_N(A) = (\xi_j(\mu(A)))_{j\le N}$ satisfies $|\mu_N|(\mathbb{T}) > 1-\varepsilon$. Consider then the harmonic function $u_N\colon\ D \to \ell_\infty^N$ that is the Poisson integral of μ_N. Note that obviously $u_N(z) = (\xi_j(u(z)))_{j\le N}$. Let $\varphi_-(t) = \liminf \|u(z)\|$ (resp. $\varphi_+(t) = \limsup \|u(z)\|$) where the limits are meant when $z \to t$ and $z \in \Gamma_\alpha(t)$. Let $v(z)$ be the Poisson integral of $|\mu| = w\cdot m$. By (3.26), on the one hand we have $\forall z \in D\ \|u(z)\| \le v(z)$, and hence $\varphi_+(t) \le \limsup v(z) = w(t)$ for a.e. t. On the other hand, we have (since $\|\xi_j\| = 1$)

$$\varphi_-(t) \ge \liminf \|u_N(z)\|$$

but by the first part of the proof

$$\int \liminf \|u_N(z)\| dm = |\mu_N|(\mathbb{T}) > 1-\varepsilon.$$

Thus we have $\varphi_- \le \varphi_+ \le w$ but also

$$\int \varphi_-\, dm > 1-\varepsilon = \int w\, dm - \varepsilon.$$

Therefore we conclude that $\varphi_- = \varphi_+ = w$ a.e. □

Let μ be a B-valued vector measure on $(\Omega, \mathcal{A})$ and let ν be a positive measure on $(\Omega, \mathcal{A})$. Recall that μ admits a Radon-Nikodým derivative (or in short a density) with respect to ν if there is a function f in $L_1(\Omega, \mathcal{A}, \nu; B)$ such that

$$\forall A \in \mathcal{A} \qquad \mu(A) = \int_A f d\nu.$$

In that case, we write simply $\mu = f.\nu$ or $d\mu = f d\nu$.

Remark 3.24. If $\mu = f.\nu$ the integral defined in Proposition 2.2 coincides with the classical Bochner integral i.e. we have

$$\forall g \in L_\infty(\Omega, \mathcal{A}, \nu) \qquad \int g d\mu = \int g f d\nu. \tag{3.52}$$

This is immediate if g is a simple function, and hence by density it remains true for all g in $L_\infty(\Omega, \mathcal{A}, \nu)$, since both sides are continuous on $L_\infty(\Omega, \mathcal{A}, \nu)$. In particular (3.52) shows that if $\mu = f.m$ on $\mathbb{T}$ then $\widehat{\mu} = \widehat{f}$ and the Poisson integral of μ coincides with the Poisson integral of f.

Theorem 3.25. *The following properties of a Banach space B are equivalent.*

(i) B has the RNP.

(ii) For all $1 < p \le \infty$, every u in $h_p(D; B)$ is the Poisson integral of a function f in $L_p(\mathbb{T}, m; B)$. In other words,

$$h^p(D; B) = \tilde{h}^p(D; B).$$

(iii) For some $1 < p \leq \infty$, the same as (ii) holds.
(iv) Every u in $h_\infty(D; B)$ admits radial limits a.e. on $\mathbb{T}$.
(v) Every u in $h_1(D; B)$ admits radial limits a.e. on $\mathbb{T}$.

Proof. (i) $\Rightarrow$ (ii). Consider u in $h^p(D; B)$, then, by Corollary 3.3, u is the Poisson integral of a B-valued bounded vector measure μ on $(\mathbb{T}, \mathcal{B})$ such that $|\mu| \ll m$. If B has the RNP, there is f in $L_1(\mathbb{T}, \mathcal{B}; B)$ such that $\mu = f.m$. By Corollary 3.3 we can assume $|\mu| \leq w.m$ with $w \in L_p(m)$. Recalling (2.2), this implies $\|f\|_B \leq w$ a.e. so that f belongs necessarily to $L_p(\mathbb{T}, m; B)$. This proves (i) $\Rightarrow$ (ii). Similarly, using Proposition 3.22, one can prove (i) $\Rightarrow$ (v). The implications (ii) $\Rightarrow$ (iii) and (v) $\Rightarrow$ (iv) are trivial, and (iii) $\Rightarrow$ (iv) follows from Proposition 3.19. It remains to prove (iv) $\Rightarrow$ (i). Assume (iv) and let $\mu \in M(\mathbb{T}, \mathcal{B}; B)$ be such that $|\mu| \leq w.m$ for some $w \in L_1(\mathbb{T}, m)$. Let

$$A_n = \{n - 1 < w \leq n\}.$$

For any integer $n > 0$, we set, for all $A \in \mathcal{B}$, $\mu_n(A) = \mu(A \cap A_n)$. Then $|\mu_n| \leq nm$. We will prove that (iv) implies that μ_n has a Radon-Nikodým derivative for any integer $n > 0$. Indeed, let u_n be the Poisson integral of μ_n. By (iv) u_n admits radial limits a.e., and by Proposition 3.19, u_n is the Poisson integral of some f_n in $L_\infty(\mathbb{T}, m; B)$. Therefore we must have $\mu_n = f_n.m$ (since these vector measures have the same Poisson integral, cf. e.g. Theorem 3.2, or use scalarization), and by (2.2) $\int \|f_n\|_B dm = \int d|\mu_n| \leq \int_{A_n} w dm$, and hence $\sum \int \|f_n\|_B dm < \infty$. Therefore, if we define $f = \sum_{n \geq 1} f_n$, we have $f \in L_1(\mathbb{T}, \mathcal{B}; B)$ and $\mu(A) = \sum \mu_n(A) = \sum \int_A f_n dm = \int_A f dm$, or equivalently $\mu = f.m$. This shows that B has the RNP relative to the single measure space $(\mathbb{T}, \mathcal{B}, m)$. By Corollary 2.14 this yields (i) in full generality. □

Proof of Theorem 3.13. We first recall Theorem 3.1 that guarantees the existence of f and μ. The case $p > 1$ follows from Corollary 3.18 with $B = \mathbb{R}$. The case $p = 1$ follows from Proposition 3.22. Indeed, we can decompose μ as $\mu = f dm + \mu^s$ with μ^s singular. Then $u = v + w$ where v (resp. w) is the Poisson integral of f (resp. μ^s), and the non-tangential limits of w vanish m-a.e. while those of v are equal to f a.e. by Corollary 3.18 applied this time with $p = 1$. □

3.5 Brownian martingales*

In this section, we will take for granted a number of facts from Brownian motion Theory. Note however that this will be used only in the rest of this chapter and in §4.5.

We start by recalling the usual definition of a standard Brownian motion in $\mathbb{R}^d$. Note that we will only use $d = 2$ for the complex case (and occasionally $d = 1$). By Brownian motion (in short BM) we mean a random process $(W_t)_{t \geq 0}$ formed of $\mathbb{R}^d$ valued variables on a probability space $(\Omega, \mathcal{A}, \mathbb{P})$ such that $t \to W_t(\omega)$ is a continuous function on $\mathbb{R}_+$ with $W_0(\omega) = 0$ for any ω, the increments $(W_{t_1}, W_{t_2} - W_{t_1}, \ldots, W_{t_n} - W_{t_{n-1}})$ are independent for any $0 < t_1 < \cdots < t_n$ and lastly for any $0 < s < t$ the distribution of

$$\frac{1}{\sqrt{t-s}}(W_t - W_s) \tag{3.53}$$

is the standard $N(0, I)$ Gaussian probability measure on $\mathbb{R}^d$.

For any $z \in \mathbb{R}^d$ we may wish to consider the process

$$W_t^z = z + W_t$$

and we may refer to it as 'a Brownian motion starting at z'. Note that our original process starts at $z = 0$ (so that $W_t^0 = W_t$).

The associated filtration is defined by $\mathcal{A}_t = \sigma(W_s \mid s < t)$. Note that by continuity this coincides with $\sigma(W_s \mid s \leq t)$.

Lemma 3.26. *Let $u\colon \mathbb{C} \to B$ be a Banach space valued harmonic function defined on the whole plane as in* (3.33) *(or* (3.34) *in the complex case). Then for any $z \in \mathbb{C}$, $(u(z + W_t))$ is a martingale indexed by* $[0, \infty)$, *which is in $L_p(B)$ for any $p < \infty$, and satisfies the condition $(A)_R$ appearing in* (1.56) *for any $R > 0$.*

Proof. It is well known that Gaussian random variables have finite moments of all order. In particular $(z + W_t)^n \in L_p$ for any $n \in \mathbb{N}$ and any $p < \infty$. Since u is assumed polynomial, this implies also $u(z + W_t) \in L_p(B)$.

By the rotational symmetry of the Gaussian distribution on $\mathbb{C}$, we have clearly (use polar coordinates) for any $t > 0$

$$\forall z \in \mathbb{C} \qquad u(z) = \mathbb{E}u(z + W_t). \tag{3.54}$$

Then for any $z \in \mathbb{C}$, $(z + W_t)_{t > 0}$ is a martingale adapted to the Brownian filtration. Indeed, for any $s < t$, the incremental independence implies

$$\mathbb{E}^{\mathcal{A}_s} u(z + W_t) = \int u(z + W_s + (W_t - W_s)(\omega'))d\mathbb{P}(\omega')$$

and hence by (3.53)

$$\mathbb{E}^{\mathcal{A}_s} u(z + W_t) = \int u(z + W_s + \sqrt{t-s}W_{t-s}(\omega'))d\mathbb{P}(\omega')$$

and by the same argument as for (3.54) (use polar coordinates)

$$\mathbb{E}^{\mathcal{A}_s} u(z + W_t) = u(z + W_s), \tag{3.55}$$

which is the announced martingale property. For the condition $(A)_R$, since u is continuous, it suffices to check that $\mathbb{E} \sup_{0 \le t \le R} \|u(z + W_t)\| < \infty$, but by (1.55) we have

$$\begin{aligned}\mathbb{E} \sup\nolimits_{0 \le t \le R} \|u(z + W_t)\| &\le (\mathbb{E} \sup\nolimits_{0 \le t \le R} \|u(z + W_t)\|^2)^{1/2} \\ &\le 2\|u(z + W_R)\|_2 < \infty. \end{aligned}$$

□

We will try not to get drawn into technical details about Brownian motion, but rather to convey the main ideas of what is going on, with only sketches of proof. The first fact that we take for granted is the very existence of BM as defined previously. To make our treatment hopefully not too painful for the unfamiliar reader, we will now set the stage more precisely.

Notation. We assume that $\Omega = C([0, \infty); \mathbb{R}^d)$. We equip this space with one of the usual metrics defining the topology of uniform convergence over compact subsets, which turns Ω into a separable Fréchet space (i.e. a complete metrizable locally convex space). We then define for any $t \ge 0$

$$\forall \omega \in \Omega \quad W_t(\omega) = \omega(t). \tag{3.56}$$

We equip Ω with its Borel σ-algebra $\mathcal{A}$ and we define $\mathcal{A}_t = \sigma(W_s \mid s \le t)$ as before. We may as well replace $\mathcal{A}$ by $\mathcal{A}_\infty = \sigma(W_s \mid s < \infty)$ (which actually coincides with it). The existence of BM is equivalent to the existence of a probability $\mathbb{P}$ on $(\Omega, \mathcal{A})$ such that the process (W_t) defined by (3.56) is a BM starting at 0 as defined earlier. In particular $\mathbb{P}$ is supported by the ωs such that $\omega(0) = 0$.

Fix $z \in \mathbb{C}$. Let $\alpha_z \colon \Omega \to \Omega$ be the mapping taking $(\omega(t))_{t \ge 0}$ to $(\omega(t) + z)_{t \ge 0}$. We denote $\mathbb{P}_z = \alpha_z(\mathbb{P})$. More precisely, for any $A \in \mathcal{A}$ we have

$$\mathbb{P}_z(A) = \mathbb{P}(\alpha_z^{-1}(A))$$

and for any integrable function of the form $\varphi(W_{t_1}, \ldots, W_{t_n})$ for some $t_1, \ldots, t_n \in [0, \infty)$ we have

$$\mathbb{E}_z \varphi(W_{t_1}, \ldots, W_{t_n}) = \mathbb{E}\varphi(z + W_{t_1}, \ldots, z + W_{t_n}).$$

Now of course $\mathbb{P}_z$ is supported by the ω's such that $\omega(0) = z$.

When Ω is equipped with $\mathbb{P}_z$, the process (W_t) becomes a BM starting at z. It has the same distribution as the process $(z + W_t)$ on $(\Omega, \mathcal{A}, \mathbb{P})$. Note that with this notation $\mathbb{P}_0 = \mathbb{P}$.

We will now concentrate on the complex valued case so we assume $d = 2$. There is a deep connection between harmonic functions and complex BM.

As we just saw in (3.55), for any harmonic function $u\colon \mathbb{C} \to \mathbb{R}$ as in (3.33), $(u(W_t))_{t\geq 0}$ is a martingale on $(\Omega, \mathcal{A}, \mathbb{P}_z)$. But since we are interested in functions defined on D we introduce the stopping time

$$T(\omega) = \inf\{t \geq 0 \mid |\omega(t)| = 1\} = \inf\{t > 0 \mid |W_t(\omega)| = 1\}.$$

Similarly for any $0 < r < 1$ we define

$$T_r(\omega) = \inf\{t \geq 0 \mid |W_t(\omega)| = r\}.$$

Note that W_n being the sum of n independent standard Gaussian variables is a.s. unbounded when $n \to \infty$. A fortiori $\{W_t(\omega) \mid t > 0\}$ is unbounded for a.a. ω. This guarantees that T (and a fortiori T_r since $T_r < T$) is a.s. finite with respect to $\mathbb{P}_z$ whenever $z \in D$. (Actually by the so-called recurrence of BM this is true for all $z \in \mathbb{C}$, but this is less obvious to check.)

Since $r \mapsto T_r$ is non-decreasing and a.s. bounded (by T), $\lim_{r\to 1} T_r(\omega)$ exists for a.a. ω, and since trivially $|W_{T_r(\omega)}| \to 1$ when $r \uparrow 1$, we must have for a.a. ω

$$\lim_{r\to 1} T_r(\omega) = T(\omega). \tag{3.57}$$

Let us return to a Banach space valued harmonic function $u\colon \mathbb{C} \to B$ defined on the whole plane as in (3.33) (or (3.34) in the complex case). Then, by Lemmas 1.53 and 3.26 for any $s \leq t < \infty$ and any stopping time S, we have

$$\mathbb{E}^{\mathcal{A}_s} u(z + W_{S\wedge t}) \overset{\text{a.s.}}{=} u(z + W_{S\wedge s}). \tag{3.58}$$

Clearly $S \wedge t \to S$ when $t \to \infty$, and hence assuming $S < \infty$ a.s. we have $u(z + W_{S\wedge t}) \to u(z + W_S)$ a.s. and if we assume that this a.s. convergence is dominated, i.e. we assume $\sup_{t>0} \|u(z + W_{S\wedge t})\| \in L_1$, then by Lebesgue's theorem the convergence holds in the L_1-sense. Thus we must have for any $0 \leq s < \infty$

$$\mathbb{E}^{\mathcal{A}_s} u(z + W_S) \overset{\text{a.s.}}{=} u(z + W_{S\wedge s}). \tag{3.59}$$

In particular, for $s = 0$, we have

$$\mathbb{E} u(z + W_S) = u(z). \tag{3.60}$$

We will freely use the following fundamental fact (for which we do not try to achieve full generality).

Proposition 3.27. *If $z \in D$ and $S = \inf\{t \mid |z + W_t| = 1\}$ then* (3.58), (3.59) *and* (3.60) *hold for any harmonic function u continuous on $\bar{D}$.*

For any continuous function $f\colon \mathbb{T} \to \mathbb{C}$ and any $z \in D$

$$\mathbb{E}_z f(W_T) = \int f \, dP^z. \tag{3.61}$$

In other words, the distribution ('law') of W_T with respect to $\mathbb{P}_z$ is exactly P^z. In particular, for $z = 0$

$$\mathbb{E}f(W_T) = \int f\,dm. \tag{3.62}$$

Proof. If $z \in D$ then the stopping time $S = \inf\{t \mid |z + W_t| = 1\}$ is finite $\mathbb{P}$-a.s. since (as we already mentioned) $t \mapsto W_t$ is $\mathbb{P}$-a.s. unbounded on $[0, \infty)$. Moreover, $z + W_{S\wedge t}$ starts and remains in D, so $\sup_{t>0} \|u(z + W_{S\wedge t})\| \le \sup \|u(\xi)\| \mid \xi \in D\} < \infty$. In particular, the preceding remarks show that (3.58), (3.59) and (3.60) hold if u is harmonic on $\mathbb{C}$. But if u is merely harmonic on D and continuous on $\bar{D}$, we know (by (ii) in Theorem 3.1) that u can be approximated uniformly on $\bar{D}$ by harmonic polynomials of the form (3.33). By continuity of all sides for the sup-norm on $\bar{D}$, (3.58), (3.59) and (3.60) can all be extended to such u's. Let f be continuous on $\mathbb{T}$ and let u be its harmonic extension inside D (i.e. its Poisson integral). Then (3.60) can be rewritten as $\mathbb{E}f(z + W_S) = u(z)$ or equivalently as $\mathbb{E}^z f(W_T) = u(z)$. But we already know $u(z) = \int f\,dP^z$ (see (3.22)). □

Remark 3.28. Note that if $z \notin D$ the stopping time S is still finite a.s. (because of the recurrence of BM). However, (3.60) is clearly wrong. It is instructive to observe that the reason lies in the lack of dominated convergence in the preceding argument for (3.59).

By the same reasoning, the distribution of W_{T_r} $(0 < r \le 1)$ is the uniform probability on the circle $\{z \mid |z| = r\}$, so that for any integrable function $f\colon \{z \mid |z| = r\} \to \mathbb{C}$ we have

$$\mathbb{E}f(W_{T_r}) = \int f(re^{it})dm(t). \tag{3.63}$$

More generally, let $\mathcal{D} \subset \mathbb{C}$ be an open domain that is conformally equivalent to D with a nice enough boundary (say a Jordan curve). Fix $z \in \mathcal{D}$. Then if $S_{\mathcal{D}}$ is the first time $z + W_t$ reaches $\partial\mathcal{D}$, the distribution of $z + W_{S_{\mathcal{D}}}$ with respect to $\mathbb{P}$ coincides with the harmonic measure $P^z_{\mathcal{D}}$. The latter is characterized as the unique probability measure on $\partial\mathcal{D}$ such that for any harmonic function v on $\mathcal{D}$ continuous on $\overline{\mathcal{D}}$ we have

$$v(z) = \int_{\partial\mathcal{D}} v(\xi)P^z_{\mathcal{D}}(d\xi).$$

Thus, if $T_{\mathcal{D}}$ is the first time W_t reaches $\partial\mathcal{D}$, we also have

$$v(z) = \mathbb{E}v(z + W_{S_{\mathcal{D}}}) = \mathbb{E}_z v(W_{T_{\mathcal{D}}}). \tag{3.64}$$

The first results of this kind go back to Kakutani [298].

Remark. In the particular case $\mathcal{D} = \mathcal{D}_r$, where $\mathcal{D}_r = \{z \in \mathbb{C} \mid |z| \le r\}$ with $0 < r \le 1$, this gives us

$$v(z) = \mathbb{E}_z v(W_{T_r}). \tag{3.65}$$

Assume $|z| < r$, say $z = se^{i\theta}$ with $0 < s < r$. Then the Poisson kernel $P_{s/r}(\theta - t)m(dt)$ coincides with $P^z_{\mathcal{D}_r}$. Indeed for any harmonic $v\colon \overline{\mathcal{D}}_r \to \mathbb{R}$ continuous on $\overline{\mathcal{D}}_r$ (say), we have

$$v(se^{i\theta}) = \int v(re^{it})P_{s/r}(\theta - t)m(dt) \tag{3.66}$$

since the right side is equal to $v_r(\frac{s}{r}e^{i\theta})$.

Combining (3.65) and (3.66) we find that for any continuous function f on $\partial\mathcal{D}_r$ we have

$$\mathbb{E}_z f(W_{T_r}) = \int f(re^{it})P_{s/r}(\theta - t)m(dt). \tag{3.67}$$

We now formulate the strong Markov property of BM (often stated simply as 'BM starts afresh at stopping times'):

Proposition 3.29. *Let T be a stopping time, assumed everywhere finite, relative to $(\mathcal{A}_t)_{t\ge 0}$. Then the process $(W_{T+t} - W_T)_{t\ge 0}$ is independent of the process $(W_{t\wedge T})_{t\ge 0}$ and has the same distribution as our original process $(W_t)_{t\ge 0}$.*

Proof. (sketch) If T is a constant time, then this reduces to the independence of the increments. If T takes only finitely many values, the verification is easy by reducing to the constant case. For a general stopping time, one can approximate T by simple stopping times as in the discretization Lemma 1.52. □

Exceptionally, in fact only in the next 10 lines, we allow ourselves a conveniently abusive notation: we use $\{\phi_t\}$ to denote a random function $t \to \phi_t$ in $\Omega = C([0,\infty); \mathbb{R}^2)$. We will use the Markov property always like this: Consider $F, G \in L_1(\Omega, \mathbb{P})$ with either F or G in L_∞. Assume that F (resp. G) depends only on the random continuous function $t \to W_{t\wedge T}$ (resp. $t \to W_{T+t}$). With our special notation, to mark the latter dependence, we will write

$$F = F\{W_{t\wedge T}\} \quad \text{and} \quad G = G\{W_{T+t}\}.$$

An explicit form of Proposition 3.29 is then:

$$\begin{aligned}\mathbb{E}(F\{W_{t\wedge T}\}G\{W_{T+t}\}) &= \mathbb{E}(F\{W_{t\wedge T}\}G\{W_T + (W_{T+t} - W_T)\}) \\ &= \mathbb{E}\left(F\{W_{t\wedge T}\}\int G\{W_T + W_t(\omega')\}\mathbb{P}(d\omega')\right) \\ &= \mathbb{E}(F\{W_{t\wedge T}\}\mathbb{E}_{W_T} G\{W_t\}).\end{aligned}$$

Equivalently, since the collection $\{W_{t\wedge T} \mid t > 0\}$ generates $\mathcal{A}_T$, this means that for any $A \in \mathcal{A}_T$ we have $\mathbb{E}(1_A G\{W_{T+t}\}) = \mathbb{E}(1_A \mathbb{E}_{W_T} G\{W_t\})$ and hence

$$\mathbb{E}^{\mathcal{A}_T} G\{W_{T+t}\} = \mathbb{E}_{W_T} G\{W_t\}. \tag{3.68}$$

This expresses in particular the Markovian character of BM: the events 'after T' depend on the past up to T only through the present, i.e. through W_T.

The next classical result, closely linking harmonic functions to martingales, is of paramount importance. Note a preliminary fact: If $p < \infty$, since any $u \in \widetilde{h}_p(D; B)$ admits radial limits a.e. (see Corollary 3.18) we can unambiguously define the random variable $u(W_T)$ (and also $u(W_{t\wedge T})$ for every $t > 0$). Indeed, this can be justified by invoking (3.62).

Proposition 3.30. *Let* $1 \le p \le \infty$. *Let* $u\colon\ D \to B$ *be a harmonic function with values in a Banach space* B. *If* $u \in h_p(D; B)$ *then the variables* $M_r = u(W_{T_r})$ $(0 \le r < 1)$ *form a martingale adapted to the filtration* $(\mathcal{A}_{T_r})_{0\le r<1}$. *Moreover*

$$\sup_{0\le r<1} \|M_r\|_p = \|u\|_{h_p(D;B)}. \tag{3.69}$$

If moreover $u \in \widetilde{h}_p(D; B)$ *and* $p < \infty$, *then* $(u(W_{t\wedge T}))_{t\ge 0}$ *is a martingale adapted to the Brownian filtration, converging a.s. and in* $L_p(B)$ *to* $u(W_T)$ *when* $t \to \infty$, *and we have a.s.*

$$\forall t \ge 0 \quad u(W_{t\wedge T}) = \mathbb{E}^{\mathcal{A}_t} u(W_T). \tag{3.70}$$

Proof. By (3.63) we have

$$\mathbb{E}|u(W_{T_r})|^p = \int |u(re^{it})|^p m(dt)$$

and (3.69) follows.

Recall that $\partial\mathcal{D}_r$ is the circle of radius r. Let $0 \le s < r < 1$. Observe that, since $s < r$, W_{T_r} depends only on $t \to W_{T_s+t}$. Indeed, this is clear: since W_t starts at $z = 0$, it reaches $\partial\mathcal{D}_s$ before $\partial\mathcal{D}_r$. Explicitly, W_{T_r} is the first point in $\partial\mathcal{D}_r$ where $t \to W_{T_s+t}$ reaches $\partial\mathcal{D}_r$. Thus (3.68) and (3.65) yield

$$\mathbb{E}^{\mathcal{A}_{T_s}} u(W_{T_r}) = \mathbb{E}_{W_{T_s}} u(W_{T_r}) = u(W_{T_s}) \tag{3.71}$$

showing that (M_r) is indeed a martingale.

As for the second part, note that we have already established this when u is continuous on $\bar{D}$. Indeed, Proposition 3.27 then tells us that $(u(W_{t\wedge T}))_{t>0}$ is a martingale, which by the continuity of u and of BM converges everywhere to $u(W_T)$ when $t \to \infty$, and is uniformly bounded by the sup-norm of u over $\bar{D}$. By dominated convergence, it also converges in $L_p(B)$. By Proposition 3.27 (which allows us to use (3.58)) we have $u(W_{t\wedge T}) = \mathbb{E}^{\mathcal{A}_t} u(W_T)$. Therefore, if

$u \in \widetilde{h}_p(D; B)$, we may apply this to u_r for any $0 < r < 1$, yielding

$$u(rW_{t\wedge T}) = \mathbb{E}^{\mathcal{A}_t} u_r(W_T) \tag{3.72}$$

but now, letting $r \to 1$, we know on the one hand, by (3.62), that $\|u_r(W_T) - u(W_T)\|_{L_p(B)} = \|u_r - u\|_{L_p(\mathbb{T},m,B)}$ and hence $u_r(W_T) \to u(W_T)$ in $L_p(B)$, and on the other hand $u(rz) \to u(z)$ for any $z \in D$, and also for a.a. $z \in \partial D$ therefore passing to the limit when $r \to 1$ in (3.72) yields (3.70). Once (3.70) is established the asserted convergences follow from the general properties of martingales: since $u(W_T) \in L_p(B)$ and $t \mapsto u(W_{t\wedge T})$ is continuous, Theorem 1.49 ensures that $u(W_{t\wedge T})$ converges a.s. and in $L_p(B)$ to $u(W_T)$ when $t \to \infty$. □

Remark. The preceding statement together with Proposition 3.19 show that for $u \in \widetilde{h}^p(D; B)$ a.e. non-tangential convergence is equivalent to convergence along almost all Brownian paths. It is tempting to naively think that the geometric reason behind this is that Brownian paths exit the disc a.s. non-tangentially, but this idea is completely wrong: the non-tangential exiting happens with zero probability. The geometric explanation seems much more delicate, see §4.7 for more on this.

Corollary 3.31. *Let $1 \le p \le \infty$. A Banach space B has the RNP iff for any u in $h_p(D; B)$ the limit $\lim_{t<T, t\uparrow T} u(W_t)$ (or equivalently $\lim_{t\to\infty} u(W_{t\wedge T})$) exists a.s. in B.*

Proof. Assume first $p > 1$. By Theorem 3.25, if B has the RNP, any such u is the Poisson integral of some f in $L_p(\mathbb{T}, m; B)$ and by the preceding Theorem we have $u(W_{t\wedge T}) \to f(W_T)$ a.s., so the desired limit exists a.e. To prove the converse, it suffices to consider the case $p = \infty$. Recall that, by (3.57), the times T_r tend to T when $r \to 1$. Then, if the limit $\lim_{t\uparrow T} u(W_t)$ exists a.s. for any $u \in h^\infty(D; B)$, so does $\lim_{r\uparrow 1} u(W_{T_r})$. Then $\varphi = \lim_{r\uparrow 1} u(W_{T_r})$ is Bochner measurable and bounded, so $\varphi \in L_\infty(B)$. We claim that φ is a function of W_T. Indeed, if B is one-dimensional, the first part of the proof makes this clear (with $\varphi = f(W_T)$), and since we may clearly assume B separable, we can use 'scalarization' to prove our claim. Thus we can write $\varphi = f(W_T)$ for some $f \in L_\infty(\mathbb{T}, m; B)$. By (3.65) we have $u(z) = \mathbb{E}_z u(W_{T_r})$ and hence in the limit when $r \to 1$, $u(z) = \mathbb{E}_z f(W_T)$. By (3.61) u is the Poisson integral of f, so that $u \in \widetilde{h}^\infty(D; B)$. Thus we conclude by the implication (iii) ⇒ (i) in Theorem 3.25 that B has the RNP.

For the case $p = 1$, any u in $h_p(D; B)$ is the Poisson integral of a bounded vector measure μ that is the sum of an absolutely continuous part μ^a and a singular one μ^s. This gives us $u(z) = u^a(z) + u^s(z)$. On the one hand, if B has the RNP, then $u^a \in \widetilde{h}_1(D; B)$ and Proposition 3.30 shows that $u^a(W_t)$ converges a.s.

when $t \uparrow T$. On the other hand, by (3.26), we have

$$\|u^s(z)\|_B \leq v(z)$$

where v is the Poisson integral of $|\mu^s|$, and in the scalar case it is known (we admit this without proof, see [176] or [26, p. 113]) that $\lim_{t\uparrow T} v(W_t) = 0$ a.s., and hence also $\lim_{t\uparrow T} u^s(W_t) = 0$ a.s. Thus we conclude that $u(W_t)$ converges a.s. when $t \uparrow T$.

The converse is obvious for $p = 1$ since $h_\infty(D; B) \subset h_1(D; B)$ and we already proved the converse for $p = \infty$. □

Remark. The 'if part' in the last corollary can be derived from the following inequality valid for any $1 \leq p < \infty$, and u in $h_p(D; B)$ and any $0 < s < r < 1$:

$$\|u_{r^2} - u_{s^2}\|_{L_p(m;B)} \leq \|u_r - u_{s^2 r^{-1}}\|_{L_p(m;B)} \leq \|u(W_{T_r}) - u(W_{T_s})\|_{L_p(\mathbb{P};B)}. \quad (3.73)$$

Indeed, by the Markov property (3.68) we have

$$\mathbb{E}^{\mathcal{A}_{T_s}} \|u(W_{T_r}) - u(W_{T_s})\|^p = \int \|u(W_{T_r}(\omega') - u(W_{T_s})\|^p d\mathbb{P}_{W_{T_s}}(\omega').$$

Assume $W_{T_s}(\omega) = se^{i\theta}$ then by (3.65) and (3.66) the last integral is equal to

$$\int \|u(re^{it}) - u(se^{i\theta})\|^p P_{s/r}(\theta - t)m(dt).$$

Averaging this with respect to θ (i.e. $W_{T_s}(\omega)$) we find

$$\mathbb{E}\|u(W_{T_r}) - u(W_{T_s})\|^p = \iint \|u(re^{it}) - u(se^{i\theta})\|^p P_{s/r}(\theta - t)m(dt)m(d\theta).$$

We now interchange the order of integrals and note that, by Jensen, for each fixed t we have

$$\int \|u(re^{it}) - u(se^{i\theta})\|^p P_{s/r}(\theta - t)m(d\theta) \geq \|u(re^{it}) - u_s * P_{s/r}(t)\|^p$$
$$= \|u_r(e^{it}) - u_{s^2/r}(e^{it})\|^p,$$

and integrating this with respect to $m(dt)$, we obtain the second inequality in (3.73). Let $f(z) = u(rz) - u((s^2/r)z)$. Since $\|f_r\|_{L_p(m;B)} \leq \|f\|_{L_p(m;B)} = \|f\|_{h_p(D;B)}$ (see (3.24)) the first inequality in (3.73) is immediate.

Complement*. We will merely sketch two alternate classical arguments for Proposition 3.27. One possibility is to define $u(z) = \mathbb{E}_z f(W_T)$ and to check that this is a bounded harmonic function on D tending to f on the boundary. Therefore it must be the Poisson integral of f, yielding (3.61).

Note that (3.62) itself is essentially obvious: By the rotational invariance of BM starting at 0, the probability distribution of $\omega \to W_T(\omega) = W_{T(\omega)}(\omega)$ with respect to $(\Omega, \mathcal{A}, \mathbb{P})$ is translation invariant on $\partial D \simeq \mathbb{T}$, so it must coincide with m. This is the basis for the second proof: We introduce the Möbius function φ_z such that $\varphi_z(0) = z$. Applying (3.62) to the composition $f(\varphi_z)$ we find

$$\mathbb{E} f \circ \varphi_z(W_T) = \int f \circ \varphi_z \, dm = \int f \, dP^z. \tag{3.74}$$

We now consider the random process $Z_t = \varphi_z(W_t)$. Note that for any $w \in D$ we have $|\varphi_z(w)| = 1$ iff $|w| = 1$. Thus if $\omega(0) = 0$, we have $T(\omega) = \inf\{t > 0 \mid |Z_t(\omega)| = 1\}$. We then invoke the so-called conformal invariance of Brownian motion. This says that the random process $(Z_{t\wedge T})$ on $(\Omega, \mathcal{A}, \mathbb{P})$ has the same distribution as the Brownian motion $(W_{t\wedge T})$ on $(\Omega, \mathcal{A}, \mathbb{P}_z)$ but with the latter run by a different 'clock'. More precisely, there is a random continuous non-decreasing function $\tau : [0, \infty) \to [0, \infty)$, tending to ∞ when $t \to \infty$, such that the process (Z_t) has the same distribution with respect to $\mathbb{P}$ as $(W_{\tau(t)})$ with respect to $\mathbb{P}_z$. But the process $(W_{\tau(t)})$ obviously reaches the unit circle at the same point as the process (W_t), so that Z_T has the same distribution with respect to $\mathbb{P}$ as (W_T) with respect to $\mathbb{P}_z$. Thus (3.74) implies (3.62).

Remark. The change of time τ is derived from the famous Ito formula involving stochastic integrals. For any bounded analytic function $\varphi : D \to \mathbb{C}$, Ito's formula (see [26, 69, 81]) tells us that

$$\varphi(W_{t\wedge T}) = \varphi(0) + \int_0^{t\wedge T} \varphi'(W_s) dW_s.$$

Let $Y_t = \int_0^{t\wedge T} \varphi'(W_s) dW_s$. The associated 'square function' is

$$V_t = \int_0^{t\wedge T} |\varphi'(W_s)|^2 ds.$$

Then if we set $\tau(t)(\omega) = \inf\{s \mid V_s \geq t\}$, the process $(Y_{\tau(t\wedge T)})$ has the same distribution as a standard BM starting at 0. See [67, p. 29] or [26, p. 75] for an intuitive argument, and [69, 81, p. 190] for a more complete account.

3.6 Notes and remarks

The results of the first part of this chapter are presented for harmonic functions with values in an arbitrary Banach space, but it has long been known that the subadditivity of the norm is all that is required for these classical results. We just replace the modulus by the norm. Thus we refer to the standard and well-known existing books such as [89] for more on this.

See Naor and Tao's recent paper for recent results (and more references) on the Hardy-Littlewood maximal inequality on a certain class of metric spaces, based on Doob's maximal inequality.

The results concerning the RNP are more specialized, but once the significance of the RNP became clear, such results must have quickly become part of the folklore. Theorem 3.25 was observed in [163].

The subject of Brownian martingales has of course a long history going back to Doob. We refer the reader to [176] and to the book [26] for more information of the martingales of the form $(u(W_{t\wedge T}))_{t>0}$, when u is harmonic. See [55] for a concise introduction to Brownian motion and stochastic calculus.

4

Analytic functions and ARNP

Unless we explicitly assume otherwise, all the Banach spaces in this chapter are over the complex scalars. Since analytic functions are a fortiori harmonic, the results of the preceding chapter can be applied to them. However, in several respects, the analytic case is better behaved. Firstly, the non-tangential (or radial) maximal inequality remains valid in L_p for $0 < p \leq 1$ when the function is analytic, and this implies numerous improvements. Secondly, for a Banach space B, the property that all B-valued bounded analytic functions on D admit radial limits a.e. is strictly more general than the same one for bounded harmonic functions, which as we saw in the preceding chapter characterizes the RNP. The property in question, called the analytic RNP (in short ARNP), is the main subject of the present chapter.

Notation. For any function φ taking values in $[-\infty, \infty]$, we will often denote $\varphi^+ = \max\{\varphi, 0\}$ and $\varphi^- = \max\{-\varphi, 0\}$, so that $\varphi = \varphi^+ - \varphi^-$. In particular, for any $x > 0$ we have $\operatorname{Log} x = \operatorname{Log}^+ x - \operatorname{Log}^- x$.

4.1 Subharmonic functions

We start by discussing the notion of subharmonicity. For convenience we choose to work with a rather strong definition of subharmonic functions, but of course this definition is equivalent to the usual one.

Definition 4.1. Let $V \subset \mathbb{C}$ be open. Let $\varphi\colon\ V \to [-\infty, \infty)$ be an upper semicontinuous function (in short u.s.c.). We will say that φ is subharmonic if for every closed disc $\overline{B}(z_0, R)$ included in V, we have for all $z \in D$

$$\varphi(z_0 + Rz) \leq \int \varphi(z_0 + Re^{it})P^z(dt). \tag{4.1}$$

The upper semi-continuity of φ guarantees that it is bounded above on compact subsets of V, so that φ^+ is integrable, and hence the integral appearing in (4.1) is unambiguously defined, although possibly equal to $-\infty$.

It is well known that this is equivalent to the apparently much weaker requirement that (4.1) be satisfied for every point $z_0 \in V$ at the centres, i.e. in the case $z = 0$. Moreover, one can only assume that (4.1) holds for a sequence of radii $R = R_n$ tending to zero. We will not use this. Of course by (3.9) (and a simple change of variable) every harmonic function is a fortiori subharmonic. We recall the following classical fact.

Proposition 4.2. *Let f be a non-vanishing analytic function in a simply connected domain V. Then there is an analytic function F such that $f = e^F$ on V.*

We can now introduce the fundamental example of a subharmonic function.

Proposition 4.3. *Let $V \subset \mathbb{C}$ be open. Let $f\colon\ V \to \mathbb{C}$ be an analytic function. Then the function*

$$\varphi(z) = \mathrm{Log}|f(z)|$$

is subharmonic on V.

Proof. By an obvious change of variables, we can reduce (4.1) to the case $z_0 = 0, R = 1$ so that $B(z_0, R) = D$. We start by the special case when f has no zero in $\overline{D}$. In that case, the function $\mathrm{Log}|f|$ is actually harmonic (and a fortiori subharmonic) in a neighbourhood of $\overline{D}$. Indeed, let V be a neighbourhood of $\overline{D}$ where f has no zero, so that there is a function F analytic in V such that $f = e^F$. Then, $\mathrm{Log}|f| = \Re(F)$ on V, and hence $\mathrm{Log}|f|$ being the real part of an analytic function, is harmonic in V. We now check the case when f may vanish in D but does not vanish on the boundary. Let $z_1, z_2, \ldots, z_n$ be the zeros of f in D repeated according to their multiplicities. Consider then the finite Blaschke product

$$B(z) = \Pi \frac{z - z_k}{1 - \bar{z}_k z},$$

and let $g = f/B$. Recall $|B| \le 1$ on D and $|B| = 1$ on ∂D. Then, g has no zero in $\overline{D}$ and satisfies

$$\forall z \in D\ |f(z)| \le |g(z)|, \quad \forall e^{it} \in \mathbb{T} \quad |g(e^{it})| = |f(e^{it})|.$$

By the first part of the proof, we have

$$\mathrm{Log}|g(z)| = \int \mathrm{Log}|g(e^{it})| P^z(dt)$$

and hence we conclude a fortiori

$$\mathrm{Log}|f(z)| \le \int \mathrm{Log}|f(e^{it})| P^z(dt).$$

Finally, assume that f has zeros on the boundary of D. We denote them say by $\{e^{i\theta_j} \mid j = 1, \ldots, m\}$, so that we can write $f = g\Pi_j(z - e^{i\theta_j})$ where g is analytic in a neighborhood of $\overline{D}$ but has no zero on the boundary of D. To conclude then it clearly suffices to show

$$\forall z \in D \quad \forall e^{i\theta} \in \mathbb{T} \qquad \mathrm{Log}|z - e^{i\theta}| \le \int \mathrm{Log}|e^{it} - e^{i\theta}| P^z(dt).$$

Or simply

$$\mathrm{Log}|z - 1| \le \int \mathrm{Log}|e^{it} - 1| P^z(dt).$$

Actually, we have equality. Indeed, by the first part of the proof we have for any $\epsilon > 0$

$$\mathrm{Log}|z - (1 + \epsilon)| = \int \mathrm{Log}|e^{it} - (1 + \epsilon)| P^z(dt),$$

and, letting ϵ tend to zero, by dominated convergence this yields

$$\mathrm{Log}|z - 1| = \int \mathrm{Log}|e^{it} - 1| P^z(dt),$$

thus completing the proof. □

We recall here a simple consequence of the classical Jensen inequality.

Proposition 4.4. *If ψ is a convex non-decreasing function on the real line and if φ is a subharmonic function on an open set $V \subset \mathbb{C}$, then the composition $\psi(\varphi)$ is again subharmonic on V. (We implicitly extend ψ by continuity at $-\infty$.)*

The proof is an obvious consequence of (4.1) and the fact that ψ is convex and non-decreasing. This gives us more examples.

Proposition 4.5. *In the same situation as in Proposition 4.3. For each $0 < p < \infty$, the functions*

$$\varphi_1(z) = |f(z)|^p \quad \text{and} \quad \varphi_2(z) = \mathrm{Log}^+|f(z)|$$

are subharmonic on V.

Proof. This follows from the preceding proposition applied to the convex non-decreasing functions $\psi_1(x) = e^{px}$ and $\psi_2(x) = x^+$, with $\varphi = \mathrm{Log}|f|$. □

With the definition we are using for subharmonicity, the next result becomes very easy.

Proposition 4.6. *Let φ be a subharmonic function on D. For $0 < r < 1$, let*

$$m(r) = \int \varphi(re^{it})dm(t).$$

Recall that, since φ is bounded above on compact subsets of D, $m(r) \in [-\infty, \infty)$. Then

$$\forall 0 \le r < s < 1 \qquad m(r) \le m(s).$$

Proof. By (4.1), applied with $z_0 = 0,\ R = s,\ z = (r/s)e^{i\theta}$, we have

$$\varphi(re^{i\theta}) \le \int \varphi(se^{it})P_{r/s}(\theta - t)dm(t), \tag{4.2}$$

and hence, after integration with respect to θ,

$$m(r) \le \int \varphi(se^{it})P_{r/s}(\theta - t)dm(t)dm(\theta) = m(s),$$

where the last equality holds because the Poisson kernel has integral 1. □

Remark. If (u_i) is a family of real valued harmonic functions and if $v = \sup_i v_i$ is u.s.c. then v clearly is subharmonic, in particular a finite supremum of harmonic functions is *always* subharmonic. For instance, if u is a harmonic function on V with values in a real Banach space B (if B is complex we consider the underlying real space structure), then $v(z) = \|u(z)\|$ is subharmonic, since it can be written as $v(z) = \sup_{\xi \in B^*, \|\xi\| \le 1} \xi(u(z))$ and v is clearly continuous hence u.s.c. By the same general idea, we have:

Proposition 4.7. *Let $V \subset \mathbb{C}$ be open. Let B be a complex Banach space. Let $f\colon V \to B$ be an analytic function. Then for each $0 < p < \infty$, the functions* $\mathrm{Log}\|f(z)\|$, $\mathrm{Log}^+\|f(z)\|$, *and* $\|f(z)\|^p$ *are subharmonic on* V.

Proof. Again we can write each of these functions as a supremum of subharmonic functions, for example (here B^* is the dual of B as a complex space) $\mathrm{Log}\|f(z)\| = \sup_{\xi \in B^*, \|\xi\| \le 1} \mathrm{Log}|\xi(f(z))|$, and each function $z \to \mathrm{Log}|\xi(f(z)|$ is subharmonic by Proposition 4.3, so the result follows from the preceding remark. □

Corollary 4.8. *For any analytic $f : D \to B$, we have for all $0 < p < \infty$ and all $0 \le r < s < 1$*

$$\int \|f(re^{it})\|^p dm(t) \le \int \|f(se^{it})\|^p dm(t). \tag{4.3}$$

Remark 4.9. By (4.2) for any analytic $f : D \to B$ we have

$$\|f(re^{i\theta})\|^p \leq \int \|f(se^{it})\|^p P_{r/s}(\theta - t)dm(t).$$

Since $|P_{r/s}(\theta - t)|$ is bounded above by a constant $\chi(r/s)$ (see (3.12)) when $0 \leq r < s$, for any z with $|z| \leq r$ we have

$$\|f(z)\|^p \leq \chi(r/s) \int \|f(se^{it})\|^p dm(t). \tag{4.4}$$

Remark 4.10. Let $\varphi\colon\ V \to [-\infty, \infty)$ be subharmonic on an open subset $V \subset \mathbb{C}$. Let $f : D \to V$ be an analytic function. Then it is well known that the composition $\varphi(f)$ is subharmonic on D (see e.g. [41, p. 45] and [42]). Thus (4.2) can be generalized to

$$\varphi(f(re^{i\theta})) \leq \int \varphi(f(se^{it}))P_{r/s}(\theta - t)dm(t).$$

4.2 Outer functions and $H^p(D)$

Although we will soon consider the B-valued case, for the moment we restrict ourselves to the complex valued case to define the classical Hardy space $H^p(D)$. Let $0 < p < \infty$. The space $H^p(D)$ is defined as the space of (complex valued) analytic functions f on D such that

$$\sup_{0 \leq r < 1} \int |f(re^{it})|^p dm(t) < \infty$$

equipped with the norm (or quasi-norm)

$$\|f\|_{H^p(D)} = \sup_{0 \leq r < 1} \left(\int |f(re^{it})|^p dm(t)\right)^{1/p}.$$

The space $H^\infty(D)$ is just the space of bounded analytic functions f on D equipped with the sup norm:

$$\|f\|_{H^\infty(D)} = \sup_{z \in D} |f(z)|.$$

Let F be a non-vanishing analytic function on D. Then there is an analytic function f so that $F = e^f$ and hence $\mathrm{Log}|F| = \mathrm{Re}(f)$ is a harmonic function. We say that F is outer if the function $\mathrm{Log}|F|$ is in $\tilde{h}_1$ (i.e. it admits radial limits a.e. and it is the Poisson integral of its radial limits). A fortiori, the function $z \mapsto |F(z)|$ admits radial and non-tangential limits a.e. As the next result shows, it is quite easy to construct 'many' outer functions. We will make extensive use of this classical fact.

Theorem 4.11. *Let w be a positive measurable function on D such that* $\text{Log}\, w \in L_1(\mathbb{T}, m)$. *Then there is an outer function F such that*

$$|F(e^{it})| = w(t) \quad a.e. \text{ on } \quad \mathbb{T}.$$

If $w \in L_p(\mathbb{T}, m)$ $(0 < p \leq \infty)$, then $F \in H^p(D)$ with $\|F\|_{H^p(D)} \leq \|w\|_p$ and F itself admits radial and non-tangential limits a.e. on $\mathbb{T}$.

Proof. Let u be the Poisson integral of $\text{Log}\, w$. Then $u \in \tilde{h}_1(D)$ and $u = \text{Log}\, w$ a.e. on $\mathbb{T}$. Let f be an analytic function such that $u = \text{Re}(f)$ and let $F = e^f$. Since $\text{Log}|F| = u$ on D, F is outer and $|F| = e^{\text{Log}|F|} = e^u = w$ a.e. on $\mathbb{T}$. For any $0 < p < \infty$, we have

$$|F(z)|^p = \exp(pu(z)) = \exp\left(p \int \text{Log}\, w(t) P^z(dt)\right)$$

and hence by Jensen's inequality

$$\leq \int \exp(p\, \text{Log}\, w(t)) P^z(dt) = \int w^p(t) P^z(dt).$$

So if $v(z)$ is the Poisson integral of w^p, we have $|F(z)|^p \leq v(z)$ and hence $\|F\|^p_{H^p(D)} \leq \|v\|_{h^1(D)} = \|w\|^p_p$. Thus $F \in H^p(D)$ with $\|F\|_{H^p(D)} \leq \|w\|_p$. Let N be any integer chosen large enough so that $Np > 1$. Then $F^{\frac{1}{N}} = \exp(f/N)$ is analytic and, since $|F^{\frac{1}{N}}|^{Np} = |F|^p$, the preceding argument shows that $F^{\frac{1}{N}} \in H^{pN}(D)$ and a fortiori $F^{\frac{1}{N}} \in h^{pN}(D; \mathbb{C})$. By Theorems 3.5 and 3.13 this implies that $z \mapsto F^{\frac{1}{N}}(z)$ admits non-tangential boundary values a.e. on $\mathbb{T}$, and hence the same is true for $z \mapsto F(z) = (F^{\frac{1}{N}}(z))^N$. □

We now describe Szegö's classical solution to the following.

Problem. Let $0 < p \leq +\infty$. For which non-negative functions $\varphi \in L_p(\mathbb{T})_+$, does there exist $f \in H^p(D)$ such that

$$\varphi = |f|?$$

The following fundamental theorem due to Szegö gives a complete answer.

Theorem 4.12. *Given $\varphi \in L_{p_+}(\mathbb{T})$ there exist $f \in H^p(D)$ such that $\varphi = |f|$ iff the following condition (called Szegö's condition) holds*

$$\int \text{Log}\, \varphi(e^{it}) dt > -\infty.$$

Proof. Necessity is proved in a more general framework in Remark 4.16. We now prove sufficiency. Assume φ satisfies Szegö's condition. Observe that necessarily $|\text{Log}\, \varphi(t)| \in L_1(m)$, because $\text{Log}^+ \varphi(t) \leq \frac{1}{p} \varphi(t)^p$, so that $\varphi \in L_p(m)$

ensures $\int \mathrm{Log}^+\varphi dm < +\infty$, while the integrability of $\mathrm{Log}^-\varphi$ is taken care of by Szegö's condition. Thus by Theorem 4.11, there is an outer function $f \in H^p(D)$ such that $\varphi = |f|$. This proves the sufficiency of Szegö's condition. □

Remark. Clearly in general the function f solving the problem in Theorem 4.12 is not unique (not even up to a constant factor). For instance, $f_0(z) = 1$ and $f_1(z) = z$ are both of modulus one on the boundary. But if the moduli agree inside the domains, then the functions have to be a multiple of each other (indeed if F, G are analytic with $|F| = |G|$ on D, then $h = F/G$ is analytic and such that $|h| = 1$ on D, thus $h = \bar{h}$ on D, so h must be constant).

Thus we obtain a certain uniqueness:

Corollary 4.13. *Given φ as in Theorem 4.12, there is a unique outer function F in $H^p(D)$ such that $|F| = \varphi$ and $F(0) > 0$.*

Proof. Indeed, the proof of Theorem 4.12 yields an outer function F. Moreover, if F, G are outer and coincide a.e. on ∂D, $\mathrm{Log}|F|$ and $\mathrm{Log}|G|$ being the Poisson integrals of their boundary values must coincide inside D. Thus we must have $|F| = |G|$ inside D. By the previous remark, F and G differ at most by a multiplicative factor of modulus one, and requiring $F(0) > 0$ determines this factor unambiguously. □

4.3 Banach space valued H^p-spaces for $0 < p \leq \infty$

Let $V \subset B$ be an open subset. Recall that a B-valued function on V is called analytic if for any $z_0 \in V$ there is $r > 0$ with $z_0 + rD \subset V$ such that for any $z \in z_0 + rD$ we can write $f(z)$ as the limit of an (absolutely) convergent series $f(z) = \sum_0^\infty x_n z^n$ with coefficients $x_n \in B$ for any $n \geq 0$. It is well known that it suffices for this that f be 'scalarly' analytic on V, i.e. such that for any $\xi \in B^*$ the $\mathbb{C}$ valued function $z \mapsto \xi(f(z))$ be analytic on V.

We first define the space $H^p(D; B)$ (or in short $H^p(B)$) as the Hardy space of all B-valued analytic functions f such that

$$\sup_{r<1}\left(\int \|f(re^{it})\|^p dm(t)\right)^{1/p} < \infty.$$

We equip it with the norm

$$\|f\|_{H^p(D;B)} = \sup_{r<1}\left(\int \|f(re^{it})\|^p dm(t)\right)^{1/p} < \infty.$$

This is a Banach space if $1 \le p \le \infty$, and only a quasi-Banach space if $0 < p < 1$. Since analyticity implies harmonicity, the functions in $H^p(D; B)$ satisfy the maximal inequalities proved in §3.3, for either $1 < p < \infty$ or $p = 1$ (see Corollary 3.11 or Theorem 3.15). However, it turns out that analyticity allows to extend the validity of (3.48) to all finite values of p including $0 < p \le 1$. This is one of the main differences between h^p-spaces and H^p-spaces.

For any f in $H^p(D; B)$, we define of course

$$\forall t \in \mathbb{T} \qquad M_\alpha f(t) = \sup_{z \in \Gamma_\alpha(t)} \|f(z)\|_B.$$

Theorem 4.14. *For any $0 < p \le \infty$, there is a constant $C_{p,\alpha}$ such that for all f in $H^p(D; B)$ we have*

$$\|M_\alpha(f)\|_{L_p(\mathbb{T},m)} \le C_{p,\alpha} \|f\|_{H^p(D;B)}. \tag{4.5}$$

A fortiori we have for some constant $C_{p,0}$

$$\| \sup_{0 \le r < 1} \|f(re^{it})\| \|_{L_p(m(dt))} \le C_{p,0} \|f\|_{H^p(D;B)}. \tag{4.6}$$

Proof. The proof is based on the subharmonicity (by Proposition 4.7) of the function u defined by

$$u(z) = \|f(z)\|^{p/2}.$$

Fix a number $s < 1$ and let $v(z) = \int u_s(e^{it}) P^z(dt)$. Note that by the subharmonicity of u

$$\|f(sz)\|^{p/2} \le v(z).$$

Moreover $v \in h^2(D)$ and clearly $\|v\|_{h^2(D)} = \|u_s\|_{L_2(\mathbb{T},m)} = \|f_s\|^{p/2}_{L_p(\mathbb{T},m;B)}$. We have by Corollary 3.11

$$\|M_\alpha(v)\|_{L_2(\mathbb{T},m)} \le A_\alpha C_2 \|v\|_{h^2(D)}.$$

Hence since $M_\alpha(v) = (M_\alpha(f_s))^{p/2}$, this yields

$$\|M_\alpha(f_s)\|^{p/2}_{L_p(\mathbb{T},m)} \le A_\alpha C_2 \|f_s\|^{p/2}_{L_p(\mathbb{T},m;B)}.$$

Taking the supremum over $s < 1$ in the preceding inequality, we obtain (4.5) with $C_{p,\alpha} = (A_\alpha C_2)^{2/p}$. □

The next statement is a key factorization theorem, allowing to reduce many questions about Banach space valued H^p-spaces to the scalar valued case.

Theorem 4.15. *Let $0 < p \le \infty$. Then every function $f \in H^p(D; B)$ can be factorized as a product $f = Fg$ with $F \in H^p(D)$ an outer function and $g \in H^\infty(D; B)$ such that*

$$\|f\|_{H^p(D;B)} = \|F\|_{H^p} \text{ and } \|g\|_{H^\infty(D;B)} = 1.$$

The non-tangential (and radial) limits $w(t) = \lim_{z \to e^{it}} \|f(z)\|$, exist a.e. and

$$\|f\|_{H^p(D;B)} = \left(\int w^p dm \right)^{1/p}. \tag{4.7}$$

Remark. From this (4.5) for $f \in H^p(D; B)$ is reduced to (and holds with the same constant $C_{p,\alpha}$ as) the complex valued case. Indeed, we have clearly $M_\alpha(f) \le M_\alpha(F)$ everywhere on $\mathbb{T}$, and hence assuming the scalar case of (4.5), we have

$$\|M_\alpha(f)\|_{L_p(\mathbb{T},m)} \le \|M_\alpha(F)\|_{L_p(\mathbb{T},m)} \le C_{p,\alpha}\|F\|_{H^p} = C_{p,\alpha}\|f\|_{H^p(B)}.$$

Proof of Theorem 4.15. We use the same notation as in §4.2. Let $f \in H^p(D; B)$. If we knew (but it is not always true) that f admits a non-tangential limit defined on $\mathbb{T}$ we could simply define F as the outer function such that $|F(t)| = \|f(t)\|$ on $\mathbb{T}$, then setting $g = f/F$ yields the desired factorization. However, although we cannot define $f(t)$ on $\mathbb{T}$, it turns out that there is a valid substitute for $\|f(t)\|$ so that this idea works in general. Here are the details. Let $w^* = M_\alpha(f)$. We claim that $\operatorname{Log} w^* \in L_1(m)$. By (4.5) we know that $w^* \in L_p(m)$. A fortiori $\operatorname{Log}^+ w^* \in L_1(m)$. We may assume that $f \not\equiv 0$ and, dividing by a suitable power of z, that $f(0) \neq 0$. Since the function Log^- is non-increasing, we have $\operatorname{Log}^- w^* \le \operatorname{Log}^- \|f_r\|$ for any $0 < r < 1$, and then the subharmonicity (by Proposition 4.7) of $\operatorname{Log}\|f\|$ implies

$$-\infty < \operatorname{Log}\|f(0)\| \le \int \operatorname{Log}\|f_r\| dm \le \int \operatorname{Log}^+ w^* dm - \int \operatorname{Log}^- \|f_r\| dm, \tag{4.8}$$

which ensures that $\operatorname{Log}^- \|f_r\|$ (and a fortiori $\operatorname{Log}^- w^*$) is integrable. This proves our claim that $\operatorname{Log} w^* \in L_1(m)$.

Let F be the outer function such that $|F| = w^*$ as in Theorem 4.11. Recall that $\operatorname{Log}|F|$ is the Poisson integral of $\operatorname{Log} w^*$. For any $0 < r < 1$ and any $z \in D$ we have

$$\operatorname{Log}\|f(rz)\| \le \int \operatorname{Log}\|f_r\| \, dP^z \le \int \operatorname{Log} w^* \, dP^z = \operatorname{Log}|F(z)|$$

and hence letting $r \to 1$ we find $\|f(z)\| \le |F(z)|$ for any $z \in D$. Thus we obtain $f = Fg$ with F outer in $H^p(D)$ and g in $H^\infty(D, B)$ but the norm estimate is not quite as announced. We need to refine the factorization to obtain that.

We view g as complex valued harmonic. Since by Proposition 3.23 (resp. Theorem 4.11) $\|g(z)\|$ (resp. $|F(z)|$) admits non-tangential limits a.e., so does $\|f(z)\|$. Let $w(z) = \lim_{r\to 1} \|f_r(z)\|_B$. Note that the a.e. convergence $\|f_r\|_B \to w$ is dominated so it holds in $L_p(m)$. Therefore $\|f\|_{H^p(D;B)} = \lim_{r\to 1} \|\|f_r\|_B\|_p = \|w\|_p$. Note also that by Fatou's lemma and by (4.8)

$$\int \mathrm{Log}^- w dm \le \liminf_{r\to 1} \int \mathrm{Log}^- \|f_r\| dm < \infty.$$

We now refine the factorization. Arguing as earlier for w^* we see that $\mathrm{Log}\, w$ is in $L_1(m)$. Let now F be the outer function such that $|F| = w$ on $\mathbb{T}$. We will show that $\|f\| \le |F|$ on D. Again, we have for any $z \in D$

$$\mathrm{Log}\|f(z)\| = \limsup_{r\to 1} \mathrm{Log}\|f(rz)\| \le \limsup_{r\to 1} \int (\mathrm{Log}^+\|f_r\| - \mathrm{Log}^-\|f_r\|)\, dP^z.$$

The convergence of $\mathrm{Log}^+\|f_r\|$ is dominated and for $\mathrm{Log}^-\|f_r\|$ we use Fatou's lemma, yielding:

$$\begin{aligned}\mathrm{Log}\|f(z)\| &\le \int \mathrm{Log}^+ w\, dP^z - \liminf_{r\to 1} \int \mathrm{Log}^- \|f_r\|\, dP^z \\ &\le \int (\mathrm{Log}^+ w - \mathrm{Log}^- w)\, dP^z\end{aligned}$$

and we conclude as announced

$$\mathrm{Log}\|f(z)\| \le \int \mathrm{Log}\, w\, dP^z = \mathrm{Log}|F(z)|.$$

Thus we find $\|f\| \le |F|$ on D, so $f = Fg$ with g in the unit ball of $H^\infty(D, B)$ and lastly by Theorem 4.11 we have $\|F\|_{H^p} \le \|w\|_p = \|f\|_{H^p(D,B)}$. So we obtain $\|F\|_{H^p}\|g\|_{H^\infty(D,B)} \le \|f\|_{H^p(D,B)}$ and since $f = Fg$ the converse is obvious. □

Remark 4.16. In the situation of the preceding Theorem, we have proved

$$\forall z \in D \qquad \mathrm{Log}\|f(z)\| \le \int_{\partial D} \mathrm{Log}\, w\, dP^z.$$

Thus the subharmonicity of $\mathrm{Log}\|f(z)\|$ on D holds roughly all the way to the boundary. In particular

$$\mathrm{Log}\|f(0)\| \le \int \mathrm{Log}\, w\, dm = \int (\mathrm{Log}^+ w - \mathrm{Log}^- w)\, dm. \tag{4.9}$$

This shows that if $f \not\equiv 0$

$$-\infty < \int \mathrm{Log}\, w\, dm.$$

Indeed, after division by a power of z we can reduce to $f(0) \ne 0$, and then (recall that $\mathrm{Log}^+ w \in L_1$ since $w \in L_p$) (4.9) ensures that $\mathrm{Log}^- w \in L_1$.

Corollary 4.17. *Let $0 < p < \infty$. Any f in $H^p(D)$ admits non-tangential limits a.e. and a fortiori radial limits a.e. If we define $f_1(e^{it}) = \lim_{r\to 1} f(re^{it})$, then $f_1 \in L_p(\mathbb{T}, m)$ (here we implicitly identify $\mathbb{T}$ to ∂D via $t \mapsto e^{it}$) and moreover if p is finite the functions f_r defined by $f_r(e^{it}) = f(re^{it})$ $(0 \le r < 1)$ converge to f_1 in $L_p(\mathbb{T}, m)$ when $r \to 1$.*

Proof. By Theorem 4.15, we can write $f = Fg$ with F outer and g in H^∞. We view g as complex valued harmonic. Since by Theorem 4.11 (resp. Theorem 3.18), F (resp. g) admits non-tangential limits a.e., so does f. By (4.5) we know that $M_\alpha(F)$ is in $L_p(\mathbb{T}, m)$, so by dominated convergence $f_1 \in L_p(\mathbb{T}, m)$ and $f_r \to f_1$ in $L_p(\mathbb{T}, m)$. □

Corollary 4.18. *For all $0 < p \le \infty$, there is a constant $C'_{p,\alpha}$ such that for all sequences $f_1, \ldots, f_n, \ldots$ in $H^p(D)$, we have (still denoting by $f_n(t)$ the non-tangential limit of f_n at t)*

$$\| \sup_{n\ge 1} M_\alpha(f_n)\|_{L_p(\mathbb{T},m)} \le C'_{p,\alpha} \| \sup_{n\ge 1} |f_n| \|_{L_p(\mathbb{T},m)}. \tag{4.10}$$

Proof. It is enough to prove this for a finite sequence $f_1, \ldots, f_N$ (with a constant independent of N). This case follows from (4.5) applied to the Banach space $B = \ell_\infty^N$ as in Corollary 3.16. Indeed, if we define $f : D \to \mathbb{C}^N$ by $f(z) = (f_1(z), \ldots, f_N(z))$, then clearly f admits radial limits a.e. and moreover $f - f_r \to 0$ in $L_p(\mathbb{T}, m; B)$ since we know the convergence for each coordinate by Corollary 4.17. We also have

$$\|f\|_{H^p(D;B)} = \lim_{r\to 1} \|f_r\|_{L_p(\mathbb{T},m;B)} \le \|f\|_{L_p(\mathbb{T},m;B)}. \tag{4.11}$$

But now $M_\alpha(f) = \sup_{n\le N} M_\alpha(f_n)$ and the radial limits of f on $\mathbb{T}$ satisfy $\|f\|_{L_p(\mathbb{T},m;B)} = \| \sup_{n\le N} |f_n| \|_{L_p(\mathbb{T},m)}$. Then, (4.10) follows from (4.5) and (4.11). □

Remark. Conversely, the reader should note that (4.10) implies (4.5) by a simple reasoning (using $f_n(z) = \xi_n(f(z))$ as earlier) so that in (4.5) the particular case of $B = \ell_\infty$ implies the general case.

Corollary 4.19. *Assume $0 < p, q, r < +\infty$ and $\frac{1}{p} = \frac{1}{q} + \frac{1}{r}$, then for all f in $H_p(D, B)$ there exists $k \in H_q(D, B)$, and $h \in H_r(D)$ such that*

$$f = kh \quad \text{and} \quad \|k\|_{H_q} \|h\|_{H_r} = \|f\|_{H_p}.$$

Proof. Let $f = Fg$ as in Theorem 4.15. Then $k = F^{p/q}g$ and $h = F^{p/r}$ gives the desired factorization. □

Remark. To illustrate the usefulness of Theorem 4.15, let σ be a measure on the unit disc D. Assume that there is a constant c such that for any $F \in H^1(D)$ we have

$$\int_D |F(z)| d\sigma(z) \le c \int_{\partial D} |F(\xi)| dm(\xi) = \|F\|_{H^1(D)}. \tag{4.12}$$

Such measures are called Carleson measures. They play a major role in connection with Carleson's proof of the famous Corona Theorem (see [33] for more on this important subject). Theorem 4.15 (applied with $p = 1$) shows that for any Banach space B and any $f \in H^1(D; B)$ we have

$$\int_D \|f(z)\| d\sigma(z) \le c \|f\|_{H^1(D;B)}. \tag{4.13}$$

Indeed, we just write $f = Fg$ and $\int_D \|f(z)\| d\sigma(z) \le \int_D |F(z)| d\sigma(z)$.

The interest of this remark, is that, by a result due to Mireille Lévy (which is but a suitable application of Hahn-Banach, see [394]), one can show that (4.13) holds for any B iff there is an operator $T : L_1(\mathbb{T}; m) \to L_1(D; \sigma)$ with $\|T\| \le c$ that coincides with the Poisson integral on analytic functions (i.e. T extends to $L_1(\mathbb{T}; m)$ the correspondence $F(\xi) \mapsto F(z)$ expressed by (4.12)). However, the mapping defined by the Poisson integral itself does *not* map $L_1(\mathbb{T}; m)$ into $L_1(D; \sigma)$.

Remark. Let $1 \le p < \infty$ and $f \in L_p(\mathbb{T}, m; B)$. We claim that the Fourier transform of f vanishes on the negative integers iff f belongs to the closure in $L_p(\mathbb{T}, m; B)$ of the subspace of all B-valued analytic polynomials restricted to $\mathbb{T}$. The 'if' part is obvious. The standard proof of the converse uses the Fejer kernel, but we may just as well use our familiar tool, namely the Poisson kernel P_r $(0 \le r < 1)$. Indeed, assume that $\widehat{f}(n) = 0$ for all $n < 0$. By Theorem 3.1, we know that, $P_r * f \to f$ in $L_p(\mathbb{T}, m; B)$ when $r \to 1$, and since $P_r * f = \sum_0^\infty \widehat{f}(n) r^n e^{int}$ we have

$$\left\| P_r * f - \sum\nolimits_0^N \widehat{f}(n) r^n e^{int} \right\|_{L_p(\mathbb{T},m;B)} \le \sum\nolimits_{n>N} r^n \|\widehat{f}(n)\|$$

and this last term tends to 0 since $r < 1$ and for all $n \in \mathbb{Z}$

$$\|\widehat{f}(n)\| \le \|f\|_{L_1(\mathbb{T},m;B)} \le \|f\|_{L_p(\mathbb{T},m;B)}.$$

This proves the 'only if' part, and the claim.

Notation. If $1 \le p < \infty$, we will denote by $\widetilde{H}^p[\mathbb{T}; B]$ the subspace of $L_p(\mathbb{T}, m; B)$ formed by all the functions f with Fourier transform vanishing on the negative integers. Equivalently, by what precedes, this is the closure in $L_p(\mathbb{T}, m; B)$ of the subspace of all B-valued analytic polynomials restricted to $\mathbb{T}$.

The latter reformulation extends to the case $0 < p < 1$: in that case, we denote by $\widetilde{H}^p[\mathbb{T}; B]$ the closed span in $L_p(\mathbb{T}, m; B)$ of the functions $\{e^{int}b \mid n \geq 0, b \in B\}$. Thus, if $0 < p < \infty$ $\widetilde{H}^p[\mathbb{T}; B]$ is the closure in $L_p(\mathbb{T}, m; B)$ of the set of all (analytic) polynomials restricted to $\mathbb{T}$.

We will denote by $\widetilde{H}^p(D; B)$ the closure in $H^p(D; B)$ of the space of B-valued analytic polynomials on D (i.e. functions of the form $P(z) = x_0 + x_1 z + \cdots x_n z^n$ with $x_k \in B$).

If $p = \infty$, we also denote by $A(D; B)$ the subspace of $H^\infty(D; B)$ of all bounded analytic functions on D that are continuously extendable to the closure of D. Equivalently, $A(D; B)$ is the closure in $H^\infty(D; B)$ of the space of all polynomials on D. We will denote by $A[\mathbb{T}; B]$ the closure in $C(\mathbb{T}; B)$ of the linear span of the functions $\{e^{int}b \mid n \geq 0, b \in B\}$.

Remark 4.20. Let $f : D \to B$ be any analytic function. Let $0 < p < \infty$. Then for any $0 \leq r < 1$, the function f_r is in $\widetilde{H}^p(D; B)$. Indeed, let $f(z) = \sum_{k \geq 0} a_k z^k$. Since the radius of convergence of this Taylor series is ≥ 1 we know that $\limsup \|a_k\|^{1/k} \leq 1$, and hence $\sum_k (r^k \|a_k\|)^p < \infty$ for any $0 \leq r < 1$. Then, if $p < 1$, by the p-triangle inequality (1.10) (which clearly holds for $E = H^p(D; B)$) we have

$$\left\| f_r - \sum\nolimits_{0 \leq k \leq n} a_k r^k z^k \right\|^p_{H^p(D;B)} \leq \sum\nolimits_{k>n} \|a_k r^k z^k\|^p_{H^p(D;B)} \leq \sum\nolimits_{k>n} (r^k \|a_k\|)^p,$$

which tends to 0 when $n \to \infty$. This shows that $f_r \in \widetilde{H}^p(D; B)$. When $p \geq 1$, $\sum_k r^k \|a_k\| < \infty$ and the usual triangle inequality yields the same conclusion.

Corollary 4.21. *Let $0 < p < \infty$. Any f in $\widetilde{H}^p(D; B)$ admits non-tangential (and radial) limits a.e. on $\partial D \simeq \mathbb{T}$. Let f_1 be the associated boundary value on $\mathbb{T}$. The correspondence $f \to f_1$ is an isometric isomorphism between $\widetilde{H}^p(D; B)$ and $\widetilde{H}^p[\mathbb{T}; B]$, and between $A(D; B)$ and $A[\mathbb{T}; B]$.*

Proof. By Theorem 2.2, the case $p \geq 1$ is already known to us more generally for harmonic functions, but for analytic functions this remains true even for $0 < p < 1$. If f is a polynomial, the existence of non-tangential (and in that case even tangential !) boundary values is obvious by continuity. Using the (non-tangential) maximal inequality, one easily extends that to any f in $\widetilde{H}^p(D; B)$. The argument for this is entirely analogous to the one used earlier for the martingale convergence Theorem 1.30 (this way to pass from a maximal inequality to an a.s. convergence is sometimes called 'the Banach principle'). Indeed, let $\epsilon > 0$ and let g be a polynomial such that $\|f - g\|_{H^p(D;B)} < \epsilon$. Define

$$\Omega_f(t) = \limsup_{z \to e^{it}\ z \in \Gamma_\alpha(t)} \|f(z) - f(e^{it})\|.$$

Then clearly $\Omega_g \equiv 0$ and $\Omega_f \le \Omega_g + \Omega_{f-g} = \Omega_{f-g}$. By the maximal inequality (cf. Theorem 4.14) we have $\|\Omega_{f-g}\|_p \le C_{p,\alpha}\|f - g\|_{L_p(\mathbb{T},m;B)} < C_{p,\alpha}\epsilon$. Therefore $\|\Omega_f\|_p \le C_{p,\alpha}\epsilon$, and we conclude, since ϵ is arbitrary, that $\Omega_f \equiv 0$.

By (4.7), the correspondence $f \to f_1$ is isometric (between $\widetilde{H}^p(D;B)$ and $\widetilde{H}^p[\mathbb{T};B]$) when restricted to analytic B-valued polynomials, so it extends by density to the whole of $\widetilde{H}^p(D;B)$. In the case $p = \infty$, $f \to f_1$ is isometric by the maximum principle. □

As the next recapitulative result shows, the converse also holds.

Theorem 4.22. *Let $f \in H^p(D;B)$, $0 < p < \infty$. The following are equivalent:*

(i) $f \in \tilde{H}^p(D;B)$.
(ii) *f admits non-tangential limits a.e.*
(iii) *f admits radial limits a.e.*
(iv) $\lim_{r,r'<1\ r,r'\to 1} \|f_r - f_{r'}\|_{L_p(m;B)} = 0$.
(v) *There is a sequence $r(n) < 1$ tending to 1 such that $f_{r(n)}$ converges in $L_p(m;B)$.*
(vi) *$f_r \to f$ in $H^p(D;B)$ when $r \to 1$.*

Moreover, if (i) holds and $p \ge 1$, f is the Poisson integral of the radial limit $t \mapsto f_1(e^{it}) = \lim_{r\to 1} f_r(e^{it})$.

Proof. (i) $\Rightarrow$ (ii) is part of the last corollary and (ii) $\Rightarrow$ (iii) is trivial. Assume (iii). By the maximal inequalities (cf. Theorem 4.14) we know $\int \sup_{r<1} \|f_r\|^p dm < \infty$. Hence by dominated convergence if f admits radial limits a.e. we have for $p < \infty$

$$\lim_{r,r'<1\ r,s\to 1} \|f_r - f_{r'}\|_{L_p(m;B)} = 0,$$

and hence (iii) $\Rightarrow$ (iv). (iv)$\Rightarrow$ (v) is obvious by the Cauchy criterion in $L_p(m;B)$. Assume (v). For any $0 < r, r' < 1$ we have by Corollary 4.8

$$\|f_r - f_{r'}\|_{H^p(D;B)} = \sup_{0\le s<1} \|(f_r)_s - (f_{r'})_s\|_{L_p(m;B)} \le \|f_r - f_{r'}\|_{L_p(m;B)}.$$

Since $H^p(D;B)$ is complete (a fact which we take for granted), if (v) holds, $f_{r(n)}$ converges in $H^p(D;B)$ when $n \to \infty$. Let $g \in H^p(D;B)$ be the limit. By (4.4), convergence in $H^p(D;B)$ implies pointwise convergence in D, and since $f(rz) \to f(z)$ for any $z \in D$, we must have $g = f$ so we conclude that (v) $\Rightarrow$ (vi). Since $f_r \in \widetilde{H}^p(D;B)$ for any $0 < r < 1$ (see Remark 4.20), (vi) $\Rightarrow$ (i). Incidentally when $p = \infty$, convergence in L_p fails in general, but f_1 is clearly Bochner measurable and bounded, or equivalently $f_1 \in L_\infty(\mathbb{T}, \mathcal{B}, m; B)$.

Lastly, if $p \geq 1$, and if $f_r \to f_1$ in $L_1(\mathbb{T}, m; B)$ we have

$$\int f_1 P^z(dt) = \int f_1(t)P_z(t)dm(t) = \lim_{r\to 1} \int f_r(t)P_z(t)dm(t) = \lim_{r\to 1} f(rz) = f(z),$$

so that f is the Poisson integral of f_1. □

Remark 4.23. In Corollary 4.19, if the function f is in $\tilde{H}_p(D; B)$ then we can find k such that $k \in \tilde{H}_q(D; B)$. Indeed, if f admits radial limits a.e. then $k = F^{-p/r} f$ also does.

We can now reformulate Corollary 4.17 with this notation:

Corollary 4.24. *Let $0 < p < \infty$. Let $\widetilde{H}^p(D) = \widetilde{H}^p(D; \mathbb{C})$. We have an isometric identity $H^p(D) = \widetilde{H}^p(D)$.*

Remark 4.25. Let $\widetilde{H^p}[\mathbb{T}] = \widetilde{H^p}[\mathbb{T}; \mathbb{C}]$. Observe that, if $p \geq 1$, by Corollary 4.21 we have an isometric identification

$$H^p(D) \simeq H^p[\mathbb{T}] = \{f \in L_p(\mathbb{T}, m; \mathbf{C}) \mid \widehat{f}(n) = 0 \quad \forall n < 0\}. \quad (4.14)$$

Note however that for $p < 1$ the Fourier transform of f does not make sense.

The following result is the famous F. and M. Riesz theorem.

Corollary 4.26. *Any measure $\mu \in M(\mathbb{T})$ such that $\widehat{\mu}(n) = 0$ for all $n < 0$ must be absolutely continuous with respect to the Lebesgue measure m.*

Proof. Indeed, let f be the complex valued harmonic function that is the Poisson integral on μ. By (3.20) f is analytic, and hence in $H^1(D)$. Let f_1 be the RN derivative of μ relative to m. On one hand, by Theorem 3.18, $f_r \to f_1$ in $L_1(m)$, in other words the measures $f_r \cdot m$ converge to $f_1 \cdot m$ in the norm of $M(\mathbb{T})$, but on the other hand, by Theorem 3.1, $f_r \cdot m$ converges weakly to μ, so we must have $\mu = f_1 \cdot m$. □

In the next section we will describe what happens to the F. and M. Riesz theorem in the Banach space valued case. For that purpose, we will need the following basic fact.

Proposition 4.27. *Let B be a complex Banach space and let μ be a bounded B-valued vector measure with Poisson integral u. Then u is analytic iff $\widehat{\mu}(n) = 0$ $\forall n < 0$. Moreover, in that case we have*

$$u(z) = \sum\nolimits_{n\geq 0} \widehat{\mu}(n)z^n, \quad u \in H^1(D; B)$$

and

$$\|u\|_{H^1(D;B)} = \|\mu\|. \quad (4.15)$$

Proof. By (3.28) we have an absolutely convergent representation for all z in D

$$u(z) = \sum_{n \geq 0} z^n \widehat{\mu}(n) + \sum_{n<0} \bar{z}^{|n|} \widehat{\mu}(n).$$

Therefore, u is analytic iff $\widehat{\mu}(n) = 0$ for all $n < 0$. Finally (4.15) follows from Theorem 3.2. □

Actually, a stronger result holds as follows:

Theorem 4.28. *Let B be an arbitrary complex Banach space. Then the Poisson integral defines an isometric isomorphism from the space of measures μ in $M(\mathbb{T}, \mathcal{B}; B)$ such that $\widehat{\mu}(n) = 0\ \forall n < 0$ onto the space $H^1(D; B)$. Moreover, every μ in $M(\mathbb{T}, \mathcal{B}; B)$ such that $\widehat{\mu}(n) = 0\ \forall n < 0$ satisfies $|\mu| \ll m$.*

Proof. By the preceding statement, it suffices for the first assertion to show that every f in $H^1(D; B)$ is the Poisson integral of a vector measure μ in $M(\mathbb{T}, \mathcal{B}; B)$, but this follows from Theorem 3.2. Moreover, by Theorem 4.14 the analyticity of f ensures that $\int \sup_{r<1} \|f_r\| dm < \infty$, therefore, going back to the proof of Theorem 3.2 and letting the function u there be our f, we find that the measure ν defined there is necessarily absolutely continuous (since $\sup_n \|u_{r_n}\|$ is integrable in this case), and since $|\mu| \leq \nu$, we obtain $|\mu| \ll m$. □

4.4 Analytic Radon-Nikodým property

Definition 4.29. A Banach space B is said to have the analytic Radon-Nikodým property (in short ARNP) if every bounded analytic function $f : D \to B$ admits radial limits a.e. that is to say, for almost every $t \in \mathbb{T}$, $f(re^{it})$ converges in B (for the norm topology) when $r \to 1$.

Theorem 4.30. *Let B be a Banach space, the following assertions are equivalent:*

(i) B has the ARNP.
(ii) For all $0 < p \leq \infty$, every f in $H^p(D; B)$ admits radial limits a.e.
(iii) For some $0 < p \leq \infty$, every f in $H^p(D; B)$ admits radial limits a.e..
(iv) Every bounded vector measure μ in $M(\mathbb{T}, \mathcal{B}; B)$ satisfying

$$\forall n < 0 \quad \widehat{\mu}(n) = \int e^{-int} \mu(dt) = 0, \tag{4.16}$$

admits a density in $L_1(\mathbb{T}, \mathcal{B}, m; B)$*, i.e. there is a function* $\varphi \in L_1(\mathbb{T}, \mathcal{B}, m; B)$ *such that for any Borel subset* $A \subset \mathbb{T}$*, we have*

$$\mu(A) = \int_A \varphi(t)dm(t). \tag{4.17}$$

Proof. We first show (i) $\Rightarrow$ (ii) $\Rightarrow$ (iii) $\Rightarrow$ (i).
(i) $\Rightarrow$ (ii). This is easy using Theorem 4.15, since every scalar function $F \in H^p(D)$ admits radial limits a.e. as we have seen (cf. Corollary 4.17).
(ii) $\Rightarrow$ (iii) and (iii) $\Rightarrow$ (i) are trivial since $H^\infty(D; B) \subset H^p(D; B)$.

To conclude the proof, it suffices to show that (iv) is equivalent to (iii) with the choice $p = 1$.

Now assume (iii) for $p = 1$. Let μ be as in (iv) and let f be its Poisson integral. Then, by Theorem 4.28, f is in $H^1(D; B)$, and admits radial limits a.e.. By Theorem 4.22, f is the Poisson integral of a function f_1 in $L_1(\mathbb{T}, m; B)$. Necessarily,

$$\mu = f_1.m$$

since for any bounded linear form ξ on B, $\xi(\mu)$ has the same Poisson integral as, and hence is equal to, $\xi(f_1).m$. Therefore (iv) holds. Conversely, assume (iv). Then let f be in $H^1(D; B)$. By Theorem 4.28, f is the Poisson integral of a measure μ satisfying the conditions in (iv), and hence admitting a density φ in $L_1(\mathbb{T}, m; B)$. Consequently (recall Remark 3.24), f is the Poisson integral of φ, hence by Corollary 3.18, for a.e. e^{it}, $f(z) \to \varphi(e^{it})$ whenever $z \to e^{it}$ non-tangentially, in particular $f(re^{it}) \to \varphi(e^{it})$ a.e., which proves that f admits radial limits a.e. □

By Theorem 4.22, this implies:

Corollary 4.31. *A Banach space B has the ARNP iff* $H^p(D, B) = \tilde{H}^p(D, B)$ *for some (or equivalently for all)* $0 < p < \infty$.

Since reflexive spaces have the RNP, all L_p-spaces a fortiori have the ARNP for $1 < p < \infty$. But in the analytic setting, as we will now show the case $p = 1$ provides a very important new example.

Theorem 4.32. *For any measure space* (Ω, μ)*, the Banach space* $B = L_1(\Omega, \mu)$ *has the ARNP.*

This is the case $p = 1, X = \mathbb{C}$ of the following more general statement:

Theorem 4.33. *If a Banach space X has the ARNP then for any* $1 \le p < \infty$ *and any measure space* (Ω, μ)*, the space* $B = L_p(\Omega, \mu; X)$ *has the ARNP.*

Proof. Consider $f = \sum_{n\geq 0} b_n z^n$ in the unit ball of $H^p(D; B)$ with $b_n \in B = L_p(\mu; X)$. Note that for any $r < 1$ (see Remark 4.20)

$$\| \sum r^n \|b_n(\cdot)\|_X \|_{L_p(\mu)} \leq \sum r^n \|b_n\|_{L_p(\mu;X)} < \infty,$$

and hence $\sum r^n \|b_n(\omega)\|_X < \infty$ for μ-almost all ω. Taking a suitable intersection, this implies that there is a measurable subset $\Omega' \subset \Omega$ such that for any $\omega \in \Omega'$ we have $\sum r^n \|b_n(\omega)\|_X < \infty$ for any $r < 1$, and hence $\sum_{n\geq 0} b_n(\omega) z^n$ converges in X for any $z \in D$. We denote $f(z)(\omega) = \sum z^n b_n(\omega)$ the X valued analytic function on D thus defined for μ-almost all ω. We claim that

$$\iint \sup_{r<1} \|f(re^{i\theta})(\omega)\|^p \, dm(\theta) d\mu(\omega) < \infty. \tag{4.18}$$

Indeed, for any $s < 1$, by Fubini we have

$$\|f_s\|^p_{L_p(m;B)} = \int \|f(se^{i\theta})(\omega)\|^p \, dm(\theta) d\mu(\omega) \leq 1.$$

Thus we can write, for μ-almost all ω, by (4.6) (applied to X instead of B)

$$\int \sup\nolimits_{r<1} \left\| \sum\nolimits_{n\geq 0} b_n(\omega) r^n s^n e^{in\theta} \right\|^p \, dm(\theta) \leq C_{p,0} \int \|f(se^{i\theta})(\omega)\|^p \, dm(\theta)$$

and hence after integration over ω

$$\iint \sup\nolimits_{r<s} \|f(re^{i\theta})(\omega)\|^p \, dm(\theta) \, d\mu(\omega) \leq C_{p,0} \|f_s\|^p_{L_p(m;B)} \leq C_{p,0}.$$

Taking the limit when $s \to 1$ this yields Claim 4.18 as announced.

Now (4.18) implies that for μ-almost all ω, $z \to f(z)(\omega)$ is in $H^p(D; X)$, and since we assume that X has the ARNP, its radial limit $\lim_{r\to 1} f(re^{i\theta})(\omega)$ exists for a.e. θ, and hence (by Fubini)

$$\lim\nolimits_{r,s\to 1} \|f(re^{i\theta})(\omega) - f(se^{i\theta})(\omega)\| = 0 \text{ for a.e. pair } (\omega, \theta) \text{ in } \Omega \times \mathbb{T}.$$

Finally, since (4.18) guarantees dominated convergence, we conclude

$$\lim_{r,s\to 1} \iint \|f_r(e^{i\theta})(\omega) - f_s(e^{i\theta})(\omega)\|^p \, d\mu(\omega) \, dm(\theta) = 0,$$

or equivalently $\lim_{r,s\to 1} \|f_r - f_s\|_{L_p(m;B)} = 0$, which implies by Theorem 4.22 that $f \in \widetilde{H}^p(D; B)$ and, by Corollary 4.31, B has the ARNP. □

Remark. Actually, the preceding statement and its proof remain valid for $p < 1$, even though $L_p(\Omega, \mu; X)$ is only a quasi-Banach space.

More generally, we have

Corollary 4.34. *A Banach lattice has the ARNP iff it does not contain an isomorphic copy of the space* c_0.

Proof. The only if part is valid in general. First observe that c_0 fails the ARNP. Indeed, consider the c_0 valued analytic function $f(z) = \sum z^k e_k$ on the unit disc. Then $f \in H^\infty(D; c_0)$, but (easy exercise) f admits radial limits nowhere on the unit circle. (It may be also instructive to prove this by considering the c_0 valued 'analytic' martingale $f_n = \sum_1^n z_k e_k$, which is bounded in $L_\infty(c_0)$ and divergent everywhere, but analytic martingales are only defined a bit further in what follows.) This example shows that c_0 fails the ARNP. Since the ARNP passes to subspaces, any B that (isomorphically) contains c_0 must fail the ARNP.

To prove the if part, let B be a Banach lattice not containing c_0. To show that B has the ARNP, we may clearly assume B separable. We claim that B semi-embeds (see Proposition 2.42) into $L_1(\Omega, \mu)$ over some probability space (Ω, μ). This follows from classical general facts about Banach lattices: First, by [54, p. 34], if $c_0 \not\subset B$ then B is order continuous, secondly, by [54, 1.a.9, p. 9] B embeds in a Banach lattice with a weak unit, lastly by [54, 1.b.14, p. 25] B is order isometric to a Köthe function space X on a probability space (Ω, μ). This means that we have an injection $X \to L_1(\Omega, \mu)$ and a subset $S \subset L_1(\Omega, \mu)_+$ such that for any $f \in X$ we have

$$\|f\| = \sup\{\int |fg| d\mu \mid g \in S\}.$$

Then by the classical Fatou lemma, the inclusion $X \to L_1(\Omega, \mu)$ is a semi-embedding (see [54, p. 30] for related details). By the proof of Proposition 2.42, the ARNP of $L_1(\Omega, \mu)$ passes to X so we conclude that B has the ARNP. □

Remark 4.35. The space $B = L_1/H^1$ over $\mathbb{T}$ gives us an example of a space failing the ARNP but not containing c_0. Indeed, let $a_n \in L_1/H^1$ denote the element associated (modulo H^1) to e^{-int}. We claim that the function $f(z) = \sum_1^\infty a_n z^n$ is in $H^\infty(D; B)$. Since it is easy to check that $\|a_n\|_B = 1$ for all $n > 0$ the function f is not in $\tilde{H}^1(D; B)$ so B fails the ARNP. Moreover, by a result due to Bourgain (see [141]) it is of cotype 2 (see §10.4) and hence certainly does not contain c_0 isomorphically. To check the claim, note that, if we denote by $Q : L_1 \to L_1/H^1$ the quotient map, we have by (3.7) $Q(P_z) = \sum_1^\infty a_n z^n$ and since $\|P_z\|_1 = 1$ we must have $\|Q(P_z)\|_B \leq 1$ for all $z \in D$.

In sharp contrast, for any reflexive subspace $Y \subset L_1$ the quotient L_1/Y has the ARNP, and actually is 2-uniformly PL-convex, which is a stronger property (see §11.7).

Remark 4.36. Let $Y \subset B$ is a subspace of a Banach space B. In general, one cannot lift an element of $\tilde{H}^p(D; B/Y)$ up to an element in $\tilde{H}^p(D; B)$ (or in $H^p(D; B)$). This lifting property is the subject of the paper [311]. For instance it holds when Y is a reflexive subspace of $B = L_1$ (see §6.8 for more on this) but it fails for $Y = H^1(\mathbb{T})$.

4.5 Hardy martingales and Brownian motion*

We now return to Brownian martingales. Our first goal is a characterization of the ARNP analogous to that in Corollary 3.31 for the RNP, but in terms of martingales obtained by composing a B-valued analytic function with BM. The main difference with the case of harmonic functions treated in the preceding chapter is the case $0 < p < 1$, which is special to analytic functions, and the case $p = 1$, which is greatly improved in the analytic case. The improvements can all be derived from the subharmonicity of $z \mapsto |f(z)|^p$, proved in Proposition 4.5. The latter produces submartingales after composition with BM.

Proposition 4.37. *Let $\varphi : D \to [-\infty, \infty)$ be a subharmonic function such that $\varphi_r \in L_1(m)$ for any $0 < r < 1$. Then for any stopping time $S \le T_r$ we have*

$$\varphi(W_S) \le \mathbb{E}^{\mathcal{A}_S} \varphi(W_{T_r}). \tag{4.19}$$

Therefore $\{\varphi(W_{T_r}) \mid 0 < r < 1\}$ is a submartingale in $L_1(\Omega, \mathcal{A}, \mathbb{P})$.

Proof. Recall that, if $|z| < r < 1$, the distribution of W_{T_r} with respect to $\mathbb{P}_z$ is given by (3.67). In particular, our assumption implies that $\varphi(W_{T_r}) \in L_1(\Omega, \mathcal{A}, \mathbb{P}_z)$ for any z with $|z| < r$. Thus by (3.67), the subharmonicity of φ (as defined in (4.1)) implies that if $|z| < r$

$$\varphi(z) \le \mathbb{E}_z \varphi(W_{T_r}),$$

and this remains trivially true if $|z| = r$ (since then $\mathbb{P}\{W_{T_r} = z\} = 1$). By the Markov property (3.68), assuming $S \le T_r$, we have

$$\mathbb{E}^{\mathcal{A}_S} \varphi(W_{T_r}) = \mathbb{E}_{W_S} \varphi(W_{T_r}).$$

Indeed, for any $\omega \in C([0, \infty); \mathbb{C})$, let $\omega'(t) = \omega(S(\omega) + t)$ for all $t > 0$. Then it is easy to check, using $S(\omega) \le T_r(\omega)$ that $\omega'(T_r(\omega')) = \omega(T_r(\omega))$. We already used an analogous argument to justify (3.71).

Therefore taking $z = W_S$ we find (4.19). Applying (4.19) with $S = T_s$ for $s < t$ shows that $\{\varphi(W_{T_r}) \mid 0 < r < 1\}$ is a submartingale with respect to the filtration $(\mathcal{A}_{T_r})$. □

Remark 4.38. In the analytic case for $f \in H^p(D; B)$, the inequality (3.73) remains valid for all values of $0 < p < \infty$: We have for any $0 < s < r < 1$

$$\|f_{r^2} - f_{s^2}\|_{L_p(m;B)} \le \|f_r - f_{s^2r^{-1}}\|_{L_p(m;B)} \le \|f(W_{T_r}) - f(W_{T_s})\|_{L_p(\mathbb{P};B)}. \quad (4.20)$$

Indeed, the same proof works, except that Jensen's inequality must be replaced by the subharmonicity of $z \mapsto \|f(re^{it}) - f(sz)\|^p$ proved in Proposition 4.7, and by (4.3).

Theorem 4.39. *Let B be any Banach space. Let $0 < p < \infty$. For any f in $H^p(D; B)$ we have (recall the notation $T = T_1$)*

$$\| \sup_{t>0} \|f(W_{t\wedge T})\| \|_p \le e^{1/p} \|f\|_{H^p(D;B)}. \quad (4.21)$$

Moreover, the following are equivalent:

(i) $f \in \widetilde{H}^p(D; B)$.
(ii) $f(W_{t\wedge T})$ *converges almost surely when* $t \to \infty$.
(iii) There is a sequence of radii $r(n) < 1$ tending to 1, such that $f(W_{T_{r(n)}})$ converges almost surely when $n \to \infty$.

Proof. Fix $1 < q < \infty$ and $0 < r < 1$. Since $\|f\|^{p/q}$ is subharmonic (see Proposition 4.7), we may apply the preceding proposition to $\varphi(z) = \|f(z)\|^{p/q}$. Let $Z_t = \varphi(W_{T_r \wedge t})$ and $M_t = \mathbb{E}^{\mathcal{A}_{T_r \wedge t}} \varphi(W_{T_r})$. By (4.19) we have $Z_t \le M_t$ and hence $\sup_{t\ge0} Z_t \le \sup_{t\ge0} M_t$. By Doob's inequality (see Remark 1.39), we have

$$\| \sup_{t\ge0} Z_t \|_q \le q' \|\varphi(W_{T_r})\|_q = q' \|f(W_{T_r})\|_p^{p/q} \le q' \|f\|_{H^p(D;B)}^{p/q}.$$

But since $\| \sup_{t\ge0} Z_t \|_q = \| \sup_{t\le T_r} \|f(W_t)\| \|_p^{p/q}$, this gives us

$$\| \sup_{t\le T_r} \|f(W_{t\wedge T})\| \|_p \le ((q')^q)^{1/p} \|f\|_{H^p(D;B)}.$$

Letting $r \to 1$, we obtain (4.21) with the constant $(q')^q$ in place of e, but since $\lim_{q\to\infty} (q')^q = e$ the announced result (4.21) follows.

Assume (i). Assume first that f extends continuously to $\bar{D}$ (for instance that f is a polynomial). Then, the Brownian paths being continuous, (ii) is obvious. Since we can approximate f in $H^p(D; B)$ by a polynomial, we can argue as in either proofs of Theorem 1.30 or Corollary 4.21 (the so-called Banach principle). Then the a.s. convergence follows easily from (4.21). Note that if f is extended a.e. to ∂D using its radial limits (see Corollary 4.21) then the same argument shows that a.s. $\lim_{t\to\infty} f(W_{t\wedge T}) = f(W_T)$, thus proving (i) $\Rightarrow$ (ii).

Assume (ii). A fortiori (recall (3.57)), $f(W_{T_r})$ converges a.s. when $r \uparrow 1$ and in particular (iii) holds.

Assume (iii). By (4.21) the convergence is dominated, therefore

$$\lim_{n,k\to\infty} \|f(W_{T_{r(n)}}) - f(W_{T_{r(k)}})\|_{L_p(\mathbb{P},B)} = 0$$

and hence by (4.20) $\lim_{n,k\to\infty} \|f_{r(n)} - f_{r(k)}\|_{L_p(m;B)} = 0$. By (v) ⇒ (i) in Theorem 4.22 this tells us that $f \in \widetilde{H}^p(D; B)$, thus proving (iii) ⇒ (i). □

Remark. Let $f \in H^p(D; B)$, with $0 < p \leq \infty$. The preceding Theorem shows that $f(W_{t\wedge T})$ converges a.s. in B (in other words f admits 'Brownian limits') iff f admits radial (or non-tangential) limits a.e. Indeed, this follows from Theorem 4.22 for $p < \infty$, and since $H^\infty(D; B) \subset H^p(D; B)$, a fortiori also for $p = \infty$.

Corollary 4.40. *Let $0 < p \leq \infty$. A Banach space B has the ARNP iff for any f in $H^p(D; B)$, $f(W_{t\wedge T})$ converges a.s. in B when $t \to \infty$. Equivalently this holds iff for any f in $H^p(D; B)$, $f(W_{T_r})$ converges a.s. when $r \uparrow 1$.*

Proof. This follows from Theorem 4.30. □

We will now give a discretized version of the last corollary, i.e. we will introduce a special class of discrete B-valued martingales (we call them Hardy martingales) and we will show that the ARNP of B is equivalent to the a.s. convergence of all B-valued Hardy martingales that are bounded in $L_p(B)$ ($0 < p \leq \infty$), in analogy with the corresponding result (Theorem 2.9) for the RNP.

Let $I = \{1, 2, \ldots\}$. Consider $\Omega = \mathbb{T}^I$ equipped with normalized Haar measure $\mathbb{P}$. For any $z \in \Omega$, we denote by z_n its n-th coordinate. Let $\mathcal{A}_n$ be the σ-algebra generated by $(z_1, \ldots, z_n)$ and $\mathcal{A}_0$ the trivial one.

Definition 4.41. A sequence $(f_n)_{n\geq 0}$ in $L_1(\Omega, \mathbb{P}; B)$ will be called a Hardy martingale if it is a martingale with respect to $(\mathcal{A}_n)_{n\geq 0}$ and if moreover for any $n \geq 1$ and any fixed $(z_1, \ldots, z_{n-1})$ the function

$$z \mapsto f_n(z_1, \ldots, z_{n-1}, z)$$

is in $\widetilde{H}^1(\mathbb{T}; B)$ i.e. its Poisson integral is in $\widetilde{H}^1(D; B)$.

We will also use the term 'Hardy martingale' for a finite sequence $(f_k)_{0\leq k\leq n}$ as earlier. This corresponds to an infinite Hardy martingale such that $f_k = f_n$ for all $k \geq n$.

Definition 4.42. We will say that a Hardy martingale (f_n) is an analytic martingale if for any n the function $z \mapsto f_n(z_1, \ldots, z_{n-1}, z)$ is a polynomial of degree 1. More explicitly this means that there is a function $\delta_n(z_1, \ldots, z_{n-1})$ such that

$$f_n(z_1, \ldots, z_{n-1}, z_n) = f_{n-1}(z_1, \ldots, z_{n-1}) + z_n\delta_n(z_1, \ldots, z_{n-1}).$$

In general Hardy martingales are better behaved than the ordinary ones, just like analytic functions are better than the harmonic ones. For instance, for

Hardy martingales, Doob's maximal inequalities remain valid for any p with $0 < p < \infty$ and not only $p > 1$:

Theorem 4.43. *Let $0 < p < \infty$. Then for any B-valued Hardy martingale (f_n) assumed bounded in $L_p(\mathbb{P}; B)$, we have*

$$\| \sup_n \|f_n\|_B \|_p \le e^{1/p} \sup_n \|f_n\|_{L_p(\mathbb{P};B)}.$$

Proof. The proof is similar in spirit to that of (4.21). We first observe that for any $1 < q < \infty$ the variables $M_n = \|f_n\|_B^{p/q}$ form a submartingale bounded in L_q. Indeed, since $z \mapsto \|f_n(z_1, \dots, z_{n-1}, z)\|_B^{p/q}$ is subharmonic (see Proposition 4.7), its integral with respect to $z \in \mathbb{T}$ is $\ge \|f_n(z_1, \dots, z_{n-1}, 0)\|_B^{p/q} = \|f_{n-1}\|_B^{p/q}$, whence the submartingale property. We then repeat the argument for (4.21). We apply Doob's inequality in L_q and let $q \to \infty$ to obtain the announced constant $e^{1/p}$. □

Corollary 4.44. *Any Banach space valued Hardy martingale converging in $L_p(\mathbb{P}; B)$ $(0 < p < \infty)$ must converge almost surely.*

Proof. The proof of the a.s. convergence in Theorem 1.30 (by the 'Banach principle') can be repeated word for word. □

Remark 4.45. Let $(f_k)_{0 \le k \le n}$ be a Hardy martingale with values in $\mathbb{C}$. Assume that the values of $(f_k)_{0 \le k \le n}$ all lie inside a compact subset of an open subset $V \subset \mathbb{C}$. Then for any analytic function $g : V \to B$ with values in a Banach space B, the composition sequence $(g(f_k))_{0 \le k \le n}$ is a B-valued Hardy martingale. The martingale property follows from the Cauchy formula. We leave the easy verification to the reader.

Theorem 4.46. *Let $0 < p < \infty$. The following properties of a Banach space are equivalent:*

- *(i) B has the ARNP.*
- *(ii) All the Hardy martingales bounded in $L_p(B)$ converge a.s. and in $L_p(B)$.*
- *(iii) All the Hardy martingales bounded in $L_\infty(B)$ converge a.s.*

The proof will be given after the next four lemmas.

Remark 4.47. Actually, by [217], the preceding are also equivalent to:

(iv) All the analytic martingales bounded in $L_\infty(B)$ converge a.s.

The idea behind this is that a Hardy martingale can be approximated by a subsequence of an analytic one. More precisely, given a finite Hardy martingale $(f_0, \dots, f_n)$ in $L_p(B)$ $(0 < p < \infty)$ and $\varepsilon > 0$, there is an extension of the underlying probability space on which an analytic martingale (g_k) can be defined together with a subsequence $k(1) < k(2) < \cdots < k(n)$ such that $g_0 = f_0$ and

$$\forall 1 \le j \le n \quad \|f_j - g_{k(j)}\|_{L_p(B)} < \varepsilon.$$

We will just give a rough outline of the argument. By perturbation we can restrict to Hardy martingales of polynomial type, and it is easy to reduce the result to the case $n = 1$ with $f_0 = 0$. Then f_1 has the same distribution as $f(W_T)$ for BM starting at 0, and by Ito's formula

$$f(W_T) = \int_0^T f'(W_s)dW_s.$$

But now, given $\delta > 0$ fixed, we can use a discretization of this stochastic integral by a finite sum of the form

$$\sum_j f'(W_{s_{j-1}})(W_{s_j} - W_{s_{j-1}})$$

where $0 < s_1 < \cdots < s_{j-1} < s_j < \cdots$ are stopping times defined by

$$s_j = \inf\{t > s_{j-1} \mid |W_t - W_{s_{j-1}}| > \delta\}.$$

Let $Z_j = \delta^{-1}(W_{s_j} - W_{s_{j-1}})$. Then the sequence (Z_j) has the same distribution as (z_j) has on $\mathbb{T}^{\mathbb{N}}$, so we have

$$\sum_j f'(W_{s_{j-1}})(W_{s_j} - W_{s_{j-1}}) = \sum_j f'\left(\delta \sum_{k<j} Z_k\right) \delta Z_j$$

and the latter is distributed like the sum of an analytic martingale. Thus letting $\delta \to 0$, we obtain the desired approximation. See [169] for complete details on the approximation argument. See [217, Prop. 6] for a different approach.

Recall that for $z \in \mathcal{D}_r = \{z \mid |z| < r\}$, the probability ('harmonic measure', or 'Jensen measure') $P^z_{\mathcal{D}_r}$ is the unique probability measure on $\partial\mathcal{D}_r$ such that any u harmonic on $\mathcal{D}_r$ and continuous on $\overline{\mathcal{D}}_r$ satisfies

$$u(z) = \int_{\partial\mathcal{D}_r} u\, dP^z_{\mathcal{D}_r}.$$

Lemma 4.48. *Fix $0 = r(0) < r(1) < \cdots < r(n) < \cdots < 1$. Let $f \in H^1(D; B)$ and let $f_n = f(W_{T_{r(n)}})$. Then $(f_n)_{n\ge 0}$ has the same distribution as a Hardy martingale.*

Proof. By Remark 4.45 it suffices to prove this for the $\mathbb{C}$ valued function $f(z) = z$, for which $f_n = W_{T_{r(n)}}$. In that case, we prove this for $(f_0, \ldots, f_n)$ by induction on n. The case $n = 0$ is trivial. Assume that there is a Hardy martingale $(\tilde{f}_0, \ldots, \tilde{f}_{n-1})$ such that $(f_0, \ldots, f_{n-1}) = (\tilde{f}_0, \ldots, \tilde{f}_{n-1})$ in distribution. Let $\mathcal{D}_n = \{z \mid |z| < r(n)\}$. By (3.68), the conditional distribution of $f_n = W_{T_{r(n)}}$

given $f_{n-1} = W_{T_{r(n-1)}}$ is equal to $P_{\mathcal{D}_n}^{W_{T_{r(n-1)}}}$. Moreover, by the Markov property, it coincides with the conditional distribution of $W_{T_{r(n)}}$ with respect to $f_0, \ldots, f_{n-1}$. For any $z \in \mathcal{D}_n$, let $\Phi_z\colon D \to \mathcal{D}_n$ be the bijective conformal mapping (essentially a Möbius transformation up to scaling) such that $\Phi_z(0) = z$. Then Φ_z transforms the uniform probability on $\partial D = \mathbb{T}$ into $P_{\mathcal{D}_n}^z$. Therefore, assuming that $\tilde{f}_0, \ldots, \tilde{f}_{n-1}$ depend on $z_1, \ldots, z_{n-1}$, if we set

$$\tilde{f}_n = \Phi_{\tilde{f}_{n-1}}(z_n)$$

then $(\tilde{f}_0, \ldots, \tilde{f}_n)$ is a Hardy martingale with the same distribution as $(f_0, \ldots, f_n)$. This completes the induction step and hence the proof. □

Remark. The same proof shows that if $f_n = u(W_{T_{r(n)}})$ for some harmonic function $u : D \to B$, then $(f_n)_{n \geq 0}$ has the same distribution as a martingale relative to $\mathbb{T}^{\mathbb{N}}$ with respect to the natural filtration.

Let $\Omega = \mathbb{T}^I$, $I = [1, 2, \ldots]$. Let $\mathcal{T}(\Omega)$ denote the set of all 'trigonometric polynomials' on Ω, i.e. the linear span of all 'monomials' of the form

$$(z_k) \mapsto z_1^{m_1} z_2^{m_2} \ldots z_n^{m_n} \qquad (n \geq 1, m_k \in \mathbb{Z}).$$

For simplicity we denote $L_p(\Omega; B)$ instead of $L_p(\Omega, \mathbb{P}; B)$.

Since $\mathcal{T}(\Omega)$ is dense in $L_p(\Omega, \mathbb{P})$, $\mathcal{T}(\Omega) \otimes B$ is dense in $L_p(\Omega; B)$. More precisely, for any $f \in L_p(\Omega; B)$ there is a $g \in L_p(\Omega; B)$ with $\|f - g\|_p < \varepsilon$ with g of the form

$$g = \sum z_1^{m_1} \ldots z_n^{m_n} b(m_1, \ldots, m_n) \tag{4.22}$$

where $b\colon \bigcup_{n \geq 1} \mathbb{Z}^n \to B$ is finitely supported. We call such a g a 'B-valued trigonometric polynomial on $\Omega = \mathbb{T}^I$'.

Consider a Hardy martingale (f_n) formed of B-valued trigonometric polynomials on $\Omega = \mathbb{T}^I$. Given an increasing sequence of integers (N_n) and $z \in \mathbb{T}$, we will 'transform' $\omega = (z_n)_{n \geq 1}$ into

$$F(\omega, z) = (z_n z^{N_n})_{n \geq 1}.$$

Since, for each fixed $z \in \mathbb{T}$, $\omega \to F(\omega, z)$ preserves the measure $\mathbb{P}$, we have

$$\forall z \in \mathbb{T}\ \forall f \in L_p(\Omega; B) \qquad \|f\|_{L_p(\Omega;B)} = \|f(F(\omega, z))\|_{L_p(\Omega;B)}. \tag{4.23}$$

Since all the variables (f_n) are finite sums of the form (4.22), it is not hard to show that the integers (N_n) can be chosen inductively so that for each n

$$z \mapsto f_n(F(\omega, z))$$

is analytic in z, i.e. its Fourier transform vanishes on the set $\mathbb{Z}_-$ of negative integers. Indeed, consider f of form

$$f(\omega) = z_1^{m_1} \dots z_n^{m_n} b \tag{4.24}$$

with $b \in B$, $m_1, \dots, m_{n-1} \in \mathbb{Z}$ and $m_n > 0$. Then

$$f(F(\omega, z)) = f(\omega) \cdot z^{m_1 N_1 + \cdots + m_n N_n} \tag{4.25}$$

and since $m_n > 0$ it is easy to choose N_n large enough (assuming $N_1, \dots, N_{n-1}$ already chosen) so that $m_1 N_1 + \cdots + m_n N_n > 0$.

By the same reasoning, we can also ensure that the spectrum (i.e. the support of the Fourier transform) of $z \mapsto df_n(F(\omega, z))$ is strictly to the right of that of all the preceding increments $z \mapsto df_j(F(\omega, z))$.

Consider now an increasing sequence of radii $(r(n))$ with $r(n) < 1$ and $r(n) \uparrow 1$. We will denote by $f_n^\omega(z)$ the analytic extension of $z \mapsto f_n(F(\omega, z))$. Similarly, we denote by $df_n^\omega(z)$ the analytic extension of $z \mapsto df_n(F(\omega, z))$. This simply extends linearly the passage from (4.24) to (4.25).

Lemma 4.49. *Let (f_n) be a Hardy martingale in $L_p(\Omega; B)$ formed by trigonometric polynomials. Let $\varepsilon_n > 0$ and $\delta_n > 0$ arbitrarily fixed. There is a choice of (N_n) and $(r(n))$ so that for any $n \geq 1$ and any $z \in \mathbb{T}$*

$$\forall j < n \qquad \|df_n^\omega(r(j)z)\|_{L_p(\Omega;B)} < \delta_n \tag{4.26}$$

$$\|f_n^\omega(z) - f_n^\omega(r(n)z)\|_{L_p(\Omega;B)} < \varepsilon_n. \tag{4.27}$$

Proof. By the preceding preliminary remarks, we can (and will) choose inductively N_n large enough so that each $z \mapsto f_n(F(\omega, z))$ is an analytic polynomial. Thus we can reduce checking (4.26) and (4.27) to the case of functions of the form (4.24), say with $\|b\| \leq 1$. Now if df_n is of that form, (4.26) is reduced to

$$|r(j)^{m_1 N_1 + \cdots + m_n N_n}| < \delta_n, \tag{4.28}$$

and since $\lim_{N \to \infty} |r(j)^{m_n N}| = 0$ (recall that since (f_n) is Hardy, $m_n > 0$) (4.28) can be ensured by choosing N_n large enough. Having chosen N_n, we now choose $r(n)$ so that (4.27) holds. Indeed, for (f_n) of the same form (4.24) with $\|b\| \leq 1$, (4.27) is reduced to

$$|1 - r(n)^{m_1 N_1 + \cdots + m_n N_n}| < \varepsilon_n$$

and since $\lim_{r \to 1} |1 - r^{m_1 N_1 + \cdots + m_n N_n}| = 0$, the desired inductive choice is possible. □

Lemma 4.50. *Fix $0 < p < \infty$. Let $\alpha = \min(p, 1)$. Assume $0 < \varepsilon_n < 1$ and $0 < \delta_n$ are such that $\varepsilon_n \to 0$ and $\sum_{j>n} \delta_j^\alpha < \varepsilon_n^\alpha$ for any $n \geq 1$. Let (N_n) and*

$(r(n))$ be increasing sequences chosen as in Lemma 4.49. Assume (f_n) bounded in $L_p(\Omega; B)$. For any ω in Ω we set

$$\forall z \in D \qquad f^{\omega}(z) = \sum\nolimits_0^{\infty} df_n(F(\omega, z)). \tag{4.29}$$

Then, for any $z \in D$, this series converges in $L_p(\Omega; B)$, $f^{\omega} \in H^{\infty}(D; L_p(\Omega; B))$ and

$$\forall n \geq 1\ \forall z \in \mathbb{T} \qquad \|f^{\omega}(r(n)z) - f_n^{\omega}(z)\|_{L_p(\Omega;B)} < 2^{1/\alpha}\varepsilon_n. \tag{4.30}$$

Proof. For simplicity we may assume $\sup_n \|f_n\|_{L_p(B)} < 1$. Each $df_n(F(\omega, z))$ is a polynomial in z and

$$f^{\omega}(r(n)z) = f_n^{\omega}(r(n)z) + \sum\nolimits_{j>n} df_j^{\omega}(r(n)z). \tag{4.31}$$

Therefore, by (4.26), this series converges in $L_p(\Omega; B)$ for any $z \in \mathbb{T}$, and by the triangle inequality in an α-normed space (see (1.10))

$$\|f^{\omega}(r(n)z) - f_n^{\omega}(r(n)z)\|_{L_p(\Omega;B)} < \left(\sum\nolimits_{j>n} \delta_j^{\alpha}\right)^{1/\alpha} < \varepsilon_n. \tag{4.32}$$

By (4.27)

$$\|f_n^{\omega}(r(n)z)\|_{L_p(\Omega;B)} \leq (\|f_n(F(\omega, z))\|_{L_p(\Omega;B)}^{\alpha} + \varepsilon_n^{\alpha})^{1/\alpha} < (1 + \varepsilon_n^{\alpha})^{1/\alpha},$$

so we obtain by the triangle inequality

$$\|f^{\omega}(r(n)z)\|_{L_p(\Omega;B)} < (1 + 2\varepsilon_n^{\alpha})^{1/\alpha}. \tag{4.33}$$

Let $\varepsilon = \sup_n \varepsilon_n$. Since this holds for any z in $\mathbb{T}$ and $r(n) \to 1$, the series (4.29) converges for any $z \in D$ and the maximum principle ensures that $z \mapsto f^{\omega}(z)$ is in $H^{\infty}(D; L_p(\Omega; B))$ with norm at most $(1 + 2\varepsilon^{\alpha})^{1/\alpha}$. Then (4.27) and (4.32) imply (4.30) by the triangle inequality. □

Lemma 4.51. *Let $0 < p < \infty$. If B has the ARNP, any Hardy martingale (f_n) bounded in $L_p(\Omega; B)$ converges in $L_p(\Omega; B)$ and a.s.*

Proof. By perturbation we may assume that each df_n is a trigonometric polynomial on Ω. Assume (f_n) bounded in $L_p(B)$. Then with the notation in Lemma 4.50, $z \mapsto f^{\omega}(z)$ is in $H^{\infty}(D; L_p(\Omega; B))$. By Theorem 4.33, $L_p(\Omega; B)$ has the ARNP and hence the sequence $\{f^{\omega}(r(n)z) \mid n \geq 0\}$ is convergent in $L_p(\mathbb{T}; L_p(\Omega; B))$. By (4.30), recalling $f_n^{\omega}(z) = f_n(F(\omega, z))$, the same holds for the sequence $\{f_n(F(\omega, z))\}$ in $L_p(\Omega \times \mathbb{T}; B)$. By (4.23) we conclude that (f_n) itself converges in $L_p(\Omega; B)$. Then by Corollary 4.44 the a.s. convergence is automatic. □

Proof of Theorem 4.46. The implication (i) ⇒ (ii) is Lemma 4.51. (ii) ⇒ (iii) is obvious. Assume (iii) and consider f in $H^{\infty}(D; B)$. By Lemma 4.48, $(f(W_{T_{r(n)}}))$

has the same distribution as a Hardy martingale bounded in $L_\infty(B)$. Therefore by (iii) it converges a.s. Equivalently, $f(W_t)$ converges a.s. when $t \uparrow T$. Therefore we conclude that (i) holds by Corollary 4.40. □

We will now describe briefly an elegant generalization of Hardy (and analytic) martingales that has the advantage that it is stable by passing to subsequences. Here we follow [217] and [162] to which we refer the reader for the proofs.

Definition 4.52. Let $V \subset B$ be an open subset. A function $\psi : V \to [-\infty, \infty)$ is called plurisubharmonic (PSH in short) if it is upper semi-continuous and its restriction to any complex line is subharmonic, i.e. for any $x, y \in B$ the function $z \mapsto \psi(x + zy)$ is subharmonic on $V \cap \{x + \mathbb{C}y\}$ assuming the latter is non-void.

Remark 4.53. It is well known ([41, p. 45] and [42]) that the PSH property of ψ implies more generally that for any analytic function f defined on an open subset of $\mathbb{C}$ and taking values in V the composition $\psi(f)$ is subharmonic.

Definition 4.54. A martingale (f_n) in $L_1(B)$ is called plurisubharmonic (in short PSH) if for any Lipschitz plurisubharmonic function $\psi : B \to \mathbb{R}$ the sequence $(\psi(f_n))$ is a submartingale. By a Lipschitz function, we mean one such that for some constant c

$$\forall x, y \in B \quad |\psi(x) - \psi(y)| \le c\|x - y\|.$$

Note that, since $\pm\psi$ is PSH when ψ is an $\mathbb{R}$-linear map, any sequence (f_n) with the preceding property must necessarily be a martingale.

For example, all analytic martingales are clearly PSH. More generally, by the preceding Remark 4.53 (see also Remark 4.10), all Hardy martingales are PSH.

Definition 4.55. A Radon probability measure μ on B such that $\int \|x\| d\mu(x) < \infty$ is called a Jensen measure if for any Lipschitz PSH function $\psi : B \to \mathbb{R}$ we have

$$\psi\left(\int x d\mu(x)\right) \le \int \psi(x) d\mu(x).$$

By definition, a PSH martingale is one such that (roughly) the conditional distribution of f_n given f_{n-1} is a Jensen measure for any $n \ge 1$.

The typical example of a Jensen measure is the image measure $F(m)$ of the Haar measure on $\mathbb{T} = \partial D$ under the boundary value function of a function $F \in \tilde{H}^1(D; B)$.

More generally, if (f_n) is a Hardy martingale with f_0 constant on $(\Omega, \mathbb{P})$, then $\mu_n = f_n(\mathbb{P})$ is Jensen for any $n \geq 1$. Note that viewing f_n as a function on $\mathbb{T}^n$ we have $\mu_n = f_n(m^n)$.

By the following beautiful results of Shangquan Bu and Walter Schachermayer, in some sense the latter example is (close to) the general case.

Theorem 4.56 ([162]). *Let μ be a Jensen measure on a complex Banach space B with barycenter x_0 and $\varepsilon > 0$. Then there is a polynomial $F : \mathbb{C} \to B$ and a measurable function φ such that $\varphi(m) = \mu$, $F(0) = x_0$ and*

$$\|F - \varphi\|_{L_1(m;B)} = \int_{\partial D} \|F(z) - \varphi(z)\| dm(z) < \varepsilon.$$

Theorem 4.57 ([162]). *Let B be a complex Banach space, $(M_n)_{n \geq 0}$ a B-valued PSH martingale and (ε_n) positive numbers. Then there is a martingale $(\hat{M}_n)$ defined on $(\mathbb{T}^{\mathbb{N}}, m^{\mathbb{N}})$ with the same distribution as the sequence (M_n), such that $\hat{M}_n$ depends only on the first n coordinates of $\mathbb{T}^{\mathbb{N}}$ (and may therefore be identified with a function on $\mathbb{T}^n$) and a Hardy martingale (F_n) such that, for every $n \geq 1$,*

$$\|(\hat{M}_n - \hat{M}_{n-1}) - (F_n - F_{n-1})\|_{L_1(\mathbb{T}^n;B)} < \varepsilon_n.$$

By Lemma 4.51, this implies

Corollary 4.58. *If B has the ARNP, any PSH martingale bounded in $L_1(B)$ converges a.s.*

4.6 B-valued h^p and H^p over the half-plane U*

Let B be a Banach space. In this section we briefly review what becomes of the preceding two chapters when one replaces the unit disc D by the upper half-plane denoted by

$$U = \{z \in \mathbb{C} \mid \Im(z) > 0\} = \{x + iy \in \mathbb{C} \mid x \in \mathbb{R},\ y > 0\}.$$

Recall that, by Riemann's conformal mapping theorem, any two simply connected open subsets of $\mathbb{C}$ (other than $\mathbb{C}$ itself), and in particular D and U, are conformally equivalent. In the case of D and U, we have an explicit conformal map

$$\Phi : D \to U \quad \text{defined by} \quad \Phi(z) = i\frac{1-z}{1+z}.$$

Moreover, the non-tangential convergence to the boundary is preserved by Φ (except at $z = -1$, which is negligible) and its inverse. Indeed, the map Φ (resp. Φ^{-1}) is actually analytic and invertible (and hence 'conformal') on $\mathbb{C} \setminus \{-1\}$

(resp. $\mathbb{C} \setminus \{-i\}$), so that Remark 3.20 applies to this case. Since we have a one to one correspondence between bounded harmonic functions on both domains, the boundary behaviour of bounded harmonic functions on U can be deduced from the case of D. Unfortunately, however, the classical Hardy spaces (h^p or H^p) are not preserved by the conformal map when $p \neq \infty$, so, although it is quite similar, the theory of h^p or H^p over U requires specific arguments, that we review without proof in this section.

Given a harmonic function $u\colon\ U \to B$ we will denote for all $y > 0$

$$\forall x \in \mathbb{R} \qquad u_y(x) = u(x + iy).$$

We now have $\mathbb{R} = \partial U$. Of course this comes equipped with the Lebesgue measure allowing us to make use of convolution with respect to the group structure of $\mathbb{R}$.

The Poisson kernel for U is defined as follows:

$$\forall y > 0 \qquad P_y(x) = \frac{1}{\pi}\frac{y}{x^2 + y^2}.$$

Note that by Hölder (with $1/p + 1/p' = 1$)

$$\|P_y\|_1 = 1, \quad \|P_y\|_\infty = (\pi y)^{-1}, \quad \|P_y\|_p \leq (\pi y)^{-1/p'} \ (1 < p < \infty). \quad (4.34)$$

In analogy with the circle, we will sometimes write if $z = x + iy$

$$P_z(t) = P_y(x - t)$$

and

$$P^z(dt) = P_y(x - t)dt.$$

According to our earlier notation, this should be denoted by P_U^z. We use here the same notation as for the Poisson kernel of D but there should be no risk of confusion.

Although the two cases are quite similar, the lack of compactness of $\overline{U}$ sometimes requires an additional technical argument to complete the half-plane version of a result first established for the disc.

If $f \in L_p(\mathbb{R}; B)$ we define its Poisson integral for all $z = x + iy$ in U by setting

$$u(z) = \int P_y(x - t)f(t)dt = P_y * f(x) = \int_{\partial U} f(t)P^z(dt).$$

Since $P_y \in L_{p'}(\mathbb{R}, dt)$ this integral converges absolutely. Let us denote by $\|\ \|_p$ the norm in $L_p(B)$.

Since $\|P_y\|_1 \leq 1$, we have for all $y > 0$

$$\|u_y\|_p = \left(\int \|u(x+iy)\|^p dx\right)^{1/p} \leq \|f\|_p.$$

Similarly, if $\mu \in M(\mathbb{R}; B)$ is a B-valued bounded vector measure on $\mathbb{R}$ we define its Poisson integral by

$$u(z) = \int P_y(x-t)\mu(dt) = P_y * \mu(x).$$

Since $P_y(x-t) = \frac{1}{\pi}\mathrm{Im}\left(\frac{1}{t-z}\right) = \frac{1}{\pi}\mathrm{Re}\left(\frac{i}{z-t}\right)$, when $B = \mathbb{C}$, the function u is the real part of an analytic function hence is harmonic in U. By scalarization, the harmonicity remains valid in the B-valued case for any B.

Actually, this Poisson integral u is well defined and harmonic in U for any (not necessarily bounded) vector measure μ such that

$$\int \frac{|\mu|(dt)}{1+t^2} < \infty.$$

Indeed, since $P_y(x-t) \leq \frac{C(x,y)}{1+t^2}$, the latter condition ensures the absolute convergence of the Poisson integrals.

Clearly we have

$$\|u_y\|_1 = \int \|u(x+iy)\|dx \leq \|\mu\|_{M(\mathbb{R};B)}.$$

Note

$$\forall y_1, y_2 > 0 \qquad P_{y_1} * P_{y_2} = P_{(y_1+y_2)}.$$

We start by the analogue of Theorem 3.1 for U.

Theorem 4.59. *Let $1 \leq p < \infty$. Consider $f \in L_p(\mathbb{R}; B)$ and let u be the Poisson integral of f. For all $y > 0$, let*

$$u_y(x) = u(x+iy).$$

Then $u_y \to f$ in $L_p(\mathbb{R}; B)$ when $y \to 0$.

Let B be a real Banach space. We denote by $h^p(U; B)$ the space of all harmonic functions $u\colon U \to B$ such that $\sup_{y>0} \int \|u(x+iy)\|^p dx < \infty$. We denote

$$\|u\|_{h^p(U;B)} = \sup_{y>0} \left(\int \|u(x+iy)\|^p dx\right)^{1/p},$$

with the usual convention when $p = \infty$. Clearly, $h^p(U; B)$ becomes a Banach space when equipped with this norm.

When u is the Poisson integral (with respect to U) of a function $f \in L_p(\mathbb{R}; B)$ as in Theorem 4.59, we have $\|u\|_{h^p(U;B)} = \|f\|_{L_p(\mathbb{R};B)}$.

Consider $u \in h^p(U; B)$, $1 \le p \le \infty$. We have for all $z = x + iy \in U$ and all $0 < s < y$

$$u(z) = \int_{\partial U} u(t+is)P^{z-is}(dt) = \int P_{y-s}(x-t)u_s(t)dt$$

and hence by (4.34)

$$\|u(z)\| \le \left(\frac{1}{\pi y}\right)^{1/p} \|u\|_{h^p(U;B)}.$$

In particular, for every $y > 0$, u is bounded and continuous on the subset $iy + \bar{U} = \{z \in \mathbb{C} \mid \Im(z) \ge y\}$, and, if $p < \infty$, $u(x+iy) \to 0$ when $y \to \infty$, uniformly over $x \in \mathbb{R}$.

We define the Fourier transform of a bounded vector measure μ on $\mathbb{R}$ by

$$\widehat{\mu}(s) = \int e^{-ist}\mu(dt)$$

and for a function f in $L_1(\mathbb{R}; B)$ we define

$$\widehat{f}(s) = \int e^{-ist}f(t)dt.$$

This definition has the well-known drawback that in the scalar case Parseval's identity becomes

$$\frac{1}{2\pi}\int |\widehat{f}(s)|^2 ds = \int |f(t)|^2 dt.$$

The Fourier transform of P_y, in analogy with the case of the disc, is given by the following formula

$$\widehat{P_y}(s) = \int e^{-ist}P_y(t)dt = e^{-y|s|}.$$

To check this note that if $s \le 0$ the function $z \to e^{-isz}$ is harmonic (actually entire) and bounded in U, and hence it is the Poisson integral of its boundary value. This yields if $s \le 0$

$$\forall z = x+iy \in U \qquad e^{-is(x+iy)} = \int e^{-ist}P_y(x-t)dt$$

hence $e^{sy} = e^{-|s|y} = \widehat{P_y}(s)$ if $s \le 0$, but $\widehat{P_y}(s) = \widehat{P_y}(-s)$ by symmetry, hence we obtain

$$\widehat{P_y}(s) = e^{-y|s|} \quad \text{for all} \quad s \text{ in } \mathbb{R}.$$

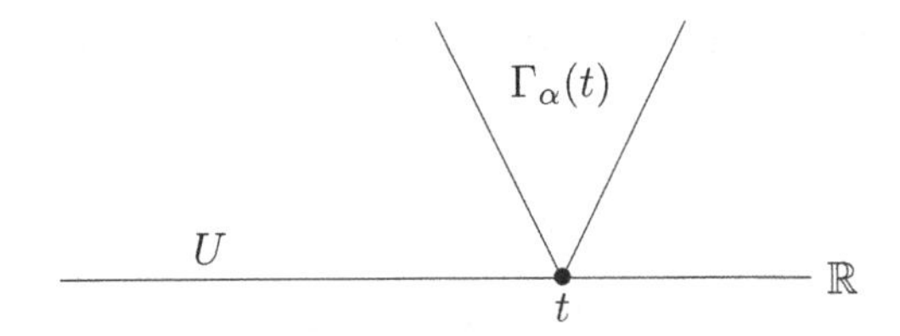

Figure 4.1. Upper half-plane.

We now turn to the non-tangential behaviour. Fix a number $\alpha > 0$. For any point $t \in \mathbb{R}$, let

$$\Gamma_\alpha(t) = \{z = x + iy\colon\ |x - t| < \alpha y\}.$$

This is a cone with vertex t and opening angle $\beta =$ Arctan α. This is analogous to (but simpler than) the Stolz domain introduced earlier for the disc.

Let u be a real valued harmonic function in U that is the Poisson integral of some function $f \in L_1\left(\frac{dt}{1+t^2}\right)$. The non-tangential maximal function of f at $t \in \mathbb{R}$ is defined, as for the unit disc, by

$$M_\alpha f(t) = \sup_{z \in \Gamma_\alpha(t)} \|u(z)\|.$$

Since the value of $\alpha > 0$ often does not really matter, we will take $\alpha = 1$ and we will denote by f^* the function $M_1 f$. We will also consider the radial maximal function, which corresponds to the case when $\alpha = 0$ and the cone $\Gamma_\alpha(t)$ degenerates to a vertical half-line. We denote the radial maximal function as

$$M_{\mathrm{rad}} f(t) = \sup_{y>0} \|u(t + iy)\|.$$

Clearly,

$$M_{\mathrm{rad}} f(t) \le M_\alpha f(t).$$

Let $1 < p \le +\infty$ and $\alpha > 0$. Then there is $C_{p,\alpha} > 0$ such that for any $u \in h_p(U; B)$ we have

$$\|M_\alpha f\|_{L_p} \le C_{p,\alpha} \|u\|_{h_p(U;B)}. \tag{4.35}$$

As before, this reduces to the scalar case by scalarization, and in any case, it follows from the Hardy-Littlewood maximal inequality, since the analogue of Theorem 3.10 remains valid for U.

We denote by $\tilde{h}^p(U; B) \subset h^p(U; B)$ $(1 \le p < \infty)$ the subspace formed of those u such that u_y converges in $L_p(B)$ to a function $f \in L_p(B)$ of which u is the Poisson integral. The Poisson integral gives us an isometric isomorphism

$$\tilde{h}^p(U;B) \simeq L_p(B),$$

and when $p = 1$, by the same proof as for Theorem 3.2, we have

$$h^1(U;B) \simeq M(\mathbb{R};B).$$

Here we may view $L_1(B) \subset M(\mathbb{R};B)$ in the usual way $f \mapsto fdt$.

The following results can be proved by the same pattern of proof as for the unit disc. We skip the details.

For any $f \in L_p(B)$ $(1 \le p < \infty)$ with Poisson integral u, we have for a.e. $t \in \mathbb{R}$

$$f(t) = \lim_{z \to t,\ z \in \Gamma_\alpha(t)} u(z).$$

Conversely, let $u \in h^p(U;B)$ $(1 < p < \infty)$. Then, by dominated convergence, $u \in \tilde{h}^p(U;B)$ iff it admits non-tangential limits a.e. and iff it admits radial limits a.e.

The space B has the RNP iff for some (or equivalently for all) $1 < p < \infty$ we have $h^p(U;B) = \tilde{h}^p(U;B)$. For the case $p = 1$, B has the RNP iff any $u \in h^1(U;B)$ admits non-tangential (or radial) limits a.e.

Let B be a complex Banach space. We denote by $H^p(U;B)$ the subspace of $h^p(U;B)$ formed of all analytic functions on U, equipped with the induced norm. Moreover we denote

$$\tilde{H}^p(U;B) = \tilde{h}^p(U;B) \cap H^p(U;B).$$

The space $\tilde{H}^1(U;B)$ can be identified (via the Poisson integral) with the subspace of $L_1(\mathbb{R};B)$ formed of all f such that $\widehat{f}(t) = 0$ for all $t < 0$. A similar statement holds for $1 < p < \infty$ but the definition of the Fourier transform becomes more technical, so we skip this.

For $u \in H^p(U;B)$, the maximal inequality (4.35) now remains valid for $0 < p \le 1$. Therefore, assuming $u \in H^p(U;B)$, then $u \in \tilde{H}^p(U;B)$ iff it admits non-tangential (or radial) limits at a.e. point on $\mathbb{R}$. Moreover, B has the ARNP iff for some (or equivalently all) $0 < p < \infty$ we have $H^p(U;B) = \tilde{H}^p(U;B)$.

It is natural to compare this again with convergence along Brownian paths. Let $y > 0$. We consider complex valued Brownian motion (W_t) starting from some point $z \in U$ (as defined in (3.56)) and we define the stopping times $T_y(\omega) = \inf\{t > 0 \mid \Im W_t(\omega) = y\}$, and $T(\omega) = \inf\{t > 0 \mid \Im W_t(\omega) = 0\}$. Then T (resp. T_y) is the exit time from U (resp. from $iy + U \subset U$). Then, in the preceding situation, $u \in \tilde{H}^p(U;B)$ iff $u(W_t(\omega))$ converges in B when $t \uparrow T(\omega)$ for a.a. path ω starting at z.

4.7 Further complements*

We cannot resist the temptation to at least describe several beautiful complements about Brownian motion and harmonic functions.

One of them, due to Fefferman-Stein is that, for the non-tangential maximal function of a harmonic function $u : D \to B$ to be in L_p $(0 < p \leq \infty)$, it actually suffices that its radial maximal function be in L_p. Thus, for any $\alpha > 0$, there is a constant $c_{p,\alpha}$ such that for any such u and any Banach space B we have

$$\|M_\alpha u\|_p \leq c_{p,\alpha}\|M_{rad}u\|_p. \tag{4.36}$$

Note that this fails in dimensions higher than 2.

The idea behind the proof is a curious fact. Although it is well known that $z \mapsto \|u(z)\|^p$ is not necessarily subharmonic when $0 < p < 1$, the following substitute holds: There is a constant C_p such that for any disc D_z with centre z, such that u is harmonic in a neighborhood of $\bar{D}_z$, we have

$$\|u(z)\|^p \leq C_p\mu(D_z)^{-1}\int_{D_z}\|u(\xi)\|^p\mu(d\xi)$$

where μ denotes the Lebesgue measure on $\mathbb{C} = \mathbb{R}^2$. For (4.36) as well as for this fact, the scalar case implies immediately the Banach valued case, simply using $\|x\| = \sup\{|x^*(x)| \mid x^* \in B^*,\ \|x^*\| \leq 1\}$. For a proof in the scalar case, we recommend [49, p. 253].

Another beautiful result is the Burkholder-Gundy-Silverstein distributional inequalities between the non-tangential maximal function and its Brownian analogue [176]. Incidentally, in what remains a historical breakthrough, they gave the first proof that a harmonic function with integrable non-tangential maximal function must be the real part of a function in $H^1(D)$. This later turned out to be a crucial stepping stone in the Fefferman-Stein program [222]. Thus for a few months around 1970 probabilists were ahead of analysts, but the latter caught up and found an analytic proof of the same fact. Nevertheless, this certainly changed for good the status of probability methods in the area. We will now outline the main ideas from [176].

One important ingredient is the Burkholder-Davis-Gundy (BDG in short) inequalities in the continuous case for Brownian martingales. Let $0 < p < \infty$. For any real valued harmonic u we set $M_t = u(W_{t\wedge T})$ and $M^*u = \sup_{t<T}|M_t|$. The BDG-inequalities say that the norm (or quasi-norm for $p < 1$) $\|M^*u\|_p$ is equivalent to $\|S(u)\|_p$ where $S(u)$ is the 'square function' defined as the integral

$$S(u) = \left(\int_0^T |\nabla u(W_t)|^2 dt\right)^{1/2}.$$

When $p \geq 1$, this equivalence is but the continuous analogue of inequalities (5.8) and (5.11). Here we restrict to the real valued case. For the B-valued case, the UMD property of the Banach space B is required and $S(u)$ must be changed (see §5.1 and §6.4). The main discovery in [176] is the equivalence of $\|M^*u\|_p$ and the L_p-norm of the non-tangential maximal function namely $\|M_\alpha u\|_p$ (see Corollary 4.61), together with the observation that

$$S(u) = S(\tilde{u}),$$

the latter simply because the Cauchy-Riemann equations

$$\partial u/\partial x = \partial \tilde{u}/\partial y \quad \partial u/\partial y = -\partial \tilde{u}/\partial x \tag{4.37}$$

imply that

$$|\nabla u| = |\nabla \tilde{u}|.$$

Indeed, combined together, these facts yield

$$\|M_\alpha \tilde{u}\|_p \simeq \|S(\tilde{u})\|_p = \|S(u)\|_p \simeq \|M_\alpha u\|_p. \tag{4.38}$$

In particular, if $\|M_\alpha u\|_p < \infty$, then $f = u + i\tilde{u} \in H^p(D)$, or equivalently

$$\|M_\alpha u\|_p < \infty \Rightarrow u \in \Re(H^p(D)),$$

and of course the converse implication also holds by (4.5) since

$$\max\{M_\alpha u, M_\alpha \tilde{u}\} \leq M_\alpha f \leq M_\alpha u + M_\alpha \tilde{u}.$$

The other major result from [176] is a comparison between the distributions of the non-tangential maximal function $M_\alpha u$ of any harmonic function $u : D \to B$ (or $u : U \to B$) and its Brownian maximal function $\sup_t \|u(W_{t\wedge T})\|$.

The proof they give for the scalar case remains valid identically for the Banach space valued case.

Theorem 4.60. *For any $0 < \alpha < \infty$ there is a positive constant c_α such that for any Banach space B and any harmonic function $u : D \to B$ we have*

$$\forall \lambda > 0 \quad c_\alpha^{-1} m\{M_\alpha u > \lambda\} \leq \mathbb{P}_0\{\sup_t \|u(W_{t\wedge T})\| > \lambda\} \leq c_\alpha m\{M_\alpha u > \lambda\}.$$

In the case of $u : U \to B$, a similar equivalence holds with

$$\mathbb{P}_0\{\sup_t \|u(W_{t\wedge T})\| > \lambda\}$$

replaced by the corresponding expression for Brownian motion 'falling from heaven'

$$\lim_{y\to\infty} \uparrow \int_{-\infty}^{\infty} \mathbb{P}_{x+iy}\{\sup_t \|u(W_{t\wedge T})\| > \lambda\} dx.$$

Using (1.3), we find:

Corollary 4.61. *For any $0 < p < \infty$, there is a constant $c'_p > 0$ such that for any B and any harmonic $u: D \to B$ we have*

$$c'^{-1}_p \|M_\alpha u\|_p \leq \| \sup_t \|u(W_{t\wedge T})\| \|_p \leq c'_p \|M_\alpha u\|_p.$$

See Bañuelos and Moore's [3] book for laws of the iterated logarithm and related refinements about the boundary behaviour of scalar valued harmonic functions.

In a similar direction, the beautiful work of Jean Brossard [148] (see also [26] for a detailed account) comparing non-tangential convergence at a point $\theta \in \partial D$ with convergence along Brownian paths exiting the disc at θ should be mentioned. Several results of this kind remain valid for a Banach space valued harmonic function on D or U. For instance, the proof given in [26, pp. 113–116] that the set of points $\theta \in \partial U$ such that $u(z) \to u(\theta)$ non-tangentially contains the set of θs such that $u(W_{t\wedge T}) \to u(\theta)$ for a.a. Brownian path conditioned to exit the disc at θ remains valid. In the converse direction, for any fixed $0 < \alpha < \infty$, the proof given in [148, Prop. 1] or [26, pp. 109–110] that the set of θs such that $\{u(W_{t\wedge T}) \mid t < T\}$ is bounded for a.a. Brownian path conditioned to exit the disc in θ contains (up to a negligible set) the set of θ's where $\sup_{z\in\Gamma_\alpha(\theta)} \|u(z\| < \infty$ also remains valid. Note however that the proofs based on the fact that $u(W_{t\wedge T})$ has the same paths as Brownian motion with a new 'clock' (see [26, p. 75]) do *not* extend to the Banach space valued case.

4.8 Notes and remarks

We only included the basic notions on subharmonicity, mainly for the function $z \to \mathrm{Log}\|f(z)\|$ when f is analytic and its consequences for outer functions. We refer the reader to Garnett's [33] (which considerably influenced our general presentation of this chapter) for a deeper view. Here again many classical results about H^p-spaces of analytic functions can be extended without effort to the Banach space valued case, so we can only give classical references such as Zygmund's all time classic [99] or the more recent monographs [25, 28, 39, 40, 48].

The ARNP was introduced by Bukhvalov and Danilevich in [163], where Theorems 4.30 and 4.32 appear, as well as the fact that a Banach lattice not containing c_0 has the ARNP. The main source for §4.5 is Edgar's paper [217]. See Bu and Schachermayer's [162] and Garling's [233] for more on this theme.

For the refinements in §4.7 we already gave detailed references, mainly to Durrett's book [26] and Jean Brossard's work [148].

Various complements on the ARNP appear in papers by Ghoussoub, Maurey and Lindenstrauss [243, 249]. In particular, the complex analogue of dentability and of the KMP are considered there.

Shangquan Bu in [153, 155] studied the ARNP for subsets $C \subset B$ of a complex Banach space B. A non-empty subset $C \subset B$ is said to have the ARNP if for each function $f : C \to \mathbb{R}$ bounded above, and each $\varepsilon > 0$, there is a Lipschitz plurisubharmonic function φ on B of Lipschitz constant 1 such that $\{x \in C \mid f(x) + \varphi(x) > 0\}$ is non-empty and of diameter less than ε. This is motivated by the following result from [243]: The space B has the ARNP iff every non-empty bounded subset of B has the ARNP in the preceding sense.

The fact that non-commutative L_1-spaces have the ARNP is proved in [261], together with some applications to the complex interpolation method (see §8.2 for background on the latter).

More results appear in [151, 152, 154, 156–159]. In particular, Bu investigated the possible analogue for the ARNP of the fact that non-RNP Banach spaces contain bounded δ-separated trees (see Theorem 2.9). Consider a Banach space B. In [159] Bu introduces the subset $G_B \subset H^\infty(D, B)$ formed of all functions $f \in H^\infty(D, B)$ such that $\exists \delta > 0$ so that for almost all $z \in \partial D$ we have

$$\limsup_{r,s\uparrow 1} \|f(rz) - f(sz)\| \geq \delta.$$

He then proves that if (and only if) B fails the ARNP then G_B is dense in $H^\infty(D, B)$, and a fortiori is non-void. This is the analogue for radial convergence of a bounded δ-separated tree. In [156] he studies an analytic analogue of the Krein-Milman property and compares it to the ARNP. In [158] he shows that B (assumed separable) has the ARNP iff for some $1 \leq p < \infty$ the space consisting of all L_p-bounded B-valued analytic martingales is separable. This is inspired by [199], where Daher proves that B (assumed separable) has the RNP (resp. ARNP) iff for some (and hence for all) $1 < p < \infty$ (resp. $1 \leq p < \infty$) the space $h^p(D; B)$ (resp. $H^p(D; B)$) is separable. In [150, 215] it is proved that the ARNP of B passes to the projective tensor product of B with L_p ($1 \leq p < \infty$). See [214] for results connecting the ARNP with representability of linear operators from L_1 to B.

In [104], Aurich introduced the following generalization of the ARNP: A Banach space B is said to have property ($*$) if for any f in $H^\infty(D; B)$ there is at least one boundary point e^{it} for which $\lim_{r\uparrow 1} f(re^{it})$ exists. For Banach lattices this is equivalent to the ARNP. See [104, 105] for more details on this property. In [160] it is proved that property ($*$) and the ARNP are equivalent for Banach spaces isomorphic to their squares. Apparently the general case remains open.

Incidentally, the radial convergence in the complex valued case can be sometimes considerably strengthened : for instance, in answer to a question of Rudin, Bourgain proves in [140] that for any function F in $H^\infty(D)$ the set of all $z \in \partial D$ such that

$$\int_0^1 |F'(rz)|dr < \infty$$

(i.e. such that $r \mapsto F(rz)$ is of bounded variation) is of Hausdorff dimension 1. In particular, quite surprisingly, this set is always non-empty although in some case it is Lebesgue-negligible.

5

The UMD property for Banach spaces

The abbreviation UMD stands for unconditionality of martingale differences. A series $\sum x_n$ in a Banach space B is said to converge unconditionally if the sequence $\sum_1^n \xi_k x_k$ converges for any choice of signs $\xi_k = \pm 1$.

A sequence (x_n) is said to be an unconditional basis for B if any $x \in B$ admits a unique representation as an unconditionally convergent series $x = \sum \alpha_k x_k$ with scalar coefficients (α_k). We say that it is an unconditional basic sequence if it is an unconditional basis for its closed linear span. In that case the unconditionality constant of (x_n) is defined as the smallest constant C such that for any scalar sequence (α_k) and any integer $N \geq 1$ we have

$$\sup_{|\xi_k|=1} \|\sum\nolimits_1^N \xi_k \alpha_k x_k\| \leq C \|\sum\nolimits_1^N \alpha_k x_k\|.$$

This chapter is devoted to the Banach spaces B that possess the unconditionality of martingale differences (UMD in short) property, i.e. those B such that for some (or all) $1 < p < \infty$ the difference sequences (df_n) of martingales in $L_p(B)$ form an unconditional basic sequence.

5.1 Martingale transforms (scalar case): Burkholder's inequalities

We start by a quick review of martingale inequalities in the scalar case. These will be derived as corollaries of the Banach space valued case, to be proved later on in this chapter.

Let $(M_n)_{n\geq 0}$ be a scalar valued martingale on a filtration $(\mathcal{A}_n)_{n\geq 0}$. We will always set $dM_0 = M_0$ (or we make the convention that $M_{-1} \equiv 0$) and

$$\forall n \geq 1 \qquad dM_n = M_n - M_{n-1}.$$

When there is no ambiguity, we will often denote by $\mathbb{E}_n$ the conditional expectation relative to $\mathcal{A}_n$. Moreover we will sometimes say n-measurable instead of $\mathcal{A}_n$-measurable.

Let $(\varphi_n)_{n\ge 0}$ be a sequence of random variables that we merely assume to be *adapted* to $(\mathcal{A}_n)_{n\ge 0}$, i.e. we assume that φ_n is $\mathcal{A}_n$-measurable for each $n \ge 0$. Let $\widetilde{M}_0 = \varphi_0 M_0$ and

$$\forall n \ge 1 \qquad \widetilde{M}_n = \varphi_0 M_0 + \sum\nolimits_1^n \varphi_{n-1}\, dM_n.$$

Then $\widetilde{M}$ is a martingale and the correspondence $M \to \widetilde{M}$ is called a martingale transform.

Recall that an adapted sequence (ψ_n) is called *predictable* if ψ_n is $\mathcal{A}_{n-1}$-measurable for each $n \ge 1$. For $(\widetilde{M}_n)$ to be a martingale, it is crucial that (ψ_n) defined by $\psi_n = \varphi_{n-1}$ (and say $\psi_0 = 0$) be predictable.

The key property about these transforms is that, if $\sup_n \|\varphi_n\|_\infty < \infty$, then $M \to \widetilde{M}$ is bounded on L_p for all $1 < p < \infty$ and is of weak type (1-1) (as defined in §5.16). This is due to Burkholder as well as the corresponding inequalities: for each $1 \le p < \infty$ there is a constant β_p such that

$$\sup \|\widetilde{M}_n\|_p \le \beta_p \sup \|\varphi_n\|_\infty \sup \|M_n\|_p \quad \text{if } 1 < p < \infty, \tag{5.1}$$

and if $p = 1$:

$$\sup_{\lambda>0} \lambda \mathbb{P}(\sup |\widetilde{M}_n| > \lambda) \le \beta_1 \sup \|\varphi_n\|_\infty \sup \|M_n\|_1. \tag{5.2}$$

For the proof, see Theorem 5.21 and Corollary 5.16.

By Doob's maximal inequality, (5.1) implies that, if $\sup \|\varphi_n\|_\infty \le 1$, we have

$$(\mathbb{E} \sup |\widetilde{M}_n|^p)^{1/p} \le \beta'_p (\mathbb{E} \sup |M_n|^p)^{1/p}, \tag{5.3}$$

where $\beta'_p = p'\beta_p$. In this form, (5.3) remains valid when $p = 1$. Namely, there is a constant β'_1 such that

$$\mathbb{E} \sup |\widetilde{M}_n| \le \beta'_1 \mathbb{E} \sup |M_n|. \tag{5.4}$$

This and (5.11) are known as B. Davis's inequality. See Corollary 5.37.

This is already of interest when each of the variables φ_n is constant and in that special case (5.1) expresses the fact that the sequence $(dM_n)_{n\ge 0}$ is an *unconditional* basic sequence in L_p, i.e. the convergence in L_p of the series $\sum dM_n$ is automatically unconditional. Let $\varepsilon = (\varepsilon_n)_n$ be a fixed choice of signs, i.e. $\varepsilon_n = \pm 1$. Then (5.1) implies for any (M_n) converging in L_p

$$\left\| \sum \varepsilon_n\, dM_n \right\|_p \le \beta_p \left\| \sum dM_n \right\|_p. \tag{5.5}$$

Replacing dM_n by $\varepsilon_n\, dM_n$ in (5.5) we find the reverse inequality

$$\left\|\sum dM_n\right\|_p \le \beta_p \left\|\sum \varepsilon_n\, dM_n\right\|_p. \tag{5.6}$$

Let ν denote as before the uniform probability on $\{-1,1\}^{\mathbb{N}}$. Recall the classical Khintchin inequalities: For any $0 < p < \infty$ there are constants $A_p > 0$ and $B_p > 0$ such that for any sequence $x = (x_n)$ in ℓ_2 we have

$$A_p\left(\sum |x_n|^2\right)^{1/2} \le \left(\int \left|\sum x_n\varepsilon_n\right|^p d\nu(\varepsilon)\right)^{1/p} \le B_p\left(\sum |x_n|^2\right)^{1/2}. \tag{5.7}$$

Note that $A_p = 1$ when $p \ge 2$ and $B_p = 1$ when $p \le 2$ are the obvious cases since the $L_p(\nu)$-norm is monotone increasing in p and $\|\sum x_n\varepsilon_n\|_{L_2(\nu)} = \left(\sum |x_n|^2\right)^{1/2}$.

Then if we integrate (5.5) and (5.6) (after raising to the p-th power) we find Burkholder's inequalities

$$A_p\beta_p^{-1}\|S\|_p \le \left\|\sum dM_n\right\|_p \le \beta_p B_p\|S\|_p \tag{5.8}$$

where S is the so-called square function defined by

$$S = \left(|M_0|^2 + \sum\nolimits_1^\infty |dM_n|^2\right)^{1/2}. \tag{5.9}$$

A similar argument can be applied to (5.2) (using (5.116)) and it yields a constant β_1' such that

$$\sup\nolimits_{\lambda>0} \lambda\mathbb{P}(S > \lambda) \le \beta_1' \sup \|M_n\|_1. \tag{5.10}$$

Moreover, using the Khintchin inequality (5.7) for $p = 1$, and also (1.43), we find

$$A_1(\beta_1')^{-1}\mathbb{E}S \le \mathbb{E}\sup|M_n| \le 2\beta_1'\mathbb{E}S. \tag{5.11}$$

Remark 5.1.* The Khintchin inequalities have well-known analogues for lacunary Fourier series of the form $\sum_{n\ge0} x_n e^{itk(n)}$ where $\inf_n k(n+1)/k(n) > 1$. When $k(n) = 2^n$, the lacunary case can be derived from a martingale inequality: Indeed, in the situation of Remark 1.21 if we apply (5.8) to the (reverse) martingale (f_{-k}) defined in (1.23) with scalar coefficients (x_k) in ℓ_2 (we first restrict to $-N \le -k \le 0$ and let $N \to \infty$) we find

$$A_p\beta_p^{-1}\left(\sum |x_n|^2\right)^{1/2} \le \left(\int_{\mathbb{T}} \left|\sum x_n z^{q^n}\right|^p dm(z)\right)^{1/p} \le B_p\beta_p\left(\sum |x_n|^2\right)^{1/2}.$$

More sophisticated inequalities, e.g. for functions of the form $\sum z^{q^n} z^{\ell^k} x_{n,k}$ where q and ℓ are distinct primes, appear in [258].

5.2 Square functions for B-valued martingales: Kahane's inequalities

In the Banach space valued case, the square function $\sup_n \left(\sum_0^n |d_k|^2\right)^{1/2}$ must be replaced by

$$\sup_n \left(\int_\Delta \left\| \sum\nolimits_0^n \varepsilon_k d_k \right\|_B^2 dv \right)^{1/2} \tag{5.12}$$

where

$$\Delta = \{-1, 1\}^{\mathbb{N}}, \qquad \nu = \otimes(\delta_1 + \delta_{-1})/2 \tag{5.13}$$

and $\varepsilon_n : \Delta \to \{-1, 1\}$ denotes the index n coordinate.

When B is a Hilbert space, for all x_k in B we have

$$\int \left\| \sum\nolimits_0^n \varepsilon_k x_k \right\|^2 dv = \sum\nolimits_0^n \|x_k\|^2$$

and hence we recover the square function, but in general this is not possible and we must work with (5.12). We will show in the next section that for the Banach spaces with the UMD property, the Burkholder inequality remains valid when the square function is replaced by (5.12).

This motivates a preliminary study of averages such as (5.12) in a general Banach space when d_k are constant.

Theorem 5.2 (Kahane). *For any $0 < p < q < \infty$ there is a constant $K(p, q)$ such that for any Banach space B and any finite subset $x_1, \ldots, x_n$ in B we have*

$$\left\| \sum \varepsilon_k x_k \right\|_{L_q(B)} \le K(p, q) \left\| \sum \varepsilon_k x_k \right\|_{L_p(B)}. \tag{5.14}$$

In particular $\left\| \sum \varepsilon_k x_k \right\|_{L_2(B)}$ is equivalent to $\left\| \sum \varepsilon_k x_k \right\|_{L_p(B)}$ for any $0 < p < \infty$.

Note that Khintchin's (5.7) is a consequence of Kahane's (5.14) in the case $B = \mathbb{C}$: Indeed, we find $B_p \le K(2, p)$ for $p \ge 2$ and $A_p \le K(p, 2)$ for $p \le 2$.

Remark 5.3. Consider the Banach space $\tilde{B}$ formed of all sequences $x = (x_n)$ of elements of B such that $\sup_n \| \sum_0^n x_k \| < \infty$, equipped with the norm

$$\|x\|_{\tilde{B}} = \sup_n \left\| \sum\nolimits_{k=0}^n x_k \right\|.$$

Then applying Kahane's inequality to the Banach space $\tilde{B}$ we immediately get that, if we denote $S^* = \sup_n \| \sum_{k=0}^n \varepsilon_k x_k \|$, then for any sequence (x_n) in B we have

$$\|S^*\|_q \le K(p, q) \|S^*\|_p.$$

We will base the proof of Kahane's Theorem on the classical hypercontractive inequality on 2-point space made famous by Nelson and Beckner [115] (but first proved in [128]), as follows:

Theorem 5.4. *Let $1 < p < q < \infty$. Let $\xi = ((p-1)/(q-1))^{1/2}$. Let B be an arbitrary Banach space. Then*

$$\forall x, y \in B \quad \left(\frac{\|x+\xi y\|^q + \|x-\xi y\|^q}{2}\right)^{1/q} \leq \left(\frac{\|x+y\|^p + \|x-y\|^p}{2}\right)^{1/p}.$$

Proof. Let $\Omega = \{-1, 1\}, \mathbb{P} = (\delta_1 + \delta_{-1})/2$. Let $\varepsilon_1\colon \ \Omega \to \{-1, 1\}$ be the identity map. The proof actually reduces to the case $B = \mathbb{R}$. Indeed, let $T\colon L_p(\Omega, \mathbb{P}) \to L_q(\Omega, \mathbb{P})$ be the operator defined by $T1 = 1, T\varepsilon_1 = \xi\varepsilon_1$. Then $T \geq 0$ ($T =$ convolution by $1 + \xi\varepsilon_1$ and $1 + \xi\varepsilon_1 \geq 0$!). Thus the passage from $\mathbb{R}$ to a general B follows from Proposition 1.6. For the case $B = \mathbb{R}$, we refer to the appendix of §5.12. □

For any finite subset $S \subset \mathbb{N}$, let w_S be the so-called Walsh function defined on $\Delta = \{-1, 1\}^{\mathbb{N}}$ by

$$w_S = \prod\nolimits_{n\in S} \varepsilon_n.$$

Corollary 5.5. *Let $1 < p < q < \infty$ and let $\xi = ((p-1)/(q-1))^{1/2}$ as before with B arbitrary. Then for any family $\{x_S \mid S \subset \{1, \ldots, n\}\}$ in B we have*

$$\left\|\sum \xi^{|S|} w_S x_S\right\|_{L_q(B)} \leq \left\|\sum w_S x_S\right\|_{L_p(B)}. \tag{5.15}$$

In particular for any $x_1, \ldots, x_n$ in B

$$\left\|\sum \varepsilon_k x_k\right\|_{L_q(B)} \leq ((q-1)/(p-1))^{1/2} \left\|\sum \varepsilon_k x_k\right\|_{L_p(B)}. \tag{5.16}$$

Proof. The proof is based on the following elementary observation: let

$$T_1\colon\ L_p(\Omega_1, m_1; B) \to L_q(\Omega_1', m_1'; B)$$

and

$$T_2\colon\ L_p(\Omega_2, m_2; B) \to L_q(\Omega_2', m_2'; B)$$

be two operators with norms ≤ 1. Then, if $q \geq p$, the operator $T_1 \otimes T_2$: $L_p(m_1 \times m_2; B) \to L_q(m_1' \times m_2'; B)$ also has norm ≤ 1. (To check this one uses the classical Hölder-Minkowski inequality that says that we have a norm 1 inclusion $L_p(m; L_q(m')) \subset L_q(m'; L_p(m))$, see §5.13.)

It follows from this observation, by iteration, that $T \otimes T \otimes \cdots \otimes T$ (n times) has norm ≤ 1 from $L_p(B)$ to $L_q(B)$, and since this operator multiplies $w_S x_S$ by $\xi^{|S|}$ we obtain (5.15) and hence (5.16). □

Proof of Kahane's Theorem. The preceding corollary already covers the case $1 < p < q < \infty$ with $K(p,q) = ((q-1)/(p-1))^{1/2}$. In particular if we set $f(\cdot) = \| \sum \varepsilon_k(\cdot) x_k \|$, we have proved for $1 < p < q < \infty$

$$\|f\|_q \le K(p,q)\|f\|_p. \tag{5.17}$$

Let $0 < r < 1 < p < q$. We will extrapolate the validity of (5.17) from (p,q) to (r,q). Define $0 < \theta < 1$ by the identity $\frac{1}{p} = \frac{1-\theta}{q} + \frac{\theta}{r}$. Then by Hölder we find

$$\|f\|_q \le K(p,q)\|f\|_q^{1-\theta}\|f\|_r^{\theta}$$

hence after division by $\|f\|_q^{1-\theta}$, we obtain

$$\|f\|_q \le K(p,q)^{1/\theta}\|f\|_r,$$

which yields $K(r,q) \le K(p,q)^{1/\theta}$. □

We might as well record here an elementary 'contraction principle': Assume $1 \le p \le \infty$. Let B be an arbitrary Banach space and let $x_1, \ldots, x_n \in B$. Then for all $\alpha_k = \pm 1$

$$\left\| \sum \alpha_k \varepsilon_k x_k \right\|_{L_p(B)} = \left\| \sum \varepsilon_k x_k \right\|_{L_p(B)}, \tag{5.18}$$

and $\forall \alpha_1, \ldots, \alpha_n \in \mathbb{R}$

$$\left\| \sum \alpha_k \varepsilon_k x_k \right\|_{L_p(B)} \le \sup |\alpha_k| \left\| \sum \varepsilon_k x_k \right\|_{L_p(B)}; \tag{5.19}$$

moreover $\forall \beta_1, \ldots, \beta_n \in \mathbb{C}$

$$\left\| \sum \beta_k \varepsilon_k x_k \right\|_{L_p(B)} \le 2 \sup |\beta_k| \left\| \sum \varepsilon_k x_k \right\|_{L_p(B)}. \tag{5.20}$$

To verify this, note that by convexity the supremum of the left side of (5.19) over all (α_k) in $\mathbb{R}^n$ with $\sup |\alpha_k| \le 1$ is attained on an extreme point, i.e. an element of $\{-1, 1\}^n$, for which (5.19) becomes an equality since $(\alpha_k \varepsilon_k)$ and (ε_k) have the same distribution. This proves (5.19) and (5.18). To verify (5.20), simply write $\beta_k = \alpha'_k + i\alpha'_k$ and use the triangle inequality.

The meaning of the next lemma is, roughly, to extrapolate (5.17) from (p,q) to $(0,q)$.

Lemma 5.6. *Let $0 < p < q < \infty$. Let $\mathcal{F}$ be a subset of $L_q(\Omega, \mathcal{A}, \mathbb{P})$. Assume that there is $C > 0$ such that*

$$\forall f \in \mathcal{F} \qquad \|f\|_q \le C\|f\|_p.$$

Then there are $\delta > 0$ and $R > 0$ such that

$$\forall f \in \mathcal{F} \qquad \mathbb{P}(|f| > R\|f\|_q) \ge \delta.$$

Proof. Let r be such that $p^{-1} = q^{-1} + r^{-1}$. Replacing f by $f\|f\|_q^{-1}$ we may assume that $\|f\|_q = 1$ for all f in $\mathcal{F}$. By Hölder's inequality for any $R > 0$ we have

$$\|f 1_{\{|f|>R\}}\|_p \leq (\mathbb{P}(|f| > R))^{1/r}.$$

Hence we can write, assuming $p \geq 1$, that

$$\begin{aligned} 1 = \|f\|_q \leq C\|f\|_p &\leq C\|f 1_{\{|f|\leq R\}}\|_p + C\|f 1_{\{|f|>R\}}\|_p \\ &\leq CR + C(\mathbb{P}(|f| > R))^{1/r}. \end{aligned}$$

Thus if we choose $R = (2C)^{-1}$ and $\delta = (2C)^{-r}$ we obtain the announced result. The case $p < 1$ is similar using (1.10). □

Let f_n and f be B-valued random variables. Recall that, by definition, f_n converges to f in probability if

$$\forall \varepsilon > 0 \qquad \mathbb{P}(\|f_n - f\| > \varepsilon) \to 0 \quad \text{when} \quad n \to \infty.$$

This convergence is in general strictly weaker than a.s. convergence. However, by Corollary 1.42, it is equivalent for sums of *independent* random variables, in particular for the sums considered in Theorem 5.7.

The corresponding topology is the natural one on the topological vector space $L_0(B)$ of B-valued Bochner measurable functions. The preceding lemma is an extrapolation principle: If the L_q-topology on a linear space $\mathcal{F}$ coincides with the L_p-topology for *some* $p < q$, then it also coincides with the topology of convergence in probability (i.e. the L_0-topology).

In particular, we obtain

Theorem 5.7.

(i) *Let (α_n) be a scalar sequence. Then $\sum_0^\infty \varepsilon_n \alpha_n$ converges in probability iff $\sum |\alpha_n|^2 < \infty$, and then it converges a.s. and in L_p for all $p < \infty$.*

(ii) *Let (x_n) be a sequence in a Banach space B. Then the series $\sum_0^\infty \varepsilon_n x_n$ converges in probability iff it converges in $L_p(B)$ for some $0 < p < \infty$. Then it converges a.s. and in $L_p(B)$ for all $0 < p < \infty$.*

Proof.

(i) By the Khintchin inequalities and the preceding lemma, the L_p- and L_0-topologies coincide on the span of $\{\varepsilon_n\}$ for any $0 < p < \infty$. Then the a.s. convergence follows either from Theorem 1.40 or from the martingale convergence theorem since $f_n = \sum_0^n \varepsilon_k \alpha_k$ is a martingale and we may choose $p > 1$.

(ii) Same argument as for (i) but using the Kahane inequalities instead of the Khintchin ones.

□

Some applications.

(i) Let $0 < r < \infty$. Let (x_n) be a sequence in the Banach (or quasi-Banach) space $B = L_r(\Omega, m)$ over a measure space. Then the series $\sum \varepsilon_n x_n$ converges a.s. in B iff

$$\int \left(\sum |x_n|^2\right)^{r/2} dm < \infty$$

or equivalently iff $(\sum |x_n|^2)^{1/2} \in B$. Indeed, choosing $p = r$ in the last Theorem this is an easy consequence.

(ii) In particular, if $B = \ell_r$, with canonical basis (e_k) and if for each n, we set $x_n = \sum x_n(k) e_k$, then $\sum \varepsilon_n x_n$ converges a.s. in B iff

$$\sum_k \left(\sum_n |x_n(k)|^2\right)^{r/2} < \infty.$$

(iii) Let (a_n) be a scalar sequence indexed by $\mathbb{Z}$. Consider the *formal* Fourier series $\sum_{n\in\mathbb{Z}} a_n e^{int}$. Let B be a Banach space of functions over the circle group $\mathbb{T}$, such as for instance the space of continuous functions $C(\mathbb{T})$ or the space $L_r(\mathbb{T}, m)$ $(0 < r < \infty)$ with respect to the normalized Haar measure m. Note that in both cases the Fourier transform of an element of B determines the element. By convention, we will write $\sum_{n\in\mathbb{Z}} a_n e^{int} \in B$ if there is an element $f \in B$ with Fourier transform equal to (a_n), i.e. such that $\forall n \in \mathbb{Z}\ \hat{f}(n) = a_n$. Then, $\sum_{n\in\mathbb{Z}} \varepsilon_n a_n e^{int} \in L_r(\mathbb{T}, m)$ for almost all choice of signs (ε_n) iff $\sum |a_n|^2 < \infty$. Note that the condition we find does not depend on r, which is surprising at first glance.

(iv) With the same notation, $f = \sum_{n\in\mathbb{Z}} \varepsilon_n a_n e^{int} \in C(\mathbb{T})$ for almost all choice of signs (ε_n) iff a.s. the partial sums of the random Fourier series

$$\sum_{|n|\le N} \varepsilon_n a_n e^{int}$$

converge uniformly over $\mathbb{T}$ when $N \to \infty$; moreover, this holds iff both (unilateral) series $\sum_0^N \varepsilon_n a_n e^{int}$ and $\sum_{-1}^{-N} \varepsilon_n a_n e^{int}$ converge a.s. uniformly over $\mathbb{T}$. By the preceding theorem, we then have for any $p < \infty$

$$\mathbb{E} \sup_{t\in\mathbb{T}} \left|\sum_{|n|>N} \varepsilon_n a_n e^{int}\right|^p \to 0.$$

Let $f_+ = \sum_{n\ge 0} \varepsilon_n a_n e^{int}$ and $f_- = \sum_{n<0} \varepsilon_n a_n e^{int}$. Observe that $f = f_+ + f_-$ has the same distribution as $f = f_+ - f_-$, and hence: $f \in B$ a.s. iff both $f_+ \in B$ a.s. and $f_- \in B$ a.s. Then, the last point (iv) follows from (the Ito-Nisio) Theorem 1.43, taking for D the countable collection of all measures μ such that $|\mu|(\mathbb{T}) < 1$ with finitely supported Fourier transform taking values in any fixed dense countable subset of $\mathbb{C}$, say in $\mathbb{Q} + i\mathbb{Q}$.

5.3 Definition of UMD

The central idea to prove the Burkholder inequalities is usually described as 'extrapolation'. Schematically, the main point in the scalar case is:

- the L_2-case is obvious by the orthogonality of martingale differences,
- the L_p-case can be deduced from the L_2-one by extrapolation. The basic principles of extrapolation go back to [175].

There are several ways to implement the extrapolation technique. In the Banach space valued case, the Burkholder-Gundy extrapolation can be adapted to show that the unconditionality of B-valued martingales in $L_p(B)$ does not depend on the value of $1 < p < \infty$. In the scalar valued case, i.e. when $B = \mathbb{R}$ or $\mathbb{C}$, we recover Burkholder's inequalities since the case $p = 2$ is obvious.

Definition 5.8. Let $1 < p < \infty$. A Banach space B is called UMD$_p$ if there is a constant C such that for any finite martingale (f_n) in $L_p(B)$ we have for any choice of signs $\xi_n = \pm 1$

$$\sup_n \left\| \sum\nolimits_0^n \xi_k df_k \right\|_{L_p(B)} \le C \sup_n \|f_n\|_{L_p(B)}. \tag{5.21}$$

Then (5.21) holds for any martingale (f_n) converging in $L_p(B)$. We will denote the best C in (5.21) by $C_p(B)$. We will say that B is UMD if this holds for some $1 < p < \infty$ (we will see that it then holds for all $1 < p < \infty$).

Clearly, any Hilbert space is UMD$_2$.

By an elementary duality argument one easily checks:

Proposition 5.9. *B is UMD$_p$ iff B^* is UMD$_{p'}$ with $\frac{1}{p} + \frac{1}{p'} = 1$. Moreover, we have*

$$C_{p'}(B^*) = C_p(B). \tag{5.22}$$

Let $x = \{x_n\}$ be a sequence in a Banach space B. We define

$$R(x) = \sup_n \left\| \sum\nolimits_0^n \varepsilon_k x_k \right\|_{L_2(\Delta,\nu;B)}.$$

Let (f_n) be any martingale on $(\Omega, \mathcal{A}, \mathbb{P})$ that is bounded in $L_p(B)$. For any $\omega \in \Omega$, we define

$$R_{df}(\omega) = R(\{df_n(\omega) \mid n \ge 0\})$$

or equivalently

$$R_{df}(\omega) = \sup_n \left\| \sum\nolimits_0^n \varepsilon_k df_k(\omega) \right\|_{L_2(\Delta,\nu;B)}$$

where the L_2-norm is with respect to the variables (ε_n) defined on (Δ, ν).

The B-valued version of Burkholder's inequalities reads as follows:

Proposition 5.10. *B is UMD$_p$ iff there are positive constants C_1, C_2 such that for any martingale (f_n) converging in $L_p(B)$ we have*

$$C_1^{-1}\|R_{df}\|_p \le \sup_n \|f_n\|_{L_p(B)} \le C_2\|R_{df}\|_p. \tag{5.23}$$

Proof. Fix a choice of signs $\xi_n = \pm 1$. Let $g_n = \sum_0^n \xi_k df_k$. Note the pointwise equality (recall (5.18))

$$R_{dg} = R_{df}.$$

The latter immediately implies the 'if-part'. Conversely, assume B UMD$_p$. Then we have (5.21). But actually, applying (5.21) to g in place of f, we also find the converse inequality

$$\sup_n \|f_n\|_{L_p(B)} \le C \sup_n \left\|\sum\nolimits_0^n \xi_k df_k\right\|_{L_p(B)}.$$

Assume for simplicity that (f_n) is a finite martingale. Then if we elevate to the p-th power and average both (5.21) and its converse over all choices of signs, we obtain (5.23); note that we use Kahane's inequalities (Theorem 5.2) to replace the L_p-norm over the signs by the L_2-norm, i.e. by R_{df}. □

As before, let $(\varphi_n)_{n\ge 0}$ be a sequence of scalar valued r.v.'s, adapted to a filtration $(\mathcal{A}_n)_{n\ge 0}$. Let $(f_n)_{n\ge 0}$ be a B-valued martingale relative to $(\mathcal{A}_n)_{n\ge 0}$. Just as in the scalar case, the sequence defined by

$$\tilde{f}_n = \varphi_0 f_0 + \sum\nolimits_1^n \varphi_{k-1}(f_k - f_{k-1})$$

forms a martingale, called a 'martingale transform' of $(f_n)_{n\ge 0}$.

Proposition 5.11. *If (5.23) holds for all $(f_n)_{n\ge 0}$ adapted to $(\mathcal{A}_n)_{n\ge 0}$ and if (φ_n) are real valued, then with the preceding notation we have*

$$\sup_{n\ge 0} \|\tilde{f}_n\|_{L_p(B)} \le C_1C_2 \sup_{n\ge 0} \|\varphi_n\|_\infty \sup_{n\ge 0} \|f_n\|_{L_p(B)}.$$

If the (φ_n)s are complex valued, this holds with $2C_1C_2$ instead of C_1C_2.

Proof. By (5.19) we have for almost all ω

$$R_{d\tilde{f}}(\omega) \le \sup_{n\ge 0} \|\varphi_n\|_\infty R_{df}(\omega),$$

so the announced inequality follows from (5.23). In the complex case, we use (5.20) instead. □

Actually, the preceding result can be made more precise in the UMD case.

Proposition 5.12. *If B is UMD$_p$, then Proposition 5.11 holds with C_1C_2 replaced by the UMD$_p$ constant of B.*

Proof. Consider first the dyadic case, i.e. we take $\Omega = \{-1, 1\}^{\mathbb{N}_*}$, with $\varepsilon_k\colon \Omega \to \{-1, 1\}$ the k-th coordinate for $k = 1, 2, \ldots$ and we set $\mathcal{A}_0 = \{\phi, \Omega\}$ and $\mathcal{A}_n = \sigma(\varepsilon_1, \ldots, \varepsilon_n)$ for all $n \geq 1$. For notational convenience, we will use the Rademacher functions defined in §1.4, i.e. we take $\Omega = [0, 1]$ and $\varepsilon_n = r_n$. In that case we have $\mathcal{A}_n = \mathcal{B}_{2^n}$ where $(\mathcal{B}_k)$ is the Haar filtration (see §1.4). Consider then a martingale transform $\tilde{f}_n = \varphi_0 f_0 + \sum_1^n \varphi_{k-1}(f_k - f_{k-1})$ associated to the predictable family (φ_{k-1}) with respect to the filtration $(\mathcal{A}_n)$. Let then $F_k = \mathbb{E}(f_n|\mathcal{B}_k)$ for $k = 1, \ldots, 2^n$, in particular so that $F_{2^n} = f_n$. We have then

$$\tilde{f}_n - \tilde{f}_{n-1} = \sum\nolimits_{2^{n-1}<k\leq 2^n} \varphi_{n-1}(F_k - F_{k-1}).$$

But now φ_{n-1} is constant on the support of $F_k - F_{k-1}$ for each $2^{n-1} < k \leq 2^n$. Indeed, by (1.21), $F_k - F_{k-1}$ and h_k have the same support and the functions $r_1, \ldots, r_{n-1}$ are all constant on that support if $2^{n-1} < k \leq 2^n$. Therefore, with respect to the filtration $(\mathcal{B}_k)$, $\tilde{f}_n$ appears as a martingale transform relative to constant multipliers (and not only predictable ones). This shows that, in the dyadic case, Proposition 5.11 holds with C_1C_2 replaced by the UMD$_p$ constant of B.

The general case can be easily reduced to the case when φ_n and f_n are all simple functions. Then using the viewpoint explained in Remark 1.19, it is easy to extend the preceding proof from the dyadic case to the general one. One replaces the original filtration $(\mathcal{T}_n)$ described in Remark 1.19 by one $(\mathcal{C}_k)$ where, for each k, at most a single atom in $\mathcal{C}_k$ is partitioned into sets of $\mathcal{C}_{k+1}$, and moreover $\mathcal{T}_n = \mathcal{C}_{K(n)}$ for any $n \geq 0$ (here $K(0) \leq K(1) \leq \cdots$). Then for any sequence (φ_n) of simple functions adapted to the original filtration $(\mathcal{T}_n)$, with $\|\varphi_n\|_\infty \leq 1$ for all n, we claim there is a sequence of constants (ξ_k) with $|\xi_k| \leq 1$ for all k such that for any $f \in L_1(B)$ we have

$$\sum \varphi_{n-1}(\mathbb{E}^{\mathcal{T}_n} - \mathbb{E}^{\mathcal{T}_{n-1}})(f) = \sum \xi_k(\mathbb{E}^{\mathcal{C}_k} - \mathbb{E}^{\mathcal{C}_{k-1}})(f).$$

Indeed, let $K(n-1) < k \leq K(n)$, if $\mathcal{C}_k = \mathcal{C}_{k-1}$ we may choose ξ_k arbitrarily, say $\xi_k = 1$. If $\mathcal{C}_k$ has one more atom A that $\mathcal{C}_{k-1}$, let A' be the atom of $\mathcal{C}_{K(n-1)} = \mathcal{T}_{n-1}$ such that $A \subset A'$ (A' is the 'parent' of A). Then φ_{n-1} is equal to a constant, say ξ_k, on A' and $(\mathbb{E}^{\mathcal{C}_k} - \mathbb{E}^{\mathcal{C}_{k-1}})(f)$ is supported by A' so we have $\varphi_{n-1}(\mathbb{E}^{\mathcal{C}_k} - \mathbb{E}^{\mathcal{C}_{k-1}})(f) = \xi_k(\mathbb{E}^{\mathcal{C}_k} - \mathbb{E}^{\mathcal{C}_{k-1}})(f)$. Since $\mathbb{E}^{\mathcal{T}_n} - \mathbb{E}^{\mathcal{T}_{n-1}} = \sum_{K(n-1)<k\leq K(n)} \mathbb{E}^{\mathcal{C}_k} - \mathbb{E}^{\mathcal{C}_{k-1}}$ we obtain the announced claim.

Since we now assume (5.21) for any filtration we can pass from constant multipliers for $(\mathcal{C}_k)$ to predictable ones for $(\mathcal{T}_n)$. □

We now come to the first basic result on UMD spaces.

Theorem 5.13. *Consider a Banach space B. Then for any $1 < q < p < \infty$, B is UMD_p iff it is UMD_q and we have positive constants $\alpha(p,q)$ and $\beta(p,q)$ depending only on p and q such that*

$$\alpha(p,q)^{-1}C_p(B) \leq C_q(B) \leq \beta(p,q)C_p(B).$$

Moreover, if p remains fixed and $q \to 1$ then $\beta(p,q) = O((q-1)^{-1})$ and $\alpha(p,q)$ remains bounded, while if q remains fixed and $p \to \infty$ then $\alpha(p,q) = O(p)$ and $\beta(p,q)$ remains bounded.

We will give several proofs of this statement. Each one is of independent interest. The first one is based on Gundy's decomposition. Its advantage is to establish Theorem 5.13 without requiring any auxiliary result. The second and third ones are based on extrapolation in the Burkholder-Gundy tradition. This technique is spectacularly efficient for dyadic martingales, or for martingales for which the lengths of the increments are predictable. For short we will use the term 'calibrated' for the latter:

Definition 5.14. A Banach space valued martingale (f_n) such that for each $n \geq 1$ the increment $\|f_n - f_{n-1}\|$ is $\mathcal{A}_{n-1}$-measurable will be called calibrated.

However, this route to Theorem 5.13 requires an argument to show that calibrated martingales suffice. One can either use a control of the jumps, as in the Burgess Davis decomposition, or Maurey's Theorem that the unconditionality of *dyadic* B-valued martingales actually implies that of all of them.

Examples of calibrated martingales include the dyadic ones and the analytic ones.

5.4 Gundy's decomposition

In our first proof of Theorem 5.13, we will use Gundy's decomposition of martingales. This is a martingale analogue of the classical Calderón-Zygmund decomposition (see [86]).

Theorem 5.15. *Let B be a Banach space. Let $(f_n)_{n\geq 0}$ be a martingale adapted to $(\mathcal{A}_n)_{n\geq 0}$, and converging in $L_1(B)$ to a limit f with $\|f\|_{L_1(B)} \leq 1$. Then for any $\lambda > 0$ there is a decomposition*

$$f = a + b + c$$

with $a, b, c \in L_1(B)$ such that:

(i) $\|a\|_{L_1(B)} \leq 2$ *and* $\mathbb{P}(\sup_n \|da_n\| \neq 0) \leq 3\lambda^{-1}$,
(ii) $\left\| \sum \|db_n\| \right\|_1 \leq 4$,
(iii) $\|c\|_{L_\infty(B)} \leq 2\lambda$ *and* $\|c\|_{L_1(B)} \leq 5$.

Note that (iii) implies for any $1 < p < \infty$

$$\|c\|_{L_p(B)} \le 5^{1/p}(2\lambda)^{1-1/p}. \tag{5.24}$$

Proof. We follow Gundy's original proof closely. Recall that by convention we denote $\mathbb{E}_n$ instead of $\mathbb{E}^{\mathcal{A}_n}$, so that $f_n = \mathbb{E}_n f$ and $df_n = f_n - f_{n-1}$ $\forall n \ge 1$ and $df_0 = f_0$. Let

$$r = \inf\{n \ge 0 \mid \|f_n\| > \lambda\},$$

with the convention $\inf \phi = \infty$. Let $v_n = \|df_n\| \cdot 1_{\{r=n\}}$. Then, with the same convention, let

$$s = \inf\left\{n \ge 0 \mid \sum\nolimits_{k=0}^{n} \mathbb{E}_k(v_{k+1}) > \lambda\right\}.$$

Finally let

$$T = r \wedge s.$$

Clearly r, s and T are stopping times. Let $a = f - f_T$ so that $a_n = f_n - f_{T\wedge n}$. Since $f_T = \mathbb{E}^{\mathcal{A}_T} f$, we clearly have $\|a\|_{L_1(B)} \le 2$. Moreover, obviously $T = \infty$ implies $da_n = 0$ so

$$\{\sup \|da_n\| \neq 0\} = \bigcup\nolimits_n \{da_n \neq 0\} \subset \{T < \infty\} = \{s < \infty\} \cup \{r < \infty\}$$

and hence

$$\mathbb{P}\{\sup \|da_n\| \neq 0\} \le \mathbb{P}(r < \infty) + \mathbb{P}(s < \infty). \tag{5.25}$$

Now by Doob's inequality (see Theorem 1.25)

$$\mathbb{P}(r < \infty) = \mathbb{P}(\sup\nolimits_n \|f_n\| > \lambda) \le \lambda^{-1} \tag{5.26}$$

and also

$$\begin{aligned}\mathbb{P}(s < \infty) = \mathbb{P}\left(\sum\nolimits_{k=0}^{\infty} \mathbb{E}_k(v_{k+1}) > \lambda\right) &\le \lambda^{-1} \sum\nolimits_0^{\infty} \mathbb{E}(\mathbb{E}_k(v_{k+1})) \\ &= \lambda^{-1} \sum\nolimits_1^{\infty} \mathbb{E} v_k.\end{aligned}$$

But now

$$\mathbb{E} v_k = \mathbb{E}(\|df_k\| 1_{\{r=k\}})$$

and $r = k$ implies $\|f_{k-1}\| \le \lambda < \|f_k\|$ hence $\|df_k\| \le \|f_k\| + \|f_{k-1}\| \le 2\|f_k\|$. This implies

$$\mathbb{E} v_k \le \mathbb{E}(2\|f_k\| 1_{\{r=k\}}) = 2\mathbb{E}\|\mathbb{E}_k(f 1_{\{r=k\}})\|$$

and, by Jensen, $\|\mathbb{E}_k(f 1_{\{r=k\}})\| \le \mathbb{E}_k(\|f\| 1_{\{r=k\}})$, therefore

$$\mathbb{E} v_k \le 2\mathbb{E}(\|f\| 1_{\{r=k\}}) \quad \text{and} \quad \sum \mathbb{E} v_k \le 2. \tag{5.27}$$

This implies

$$\mathbb{P}(s < \infty) \le \lambda^{-1} \sum\nolimits_0^\infty \mathbb{E} v_k \le 2\lambda^{-1} \|f\|_{L_1(B)} \le 2\lambda^{-1}. \tag{5.28}$$

Thus combining (5.25), (5.26) and (5.28) we obtain (i).

We now turn to the decomposition $f - a = b + c$. We will define b and c through their increments db_n and dc_n. Note that $f - a = f_T$ so we must have a priori

$$b_n + c_n = f_{T \wedge n},$$

which will guarantee that

$$\|b + c\|_{L_1(B)} \le 1. \tag{5.29}$$

Also

$$f_{T\wedge n} - f_{T \wedge (n-1)} = df_n 1_{\{n \le T\}} = df_n \cdot 1_{\{n \le r\}} \cdot 1_{\{n \le s\}} = \gamma_n + \delta_n$$

where

$$\gamma_n = df_n \cdot 1_{\{n<r\}} 1_{\{n \le s\}}$$
$$\delta_n = df_n \cdot 1_{\{n=r\}} 1_{\{n \le s\}}.$$

Obviously since $(f_{T\wedge n})$ is a martingale we have

$$\mathbb{E}_{n-1}(\gamma_n + \delta_n) = 0 \qquad \forall n \ge 1$$

so we can define $db_0 = \delta_0$, $dc_0 = \gamma_0$ and for all $n \ge 1$

$$db_n = \delta_n - \mathbb{E}_{n-1}(\delta_n)$$
$$dc_n = \gamma_n + \mathbb{E}_{n-1}(\delta_n).$$

Since $\mathbb{E}_{n-1}(\delta_n) = -\mathbb{E}_{n-1}(\gamma_n)$ these are indeed martingale differences.

Note that, by Jensen,

$$\mathbb{E} \sum \|db_n\| \le 2\mathbb{E} \sum \|\delta_n\| \le 2\mathbb{E} \sum |v_n|$$

and hence by (5.27) we have (ii).

We now turn to (iii). First note that by (5.29), (ii) and the triangle inequality we have

$$\|c\|_{L_1(B)} \le 5.$$

Finally, $\sum \gamma_n = \sum_{n \le s} df_n 1_{\{n<r\}} = f_{(r-1)\wedge s}$ if $r \ge 1$ and $\sum \gamma_n = 0$ if $r = 0$, so that by definition of r

$$\left\| \sum \gamma_n \right\|_{L_\infty(B)} \le \lambda. \tag{5.30}$$

Moreover, since $\{n \le s\}$ is $(n-1)$-measurable

$$\left\|\sum\nolimits_{n\ge 1} \mathbb{E}_{n-1}(\delta_n)\right\| = \left\|\sum\nolimits_{n\le s} \mathbb{E}_{n-1}(df_n 1_{\{n=r\}})\right\|$$

and hence by Jensen this is

$$\le \sum\nolimits_{n\le s} \mathbb{E}_{n-1}(\|df_n\| \cdot 1_{\{n=r\}}) = \sum\nolimits_{k<s} \mathbb{E}_k(\|df_{k+1}\| 1_{\{r=k+1\}}),$$

which by definition of s is $\le \lambda$. Thus we conclude $\left\| \sum_{n\ge 1} \mathbb{E}_{n-1}(\delta_n)\right\|_{L_\infty(B)} \le \lambda$ and (iii) follows from (5.30) by the triangle inequality. □

Corollary 5.16. *Assume that B is UMD$_p$ for some $1 < p < \infty$. Then all martingale transforms are of weak type (1-1) in the sense of §5.16. More precisely there is a constant δ_p (depending only on p) such that for all martingales $(f_n)_{n\ge 0}$ bounded in $L_1(B)$ and for all choices of signs $\varepsilon_n = \pm 1$ the transformed martingale $\tilde{f}_n = \sum_0^n \varepsilon_k df_k$ satisfies*

$$\sup\nolimits_{\lambda>0} \lambda \mathbb{P}(\sup\nolimits_{n\ge 0} \|\tilde{f}_n\| > \lambda) \le \delta_p C_p(B) \sup_{n\ge 0} \|f_n\|_{L_1(B)}. \tag{5.31}$$

More generally, the same holds when

$$\tilde{f}_n = \sum\nolimits_0^n \varphi_{k-1} df_k$$

and $(\varphi_n)_{n\ge 0}$ is an adapted sequence of scalar valued variables such that

$$\forall n \ge 0 \qquad \|\varphi_n\|_\infty \le 1,$$

with the usual convention $\varphi_{-1} \equiv 0$.

Proof. By homogeneity, we may assume $\|f\|_{L_1(B)} \le 1$. Let $t = C_p(B)\lambda$. We have $\tilde{f}_n = \tilde{a}_n + \tilde{b}_n + \tilde{c}_n$ and hence (since $t \ge \lambda$)

$$\mathbb{P}(\sup \|\tilde{f}_n\| > 3t) \le \mathbb{P}(\sup \|\tilde{a}_n\| > \lambda) + \mathbb{P}(\sup \|\tilde{b}_n\| > \lambda) + \mathbb{P}(\sup \|\tilde{c}_n\| > t). \tag{5.32}$$

We estimate each term on the right-hand side separately: since $\sup \|\tilde{a}_n\| > \lambda$ implies $\sup_n \|da_n\| \ne 0$ we have by (i)

$$\mathbb{P}(\sup \|\tilde{a}_n\| > \lambda) \le 3\lambda^{-1}.$$

By Chebyshev's inequality, since $\sup \|\tilde{b}_n\| \le \sum \|db_n\|$

$$\mathbb{P}(\sup \|\tilde{b}_n\| > \lambda) \le \lambda^{-1} \sum \mathbb{E}\|db_n\| \le 4\lambda^{-1}.$$

Finally, by (5.24), UMD$_p$ and Doob's inequality, we have

$$\left\| \sup \|\tilde{c}_n\| \right\|_p \le p' C_p(B) 5^{1/p} \lambda^{1-1/p}$$

and hence by Chebyshev again

$$\mathbb{P}(\sup \|\tilde{c}_n\| > t) \le (p' 5^{1/p})^p \lambda^{-1},$$

so by (5.32) we obtain

$$\mathbb{P}(\sup_{n\ge 0} \|\tilde{f}_n\| > 3C_p(B)\lambda) \le (7 + 5(p')^p)\lambda^{-1},$$

which (by homogeneity) implies (5.31) with the constant $\delta_p = 3(7 + 5(p')^p)$.

For the more general case of predictable multipliers (φ_n), the same argument works using Proposition 5.11. □

Remark 5.17. We note here for future use that when $p \to 1$ (resp. $p \to \infty$) we have $\delta_p = O(p')$ (resp. $\delta_p = O(1)$ since $(p')^p \to e$).

Remark 5.18. Note that (5.31) implies a fortiori that there is a constant $C_{1,\infty}(B)$ such that for any martingale (f_n) and any $\varepsilon_n = \pm 1$ we have

$$\sup_{n\ge 0}\sup_{\lambda>0} \lambda \mathbb{P}(\|\tilde{f}_n\| > \lambda) \le C_{1,\infty}(B) \sup_{n\ge 0} \|f_n\|_{L_1(B)}. \tag{5.33}$$

Corollary 5.19. *In the scalar case (i.e. $B = \mathbb{R}$ or $\mathbb{C}$) we find (5.31) with $C \le 81$. Moreover, for any martingale $(f_n)_{n\ge 0}$ bounded in L_1 we have*

$$\sup_{\lambda>0} \lambda \mathbb{P}\left(\left(\sum |df_n|^2\right)^{1/2} > \lambda\right) \le 81 \sup_{n\ge 0} \|f_n\|_1.$$

More generally, if B is a Hilbert space we find

$$\sup_{\lambda>0} \lambda \mathbb{P}\left(\left(\sum \|df_n\|^2\right)^{1/2} > \lambda\right) \le 81 \sup_{n\ge 0} \|f_n\|_{L^1(B)}. \tag{5.34}$$

Proof. The first assertion is clear since $C_2(\mathbb{R}) = C_2(\mathbb{C}) = 1$ and $\delta_2 = 81$. To prove the second one, one simply observes that

$$\left\| \left(\sum |df_n|^2\right)^{1/2} \right\|_2 = \|f\|_2$$

so that the same argument applies when we substitute $S(f) = \left(\sum |df_n|^2\right)^{1/2}$ to $\sup_{n\ge 1} |\tilde{f}_n|$: Indeed, we first write $S(f) \le S(a) + S(b) + S(c)$, $S(b) \le \sum |db_n|$ and $\|S(c)\|_2 = \|c\|_2$, then we can argue as in the proof of Corollary 5.16. □

Remark 5.20. Note that (5.34) only requires that B is UMD and of cotype 2. See §10.4 for more on this topic.

Theorem 5.21 (Burkholder's inequalities)**.** *For any $1 < p < \infty$, there is a positive constant β_p such that, for any scalar martingale (M_n) in L_p and for any* *predictable uniformly bounded scalar sequence* (φ_n), *we have*

$$\sup_n \left\| \sum_0^n \varphi_k dM_k \right\|_p \le \beta_p \sup_n \|M_n\|_p \sup_n \|\varphi_n\|_\infty.$$

Let S *be the square function defined by* (5.9). *There are positive constants* a_p *and* b_p *such that any scalar martingale* (M_n) *in* L_p *satisfies*

$$a_p^{-1}\|S\|_p \le \sup_n \|M_n\|_p \le b_p\|S\|_p. \tag{5.35}$$

Proof. By homogeneity we may assume $\sup_n \|\varphi_n\|_\infty \le 1$. Let T_φ be the transformation taking $f \in L_2$ to $\sum \varphi_n df_n$. Clearly, by Parseval, $\|T_\varphi : L_2 \to L_2\| \le 1$. A fortiori, T_φ is of weak type (2-2). By Corollary 5.16 applied to $B = \mathbb{C}$, T_φ is of weak type (1-1), hence by Marcinkiewicz Theorem 5.78, for any $1 < p < 2$, there is β_p so that $\|T_\varphi : L_p \to L_p\| \le \beta_p$. By duality, since T_φ is (essentially) self-adjoint we have $\|T_\varphi : L_{p'} \to L_{p'}\| \le \beta_p$ for any $p' > 2$. This establishes the first assertion. To prove the second one, we first fix a choice of signs $\varepsilon = (\varepsilon_n)$ and we use $\varphi_n = \varepsilon_n$. Let $T_\varepsilon M_n = \sum_0^n \varepsilon_k dM_k$. The first assertion gives us $\|T_\varepsilon M_n\|_p \le \beta_p\|M_n\|_p$ but since $T_\varepsilon(T_\varepsilon M_n) = M_n$ we have by iteration $\|M_n\|_p \le \beta_p\|T_\varepsilon M_n\|_p$. Therefore,

$$(\beta_p)^{-1}\|T_\varepsilon M_n\|_p \le \|M_n\|_p \le \beta_p\|T_\varepsilon M_n\|_p.$$

But if we now integrate with respect to ε and use the Khintchin inequalities (5.7), letting $S_n = \left(\sum_o^n |dM_k|^2\right)^{1/2}$ we find

$$A_p(\beta_p)^{-1}\|S_n\|_p \le \|M_n\|_p \le \beta_p B_p\|S_n\|_p,$$

and we conclude by taking the supremum over n.

Note that, for $1 < p < 2$, we can also deduce the square function inequality $\|S\|_p \le b_p \sup_n \|M_n\|_p$ from Corollary 5.19 by the sublinear version of Marcinkiewicz Theorem (see Remark 5.79). □

Proof of Theorem 5.13. Assume B UMD$_p$. Then by Corollary 5.16 and by the Marcinkiewicz interpolation Theorem 5.78, B is UMD$_q$ for all $1 < q < p < \infty$. Moreover, we have

$$C_q(B) \le K(1, p, q)\delta_p{}^{1-p'/q'} C_p(B).$$

Thus we obtain half of the desired equivalence with $\beta(p, q) = K(1, p, q)\delta_p{}^{1-p'/q'}$. But now, conversely if B is UMD$_q$ then by (5.22) B^* is UMD$_{q'}$ with $1 < p' < q' < \infty$ and hence we may repeat the preceding argument for B^*, and obtain that B^* is UMD$_{p'}$, or equivalently B is UMD$_p$, with

$$C_p(B) = C_{p'}(B^*) \le \beta(q', p')C_{q'}(B^*) = \beta(q', p')C_q(B).$$

Thus we obtain Theorem 5.13 with $\alpha(p, q) = \beta(q', p')$. Moreover, by (5.120), if p remains fixed, when $q \to 1$ (i.e. $q' \to \infty$) we find $\beta(p, q) = O((q-1)^{-1})$ and using Remark 5.17 we obtain $\beta(q', p') = O(1)$. □

We now give the basic examples of UMD spaces.

Corollary 5.22. *All Hilbert spaces (and in particular $\mathbb{R}$ or $\mathbb{C}$) are UMD. Let (S, Σ, m) be an arbitrary measure space and let $1 < p < \infty$. Then the Banach space $B = L_p(S, \Sigma, m)$ is UMD. More generally, if B is any UMD space, then the space $L_p(S, \Sigma, m; B)$ is UMD.*

Proof. Clearly all Hilbert spaces are UMD$_2$ and hence UMD$_p$ for all $1 < p < \infty$. By Fubini's theorem, it is easy to see that, if B is UMD$_p$, then the space $L_p(S, \Sigma, m; B)$ is UMD$_p$. The case $B = \mathbb{C}$ corresponds to the first assertion for $\mathbb{R}$ or $\mathbb{C}$. Since UMD$_p$ does not depend on p, this proves all the assertions. □

5.5 Extrapolation

We will now give a different approach to Theorem 5.13 based on 'extrapolation'. This is particularly efficient in the dyadic or calibrated case (see Definition 5.14).

We start by a rather general version of the extrapolation principle.

Lemma 5.23. *Let $(v_n)_{n\geq 0}$ and $(w_n)_{n\geq 0}$ be adapted sequences of non-negative random variables, converging a.s. to limits denoted by v_∞ and w_∞. Fix $p > 0$. Assume that for any stopping time $T\colon\ \Omega \to \mathbb{N} \cup \{\infty\}$ we have*

$$\|1_{\{T>0\}} v_T\|_p \leq \|1_{\{T>0\}} w_T\|_p. \tag{5.36}$$

Moreover, assume that there is an adapted non-negative sequence $(\psi_n)_{n\geq 0}$ (which is said to 'control the jumps') such that

$$\forall n \geq 0 \qquad w_{n+1} - w_n \leq \psi_n.$$

Let $w^* = \sup_n w_n$ and $\psi^* = \sup_n \psi_n$. Then for any $t > 0$

$$\mathbb{P}\{v_\infty > t\} \leq t^{-p}\mathbb{E}(t^p \wedge w^{*p}) + \mathbb{P}\{w^* + \psi^* > t\}, \tag{5.37}$$

and hence for any $0 < q < p$

$$\mathbb{E}v_\infty^q \leq (p/(p-q))\mathbb{E}w^{*q} + \mathbb{E}(w^* + \psi^*)^q. \tag{5.38}$$

Proof. Let $T = \inf\{n \mid w_n + \psi_n > t\}$. Note that on $\{T > 0\}$ we have

$$w_T \leq w_{T-1} + \psi_{T-1} \leq t,$$

and hence

$$1_{\{T>0\}}w_T \le t \wedge w^*. \tag{5.39}$$

We have obviously

$$\begin{aligned}\mathbb{P}(v_\infty > t) &\le \mathbb{P}(v_T > t, T = \infty) + \mathbb{P}(v_\infty > t, T < \infty)\\ &\le \mathbb{P}(v_T > t, T > 0) + \mathbb{P}(T < \infty)\\ &\le t^{-p}\mathbb{E}(1_{\{T>0\}}v_T^p) + \mathbb{P}(\sup(w_n + \psi_n) > t)\end{aligned}$$

and hence by (5.36) and (5.39)

$$\begin{aligned}&\le t^{-p}\mathbb{E}(1_{\{T>0\}}w_T^p) + \mathbb{P}(w^* + \psi^* > t)\\ &\le t^{-p}\mathbb{E}(t^p \wedge w^{*p}) + \mathbb{P}(w^* + \psi^* > t),\end{aligned}$$

which proves (5.37). Then, using $\mathbb{E}v_\infty^q = \int_0^\infty qt^{q-1}\mathbb{P}\{v_\infty > t\}dt$ and also $1 \wedge (w^*/t)^p = 1_{\{w^*>t\}} + (w^*/t)^p 1_{\{w^*\le t\}}$ we obtain (5.38) by an elementary computation. □

Second Proof of Theorem 5.13. Let $1 < q < p < \infty$. Assume B UMD$_p$. Consider a finite dyadic martingale (f_n) and a fixed choice of signs (ε_n). We will apply Lemma 5.23. Let $\tilde{f}_\infty = \sum \varepsilon_n df_n$, $\tilde{f}_n = \mathbb{E}_n(\tilde{f}_\infty)$ and let T be a stopping time. We set $v_n = \|\tilde{f}_n\|_B$ and $w_n = C_p(B)\|f_n\|_B$. By (5.21) applied to the martingale $(1_{\{T>0\}}f_{n\wedge T})$, we have

$$\|1_{\{T>0\}}v_T\|_p = \|1_{\{T>0\}}\tilde{f}_T\|_{L_p(B)} \le C_p(B)\|1_{\{T>0\}}f_T\|_{L_p(B)} = \|1_{\{T>0\}}w_T\|_p$$

and hence (5.36) holds. For dyadic (or calibrated) martingales, $\|df_{n+1}\|_B$ is n-measurable, so we can take simply $\psi_n = C_p(B)\|df_{n+1}\|_B$. Note that $w^* + \psi^* \le 3C_p(B)f^*$. By Doob's maximal inequality, if $1 < q < p$, (5.38) implies

$$\|\tilde{f}_\infty\|_{L_q(B)} = \|v_\infty\|_q \le C_p(B)\left(3 + (p/(p-q))\right)^{1/q} q'\|f\|_{L_q(B)}.$$

This shows B is UMD$_q$ with $\beta(p,q) \le \left(3 + (p/(p-q))\right)^{1/q} q'$. By duality (see (5.22)), the preceding argument applied to B^* shows B^* is UMD$_{q'}$ for $1 < q' < p' < \infty$, and hence that B also is UMD$_q$ for any $1 < p < q < \infty$. Again, by (5.22), $\alpha(p,q) = \beta(q', p')$ and the required growth holds when $q \to 1$ with p fixed. All this is restricted to the dyadic filtration, but it is known (see §5.8), that it suffices. □

Remark 5.24. Assume B UMD$_p$ again. The preceding argument shows that, for any q with $0 < q < p$, there is a constant $D(q,p)$ such that for any dyadic B-valued martingale (or any calibrated one) we have

$$\|\sup_n \|\tilde{f}_n\|_B\|_q \le D(q,p)C_p(B)\|\sup_n \|f_n\|_B\|_q.$$

Indeed, this follows easily from Lemma 5.23 setting $v_n = \sup_{k\le n} \|\tilde{f}_k\|_B$ and $w_n = \sup_{k\le n} \|f_k\|_B$. This can be applied to the martingales (f_n) on $\mathbb{T}^{\mathbb{N}}$ with differences of the form

$$df_n(z_1, \dots, z_n) = z_n\varphi_{n-1}(z_1, \dots, z_{n-1}).$$

The latter are called 'analytic' martingales in §4.5.

For any scalar or Banach space valued random variable F, we set

$$\|F\|_{p,\infty} = (\sup_{c>0} c^p \mathbb{P}\{\|F\| > c\})^{1/p}.$$

Remark 5.25. Lemma 5.23 clearly still holds if we replace, in the assumption (5.36), the L_p-norm of $1_{\{T>0\}}v_T$ by $\|1_{\{T>0\}}v_T\|_{p,\infty}$. Indeed, this suffices to prove (5.37).

We now turn to a slightly more involved (but also more powerful) extrapolation result. This has the surprising feature that the starting assumption required is particularly weak: we can deduce strong-type inequalities in L_q for all values $0 < q < \infty$ from a rather weak one for a single value $q = p$.

Let B and B_1 be Banach spaces. Let $(\mathcal{A}_n)_{n\ge 0}$ be an arbitrary filtration. Let $(w_n)_{n\ge 0}$ (resp. $(v_n)_{n\ge 0}$) be B-valued (resp. B_1 valued) adapted sequences of random variables, converging a.s. to a limit denoted by w_∞ (resp. v_∞). We will denote as usual

$$w_n^* = \sup_{0\le k\le n} \|w_k\| \quad \text{and} \quad w^* = \sup_{0\le k<\infty} \|w_k\|.$$

We will assume that there is an adapted non-negative (jump control) sequence $(\psi_n)_{n\ge 0}$ such that

$$\forall n \ge 0 \qquad \max\{\|w_{n+1} - w_n\|, \|v_{n+1} - v_n\|\} \le \psi_n, \tag{5.40}$$

$$\|v_0\| \le \|w_0\| \tag{5.41}$$

We have then

Lemma 5.26. *Let $0 < p < \infty$. Let $0 < \delta < 1$ and $\beta > 1 + \delta$. With the preceding assumptions* (5.40), (5.41) *if for any pair $R \le T$ of stopping times (such that $\mathbb{E}\|w_T - w_R\|^p < \infty$) we have*

$$\|v_T - v_R\|_{p,\infty} \le \|w_T - w_R\|_p, \tag{5.42}$$

then for any $\lambda > 0$ we have

$$\mathbb{P}\{v^* > \beta\lambda,\ w^* \vee \psi^* \le \delta\lambda\} \le \alpha\mathbb{P}\{v^* > \lambda\}, \tag{5.43}$$

where

$$\alpha = (3\delta(\beta - 1 - \delta)^{-1})^p.$$

Moreover, for any $0 < q < \infty$ *there is a constant* $\chi_{p,q}$*, depending only on* p *and* q*, which is* $O(q)$ *when* $q \to \infty$ *while* p *remains fixed, for which we have*

$$\|v^*\|_q \le \chi_{p,q}\|w^* \vee \psi^*\|_q. \tag{5.44}$$

Proof. To prove this, we define stopping times R, S, T by

$$R = \inf\{n \mid \|v_n\| > \lambda\},$$

$$S = \inf\{n \mid \|v_n\| > \beta\lambda\},$$

$$T' = \inf\{n \mid \max\{\|w_n\|, \psi_n\} > \delta\lambda\},$$

with the usual convention $\inf \phi = \infty$. Note that $R \le S$. We define

$$T = (S \wedge T') \vee R$$

so that $R \le T$. Let $F = w_T - w_R$, $G = v_T - v_R$.

Note that the left-hand side of (5.43) is equal to $\mathbb{P}\{S < \infty,\ T' = \infty\}$.

If we apply (5.42) to G and F we find for any $c > 0$

$$\mathbb{P}\{\|G\| > c\} \le c^{-p}\mathbb{E}\|F\|^p. \tag{5.45}$$

Note that $\{v^* \le \lambda\} = \{R = \infty\} \subset \{F = 0\}$ and hence

$$\{F \ne 0\} \subset \{v^* > \lambda\}.$$

We claim that

$$\|F\| \le 3\delta\lambda \text{ on } \{F \ne 0\}.$$

Indeed, this is clear on $\{F \ne 0\} \cap \{T' = \infty\}$ since on that set we even have $\|F\| \le \|w_T\| + \|w_R\| \le 2\delta\lambda$.

Consider now the set $\{F \ne 0\} \cap \{T' < \infty\}$. If $F \ne 0$ then necessarily $R < T$ and hence $T = S \wedge T'$, $R < T'$ and $S \wedge T' > 0$. Thus, on the set $\{F \ne 0\} \cap \{T' < \infty\}$, we have $T = S \wedge T' < \infty$ and

$$\|F\| = \|w_{S\wedge T'} - w_R\| \le \|w_{S\wedge T'} - w_{S\wedge T'-1}\| + \|w_{S\wedge T'-1}\| + \|w_R\|.$$

Now if $R < T'$ we have $\|w_R\| \le \delta\lambda$. Moreover, by definition of T', if $0 < n \le T' < \infty$, we have $\|w_n - w_{n-1}\| \le \psi_{n-1} \le \delta\lambda$ and $\|w_{n-1}\| \le \delta\lambda$, and hence $\|w_{S\wedge T'} - w_{S\wedge T'-1}\| \le \delta\lambda$ and $\|w_{S\wedge T'-1}\| \le \delta\lambda$. This yields $\|F\| \le 3\delta\lambda$ pointwise, proving the claim.

Thus we obtain

$$\mathbb{E}\|F\|^p \le \mathbb{P}\{v^* > \lambda\}(3\delta\lambda)^p. \tag{5.46}$$

Note that $\{w^* \vee \psi^* \le \delta\lambda\} = \{T' = \infty\}$. Therefore

$$\{v^* > \beta\lambda,\ w^* \vee \psi^* \le \delta\lambda\} = \{S < \infty, T' = \infty\}. \tag{5.47}$$

But now if $c = \lambda(\beta - 1 - \delta)$ we have

$$\{S < \infty, T' = \infty\} \subset \{\|G\| > c\}.$$

Indeed, observe first that by (5.41) $\|v_0\| \le \|w_0\| \le \delta\lambda \le \lambda$ on $\{T' = \infty\}$ and hence $R > 0$ on $\{T' = \infty\}$. Then, recalling $R \le S$ and $\|v_R - v_{R-1}\| \le \psi_{R-1}$, we have

$$\|v_S - v_R\| \ge \|v_S\| - \|v_R\| \ge \|v_S\| - \|v_{R-1}\| - \psi_{R-1}$$

and by definition of S (resp. R), $S < \infty$ (resp. $0 < R < \infty$) implies $\|v_S\| > \beta\lambda$ (resp. $\|v_{R-1}\| \le \lambda$) and moreover we have $\psi_{R-1} \le \psi^* \le \delta\lambda$ on $\{T' = \infty\}$. Thus

$$\|v_S\| - \|v_{R-1}\| - \psi_{R-1} \ge \beta\lambda - \lambda - \delta\lambda = c \text{ on } \{S < \infty, T' = \infty\}.$$

Therefore, since $G = v_S - v_R$ on $\{T' = \infty\}$, we obtain as announced

$$\{S < \infty, T' = \infty\} \subset \|G\| > c\}.$$

By (5.45), (5.46) and (5.47), since $c^{-p}(3\delta\lambda)^p = \alpha$, we obtain Claim 5.43.

We now come to the extrapolation trick, for which it is crucial to observe that α can be made as small as we wish by choosing δ small enough. From Claim 5.43 we obtain an estimate that, at first glance seems like it is going in the wrong direction, namely since

$$\mathbb{P}\{v^* > \beta\lambda\} = \mathbb{P}\{v^* > \beta\lambda, w^* \wedge \psi^* > \delta\lambda\} + \mathbb{P}\{v^* > \beta\lambda, w^* \wedge \psi^* \le \delta\lambda\}$$

we have a fortiori

$$\mathbb{P}\{v^* > \beta\lambda\} \le \mathbb{P}\{w^* > \delta\lambda\} + \alpha\mathbb{P}\{v^* > \lambda\}.$$

We will now use the elementary Lemma 5.27: We simply observe that when $\beta > 1$ is fixed and $\delta \to 0$ we have $\alpha \to 0$, so we can find δ small enough so that $\beta^q\alpha < 1$, and then (5.51) gives us (5.44) with $\chi_{p,q} \le (\beta/\delta)(1 - \beta^q\alpha)^{-1/q}$. However, to check that $\chi_{p,q}$ grows like q we need a more refined choice of the parameters. Fix $p > 0$ and assume $q > 2$. We set $\beta = 1 + \delta + 1/q$ and $\delta = \delta_0/q$. We have

$$\beta^q\alpha = \left(1 + \frac{1+\delta_0}{q}\right)^q (3\delta_0)^p \le e^{1+\delta_0}(3\delta_0)^p.$$

Thus we can select and fix $\delta_0 < 1$ so that $e^{1+\delta_0}(3\delta_0)^p = 1/2$ and then we find

$$\chi_{p,q} \le (\beta/\delta)(1-\beta^q\alpha)^{-1/q} \le 6q/\delta_0. \tag{5.48}$$

□

Lemma 5.27. *Let $Y, X \ge 0$ be random variables. Assume that there are positive constants β, δ, α so that for any $\lambda > 0$*

$$\mathbb{P}\{Y > \beta\lambda\} \le \mathbb{P}\{X > \delta\lambda\} + \alpha\mathbb{P}\{Y > \lambda\}. \tag{5.49}$$

Then for any $0 < q < \infty$ we have

$$\mathbb{E}Y^q \le (\beta/\delta)^q\mathbb{E}X^q + \beta^q\alpha\mathbb{E}Y^q, \tag{5.50}$$

and if we assume $\beta^q\alpha < 1$ we have

$$\|Y\|_q \le (\beta/\delta)(1-\beta^q\alpha)^{-1/q}\|Y\|_q. \tag{5.51}$$

Proof. If we multiply both sides of (5.49) by $q\lambda^{q-1}$ and integrate over $\lambda \in \mathbb{R}_+$, we obtain

$$\mathbb{E}(Y/\beta)^q \le \mathbb{E}(X/\delta)^q + \alpha\mathbb{E}Y^q,$$

which is the same as (5.50). Then (5.51) is immediate. □

Remark 5.28. Typically, we will use Lemma 5.26 under the additional assumption that there is a numerical constant $C \ge 1$ such that

$$\psi^* \le Cw^*. \tag{5.52}$$

In that case, (5.44) yields

$$\|v^*\|_q \le C\chi_{p,q}\|w^*\|_q. \tag{5.53}$$

Remark 5.29. Let

$$(w_T - w_R)^* = \sup_{n\ge 0}\|w_{T\wedge n} - w_{R\wedge n}\|.$$

If we replace (5.42) by the following

$$\|v_T - v_R\|_{p,\infty} \le \|(w_T - w_R)^*\|_p, \tag{5.54}$$

the proof remains valid and the same conclusion holds. The reason is that on the one hand $(w_T - w_R)^* = 0$ on the set $\{R = \infty\}$ and on the other hand, by the same argument, $(w_T - w_R)^* \le 3\delta\lambda$ on its support.

Actually, we could weaken the assumption further, by replacing $\|(w_T - w_R)^*\|_p$ by $p^{-1}\|(w_T - w_R)^*\|_{p,1}$ where $\| \; \|_{p,1}$ is the Lorentz space quasi-norm (see (8.57) for the definition). Indeed, if f is the indicator function of a set A, we have $p^{-1}\|f\|_{p,1} = \mathbb{P}(A)^{1/p}$, so we still have

$$p^{-1}\|(w_T - w_R)^*\|_{p,1} \leq 3\delta\lambda\,(\mathbb{P}\{R < \infty\})^{1/p}.$$

In the language of interpolation theory, this shows that a 'very weak type (p, p)' assumption for a single p suffices to imply a 'strong (q, q)' bound such as (5.53) for all q, which is a rather surprising fact.

Let (f_n), (g_n) be B-valued martingales relative to a filtration $(\mathcal{A}_n)$. For simplicity we assume that both are finite i.e. for some $N > 0$ we have $f_n = f_N$ for all $n > N$ and similarly for g_n (thus we may write $f_\infty = f_N$ and $g_\infty = g_N$). We denote as usual

$$f_n^* = \sup_{0\leq k\leq n} \|f_k\|_B \quad \text{and} \quad f^* = \sup_{0\leq k<\infty} \|f_k\|_B.$$

Theorem 5.30. *We assume that (f_n) is a B-valued calibrated martingale and that (g_n) satisfies (with the convention $df_0 = f_0$)*

$$\forall n \geq 0, \forall \omega \in \Omega \quad \|dg_n(\omega)\| \leq \|df_n(\omega)\|.$$

Let $0 < p < \infty$ and let $C \geq 1$ be a constant. If for any pair $R \leq T$ of stopping times we have

$$\|g_T - g_R\|_{p,\infty} \leq C\|f_T - f_R\|_p, \tag{5.55}$$

then for any $0 < q < \infty$ we have

$$(\mathbb{E}g^{*q})^{1/q} \leq \chi_{p,q} C(\mathbb{E}f^{*q})^{1/q}. \tag{5.56}$$

Proof. We may restrict to $f_0 = 0$. Let $v_n = g_n/C$, $w_n = f_n$ and $\psi_n = \|df_{n+1}\|$. Then (5.40), (5.41) and (5.42) obviously hold under the present assumptions. Thus Lemma 5.26 yields the conclusion. □

In the dyadic case, the equivalence of UMD_p and UMD_q can be extended to all $0 < p, q < \infty$, but using the L_p-norms of maximal functions, as follows:

Theorem 5.31. *Let B be a Banach space. Consider for $0 < p < \infty$ the following property:*
$(B)_p$: There is a constant C such that for any B-valued dyadic martingale (f_n) and any martingale transform $\tilde{f}_n = \sum_{k\leq n} \varepsilon_k df_k$ ($\varepsilon_k \in \pm 1$) we have

$$(\mathbb{E}\sup_n \|\tilde{f}_n\|^p)^{1/p} \leq C(\mathbb{E}\sup_n \|f_n\|^p)^{1/p}.$$

Then

$$\forall 0 < p, q < \infty \quad (B)_p \Leftrightarrow (B)_q.$$

Moreover, for $(B)_p$ to hold, it suffices that the weak-type version holds, more precisely it suffices that there exists C' such that for any (f_n) and (ε_n) we have

$$\sup_n (\sup_{c>0} c^p \mathbb{P}\{\|\tilde{f}_n\| > c\})^{1/p} \le C'(\mathbb{E} \sup_n \|f_n\|^p)^{1/p}. \tag{5.57}$$

Proof. Let $(g_n) = (\tilde{f}_n)$. We will restrict all the inequalities under consideration to finite martingales of length N with N arbitrarily large. Fix $N > 0$. We assume that $f_n = f_N$ for all $n \ge N$. Assume (5.57). Then for any pair $R \le T$ of stopping times we have

$$\|g_T - g_R\|_{p,\infty} \le C'\|(f_T - f_R)^*\|_p. \tag{5.58}$$

By Remark 5.29 this implies $(B)_q$ for any $0 < q < \infty$. Since $(B)_p \Rightarrow$ (5.57), this shows $(B)_p \Rightarrow (B)_q$ for any $0 < p, q < \infty$. □

Third Proof of Theorem 5.13. We will again anticipate (see §5.8): We admit that the UMD property and the UMD constant are unchanged if we restrict them to the dyadic case. Let $g_n = \tilde{f}_n = \sum_{0\le k\le n} \varepsilon_k df_k$ be the transform of a finite dyadic martingale. Consider any pair $R \le T$ of stopping times, and let $h = f_T - f_R$. Then we have $g_T - g_R = \tilde{h}$. Thus, when $1 < p < \infty$, if B is UMD_p all B-valued dyadic martingale transforms a fortiori must satisfy (5.55) with $C \le C_p(B)$. Then Theorem 5.30, together with Doob's inequality (1.38), implies that for any $1 < q < \infty$

$$C_q(B) \le \chi_{p,q} q' C_p(B).$$

It follows that $\beta(p,q) \le \chi_{p,q} q'$ and $\alpha(p,q) \le \chi_{q,p} p'$. By (5.48) (recalling (5.22)) we obtain the announced growth for $\beta(p,q)$ and $\alpha(p,q)$. □

5.6 The UMD_1 property: Burgess Davis decomposition

The following classical decomposition due to Burgess Davis is very useful to control the 'jumps' of a martingale when a priori their length is not predictable.

Theorem 5.32. *A general B-valued martingale $(f_n)_{n\ge 0}$ can be decomposed as a sum*

$$f_n = h_n + g_n$$

with $h_0 = 0$, so that almost surely for all $n \ge 1$

$$\|dg_n\| \le 4 \sup_{0\le k\le n-1} \|df_k\|,$$

and for any $p \ge 1$

$$\left\| \sum_1^\infty \|dh_n\| \right\|_p \le (2 + 2p) \, \| \sup_n \|df_n\| \, \|_p. \tag{5.59}$$

Proof. Let $\delta_n = \sup_{0\le k\le n} \|df_k\|$ and $\delta_\infty = \sup_{k\ge 0} \|df_k\|$. By convention we set $\delta_{-1} = 0$. We define h and g via their increments by setting $h_0 = 0$, $g_0 = f_0$ and for all $n \ge 1$

$$dh_n = df_n \cdot 1_{\{\delta_n > 2\delta_{n-1}\}} - \mathbb{E}_{n-1}(df_n \cdot 1_{\{\delta_n > 2\delta_{n-1}\}}$$

and

$$dg_n = df_n \cdot 1_{\{\delta_n \le 2\delta_{n-1}\}} - \mathbb{E}_{n-1}(df_n \cdot 1_{\{\delta_n \le 2\delta_{n-1}\}}.$$

Then $\mathbb{E}_{n-1}(df_n) = 0$ implies $df_n = dh_n + dg_n$. Moreover, we have clearly $\|dg_n\| \le 2\delta_{n-1} + \mathbb{E}_{n-1}(2\delta_{n-1}) = 4\delta_{n-1}$.

Note that when $\delta_n > 2\delta_{n-1}$ then $\|df_n\| \le \|df_n\| + (\delta_n - 2\delta_{n-1}) \le 2(\delta_n - \delta_{n-1})$ so that

$$1_{\{\delta_n > 2\delta_{n-1}\}} \|df_n\| \le 2(\delta_n - \delta_{n-1}).$$

Therefore we have pointwise

$$\|dh_n\| \le 2(\delta_n - \delta_{n-1}) + \mathbb{E}_{n-1}(2(\delta_n - \delta_{n-1})).$$

Then $\| \sum \delta_n - \delta_{n-1} \|_p = \|\delta_\infty\|_p$, and by the dual to Doob's inequality (namely Theorem 1.26) we have

$$\left\| \sum \mathbb{E}_{n-1}(\delta_n - \delta_{n-1}) \right\|_p \le p \left\| \sum \delta_n - \delta_{n-1} \right\|_p = p\|\delta_\infty\|_p,$$

and hence (5.59) follows from the triangle inequality. □

By the triangle inequality, the following statement is immediate.

Corollary 5.33. *Let $f_n^* = \sup_{k\le n} \|f_k\|$. If we assume $\mathbb{E} \sup_n \|f_n\| < \infty$, the preceding decomposition $f_n = h_n + g_n$ with $h_0 = 0$ is such that almost surely*

$$\forall n \ge 1 \quad \|dg_n\| \le 8\, f_{n-1}^*, \tag{5.60}$$

and $\sum_1^\infty \mathbb{E}\|dh_n\| \le 8\, \mathbb{E} \sup_n \|f_n\|$. More generally, for any $p \ge 1$ we have

$$\left\| \sum\nolimits_1^\infty \|dh_n\| \right\|_p \le (4 + 4p)\, \|\sup \|f_n\|\,\|_p \tag{5.61}$$

and hence

$$\|\sup \|g_n\|\,\|_p \le (5 + 4p)\, \|\sup \|f_n\|\,\|_p. \tag{5.62}$$

Theorem 5.34. *Let B be a UMD Banach space. Let $C = 70\, C_2(B)$. Then for all filtrations $(\mathcal{A}_n)_{n\ge 0}$ and all choices of signs $\varepsilon_n = \pm 1$ we have for all martingales $(f_n)_{n\ge 0}$*

$$\mathbb{E} \sup_n \left\| \sum\nolimits_0^n \varepsilon_k df_k \right\| \le C \mathbb{E} \sup_n \|f_n\|. \tag{5.63}$$

Proof. We will use Corollary 5.33. Let $\tilde{f}_n = \sum_{k \le n} \varepsilon_k df_k$ and let $\tilde{f}_n^* = \sup_{k \le n} \|\tilde{f}_k\|$ and $\tilde{f}^* = \sup_n \tilde{f}_n^*$, and similarly for (g_n) and (h_n). By the UMD property and Doob's maximal inequality, we have for any stopping time T (note that $(1_{\{T>0\}}\tilde{g}_{n \wedge T})$ is a martingale)

$$\|\tilde{g}_T^* 1_{\{T>0\}}\|_2 \le 2\|\tilde{g}_T 1_{\{T>0\}}\|_{L_2(B)} \le 2\, C_2(B)\|g_T 1_{\{T>0\}}\|_{L_2(B)}.$$

By the triangle inequality we have on one hand

$$\|\tilde{f}^*\|_1 \le \sum \|dh_n\|_{L^1(B)} + \|\tilde{g}^*\|_1 \le 8\|f^*\|_1 + \|\tilde{g}^*\|_1.$$

On the other hand, let us apply (5.38) in the case $q = 1$ $p = 2$ with $v_n = \tilde{g}_n^*$, $w_n = 2\, C_2(B) g_n^*$ and $\psi_n = 2\, C_2(B)(8 f_n^*)$. Note that (5.60) implies $g_{n+1}^* - g_n^* \le \|g_{n+1}\| - \|g_n\| \le \|g_{n+1} - g_n\| \le 8\, f_n^*$, and hence $w_{n+1} - w_n \le \psi_n$ for all $n \ge 0$. Then we find

$$\|\tilde{g}^*\|_1 \le 16\, C_2(B)\|f^*\|_1 + 6\, C_2(B)\|g^*\|_1$$

hence, using $\|g^*\|_1 \le \|f^*\|_1 + \|h^*\|_1 \le 9\|f^*\|_1$, we obtain the announced result after some arithmetic. □

Remark 5.35. Conversely, any space B satisfying (5.63) for some C must be UMD. More generally, for $\varepsilon = (\varepsilon_n) \in \{-1, 1\}^{\mathbb{N}}$, let us denote $T_\varepsilon(f) = \sum_0^\infty \varepsilon_n df_n$. If for some $r > 0$ and C we have for any such ε and $f \in L_\infty(B)$

$$\|T_\varepsilon(f)\|_r \le C\|f\|_{L_\infty(B)}, \tag{5.64}$$

then B is UMD. Let us briefly sketch the argument. Unfortunately, we will use several results, notably on type p, discussed only later on. Arguing as in Corollary 5.39, we observe that if B satisfies (5.64), then B cannot contain ℓ_1^n's uniformly and hence is of type p for some $p > 1$ by Theorem 10.45. Let $s = r/(r + 1)$. By the Gundy decomposition in Theorem 5.15, (5.64) implies that for some C' we have $\|T_\varepsilon(f)\|_{s,\infty} \le C'\|f\|_{L_1(B)}$. A fortiori, $\|T_\varepsilon(f)\|_{s,\infty} \le C'\|f\|_{L_p(B)}$. By the reverse Hölder principle described in Proposition 5.77 applied with any $q < s$ (note that we may restrict to symmetric f's), this implies for some constant c, using the type p inequality, an estimate $\|T_\varepsilon(f)\|_{p,\infty} \le c\|f\|_{L_p(B)}$. Since B is of type 1, for some c' we also have $\|T_\varepsilon(f)\|_{1,\infty} \le c'\|f\|_{L_1(B)}$. Then by the Marcinkiewicz Theorem 5.78 we conclude that B is UMD$_t$ for any $1 < t < p$.

One more Proof of Theorem 5.13. This is merely a variant of the second proof that avoids the restriction to dyadic martingales by making use of the B. Davis decomposition. Let $1 < q < \infty$. Assume B UMD$_p$. Consider a finite martingale (f_n) in $L_q(B)$ and let (g_n) be as in Lemma 5.33. Note that (g_n) is also a finite martingale. We may assume for simplicity $f_0 = g_0 = 0$. Fix a choice of signs $\varepsilon = (\varepsilon_n)$. Let $g_\infty = \sum_1^\infty \varepsilon_k dg_k$, $\tilde{g}_n = \sum_1^n \varepsilon_k dg_k$ and let $\tilde{g}^* = \sup_n \|\tilde{g}_n\|$,

$\tilde{g}_n^* = \sup_{k \le n} \|\tilde{g}_k\|$ and $g_n^* = \sup_{k \le n} \|g_k\|$. Since B is UMD$_p$, by (5.21) and Doob's inequality, we have

$$\|\tilde{g}^*\|_p \le p' C_p(B) \|g_\infty\|_{L_p(B)}.$$

Since this also holds for all the stopped martingales $(1_{\{T>0\}} g_{n \wedge T})$ for any stopping time T, we have

$$\|1_{\{T>0\}} \tilde{g}_T^*\|_p \le p' C_p(B) \|1_{\{T>0\}} g_T\|_{L_p(B)},$$

and a fortiori if we set $v_n = \tilde{g}_n^*$ we have

$$\|1_{\{T>0\}} v_T\|_p \le p' C_p(B) \|1_{\{T>0\}} g_T\|_{L_p(B)}.$$

We will use Lemma 5.23 with $w_n = p' C_p(B) g_n^*$ and $\psi_n = p' C_p(B)(8 f_n^*)$. Therefore by (5.38), we have for any $1 \le q < p$

$$\|\tilde{g}^*\|_q \le p' C_p(B)[8\|f^*\|_q + (1 + (p/(p-q))^{1/q}) \|g^*\|_q]$$

and hence by (5.62)

$$\begin{aligned} \|\tilde{g}^*\|_q &\le p' C_p(B)(8 + (1 + (p/(p-q))^{1/q})(5 + 4q)) \|f^*\|_q \\ &= C_p(B) C(p,q) \|f^*\|_q \end{aligned}$$

Finally, since we have trivially

$$\tilde{f}^* \le \tilde{g}^* + \sum \|dh_n\|,$$

recalling (5.61) and assuming $1 \le q < p$ we obtain

$$\begin{aligned} \|\tilde{f}^*\|_q &\le \|\tilde{g}^*\|_q + \left\| \sum \|dh_n\| \right\|_q \\ &\le (C_p(B) C(p,q) + 8q) \|f^*\|_q. \end{aligned}$$

Now, when $1 < q < p$, Doob's maximal inequality yields

$$\|\tilde{f}\|_{L_q(B)} \le \|\tilde{f}^*\|_q \le q'(C_p(B) C(p,q) + 8q) \|f\|_{L_q(B)},$$

and hence

$$C_q(B) \le q' C_p(B) C(p,q) + 8q.$$

This shows that UMD$_p \Rightarrow$ UMD$_q$. The converse is proved by duality as in the second proof. □

Definition. A Banach space B is called UMD$_1$ if there is a constant C such that for any martingale in $L_1(B)$ we have for any choice of signs $\varepsilon_n = \pm 1$

$$\mathbb{E} \sup_n \left\| \sum\nolimits_0^n \varepsilon_k df_k \right\|_B \le C \mathbb{E} \sup_n \|f_n\|_B. \tag{5.65}$$

We will denote the best C in (5.65) by $C_1(B)$.

The preceding (or Theorem 5.34) shows that for any $p \neq 1$ $\text{UMD}_p \Rightarrow \text{UMD}_1$. (Just take $q = 1$ in the preceding 'third' proof, and stop the proof before the last step.) The converse, namely $\text{UMD}_1 \Rightarrow \text{UMD}$ is also true by the preceding Remark 5.35.

Here is the analogue of Proposition 5.10 for the case $p = 1$:

Proposition 5.36. *A Banach space B is* UMD_1 *(or equivalently UMD) iff there are constants C_1' and C_2' such that for any martingale (f_n) in $L_1(B)$ we have*

$$(C_1')^{-1}\mathbb{E}R_{df} \le \mathbb{E}\sup_n \|f_n\|_B \le C_2'\mathbb{E}R_{df}.$$

Here we recall that

$$R_{df}(\omega) = \sup_n \left\| \sum_0^n \varepsilon_k df_k(\omega) \right\|_{L_2(\Delta,\nu;B)}.$$

Proof. Since $\tilde{\tilde{f}} = f$, we may apply (5.65) with $\tilde{f}$ in place of f and we obtain

$$C^{-1}\mathbb{E}\sup \|f_n\| \le \mathbb{E}\sup \left\| \sum_0^n \varepsilon_k df_k \right\| \le C\mathbb{E}\sup \|f_n\|.$$

After averaging over the choices of signs $\varepsilon = (\varepsilon_n)$, this becomes

$$C^{-1}\mathbb{E}\sup \|f_n\| \le \mathbb{E}\Phi \le C\mathbb{E}\sup \|f_n\| \tag{5.66}$$

where $\Phi(\omega) = \int \sup_n \| \sum_0^n \varepsilon_k df_k(\omega)\|_B d\nu$. By Doob's inequality (or by (1.43) with $\sqrt{2}$ in place of 2)

$$\Phi(\omega) \le \| \sup_n \| \sum_0^n \varepsilon_k df_k(\omega)\|_B\|_{L_2(\nu)} \le 2R_{df}(\omega),$$

and by Kahane's inequality (see Remark 5.2), we have $R_{df} \le K(2,1)\Phi$. Thus Φ being equivalent to R_{df}, the proposition follows from (5.66). □

Corollary 5.37. *In the scalar (or Hilbert space) valued case there is a positive constant β_1' such that for any scalar martingale (M_n) and any predictable sequence (φ_n) with $\|\varphi_n\|_\infty \le 1$ we have*

$$\mathbb{E}\sup_n \left| \sum_0^n \varphi_k dM_k \right| \le \beta_1' \mathbb{E}\sup |M_n|.$$

Let S be the square function defined by (5.9). *There are constants a_1', b_1' such that for any martingale (M_n) in L_1 we have*

$$(a_1')^{-1}\|S\|_1 \le \mathbb{E}\sup |M_n| \le b_1'\|S\|_1. \tag{5.67}$$

Remark. A fortiori, there is a constant b_1 such that

$$\sup \mathbb{E}|M_n| \le b_1\|S\|_1. \tag{5.68}$$

This partially extends (5.35) to $p = 1$.

5.7 Examples: UMD implies super-RNP

Consider again the martingale $M_n = \prod_1^n (1 + \varepsilon_k)$ on $\Delta = \{-1, 1\}^{\mathbb{N}}$ with respect to the filtration $\mathcal{A}_n = \sigma(\varepsilon_1, \ldots, \varepsilon_n)$. We set $M_0 = 1$ and let $\mathcal{A}_0$ be the trivial σ-algebra. Note that $M_k = 2^k 1_{\{\varepsilon_1 = \ldots = \varepsilon_k = 1\}}$ and $dM_k = \varepsilon_k M_{k-1}$ for all $k \geq 1$. In particular

$$\forall n \geq 0 \quad \|M_n\|_1 = \mathbb{E} M_n = 1.$$

Let $\Omega_0 = \{\varepsilon_1 = -1\}$, $\Omega_k = \{\varepsilon_1 = \cdots = \varepsilon_k = 1, \varepsilon_{k+1} = -1\}$ for all $0 < k < n$ and lastly

$$\Omega_n = \{\varepsilon_1 = \varepsilon_2 = \cdots = \varepsilon_n = 1\}.$$

Note that $\Omega_0, \Omega_1, \ldots, \Omega_n$ form a partition of our probability space Δ. We have for any $n \geq 1$

$$\sup\nolimits_{0 \leq k \leq n} |M_k| = 1_{\Omega_0} + \sum\nolimits_{0<k<n} 1_{\Omega_k} 2^k + 1_{\Omega_n} 2^n$$

and hence

$$\mathbb{E} \sup\nolimits_{0 \leq k \leq n} |M_k| = 2^{-1} + \sum\nolimits_{0<k<n} 2^{-k-1} 2^k + 1 = n/2 + 1.$$

Let

$$S_n = (|M_0|^2 + |dM_1|^2 + \cdots + |dM_n|^2)^{1/2} = (1 + |M_0|^2 + \cdots + |M_{n-1}|^2)^{1/2}.$$

We have

$$\begin{aligned} S_n = \; & 1_{\Omega_0} 2^{1/2} + \sum\nolimits_{0<k<n} 1_{\Omega_k} (1 + 1 + 2^2 + \cdots + 2^{2k})^{1/2} \\ & + 1_{\Omega_n} (1 + 1 + 2^2 + \cdots + 2^{2n-2})^{1/2} \end{aligned}$$

and hence

$$\mathbb{E} S_n = 2^{-1/2} + \sum\nolimits_{0<k<n} 2^{-k-1} \left(1 + \frac{2^{2k+2} - 1}{3}\right)^{1/2} + 2^{-n} \left(1 + \frac{2^{2n} - 1}{3}\right)^{1/2},$$

which shows that there is $\alpha > 0$ independent of n such that

$$n/\alpha \leq \mathbb{E} S_n \leq \alpha n.$$

As a consequence, we may infer that

$$nA_1/\alpha \leq \sup_{\xi_k = \pm 1} \left\| \sum\nolimits_0^n \xi_k dM_k \right\|_1. \tag{5.69}$$

Indeed, by Khintchin's inequality (5.7), we have for any ω

$$A_1 S_n(\omega) \leq \int \left| \sum\nolimits_0^n \xi_k dM_k(\omega) \right| d\nu(\xi)$$

and hence after integration in ω

$$A_1 \mathbb{E} S_n \le \int \left\| \sum\nolimits_0^n \xi_k dM_k \right\|_1 d\nu(\xi)$$

which obviously implies (5.69).

In particular this shows that the inequality $\|S\|_p \le a_p \sup_n \|M_n\|_p$ (see Theorem 5.21), as well as Doob's maximal inequality do not remain valid for $p = 1$. We will now show that the spaces $\ell_1, c_0, L_1, C_{[0,1]}$ and L_∞ all fail UMD.

The proof is based on the following:

Proposition 5.38. *For each* $1 < p < \infty$, *there is* $\delta > 0$ *such that the* UMD$_p$-*constant of* ℓ_1^N *satisfies*

$$\forall N \ge 1 \qquad C_p(\ell_1^N) \ge \delta \operatorname{Log}(N). \tag{5.70}$$

Proof. It suffices to show this for $N = 2^n$. Consider $B = L_1(\Delta, \nu)$ and let (f_n) be the B-valued martingale defined by $f_n(\omega) = \prod_1^n (1 + \varepsilon_k(\omega)\varepsilon_k)$. Note that by the translation invariance of ν (indeed ν is the Haar measure on $\{-1, 1\}^{\mathbb{N}}$) we have for any fixed ω

$$\|f_n(\omega)\|_B = \|M_n\|_1 \tag{5.71}$$

where (M_n) is the scalar valued martingale in the previous paragraph. Similarly, for any choice of signs $\xi = (\xi_n)_n$ we have for any ω

$$\left\| \sum\nolimits_0^n \xi_k df_k(\omega) \right\|_B = \left\| \sum\nolimits_0^n \xi_k dM_k \right\|_1 . \tag{5.72}$$

Now observe that $(f_1, \ldots, f_n)$ actually takes values in a subspace of B that is isometric to $\ell_1^{2^n}$, namely the subspace generated by the indicating functions of the 2^n disjoint atoms of $\mathcal{A}_n = \sigma(\varepsilon_1, \ldots, \varepsilon_n)$. Thus, we have

$$\left\| \sum\nolimits_0^n \xi_k df_k \right\|_{L_p(B)} \le C_p(\ell_1^{2^n}) \left\| \sum\nolimits_0^n df_k \right\|_{L_p(B)}$$

and hence by (5.71) and (5.72)

$$\left\| \sum\nolimits_0^n \xi_k dM_k \right\|_1 \le C_p(\ell_1^{2^n}) \|M_n\|_1 \le C_p(\ell_1^{2^n}),$$

which implies by (5.69)

$$n A_{1/\alpha} \le C_p(\ell_1^{2^n}).$$

□

Remark. Note that by (5.72) we have for any fixed $\omega \in \Omega$

$$R_{df}(\omega) = \left(\int_\Delta \left\| \sum\nolimits_0^n \xi_k dM_k \right\|_1^2 d\nu(\xi) \right)^{1/2},$$

and hence by Jensen's inequality and by (5.7) $A_1\mathbb{E}S_n \le R_{df}(\omega) \le \mathbb{E}S_n$. Thus, we obtain for any ω

$$(A_1/\alpha)n \le R_{df}(\omega) \le \alpha n.$$

Remark. Note that the constant $C_p(\ell_1^N)$ grows at most like $\operatorname{Log} N$. Indeed, let $1 < q < \min\{p, 2\}$, since ℓ_1^N is $N^{1/q'}$-isomorphic to ℓ_q^N, we have $C_p(\ell_1^N) \le N^{1/q'} C_p(\ell_q^N)$ and, by Theorem 5.13, $C_p(\ell_q^N) \le \alpha(p,q) C_q(\ell_q^N)$, but $\alpha(p,q)$ remains bounded when $q \to 1$, so for some constant γ independent of q we have

$$C_p(\ell_1^N) \le \gamma N^{1/q'} C_q(\ell_q^N).$$

Moreover, assuming the scalars are $\mathbb{C}$, for some constant $\beta > 0$ we have a bound

$$C_q(\ell_q^N) = C_q(\mathbb{C}) \le \beta(2,q) C_2(\mathbb{C}) = \beta(2,q) \le \beta q (q-1)^{-1} = \beta q'.$$

Taking $q' = \operatorname{Log} N$ (and assuming $N \ge e^2$) it follows that

$$C_p(\ell_1^N) \le e\gamma\beta \operatorname{Log} N.$$

Remark. We remind the reader that, by Theorem 5.13, the UMD$_2$ and UMD$_p$ constants are equivalent for any $1 < p < \infty$, so when discussing the growth as a function of N, we may use any $1 < p < \infty$.

Bourgain [135] extended the preceding logarithmic growth to any N-dimensional space with a 1-symmetric basis, assuming finite cotype, and he proved a $(\operatorname{Log} N)^2$ bound without the latter assumption. Thus, it is tempting to think that the UMD constant of finite-dimensional spaces grows slowly with respect to the dimension, but this guess is not correct even for spaces with 1-unconditional basis. See Bourgain's [135] or the striking recent examples by Y. Qiu in [405]. The UMD$_2$ constant of an N-dimensional space is obviously majorized by its distance to ℓ_2^N, which is at most $\sqrt{N}$ (cf. e.g. [79, p. 27]). The example in [135, p. 48] has a UMD$_2$ constant $\ge cN^\theta$ with $\theta = (\operatorname{Log}_2(3/2))/2 = 0.29\ldots$. It is an open problem whether the UMD$_2$ constant of an N-dimensional space can be $\approx \sqrt{N}$.

We will use freely the obvious observation that if F is c-isomorphic to a subspace $E \subset B$ of a UMD space B, then F itself is UMD and $C_p(F) \le cC_p(B)$. We say that a Banach space B contains ℓ_1^ns uniformly (equivalently, in the terminology of §11.1, ℓ_1 is finitely representable in B) if for any $n \ge 1$ and $\varepsilon > 0$ there is a subspace $E \subset B$ that is $(1+\varepsilon)$-isomorphic to ℓ_1^n. We then have

Corollary 5.39. *The spaces ℓ_1, L_1 (and also $c_0, \ell_\infty, C([0,1])$ and $L_\infty[0,1]$) all fail the UMD property. More generally, any Banach space B that contains ℓ_1^ns uniformly must fail UMD.*

Proof. The last assertion is an obvious consequence of the proposition. Then the other assertions all follow since each of the spaces listed contains ℓ_1^ns uniformly. Indeed, note that $\ell_1^n \subset \ell_\infty^{2^n}$ isometrically (at least in the real case, in the complex case the embedding is only 2-isomorphic, as in (5.20)). Alternatively, one can use duality (we have $C_p(\ell_1^N) = C_{p'}(\ell_\infty^N)$) to deduce from the proposition that $c_0, \ell_\infty, C([0,1])$ and $L_\infty([0,1])$ fail UMD$_{p'}$. □

Remark 5.40. We will see in Corollary 10.47 and Remark 10.48 that UMD implies reflexivity and even super-reflexivity (but not conversely). A fortiori, UMD implies the RNP (and even the super-RNP), but we will now give a quick direct proof of that. Assume by contradiction that B fails the RNP. Then, for some $\delta > 0$, B contains a δ-separated tree in its unit ball, by Theorem 2.9. This produces a fortiori, for any n, a martingale $f_1, \ldots, f_n$ in the unit ball of $L_p(B)$ $(1 < p < \infty)$ such that $\|df_k\|_{L_p(B)} = \|f_k - f_{k-1}\|_{L_p(B)} \geq \delta$ for all $1 \leq k \leq n$. This implies that B fails UMD_p. Indeed, the UMD_p property implies that for any $\xi_k = \pm 1$

$$\left\|\sum \xi_k df_k\right\|_{L_p(B)} \leq C_p(B),$$

and hence by the contraction principle (5.19) for any $x \in \mathbb{R}^n$ we have

$$\left\|\sum x_k df_k\right\|_{L_p(B)} \leq C_p(B) \sup_{1\leq k\leq n} |x_k|.$$

In the converse direction, we have obviously $\|x_k df_k\|_{L_p(B)} \leq 2\|\sum x_k df_k\|_{L_p(B)}$ for any k, therefore we obtain

$$(\delta/2) \sup_{1\leq k\leq n} |x_k| \leq \left\|\sum x_k df_k\right\|_{L_p(B)} \leq C_p(B) \sup_{1\leq k\leq n} |x_k|.$$

Thus, letting $c = 2C_p(B)/\delta$, the space $L_p(B)$ (viewed as a real Banach space) contains a subspace $E(n)$ that is c-isomorphic to ℓ_∞^n (in fact it even contains an isomorphic copy of the space c_0). But this contradicts the UMD_p property since we have seen by (5.22) and (5.70) that $C_p(\ell_\infty^n) = C_{p'}(\ell_1^n) \to \infty$, and we must have $C_p(\ell_\infty^n) \leq cC_p(E(n)) \leq cC_p(L_p(B)) = cC_p(B)$. This contradiction shows, as promised, that UMD implies the RNP.

Actually, we will see in §11.1 that any UMD space admits an equivalent uniformly convex norm, which is a much stronger implication.

5.8 Dyadic UMD implies UMD

We wish to show that we may restrict ourselves to the dyadic filtration in the definition of UMD spaces. The proof (from [343]) is based on a result of independent interest, asserting that for any $1 \leq p < \infty$, any finite B-valued martingale $(f_0, \ldots, f_N)$ (on a large enough probability space) is a perturbation in $L_p(B)$ of a martingale $(f'_0, \ldots, f'_N)$ that has the same distribution as a subsequence of a dyadic martingale. In particular, $(f'_0, \ldots, f'_N)$ is formed of simple functions. Thus our first goal is to show that we can indeed do this with simple functions, as follows:

Lemma 5.41. *Let $(f_n)_{n\geq 0}$ be a martingale in $L_p(\Omega, \mathcal{A}, \mathbb{P}; B)$ $(1 \leq p < \infty)$. Let $\varepsilon > 0$. Then there is a martingale $(f'_n)_{n\geq 0}$ formed of simple functions such that*

$$\forall n \geq 0 \qquad \|f_n - f'_n\|_{L_p(B)} < \varepsilon.$$

Proof. Let $\delta = \varepsilon/2$. Fix $\delta_n > 0$ with $\sum \delta_n < \delta$. Let $\mathcal{A}_n = \sigma(f_0, \ldots, f_n)$. For each $n \geq 0$, let F_n be an $\mathcal{A}_n$-measurable simple function such that

$$\|df_n - F_n\|_{L_p(B)} < \delta_n.$$

Let $\mathcal{B}_n = \sigma(F_0, \ldots, F_n)$. Note $\mathcal{B}_n \subset \mathcal{A}_n$ and $\mathcal{B}_n$ is finite since the F_k's are simple functions. Let $f'_n = F_0 + \sum_1^n F_k - \mathbb{E}^{\mathcal{B}_{k-1}} F_k$. Note that since $\mathbb{E}^{\mathcal{B}_{n-1}}(df_n) = 0$ for all $n \geq 1$

$$\|df_n - df'_n\|_{L_p(B)} \leq \|df_n - F_n\|_{L_p(B)} + \|E^{\mathcal{B}_{n-1}}(F_n - df_n)\|_{L_p(B)} < 2\delta_n$$

and hence

$$\|f_n - f'_n\|_{L_p(B)} \leq 2\sum\nolimits_0^n \delta_k < 2\delta.$$

Since $\mathcal{B}_n$ is finite, f'_n is a simple function and finally $(f'_n)_{n\geq 0}$ is clearly a martingale, so the result follows. □

We advise the reader to review Remark 1.18 before reading further.

Recall (see §1.4) that the dyadic filtration on $([0, 1], dt)$ is defined by $\mathcal{A}_0 = (\emptyset, [0, 1])$ and $\mathcal{A}_m$ is generated by the partition (up to negligible intersections) into the 2^m intervals

$$[(j-1)2^{-m}, j2^{-m}] \qquad j = 1, 2, 3, \ldots, 2^m.$$

Let $\mathcal{B}_0 \subset \cdots \subset \mathcal{B}_N$ be a filtration with length N formed of finite σ-algebras on $(\Omega, \mathbb{P})$, with $\mathcal{B}_0$ trivial. We denote $\beta = (\mathcal{B}_0, \cdots, \mathcal{B}_N)$. Let $\{a(i_1) \mid 1 \leq i_1 \leq N(0)\}$ be the set of atoms of $\mathcal{B}_1$. Then each $a(i_1)$ is the disjoint union of a family of $\mathcal{B}_2$-atoms $\{a(i_1, i_2) \mid 1 \leq i_2 \leq N(i_1)\}$, and so on, so that each atom $a(i_1, \ldots, i_k)$ of $\mathcal{B}_k$ is the disjoint union of a family of $\mathcal{B}_{k+1}$-atoms $a(i_1, \ldots, i_{k+1}) \mid 1 \leq i_{k+1} \leq N(i_1, \ldots, i_k)\}$.

This indexation simply boils down to identifying β with a rooted tree, with root 0, of which each vertex x is equipped with a weight $w_x > 0$ such that w_x is equal to the sum $\sum w_y$ over all the successors y of x and $w_0 = 1$.

Let $\beta' = (\mathcal{B}'_0, \ldots, \mathcal{B}'_N)$ be another filtration of length N of finite σ-algebras on $(\Omega', \mathbb{P}')$. We say that β' is isomorphic to β if there is a measure preserving bijection $\pi : \mathcal{B}_N \to \mathcal{B}'_N$ that is a Boolean isomorphism, such that its restriction induces a Boolean isomorphism $\pi : \mathcal{B}_k \to \mathcal{B}'_k$ for all $k = 0, \dots, N$. By a Boolean isomorphism $\pi : \mathcal{B} \to \mathcal{B}'$, we mean a bijection such that $\pi(A \cap B) = \pi(A) \cap \pi(B)$, $\pi(\Omega) = \Omega'$ and $\pi(\Omega \setminus A) = \pi(\Omega' \setminus \pi(A))$ for any $A, B \in \mathcal{B}$.

When this holds, we can define $\psi : L_\infty(\Omega, \mathcal{B}_N, \mathbb{P}) \to L_\infty(\Omega', \mathcal{B}'_N, \mathbb{P}')$ as the unique linear map such that

$$\psi(1_A) = 1_{\pi(A)}$$

for any atom $A \in \mathcal{B}_N$. Clearly $\pi(A)$ is an atom in $\mathcal{B}'_N$, and we have $\psi(xy) = \psi(x)\psi(y)$ for any $x, y \in L_\infty(\Omega, \mathcal{B}_N, \mathbb{P})$. Then, it is easy to check that

$$\forall k = 0, \dots, N \quad \mathbb{E}^{\mathcal{B}'_k}\psi = \psi\mathbb{E}^{\mathcal{B}_k}. \tag{5.73}$$

Let $1 \le p \le \infty$. Consider a Banach space B. Since $\mathbb{P}'(\pi(A)) = \mathbb{P}(A)$ for any $A \in \mathcal{B}_N$, ψ extends to an isometric isomorphism

$$\widetilde{\psi} : L_p(\Omega, \mathcal{B}_N, \mathbb{P}; B) \to L_p(\Omega', \mathcal{B}'_N, \mathbb{P}'; B)$$

that takes $L_p(\Omega, \mathcal{B}_k, \mathbb{P}; B)$ to $L_p(\Omega', \mathcal{B}'_k, \mathbb{P}'; B)$ for any $k = 0, \dots, N$. Thus, by (5.73), it defines an isometric isomorphism from the space of β-martingales to that of β'-martingales, when these are equipped with the norms induced by $L_p(\mathcal{B}_N, \mathbb{P}; B)$ and $L_p(\mathcal{B}'_N, \mathbb{P}'; B)$. Moreover, $\widetilde{\psi}$ commutes with all martingale transforms. More explicitly, for any scalars ξ_k we have

$$\sum\nolimits_0^N \xi_k \widetilde{\psi}(df_k) = \widetilde{\psi}\left(\sum\nolimits_0^N \xi_k df_k\right). \tag{5.74}$$

Remark 5.42. Two length N filtrations β, β' are isomorphic iff there is a common indexation of their respective atoms (with the same generating degrees $\{N(i_1, \dots, i_k)\}$) $a(i_1, \dots, i_k)$ and $a'(i_1, \dots, i_k)$ such that

$$\mathbb{P}(a(i_1, \dots, i_k)) = \mathbb{P}'(a'(i_1, \dots, i_k))$$

for any $k = 0, \dots, N$ and any $i_1, \dots, i_k$.

Indeed, setting $a'(i_1, \dots, i_k) = \pi(a(i_1, \dots, i_k))$ produces the desired common indexation. Conversely, given the common indexation, this defines π.

Remark 5.43. Clearly, any β is isomorphic to one β' on the Lebesgue interval, such that all the atoms of $\mathcal{B}'_k$ are intervals for any $k = 0, \dots, N$.

The conventional way to index the sets of indices for β' is to order them from left to right. More precisely, this is done as follows: there are $0 < t_1 < t_2 < \cdots < 1$ such that (up to negligible sets) $a'(1) = [0, t_1)$, $a'(2) = [t_2, t_3)$ and so on. Moreover, for each fixed $k = 1, \dots, N$ and each fixed $i_1, \dots, i_k$ so that $a'(i_1, \dots, i_k) = [a, b)$similarly there are $s_1 < s_2 < \cdots$ (depending on $(i_1, \ldots, i_k))$ such that $a'(i_1, \ldots, i_k, 1) = [a, s_1), a'(i_1, \ldots, i_k, 2) = [s_1, s_2), \ldots$.

In the sequel, it will be convenient to refer to this indexation as the 'conventional indexation' of β'.

Let $\varepsilon > 0$. Let β' be another length N filtration with the same indexation and the same generating degrees as β. We denote its atoms by $\{a'(i_1, \ldots, i_k)\}$. We will say that β, β', assumed both defined on the same probability space $(\Omega, \mathbb{P})$, (e.g. the Lebesgue interval), are ε-close (or that β' is ε-close to β) if for any k and any $i_1, \ldots, i_k$ we have

$$\mathbb{P}(a(i_1, \ldots, i_k)\Delta a'(i_1, \ldots, i_k)) < \varepsilon, \tag{5.75}$$

where $a\Delta b$ means the symmetric difference $(a \cup b) \setminus (a \cap b)$. Obviously we have

$$\mathbb{P}(a\Delta b) < \varepsilon \Rightarrow |\mathbb{P}(a) - \mathbb{P}(b)| < \varepsilon. \tag{5.76}$$

Lemma 5.44. *Fix N and β of length N as earlier. Then there is a function $\delta(\varepsilon)$ that is $o(\varepsilon)$ when $\varepsilon \to 0$ such that, whenever β' is ε-close to β, on the same $(\Omega, \mathbb{P})$, the following holds: For any $f \in L_p(\mathcal{B}_N, \mathbb{P}; B)$ with $\|f\|_{L_p(\mathcal{B}_N, \mathbb{P}; B)} = 1$*

$$\sup_{0 \le k \le N} \|\mathbb{E}^{\mathcal{B}_k}(f) - \mathbb{E}^{\mathcal{B}'_k}(f)\|_{L_p(\mathbb{P}; B)} \le \delta(\varepsilon).$$

Proof. First observe that, by the normalization of f, since β is fixed, there is a constant C_β (depending only on β) such that $\|f\|_\infty \le C_\beta$. Then

$$\mathbb{E}^{\mathcal{B}_k}(f) - \mathbb{E}^{\mathcal{B}'_k}(f) = \sum\nolimits_{i_1, \cdots, i_k} \gamma(i_1, \ldots, i_k)(f) \tag{5.77}$$

where $\gamma(i_1, \ldots, i_k)(f)$ stands for

$$\frac{1_{a(i_1, \cdots, i_k)}}{\mathbb{P}(a(i_1, \ldots, i_k))} \mathbb{E}(f 1_{a(i_1, \cdots, i_k)}) - \frac{1_{a'(i_1, \cdots, i_k)}}{\mathbb{P}(a'(i_1, \ldots, i_k))} \mathbb{E}(f 1_{a'(i_1, \cdots, i_k)}).$$

Note that by (5.75) and (5.76) there is $\delta_1(\varepsilon) = o(\varepsilon)$ such that

$$\left\| \frac{1_{a(i_1, \cdots, i_k)}}{\mathbb{P}(a(i_1, \ldots, i_k))} - \frac{1_{a'(i_1, \cdots, i_k)}}{\mathbb{P}(a'(i_1, \ldots, i_k))} \right\|_p \le \delta_1(\varepsilon)$$

and also

$$\begin{aligned} \|\mathbb{E}(f 1_{a(i_1, \cdots, i_k)}) - \mathbb{E}(f 1_{a'(i_1, \cdots, i_k)})\| &\le \|f\|_\infty \mathbb{P}(a(i_1, \ldots, i_k)\Delta a'(i_1, \ldots, i_k)) \\ &\le C_\beta \varepsilon. \end{aligned}$$

Using these bounds to majorize (5.77), it is easy to conclude. □

A positive number $w > 0$ will be called dyadic if there are integers $k > 0$ and $m \ge 0$ such that $w = 2^{-m}k$.

We will say that a length N filtration β as before is conditionally dyadic if all the numbers $\{\mathbb{P}(a(i_1) \mid i_1 \le N(0)\}$ and, for any $k = 1, \ldots, N$, all the numbers (conditional probabilities)

$$\frac{\mathbb{P}(a(i_1, \ldots, i_k, i_{k+1}))}{\mathbb{P}(a(i_1, \ldots, i_k))}$$

are dyadic.

Remark 5.45. Let $w_i > 0$ $(1 \le i \le n)$ such that $\sum_1^n w_i = 1$. Then for any $\varepsilon > 0$ there are dyadic numbers w_i' such that $\sum w_i' = 1$ and $\sup |w_i - w_i'| < \varepsilon$. This is elementary: Just approximate from what follows all but one of the w_i's by a dyadic number w_i', then the last one is determined by the condition $\sum_1^n w_i = 1$.

Lemma 5.46. *Let β be a length N filtration of finite σ-algebras on* $[0, 1]$, *such that all the atoms appearing are intervals as in Remark 5.43. Then for any $\varepsilon > 0$ there is a conditionally dyadic β' that is ε-close to β.*

Proof. Assume that the common indexation of β and β' is the conventional one. Then, assuming ε small enough, (5.75) simply implies that the extremities of the interval $a(i_1, \ldots, i_k)$ are at distance $< \varepsilon$ to those of $a'(i_1, \ldots, i_k)$ in $[0, 1]$. Conversely, their being at distance $< \varepsilon/2$ implies (5.75). Thus we can reduce the proximity of β' to β to that of the respective endpoints, as long as we use the conventional indexation for both β and β'.

Let us check the lemma by induction on N. The case $N = 0$ is trivial. Assume the lemma proved for a certain N for any $\varepsilon > 0$, with common conventional indexation, and let us prove the same for $N + 1$.

Fix $i_1, \ldots, i_N$. Let $w_i = \mathbb{P}(a(i_1, \ldots, i_N, i)(\mathbb{P}(a(i_1, \ldots, i_N))^{-1}$, and let $n = N(i_1, \cdots, i_N)$. By Remark 5.45 there are dyadic numbers w_i' with $\sum w_i' = 1$ such that $|w_i - w_i'| < \varepsilon$ for all i. Clearly this implies that there is a partition of $a'(i_1, \ldots, i_N)$ in disjoint subintervals $a'(i_1, \ldots, i_N, i)$ of length $w_i'|a'(i_1, \ldots, i_N)|$. Moreover, we can clearly ensure that the end points of $a'(i_1, \ldots, i_N, i)$ are close to those of $a(i_1, \ldots, i_N, i)$. More precisely, we may assume that $a(i_1, \ldots, i_N) = (a, b]$ and $a'(i_1, \ldots, i_N) = (a', b']$ with $\max\{|a - a'|, |b - b'|\} < \varepsilon$. Let $c = b - a$ and $c' = b' - a'$. Then the end points determining the partition of $a(i_1, \ldots, i_N)$ (resp. $a'(i_1, \ldots, i_N)$) are $a, a + cw_1, a + c(w_1 + w_2), \ldots, a + c(\sum w_i)$ (resp. $a', a' + c'w_1', a' + c'(w_1' + w_2'), \ldots, a' + c'(\sum w_i')$). In other words, if we denote $a(i_1, \ldots, i_N, i) = (a_i, b_i]$ and $a'(i_1, \ldots, i_N, i) = (a_i', b_i']$, we find

$$\max\{|a_i - a_i'|, |b_i - b_i'|\} \le |a - a'| + n \sup |cw_i - c'w_i'| \le \varepsilon + 3n\varepsilon.$$

Then the σ-algebra $\mathcal{B}'_{N+1}$ generated by $\{a'(i_1, \dots, i_N, i_{N+1})\}$ is conditionally dyadic since the ratios $|a'(i_1, \dots, i_N, i)||a'(i_1, \dots, i_N)|^{-1} = w'_i$ are all dyadic. Thus the resulting $(\mathcal{B}'_0, \dots, \mathcal{B}'_{N+1})$ is conditionally dyadic, it is $\varepsilon(3n+1)$-close to $(\mathcal{B}_0, \dots, \mathcal{B}_{N+1})$, and its indexation is conventional. Since $\varepsilon > 0$ is arbitrary, this completes the induction step, and hence the proof. □

The last lemma, where we embed β into a subsequence of the dyadic filtration $(\mathcal{A}_n)$, is a bit more delicate than the preceding two elementary ones:

Lemma 5.47. *Let $\beta = (\mathcal{B}_0, \dots, \mathcal{B}_N)$ be a conditionally dyadic filtration of length N of finite σ-algebras on* $[0, 1]$. *We assume all the atoms are intervals. Then there are $0 = m(0) \le m(1) \le \cdots m(N)$ such that $\mathcal{B}_k \subset \mathcal{A}_{m(k)}$ and moreover*

$$\mathbb{E}^{\mathcal{B}_k} = \mathbb{E}^{\mathcal{A}_{m(k)}}\mathbb{E}^{\mathcal{B}_{k+1}} \tag{5.78}$$

for any $k = 0, \dots, N-1$. A fortiori, we have

$$\mathbb{E}^{\mathcal{B}_k} = \mathbb{E}^{\mathcal{A}_{m(k)}}\mathbb{E}^{\mathcal{B}_N} \tag{5.79}$$

for any $k = 0, \dots, N$.

Proof. We first verify that (5.78) implies (5.79). Indeed, if $k = N$, (5.79) follows from $\mathcal{B}_N \subset \mathcal{A}_{m(N)}$, and if $k = N-1$ from (5.78). Now if $k+1 < N$, (5.78) yields

$$\mathbb{E}^{\mathcal{B}_k} = \mathbb{E}^{\mathcal{A}_{m(k)}}\mathbb{E}^{\mathcal{B}_{k+1}} = \mathbb{E}^{\mathcal{A}_{m(k)}}\mathbb{E}^{\mathcal{A}_{m(k+1)}}\mathbb{E}^{\mathcal{B}_{k+2}} = \mathbb{E}^{\mathcal{A}_{m(k)}}\mathbb{E}^{\mathcal{B}_{k+2}} = \cdots = \mathbb{E}^{\mathcal{A}_{m(k)}}\mathbb{E}^{\mathcal{B}_N},$$

showing that (5.78) implies (5.79) (incidentally the converse is also true).

We will prove this lemma by induction on N. The case $N = 0$ is trivial. Assume the lemma known up to N and let us prove it for $N+1$. We use the conventional indexation for the atoms of $\mathcal{B}_N$. Consider an atom $a(i_1, \dots, i_N)$ of $\mathcal{B}_N$. As earlier we set $w_i = |a(i_1, \dots, i_N, i)||a(i_1, \dots, i_N)|^{-1}$, but now we assume all w_i's dyadic. Since (by the induction hypothesis) $\mathcal{B}_N \subset \mathcal{A}_{m(N)}$ we know $a(i_1, \dots, i_N) \in \mathcal{A}_{m(N)}$, and hence there is a family $\{I^t \mid t \in \mathcal{F}(i_1, \dots, i_N)\}$ of atoms of $\mathcal{A}_{m(N)}$, such that $a(i_1, \dots, i_N) = \cup\{I^t \mid t \in \mathcal{F}(i_1, \dots, i_N)\}$. Let m be such that (using a common denominator) $w_i = 2^{-m}k_i$ for some integer k_i. By the same (common denominator) argument we may also choose m common to all $a(i_1, \dots, i_N)$, so that m does not depend on $i_1, \dots, i_N$.

Recall that each atom I^t of $\mathcal{A}_{m(N)}$ is the union of 2^m disjoint subintervals in $\mathcal{A}_{m(N)+m}$ of length $2^{-m}|I^t|$. Clearly we may find a partition of I^t in $\mathcal{A}_{m(N)+m}$, which we denote by $\{I^t(i_{N+1}) \mid i_{N+1} \le N(i_1, \dots, i_N)\}$, such that

$$\forall i \le N(i_1, \dots, i_N) \quad \frac{|I^t(i)|}{|I^t|} = w_i.$$

Then we set $m(N+1) = m(N) + m$ and for all $i \le N(i_1, \dots, i_N)$

$$a(i_1, \dots, i_N, i) = \cup\{I^t(i) \mid t \in \mathcal{F}(i_1, \dots, i_N)\}.$$

Recall that $I^t(i) \subset I^t$ and I^t is an atom of $\mathcal{A}_{m(N)}$ so that

$$\mathbb{E}^{\mathcal{A}_{m(N)}}(1_{I^t(i)}) = 1_{I^t}\frac{|I^t(i)|}{|I^t|} = 1_{I^t} w_i. \tag{5.80}$$

Then the σ-algebra $\mathcal{B}_{N+1}$ generated by $\{a(i_1, \dots, i_N, i_{N+1})\}$ is included in $\mathcal{A}_{m(N+1)}$, and we claim that

$$\mathbb{E}^{\mathcal{B}_N} = \mathbb{E}^{\mathcal{A}_{m(N)}}\mathbb{E}^{\mathcal{B}_{N+1}}. \tag{5.81}$$

Indeed, it suffices to verify this on the indicator $f = 1_{a(i_1,\cdots,i_N,i)}$ of an atom of $\mathcal{B}_{N+1}$, for which of course $\mathbb{E}^{\mathcal{B}_{N+1}}(f) = f$. But then by (5.80)

$$\mathbb{E}^{\mathcal{A}_{m(N)}}(f) = \sum_{t\in\mathcal{F}(i_1,\cdots,i_N)} \mathbb{E}^{\mathcal{A}_{m(N)}}(1_{I^t(i)}) = \sum_{t\in\mathcal{F}(i_1,\cdots,i_N)} 1_{I^t} w_i = 1_{a(i_1,\cdots,i_N)} w_i$$

and

$$\mathbb{E}^{\mathcal{B}_N}(f) = 1_{a(i_1,\cdots,i_N)}|a(i_1, \dots, i_N, i)||a(i_1, \dots, i_N)|^{-1} = 1_{a(i_1,\cdots,i_N)} w_i.$$

This completes the induction step, and hence the proof. □

Remark. Conversely, (5.78) clearly implies that β is conditionally dyadic.

Proposition 5.48. *Let $(f_0, \dots, f_N)$ be a finite martingale in $L_p([0,1], dt; B)$ formed of simple functions. Then for any $\varepsilon > 0$ and $1 \le p < \infty$ there are integers $0 = m(0) \le m(1) \le \cdots m(N)$ and an $\mathcal{A}_{m(N)}$-measurable $f'_N \in L_p([0,1], dt; B)$ such that the martingale $f'_k = \mathbb{E}^{\mathcal{A}_{m(k)}} f'_N$ satisfies*

$$\sup\nolimits_{0\le k\le N} \|f_k - f'_k\|_{L_p(\mathbb{P};B)} \le \varepsilon.$$

Proof. Let $\mathcal{B}_k = \sigma(f_0, \dots, f_k)$. By Remark 5.43 we may assume that all the atoms of $(\mathcal{B}_0, \dots, \mathcal{B}_N)$ are intervals. By Lemmas 5.46 and 5.44 we can find a conditionally dyadic filtration $(\mathcal{B}'_0, \dots, \mathcal{B}'_N)$ such that the martingale (f'_k) defined by $f'_k = \mathbb{E}^{\mathcal{B}'_k} f_N$ satisfies $\sup_{0\le k\le N} \|f_k - f'_k\|_{L_p(\mathbb{P};B)} < \varepsilon$. By Lemma 5.47, there are $0 = m(0) \le m(1) \le \cdots m(N)$ such that $f'_k = \mathbb{E}^{\mathcal{A}_{m(k)}} f'_N$ for any $k = 1, \dots, N$. This completes the proof. □

Theorem 5.49. *Let $1 < p < \infty$. To compute the UMD_p constant of a Banach space B, we may restrict ourselves to martingale differences relative to the dyadic filtration on [0,1], i.e. the unconditionality constant of the dyadic case in $L_p(B)$ dominates that of any martingale difference sequence.*

Proof. Assume we know that

$$\left\|\sum \varepsilon_n df_n\right\|_{L^p(B)} \leq C\left\|\sum df_n\right\|_{L^p(B)} \tag{5.82}$$

for any $\varepsilon_n = \pm 1$ and any finite dyadic martingale (f_n) relative to the dyadic filtration $(\mathcal{A}_n)$. Then, by an obvious blocking argument, the same still holds for martingales of arbitrary finite length N relative to a subsequence $(\mathcal{A}_{m(k)})$ $(0 = m(0) \leq \cdots \leq m(k) \leq \cdots)$ of the dyadic filtration. Thus (5.82) holds for any (f'_k) of the form $f'_k = \mathbb{E}^{\mathcal{A}_{m(k)}} f'_N$ for $0 \leq k \leq N$. By Proposition 5.48, (5.82) holds for any martingale $(f_0, \ldots, f_N)$ formed of simple functions on $[0, 1]$, and hence by Remark 5.43 and (5.74) on any probability space, and lastly by Lemma 5.41 for any martingale $(f_0, \ldots, f_N)$. □

5.9 The Burkholder-Rosenthal inequality

We now turn to what we call the Burkholder-Rosenthal inequality, because Burkholder apparently was inspired by Rosenthal's discovery of this inequality for sums of independent random variables.

Let $(f_n)_{n\geq 0}$ be a scalar (or Hilbert space valued) martingale in L_2. We will denote by $\sigma(f)$ the 'conditioned square function', namely

$$\sigma(f) = \left(\|f_0\|^2 + \sum\nolimits_1^\infty \mathbb{E}_{n-1}\|df_n\|^2\right)^{1/2}. \tag{5.83}$$

We will also denote

$$d^*(f) = \sup_{n\geq 0} \|df_n\|.$$

We have then

Theorem 5.50 (Burkholder-Rosenthal inequality). *For any $2 \leq p < \infty$, there are positive constants α'_p, β'_p such that any scalar or Hilbert space valued martingale $(f_n)_{n\geq 0}$ in L_p satisfies*

$$\alpha'_p(\|\sigma(f)\|_p + \|d^*(f)\|_p) \leq \sup_{n\geq 0} \|f_n\|_p \leq \beta'_p[\|\sigma(f)\|_p + \|d^*(f)\|_p] \tag{5.84}$$

Proof. We will prove this in the scalar case only. The Hilbert space case is identical. For short, we will write σ and d^* instead of $\sigma(f)$ and $d^*(f)$. By convention, we set $\mathbb{E}_{-1}|d_0|^2 = |d_0|^2$. Recall that we have by (5.35)

$$a_p\|S\|_p \leq \sup \|f_n\|_p \leq b_p\|S\|_p. \tag{5.85}$$

Since $p/2 \geq 1$, by Theorem 1.26 (the dual to Doob's inequality) we have on one hand

$$\left\| \sum \mathbb{E}_{n-1}|d_n|^2 \right\|_{p/2} \leq (p/2) \left\| \sum |d_n|^2 \right\|_{p/2}$$

therefore

$$\|\sigma\|_p \leq (p/2)^{1/2}\|S\|_p.$$

On the other hand, by Doob's (1.30) we have

$$\|f^*\|_p \leq p' \sup \|f_n\|_p;$$

since $d^* \leq 2f^*$, this last inequality implies

$$\|d^*\|_p \leq 2p' \sup \|f_n\|_p.$$

Therefore we obtain

$$\|\sigma\|_p + \|d^*\|_p \leq ((p/2)^{1/2}a_p^{-1} + 2p') \sup \|f_n\|_p.$$

For the other side, we will estimate

$$S^2 - \sigma^2 = \sum\nolimits_1^\infty |d_n|^2 - \mathbb{E}_{n-1}|d_n|^2.$$

Since $d'_n = |d_n|^2 - \mathbb{E}_{n-1}|d_n|^2$ are martingale differences, we have by (5.35) and (5.68)

$$\|S^2 - \sigma^2\|_{p/2} \leq b_{p/2} \left\| \left(\sum |d'_n|^2 \right)^{1/2} \right\|_{p/2} \leq b_{p/2}(\mathrm{I} + \mathrm{II}),$$

where $\mathrm{I} = \left\| \left(\sum |d_n|^4\right)^{1/2} \right\|_{p/2}$ and

$$\mathrm{II} = \left\| \left(\sum (\mathbb{E}_{n-1}|d_n|^2)^2 \right)^{1/2} \right\|_{p/2} \leq \left\| \sum \mathbb{E}_{n-1}|d_n|^2 \right\|_{p/2} = \|\sigma\|_p^2.$$

But now $\left(\sum |d_n|^4\right)^{1/4} \leq (Sd^*)^{1/2}$ and hence by Hölder

$$\mathrm{I} = \left\| \left(\sum |d_n|^4 \right)^{1/4} \right\|_p^2 \leq \|(Sd^*)^{1/2}\|_p^2 \leq \|S\|_p \|d^*\|_p.$$

Therefore by the arithmetic/geometric mean inequality for any $t > 0$

$$\sqrt{\mathrm{I}} \leq 2^{-1}(t\|S\|_p + t^{-1}\|d^*\|_p).$$

Recapitulating, this gives us since $\sqrt{\mathrm{I} + \mathrm{II}} \leq \sqrt{\mathrm{I}} + \sqrt{\mathrm{II}}$

$$\begin{aligned} \||S^2 - \sigma^2|^{1/2}\|_p &\leq (b_{p/2})^{1/2}(\sqrt{\mathrm{I}} + \sqrt{\mathrm{II}}) \\ &\leq (b_{p/2})^{1/2}(2^{-1}t\|S\|_p + 2^{-1}t^{-1}\|d^*\|_p + \|\sigma\|_p). \end{aligned}$$

But now

$$S = \sqrt{S^2} \le |S^2 - \sigma^2|^{1/2} + \sigma$$

therefore $\|S\|_p \le (b_{p/2})^{1/2}(2^{-1}t\|S\|_p + 2^{-1}t^{-1}\|d^*\|_p + \|\sigma\|_p) + \|\sigma\|_p$. Thus, if we choose t so that $(b_{p/2})^{1/2}t = 1$ we find

$$\|S\|_p \le 2^{-1}\|S\|_p + 2^{-1}b_{p/2}\|d^*\|_p + (b_{p/2})^{1/2}\|\sigma\|_p + \|\sigma\|_p,$$

which implies $\|S\|_p \le b_{p/2}\|d^*\|_p + 2(b_{p/2})^{1/2}\|\sigma\|_p + 2\|\sigma\|_p$, so that we obtain the desired inequality with $\beta'_p = \max\{b_{p/2}, 2(b_{p/2})^{1/2} + 2\}$. □

Lemma 5.51. *Let $2 \le p \le \infty$. Any scalar (or Hilbert space) valued martingale $(f_n)_{n\ge 0}$ satisfies*

$$\left(\sum\nolimits_0^\infty \|df_n\|_p^p\right)^{1/p} \le 2^{1/p'}\|f\|_p$$

Proof. Consider f in L_∞. Let $f_n = \mathbb{E}_n f$. We have trivially both

$$\left(\sum\nolimits_0^\infty \|df_n\|_2^2\right)^{1/2} \le \|f\|_2 \quad \text{and} \quad \sup_n \|df_n\|_\infty \le 2\|f\|_\infty.$$

Therefore the inequality follows by the Riesz interpolation Theorem, stated as Corollary 8.16 in this volume (see §8.2 for more on complex interpolation). □

Let us denote for (f_n) as in (5.83)

$$\sigma_p(f) = \left(\|f_0\|^p + \sum\nolimits_1^\infty \|df_n\|^p\right)^{1/p}.$$

Note that

$$\|\sigma_p(f)\|_p = \left(\sum\nolimits_0^\infty \|df_n\|_p^p\right)^{1/p}.$$

Then the following variant of the Burkholder-Rosenthal inequality is particularly useful:

Theorem 5.52. *Let $2 \le p < \infty$. Let $\alpha''_p = 2^{-1}\min(\alpha'_p, 2^{-1/p'})$. Any scalar (or Hilbert space) valued martingale $(f_n)_{n\ge 0}$ satisfies*

$$\alpha''_p(\|\sigma(f)\|_p + \|\sigma_p(f)\|_p) \le \sup \|f_n\|_p \le \beta'_p(\|\sigma(f)\|_p + \|\sigma_p(f)\|_p). \quad (5.86)$$

Proof. The first inequality follows from Lemma 5.51. Moreover, we have trivially $d^*(f) \le \sigma_p(f)$ and hence

$$\|d^*(f)\|_p \le \|\sigma_p(f)\|_p.$$

Thus we obtain the second inequality from (5.84). □

Remark 5.53. In [295, §6], the following dual form of (5.86) is proved for any $1 < p \le 2$: The norm $\sup \|f_n\|_p$ is equivalent to the following one:

$$\inf\{\|\sigma(g)\|_p + (\sum \|dh_n\|_p^p)^{1/p}\}$$

where the inf runs over all possible decompositions $df_n = dg_n + dh_n$ in L_p.

Corollary 5.54. *Let (Y_n) be independent random variables in L_p, $2 < p < \infty$, with mean zero, i.e. $\mathbb{E}Y_n = 0$ for all n. Then the series $f = \sum Y_n$ converges in L_p iff both $\sum \|Y_n\|_2^2 < \infty$ and $\sum \|Y_n\|_p^p < \infty$. Moreover, we have*

$$\alpha_p'' \left[\left(\sum \|Y_n\|_2^2\right)^{1/2} + \left(\sum \|Y_n\|_p^p\right)^{1/p}\right]$$
$$\le \|f\|_p \le \beta_p' \left[\left(\sum \|Y_n\|_2^2\right)^{1/2} + \left(\sum \|Y_n\|_p^p\right)^{1/p}\right].$$

Proof. Let $\mathcal{A}_n = \sigma(Y_0, Y_1, \ldots, Y_n)$. Clearly, since the (Y_n)s are independent, we have $\mathbb{E}^{\mathcal{A}_{n-1}}|Y_n|^2 = \mathbb{E}|Y_n|^2$, hence if $f = \sum Y_n$ (i.e. $df_n = Y_n$), we have $\sigma(f) = (\sum \|Y_n\|_2^2)^{1/2}$ and $\|\sigma_p(f)\|_p = (\sum \|Y_n\|_p^p)^{1/p}$. Thus the result follows from (5.86). □

Corollary 5.55. *Let $(\Omega, \mathcal{A}, \mathbb{P})$ be a probability space. Let $(\mathcal{C}_n)$ be a sequence of independent σ-subalgebras of $\mathcal{A}$. Let $\Sigma_p \subset L_p(\Omega, \mathcal{A}, \mathbb{P})$ be the closure of the linear space of all the finite sums $\sum Y_n$ with $Y_n \in L_p$, $\mathcal{C}_n$-measurable and with $\mathbb{E}Y_n = 0$ for all n. Then the orthogonal projection $Q\colon\ L_2 \to \Sigma_2$, defined by*

$$\forall f \in L_2 \qquad Q(f) = \sum (\mathbb{E}^{\mathcal{C}_n}(f) - \mathbb{E}(f))$$

is bounded on L_p for all $1 < p < \infty$.

Proof. By duality, it suffices to show this for $2 < p < \infty$. Let $f \in L_p$ and let $\mathcal{A}_n = \sigma(\mathcal{C}_0, \mathcal{C}_1, \ldots, \mathcal{C}_n)$. As usual we set $df_n = \mathbb{E}_n f - \mathbb{E}_{n-1} f$. We may as well assume $\mathcal{A} = \mathcal{A}_\infty$. Then Σ_p clearly coincides with the set of all f in L_p such that df_n is $\mathcal{C}_n$-measurable for all n, and we have $Qf = \sum \mathbb{E}^{\mathcal{C}_n} df_n$ for all f in L_2. Assuming $p > 2$, we have

$$\left(\sum \|\mathbb{E}^{\mathcal{C}_n} df_n\|_2^2\right)^{1/2} \le \left(\sum \|df_n\|_2^2\right)^{1/2} = \|f\|_2 \le \|f\|_p,$$

and by interpolation between the cases $p = \infty$ and $p = 2$ (as in Lemma 5.51) we have

$$\left(\sum \|\mathbb{E}^{\mathcal{C}_n} df_n\|_p^p\right)^{1/p} \le 2^{1/p'} \|f\|_p. \tag{5.87}$$

Therefore, by Corollary 5.54 we find

$$\|Qf\|_p \le \beta_p' \left[\left(\sum \|\mathbb{E}^{\mathcal{C}_n} df_n\|_2^2\right)^{1/2} + \left(\sum \|\mathbb{E}^{\mathcal{C}_n} df_n\|_p^p\right)^{1/p}\right] \le \beta_p'(1 + 2^{1/p'})\|f\|_p,$$

which means $\|Q\colon\ L_p \to L_p\| \le \beta_p'(1 + 2^{1/p'})$. □

Corollary 5.56. *Let $p \geq 2$. Let (Y_n) be a sequence of independent mean zero random variables in $L_p(\Omega, \mathcal{A}, \mathbb{P})$ with $\|Y_n\|_p = 1$. Let $w_n = \|Y_n\|_2$ and $w = (w_n)$. Let $x = (x_n)$ be a scalar sequence. Then the series $\sum x_n Y_n$ converges in L_p iff both $\sum w_n^2 |x_n|^2 < \infty$ and $\sum |x_n|^p < \infty$. Let $X_{p,w}$ be the space of all such sequences with norm $\|x\|_{p,w} = \left(\sum w_n^2 |x_n|^2\right)^{1/2} + \left(\sum |x_n|^p\right)^{1/p}$. We have then*

$$\alpha''_p \|x\|_{p,w} \leq \left\| \sum x_n Y_n \right\|_p \leq \beta'_p \|x\|_{p,w}.$$

Therefore, as a Banach space, the span in L_p of (Y_n) depends only on $w = (w_n)$.

Proof. This is immediate from (5.86) (see Corollary 5.54). □

Corollary 5.57. *Let $p \geq 2$. Let $w_n > 0$. Let (Y_n) be a sequence of independent symmetric random variables with $\|Y_n\|_2 = w_n$, $\|Y_n\|_p = 1$ and such that, for each n, $|Y_n|$ has only one non-zero value. Then the orthogonal projection P onto the closed span of (Y_n) in L_2 is bounded on L_p. Consequently, the space $X_{p,w}$ is isomorphic to a complemented subspace of L_p.*

Proof. An elementary calculation shows that, since $|Y_n|$ is a multiple of an indicator function we have

$$\|Y_n\|_p \|Y_n\|_{p'} = \|Y_n\|_2^2,$$

and hence, since $\|Y_n\|_p = 1$, $\|Y_n\|_{p'} = \|Y_n\|_2^2$. Let C_n be the σ-algebra generated by Y_n. Let Q be as in Corollary 5.55. Note that $\langle f, Y_n \rangle = \langle \mathbb{E}^{C_n}(f) - \mathbb{E}(f), Y_n \rangle$ for all n. We have

$$\forall f \in L_2 \qquad Pf = \sum \|Y_n\|_2^{-2} \langle f, Y_n \rangle Y_n = \sum \|Y_n\|_2^{-2} \langle Qf, Y_n \rangle Y_n.$$

We have clearly, on one hand $\|Pf\|_2 \leq \|f\|_2 \leq \|f\|_p$, and on the other one

$$\sum |\|Y_n\|_2^{-2} \langle f, Y_n \rangle|^p = \sum \|Y_n\|_2^{-2p} |\langle \mathbb{E}^{C_n}(f) - \mathbb{E}(f), Y_n \rangle|^p.$$

By Hölder, the latter is

$$\begin{aligned} &\leq \sum (\|Y_n\|_2^{-2} \|Y_n\|_{p'})^p \|\mathbb{E}^{C_n}(f) - \mathbb{E}(f)\|_p^p = \sum \|\mathbb{E}^{C_n}(f) - \mathbb{E}(f)\|_p^p \\ &\leq (2^{1/p'} \|f\|_p)^p, \end{aligned}$$

where at the last step we used (5.87); therefore Corollary 5.56 yields

$$\|Pf\|_p \leq \beta'_p (1 + 2^{1/p'}) \|f\|_p.$$ □

Remark 5.58. In the preceding statement, the dual of the space $X_{p,w}$ can be identified with the closed span in $L_{p'}$ of the variables (Y_n). Since $X_{p,w}$ is the intersection of ℓ_p with a weighted ℓ_2-space, its dual is the sum of the respective duals. It follows that, the series $\sum x_n Y_n \|Y_n\|_{p'}^{-1}$ converges in $L_{p'}$ iff (x_n)

admits a decomposition of the form $x_n = a_n + b_n$ with both $\sum |a_n|^{p'} < \infty$ and $\sum |w_n^{-1} b_n|^2 < \infty$, and the corresponding norms are equivalent.

Remark 5.59. Fix $p > 2$. Let $q = 2p/(p-2)$ so that $1/2 = 1/p + 1/q$. By Hölder, we have

$$\left(\sum w_n^2 |x_n|^2\right)^{1/2} \leq \left(\sum |x_n|^p\right)^{1/p} \left(\sum |w_n|^q\right)^{1/q},$$

so that, on one hand, if $\sum |w_n|^q < \infty$, then $X(p, w) = \ell_p$ and on the other hand, if $\inf w_n > 0$, obviously $X(p, w) = \ell_2$. Now if $w = (w_n)$ splits as the disjoint union of a sequence such that $\sum |w_n|^q < \infty$ and one such that $\inf w_n > 0$, then $X(p, w)$ is isomorphic to $\ell_p \oplus \ell_2$. If none of these three cases happens, w must satisfy both $\liminf w_n = 0$ and $\sum_{n:w_n<\varepsilon} |w_n|^q = \infty$ for any $\varepsilon > 0$. Rosenthal proved that the resulting space $X(p, w)$ is actually independent of w up to isomorphism, as long as w satisfies the latter requirements. More precisely, if w and w' are two sequences both satisfying these, then $X(p, w)$ and $X(p, w')$ are isomorphic to the same Banach space, which therefore can be denoted simply by X_p.

Historically, this space was the first example of a genuinely new $\mathcal{L}_p$-space in the sense of [332], (for $1 < p \neq 2 < \infty$ this means simply that it is isomorphic to a complemented subspace of L_p but not isomorphic to Hilbert space) one that was not obtained by direct sums from the classical examples ℓ_2, ℓ_p or L_p. Shortly after that breakthrough, uncountably many examples of $\mathcal{L}_p$-spaces were produced in [145].

5.10 Stein inequalities in UMD spaces

Bourgain [137] observed that the UMD property of a Banach space B implies a certain B-valued version of Stein's inequality. In its most classical form, Stein's inequality is as follows. Consider a filtration $(\mathcal{A}_n)_{n\geq 0}$ on a probability space $(\Omega, \mathbb{P})$ and let $1 < p < \infty$. Then for *any* sequence $(F_n)_{n\geq 0}$ in L_p we have

$$\left\| \left(\sum |\mathbb{E}_n F_n|^2\right)^{1/2} \right\|_p \leq C(p) \left\| \left(\sum |F_n|^2\right)^{1/2} \right\|_p \tag{5.88}$$

where $C(p)$ is a constant depending only on p.

When $p = 1$ this is no longer valid.

As usual in the B-valued case, the 'square function' $\left(\sum |F_n|^2\right)^{1/2}$ must be replaced by an average of $\left\|\sum \varepsilon_n F_n\right\|_B$ over all signs $\varepsilon = (\varepsilon_n)$. In particular, Bourgain proved that, if B is UMD, if $F = (F_n)_{n\geq 0}$ is an *arbitrary* sequence in $L_p(\Omega, \mathbb{P}; B)$ this substitution leads to the following far reaching generalization of Stein's inequality, where the constant $C_p(B)$ is precisely the UMD$_p$ constant of B.

Theorem 5.60. *Assume B UMD. Let $1 < p < \infty$ and let $F = (F_n)_{n\geq 0}$ be an arbitrary sequence in $L_p(\Omega, \mathbb{P}; B)$. If the series $\sum \varepsilon_n F_n$ converges in the space $L_p(\Delta \times \Omega, \nu \times \mathbb{P}; B)$, then $\sum \varepsilon_n \mathbb{E}_n(F_n)$ also converges in $L_p(\Delta \times \Omega, \nu \times \mathbb{P}; B)$ and satisfies*

$$\left\|\sum \varepsilon_n \mathbb{E}_n(F_n)\right\|_{L_p(d\nu\times dP;B)} \leq C_p(B) \left\|\sum \varepsilon_n F_n\right\|_{L_p(d\nu\times dP;B)}. \tag{5.89}$$

Proof. Consider as usual $\Delta = \{-1, +1\}^{\mathbb{N}}$ equipped with the filtration

$$\mathcal{B}_n = \tau(\varepsilon_0, \varepsilon_1, \ldots, \varepsilon_n).$$

Then we define a filtration $(\mathcal{C}_n)_{n\geq 0}$ on $\Omega \times \Delta$ by setting

$$\mathcal{C}_{2j} = \mathcal{A}_j \otimes \mathcal{B}_j$$
$$\mathcal{C}_{2j+1} = \mathcal{A}_{j+1} \otimes \mathcal{B}_j.$$

Note that this is an increasing filtration. Now consider $f \in L_p(\Omega \times \Delta; B)$ defined by

$$f = \sum\nolimits_{n\geq 0} F_n \varepsilon_n.$$

We will apply (5.21) to the martingale

$$f_n = \mathbb{E}^{\mathcal{C}_n}(f).$$

Note that we have

$$f_{2j} = \sum\nolimits_{n\leq j} \mathbb{E}_j(F_n)\varepsilon_n$$

and

$$f_{2j+1} = \sum\nolimits_{n\leq j} \mathbb{E}_{j+1}(F_n)\varepsilon_n.$$

This implies that the increments are of two kinds: On one hand $df_{2j+1} = \sum_{n\leq j} d(F_n)_{j+1}\varepsilon_n$ and on the other one $df_{2j} = \mathbb{E}_j(F_j)\varepsilon_j$.

Thus, by the definition of UMD$_p$ as in (5.21), we find

$$\left\|\sum \mathbb{E}_j(F_j)\varepsilon_j\right\|_{L_p(B)} = \left\|\sum df_{2j}\right\|_{L_p(B)} \leq C_p(B)\|f\|_{L_p(B)}, \tag{5.90}$$

because $\|\sum df_{2j}\|_{L_p(B)} \leq \sup_{\xi_k=\pm 1} \|\sum \xi_k df_k\|_{L_p(B)}$. □

Remark. It seems to be an open question whether conversely any B satisfying (5.89) for some constant must be UMD.

When B is isomorphic to a Hilbert space (and in some sense only then, see [315]), then $\left(\int \|\sum \varepsilon_n x_n\|^2 d\nu\right)^{1/2}$ is equivalent to $\left(\sum \|x_n\|^2\right)^{1/2}$, but in a general Banach space these two ways to measure the 'quadratic variation' of a sequence are quite different. For instance, Bourgain's version of the Stein inequality implies at once the following extensions where ℓ_2 is replaced by ℓ_q $(1 < q < \infty)$ or any UMD space with unconditional basis:

Corollary 5.61. *Let B be a UMD Banach space equipped with an unconditional basis (e_n). Then, in the situation of* (5.88), *we have for any* $1 < p < \infty$

$$\sup_N \left\|\sum_1^N \mathbb{E}_n(F_n)e_n\right\|_{L_p(B)} \le C_p(B) \sup_N \left\|\sum_1^N F_n e_n\right\|_{L_p(B)}.$$

This holds in particular, for $B = \ell_q$ for any $1 < q < \infty$.

Proof. By the unconditionality of (e_n) we have

$$\left\|\sum_1^N F_n e_n\right\|_{L_p(\mathbb{P};B)} = \|\sum_1^N \varepsilon_n F_n e_n\|_{L_p(\nu\times\mathbb{P};B)}.$$

Thus the corollary is an immediate consequence of (5.89). □

Remark 5.62. In [136], Bourgain uses Muckenhoupt's theory of A_p-weights to prove that in the situation of Corollary 5.61, there is a constant c_p such that for any $F = \sum F_n e_n \in L_p(B)$ we have

$$\left\|\sum F_n^* e_n\right\|_{L_p(B)} \le c_p \left\|\sum F_n e_n\right\|_{L_p(B)},$$

$$\left\|\sum S(F_n) e_n\right\|_{L_p(B)} \le c_p \left\|\sum F_n e_n\right\|_{L_p(B)},$$

where F_n^* (resp. $S(F_n)$) is the maximal (resp. square) function of the scalar valued variable F_n. Note that the second inequality is also an easy consequence of Maurey's version of Khintchin's inequality for Banach lattices described in Theorem 10.51.

See [129, 326] for a discussion of Doob's maximal inequality with weights, in connection with the preceding remark when $B = \ell_q$ $(q \neq p)$.

5.11 Burkholder's geometric characterization of UMD space

It seemed natural to search for a natural, hopefully 'geometric', characterization of the UMD property. One first candidate that appeared is 'super-reflexivity'. This is equivalent to uniform convexity up to an equivalent renorming. Indeed, as we explain in §10.4, UMD implies super-reflexive. It is natural to wonder whether the converse holds but this is not so (see §10.4 for more information) and actually, the class of UMD spaces seems much smaller. This seemed to kill the hopes for a nice characterization. However, in [167], Burkholder found a somewhat geometric condition, equivalent to the UMD property, which he named ζ-convexity.

Definition 5.63. A Banach space B is called ζ-convex if there is a function $u\colon B \times B \to \mathbb{R}$ that is biconvex (i.e. separately convex in each of the two variables), satisfying $u(0,0) > 0$ and such that

$$\forall x, y \in B \qquad u(x,y) \le \|x+y\| \quad \text{whenever} \quad \max\{\|x\|, \|y\|\} \ge 1. \tag{5.91}$$

Burkholder's initial definition of ζ-convexity was slightly different, but he later proposed in [171] a more convenient (but equivalent) formulation involving a function u (as in the preceding definition) rather than his original symmetric biconvex function ζ satisfying (5.91) only when $\|x\| \le 1 \le \|y\|$. Note that it is easy to see that if B is ζ-convex there is a largest biconvex function satisfying (5.91). Indeed, the supremum of the family of all such functions belongs to the family. We will denote it by ζ_B.

His initial main result can be summarized as follows:

Theorem 5.64. *A Banach space B is UMD iff it is ζ-convex. More precisely, the best constant $C_{1,\infty}(B)$ in (5.33) satisfies $\frac{2}{\zeta_B(0,0)} \le C_{1,\infty}(B) \le \frac{4}{\zeta_B(0,0)}$.*

Proof. We skip the proof of the upper bound, which can be found in [167] or in [72]. The lower bound follows immediately from Lemma 5.66. Indeed, the latter lemma can be rephrased as saying that $\frac{2}{\zeta_B(0,0)}$ is the smallest C for the inequality

$$\forall \varepsilon_k = \pm 1 \quad \inf\nolimits_\omega \left\| \sum\nolimits_1^n \varepsilon_k df_k(\omega) \right\| \le C \|f_\infty\|_{L_1(B)},$$

and it is obvious that this holds if $C = C_{1,\infty}(B)$. □

In the Hilbert space case, when $B = H$, we have $\zeta_H(x,y) = \|x+y\|$ if $\max\{\|x\|, \|y\|\} \ge 1$ and $\zeta_H(x,y) = [\|x+y\|^2 + (1 - \|x\|^2)((1 - \|y\|^2)]^{1/2}$ otherwise. Thus $\zeta_H(0,0) = 1$.

Remark 5.65. The equality $C_{1,\infty}(B) = \frac{2}{\zeta_B(0,0)}$ for an arbitrary B seems to be still open. If the function ζ_B also satisfies $\zeta_B(x,-x) \le \zeta_B(0,0)$ for all $x \in B$, then the latter equality holds. It holds in particular when $B = H$, so we have

$C_{1,\infty}(H) = 2$. See [167]. Moreover, $C_{1,\infty}(B) = 2$ iff $\zeta_B(0, 0) = 1$ and this happens iff B is isometric to a Hilbert space. This remarkable characterization of Hilbert space was proved in [168] (see also [319, 320]).

Let us say that a function v is biconcave if $-v$ is biconvex.

We will need the notion of zigzag martingale: a martingale $Z_n = (X_n, Y_n)$ $(n \geq 0)$ with values in $B \times B$ will be called zigzag if it is such that for any $n \geq 1$ either $dX_n = 0$ or $dY_n = 0$. The interest of this notion is that if (g_n) is a $\pm$-transform of (f_n) (meaning by this $g_n = g_0 + \sum_1^n \varepsilon_k df_k$ with $\varepsilon_k = \pm 1$) then the martingale $Z_n = (f_n + g_n, f_n - g_n)$ is zigzag, and the converse is also immediate. A priori, we consider martingales with respect to any filtration, on any probability space, but we may clearly restrict to the Lebesgue interval. If (Z_n) is formed of simple functions (resp. dyadic) we say that it is a simple (resp. dyadic) zigzag martingale.

Let $Z(x, y)$ denote the set of simple zigzag martingales $Z_n = (X_n, Y_n)$ $(n \geq 0)$ such that $X_0 = x$, $Y_0 = y$.

Lemma 5.66. *For any $x, y \in B$ we have*

$$\zeta_B(x, y) = \inf \mathbb{E}\|X_\infty + Y_\infty\| \tag{5.92}$$

where the infimum runs over all simple zigzag martingales (X_n, Y_n) in $Z(x, y)$ such that $\|X_\infty - Y_\infty\| \geq 2$ almost surely. Moreover, the infimum remains the same if we restrict it to dyadic zigzag martingales.

Sketch. Let $w(x, y)$ be the right-hand side of (5.92). Since $\|X_\infty - Y_\infty\| \geq 2$ implies $\max\{\|X_\infty\|, \|Y_\infty\|\} \geq 1$, for any martingale (X_n, Y_n) as in (5.92) we have $\zeta_B(X_\infty, Y_\infty) \leq \|X_\infty + Y_\infty\|$ and hence $\mathbb{E}\zeta_B(X_\infty, Y_\infty) \leq \mathbb{E}\|X_\infty + Y_\infty\|$. But then, by (1.20), the biconvexity and the zigzag property imply $\mathbb{E}\zeta_B(X_{n-1}, Y_{n-1}) \leq \mathbb{E}\zeta_B(X_n, Y_n)$ for any $n \geq 1$, and hence

$$\zeta_B(x, y) = \mathbb{E}\zeta_B(X_0, Y_0) \leq \mathbb{E}\zeta_B(X_\infty, Y_\infty) \leq \mathbb{E}\|X_\infty + Y_\infty\|.$$

This shows $\zeta_B(x, y) \leq w(x, y)$. For the converse we claim that w itself is a biconvex function. Taking the claim for granted note that $w(x, y) \leq \|x + y\|$ if $\|x - y\| \geq 2$, as can be seen by taking $(X_n, Y_n) = (x, y)$ for all n. Then if $\max\{\|x\|, \|y\|\} \geq 1$ and $\|x - y\| < 2$, we may assume (say) $\|y\| \geq 1$ and hence $w(-y, y) \leq 0$. The conditions $\|x - y\| < 2$ and $\|y\| \geq 1$ guarantee $x + y \neq 0$ and hence when $t > 0$ is large enough we have $\| - 2y + t(x + y)\| \geq 2$ and hence $w(-y + t(x + y), y) \leq t\|x + y\|$. Thus we may write

$$w(x, y) \leq (1 - t^{-1})w(-y, y) + t^{-1}w(-y + t(x + y), y) \leq \|x + y\|.$$

By the maximality of ζ_B we conclude that $\zeta_B = w$. To check the claim, we will show that $x \mapsto w(x, y)$ is convex when y is fixed. The proof for the other side is identical. Let $Z^{(2)}(x, y) \subset Z(x, y)$ denote the subset formed of the simple zigzag martingales such that a.s. $\|X_\infty - Y_\infty\| \geq 2$. Assume $x = (x_{-1} + x_1)/2$. Fix $\delta > 0$. Let $Z_n^j = (X_n^j, Y_n^j)$ $(j = -1, 1)$ be in $Z^{(2)}(x_j, y)$ on the Lebesgue interval such that $\mathbb{E}\|X_\infty^j + Y_\infty^j\| < w(x_j, y) + \delta$. We can then form a martingale (X_n, Y_n) in $Z^{(2)}(x, y)$ (obtained by 'splicing') as follows. This is the martingale that starts $(n = 0)$ at (x, y) then moves $(n = 1)$ either to (x_{-1}, y) or (x_1, y) with probability $1/2$, and then, after having moved to (x_j, y) it continues $(n \geq 2)$ along the path of $(Z_{n-1}^j)_{n \geq 2}$. It is easy to make this precise, we do not spell out the details (the details are similar to those given later on in the proof of (ii) $\Rightarrow$ (i) in Theorem 10.6). Then we have

$$\begin{aligned} w(x, y) &\leq \mathbb{E}\|X_\infty + Y_\infty\| = (\mathbb{E}\|X_\infty^{-1} + Y_\infty^{-1}\| + \mathbb{E}\|X_\infty^1 + Y_\infty^1\|)/2 \\ &\leq (w(x_{-1}, y) + w(x_1, y))/2 + \delta. \end{aligned}$$

Letting $\delta \to 0$, this shows that $x \mapsto w(x, y)$ is midconvex. Using a biased variable $= \pm 1$ with unequal probability, the same argument shows that it is convex. But this is somewhat irrelevant because it is easy to check that $x \mapsto w(x, y)$ is locally bounded above (say by $4 + \|x + y\|$) and for such functions midconvex implies convex (cf. [82, p. 215]). This observation also proves the last assertion, if we restrict the definition of w to dyadic martingales, we again obtain a bi-midconvex function but, since it is automatically biconvex, it must be the same. □

Theorem 5.67. *Let $1 < p < \infty$ and let $C \geq 1$. A Banach space B is UMD$_p$ with $C_p(B) \leq C$ iff the function $v : B \times B \to \mathbb{R}$ defined by*

$$V(x, y) = \|(x - y)/2\|^p - C^p\|(x + y)/2\|^p$$

admits a biconcave majorant U on $B \times B$.

Proof. Assume $V \leq U$ with U biconcave. We may replace U by the function U_0 defined by

$$U_0(x, y) = \inf_{t \neq 0} |t|^{-p} U(tx, ty).$$

Then it is easy to check that U_0 is still a biconcave majorant of V, and that $U_0(0, 0) \leq 0$. Then, just like earlier for the biconvexity of u, by (1.20), for any martingale $Z_n = (X_n, Y_n)$ in $Z(x, y)$ the biconcavity of U_0 implies $\mathbb{E}U_0(X_n, Y_n) \leq \mathbb{E}U_0(X_{n-1}, Y_{n-1})$ for any $n \geq 1$ and hence $\mathbb{E}U_0(X_\infty, Y_\infty) \leq \mathbb{E}U_0(X_0, Y_0) = U_0(x, y)$. Choosing $x = y = 0 = f_0 = g_0$ and $Z_n = (f_n + g_n, f_n - g_n)$ with

$g_n = \sum_1^n \varepsilon_k df_k$ with $\varepsilon_k = \pm 1$, we find

$$\mathbb{E}V(X_\infty, Y_\infty) \leq \mathbb{E}U_0(X_\infty, Y_\infty) \leq U_0(0,0) \leq 0,$$

which implies $\|g_\infty\|_{L_p(B)} \leq C\|f_\infty\|_{L_p(B)}$. In other words, $C_p(B) \leq C$.

Conversely, assume that $C_p(B) \leq C$. Let

$$U(x,y) = \sup \mathbb{E}V(X_\infty, Y_\infty)$$

where the supremum runs over all martingales $Z_n = (X_n, Y_n)$ in $Z(x,y)$. Obviously $V(x,y) \leq U(x,y)$. Then $C_p(B) \leq C$ implies $U(0,0) \leq 0$ and even $U(x,x) \leq 0$ for any $x \in B$. By the same 'splicing' argument as in the proof of Lemma 5.66, we can check that U is biconcave. A priori, at this stage, U might take the value ∞. But, by the biconcavity, for any $x, y \in B$ $(1/2)(U(x,y) + U(x, 2x-y)) \leq U(x,x) \leq 0$, therefore $U(x,y) < \infty$. This shows that U is a biconcave majorant of V. □

Remark 5.68. As we already mentioned, by classical results (cf. [82, p. 215]) if a function f on B is midconvex (i.e. such that $f((x+y)/2) \leq (f(x)+f(y))/2$ for any $x, y \in B$), and if it is bounded above on an open ball, then it is continuous and hence convex in the usual sense. Thus if the function $-U$ in Theorem 5.67 is only separately midconvex in each variable, it is automatically (continuous and) convex and hence U is (continuous and) biconcave. This observation provides a simpler proof that $C_p(B)$ is attained when we restrict to dyadic martingales (see Theorem 5.49). Indeed, we just note that, if we restrict $Z(x,y)$ to dyadic martingales, the dyadic version of $C_p(B) \leq C$ already implies that V has a biconcave majorant.

The following result from [302] is closely connected to Burkholder's characterization but the meaning of the condition is somewhat easier to grasp, since it involves only the rather standard class of (differences of) convex continuous functions.

Theorem 5.69. *Let B be a real Banach space. We set $X = B \oplus B^*$. Then B is UMD iff the function $\varphi\colon X \to \mathbb{R}$ defined for all $x = (b, b^*) \in X$ by*

$$\varphi(b, b^*) = b^*(b)$$

is the difference of two convex continuous functions on X.

Remark 5.70. Let X be a Banach space. Then $\varphi\colon X \to \mathbb{R}$ is the difference of two convex continuous functions on X iff there is a convex continuous $\psi\colon X \to \mathbb{R}$ such that $\psi \pm \varphi$ are both convex and continuous. Indeed, if $\psi \pm \varphi$ are convex and continuous then $\psi = \varphi_1 - \varphi_2$ with $\varphi_1 = (\psi + \varphi)/2$, $\varphi_2 = (\psi - \varphi)/2$. Conversely, if $\varphi = \varphi_1 - \varphi_2$ with φ_1, φ_2 convex continuous then if $\psi = \varphi_1 + \varphi_2$, ψ is convex continuous and both $\psi + \varphi$ and $\psi - \varphi$ are also convex and continuous. □

In the rest of this section we set $\Omega = \{-1, 1\}^{\mathbb{N}_*}$, equipped with its usual probability denoted here by $\mathbb{P}$, let $\varepsilon_k\colon \Omega \to \{-1, 1\}$ be the k-th coordinate for $k = 1, 2, \ldots$. We set $\mathcal{A}_0 = \{\phi, \Omega\}$ and $\mathcal{A}_n = \sigma(\varepsilon_1, \ldots, \varepsilon_n)$ for all $n \geq 1$. For any $f \in L_1(X)$, as usual we denote $f_n = \mathbb{E}_n f = \mathbb{E}^{\mathcal{A}_n}(f)$, and $df_n = f_n - f_{n-1}$.

We will use the following.

Lemma 5.71. *Let X be any Banach space. Let $V\colon X \to X^*$ be a bounded linear operator. Assume that there is a constant C such that for all finite X valued dyadic martingale (f_n) such that $f_0 = 0$ with limit $f = \sum df_n$, we have*

$$\sum\nolimits_1^\infty \mathbb{E}|V(df_n)(df_n)| \leq C\|f\|^2_{L_\infty(X)}. \tag{5.93}$$

Then there is a constant C' such that for all such (f_n) we actually have

$$\sum\nolimits_1^\infty \mathbb{E}|V(df_n)(df_n)| \leq C'\mathbb{E}\|f\|^2_X. \tag{5.94}$$

Proof. Fix $k \geq 0$. For simplicity we set $d_n = df_n$. We first show that

$$\mathbb{E}_k \sum\nolimits_{n>k} |V(d_n)(d_n)| \leq 4C\|f\|^2_{L_\infty(B)}. \tag{5.95}$$

To check this, we will apply (5.93) to $f - f_k$. We fix $(\varepsilon_1, \ldots, \varepsilon_k)$, and let

$$\forall \omega \in \{-1, 1\}^{\mathbb{N}_*} \quad F(\omega) = f(\varepsilon_1, \ldots, \varepsilon_k, \omega) - f_k(\varepsilon_1, \ldots, \varepsilon_k).$$

Applying (5.93) to F with $(\varepsilon_1, \ldots, \varepsilon_k)$ fixed we find

$$\sum\nolimits_{n>k} \mathbb{E}_\omega |V(d_n)(d_n)|(\varepsilon_1, \ldots, \varepsilon_k, \omega)$$
$$\leq C \sup\nolimits_\omega \|f(\varepsilon_1, \ldots, \varepsilon_k, \omega) - f_k(\varepsilon_1, \ldots, \varepsilon_k)\|^2,$$

which implies (5.95). Furthermore, for any stopping time T_0, this implies

$$\int_{\{T_0=k\}} \sum\nolimits_{n>T_0} |V(d_n)(d_n)| \leq C\mathbb{P}\{T_0 = k\}\|f - f_{T_0}\|^2_{L_\infty(X)},$$

so that summing over k we find

$$\mathbb{E}\left(1_{\{T_0<\infty\}} \sum\nolimits_{n>T_0} |V(d_n)(d_n)|\right) \leq C\mathbb{P}\{T_0 < \infty\}\|f - f_{T_0}\|^2_{L_\infty(X)}. \tag{5.96}$$

Now let $T_1 \geq T_0$ be another stopping time. Replacing f by f_{T_1} in (5.96) we find

$$\mathbb{E}\left(1_{\{T_0<\infty\}} \sum\nolimits_{T_0<n\leq T_1} |V(d_n)(d_n)|\right) \leq C\mathbb{P}\{T_0 < \infty\}\|f_{T_1} - f_{T_0}\|^2_{L_\infty(X)}. \tag{5.97}$$

We will now prove (5.94). We may assume by homogeneity that $\mathbb{E}\|f\|^2_X = 1$. We define $T_0 = \inf\{n \geq 0 \mid \|f_n\| + \|d_{n+1}\| > 1\}$ and for all $m \geq 1$

$$T_m = \inf\{n > T_{m-1} \mid \|f_n - f_{T_{m-1}}\| + \|d_{n+1}\| > 2^m\}.$$

We have then

$$\mathbb{E}\sum_1^\infty |V(d_n)(d_n)| = \mathrm{I} + \mathrm{II}$$

where $\mathrm{I} = \mathbb{E}\sum_{n\le T_0} |V(d_n)(d_n)|$ and $\mathrm{II} = \sum_{m\ge 1}\mathbb{E}\sum_{T_{m-1}<n\le T_m} |V(d_n)(d_n)|$. But by (5.97) we have

$$\mathrm{II} \le \sum_{m\ge 1} C\mathbb{P}\{T_{m-1} < \infty\}\|f_{T_m} - f_{T_{m-1}}\|^2_{L_\infty(X)}. \tag{5.98}$$

$$\text{Since } \|f_{T_m} - f_{T_{m-1}}\| \le 2^m \text{ if } T_m > T_{m-1}$$
$$(\text{because } \|f_{T_m} - f_{T_{m-1}}\| \le \|f_{T_m - 1} - f_{T_{m-1}}\| + \|d_{T_m}\| \le 2^m)$$

and also $f_{T_m} - f_{T_{m-1}} = 0$ if $T_m = T_{m-1}$, we have $\|f_{T_m} - f_{T_{m-1}}\|_{L_\infty(X)} \le 2^m$. Moreover

$$\begin{aligned}\mathbb{P}\{T_{m-1} < \infty\} &\le \mathbb{P}\{\sup_{n,k}\|f_n - f_k\| + \|d_{n+1}\| > 2^{m-1}\}\\ &\le \mathbb{P}\{4\sup\|f_n\| > 2^{m-1}\}\end{aligned}$$

so (5.98) implies (setting $f^* = \sup\|f_n\|$)

$$\mathrm{II} \le C\sum_{m\ge 1}\mathbb{P}\{4f^* > 2^{m-1}\}2^{2m} \le C'\mathbb{E}f^{*2} \le 4C'\mathbb{E}\|f\|_X^2 \le 4C',$$

where at the last step we used Doob's inequality. In addition, by (5.93) applied with f_{T_0} in place of f we have $\mathrm{I} \le C\|f_{T_0}\|^2_{L_\infty(X)}$ and if $T_0 \ge 1$ we again have $\|f_{T_0}\| \le \|f_{T_0 - 1}\| + \|d_{T_0}\| \le 1$ while if $T_0 = 0$ we have $f_{T_0} = 0$. Therefore we find $\mathrm{I} \le C$ and we conclude $\mathbb{E}\sum_1^\infty |V(d_n)(d_n)| \le C + 4C'$. By homogeneity this proves the announced result. □

Proof of Theorem 5.69. Let us equip the space $X = B \oplus B^*$ with the norm $\|(b, b^*)\| = (\|b\|^2 + \|b^*\|^2)^{1/2}$. Note $X^* = B^* \oplus B^{**}$ and $B \subset B^{**}$ so that $B^* \oplus B \subset X^*$ isometrically. Note for further reference that

$$\forall x = (b, b^*) \in X \quad |\varphi(x)| = |b^*(b)| \le \|x\|^2. \tag{5.99}$$

We define $V\colon\ X \to X^*$ as the unique self-adjoint linear map such that $V(x)(x) = \varphi(x)$ for all x in X. Note that $V(b, b^*) = (b^*, b)$ and V is isometric. Equivalently, this definition means

$$\forall x, y \in X \quad V(x)(y) = V(y)(x) = (\varphi(x + y) - \varphi(x - y))/4. \tag{5.100}$$

If $x = (b, b^*)$ and $y = (c, c^*)$, we have $V(x)(y) = (b^*(c) + c^*(b))/2$. Note that $\|V\| \le 1$ (actually $\le 1/2$). Assume that B and hence X is UMD. We claim that if C is the UMD_2 constant of X then any finite X valued dyadic martingale (f_n) in $L_2(X)$ satisfies

$$\sum_1^\infty \mathbb{E}|V(d_n)(d_n)| \le C\mathbb{E}\|f\|_X^2.$$

Indeed, since $V(d_n)(d_n)$ is predictable (recall that in the dyadic case $d_n = \varepsilon_n \varphi_{n-1}$ with φ_{n-1} $\mathcal{A}_{n-1}$-measurable), the random variable $\xi_n = \text{sign}(V(d_n)(d_n))$ is $\mathcal{A}_{n-1}$-measurable (=predictable) so that we can write

$$\sum\nolimits_1^n \mathbb{E}|V(d_n)(d_n)| = \mathbb{E} \sum\nolimits_1^\infty \xi_n V(d_n)(d_n) = \mathbb{E}\left(V\left(\sum\nolimits_1^\infty \xi_n d_n\right)\left(\sum\nolimits_0^\infty d_n\right)\right)$$

and hence recalling Proposition 5.12

$$\begin{aligned} &\le \|V\| \mathbb{E}\left(\left\|\sum\nolimits_1^\infty \xi_n d_n\right\| \left\|\sum\nolimits_0^\infty d_n\right\|\right) \\ &\le \|V\| \left\|\sum\nolimits_1^\infty \xi_n d_n\right\|_{L_2(X)} \left\|\sum\nolimits_0^\infty d_n\right\|_{L_2(X)} \\ &\le C\|f\|^2_{L_2(X)}. \end{aligned}$$

This proves our claim.

We now define for any x in X

$$\psi(x) = \inf\left\{C\mathbb{E}\|f\|_X^2 - \sum\nolimits_1^\infty |\mathbb{E}V(df_n)(df_n)|\right\}$$

where the infimum runs over all finite dyadic martingales (f_n) with $f_0 = x$. Note in passing that $\psi(-x) = \psi(x)$ and also, by the claim

$$\forall x \in X \quad 0 \le \psi(x) \le C\|x\|^2. \tag{5.101}$$

Indeed, the upper bound follows by considering f_n such that $f_n = x$ for all n.

We will now show that for any y in X

$$|V(y)(y)| \le 2^{-1}(\psi(x+y) + \psi(x-y)) - \psi(x). \tag{5.102}$$

Let $\varepsilon > 0$ and let (f_n), (g_n) be such that $f_0 = x + y$, $g_0 = x - y$ and

$$\begin{aligned} C\mathbb{E}\|f\|^2 - \sum\nolimits_1^\infty \mathbb{E}|V(df_n)(df_n)| &< \psi(x+y) + \varepsilon \\ C\mathbb{E}\|g\|^2 - \sum\nolimits_1^\infty \mathbb{E}|V(dg_n)(dg_n)| &< \psi(x-y) + \varepsilon. \end{aligned}$$

We then define a dyadic martingale F_n by setting $F_0 = x$, $F_1 = x + \varepsilon_1 y$ and then for $n > 1$ $F_n(\varepsilon_1, \varepsilon_2, \ldots, \varepsilon_n) = f(\varepsilon_2, \ldots, \varepsilon_n)$ if $\varepsilon_1 = 1$ and $= g(\varepsilon_2, \ldots, \varepsilon_n)$ if $\varepsilon_1 = -1$. We then find since $|V(dF_1)(dF_1)| = |V(y)(y)|$

$$\begin{aligned} \psi(x) &\le C\mathbb{E}\|F\|^2 - \sum\nolimits_1^\infty \mathbb{E}|V(dF_n)(dF_n)| \\ &< 2^{-1}(\psi(x+y) + \psi(x-y)) + \varepsilon - |V(y)(y)| \end{aligned}$$

and hence we obtain (5.102). But then (5.102) can be rewritten as

$$\begin{aligned} |V(y)(y)| &= |2^{-1}(\varphi(x+y) + \varphi(x-y)) - \varphi(x)| \\ &\le 2^{-1}(\psi(x+y) + \psi(x-y)) - \psi(x). \end{aligned} \tag{5.103}$$

This implies that the functions $\chi = \psi \pm \varphi$ are both convex, or more precisely midconvex (meaning by this that $\chi(x) \le 2^{-1}(\chi(x+y) + \chi(x-y))$ for all x, y), but by (5.99) and (5.101), it follows that $\psi \pm \varphi$ are bounded on bounded sets and hence by classical results (cf. [82, p. 215]) they are actually both convex and continuous. By Remark 5.70 this completes the proof of the 'only if' part.

Conversely, assume that φ is the difference of two convex continuous functions. By Remark 5.70 there is ψ convex continuous such that $\psi \pm \varphi$ is convex and hence (5.103) holds. Let (f_n) be a finite X valued dyadic martingale, and let $d_n = df_n$. Applying (5.103) with $x = f_{n-1}$, $y = d_n$ we find for all $n \ge 1$

$$|V(d_n)(d_n)| \le \mathbb{E}_{n-1}(\psi(f_n) - \psi(f_{n-1}))$$

and hence after integration

$$\sum\nolimits_1^\infty \mathbb{E}|V(d_n)(d_n)| \le \mathbb{E}(\psi(f) - \psi(0)).$$

Assume $f_0 = 0$. Since ψ is continuous, there is $r > 0$ such that $\|x\| \le r$ implies $|\psi(x) - \psi(0)| \le 1$. Therefore if $\|f\|_{L_\infty(X)} \le r$ we find $\sum_1^\infty \mathbb{E}|V(d_n)(d_n)| \le 1$. By homogeneity, this implies that

$$\sum\nolimits_1^\infty \mathbb{E}|V(d_n)(d_n)| \le (1/r)^2 \|f\|^2_{L_\infty(X)}, \tag{5.104}$$

and hence, by Lemma 5.71, we obtain (5.94). Let (g_n) be another finite dyadic X valued martingale. Let $d'_n = dg_n$. Clearly by polarization

$$V(d_n)(d'_n) = 4^{-1}(V(d_n + d'_n)(d_n + d'_n) - V(d_n - d'_n)(d_n - d'_n))$$

and hence (5.94) implies

$$\sum\nolimits_1^\infty \mathbb{E}|V(d_n)(d'_n)| \le (C'/4)(\|f+g\|^2_{L_2(X)} + \|f-g\|^2_{L_2(X)}). \tag{5.105}$$

Now let $\xi_n = \pm 1$ be arbitrary signs. Let $\tilde{f} = \sum \xi_n d_n$. Since $V\colon X \to X^*$ is isometric, we have

$$\|\tilde{f}\|_{L_2(X)} = \|V(\tilde{f})\|_{L_2(X^*)}, = \sup\{\mathbb{E}|V(\tilde{f})(g)| \mid g \in B_{L_2(X)}\}. \tag{5.106}$$

But for $g \in B_{L_2(X)}$ we have

$$\mathbb{E}V(\tilde{f})(g) = \mathbb{E}\sum\nolimits_1^\infty V(d\tilde{f}_n)(dg_n) = \mathbb{E}\sum\nolimits_1^\infty \xi_n V(d_n)(dg_n)$$

and hence by (5.105) if $g, f \in B_{L_2(X)}$

$$|\mathbb{E}V(\tilde{f})(g)| \le \mathbb{E}\sum\nolimits_1^\infty |V(d_n)(dg_n)| \le 2C'.$$

Thus, by (5.106), we obtain $\|\tilde{f}\|_{L_2(X)} \le 2C'$. By homogeneity, $\|\tilde{f}\|_{L_2(X)} \le 2C'\|f\|_{L_2(X)}$ for all f. In other words, $C_2(X) \le 2C'$. A fortiori $C_2(B) \le 2C'$. □

5.12 Appendix: hypercontractivity on $\{-1, 1\}$

In this appendix, we reproduce the proof given (already as an appendix) in Beckner's paper [115] for Theorem 5.4 in the case $B = \mathbb{R}$.

Fix $x, y \in \mathbb{R}$. Let

$$f_{x,y}(p) = \left(\frac{\left|x + \frac{1}{\sqrt{p-1}} y\right|^p + \left|x - \frac{1}{\sqrt{p-1}} y\right|^p}{2} \right)^{\frac{1}{p}}.$$

We need to show that $p \mapsto f_{x,y}(p)$ is monotone decreasing on $(1, \infty)$. Assume that we are given a pair $p < q$ such that $f_{x,y}(p) \leq f_{x,y}(q)$ for any $x, y \in \mathbb{R}$. Then a simple duality argument shows that $f_{x,y}(q') \leq f_{x,y}(p')$ for any $x, y \in \mathbb{R}$, where p', q' denote as usual the conjugate exponents.

This shows that it suffices to settle the case $1 < p < q \leq 2$. Moreover, it is easy to see that it suffices to consider the case when $x = 1$ and $y > 0$. In other words, it suffices to check that the function $f_{1,y}$ is monotone decreasing on $(1, 2)$ for any fixed $y > 0$. The monotonicity of this function is equivalent to the inequality

$$\left\{ \frac{\left|1 + \sqrt{\frac{p-1}{q-1}} y\right|^q + \left|1 - \sqrt{\frac{p-1}{q-1}} y\right|^q}{2} \right\}^{1/q} \leq \left\{ \frac{|1 + y|^p + |1 - y|^p}{2} \right\}^{1/p}$$

for $1 < p \leq q \leq 2$. Suppose we have proved this inequality for $0 < y \leq 1$. Then we can obtain the case $y > 1$ from this result. For observe that if $0 < w \leq 1$, we have $|w \pm t|^2 \leq |1 \pm tw|^2$ for any $0 < t < 1$, and hence

$$\left| w \pm \sqrt{\frac{p-1}{q-1}} \right|^2 \leq \left| 1 + \sqrt{\frac{p-1}{q-1}} w \right|^1;$$

so we have

$$\left\{ \frac{\left|w + \sqrt{\frac{p-1}{q-1}}\right|^q + \left|w - \sqrt{\frac{p-1}{q-1}}\right|^q}{2} \right\}^{1/q} \leq \left\{ \frac{\left|1 + \sqrt{\frac{p-1}{q-1}} w\right|^q + \left|1 - \sqrt{\frac{p-1}{q-1}} w\right|^q}{2} \right\}^{1/q}$$

$$\leq \left\{ \frac{|1 + w|^p + |1 - w|^p}{2} \right\}^{1/p}.$$

Now dividing through both sides of this equation by a factor w and setting $y = 1/w$, we obtain the preceding inequality for the case $y > 1$. Thus we need only to show the inequality

$$\left\{\frac{\left|1+\sqrt{\frac{p-1}{q-1}}y\right|^q+\left|1-\sqrt{\frac{p-1}{q-1}}y\right|^q}{2}\right\} \le \left\{\frac{|1+y|^p+|1-y|^p}{2}\right\}^{1/p}$$

for the restricted case $0 < y \le 1$ and $1 < p \le q \le 2$. With the use of the binomial expansion, this inequality is equivalent to

$$\left[\sum\nolimits_{k=0}^{m}\binom{q}{2k}\left(\tfrac{p-1}{q-1}\right)^k y^{2k}\right]^{p/q} \le \sum\nolimits_{k=0}^{\infty}\binom{p}{2k}y^{2k},$$

and for $1 < p \le q \le 2$ the binomial coefficients $\binom{p}{2k}$ and $\binom{q}{2k}$ are both positive, and in addition

$$\frac{p}{q}\binom{q}{2k}\left(\frac{p-1}{q-1}\right)^k \le \binom{p}{2k}.$$

Using the elementary result that for $0 < \lambda \le 1$ and $x > 0$

$$(1+x)^\lambda \le 1+\lambda x,$$

we have

$$\begin{aligned}\left[1+\sum\nolimits_{k=1}^{\infty}\binom{q}{2k}\left(\frac{p-1}{q-1}\right)^k y^{2k}\right]^{p/q} &\le 1+\frac{p}{q}\sum\nolimits_{k=1}^{\infty}\binom{q}{2k}\left(\frac{p-1}{q-1}\right)^k y^{2k}\\ &\le 1+\sum\nolimits_{k=1}^{\infty}\binom{p}{2k}y^{2k}.\end{aligned}$$

5.13 Appendix: Hölder-Minkowski inequality

For further reference, we wish to review here a classical set of inequalities usually referred to as 'the Hölder-Minkowski inequality'. Let $0 < q \le p \le \infty$ and let $(\Omega, \mathcal{A}, m)$ be any measure space. Consider a sequence (x_n) in $L_p(\Omega, \mathcal{A}, m)$. Then

$$\left\|\left(\sum |x_n|^q\right)^{1/q}\right\|_p \le \left(\sum \|x_n\|_p^q\right)^{1/q}. \tag{5.107}$$

Indeed, this is an easy consequence of the fact (since $p/q > 1$) that $L_{p/q}$ is a normed space. In particular, when $q = 1$ we find

$$\left\|\sum |x_n|\right\|_p \le \sum \|x_n\|_p$$

that is but the triangle inequality in L_p. If $0 < p \le q \le \infty$, the inequality is reversed: we have

$$\left\| \left(\sum |x_n|^q \right)^{1/q} \right\|_p \ge \left(\sum \|x_n\|_p^q \right)^{1/q}. \tag{5.108}$$

In particular, when $q = \infty$, we find simply the obvious inequality

$$\| \sup_n |x_n| \|_p \ge \sup_n \|x_n\|_p.$$

One way to check (5.108) is to set $r = q/p$, $r' = r(r-1)^{-1}$ and $y_n = |x_n|^p$. Then (5.108) is the same as

$$\left\| \left(\sum |y_n|^r \right)^{1/r} \right\|_1 \ge \left(\sum \|y_n\|_1^r \right)^{1/r}$$

that is easy to derive from

$$\left(\sum |y_n|^r \right)^{1/r} = \sup \left\{ \sum \alpha_n |y_n| \mid \alpha_n \ge 0 \sum |\alpha_n|^{r'} \le 1 \right\}.$$

Indeed, we find

$$\int \left(\sum |y_n|^r \right)^{1/r} \ge \sup_{\sum |\alpha_n|^{r'} \le 1} \int \sum |\alpha_n| \, |y_n| = \left(\sum \|y_n\|_1^r \right)^{1/r}.$$

In its simplest form (5.107) and (5.108) reduce to: $\forall x, y \in L_p$

$$\|(|x|^q + |y|^q)^{1/q}\|_p \le (\|x\|_p^q + \|y\|_p^q)^{1/q} \quad \text{if} \quad p \ge q$$

$$\|(|x|^q + |y|^q)^{1/q}\|_p \ge (\|x\|_p^q + \|y\|_p^q)^{1/q} \quad \text{if} \quad p \le q.$$

It is easy to see that actually the preceding inequalities imply conversely (5.107) and (5.108), by iteration.

In the opposite direction, one can easily deduce from (5.107) and (5.108) the following refinements of (5.107) and (5.108). Let $(\Omega', \mathcal{A}', m')$ be another measure space. Consider a measurable function $F\colon\ \Omega \times \Omega' \to \mathbb{R}$. Then (5.107) and (5.108) become

$$\|F\|_{L_p(m;L_q(m'))} \le \|F\|_{L_q(m';L_p(m))} \quad \text{if} \quad p \ge q \tag{5.109}$$

$$\|F\|_{L_p(m;L_q(m'))} \ge \|F\|_{L_q(m';L_p(m))} \quad \text{if} \quad p \le q. \tag{5.110}$$

The latter two inequalities reduce to the same one if one exchanges p and q: We have a norm 1 inclusion $L_q(m'; L_p(m)) \subset L_p(m; L_q(m'))$ when $p \ge q$.

Essentially the same proof as for (5.107) and (5.108) establishes (5.109) and (5.110). One can also deduce the latter from (5.107) and (5.108) by a simple argument, approximating integrals by finite sums. Note that (5.107) and (5.108) correspond to $\Omega' = \mathbb{N}$ equipped with the counting measure $m' = \sum \delta_n$.

5.14 Appendix: basic facts on weak-L_p

Let (Ω, m) be a measure space. Let $0 < p < \infty$. For any $Z \in L_0(\Omega, \mathcal{A}, m)$, let

$$\|Z\|_{p,\infty} = \left(\sup_{t>0} t^p m(\{|Z| > t\})\right)^{1/p}.$$

The following inequality is immediate

$$\forall x, y \in L_0(\Omega, \mathcal{A}, m) \quad \| |x| \vee |y| \|_{p,\infty} \le (\|x\|_{p,\infty}^p + \|y\|_{p,\infty}^p)^{1/p}.$$

More generally, for any sequence $x_n \in L_0(\Omega, \mathcal{A}, m)$

$$\| \sup_n |x_n| \|_{p,\infty} \le (\sum\nolimits_n \|x_n\|_{p,\infty}^p)^{1/p}. \tag{5.111}$$

A fortiori there is a constant c such that $\|x + y\|_{p,\infty} \le c(\|x\|_{p,\infty} + \|y\|_{p,\infty})$.

We denote by $L_{p,\infty}(\Omega, \mathcal{A}, m)$ (or briefly $L_{p,\infty}$ if there is no ambiguity) the space of those Z such that $\|Z\|_{p,\infty} < \infty$, and we equip it with the quasi-norm $Z \mapsto \|Z\|_{p,\infty}$. The space $L_{p,\infty}$ is called weak-L_p because it contains L_p and we have obviously

$$\forall Z \in L_p \quad \|Z\|_{p,\infty} \le \|Z\|_p.$$

Moreover, if m is a probability (or is finite) then $L_{p,\infty} \subset L_q$ for any $0 < q < p$, and there is a positive constant $c = c(p, q)$ such that

$$\|Z\|_q \le c\|Z\|_{p,\infty}. \tag{5.112}$$

Indeed, this is easy to check using (1.3) (replace p by q in (1.3)).

More generally, given a Banach (or quasi-Banach) space B we denote by $L_{p,\infty}(\Omega, \mathcal{A}, m; B)$ (or simply $L_{p,\infty}(B)$) the space of those $f \in L_0(\Omega, m; B)$ such that $\omega \mapsto \|f(\omega)\|_B$ is in $L_{p,\infty}$ and we equip it with the quasi-norm

$$\|f\|_{L_{p,\infty}(B)} = \left(\sup_{t>0} t^p m\{\|f\|_B > t\}\right)^{1/p}.$$

When $p = 1$ (arguably the most important case) and also when $0 < p < 1$, the quasi-norm $Z \mapsto \|Z\|_{p,\infty}$ is *not* equivalent to a norm (unless $L_0(\Omega, \mathcal{A}, m)$ is finite-dimensional). But it is so when $p > 1$:

Proposition 5.72. *Assume $p > 1$ and let $p' = p/(p-1)$. Then for any $Z \in L_{p,\infty}$ we have*

$$\|Z\|_{p,\infty} \le \sup_{E \in \mathcal{A}} \left\{ m(E)^{-1/p'} \int_E |Z| dm \right\} \le p' \|Z\|_{p,\infty}. \tag{5.113}$$

In particular, $Z \mapsto \|Z\|_{p,\infty}$ is equivalent to a norm, the middle term in (5.113).

Proof. Since this will be proved in §8.5 in the wider context of Lorentz spaces, we only indicate a quick direct proof. The lower bound in (5.113) is obtained by choosing $E = \{|Z| > t\}$. For the upper bound, we use $\int_E |Z|dm = \int_0^\infty m(\{|Z| > t\} \cap E)dt$. Then, assuming $\|Z\|_{p,\infty} \le 1$, we may write for any $s > 0$

$$\int_E |Z|dm \le sm(E) + \int_s^\infty m(\{|Z| > t\} \cap E)dt \le sm(E) + \frac{1}{p-1}s^{1-p},$$

and with the optimal choice of s, this yields $\int_E |Z|dm \le p'm(E)^{1/p'}$. □

Corollary 5.73. *Let $0 < r < 1$. Then for any sequence (x_n) in $L_{1,\infty}$ we have*

$$\left\| \left(\sum |x_n|^r \right)^{1/r} \right\|_{1,\infty} \le (1-r)^{-1/r} \left(\sum \|x_n\|_{1,\infty}^r \right)^{1/r}. \tag{5.114}$$

More generally, let $(\Omega', \mathcal{A}', m')$ be another measure space. Then for any measurable function S on the product $\Omega \times \Omega'$, we have

$$\left\| \left(\int |S(\cdot,\omega')|^r dm'(\omega') \right)^{1/r} \right\|_{1,\infty} \le (1-r)^{-1/r} \left(\int \|S(\cdot,\omega')\|_{1,\infty}^r dm'(\omega') \right)^{1/r}. \tag{5.115}$$

Proof. Let $p = 1/r$. Note that for any Z we have $\|Z\|_{p,\infty} = \||Z|^p\|_{1,\infty}^{1/p}$. Let $y_n = |x_n|^r$. By (5.113) we have $\|\sum |y_n|\|_{p,\infty} \le p' \sum \|y_n\|_{p,\infty}$, but $\|\sum |y_n|\|_{p,\infty} = \|(\sum |x_n|^r)^{1/r}\|_{1,\infty}^r$ and $\sum \|y_n\|_{p,\infty} = \sum \|x_n\|_{1,\infty}^r$, and (5.114) follows. When S is a step function, (5.115) is essentially the same as (5.114). The general case follows by a routine approximation argument. □

Remark. In analogy with the Hölder-Minkowski inequality (5.109), the proof of (5.115) (with p/r in place of $1/r$) actually shows that, for any $0 < r < p < \infty$, we have a bounded inclusion

$$L_r(m'; L_{p,\infty}(m)) \subset L_{p,\infty}(m; L_r(m')).$$

Corollary 5.74. *Let $0 < r < 1$. Then for any finite sequence (x_n) in $L_{1,\infty}$ we have*

$$A_r \left\| (\sum |x_n|^2)^{1/2} \right\|_{1,\infty} \le (1-r)^{-1/r} \left(\mathbb{E} \left\| \sum \varepsilon_n x_n \right\|_{1,\infty}^r \right)^{1/r}, \tag{5.116}$$

where A_r is the constant appearing in Khintchin's inequality (5.7).

Proof. We apply (5.115) with $(\Omega', m') = (\Delta, \nu)$ and $S(\cdot, \omega') = \sum \varepsilon_n(\omega')x_n(\cdot)$. Then, using (5.7), (5.116) follows. □

5.15 Appendix: reverse Hölder principle

The classical Hölder inequality implies that for any measurable function $Z \ge 0$ on a probability space and any $0 < q < p < \infty$ we have $\|Z\|_q \le \|Z\|_p$. By the 'reverse Hölder principle' we mean the following two statements (closely related to [164]) in which the behaviour of Z in L_q controls conversely its belonging to weak-L_p. Our first principle corresponds roughly to the case $q = 0$.

Proposition 5.75. *Let $0 < p < \infty$. For any $0 < \delta < 1$ and any $R > 0$ there is a constant $C_p(\delta, R)$ such that the following holds. Consider a random variable $Z \geq 0$ and a sequence $(Z^{(n)})_{n\geq 0}$ of independent copies of Z. We have then*

$$\sup_{N\geq 1} \mathbb{P}\left\{\sup_{n\leq N} N^{-1/p} Z^{(n)} > R\right\} \leq \delta \Rightarrow \|Z\|_{p,\infty} \leq C_p(\delta, R). \quad (5.117)$$

Proof. Assume $\mathbb{P}\{N^{-1/p}\sup_{n\leq N} Z^{(n)} > R\} \leq \delta$ for all $N \geq 1$. By independence of $Z^{(1)}, Z^{(2)}, \ldots$ we have

$$\mathbb{P}\left\{\sup_{n\leq N} Z^{(n)} \leq RN^{1/p}\right\} = (\mathbb{P}\{Z \leq RN^{1/p}\})^N,$$

therefore $\mathbb{P}\{Z \leq RN^{1/p}\} \geq (1-\delta)^{1/N}$ and hence

$$\mathbb{P}\{Z > RN^{1/p}\} \leq 1 - (1-\delta)^{1/N} \leq c_1(\delta)N^{-1}.$$

Consider $t > 0$ and $N \geq 1$ such that $RN^{1/p} < t \leq R(N+1)^{1/p}$. We have

$$\mathbb{P}\{Z > t\} \leq c_1(\delta)N^{-1} \leq c_2(\delta, R)t^{-p}.$$

Since we trivially have $\mathbb{P}\{Z > t\} \leq 1$ if $t \leq R$, we obtain as announced

$$\|Z\|_{p,\infty} \leq (\max\{R, c_2(\delta, R)\})^{1/p}.$$

□

Corollary 5.76. *For any $0 < q < p < \infty$ there is a constant $R(p, q)$ such that for any Z as in Proposition 5.75 we have*

$$\|Z\|_{p,\infty} \leq R(p, q) \sup_{N\geq 1} \|N^{-\frac{1}{p}} \sup_{n\leq N} Z^{(n)}\|_q. \quad (5.118)$$

Proof. By homogeneity we may assume $\sup_{N\geq 1} \|N^{-1/p}\sup_{n\leq N} Z^{(n)}\|_q \leq 1$. Then $\mathbb{P}\{N^{-1/p}\sup_{n\leq N} Z^{(n)} > \delta^{-1/q}\} \leq \delta$, so by Proposition 5.75 with $R = \delta^{-1/q}$ and (say) $\delta = 1/2$ we obtain (5.118). □

Remark. Conversely by (5.111) and (5.112) $\sup_{N\geq 1} \|N^{-1/p}\sup_{n\leq N} Z^{(n)}\|_q \leq c \sup_{N\geq 1} \|N^{-1/p}\sup_{n\leq N} Z^{(n)}\|_{p,\infty} \leq c\|Z\|_{p,\infty}$. Thus, using $q = 1$, we find an alternate proof that $Z \mapsto \|Z\|_{p,\infty}$ is equivalent to a norm when $p > 1$.

The following Banach space valued version of the 'principle' is very useful. Let B be an arbitrary Banach space and let $f\colon\ \Omega \to B$ be a B-valued random variable. We will denote again by $f^{(1)}, f^{(2)}, \ldots$ a sequence of independent copies of the variable f.

Proposition 5.77. *For any $1 \leq q < p < \infty$ there is a constant $R'(p, q)$ such that any f in $L_q(B)$ with $\mathbb{E}(f) = 0$ satisfies*

$$\|f\|_{L^{p,\infty}(B)} \leq R'(p, q) \sup_{N\geq 1} N^{-1/p}\|f^{(1)} + \cdots + f^{(N)}\|_{L_q(B)}.$$

Moreover, this also holds for $0 < q < p < \infty$ if we assume f symmetric.

Proof. Assume $q \geq 1$ and $N^{-1/p}\|f^{(1)} + \cdots + f^{(N)}\|_{L_q(B)} \leq 1$ for all $N \geq 1$. By Corollary 1.41 we have

$$\left\| \sup_{1 \leq n \leq N} N^{-1/p}\|f^{(1)} + \cdots + f^{(n)}\|_B \right\|_q \leq 2^{1+1/q}$$

and hence by the triangle inequality

$$\left\| \sup_{1 \leq n \leq N} N^{-1/p}\|f^{(n)}\|_B \right\|_q \leq 2^{2+1/q}.$$

Therefore we conclude by Corollary 5.76 applied to $Z(\cdot) = \|f(\cdot)\|_B$.

If $0 < q < 1$ and f is symmetric, the same argument works but using (1.43). □

5.16 Appendix: Marcinkiewicz theorem

In the next statement, it will be convenient to use the following terminology. Let X, Y be Banach spaces, let (Ω, m), (Ω', m') be measure spaces and let $T\colon L_p(m; X) \to L_0(m'; Y)$ be a linear operator. We say that T is of weak type (p, p) with constant C if we have for any f in $L_p(m; X)$

$$\|Tf\|_{p,\infty} = (\sup_{t>0} t^p m'(\|Tf\| > t))^{1/p} \leq C\|f\|_{L_p(X)}.$$

We say that T is of strong type (p, p) if it bounded from $L_p(X)$ to $L_p(Y)$. We invoke on numerous occasions the following famous classical result due to Marcinkiewicz. Although we will prove a more general result in Chapter 8, we include a quick direct proof here for the convenience of the reader, in case he/she is reluctant to go into general interpolation theory.

Theorem 5.78 (Marcinkiewicz). *Let $0 < p_0 < p_1 \leq \infty$. In the preceding situation, assume that T is both of weak type (p_0, p_0) with constant C_0 and of weak type (p_1, p_1) with constant C_1. Then for any $0 < \theta < 1$, T is of strong type (p_θ, p_θ) with $p_\theta^{-1} = (1-\theta)p_0^{-1} + \theta p_1^{-1}$, and moreover we have*

$$\|T\colon L_{p_\theta}(X) \to L_{p_\theta}(Y)\| \leq K(p_0, p_1, p)C_0^{1-\theta}C_1^{\theta}$$

where $K(p_0, p_1, p)$ is a constant depending only on p_0, p_1, p.

Proof. Let $f \in L_{p_0}(X) \cap L_{p_1}(X)$. Consider a decomposition $f = f_0 + f_1$ with

$$f_0 = f \cdot 1_{\{\|f\| > \gamma\lambda\}} \quad \text{and} \quad f_1 = f \cdot 1_{\{\|f\| \leq \gamma\lambda\}},$$

where $\gamma > 0$ and $\lambda > 0$ are fixed. We have by our assumptions

$$m'(\|T(f_0)\| > \lambda) \le (C_0\lambda^{-1})^{p_0} \int_{\{\|f\|>\gamma\lambda\}} \|f\|^{p_0}\, dm$$

$$m'(\|T(f_1)\| > \lambda) \le (C_1\lambda^{-1})^{p_1} \int_{\{\|f\|\le\gamma\lambda\}} \|f\|^{p_1}\, dm$$

hence since $\|T(f)\| \le \|T(f_0)\| + \|T(f_1)\|$

$$m'(\|T(f)\| > 2\lambda) \le C_0^{p_0}\lambda^{-p_0} \int_{\|f\|>\gamma\lambda} \|f\|^{p_0} dm + C_1^{p_1}\lambda^{-p_1} \int_{\|f\|\le\gamma\lambda} \|f\|^{p_1} dm. \tag{5.119}$$

Let $p = p_\theta$. If we now multiply (5.119) by $2^p p\lambda^{p-1}$ and integrate with respect to λ, using

$$\int_{\{\|f\|>\gamma\lambda\}} \lambda^{p-p_0-1} d\lambda = (p-p_0)^{-1}(\|f\|/\gamma)^{p-p_0}$$

and

$$\int_{\{\|f\|\le\gamma\lambda\}} \lambda^{p-p_1-1} d\lambda = (p_1-p)^{-1}(\|f\|/\gamma)^{p-p_1},$$

we find

$$\int \|T(f)\|^p\, dm' \le \frac{2^p p C_0^{p_0}\gamma^{p_0-p}}{p-p_0} \int \|f\|^p\, dm + \frac{2^p p C_1^{p_1}\gamma^{p_1-p}}{p_1-p} \int \|f\|^p\, dm.$$

Hence, we obtain the estimate

$$\|T\colon L_p(X) \to L_p(Y)\| \le \frac{2p^{1/p}C_0^{p_0/p}\gamma^{(p_0-p)/p}}{(p-p_0)^{1/p}} + \frac{2p^{1/p}C_1^{p_1/p}\gamma^{(p_1-p)/p}}{(p_1-p)^{1/p}},$$

so that choosing γ so that

$$C_0^{p_0}\gamma^{p_0-p} = C_1^{p_1}\gamma^{p_1-p}$$

we finally find the announced result with

$$K(p_0, p_1, p) = 2p^{1/p}(p-p_0)^{-1/p} + 2p^{1/p}(p_1-p)^{-1/p}. \tag{5.120}$$

□

Remark 5.79. It is fairly obvious and well known that the preceding proof remains valid for 'sublinear' operators. Indeed, all that we need for the operator T is the pointwise inequalities

$$\|T(f_0 + f_1)\|_B \le \|T(f_0)\|_B + \|T(f_1)\|_B$$

for any pair f_0, f_1 in $L_{p_0}(X) \cap L_{p_1}(X)$, and also the positive homogeneity, i.e.

$$\forall\lambda \ge 0,\ \forall f \in L_{p_0}(X) \cap L_{p_1}(X)\ \ \|T(\lambda f)\|_B = \lambda\|T(f)\|_B.$$

5.17 Appendix: exponential inequalities and growth of L_p-norms

In many cases the growth of the L_p-norms of a function when $p \to \infty$ can be advantageously reformulated in terms of its exponential integrability. This is made precise by the following elementary and well-known lemma.

Lemma 5.80. *Fix a number $a > 0$. The following properties of a random variable $f \geq 0$ are equivalent:*

(i) $\sup_{p\geq 1} p^{-1/a}\|f\|_p < \infty$
(ii) *There is a number t such that $\mathbb{E}\exp|f/t|^a \leq e$.*

Moreover, let

$$\|f\|_{\exp L^a} = \inf\{t \geq 0 \mid \mathbb{E}\exp|f/t|^a \leq e\}.$$

There is a constant C such that for any $f \geq 0$ we have

$$C^{-1}\sup_{p\geq 1} p^{-1/a}\|f\|_p \leq \|f\|_{\exp L^a} \leq C\sup_{p\geq 1} p^{-1/a}\|f\|_p.$$

Proof. Assume that the supremum in (i) is ≤ 1. Then

$$\mathbb{E}\exp|f/t|^a = 1 + \sum\nolimits_1^\infty \mathbb{E}|f/t|^{an}(n!)^{-1} \leq 1 + \sum\nolimits_1^\infty (an)^n t^{-an}(n!)^{-1}$$

hence by Stirling's formula for some constant C

$$\leq 1 + C\sum\nolimits_1^\infty (an)^n t^{-an} n^{-n} e^n = 1 + C\sum\nolimits_1^\infty (at^{-a}e)^n$$

from which it becomes clear (since $1 < e$) that (i) implies (ii). Conversely, if (ii) holds we have a fortiori for all $n \geq 1$

$$(n!)^{-1}\|f/t\|_{an}^{an} \leq \mathbb{E}\exp|f/t|^a \leq e$$

and hence

$$\|f\|_{an} \leq e^{\frac{1}{an}}(n!)^{\frac{1}{an}} t \leq e^{\frac{1}{a}} n^{\frac{1}{a}} t = (an)^{\frac{1}{a}} t (e/a)^{1/a},$$

which gives $\|f\|_p \leq p^{1/a} t (e/a)^{1/a}$ for the values $p = an$, $n = 1, 2, \ldots$. One can then easily interpolate (using Hölder's inequality) to obtain (i). The last assertion is now a simple recapitulation left to the reader. □

5.18 Notes and remarks

The inequalities (5.1) and (5.2) were obtained in a 1966 paper by Burkholder. We refer the reader to the classical papers [165] and [175] for more on this. See also the book [34]. The best constant β_p in (5.1) is equal to $p^* - 1$ where $p^* = \max\{p, p'\}$ and $1 < p < \infty$. It is also the best constant when we restrict to constant multipliers (φ_n). Thus the unconditionality constant of the Haar system in L_p is also equal to $p^* - 1$, and we have

$$C_p(L_p) = p^* - 1.$$

More generally, the inequality $\sup \|\widetilde{M}_n\|_p \le (p^* - 1) \sup \|M_n\|_p$ holds for any pair of martingales with values in a Hilbert space H such that

$$\|\widetilde{M}_0\|_H \le \|M_0\|_H \quad \text{and} \quad \|\widetilde{M}_n - \widetilde{M}_{n-1}\|_H \le \|M_n - M_{n-1}\|_H$$

(pointwise) for all $n \ge 1$. These are called 'differentially subordinate' by Burkholder. Of course this implies that for any B with $\dim(B) \ge 1$ we have

$$C_p(B) \ge p^* - 1,$$

with equality in the Hilbert space case.

Incidentally, the equality $C_2(B) = 1$ implies that B is isometric to Hilbert space. Indeed, this implies that the Gaussian (or Rademacher) K-convexity constant as defined in [79, p. 20] is also $= 1$, then [79, Theorem 3.11] implies the desired isometry, for any finite-dimensional subspace, which is enough.

For (5.8) the best constants are also partially known: We have for any $1 < p < \infty$

$$(p^* - 1)^{-1} \|S\|_p \le \|f\|_p \le (p^* - 1)\|S\|_p, \tag{5.121}$$

and $\|f\|_p \le (p-1)\|S\|_p$ is optimal for $2 \le p < \infty$ while $\|f\|_p \ge (p-1)\|S\|_p$ is optimal for $1 < p \le 2$, but the best constants in the remaining cases are not known. Their order of magnitude, however is known: On one hand there is a universal constant c valid for all $1 \le p \le 2$ so that $\|f\|_p \le c\|S\|_p$ (and for $p = 1$ the best c is equal to 2, see Osękowski's book [72, Theorem 8.7, p. 413]), and on the other hand $\|S\|_p \le 2\sqrt{p}\|f\|_p$ for all $p \ge 2$ (see [172, p. 89] and references there).

By Doob's inequality, when $p \ge 2$, (5.121) implies $\|f^*\|_p \le p\|S\|_p$, and the constant p is still best possible in the latter.

We now turn to the weak-type inequalities for $p = 1$. The best constant β_1 in (5.2) is equal to 2, even in the differentially subordinate case. Moreover, only the B's that are isometric to a Hilbert space can satisfy the optimal inequality (see Remark 5.17) $\sup_n \sup_{\lambda>0} \lambda\mathbb{P}\{\|\tilde{f}_n\| > \lambda\} \le 2 \sup_n \|f_n\|_{L_1(B)}$.

The best constant β_1' in (5.10) is equal to $\sqrt{e}$. This is due to Cox [191].

As noted by Burkholder, by Kwapień's theorem in [315], the differentially subordinate extensions of inequalities such as (5.1) or (5.2) in the Banach valued case can hold *only if* the space is isomorphic to a Hilbert space.

For all the preceding sharp inequalities we refer the reader to [165, 170, 173, 174], to [34] and also to Osękowski's more recent book [72]. We recommend Burkholder's surveys [172, 174], and the collection of his selected works in [14].

The best constants in the Khintchin inequalities are known: see [260, 430]. Szarek [430] proved that $A_1 = 2^{-1/2}$. More generally, let γ_p be the L_p-norm of a standard Gaussian distribution (with mean zero and variance 1). It is well known that

$$\gamma_p = 2^{1/2}\left(\Gamma((p+1)/2)/\sqrt{\pi}\right)^{1/p} \quad 0 < p < \infty.$$

Let $p_0 = 1.87\ldots$ be the unique solution in the interval $]1, 2[$ of the equation $2^{1/2-1/p} = \gamma_p$ (or explicitly $\Gamma((p+1)/2) = \sqrt{\pi}/2$), then Haagerup (see [260]) proved:

$$A_p = 2^{1/2-1/p} \quad 0 < p \le p_0, \tag{5.122}$$

$$A_p = \gamma_p \quad p_0 \le p \le 2, \tag{5.123}$$

$$B_p = \gamma_p \quad 2 \le p < \infty. \tag{5.124}$$

The lower bounds $A_p \ge \max\{\gamma_p, 2^{1/2-1/p}\}$ for $p \le 2$ and $B_p \ge \gamma_p$ for $p \ge 2$ are easy exercises (by the Central Limit Theorem).

For Kahane's inequalities, some of the optimal constants are also known, in particular (see [318]), if $0 < p \le 1 \le q \le 2$, we have $K(p,q) = 2^{\frac{1}{p}-\frac{1}{q}}$.

Kahane's inequalities follow from the results in the first edition of [44]. The idea to derive them from the 2-point hypercontractive inequality is due to C. Borell. See [110, 256, 336] for generalizations.

The property UMD was introduced by B. Maurey and the author (see [343]), together with the author's observation that Burkholder's ideas could be extended to show that $\mathrm{UMD}_p \Leftrightarrow \mathrm{UMD}_q$ for any $1 < p, q < \infty$ (Theorem 5.13). It was also noted initially that UMD_p implies super-reflexivity (and a fortiori reflexivity), but not conversely. See Chapter 11 for more on this. Proposition 5.12 appears in [167, Theorem 2.2]. The Gundy decomposition appearing in Theorem 5.15 comes from [257].

The extrapolation principle appearing in §5.5 (sometimes called 'good λ-inequality') is based on the early ideas of Burkholder and Gundy ([175]), but our presentation in Lemma 5.23 was influenced by the refinements from [325]. As for Lemma 5.26 and the other statements in §5.5, our source is Burkholder in [165, 174].

§5.6 is a simple adaptation to the B-valued case of Burgess Davis classical results from [201]. The examples presented in §5.7 are sort of 'folkloric'.

§5.8 is due to B. Maurey [343]. Note that if a UMD Banach space B has an unconditional basis (e_n), then the functions $t \mapsto h_k(t)e_n$, form an unconditional basis for $L_p([0,1];B)$. We leave the proof as an exercise (hint: use Remark 10.53). Conversely, Aldous proved that if $L_p([0,1];B)$ has an unconditional basis then B is necessarily UMD, see [100] for details.

The Burkholder-Rosenthal inequality in Theorem 5.50 appears in [165]. It was preceded by Rosenthal's paper [415] from which Corollaries 5.54 to 5.56 are extracted. §5.10 is due to Bourgain [137], but the original Stein inequality comes from [85].

§5.11 is motivated by Burkholder's characterization of UMD spaces in terms of ζ-convexity (Theorem 5.64), for which we refer to [167, 169, 171, 173]. Theorem 5.69 is due to Kalton, Konyagin and Vesely [302]. Its proof is somewhat similar to the renorming of super-reflexive spaces proved earlier in [387] but presented later on in this volume (see Chapters 10 and 11). In sharp contrast, Burkholder's proof that UMD implies zeta convexity is based on the weak-type (1,1) bound (5.31). An excellent detailed presentation of Burkholder's method is given in Osękowski's book [72].

6

The Hilbert transform and UMD Banach spaces

6.1 Hilbert transform: HT spaces

The close connection between martingale transforms and the Hilbert transform or more general 'singular integrals' was noticed very early on. In this context, the classical Calderón-Zygmund (CZ in short) decomposition is a fundamental tool (see e.g. [86]). (We saw its martingale counterpart, the Gundy decomposition, in the preceding chapter.) Thus it is not surprising that in the Banach space valued case, the two kinds of transforms are bounded for exactly the same class of Banach spaces. The goal of this chapter is the proof of this equivalence.

Definition 6.1. Let $1 < p < \infty$. Let m denote the normalized Haar measure on the torus $\mathbb{T}$. A Banach space B is called HT_p if there is a constant C such that for any finitely supported function $x\colon \mathbb{Z} \to B$ we have

$$\left(\int \|\sum_{n>0} z^n x_n - \sum_{n<0} z^n x_n\|^p dm(z)\right)^{1/p} \le C\left(\int \|\sum_{n\in\mathbb{Z}} z^n x_n\|^p dm(z)\right)^{1/p}.$$

We will denote by $C_p^{HT}(B)$ the best constant C for which this holds.

The Hilbert transform on the circle $\mathbb{T}$ is the transformation

$$H^{\mathbb{T}}\colon \sum_{n\in\mathbb{Z}} z^n x_n \to (-i)\left(\sum_{n>0} z^n x_n - \sum_{n<0} z^n x_n\right). \tag{6.1}$$

Thus $C_p^{HT}(B)$ is equal to its norm acting on $L_p(\mathbb{T}, m; B)$.

Clearly any Hilbert space is an HT space, and $C_2^{HT}(B) = 1$ if B is isometric to a Hilbert space. Actually, we will show that HT and UMD are identical notions.

Note that we have obviously by duality for any B

$$C_2^{HT}(B^*) = C_2^{HT}(B), \tag{6.2}$$

and more generally if $1/p + 1/p' = 1$

$$C_{p'}^{HT}(B^*) = C_p^{HT}(B). \tag{6.3}$$

Remark 6.2. One can often replace the Hilbert transform by the orthogonal projection $P_0 : L_2(\mathbb{T}, m) \to H^2[\mathbb{T}]$, which is usually referred to as 'the Riesz projection'. More generally, let P_N denote the orthogonal projection from $L_2(\mathbb{T}, m)$ onto the closed span of $\{z^n | n \geq N\}$ (or equivalently from $L_2(\mathbb{T}, m)$ onto $z^N H^2[\mathbb{T}]$).

We claim that B is HT_p iff P_0 is bounded on $L_p(\mathbb{T}, m; B)$ and in the latter case, P_0 defines a bounded linear projection from $L_p(\mathbb{T}, m; B)$ onto $\widetilde{H}^p[\mathbb{T}; B]$. Indeed, for any $N \in \mathbb{Z}$, we have $z^N P_0(z^{-N} \sum_{n\in\mathbb{Z}} z^n x_n) = \sum_{n\geq N} z^n x_n$, and hence for any N

$$\|P_N \colon L_p(\mathbb{T}, m; B) \to L_p(\mathbb{T}, m; B)\| = \|P_0 \colon L_p(\mathbb{T}, m; B) \to L_p(\mathbb{T}, m; B)\|.$$

Then, if we let

$$C_p^H(B) = \|P_0 \colon L_p(\mathbb{T}, m; B) \to L_p(\mathbb{T}, m; B)\|,$$

by (6.1) we have $P_0 f = 2^{-1}(iH^{\mathbb{T}} + I)(f) + 2^{-1} \int f dm$ and hence

$$2C_p^H(B) - 2 \leq C_p^{HT}(B) \leq 2C_p^H(B). \tag{6.4}$$

Theorem 6.3. *Let B be a Banach space such that $C_p^{HT}(B) < \infty$ for some $1 < p < \infty$. Then $C_p^{HT}(B) < \infty$ for all $1 < p < \infty$. Moreover there is an absolute numerical constant K such that for any $1 < p < \infty$ and any B we have*

$$C_2^{HT}(B) \leq K C_p^{HT}(B). \tag{6.5}$$

Proof. The idea of the proof goes back to Calderón and Zygmund (CZ in short) but it was apparently Jacob Schwartz [425] who first observed that the CZ-idea remains valid in the Banach space valued case (and even for operator valued singular integral kernels). Since (6.5) plays some role in the sequel, we sketch the proof. The key point is to show that if $C_p^{HT}(B) < \infty$ then $H \otimes Id_B$ is of weak-type (1,1) on $L_1(T; B)$. This is where the CZ decomposition is used. This yields

$$\|H \colon L_1(B) \to L_{1\infty}(B)\| \leq K_1 C_p^{HT}(B)$$

where K_1 is a numerical constant. Then, by the Marcinkiewicz interpolation theorem (see the appendix to this chapter), if $1 < 2 < p$, we obtain

$$C_2^{HT}(B) \leq K C_p^{HT}(B)$$

where K is a numerical constant, independent of p and B. Moreover, if $1 < p < 2$, replacing B by B^* and p by $p' \geq 2$, we obtain $C_2^{HT}(B) = C_2^{HT}(B^*) \leq KC_{p'}^{HT}(B^*) = KC_p^{HT}(B)$. □

In Definition 6.1 the space B is implicitly assumed to be a complex Banach space. However, the Hilbert transform originally arose in the study of real valued harmonic functions, where the 'conjugate' function $\tilde{u}$ of a real valued harmonic function u on D is defined as the unique harmonic function on D with $\tilde{u}(0) = 0$ such that the ($\mathbb{C}$ valued) function $u + i\tilde{u}$ is analytic on D. One defines similarly the conjugate function on the upper half-plane U, but the condition $\tilde{u}(0) = 0$ is replaced by $\lim_{y\to\infty} \tilde{u}(x + iy) = 0$.

In the case of D since we know that $P_z(t) = \Re\frac{e^{it}+z}{e^{it}-z}$ and the imaginary part vanishes when $z = 0$, if u is the Poisson integral of $f \in L_1(\mathbb{T})$ we must have for any $z = re^{i\theta} \in D$

$$u + i\tilde{u}(z) = \int \frac{e^{it} + z}{e^{it} - z} f(t)dm(t), \tag{6.6}$$

and hence

$$\tilde{u}(z) = \int \Im\left(\frac{e^{it} + z}{e^{it} - z}\right) f(t)dm(t) = \int \frac{2r\sin(\theta - t)}{1 - 2r\cos(\theta - t) + r^2} f(t)dm(t).$$

Let us denote by $\tilde{f}$ the boundary values (say the radial limits) of $\tilde{u}$ assuming they exist a.e., for instance when f is a trigonometric polynomial.

When $u(z) = \Re(z^n)$ (resp. $u(z) = \Im(z^n)$) with $n > 0$ we have obviously $\tilde{u}(z) = \Im(z^n)$ (resp. $\tilde{u}(z) = -i\Re(z^n)$) and for $n = 0$ we must have $\tilde{u} = 0$. From this it is easy to deduce that

$$\tilde{f} = H^{\mathbb{T}} f.$$

Since $H^{\mathbb{T}}$ is bounded on $L_2(\mathbb{T})$, this shows that $u \mapsto \tilde{u}$ is bounded on $h^2(D)$, since the latter can be identified with $L_2(\mathbb{T}; \mathbb{R})$.

When f is smooth enough, in particular if it is a trigonometric polynomial, we can rewrite $\tilde{f}(\theta)$ as a 'principal value integral', i.e. we have

$$\tilde{f}(\theta) = \lim_{\varepsilon\to 0} \int_{|\theta - t| > \varepsilon} \cot(\frac{\theta - t}{2}) f(t)dm(t), \tag{6.7}$$

which is traditionally written

$$\tilde{f}(\theta) = p.v. \int \cot\left(\frac{\theta - t}{2}\right) f(t)dm(t),$$

to emphasize that the integral is not absolutely convergent.

More generally, we will verify (6.7) for any f such that, for some constants c and $\delta > 0$, we have $|f(t) - f(\theta)| \leq c|t - \theta|^\delta$ for all $\theta, t \in \mathbb{T}$. Indeed, since the integrand is odd, $\int_{|\theta-t|>\varepsilon} \cot(\frac{\theta-t}{2})dm(t) = 0$ and hence we have for any $\theta \in \mathbb{T}$

$$\int_{|\theta-t|>\varepsilon} \cot\left(\frac{\theta-t}{2}\right) f(t)dm(t) = \int_{|\theta-t|>\varepsilon} \cot\left(\frac{\theta-t}{2}\right)(f(t)-f(\theta))dm(t),$$

and now since $|\cot(\frac{\theta-t}{2})(f(t)-f(\theta))| \in O(|\theta-t|^{\delta-1})$ we have absolute convergence and the limit in (6.7) exists for any $\theta \in \mathbb{T}$. Furthermore, again the oddness of the integrand implies

$$\tilde{u}(re^{i\theta}) = \int \frac{2r\sin(\theta-t)}{1-2r\cos(\theta-t)+r^2}(f(t)-f(\theta))dm(t)$$

and hence by dominated convergence

$$\tilde{f}(\theta) = \lim_{r\to 1} \tilde{u}(re^{i\theta}) = \int \cot\left(\frac{\theta-t}{2}\right)(f(t)-f(\theta))dm(t),$$

which is the same as $\lim_{\varepsilon\to 0}\int_{|\theta-t|>\varepsilon} \cot\left(\frac{\theta-t}{2}\right) f(t)dm(t)$. This proves (6.7) (and also that the radial limits exist for any θ) for any such f. Moreover $\tilde{f}$ is bounded. Let $g \in L_1(\mathbb{T}, m)$. We have

$$\int g(\theta)\tilde{f}(\theta)dm(\theta) = \int\int g(\theta)\cot\left(\frac{\theta-t}{2}\right)(f(t)-f(\theta))dm(t)dm(\theta),$$

and hence we note for further reference that if f, g are supported on disjoint compact sets (so that $g(\theta)f(\theta)$ vanishes identically), this implies

$$\int g(\theta)\tilde{f}(\theta)dm(\theta) = \int\int \cot\left(\frac{\theta-t}{2}\right) g(\theta)f(t)dm(t)dm(\theta),$$

which we view as expressing that the transformation $f \mapsto \tilde{f}$ admits $\cot(\frac{\theta-t}{2})$ as its kernel.

Given a real Banach space $B_{\mathbb{R}}$, we may view it as embedded isometrically in a complex one B such that $B = B_{\mathbb{R}} + iB_{\mathbb{R}}$. Then we can define the conjugate function $\tilde{u}$ for any $u \in h^p(D; B)$ that is a $B_{\mathbb{R}}$ valued polynomial of the form (3.33), and similarly for $\tilde{f}$. With this definition it becomes clear, by Theorem 3.4, that B is HT_p $(1 < p < \infty)$ iff the conjugation $u \mapsto \tilde{u}$ extends to a bounded linear map on $\widetilde{h}^p(D; B)$.

We will return to the theme of conjugate functions in §7.1.

A parallel notion of Hilbert transform and property HT can be introduced with $\mathbb{T}$ and D replaced by $\mathbb{R}$ and the upper half-plane U, but the resulting property HT is equivalent with the same constant. Since this equivalence will be needed in the sequel, we wish to prove this now, based on classical ideas. We will denote by $H^{\mathbb{R}}$ the Hilbert transform on $\mathbb{R}$. This is defined as the operator on $L_2(\mathbb{R})$ that acts by multiplication of the Fourier transform by the function

$$\forall x \in \mathbb{R} \quad \varphi(x) = -i\,\mathrm{sign}(x).$$

This means that for any test function $f \in L_2(\mathbb{R})$ with Fourier transform

$$\widehat{f}(y) = \int f(x)e^{-ixy}dx,$$

the image $g = H^{\mathbb{R}}(f)$ is determined by the identity

$$\widehat{g} = \varphi \widehat{f}. \tag{6.8}$$

When f is smooth enough, say if it is C^∞ with compact support, it is well known that we can rewrite $H^{\mathbb{R}} f(x)$ as a 'principal value integral', i.e. we have at any point x

$$H^{\mathbb{R}} f(x) = \frac{1}{\pi} \lim_{\varepsilon \to 0} \int_{|x-t|>\varepsilon} \frac{1}{x-t} f(t)dt = \frac{1}{\pi} p.v. \int \frac{1}{x-t} f(t)dt. \tag{6.9}$$

We may then observe that since the kernel $\frac{1}{x-t} 1_{[-1,1]}(x-t)$ is odd, as earlier for the circle, we have

$$\int_{|x-t|>\varepsilon} \frac{1}{x-t} f(t)dt = \int_{|x-t|>\varepsilon} \frac{1}{x-t} (f(t) - f(x)1_{[-1,1]}(x-t))dt$$

and hence, since f is differentiable at x, we can write $H^{\mathbb{R}} f(x)$ as an absolutely convergent integral, namely

$$H^{\mathbb{R}} f(x) = \frac{1}{\pi} \int \frac{1}{x-t} (f(t) - f(x)1_{[-1,1]}(x-t))dt.$$

Then if g is a bounded measurable function with compact support we have

$$\int g(x) H^{\mathbb{R}} f(x)dx = \int \int g(x) \frac{1}{\pi} \frac{1}{x-t} (f(t) - f(x)1_{[-1,1]}(x-t))dtdx,$$

and if we assume that f and g are supported on disjoint compact sets, so that $f(x)g(x) = 0$ for all x, we find

$$\int g(x) H^{\mathbb{R}} f(x)dx = \int \int \frac{1}{\pi} \frac{1}{x-t} g(x) f(t)dtdx, \tag{6.10}$$

and this integral is absolutely convergent. This identity (6.10) holds for any pair f, g of C^∞ functions with disjoint compact supports but, by density, it clearly remains true for any pair $f, g \in L_2(\mathbb{R})$ supported on disjoint compact sets. This expresses the fact that $\frac{1}{\pi} \frac{1}{x-t}$ is the kernel associated to the (singular integral) operator $H^{\mathbb{R}}$.

Remark 6.4. Let $K(x, t)$ be a measurable function on $\mathbb{R} \times \mathbb{R}$ bounded on $\{(x, t) \mid |x - t| > \delta\}$ for any $\delta > 0$. By a repeated application of Lebesgue's

Theorem (3.35) we have for almost all $(x', t') \in \mathbb{R} \times \mathbb{R}$

$$K(x', t') = \lim_{\delta_1 \to 0} \lim_{\delta_2 \to 0} \frac{1}{2\delta_1} \frac{1}{2\delta_2} \int_{x'-\delta_1}^{x'+\delta_1} \int_{t'-\delta_2}^{t'+\delta_2} K(x, t) dx dt.$$

Therefore, if there is an operator $T : L_2(\mathbb{R}) \to L_2(\mathbb{R})$ admitting K as its kernel, *i.e.* such that $\int gT(f) = \int g(x)K(x,t)f(t)dxdt$ for any pair $f, g \in L_2(\mathbb{R})$ supported on disjoint compact sets, then K is uniquely determined by T on the complement of a negligible set. (Note that the diagonal is negligible in $\mathbb{R} \times \mathbb{R}$.)

We will use the following very simple characterization of the Hilbert transform on $\mathbb{R}$.

For any $r > 0$ let us denote by $D_r\colon L_2(\mathbb{R}) \to L_2(\mathbb{R})$ the (isometric) dilation operator D_r defined by

$$D_r f(x) = f(x/r) r^{-1/2}.$$

Let us denote by $\mathcal{S}\colon L_2(\mathbb{R}) \to L_2(\mathbb{R})$ the (isometric) symmetry defined by

$$\mathcal{S} f(x) = f(-x).$$

Theorem 6.5. *Let $T\colon L_2(\mathbb{R}) \to L_2(\mathbb{R})$ be an operator on $L_2(\mathbb{R})$ commuting with all translations, with all dilations $\{D_r \mid r > 0\}$ and such that $T\mathcal{S} = -T\mathcal{S}$. Then T is a multiple of the Hilbert transform $H^{\mathbb{R}}$.*

Proof. Since T commutes with translations, it must be a 'convolution' or equivalently there is an associated multiplier φ in $L_\infty(\mathbb{R})$ such that

$$\widehat{Tf} = \varphi \hat{f}.$$

Note that for any $r > 0$

$$\forall y \in \mathbb{R} \qquad \widehat{D_r f}(y) = \hat{f}(ry) r$$

so that $\widehat{D_r T f}(y) = \hat{f}(ry)\varphi(ry)r$ and $\widehat{TD_r f}(y) = \varphi(y)\hat{f}(ry)r$. Thus if $TD_r = D_r T$, we must have $\varphi(ry) = \varphi(y)$ for all $r > 0$, which clearly implies that φ is constant (almost everywhere) both on $(0, \infty)$ and on $(-\infty, 0)$. In particular, after a negligible correction, we may assume $\varphi = \varphi(1)$ (resp. $\varphi = \varphi(-1)$) on $(0, \infty)$ (resp. $(-\infty, 0)$). Finally, since $\widehat{\mathcal{S}f}(y) = \hat{f}(-y)$, $T\mathcal{S} = -T\mathcal{S}$ implies $\varphi(-y) = -\varphi(y)$. Thus we conclude that

$$\varphi(y) = \varphi(1)\text{sign}(y) = c \cdot (-i \text{ sign}(y))$$

with $c = i\varphi(1)$, completing the proof. □

Remark 6.6. Assuming more regularity of the kernel, a second proof can be derived from the following observations. Let $(x, t) \mapsto K(x, t)$ be a function defined for $x \neq t$. Suppose that $K(x, t) = K(x + s, t + s)$ for any $s \in \mathbb{R}$

(invariance by translation), that $K(x,t) = \rho^{-1}K(x/\rho, t/\rho)$ for any $\rho > 0$ (invariance by dilation) and $K(x,t) = -K(t,x)$. This implies that

$$K(x,t) = (x-t)^{-1}K(1,0).$$

Indeed, taking $\rho = x - t$, we have $K(x,t) = K(x-t,0) = (x-t)^{-1}K(1,0)$ when $x > t$, and again $K(x,t) = -K(t,x) = (x-t)^{-1}K(1,0)$ when $x < t$.

We will say that a Banach space B is HT_p on $\mathbb{R}$ with constant C if the operator $H^{\mathbb{R}} \otimes Id_B$ extends to a bounded operator on $L_p(\mathbb{R}; B)$ with norm $= C$.

We will show that this holds iff B is HT_p on $\mathbb{T}$. That HT_p on $\mathbb{R}$ implies HT_p on $\mathbb{T}$ follows easily from the fact that the respective kernels have the same singularity at $x = 0$, namely we have $\cot\frac{x}{2} \approx \frac{2}{x}$ when $x \to 0$, see Remark 6.9 for more details.

We will use the following basic fact in §6.3.

Proposition 6.7. *If a Banach space B is HT_p on $\mathbb{R}$ with constant C then B is HT_p (on $\mathbb{T}$) with constant $C_p^{HT}(B) \le C$.*

Proof. Since the reduction to this case is easy, we assume B finite-dimensional. Let $f = \sum e^{int}x_n \in L_p(\mathbb{T}; B)$ and $\xi = \sum e^{int}y_n \in L_{p'}(\mathbb{T}; B^*)$ be trigonometric polynomials. Since

$$\left|\int \langle \xi(-t), (H^{\mathbb{T}}f)(t)\rangle dm(t)\right| = \left|\sum \langle y_n, x_n\rangle \varphi(n)\right|,$$

it suffices to show that

$$\left|\sum \langle y_n, x_n\rangle \varphi(n)\right| \le C\|f\|_{L_p(\mathbb{T};B)}\|\xi\|_{L_{p'}(\mathbb{T};B^*)}. \tag{6.11}$$

Let $\varepsilon > 0$. Consider the Gaussian kernel on $\mathbb{R}$

$$g_\varepsilon(x) = (2\pi\varepsilon)^{-1/2}e^{-x^2/2\varepsilon}.$$

Then $\widehat{g_\varepsilon}(x) = e^{-\varepsilon x^2/2}$, $\int g_\varepsilon(x)dx = 1$, $g_\varepsilon * g_{\varepsilon'} = g_{\varepsilon+\varepsilon'}$ $(\forall \varepsilon' > 0)$.

Recall the Fourier inversion and self-duality formulae

$$\forall\, \psi, \chi \in L_2(\mathbb{R}) \quad \widehat{\widehat{\psi}}(x) = 2\pi\psi(-x) \quad \text{and} \tag{6.12}$$
$$\int \widehat{\psi}(x)\widehat{\chi}(-x)dx = 2\pi \int \psi(x)\chi(x)dx.$$

We now wish to replace f and ξ by (2π-periodic) functions of a real variable (instead of functions on $\mathbb{T}$), so we denote by $\underline{f}$ and $\underline{\xi}$ the corresponding functions, defined by:

$$\forall x \in \mathbb{R} \quad \underline{f}(x) = \sum e^{-inx}x_n \quad \text{and} \quad \underline{\xi}(x) = \sum e^{-inx}y_n.$$

Note that $x \mapsto e^{-inx}$ is the Fourier transform of the Dirac measure δ_n.

Then, the functions $\psi_\varepsilon = (1/2\pi)\underline{f}\ \widehat{g_\varepsilon^{1/p'}}$ and $\chi_\varepsilon = (1/2\pi)\underline{\xi}\ \widehat{g_\varepsilon^{1/p}}$ admit Fourier transforms respectively

$$\widehat{\psi_\varepsilon} = \widehat{\underline{f}} * g_\varepsilon^{1/p'} = \sum x_n \delta_n * g_\varepsilon^{1/p'}$$

$$\widehat{\chi_\varepsilon} = \widehat{\underline{\xi}} * g_\varepsilon^{1/p} = \sum \xi_m \delta_m * g_\varepsilon^{1/p}$$

and satisfy (note the special reverse choice of p, p' here)

$$\left| \int \langle \chi_\varepsilon(-x), H^{\mathbb{R}}(\psi_\varepsilon)(x) \rangle dx \right| \leq C \|\psi_\varepsilon\|_{L_p(\mathbb{R};B)} \|\chi_\varepsilon\|_{L_{p'}(\mathbb{R};B^*)}. \tag{6.13}$$

We plan to show that when $\varepsilon \to 0$, (6.13) tends to (6.11). By (6.12), we have

$$\left| \int \langle \chi_\varepsilon(-x), H^{\mathbb{R}}(\psi_\varepsilon)(x) \rangle dx \right| = (1/2\pi) \left| \int \varphi(y) \langle \widehat{\chi_\varepsilon}(y), \widehat{\psi_\varepsilon}(y) \rangle dy \right|. \tag{6.14}$$

Note

$$\int |\widehat{g_\varepsilon^{1/p'}}(y)|^p dy = \int \sqrt{2\pi\varepsilon}(p')^{p/2} e^{-pp'\varepsilon y^2/2} dy = 2\pi (p')^{p/2}/\sqrt{pp'}.$$

By the form of the Gaussian kernel, since $\underline{f}$ is 2π-periodic, we have

$$\lim_{\varepsilon \to 0} \left| \int \|\underline{f}\|^p |\widehat{g_\varepsilon^{1/p'}}|^p dy - \int_0^{2\pi} \|f(y)\|^p \frac{dy}{2\pi} \int |\widehat{g_\varepsilon^{1/p'}}(y)|^p dy \right| = 0.$$

Therefore

$$\lim_{\varepsilon \to 0} \|\psi_\varepsilon\|_{L_p(\mathbb{R};B)} = (2\pi)^{-1/p'} \sqrt{p'} (\sqrt{pp'})^{-1/p} \|f\|_{L_p(\mathbb{T};B)}$$

and similarly

$$\lim_{\varepsilon \to 0} \|\chi_\varepsilon\|_{L_{p'}(\mathbb{R};B^*)} = (2\pi)^{-1/p} \sqrt{p} (\sqrt{pp'})^{-1/p'} \|\xi\|_{L_{p'}(\mathbb{T};B^*)},$$

which together gives us

$$2\pi \lim_{\varepsilon \to 0} \|\psi_\varepsilon\|_{L_p(\mathbb{R};B)} \|\chi_\varepsilon\|_{L_{p'}(\mathbb{R};B^*)} = \|f\|_{L_p(\mathbb{T};B)} \|\xi\|_{L_{p'}(\mathbb{T};B^*)}.$$

As for the other side, we have

$$\int \varphi(y) \langle \widehat{\chi_\varepsilon}(y), \widehat{\psi_\varepsilon}(y) \rangle dy = \sum_{n,m} \langle x_n, y_m \rangle a^\varepsilon_{n,m}$$

with

$$a^\varepsilon_{n,m} = \int \varphi(y) (\delta_m * g_\varepsilon^{1/p})(y) (\delta_n * g_\varepsilon^{1/p'})(y) dy,$$

but, distinguishing the cases $n = m$ and $n \neq m$, we have (easy verification)

$$\lim_{\varepsilon\to 0} a^{\varepsilon}_{n,n} = \lim_{\varepsilon\to 0} \int \varphi(y+n)g_{\varepsilon}(y)dy = \varphi(n) \qquad \forall n \in \mathbb{Z}$$

$$\lim_{\varepsilon\to 0} |a^{\varepsilon}_{n,m}| \leq \|\varphi\|_{\infty} \lim_{\varepsilon\to 0} \int g_{\varepsilon}^{1/p}(y+n-m)g_{\varepsilon}^{1/p'}(y) = 0 \quad \forall n \neq m \in \mathbb{Z}.$$

Thus, taking the limit of (6.13) when $\varepsilon \to 0$, and recalling (6.14) we obtain (6.11). □

For the sake of completeness, we now choose, among many available possibilities, a quick argument to show that the converse is also true and the constants in Proposition 6.7 are actually equal.

Proposition 6.8. *If a Banach space B is HT_p (on $\mathbb{T}$) then it is HT_p on $\mathbb{R}$ with constant $C \leq C_p^{HT}(B)$.*

Proof. The idea of the proof is that the kernel $K_{\mathbb{T}}(t,\theta) = (2\pi)^{-1}\cot(\frac{t-\theta}{2})$ (resp. $K_{\mathbb{R}}(t,\theta) = \pi^{-1}\frac{1}{t-\theta}$) of $H^{\mathbb{T}}$ (resp. $H^{\mathbb{R}}$) are such that

$$\lim_{s\to 0} sK_{\mathbb{T}}(st, s\theta) = K_{\mathbb{R}}(t,\theta).$$

Heuristically, the real line behaves like a circle of radius $1/s \to \infty$.

Again we may assume B finite-dimensional. Let $R > 0$. Let $f, g \in L_2(\mathbb{R})$ with support in $[-R, R]$. Then for any $0 < r < \pi/R$ we define a function $\Xi_r f \in L_2(\mathbb{T}, m))$ by setting

$$\forall t \in [-\pi,\pi] \quad \Xi_r f(e^{it}) = D_r f(t) = f(t/r)r^{-1/2}.$$

Note that $[-rR, rR] \subset [-\pi,\pi]$ and hence $\|\Xi_r f\|^2_{L_2(\mathbb{T},m)} = (1/2\pi)\|f\|^2_{L_2(\mathbb{R})}$. We then consider the operator T defined by

$$\langle Tf, g\rangle = \lim_{r\to 0,\mathcal{U}} (\operatorname{Log} r)^{-1} \int_0^r 2\pi \langle H^{\mathbb{T}} \Xi_s f, \Xi_s g\rangle ds/s$$

where the limit is with respect to an utrafilter $\mathcal{U}$ refining the net of the convergence $r \to 0$. It is easy to check that the assumptions of Theorem 6.5 hold so that T is a multiple of $H^{\mathbb{R}}$. We claim that actually $T = H^{\mathbb{R}}$. Indeed, if we assume that f, g are supported in disjoint compact sets, the same is true for $\Xi_s f, \Xi_s g$ for $s > 0$ small enough and hence

$$\begin{aligned} 2\pi \langle H^{\mathbb{T}} \Xi_s f, \Xi_s g\rangle &= (2\pi)^{-1} \int_{\mathbb{R}} \cot\left(\frac{t-\theta}{2}\right) f(t/s)g(\theta/s)dt d\theta/s \\ &= (2\pi)^{-1} \int_{\mathbb{R}} s\cot\left(\frac{s(t-\theta)}{2}\right) f(t)g(\theta)dtd\theta \\ &\to \int_{\mathbb{R}} \pi^{-1}\frac{1}{t-\theta} f(t)g(\theta)dtd\theta \text{ when } s \to 0. \end{aligned}$$

Thus, by (6.9), we find

$$\langle Tf, g\rangle = \langle H^{\mathbb{R}} f, g\rangle,$$

proving our claim that $T = H^{\mathbb{R}}$. Now for $f \in L_p(\mathbb{R}; B)$ and $g \in L_{p'}(\mathbb{R}; B^*)$, since $\langle (T \otimes Id)f, g\rangle$ is a limit of averages of $2\pi \langle H^{\mathbb{T}} \Xi_s f, \Xi_s g\rangle$ and since when s is small enough

$$\|\Xi_s f\|_{L_p(\mathbb{T};m,B)} \|\Xi_s g\|_{L_{p'}(\mathbb{T},m;B^*)} = (2\pi)^{-1} \|f\|_{L_p(\mathbb{R};B)} \|g\|_{L_{p'}(\mathbb{R};B^*)}$$

we find

$$\begin{aligned} |\langle (T \otimes Id)f, g\rangle| &\le C_p^{HT} 2\pi \limsup_{s\to 0} \|\Xi_s f\|_{L_p(\mathbb{T};m,B)} \|\Xi_s g\|_{L_{p'}(\mathbb{T},m;B^*)} \\ &= C_p^{HT} \|f\|_{L_p(\mathbb{R};B)} \|g\|_{L_{p'}(\mathbb{R};B^*)}. \end{aligned}$$

Thus, we obtain as announced $\|H^{\mathbb{R}} \otimes Id\|_{B(L_p(\mathbb{R};B))} \le C_p^{HT}$. □

Remark 6.9. Let B be any Banach space and let $1 \le p \le \infty$. Let $G = \mathbb{T}$ (or $G = \mathbb{R}$ or any other locally compact Abelian group equipped with Haar measure). Then for any $f \in L_p(G; B)$ and any $\varphi \in L_1(G)$ the convolution $f * \varphi$ is in $L_p(G; B)$ and we have

$$\|f * \varphi\|_{L_p(G;B)} \le \|f\|_{L_p(G;B)} \|\varphi\|_{L_1(G)}.$$

In other words, the convolution by any $\varphi \in L_1(G)$ defines a bounded operator on $L_p(G; B)$. This is easy to check using Jensen's inequality (1.4).

As a consequence, if the singular kernels of two 'principal value' convolution operators T_1, T_2 such as (6.7) on $\mathbb{T}$ (or (6.9) on $\mathbb{R}$) differ by a function $\varphi \in L_1(G)$, then T_1, T_2 will be bounded on $L_p(G; B)$ for the same class of Banach spaces B.

We can use this idea to give a simpler proof that if B is HT_p on $\mathbb{R}$ then it is also HT_p on $\mathbb{T}$. Indeed, by restricting $H^{\mathbb{R}}$ to functions supported on $[-\pi, \pi]$, if B is HT_p on $\mathbb{R}$, it is easy to see that the 'principal value' integral

$$T_1(f)(\theta) = \frac{1}{\pi} p.v. \int_{-\pi}^{\pi} \frac{1}{\theta - t} f(t)dt$$

is bounded on $L_p([-\pi, \pi]; B)$, or equivalently on $L_p(\mathbb{T}; B)$. But since

$$\varphi(\theta) = \frac{1}{\pi}\frac{1}{\theta} - \frac{1}{2\pi} \cot\left(\frac{\theta}{2}\right) \in L_1([-\pi, \pi], d\theta)$$

we conclude that $T_2(f)(\theta) = \frac{1}{2\pi} p.v. \int_{-\pi}^{\pi} \cot(\frac{\theta-t}{2}) f(t)dt$ is also bounded on $L_p(\mathbb{T}; B)$, and hence B is HT_p on $\mathbb{T}$.

6.2 Bourgain's transference theorem: HT implies UMD

The main result of this section is due to J. Bourgain [134]. It allows for transplanting certain Fourier multipliers from $\mathbb{T}$ to $\mathbb{T}^{\mathbb{N}}$. Let $G = \mathbb{T}^{\mathbb{N}}$ and let $\Gamma = \mathbb{Z}^{(\mathbb{N})}$ be its dual group (one often denotes this by $\Gamma = \hat{G}$ or $G = \hat{\Gamma}$) identified with the set of all the continuous characters on G. Here $\mathbb{Z}^{(\mathbb{N})}$ denotes the set of all sequences of integers $(n_k)_{k\ge 0}$ in $\mathbb{Z}^{\mathbb{N}}$ such that $\sum_0^\infty |n_k| < \infty$. Then, to any such $n = (n_k)_{k\ge 0}$ is associated the function $\gamma_n : \mathbb{T}^{\mathbb{N}} \to \mathbb{T}$ defined by

$$\forall z = (z_k) \in \mathbb{T}^{\mathbb{N}} \quad \gamma_n(z) = \prod\nolimits_{k\ge 0} z_k^{n_k}.$$

This gives us a one-to-one correspondence between $\mathbb{Z}^{(\mathbb{N})}$ and the multiplicative group formed of all continuous characters, which is also a group isomorphism. Then $\{\gamma_n \mid n \in \mathbb{Z}^{(\mathbb{N})}\}$ forms an orthonormal basis of $L_2(G;\mu)$, where μ is the normalized Haar measure on G.

The set $\Gamma = \mathbb{Z}^{(\mathbb{N})}$ can be totally ordered by the (reverse) lexicographic order, which can be defined like this: a non-zero element $n = (n_k)_{k\ge 0}$ is > 0 iff there is an integer K such that $n_K > 0$ and $n_k = 0 \; \forall\, k > K$. Equipped with this order structure, Γ becomes an ordered group for which we can develop Harmonic Analysis on the model of the group $\mathbb{Z}$. This is entirely classical. See [84, Chapter 8] for details. In particular, we can define the space $H^p(G)$ as the closed span in $L_p(G,\mu)$ of the characters in $\Gamma_+ = \{n > 0\} \cup \{0\}$. Consider $n = (n_k)_{k\ge 0}$ and $m = (m_k)_{k\ge 0}$ in Γ. We set

$$\langle n, m\rangle = \sum_{k=0}^{\infty} n_k m_k.$$

We will show that several Fourier multipliers can be transplanted from $\mathbb{Z}$ to Γ. The combinatorial key behind Bourgain's theorem is the following useful fact.

Lemma 6.10. *For any finite subset $A \subset \Gamma$ there is an $m \in \Gamma$ such that*

$$\forall n \in A \setminus \{0\} \quad sign(\langle n, m\rangle) = sign(n).$$

where $sign(n) = \pm$ depending whether $n > 0$ or $n < 0$ in Γ (just like in $\mathbb{Z}$). More generally, for any sequence of signs $(\varepsilon_k)_{k\ge 0}$ there is $m \in \Gamma$ such that

$$\forall n \in A \setminus \{0\} \quad sign(\langle(\varepsilon_k n_k), m\rangle) = sign((\varepsilon_k n_k)).$$

Proof. The second assertion follows from the first one applied to $A' = \{(\varepsilon_k n_k) \mid n \in A\}$, so it suffices to prove the first one.

Let $W_K = \{n \in \Gamma \mid n_K \neq 0,\ n_j = 0\ \forall j > K\}$ be the set of 'words' of length exactly $K+1$. The idea is simply that for any $n \in W_K$ we have

$$\lim_{m_K \to \infty} \operatorname{sign}(\langle n, m \rangle) = \operatorname{sign}(n_K).\infty = \operatorname{sign}(n).\infty.$$

By induction on K we will prove that for any $A \subset W_0 \cup \cdots \cup W_K$ there is $m \in W_0 \cup \cdots \cup W_K$ such that

$$\forall n \in A \quad \operatorname{sign}(\langle n, m \rangle) = \operatorname{sign}(n). \tag{6.15}$$

The case $K = 0$ is trivial. Assume that we have proved this for any $A \subset W_0 \cup \cdots \cup W_{K-1}$ and let us prove it for $A \subset W_0 \cup \cdots \cup W_K$. By the induction hypothesis there is $(m_0, \ldots, m_{K-1}, 0, \ldots)$ such that $\operatorname{sign}(\langle n, m \rangle) = \operatorname{sign}(n)$ for any $n \in A \cap (W_0 \cup \cdots \cup W_{K-1})$. Consider now any $n \in A \cap W_K$ so that $n_K \neq 0$. Then choosing $m_K > 0$ large enough ensures that $\operatorname{sign}(\langle n, m \rangle) = \operatorname{sign}(n)$, and since A is finite, we can achieve this for any $n \in A \cap W_K$. Then $m = (m_0, \ldots, m_{K-1}, m_K, 0, \ldots)$ satisfies (6.15), completing the induction argument. □

Remark. A look at the preceding proof shows that if we are given in advance any (large) number N we can find $m \in \Gamma$ such that moreover $|\langle n, m \rangle| > N$ for any $n \in A$. Let $Z_{A,N}$ be the set of suitable ms. Since all these sets $Z_{A,N}$ are non-void, they generate a filter $\mathcal{Z}$ (or a net) on Γ such that for any non-zero n in Γ we have

$$\lim_{m,\mathcal{Z}} \langle m, n \rangle = \operatorname{sign}(n) \cdot \infty$$

where $\operatorname{sign}(n) = \pm$ depending whether $n > 0$ or $n < 0$. This net has a fortiori the following property: for any integer k and any $n(1) < n(2) < \cdots < n(k)$ in Γ, there is an m such that, in $\mathbb{Z}$, we have

$$\langle m, n(1) \rangle < \langle m, n(2) \rangle < \cdots < \langle m, n(k) \rangle.$$

Given a bounded function $\varphi\colon\ \Gamma \to \mathbb{C}$ on a discrete group Γ (we refer to φ as a 'multiplier'), we will always denote by $M_\varphi\colon\ L_2(\widehat{\Gamma}) \to L_2(\widehat{\Gamma})$ the corresponding multiplier operator on $L_2(\widehat{\Gamma})$ defined by $M_\varphi \gamma = \varphi(\gamma)\gamma$ for any $\gamma \in \Gamma$. Here $G = \widehat{\Gamma}$ is the (compact) dual of Γ equipped with its normalized Haar measure. When it is bounded, we denote by $M_\varphi\colon\ L_p(G; B) \to L_p(G; B)$ the operator that extends $M_\varphi \otimes Id_B$. As usual, we denote $L_p(G, \mu; B)$ simply by $L_p(G; B)$.

The next result is easy to obtain by a classical transference argument.

Lemma 6.11. *Let B be a Banach space. Let $1 \leq p < \infty$. Let $\varphi\colon\ \mathbb{Z} \to \mathbb{C}$ be a multiplier. For any m in Γ we define the multiplier $\varphi_m\colon\ \Gamma \to \mathbb{C}$ by*

$$\forall\, n \in \Gamma \qquad \varphi_m(n) = \varphi(\langle n, m \rangle).$$

We have then:

$$\|M_{\varphi_m}\colon\ L_p(G;B)\to L_p(G;B)\|\le\|M_\varphi\colon\ L_p(\mathbb{T},B)\to L_p(\mathbb{T},B)\|.$$

Proof. By homogeneity we may assume $\|M_\varphi\colon\ L_p(\mathbb{T},B)\to L_p(\mathbb{T},B)\|=1$. By Fubini we also have $\|M_\varphi\colon\ L_p(\mathbb{T};L_p(G;B))\to L_p(\mathbb{T};L_p(G;B))\|=1$. Fix $m\in\Gamma$. For any $z=(z_k)\in\Gamma$ and any $w\in\mathbb{T}$ we denote

$$w.z=(w^{m_k}z_k)_{k\ge0}.$$

To any $f\in L_p(G;B)$ we associate $F\in L_p(\mathbb{T},L_p(G;B))$ defined by

$$F(w)(z)=f(w.z).$$

We may assume that f is a finite sum of the form $f=\sum_n x_n\gamma_n$. Note that $\gamma_n(w.z)=w^{\langle n,m\rangle}\gamma_n(z)$, and hence

$$M_\varphi F(w)(z)=M_{\varphi_m}f(w.z).$$

We have then by the translation invariance of the $L_p(G;B)$-norm

$$\|M_{\varphi_m}f\|_{L_p(G;B)}=\|M_\varphi F\|_{L_p(\mathbb{T};L_p(G;B))}\le\|F\|_{L_p(\mathbb{T};L_p(G;B))}=\|f\|_{L_p(G;B)},$$

which means that $\|M_{\varphi_m}\colon\ L_p(G;B)\to L_p(G;B)\|\le1$. □

Remark 6.12. An analogous result holds for multipliers of weak-type (1,1). We have:

$$\|M_{\varphi_m}\colon\ L_1(G;B)\to L_{1,\infty}(G;B)\|\le\|M_\varphi\colon\ L_1(\mathbb{T},B)\to L_{1,\infty}(\mathbb{T},B)\|.$$

Indeed, we can use the elementary fact that for any function f on a product of measure spaces such as $G\times\mathbb{T}$ we have

$$\|f\|_{L_{1,\infty}(G\times\mathbb{T})}\le\int\|f(x,\cdot)\|_{L_{1,\infty}(\mathbb{T})}\mu(dx).$$

To state the main point of this section, we need some specific notation.

We will apply Lemma 6.11 to the multiplier $\varphi\colon\ \mathbb{Z}\to\mathbb{C}$ defined by $\varphi(0)=0$ and $\varphi(n)=\ \mathrm{sign}(n)$. Up to a factor i, The corresponding operator on $L_2(\mathbb{T})$ is the Hilbert transform $H^{\mathbb{T}}$, namely $M_\varphi=iH^{\mathbb{T}}$.

We equip the space (G,μ_G) with the filtration $(\mathcal{A}_k)_{k\ge0}$ defined by

$$\mathcal{A}_k=\sigma(z_0,z_1,\ldots,z_k)$$

where $z_k\colon\ G\to\mathbb{T}$ denotes the k-th coordinate on G.

Given a Banach space B, any element f in $L_1(G;B)$ be written as a convergent series $f=\sum_{k\ge0}df_k$ of martingale differences with respect to this filtration. A martingale $(f_n)_{n\ge0}$ is called a Hardy martingale if for any $n\ge0$ the variable

f_n depends 'analytically' on the 'last' variable z_n; equivalently this means that each f_n belongs to the closure of $H^1(G) \otimes B$ in $L_1(G; B)$. We will denote by

$$H_k\colon\ L_2(G) \to L_2(G)$$

the Hilbert transform acting on the k-th variable only. So that if $L_2(G)$ is identified with $\bigotimes_{k\geq 0} L_2(\mathbb{T})$ then H_k corresponds to the transformation

$$I \otimes \cdots \otimes I \otimes H \otimes I \otimes \cdots$$

with $H^{\mathbb{T}}$ sitting at the coordinate of index k. We can now state Bourgain's transference theorem:

Theorem 6.13. *Let $1 < p < \infty$. Let B be any Banach space such that the Hilbert transform $H^{\mathbb{T}}$ is bounded on $L_p(\mathbb{T}, B)$. Recall $C_p^{HT}(B) = \|H^{\mathbb{T}}\colon\ L_p(B) \to L_p(B)\|$. Then for any choice of signs $\varepsilon_k = \pm 1$ and for any finite martingale $f = \int f d\mu + \sum_{k\geq 0} df_k$ in $L_p(G; B)$ we have*

$$\left\| \sum\nolimits_{k\geq 0} \varepsilon_k H_k df_k \right\|_{L_p(B)} \leq C_p^{HT}(B) \|f\|_{L_p(B)}. \tag{6.16}$$

Proof. We may assume that the Fourier transform of f is supported on a finite subset $A \subset \Gamma$. Recall that we set $\varphi(0) = 0$ and $\varphi(n) = \text{sign}(n)$ for any $n \in \mathbb{Z}$. Let us denote simply $\varepsilon n = (\varepsilon_k n_k)$ for any $n \in \Gamma$ and any $\varepsilon = (\varepsilon_k) \in \{-1, 1\}^{\mathbb{N}}$. By Lemma 6.10, we can find $m \in \Gamma$ such that for any $n \in A$ we have

$$\varphi_{\varepsilon m}(n) = \text{sign}(\varepsilon n).$$

More precisely, if $n = (n_k)_{k\geq 0}$ with $n_K \neq 0$ and $n_k = 0\ \forall\, k > K$, we have

$$\varphi_{\varepsilon m}(n) = \varepsilon_K\ \text{sign}(n_K).$$

Thus $\varphi_{\varepsilon m}(0) = 0$ and

$$M_{\varphi_{\varepsilon m}}[df_K] = i\varepsilon_K H_K df_K.$$

Thus the theorem now follows from Lemma 6.11, since $M_\varphi = iH^{\mathbb{T}}$. □

Corollary 6.14. *In the situation of Theorem 6.13, let P_+ denote the orthogonal projection from $L_2(G)$ onto the subspace $H^2[G]$ formed of all f such that the associated martingale (f_n) is Hardy. Then for any finite sum $f = \int f d\mu + \sum_{k\geq 0} df_k$ in $L_p(G; B)$ we have*

$$\|(P_+ \otimes Id_B)(f)\|_{L_p(B)} \leq (2^{-1} C_p^{HT}(B) + 1)\|f\|_{L_p(B)}.$$

In other words, if we denote by $\tilde H^p[G;B] \subset L_p(G;B)$ the closure of $H^p[G] \otimes B$ in $L_p(G;B)$, then P_+ is a bounded linear projection from $L_p(G;B)$ to $\tilde H^p(G;B)$ of norm at most $2^{-1}C_p^{HT}(B)+1$.

Proof. Just like on $\mathbb{T}$ (see (6.4)), this follows from the identity

$$\begin{aligned} P_+(f) &= \int f d\mu + \sum 2^{-1}(1+iH_k)df_k \\ &= 2^{-1}\int f d\mu + 2^{-1}f + 2^{-1}i\sum H_k df_k, \end{aligned}$$

from which we deduce $\|P_+(f)\|_{L_p(B)} \le (2^{-1}C_p^{HT}(B)+1)\|f\|_{L_p(B)}$. □

Corollary 6.15. *In the situation of Theorem 6.13, let $A \subset \mathbb{N}$ be a subset such that for each n in A, the variable df_n is analytic in the variable z_n. We have then for all f in $L_p(B)$*

$$\left\| \sum_{n\in A} df_n \right\|_{L_p(B)} \le C_p^{HT}(B)\|f\|_{L_p(B)}.$$

Proof. Note that $iH_k\, df_k = df_k$ for all k in A. By Theorem 6.13 we have

$$\left\| \sum_{k\in A} df_k \pm \sum_{k\notin A} iH_k\, df_k \right\|_{L_p(B)} \le C_p^{HT}(B)\|f\|_{L_p(B)}.$$

Hence the first part follows from the triangle inequality. □

The next statement is the result for which Bourgain invented the preceding transference 'trick'.

Corollary 6.16. *For any Banach space B and any $1 < p < \infty$ we have*

$$C_p(B) \le C_p^{HT}(B)^2.$$

In particular, HT implies UMD.

Proof. We may restrict consideration to $f \in L_p(B)$ with vanishing mean so that $f_0 = 0$. Let $g = \sum_{k\ge 0} \varepsilon_k H_k df_k$. By (6.16) we have $\|g\|_{L_p(B)} \le C_p^{HT}(B)\|f\|_{L_p(B)}$. By (6.16) applied to g, with all the signs ε_k equal to 1, we find

$$\left\| \sum\nolimits_{k\ge 0} H_k dg_k \right\|_{L_p(B)} \le C_p^{HT}(B)\|g\|_{L_p(B)}.$$

But since $H_k dg_k = \varepsilon_k H_k^2 df_k = -\varepsilon_k df_k$ for any $k \ge 0$, (we can include $k = 0$ because $f_0 = g_0 = 0$) we obtain $\|\sum_{k\ge 0}\varepsilon_k df_k\|_{L_p(B)} \le C_p^{HT}(B)^2\|f\|_{L_p(B)}$. □

In the sequel, it will be useful for us to *avoid* squaring the constant p. This is possible if one restricts the UMD property to Hardy martingales. More precisely, if we denote by $C_p^a(B)$ the smallest constant C such that (5.21) holds for any B-valued Hardy martingale $f = (f_n)_{n\geq 0}$ in $L_p(G; B)$, then we have

$$C_p^a(B) \leq C_p^{HT}(B).$$

Indeed, when (f_n) is a Hardy martingale, $H_k(df_k) = df_k$, hence this follows immediately from Theorem 6.13.

See §6.7 for more on this.

However, the following seems to be still open.

Problem 6.17. Is there an absolute constant K such that $C_p(B) \leq KC_p^{HT}(B)$ for any $1 < p < \infty$ and any B?

6.3 UMD implies HT

The main result of this section is:

Theorem 6.18 ([169]). *Let $1 < p < \infty$ and let B be a UMD$_p$ Banach space. Then the Hilbert transform on $\mathbb{R}$ is bounded on $L_p(\mathbb{R}; B)$.*

Combined with Corollary 6.16 this yields:

Theorem 6.19. *UMD and HT are equivalent properties for Banach spaces.*

The traditional way to prove that UMD implies HT is to use Brownian motion as in the next section (see [169] in turn fundamentally based on [176]). We will use instead an idea in [378]. For that purpose, we first need to introduce some specific notation. Let $I = [0, 1)$. We denote

$$\mathcal{D}_0^{0,1} = \{k + [0, 1) \mid k \in \mathbb{Z}\}.$$

Then $\mathcal{D}_0^{0,1}$ constitutes a disjoint partition of $\mathbb{R}$ into intervals of equal measure 1 with 0 as origin (i.e. 0 is the endpoint of some interval). Similarly, we denote for each n in $\mathbb{Z}$

$$\mathcal{D}_n^{0,1} = \{2^n k + [0, 2^n) \mid k \in \mathbb{Z}\},$$

so that $\mathcal{D}_n^{0,1}$ is a partition of $\mathbb{R}$ into intervals of equal measure 2^n admitting 0 as origin. Equivalently, $\mathcal{D}_n^{0,1}$ is obtained from $\mathcal{D}_0^{0,1}$ by a dilation of ratio 2^n.

We set

$$\mathcal{D}^{0,1} = \bigcup_{n\in\mathbb{Z}} \mathcal{D}_n^{0,1}$$

for each I in $\mathcal{D}^{0,1}$, we denote by I_+ the left half of I and by I_- the right half, so that, for instance, if $I = [0, 1)$ we have $I_+ = [0, \frac{1}{2})$ and $I_- = [\frac{1}{2}, 1)$. (Note that actually the endpoints are irrelevant because they are negligible for all present measure theoretic purposes.)

Remark 6.20. Let $\mathcal{C}_n$ be the σ-algebra on $\mathbb{R}$ generated by the partition $\mathcal{D}_n^{0,1}$. The family $(\mathcal{C}_n)_{n\in\mathbb{Z}}$ gives us an example of a filtration on an infinite measure space, namely $(\mathbb{R}, dx)$, for which the conditional expectation makes perfectly good sense (see Remark 1.13). Thus martingale theory tells us that for any $f \in L_p(\mathbb{R}, dx)$ $(1 \le p < \infty)$ the sequence $f_n = \mathbb{E}^{\mathcal{C}_n} f$ is a martingale indexed by $\mathbb{Z}$ such that $f_n \to f$ in L_p when $n \to +\infty$, and also (reverse martingale) $f_n \to 0$ in L_p when $n \to -\infty$ (because 0 is the only constant function in L_p). But actually, although rather elegant, this viewpoint is not so important because the main issues we address are all local, and locally $\mathbb{R}$ can be treated as a bounded interval and hence 'at the bottom' as a probability space.

We will denote

$$\forall I \in \mathcal{D}^{0,1} \qquad h_I = |I|^{-1/2}(1_{I_+} - 1_{I_-}),$$

so that h_I is positive on I_+ and negative on I_- (curiously this rather natural convention is opposite to the one made in [378]).

It is well known that $\{h_I \mid I \in \mathcal{D}^{0,1}\}$ forms an orthonormal basis of $L_2(\mathbb{R})$.

The fundamental operator that we will study is the linear operator $T^{0,1}$: $L_2(\mathbb{R}) \to L_2(\mathbb{R})$ defined as follows:

$$\forall f \in L_2(\mathbb{R}) \qquad T^{0,1}(f) = \sum \langle f, h_I \rangle 2^{-1/2}(h_{I_+} - h_{I_-}).$$

Note that

$$T^{0,1} h_I = 2^{-1/2}(h_{I_+} - h_{I_-})$$

and it is easy to see that the family $\{2^{-1/2}(h_{I_+} - h_{I_-}) \mid I \in \mathcal{D}^{0,1}\}$ is orthonormal in $L_2(\mathbb{R})$. Therefore, $T^{0,1}$ is an isometry on $L_2(\mathbb{R})$ into itself.

We will first show that if (and only if) B is UMD then $T^{0,1}$ is bounded on $L_2(\mathbb{R}; B)$. This is rather easy:

Lemma 6.21. *For each $1 < p < \infty$, $T^{0,1}$ is bounded on $L_p(\mathbb{R})$. More generally, if B is UMD$_p$ then $T^{0,1}$ is bounded on $L_p(\mathbb{R}; B)$.*

Proof. Consider f in $L_2(\mathbb{R}; B)$. We may assume (by density) that f is a sum $f = \sum_{I\in\mathcal{D}^{0,1}} x_I h_I$ with only finitely many non-zero x_Is. We define $T^{0,1}(f) = \sum x_I T^{0,1}(h_I)$. Note: according to our previous notation (see Proposition 1.6) we should denote this by $\widetilde{T^{0,1}}(f)$, but for simplicity we choose to abuse the notation here.

Let us first assume that $f = \sum_{I \subset I_0} x_I h_I$ where $I_0 = [0, 1)$. Then with an appropriate reordering the series $f = \sum x_I h_I$ appears as a sum of dyadic martingale differences, to which the Burkholder inequalities can be applied.

Assume $B = \mathbb{C}$. Then the square function $S(f)$ of the corresponding martingale is given by

$$S(f)^2 = \sum_I |x_I|^2 |h_I|^2 = \sum_I |I|^{-1} |x_I|^2 1_I.$$

Let $g = T^{0,1}(f) = \sum x_I 2^{-1/2}(h_{I_+} - h_{I_-})$. Note that g is also a sum of dyadic martingale differences and since

$$|2^{-1/2}(h_{I_+} - h_{I_-})| = |h_I| \tag{6.17}$$

we have the following pointwise equality: $S(g)^2 = S(f)^2$. By Theorem 5.21 we have

$$\|g\|_p \le b_p \|S(g)\|_p = b_p \|S(f)\|_p \le b_p a_p \|f\|_p,$$

which proves that $\|T^{0,1}\colon\ L_p(I_0) \to L_p(I_0)\| \le a_p b_p$.

Now, if B is UMD$_p$ and $x_I \in B$, we can argue similarly: indeed, by (6.17) and by (5.19) we have the following pointwise equality

$$R(\{x_I 2^{-1/2}(h_{I_+} - h_{I_-})\}) = R(\{x_I h_I\}),$$

and hence by (5.23) we have

$$\|g\|_{L_p(B)} = \|T^{0,1}(f)\|_{L_p(B)} \le C_1 C_2 \|f\|_{L_p(B)}.$$

Now in the general case, when $f = \sum x_I h_I$ with only finitely many non-zero x_Is, we may clearly decompose f as a sum $f = f^1 + f^2$ so that there are intervals $I^1 \subset (-\infty, 0]$ and $I^2 \subset [0, +\infty)$ in $\mathcal{D}^{0,1}$ such that $f^1 = \sum_{I \subset I^1} x_I h_I$ and $f^2 = \sum_{I \subset I^2} x_I h_I$. Since we have $\|f\|^p_{L_p(B)} = \|f^1\|^p_{L_p(B)} + \|f^2\|^p_{L_p(B)}$, it suffices to majorize separately $\|T^{0,1}(f^1)\|_{L_p(B)}$ and $\|T^{0,1}(f^2)\|_{L_p(B)}$. But for each of them we may argue exactly as before, replacing I_0 in the preceding by I^1 or I^2. □

We will use Theorem 6.5. Recall that, for any $r > 0$, we denote by $D_r\colon\ L_2(\mathbb{R}) \to L_2(\mathbb{R})$ the (isometric) dilation operator D_r defined by

$$D_r f(x) = f(x/r) r^{-1/2}.$$

We now introduce the family of intervals $\mathcal{D}^{\alpha,r}$ obtained from $\mathcal{D}^{0,1}$ by dilating the intervals in $\mathcal{D}^{0,1}_0$ from measure 1 to measure r and translating the origin from 0 to α. In other words the intervals in $\mathcal{D}^{\alpha,r}$ are the images of those in $\mathcal{D}^{0,1}$ under the transformation $x \mapsto rx + \alpha$. They are of length $r2^n$ ($n \in \mathbb{Z}$).

Clearly the collection $\{h_I \mid I \in \mathcal{D}^{\alpha,r}\}$ is still an orthonormal basis in $L_2(\mathbb{R})$. Let us denote $T^{\alpha,r}$ the operator we obtain from $T^{0,1}$ after we replace $\mathcal{D}^{0,1}$

by $\mathcal{D}^{\alpha,r}$. More precisely

$$T^{\alpha,r}(f) = \sum\nolimits_{I\in\mathcal{D}^{\alpha,r}} \langle f, h_I\rangle (h_{I_+} - h_{I_-})2^{-1/2}.$$

Obviously, $T^{\alpha,r}$ is an isometry on $L_2(\mathbb{R})$. Let G be the group $\mathbb{R}\times\mathbb{R}_+$ where $\mathbb{R}$ is equipped with addition and $\mathbb{R}_+ = \{r > 0\}$ is equipped with multiplication. The integral of a (nice enough) function $F\colon\ (\alpha, r) \to F(\alpha, r)$ with respect to Haar measure on G is given by

$$\int_{-\infty}^{\infty}\int_{0}^{\infty} F(\alpha, r)\, d\alpha dr/r.$$

But since it is an infinite measure, we need to use instead an invariant mean Φ on G. By definition this is a translation invariant positive linear form on $L_\infty(G)$ such that $\Phi(1) = 1$. So Φ looks like a probability but it is not unique and it defines only an additive (and not σ-additive) set function on the Borel subsets of $\mathbb{R}$. Nevertheless, for any F in $L_\infty(G)$, abusing the notation, we will denote $\Phi(F)$ by

$$F \to \int F(\alpha, r)\, d\Phi(\alpha, r).$$

Note that G being commutative is amenable. We will use a specific choice of Φ as follows: we choose non-trivial ultrafilters $\mathcal{U}$ on $\mathbb{R}$ and $\mathcal{V}$ on $\mathbb{R}_+$ and we set (actually, we will show that, for the specific F to which we apply this, the true limits exist so that we can avoid using ultrafilters)

$$\int F(\alpha, r) d\Phi(\alpha, r) = \lim_{\substack{R\to\infty\\ \mathcal{V}}} \frac{1}{2\,\mathrm{Log}\, R}\int_{R^{-1}}^{R}\left(\lim_{\substack{a\to\infty\\ \mathcal{U}}} \frac{1}{2a}\int_{-a}^{a} F(\alpha, r)\, d\alpha\right)\frac{dr}{r}. \tag{6.18}$$

From now on, the invariant mean Φ is defined by (6.18). This choice guarantees a supplementary invariance under symmetries as follows:

$$\int F(\alpha, r) d\Phi(\alpha, r) = \int F(-\alpha, r) d\Phi(\alpha, r). \tag{6.19}$$

We then define the operator $\mathcal{T}$ on $L_2(\mathbb{R})$ by setting

$$\forall f, g \in L_2(\mathbb{R}) \quad \langle \mathcal{T}(f), g\rangle = \int \langle T^{\alpha,r}(f), g\rangle\, d\Phi(\alpha, r). \tag{6.20}$$

Note that, since $(\alpha, r) \to \langle T^{\alpha,r}(f), g\rangle$ is in $L_\infty(G)$, this does make sense.

Let us denote by $\lambda(t)\colon\ L_2(\mathbb{R}) \to L_2(\mathbb{R})$ the (unitary) operator of translation by t.

Theorem 6.22. *The operator $\mathcal{T}$ is a non-zero multiple of the Hilbert transform on $L_2(\mathbb{R})$.*

First part of the proof. We will use Theorem 6.5. Given I in $\mathcal{D}^{0,1}$, we denote by $I^{\alpha,r}$ the interval obtained from I after first dilation of ratio r (from 0) and then translation by α, so that $|I^{\alpha,r}| = r|I|$.

Clearly $I \to I^{\alpha,r}$ is a (1-1)-correspondence. Moreover, we have obviously $h_{I^{\alpha,r}} = \lambda(\alpha)D_r h_I$ and similarly with I_+ and I_- in place of I. We have

$$T^{\alpha,r}(f) = \sum\nolimits_{I\in\mathcal{D}^{0,1}} \langle f, h_{I^{\alpha,r}}\rangle (h_{I_-^{\alpha,r}} - h_{I_+^{\alpha,r}})2^{-1/2} \tag{6.21}$$

and hence

$$T^{\alpha,r} = (\lambda(\alpha)D_r)T^{0,1}(\lambda(a)D_r)^* = \lambda(\alpha)D_r T^{0,1}D_r^{-1}\lambda(\alpha)^{-1}. \tag{6.22}$$

By the (translation) invariance of Φ, it follows immediately from this that $\mathcal{T} = \lambda(t)\mathcal{T}\lambda(t)^{-1}$ for any t in $\mathbb{R}$. Note that $\lambda(\alpha)D_r = D_r\lambda(\alpha/r)$, so that

$$T^{\alpha,r} = D_r\lambda(\alpha/r)T^{0,1}\lambda(\alpha/r)^{-1}D_r^{-1}. \tag{6.23}$$

Therefore, again by the invariance of Φ we must have $\mathcal{T} = D_r\mathcal{T}D_r^{-1}$ for any $r > 0$. Finally, observe that, for any I in $\mathcal{D}^{0,1}$, we have $h_I(-t) = -h_{-I}(t)$ but also $(-I)_+ = -I_-$ and $(-I)_- = -I_+$, so that

$$T^{0,1}\mathcal{S}h_I = -\mathcal{S}T^{0,1}h_I.$$

Therefore we have $T^{0,1}\mathcal{S} = -\mathcal{S}T^{0,1}$ and a fortiori $T^{\alpha,r}\mathcal{S} = -\mathcal{S}T^{-\alpha,r}$ for any (α, r) in G. Thus, by (6.19) we must also have $\mathcal{T}\mathcal{S} = -\mathcal{S}\mathcal{T}$. By Theorem 6.5, it follows that $\mathcal{T}$ is a multiple of the Hilbert transform on $\mathbb{R}$.

But there remains a crucial point: to check that $\mathcal{T} \neq 0$! This turns out to be delicate. We do not know a simple proof of this sticky point, avoiding the calculations of the second part of the proof, which require first a more detailed study of the kernel of $T^{\alpha,r}$. □

To complete the proof, we will essentially reproduce the argument from Stefanie Petermichl's [378], just inserting a few more details to ease the reader's task. The pictures (reproducing those from [378]) are crucial to understand what goes on. We will compute the kernel of $\mathcal{T}$ by averaging the kernels of $T^{\alpha,r}$ as in (6.20).

Remark 6.23. We will use the following well-known elementary fact (related to the theory of almost periodic functions). Let $f : \mathbb{R} \to \mathbb{R}$ be bounded and measurable. Assume that there are $r > 0$ and $c \geq 0$ such that for any integers $N \geq 0$ and $m \geq 1$

$$\sup_{\alpha\in\mathbb{R}} |f(\alpha) - f(rm2^N + \alpha)| \leq c2^{-N}. \tag{6.24}$$

Then the averages $M_R(f) = (2R)^{-1}\int_{-R}^{R} f(\alpha)d\alpha$ converge when $R \to \infty$. Of course, a fortiori, this holds if f is a periodic function with period r (in which case (6.24) holds with $c = 0$).

Indeed, fix $\varepsilon > 0$ and N such that $c2^{-N} < \varepsilon$. Let $P = r2^N$, so that we have $\sup_{\alpha\in\mathbb{R}} |f(\alpha) - f(mP+\alpha)| < \varepsilon$ for any $m \geq 1$. We have for any $k \geq 1$

$$|M_{kP}(f) - \frac{1}{P}\int_0^P f(\alpha)d\alpha| \leq (2k)^{-1}\left|\sum_{m=-k}^{k-1} \frac{1}{P}\int_0^P (f(mP+\alpha) - f(\alpha))d\alpha\right| \leq \varepsilon.$$

Moreover, for any $R \geq P$ such that $kP \leq R < (k+1)P$ we have

$$|M_R(f) - \frac{kP}{R}M_{kP}(f)| \leq \frac{P}{R}\sup_{\alpha\in\mathbb{R}} |f(\alpha)|,$$

which implies

$$\limsup_{R\to\infty} |M_R(f) - \frac{1}{P}\int_0^P f(\alpha)d\alpha| \leq \varepsilon,$$

from which the announced convergence follows easily by the Cauchy criterion.

Recall that for each α, r we defined a dyadic shift operator $T^{\alpha,r}$ by

$$(T^{\alpha,r}f)(x) = \sum_{I\in\mathcal{D}^{\alpha,r}} (f, h_I)(h_{I_+}(x) - h_{I_-}(x)).$$

Its L^2 operator norm is $\sqrt{2}$ and its representing kernel, defined for any $x \neq t$, is

$$K^{\alpha,r}(x,t) = \sum_{I\in\mathcal{D}^{\alpha,r}} h_I(t)(h_{I_+}(x) - h_{I_-}(x)). \tag{6.25}$$

Let $\mathcal{D}_n^{\alpha,r} \subset \mathcal{D}^{\alpha,r}$ denote the subcollection formed of the intervals of length $r2^n$ that form a partition of $\mathbb{R}$. Note that $\mathcal{D}^{\alpha,r} = \cup_{n\in\mathbb{Z}}\mathcal{D}_n^{\alpha,r}$. Note also that $\mathcal{D}_n^{\alpha,2r} = \mathcal{D}_{n+1}^{\alpha,r}$, and hence $\mathcal{D}^{\alpha,2r} = \mathcal{D}^{\alpha,r}$. Let

$$K_n^{\alpha,r}(x,t) = \sum\nolimits_{I\in\mathcal{D}_n^{\alpha,r}} h_I(t)(h_{I_+}(x) - h_{I_-}(x)),$$

so that (for any $x \neq t$),

$$K^{\alpha,r} = \sum\nolimits_{n\in\mathbb{Z}} K_n^{\alpha,r}.$$

Lemma 6.24. *The convergence of the sum (6.25) is uniform for $|x-t| \geq \delta$ for every $\delta > 0$. For $x \neq t$, let*

$$K_n^r(x,t) = \lim_{a\to\infty}\frac{1}{2a}\int_{-a}^{a} K_n^{\alpha,r}(x,t)\,d\alpha, \tag{6.26}$$

$$K^r(x,t) = \lim_{a\to\infty}\frac{1}{2a}\int_{-a}^{a} K^{\alpha,r}(x,t)\,d\alpha, \tag{6.27}$$

and also

$$K(x,t) = \lim_{R\to\infty} \frac{1}{2\,\mathrm{Log}\,R} \int_{1/R}^{R} K^r(x,t)\frac{dr}{r}. \tag{6.28}$$

These 3 limits exist pointwise and the convergence is bounded for $|x-t| \geq \delta$ *for every* $\delta > 0$*. In addition, each of these is a function of* $x - t$*. Moreover,*

$$K^r(x,t) = \sum\nolimits_{n\in\mathbb{Z}} K_n^r(x,t). \tag{6.29}$$

Proof. For any $x \in \mathbb{R}$, let $I_n(x)$ be the unique interval containing x in $\mathcal{D}_n^{\alpha,r}$. We have

$$|h_I(t)(h_{I_+}(x) - h_{I_-}(x))| = 1_{t\in I}1_{x\in I}\sqrt{2}|I|^{-1}.$$

In particular,

$$|K_n^{\alpha,r}(x,t)| \leq \sqrt{2}(r2^n)^{-1}. \tag{6.30}$$

Note that $I_n(x) = I_n(t) \Rightarrow |x-t| \leq |I_n(x)| = |I_n(t)| = r2^n$ and hence $\forall \alpha \in \mathbb{R}$ and $\forall r > 0$ we have

$$\sum_{I\in\mathcal{D}^{\alpha,r}} |h_I(t)(h_{I_+}(x) - h_{I_-}(x))| = \sum_{n\in\mathbb{Z}} \sqrt{2}|I_n(x)|^{-1}1_{I_n(x)=I_n(t)} \leq 2\sqrt{2}/|x-t|.$$

In particular, the sum converges absolutely and uniformly for $|x-t| \geq \delta$ for every $\delta > 0$, and we have

$$|K^{\alpha,r}(x,t)| \leq 2\sqrt{2}/|x-t|. \tag{6.31}$$

The existence of the limits is due to either the periodicity or the 'almost periodicity' in α and the (multiplicative) periodicity in r. More precisely, note that the sum defining $K_n^{\alpha,r}(x,t)$ is finite and periodic in α (with period $r2^n$). From this it is easy to show (see Remark 6.23) that the limit (6.26) exists. Furthermore, for any fixed integer $m \in \mathbb{Z}$, we have $\mathcal{D}_n^{\alpha,r} = \mathcal{D}_n^{\alpha+rm2^n,r}$, and a fortiori $\mathcal{D}_n^{\alpha,r} = \mathcal{D}_n^{\alpha+rm2^N,r}$ for any $N \geq n$, and consequently $K_n^{\alpha,r} = K_n^{\alpha+rm2^N,r}$ if $n \leq N$. Therefore

$$|K^{\alpha,r} - K^{\alpha+rm2^N,r}| \leq \sum\nolimits_{n>N} |K_n^{\alpha,r} - K_n^{\alpha+rm2^N,r}|,$$

so that by (6.30)

$$|K^{\alpha,r} - K^{\alpha+rm2^N,r}| \leq 2\sqrt{2}\sum\nolimits_{n>N}(r2^n)^{-1} = 2\sqrt{2}(r2^N)^{-1}.$$

By Remark 6.23, this implies that the limit (6.27) exists for any r and any (x,t), and by (6.31) we have

$$|K^r(x,t)| \leq 2\sqrt{2}/|x-t|. \tag{6.32}$$

Since $\mathcal{D}^{\alpha,2r} = \mathcal{D}^{\alpha,r}$, we have $K^r(x,t) = K^{2r}(x,t)$. Note that the change of variable $r = \exp s$ gives us a periodic function in s, so that Remark 6.23 again implies that the limit in R exists in (6.28).

Note that for any $\beta \in \mathbb{R}$ we have $K_n^{\alpha,r}(x-\beta, t-\beta) = K_n^{\alpha+\beta,r}(x,t)$. This allows us to use the invariance property of the 'mean' (or 'Banach limit') defining (6.26). Indeed, from this and (6.30), we deduce that $K_n^r(x-\beta, t-\beta) = K_n^r(x,t)$, and hence $K_n^r(x,t) = K_n^r(x-t, 0)$. Thus (6.26) depends only on $x-t$ and consequently also for (6.27) and (6.28).

Recall that $K_n^{\alpha,r}(x,t) = 0$ whenever $2^n < |x-t|/r$. Thus by (6.30), it is clear that (6.29) holds. □

Lemma 6.25. *We have for any $x \neq t$*

$$K(x,t) = (x-t)^{-1}K(1,0).$$

Moreover, the function $(x,t) \mapsto K(x,t)$ (defined for $x \neq t$) is the kernel of the operator $\mathcal{T}$ in the following sense: For any pair of functions $f, g \in L_2(\mathbb{R})$ assumed bounded and supported respectively in two disjoint compact sets we have

$$\langle \mathcal{T}f, g\rangle = \int K(x,t)f(t)g(x)dtdx. \tag{6.33}$$

Proof. Recall that the intervals in $\mathcal{D}^{\alpha\rho,r\rho}$ are the images of those in $\mathcal{D}^{\alpha,r}$ under the transformation $x \mapsto \rho x$. From this and (6.25), it is easy to check that $K^{\alpha,r}(x/\rho, t/\rho) = \rho K^{\alpha\rho,r\rho}(x,t)$ for any $\rho > 0$. Averaging first in α and then in r as in (6.27) and (6.28) yields first $K^r(x/\rho, t/\rho) = \rho K^{r\rho}(x,t)$ and then $K(x/\rho, t/\rho) = \rho K(x,t)$. By Remark 6.6 we conclude that

$$K(x,t) = (x-t)^{-1}K(1,0).$$

Going back to (6.20), it is now clear that $K(x,t)$ is the kernel of $\mathcal{T}$. Indeed, assuming f, g bounded on $\mathbb{R}$ and $(x,t) \mapsto f(t)g(x)$ supported in $\{(x,t) \mid |x-t| \geq \delta\}$ for some $\delta > 0$, the right-hand side of (6.33) is equal to (by (6.31) and (6.32) the convergences are dominated)

$$\lim_{R\to\infty} \frac{1}{2\,\mathrm{Log}\,R}\int_{\frac{1}{R}}^{R} \lim_{a\to\infty}\frac{1}{2a}\int_{-a}^{a} K^{\alpha,r}(x,t)f(t)g(x)dtdxd\alpha\frac{dr}{r},$$

and this is the same as

$$\lim_{\substack{R\to\infty\\ \mathcal{V}}} \frac{1}{2\,\mathrm{Log}\,R}\int_{R^{-1}}^{R}\left(\lim_{\substack{a\to\infty\\ \mathcal{U}}}\frac{1}{2a}\int_{-a}^{a}\left(\int K^{\alpha,r}(x,t)f(t)g(x)dtdx\right)d\alpha\right)dr/r,$$

$$= \lim_{\substack{R\to\infty\\ \mathcal{V}}} \frac{1}{2\,\mathrm{Log}\,R}\int_{R^{-1}}^{R}\left(\lim_{\substack{a\to\infty\\ \mathcal{U}}}\frac{1}{2a}\int_{-a}^{a}\langle T^{\alpha,r}(f), g\rangle\, d\alpha\right)dr/r = \langle \mathcal{T}(f), g\rangle,$$

which means that K is the kernel of $\mathcal{T}$. □

Proof that $\mathcal{T} \neq 0$. Let us first give a picture for $h_I(t)(h_{I_+}(x) - h_{I_-}(x))$:

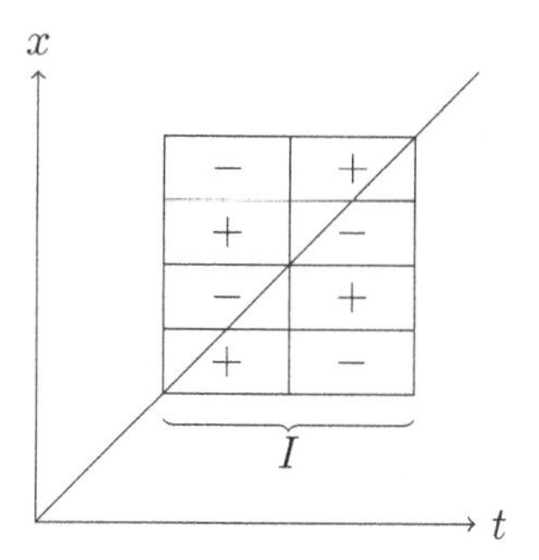

We have $h_I(t)(h_{I_+}(x) - h_{I_-}(x)) \neq 0$ if and only if the point (x, t) lies in the preceding square $I \times I$. Its value is $\pm\sqrt{2}/|I|$, where the correct sign is indicated inside the smaller rectangles. Recall that

$$K_n^r(x,t) = \lim_{a\to\infty} \frac{1}{2a} \int_{-a}^{a} K_n^{\alpha,r} \, d\alpha. \tag{6.34}$$

Our next goal is to compute it (and to show that it exists) for fixed $r > 0$ and $n \in \mathbb{Z}$ and assuming $x > t$. The picture is the following:

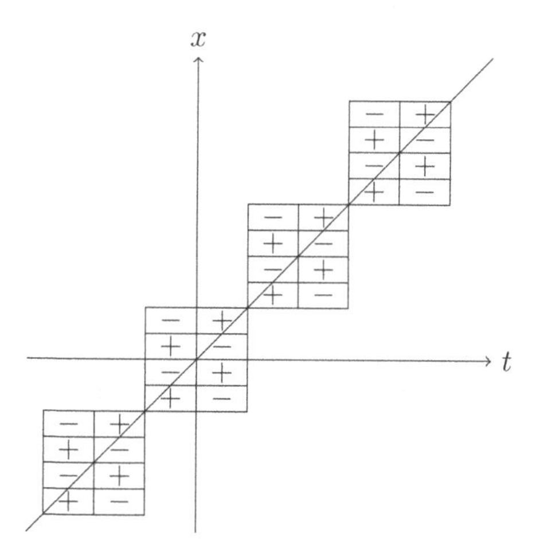

The exact position of the squares along the diagonal depends on the starting point α. The picture will repeat itself for two values of α that differ by an integer multiple of $|I|$. We compute (6.34) in (x, t) by considering the probability that (x, t) lies in any of the squares. As we already mentioned in Lemma 6.24, this only depends on $x-t$. Thus we have $K_n^r(x,t) = K_n^r(x-t,0)$ and hence setting $K_n^r(s) = K_n^r(s,0)$ we may write $K_n^r(x,t) = K_n^r(x-t)$.

- If $x-t=0$ then $K_n^r(x,t) = (1/4 - 1/4 + 1/4 - 1/4)\cdot\sqrt{2}/|I| = 0$, and similarly:
- if $x-t = |I|/4$ then $K_n^r(x,t) = 3/4\cdot\sqrt{2}/|I|$,
- if $x-t = |I|/2$ then $K_n^r(x,t) = 0$,
- if $x-t = 3|I|/4$ then $K_n^r(x,t) = -1/4\cdot\sqrt{2}/|I|$,
- if $x-t \geq |I|$ then $K_n^r(x,t) = 0$.

In between the preceding computed values, the function $K_n^r(x,t) = K_n^r(x-t)$ is piecewise linear in $x-t$, so we obtain for it the following graph, depending on n and r:

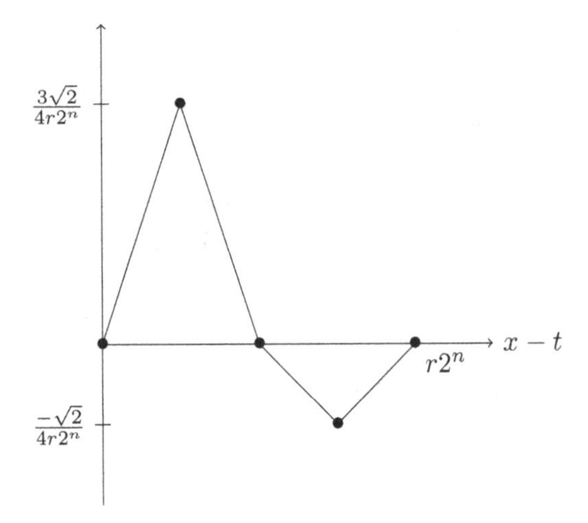

Some explanations: The value when $x-t=c$ is given by what happens in the picture when one averages the values of the kernel over the line parallel to the diagonal with equation $x-t=c$. By looking at the picture one sees that the value will change only at the values indicated, i.e. when $c=0$, $|I|/4$, $|I|/2$, $3|I|/4$ and $c \mapsto K_n^r(c)$ vanishes when $c \geq |I|$. Moreover, by elementary plane geometry, one sees that $c \mapsto K_n^r(c)$ is linear on the interval lying between any of two consecutive values among $c = 0, |I|/4, |I|/2, 3|I|/4$. Consider for instance the first interval: When $c \in [0, |I|/4]$. Let $\theta = c(|I|/4)^{-1}$. We need to compute the average of a function that (if we start from the bottom of the square in the first picture) is equal to $\sqrt{2}/|I|$ times successively $+1, -1, +1, -1$ and finally 0 and then the values repeat with period equal to

$|I|$. These successive values are the same on the diagonal except that, there (for $c = \theta = 0$), the interval with value 0 shrinks to a singleton. Now when $c = \theta(|I|/4)$ with $0 < \theta < 1$ the values to be averaged are now $+1, -1, +1, -1, 0$ with respective proportions $1/4, (1-\theta)/4, (1+\theta)/4, (1-\theta)/4, \theta/4$. So we find

$$\begin{aligned} K_n^r(c) &= (\sqrt{2}/|I|)[1/4 - (1-\theta)/4 + (1+\theta)/4 - (1-\theta)/4 + 0(\theta/4)] \\ &= (\sqrt{2}/|I|)[3\theta/4]. \end{aligned}$$

Thus we find as announced that $K_n^r(c)$ is linear in c between the values $c = 0$ and $c = |I|/4$, as indicated in the preceding graph. On the other 3 intervals, the linearity can be checked similarly by looking at the picture.
Now we compute $K^r(x,t)$.

We will compute $K^r(x,t)$ using $K_n^r(x,t)$ for different values of n and summing over $n \in \mathbb{Z}$. It suffices to compute $K^r(x,t)$ for the values $x - t = 3/4 \cdot r2^m$ and $x - t = r2^m$ ($m \in \mathbb{Z}$):

$$K^r\left(\frac{3}{4}r2^m\right) = -\frac{1}{4}\frac{\sqrt{2}}{r2^m} + \frac{3}{16}\frac{\sqrt{2}}{r2^m} + \frac{9}{64}\frac{\sqrt{2}}{r2^m}\left(1 + \frac{1}{4} + \frac{1}{16} + \cdots\right) = \frac{\sqrt{2}}{8r2^m}, \tag{6.35}$$

$$K^r(r2^m) = \frac{3}{16}\frac{\sqrt{2}}{r2^m}\left(1 + \frac{1}{4} + \frac{1}{16} + \cdots\right) = \frac{\sqrt{2}}{4r2^m}. \tag{6.36}$$

Here again a few more explanations might help. Consider for example (6.35). Let $c_m = \frac{3}{4}r2^m$. We have

$$K^r\left(\frac{3}{4}r2^m\right) = \sum\nolimits_{n\in\mathbb{Z}} K_n^r(c_m) = K_m^r(c_m) + \sum\nolimits_{n>m} K_n^r(c_m) + \sum\nolimits_{n<m} K_n^r(c_m).$$

But when $n < m$ we have $c_m \geq \frac{3}{2}r2^n > r2^n$ and hence since K_n^r is supported on $[0, r2^n]$ the sum over $n < m$ vanishes. A close look at the preceding graph then gives us

$$\begin{aligned} \sum\nolimits_{n>m} K_n^r(c_m) &= K_{m+1}^r(c_m) + \sum\nolimits_{n>m+1} K_n^r(c_m) \\ &= \frac{3}{16}\frac{\sqrt{2}}{r2^m} + \frac{9}{64}\frac{\sqrt{2}}{r2^m}\left(1 + \frac{1}{4} + \frac{1}{16} + \cdots\right). \end{aligned}$$

The justification of (6.36) is similar but simpler.

We now claim that equations (6.35) and (6.36) imply that

$$\frac{3\sqrt{2}}{32(x-t)} \leq K^r(x-t) \leq \frac{\sqrt{2}}{4(x-t)} \quad \forall r > 0. \tag{6.37}$$

Indeed, assume that $r2^{m-1} \leq x - t \leq r2^m$. Note that the function K^r restricted either to the interval $[\frac{3}{4}r2^m, r2^m]$ or to $[r2^{m-1}, \frac{3}{4}r2^m]$ is linear (since it is a sum of the functions $\{K_n^r \mid n \geq m\}$ that are each linear on this interval). From the values of $x \mapsto xK^r(x)$ at the end points, as given by (6.35) and (6.36), it follows that K^r restricted to the interval $[\frac{3}{4}r2^m, r2^m]$ (resp. $[r2^{m-1}, \frac{3}{4}r2^m]$) is of the form $K^r(x) = ax + b$ with a, b both positive (resp. both negative), and hence $x \mapsto xK^r(x)$ is monotone on both intervals. Applying this to both intervals, we obtain the claim for $r2^{m-1} \leq x - t \leq r2^m$ and hence for $x - t \geq 0$. Since K^r is odd this suffices.

The expression in Lemma 6.24 is obtained from $K^r(x-t)$ by a limit of averages in r, so it is clear from (6.37) that the constant $c = K(1)$ such that $K(x,t) = c(x-t)^{-1}$ is such that $3\sqrt{2}/32 \leq c \leq \sqrt{2}/4$. In particular $c \neq 0$, and since K is the kernel of $\mathcal{T}$, $\mathcal{T} \neq 0$. The proof of Theorem 6.22 is now complete. □

Remark 6.26. By (6.9), the equality $K(x,t) = K(1)(x-t)^{-1}$ means that $\mathcal{T} = \pi K(1) H^{\mathbb{R}}$. At the cost of a little more effort one can compute the exact value of $K(1)$. (A different way to calculate this same number appears in [379].) Here is an outline: We first compute $K_r(1)$. Let n be such that

$$r2^{n-1} \leq 1 < r2^n.$$

Equivalently we have $1/r = 2^s$ and $n - 1 = [s]$ where $[s] \in \mathbb{Z}$ denotes the largest integer $N \leq s$ in $\mathbb{Z}$. Let $0 \leq \theta < 1$ be such that

$$1 = (1-\theta)r2^{n-1} + \theta r2^n.$$

We have to distinguish $\theta \leq 1/2$ and $\theta > 1/2$, because K_r is linear on both intervals corresponding to $\theta \leq 1/2$ and $\theta > 1/2$. When $\theta \leq 1/2$ we have $1 = (1-2\theta)r2^{n-1} + 2\theta(3/4)r2^n$, and hence

$$K_r(1) = (1-2\theta)K_r(r2^{n-1}) + 2\theta K_r(((3/4)r2^n)$$

so that by (6.35) and (6.36) we find

$$K_r(1) = \sqrt{2}(r2^{n+1})^{-1}(1 - 3\theta/2).$$

Now if $\theta \geq 1/2$ we have $1 = 2(1-\theta)(3/4)r2^n + (2\theta - 1)r2^n$, and we find similarly

$$K_r(1) = \sqrt{2}(r2^{n+1})^{-1}(\theta/2).$$

Note $\theta = \frac{1-r2^{n-1}}{r2^{n-1}} = \frac{1-2^{[s]-s}}{2^{[s]-s}} = 2^{s-[s]} - 1$. So if we set $x = s - [s]$, we have $\theta = 2^x - 1$. Then $0 \leq x < 1$ and $r = 2^{-s} = 2^{-x-[s]}$. Let $a = \mathrm{Log}_2(3/2)$. Note that $\theta \leq 1/2$ iff $x \leq a$. Thus we have

$$K_r(1) = \sqrt{2}(1/4)\left(1_{\{x \leq a\}}((5/2)2^x - (3/2)2^{2x}) + 1_{\{x>a\}}((2^{2x} - 2^x)/2)\right)$$

Note that when $R \to \infty$ and $R = 2^Y$

$$(2\,\mathrm{Log}\, R)^{-1} \int_{1/R}^{R} f(r)dr/r = (2Y)^{-1} \int_{-Y}^{Y} f(2^y)dy = (2Y)^{-1} \int_{-Y}^{Y} f(2^{-y})dy$$

we apply this to $f(r) = K_r(1)$. Note that $r = 2^{-s}$ and $f(2^{-s})$ depends only on $x = s - [s]$, and hence is periodic with period 1. Thus it is clear that the average of $(2Y)^{-1} \int_{-Y}^{Y} f(2^{-y})dy$ is equal to

$$\int_0^1 f(2^{-y})dy.$$

So we conclude that

$$K(1) = \sqrt{2}(1/4)\left(\int_0^a ((5/2)2^x - (3/2)2^{2x})dx + \int_a^1 ((2^{2x} - 2^x)/2)dx\right)$$

and we find

$$K(1) = \sqrt{2}(1/8)(1/\mathrm{Log}\, 2).$$

Proof of Theorem 6.18. By (6.23), the operators $T^{\alpha,r}$ are obtained from $T^{0,1}$ by conjugation with respect to the group generated by translations and dilations. Since translations and dilations are isometric on $L_2(B)$, the norm on $L_2(B)$ of the single operator $T^{0,1}$ is the same as that of any of the operators $T^{\alpha,r}$. Thus, by Lemma 6.21, the UMD$_2$ property of B implies a uniform bound of the norm on $L_2(B)$ for the operators $T^{\alpha,r}$, and hence by averaging for the operator $\mathcal{T}$. By Theorem 6.22 and Proposition 6.7, we conclude that B is HT$_2$. The proof works just as well for $1 < p \neq 2 < \infty$ once one makes a cosmetic change of the normalization of the dilation D_r so that it remains isometric on L_p and $L_p(B)$: we just reset D_r to be equal to $D_r f(x) = f(x/r)r^{-1/p}$. □

Remark 6.27. If we replace the minus sign in (6.21) by a plus sign, then the resulting operator is even and hence the same reasoning shows that it is a scalar multiple of the identity. The reader in need of calculus exercises can verify that indeed the preceding calculations lead to $K(1) = 0$, and hence $K(x, t) = 0$ for all $x \neq t$, if the sign is changed.

Remark 6.28. Both proofs of $UMD \Rightarrow HT$ in this section and the next one produce an estimate of the form $C_2^{HT}(B) \leq cC_2(B)^2$ (for some absolute constant c). It seems to be an open problem whether this can be improved to $C_2^{HT}(B) \leq cC_2(B)$. See [240] and Remark 6.40 for more on this problem.

6.4 UMD implies HT (with stochastic integrals)*

In this section we briefly outline an alternate proof of the implication $UMD \Rightarrow HT$ based on the ideas of Burkholder, Gundy and Silverstein in [176]. The adaptation to the Banach space valued case comes from [169, 232].

To describe the main idea, we first need to relate the UMD property with a certain form of 'decoupling' inequality. Let B be an arbitrary Banach space. Consider a B-valued harmonic function $u : D \to B$ (we could restrict the discussion to a polynomial). We will use the representation of the martingale $M_t = u(W_{t\wedge T})$ given by the Ito-formula:

$$u(W_{t\wedge T}) = u(W_0) + \int_0^{t\wedge T} \nabla u(W_s).dW_s.$$

Writing $W_t = W_t^1 + iW_t^2$ and $\nabla u(x+iy) = (\frac{\partial u}{\partial x}(x+iy), \frac{\partial u}{\partial y}(x+iy))$ we can rewrite this as

$$M_t = u(W_{t\wedge T}) = u(W_0) + \int_0^{t\wedge T} \left[\frac{\partial u}{\partial x}(W_s)dW_s^1 + \frac{\partial u}{\partial y}(W_s)dW_s^2\right].$$

Assuming BM defined on $(\Omega, \mathbb{P})$ as before, we define the 'decoupled' martingale $\widehat{M}_t$ on $(\Omega \times \Omega, \mathbb{P} \times \mathbb{P})$ by setting for any $(\omega, \omega') \in \Omega \times \Omega$

$$\widehat{M}_t(\omega, \omega') = u(W_0(\omega)) + \int_0^{t\wedge T(\omega)} \nabla u(W_s(\omega)).dW_s(\omega')$$

or equivalently

$$\widehat{M}_t(\omega, \omega') = u(W_0(\omega)) + \int_0^{t\wedge T(\omega)} \left[\frac{\partial u}{\partial x}(W_s(\omega))dW_s^1(\omega') + \frac{\partial u}{\partial y}(W_s(\omega))dW_s^2(\omega')\right].$$

With this notation, the key fact to extend the proof of [176], and specifically (4.38), is the following fact:

Lemma 6.29. *If B is UMD, then for any $1 < p < \infty$ there is a constant D_p such that for any $f \in L_p(\mathbb{T}; B)$ with Poisson integral u (recall $u \in \tilde{h}^p(D; B)$) we have*

$$D_p^{-1} \sup_{t>0} \|\widehat{M}_t\|_p \le \|u\|_{h^p(D;B)} = \|f\|_{L_p(\mathbb{T};B)} \le D_p \sup_{t>0} \|\widehat{M}_t\|_p. \tag{6.38}$$

Sketch of proof. Consider a B-valued dyadic martingale (f_n) on the probability space $(\Omega, \mathbb{P}) = (D_\infty, d\nu)$ as before. We can write

$$d_n = f_n - f_{n-1} = \varepsilon_n \varphi_{n-1}(\varepsilon_1, \ldots, \varepsilon_{n-1}).$$

We define another martingale difference sequence $\widehat{d_n}$ on $(\Omega \times \Omega, \mathbb{P} \times \mathbb{P})$ by setting $\widehat{d_0} = d_0$ and for any $(\varepsilon, \varepsilon') \in \Omega \times \Omega$

$$\widehat{d_n}(\varepsilon, \varepsilon') = \varepsilon'_n \varphi_{n-1}(\varepsilon_1, \ldots, \varepsilon_{n-1}),$$

and we define $\widehat{f_n}$ by

$$\widehat{f_n} = \sum\nolimits_{0 \le k \le n} \widehat{d_k}.$$

By (5.23) and by Kahane's inequalities (Theorem 5.2) there is a constant D_p such that for any (f_n) we have

$$D_p^{-1} \sup_n \|\widehat{f_n}\|_{L_p(B)} \le \sup_n \|f_n\|_{L_p(B)} \le D_p \sup_n \|\widehat{f_n}\|_{L_p(B)}. \tag{6.39}$$

We now claim that a similar inequality holds if we replace (ε_n) by a Gaussian i.i.d. sequence (g_n). Indeed, consider now $\Omega = \mathbb{R}^{\mathbb{N}*}$ equipped with its canonical Gaussian measure $\mathbb{P}$, denote the coordinates by (g_n) and let (F_n) be a martingale on $(\Omega, \mathbb{P})$ associated to the natural filtration as in §1.4, such that the increments dF_n are of the form

$$dF_n = F_n - F_{n-1} = g_n \psi_{n-1}(g_1, \ldots, g_{n-1}).$$

We again denote $\widehat{dF_0} = dF_0$, $\widehat{dF_n}(\omega, \omega') = g_n(\omega')\varphi_{n-1}(g_1(\omega), \ldots, g_{n-1}(\omega))$, and

$$\widehat{F_n} = \sum\nolimits_{0 \le k \le n} \widehat{dF_k}.$$

By a suitable application of the central limit theorem (CLT in short), we can show that (6.39) 'automatically' implies

$$D_p^{-1} \sup_n \|\widehat{F_n}\|_{L_p(B)} \le \sup_n \|F_n\|_{L_p(B)} \le D_p \sup_n \|\widehat{F_n}\|_{L_p(B)}. \tag{6.40}$$

Indeed, we can reduce to the case when the ψ_n's are all polynomials and the CLT gives us the following approximations

$$g_1 \approx G_1 = N(1)^{-1/2}(\varepsilon_1 + \cdots + \varepsilon_{N(1)})$$
$$g_2 \approx G_2 = N(2)^{-1/2}(\varepsilon_{N(1)+1} + \cdots + \varepsilon_{N(1)+N(2)}),$$

and so on. Thus one finds

$$F_n(g_1, \ldots, g_n) \approx F_n(G_1, \ldots, G_n)$$

and similarly for $\widehat{F_n}$. Then (6.39) implies for any n

$$\begin{aligned} D_p^{-1} & \left\| \sum_0^n G_k(\omega')\psi_k(G_1(\omega), \ldots, G_{k-1}(\omega)) \right\|_{L_p(\mathbb{P}\times\mathbb{P};B)} \\ & \le \|F_n(G_1, \ldots, G_n)\|_{L_p(\mathbb{P};B)} \\ & \le D_p \left\| \sum_0^n G_k(\omega')\psi_k(G_1(\omega), \ldots, G_{k-1}(\omega)) \right\|_{L_p(\mathbb{P}\times\mathbb{P};B)} \end{aligned}$$

and in the limit we obtain Claim 6.40. Lastly, one can deduce (6.38) from (6.40) by a suitable discretization of the stochastic integrals, analogous to the one described for Lemma 4.48. □

We can now give an alternate proof that UMD implies HT. We will use the 'conjugate' harmonic function $\tilde{u}$. The latter can be defined viewing B as a real Banach space embedded isometrically into its complexification, as we did at the end of §6.1, but invoking the complexification of B is rather unpleasant. When u is a polynomial with coefficients in B of the form

$$u(z) = x_0 + \sum_{n>0} \Re(z^n)x_n + \Im(z^n)y_n$$

then u is the Poisson integral of its boundary values given by

$$f(e^{it}) = x_0 + \sum_{n>0} \cos(nt)x_n + \sin(nt)y_n,$$

and $\tilde{u}$ is given explicitly by the simple formula

$$\tilde{u}(e^{it}) = \sum_{n>0} \sin(nt)x_n - \cos(nt)y_n.$$

This also shows that if we denote by $\tilde{f}$ the boundary values of $\tilde{u}$ (so that $\tilde{u}$ is the Poisson integral of $\tilde{f}$), then we have

$$\tilde{f} = H^{\mathbb{T}} f.$$

Second proof of Theorem 6.18. Assume B has the UMD$_p$ property. Let f, u be as in the preceding lemma. For convenience, let us denote

$$\underline{S}(u)(\omega) = \left(\mathbb{E}_{\omega'} \left\| \int_0^{T(\omega)} \nabla u(W_s(\omega)).dW_s(\omega') \right\|^p \right)^{1/p}.$$

Note that if $p = 2$ and if $B = \mathbb{R}$ (or if B is a Hilbert space) then $\underline{S}(u) = S(u)$ where $S(u)$ is as in (4.38) (and the two quantities are in any case equivalent for $p \neq 2$). The key observation (as in (4.38)) is that the Cauchy-Riemann equations (4.37) imply that $\underline{S}(u) = \underline{S}(\tilde{u})$. Therefore, (6.38) yields

$$\|H^{\mathbb{T}} f\|_{L_p(\mathbb{T};B)} = \|\tilde{f}\|_{L_p(\mathbb{T};B)} \leq D_p^2 \|f\|_{L_p(\mathbb{T};B)},$$

which completes our second proof that $UMD \Rightarrow HT$.

A variant of the preceding proof can be based on the discretization in Remark 4.47, as was done initially by Burkholder in [169]. □

Remark. We refer to [52] for numerous complements on stochastic integrals and decoupling inequalities for multiple integrals. See also [59] for a general presentation of stochastic integration in Banach spaces.

6.5 Littlewood-Paley inequalities in UMD spaces

Let $x = (x_n)_{n \geq 0}$ be a sequence in a Banach space B. We will denote

$$R(x) = \sup_n \left\| \sum\nolimits_0^n \varepsilon_k x_k \right\|_{L_2(B)}$$

where $L_2(B) = L_2(\Delta, \nu; B)$ with (Δ, ν) as in (5.13).

Let $M = (M_n)_{n \geq 0}$ be a B-valued martingale with associated difference sequence $dM = (dM_n)_{n \geq 0}$. Then $R(dM)$ should be understood as a random variable. In our earlier study of the UMD property, we showed that if B is UMD then, if $1 < p < \infty$, all B-valued martingales satisfy

$$a_p \|R(dM)\|_p \leq \sup_n \|M_n\|_{L_p(B)} \leq b_p \|R(dM)\|_p$$

where a_p, b_p are positive constants depending on B. For Fourier series in the scalar case, there is an analogue of this, namely the famous Littlewood-Paley inequalities (see [30] and [85] for background). These say that for any f in $L_p(\mathbb{T})$ the dyadic partial sums

$$\Delta_n^+(f) = \sum_{2^n \leq k < 2^{n+1}} \hat{f}(k) e^{ikt} \tag{6.41}$$

$$\Delta_n^-(f) = \sum_{2^n \leq -k < 2^{n+1}} \hat{f}(k) e^{ikt} \tag{6.42}$$

satisfy for any $1 < p < \infty$ the following inequality

$$a'_p \|V_2(f)\|_p \leq \|f\|_p \leq b'_p \|V_2(f)\|_p$$

where a'_p, b'_p are positive constants and

$$V_2(f) = \left(|\hat{f}(0)|^2 + \sum |\Delta_n^+(f)|^2 + \sum |\Delta_n^-(f)|^2 \right)^{1/2}. \tag{6.43}$$

We will now show, following Bourgain [137], that any HT Banach space B satisfies an analogue of this where f is now in $L_p(\mathbb{T}; B)$ and where

$$V(f)^2 = \|\hat{f}(0)\|_B^2 + R(\{\Delta_n^+(f)\})^2 + R(\{\Delta_n^-(f)\})^2. \tag{6.44}$$

Note that in the scalar case (or in the Hilbert space case) (6.43) and (6.44) are the same.

Theorem 6.30 ([137]). *Let B be an HT and hence a UMD space. Then for any $1 < p < \infty$ there are positive constants a''_p, b''_p such that for any f in $L_p(\mathbb{T}; B)$ we have*

$$a''_p \|V(f)\|_p \leq \|f\|_{L_p(B)} \leq b''_p \|V(f)\|_p$$

where $V(f)$ is defined by (6.44) with $\Delta_n^+(f)$, $\Delta_n^-(f)$ given by (6.41) and (6.42).

As a corollary, we obtain an analogue of the classical Marcinkiewicz multiplier theorem in the B-valued case. Both the theorem and its corollary are proved further in what follows.

Corollary 6.31. *Let B be as in the Theorem. Let $\varphi\colon \mathbb{Z} \to \mathbb{C}$ be a function such that $\sup_{n\in\mathbb{Z}} |\varphi(n)| \leq 1$, and moreover*

$$\sum_{2^n \leq k < 2^{n+1}} |\varphi(k) - \varphi(k-1)| \leq 1 \text{ and } \sum_{2^n \leq k < 2^{n+1}} |\varphi(-k) - \varphi(-k+1)| \leq 1. \tag{6.45}$$

Then the Fourier multiplier

$$M_\varphi\colon\ f \to \sum \hat{f}(k)\varphi(k)e^{ikt}$$

is bounded on $L_p(B)$ for any $1 < p < \infty$, with norm less than a constant $C(p, B)$ depending only on p and B.

Remark 6.32. Let $\alpha_0, \alpha_1, \ldots, \alpha_N$ be real numbers such that

$$|\alpha_0| + \sum_1^N |\alpha_k - \alpha_{k-1}| \leq 1.$$

Let $(e_0, \ldots, e_N)$ be the canonical basis in $\mathbb{R}^{N+1}$. We have then

$$\sum\nolimits_0^N \alpha_j e_j = \alpha_0 \left(\sum_{j\geq 0} e_j\right) + (\alpha_1 - \alpha_0)\sum_{j\geq 1} + \cdots + (\alpha_N - \alpha_{N-1})\sum_{j\geq N} e_j.$$

Therefore $\alpha = (\alpha_0, \ldots, \alpha_N)$ lies in the absolutely convex hull of the set

$$\Big\{ \sum_{K\leq j\leq N} e_j \mid 0 \leq K \leq N \Big\}.$$

We will make extensive use of multipliers of bounded variation on $\mathbb{Z}$. More precisely, we will be interested in the finitely supported functions $\varphi\colon\ \mathbb{Z} \to \mathbb{R}$ such that

$$\sum |\varphi(k) - \varphi(k-1)| \leq 1. \tag{6.46}$$

Let us denote this class by $\mathcal{V}$. It follows easily from the previous remark that any element of $\mathcal{V}$ is in the absolutely convex hull of the subset $\mathcal{V}_1 \subset \mathcal{V}$ formed of the indicator functions of a bounded interval $I \subset \mathbb{Z}$.

It will be convenient to introduce some new terminology:

Definition 6.33. Let B be an arbitrary Banach space. A family of operators $\mathcal{C}$ acting on B is called R-bounded if there is a constant C such that for any n and any $x_1, \dots, x_n$ in B, and any $T_1, \dots, T_n$ in $\mathcal{C}$ we have

$$R(\{T_j x_j\}) \le CR(\{x_j\}).$$

The smallest such C is called the R-bound of $\mathcal{C}$.

If B is HT, the multipliers in $\mathcal{V}_1$ are uniformly bounded on $L_p(B)$ (by Remark 6.2) and hence by the preceding observation the same is true for the multipliers in $\mathcal{V}$. We will need the following refinement of this:

Lemma 6.34. *If B is an HT space, then the collection formed by the Fourier multipliers M_φ with $\varphi \in \mathcal{V}$ is R-bounded on $L_p(B)$, for any $1 < p < \infty$, with R-bound less than a constant $C'(p, B)$ depending only on p and B.*

Proof. Consider $f_1, \dots, f_n$ in $L_p(B)$. We must show $\exists C \forall n \forall \varphi_1, \dots, \varphi_n \in \mathcal{V}$

$$\|R(\{M_{\varphi_j} f_j\})\|_p \le C \|R(\{f_j\})\|_p. \tag{6.47}$$

Since $\mathcal{V}$ is included in the absolutely convex hull of $\mathcal{V}_1$ it clearly suffices to check this for $\varphi_j \in \mathcal{V}_1$.

Since the indicator function of an interval $[a, b)$ is the difference of the indicators of $[a, \infty)$ and $[b, \infty)$, it suffices to prove (6.47) when $\varphi_j = \psi_{K_j}$ with $\psi_K(n) = \begin{cases} 1 & \text{if } n \ge K \\ 0 & \text{otherwise,} \end{cases}$ where $\{K_j\}$ is arbitrary in $\mathbb{Z}$. By the HT property, we know that ψ_0 defines a bounded multiplier on $L_p(\mathbb{T}; B)$ (see Remark 6.2), or more generally (by Fubini) on $L_p(\mathbb{T} \times \Delta, dt \times dv; B)$. Hence we have a constant C' such that for any sequence of functions $(f_n)_{n \ge 0}$ in $L_p(\mathbb{T}; B)$ we have

$$\|R(\{M_{\psi_0} f_n\})\|_p \le C' \|R(\{f_n\})\|_p. \tag{6.48}$$

Now observe that (if we denote $z = e^{it}$ the variable in $\mathbb{T}$) for any f in $L_p(B)$

$$M_{\psi_K}(f) = z^K M_{\psi_0}(z^{-K} f). \tag{6.49}$$

Moreover, by (5.20) we have for any z in $\mathbb{C}$

$$2^{-1}R(\{f_n\}) \le R(\{z^{K_n}f_n\}) \le 2R(\{f_n\}). \tag{6.50}$$

Therefore, if $\varphi_j = \psi_{K_j}$ we find by (6.49) and (6.50)

$$R(\{M_{\varphi_j}f_j\}) \le 2R(\{M_{\psi_0}(z^{-K_j}f_j)\})$$

hence by (6.48) and by (6.50) again

$$\|R(\{M_{\varphi_j}f_j\})\|_p \le 4C'\|R(\{f_j\})\|_p. \qquad \square$$

Proof of Theorem 6.30. Let $(\mathcal{A}_n)_{n\ge 0}$ be the dyadic filtration on $\mathbb{T} \simeq [0, 2\pi)$. In what follows, any function f on $[0, 2\pi)$ will be implicitly extended to a 2π-periodic function on $\mathbb{R}$, so that its average on $\mathbb{T}$ coincides with its average on any interval of length 2π. Let $L_p(B) = L_p(\mathbb{T}, dx/2\pi; B)$. Let $I^n = [0, 2\pi 2^{-n})$ and $I^n(j) = I^n + (2\pi)2^{-n}j$. The conditional expectation $\mathbb{E}_n$ with respect to $\mathcal{A}_n$ is given by

$$E_n f(x) = \int_0^{2\pi} f(y)K_n(x, y)dy \quad \text{where}$$

$$K_n(x, y) = 2^{n-1}\pi^{-1}\sum\nolimits_0^{2^n-1} 1_{I^n(j)}(x)1_{I^n(j)}(y). \tag{6.51}$$

Since we assume B UMD, for any choice of signs $\varepsilon = (\varepsilon_n)$ the kernel $L_\varepsilon = \sum_1^\infty \varepsilon_n(K_n - K_{n-1})$ is bounded on $L_p(B)$. Thus for some C we have

$$\forall f \in L_p(B) \qquad \left\| \int L_\varepsilon(x, y)f(y)dy \right\|_{L_p(B)} \le C\|f\|_{L_p(B)}.$$

By translation invariance of Haar measure (applying the preceding to the function $y \to f(y - \theta)$), we find for any fixed θ

$$\left\| \int L_\varepsilon(x + \theta, y)f(y - \theta)dy \right\|_{L_p(B)} \le C\|f\|_{L_p(B)}.$$

Hence, after averaging over θ, we find

$$\left\| \int M_\varepsilon(x, y)f(y)dy \right\|_{L_p(B)} \le C\|f\|_{L_p(B)}$$

with $M_\varepsilon(x, y) = (2\pi)^{-1}\int_0^{2\pi} L_\varepsilon(x + \theta, y + \theta)d\theta$. We will now 'calculate' M_ε.

Clearly $M_\varepsilon(x, y) = N_\varepsilon(x - y)$ and $N_\varepsilon(\cdot) = \sum_1^\infty \varepsilon_n(\Phi_n(\cdot) - \Phi_{n-1}(\cdot))$ where $\Phi_n(x) = (2\pi)^{-1}\int_0^{2\pi} K_n(x + \theta, y + \theta)d\theta$. By (6.51), we have

$$\begin{aligned}\Phi_n(x - y) &= 2^{n-1}\pi^{-1}\sum\nolimits_0^{2^n-1}\int 1_{I^n}(x - 2\pi j2^{-n} + \theta)1_{I^n}(y - 2\pi j2^{-n} + \theta)d\theta/2\pi \\ &= 4^n(2\pi)^{-1}1_{I^n} * 1_{-I^n}(x - y),\end{aligned}$$

where the convolution is meant on $\mathbb{T}$. Hence for any k in $\mathbb{Z}$

$$\hat{\Phi}_n(k) = 4^n(2\pi)^{-1}|(1_{I^n})\hat{}(k)|^2 = 4^n(2\pi)^{-1}(\pi^2k^2)^{-1}\sin^2(k2^{-n}\pi).$$

Therefore we find

$$\hat{N}_\varepsilon(k) = 2^{-1}\pi^{-3}k^{-2}\sum\nolimits_1^\infty \varepsilon_n 4^n(\sin^2(k2^{-n}\pi) - 4^{-1}\sin^2(k2^{-n+1}\pi))$$
$$= 2^{-1}\pi^{-3}k^{-2}\sum\nolimits_1^\infty \varepsilon_n 4^n \sin^4(k2^{-n}\pi).$$

Thus we have proved that if we set

$$\sigma_n(k) = 4^n k^{-2}\sin^4(k2^{-n}\pi)$$

there is a constant C' such that

$$\|R(\{M_{\sigma_n}f\})\|_p \le C'\|f\|_{L_p(B)}. \tag{6.52}$$

We will now consider the multiplier φ_n supported by $[2^{n-2}, 2^{n-1})$ and defined by $\varphi_n(k) = \sigma_n(k)^{-1}$ if $2^{n-2} \le k < 2^{n-1}$. Clearly

$$M_{\varphi_n}M_{\sigma_n}f = \sum_{2^{n-2}\le k<2^{n-1}} \hat{f}(k)z^k = \Delta_{n-2}^+(f).$$

We claim that $\sum_k |\varphi_n(k) - \varphi_n(k-1)| \le \gamma$ for some γ independent of n. This follows by elementary calculus: Indeed, if we let $\sigma_n(t) = 4^n t^{-2}\sin^4(t2^{-n}\pi)$, or equivalently $\sigma_n(t) = \theta(t2^{-n})$ with $\theta(t) = t^{-2}\sin^4(t\pi)$, then on the interval $[2^{n-2}, 2^{n-1})$ we have bounds

$$|\sigma_n(t)| \ge \alpha \quad \text{and} \quad |\sigma_n'(t)| \le \beta 2^{-n}$$

with $\alpha > 0$ and $\beta > 0$ independent of n; from this it follows that the derivative of $t \to \sigma_n(t)^{-1}$ is bounded above by $\beta\alpha^{-2}2^{-n}$ for t in $[2^{n-2}, 2^{n-1})$. Then the mean value theorem yields $\sum_{2^n\le k<2^{n+1}} |\sigma_n(k)^{-1} - \sigma_n(k-1)^{-1}| \le \beta\alpha^{-2}$. This verifies our claim with (say) $\gamma = \alpha^{-1} + \beta\alpha^{-2}$.

Thus we find that $\gamma^{-1}\varphi_n \in \mathcal{V}$. We can now conclude the proof: since $\Delta_{n-2}^+(f) = M_{\varphi_n}M_{\sigma_n}f$, by Lemma 6.34 and by (6.52)

$$\|R(\{\Delta_n^+(f)\})\|_p \le C'(p,B)\gamma\|R(\{M_{\sigma_n}f\})\|_p \le C'(p,B)\gamma C'\|f\|_{L_p(B)}.$$

Since B is HT, we immediately deduce from this that $\exists a_p'' > 0$ such that

$$a_p''\|V(f)\|_p \le \|f\|_{L_p(B)}, \tag{6.53}$$

i.e. we obtain 'one-half' of the desired result. But then a simple duality argument shows that, if p' is the conjugate of p, (6.53) implies for any g in $L_{p'}(B^*)$

$$a_p''\|g\|_{L_{p'}(B^*)} \le \|V(g)\|_{p'} \tag{6.54}$$

Finally, since these inequalities have now been established for any UMD space, and UMD is a self-dual property, we conclude (since B^* is UMD and $B \subset (B^*)^*$) that B itself also satisfies an inequality like (6.54) for any $1 < p' < \infty$. This gives the 'missing half' of the desired result. □

Proof of Corollary 6.31. By Theorem 6.30, it suffices to show that, for some constant C, if $g = M_\varphi f$ then

$$\|V(g)\|_p \le C\|V(f)\|_p.$$

This follows from (6.47). □

6.6 The Walsh system Hilbert transform

The 'Walsh functions' are the same as the collection $\{w_A\}$ considered earlier in (5.15) on $\Delta = \{-1, 1\}^{\mathbb{N}}$, but with a special indexation and a specific order. Recall that we denote by ε_n the coordinates on Δ and we have set $w_A = \prod_{n\in A} \varepsilon_n$. For any integer $j \ge 0$, let us consider its binary expansion $j = 2^{s_0} + 2^{s_1} + \cdots + 2^{s_k}$ with $0 \le s_0 < s_1 < \cdots < s_k$. Let $A(j) = \{s_0, s_1, \ldots, s_k\}$. We then set $w_j = w_{A(j)}$, and $w_0 \equiv 1$. This sequence $\{w_j \mid j \ge 0\}$, which is obviously an orthonormal basis of $L_2(\Delta, \nu)$, is sometimes called the Fourier-Walsh system. It shares many properties of the trigonometric system $\{e^{ijt} \mid j \in \mathbb{Z}\}$ on $\mathbb{T}$.

In particular, this system is a basis of $L_p(\Delta, \nu)$ for any $1 < p < \infty$, i.e. the family of projections

$$P_N\colon\ f \to \sum\nolimits_0^N \langle f, w_j\rangle w_j$$

are uniformly bounded on $L_p(\Delta, \nu)$. More generally, we have

Theorem 6.35. *A Banach space B is UMD$_p$ iff*

$$\sup\nolimits_N \|P_N\colon\ L_p(\nu; B) \to L_p(\nu; B)\| < \infty.$$

Moreover, we have the UMD$_p$ constant $C_p(B)$ $(1 < p < \infty)$ satisfies

$$\sup_N \|P_N\colon L_p(\nu; B) \to L_p(\nu; B)\| \le C_p(B) \le 2\sup_N \|P_N\colon L_p(\nu; B) \to L_p(\nu; B)\|. \tag{6.55}$$

Proof. Let $N = 2^{s_0} + 2^{s_1} + \cdots + 2^{s_k}$ with $0 \le s_0 < s_1 < \cdots < s_k$. Consider f in $L_p(\nu; B)$. Let $\mathcal{B}_n = \sigma(\varepsilon_0, \ldots, \varepsilon_n)$, $f_n = \mathbb{E}^{\mathcal{B}_n} f$ and $d_n(f) = f_n - f_{n-1}$. Note that $L_2(\mathcal{B}_s) = \overline{\mathrm{span}}[w_k \mid 0 \le k < 2^{s+1}]$ and

$$L_2(\mathcal{B}_s) \ominus L_2(\mathcal{B}_{s-1}) = \overline{\mathrm{span}}[w_k \mid 2^s \le k < 2^{s+1}\}.$$

The following key identity is not hard to verify (left to the reader).

$$w_N P_N(w_N f) = \sum\nolimits_0^k d_{s_j}(f).$$

Therefore, if B is UMD$_p$, we have

$$\|P_N(f)\|_{L_p(B)} = \|w_N P_N(w_N w_N f)\|_{L_p(B)} = \left\|\sum\nolimits_0^k d_{s_j}(w_N f)\right\|_{L_p(B)}$$

$$\leq C_p(B)\left\|\sum\nolimits_{s\geq 0} d_s(w_N f)\right\|_{L_p(B)} = C_p(B)\|w_N f\|_{L_p(B)} = C_p(B)\|f\|_{L_p(B)}.$$

Conversely, estimating separately the subsets $\{j \mid \xi_j = 1\}$ and $\{j \mid \xi_j = -1\}$ we find that for any choice of signs (ξ_j) we have

$$\|\sum \xi_s df_s\|_{L_p(B)} \leq 2\sup \|\sum\nolimits_0^k d_{s_j}(f)\|_{L_p(B)}$$

where the sup runs over all possible $0 \leq s_0 < s_1 < \cdots < s_k$, and hence, since $f \mapsto w_N P_N(w_N f)$ has the same norm as P_N, we obtain the converse direction $C_p(B) \leq 2\sup_N \|P_N : L_p(B) \to L_p(B)\|$. □

In view of Remark 6.2, (6.55) seems like evidence that the answer to Problem 6.17 is positive.

6.7 Analytic UMD property*

Given the special properties enjoyed by Banach space valued analytic functions among the harmonic ones, it is natural to consider what becomes of the UMD property if one restricts to analytic or Hardy martingales.

Definition 6.36. Let $0 < p < \infty$. A Banach space B is called AUMD$_p$ if there is a constant C such that for any Hardy martingale (f_n) converging in $L_p(B)$ we have for any choice of signs $\varepsilon_n = \pm 1$

$$\sup_n \left\|\sum\nolimits_0^n \varepsilon_k df_k\right\|_{L_p(B)} \leq C \sup_n \|f_n\|_{L_p(B)}. \tag{6.56}$$

Let $C_p^a(B)$ be the smallest constant C for which this holds.

Obviously, UMD$_p$ implies AUMD$_p$ for $p > 1$ and $C_p^a(B) \leq C_p(B)$. We will content ourselves of a summary of most of what is known on this property.

Let $0 < p < \infty$. Recall that for Hardy martingales (f_n), we have (see Theorem 4.43)

$$\|\sup \|f_n\|_B\|_p \leq e^{1/p} \sup \|f_n\|_{L_p(B)} \tag{6.57}$$

i.e. the quantities $\sup \|f_n\|_{L_p(B)}$ and $\|\sup \|f_n\|_B\|_p$ are equivalent. In particular, UMD$_1$ as defined in §5.6 implies AUMD$_1$.

Note that restricting to analytic martingales leads to an equivalent definition. Indeed, since Hardy martingales are perturbations of subsequences of analytic ones (see Remark 4.47), the unconditionality of the latter implies that of the former.

Since we can restrict to analytic martingales, by Remark 5.24, we have $\mathrm{AUMD}_p \Rightarrow \mathrm{AUMD}_q$ for any $0 < q \le p$. The converse is also true. We choose to prove it following Burkholder's idea in [174] already used in §5.8 for Theorem 5.30.

Theorem 6.37. *Let B be a Banach space. Consider for $0 < p < \infty$ the following property:*
$(AB)_p$: There is a constant C such that for any B-valued analytic martingale (f_n) and any martingale transform $\tilde{f}_n = \sum_{k \le n} \varepsilon_k df_k$ $(\varepsilon_k \in \pm 1)$ we have

$$(\mathbb{E} \sup_n \|\tilde{f}_n\|^p)^{1/p} \le C(\mathbb{E} \sup_n \|f_n\|^p)^{1/p}.$$

Then $(AB)_p \Leftrightarrow AUMD_p$, and $\forall\, 0 < p, q < \infty \quad (AB)_p \Leftrightarrow (AB)_q$.

Moreover, for $(AB)_p$ to hold, it suffices that the weak-type version holds, more precisely it suffices that there exists C' such that for any (f_n) and (ε_n) we have

$$\forall n \ge 0 \quad (\sup_{c>0} c^p \mathbb{P}\{\|\tilde{f}_n\| > c\})^{1/p} \le C'(\mathbb{E}\|\tilde{f}_n\|^p)^{1/p}. \tag{6.58}$$

Furthermore, with the notation in Theorem 5.30, we have $C_q^a(B) \le e^{1/q}\chi_{p,q}C'$ and

$$C_q^a(B) \le e^{1/q}\chi_{p,q}C_p^a(B).$$

Proof. As we already mentioned $(AB)_p \Leftrightarrow AUMD_p$ follows from Remark 4.47. For $(AB)_p \Leftrightarrow (AB)_q$, the argument is similar to the third proof of Theorem 5.13 given at the end of §5.5: Once we observe that analytic martingales are stable under stopping, this is an immediate consequence of Theorem 5.30 and (6.57). □

The main novelty compared with the 'ordinary' UMD is this:

Proposition 6.38. *The spaces $B = \mathbb{C}$ and $B = L_1$ are $AUMD_1$. More generally if a Banach space B is $AUMD_1$, then for any measure space (Ω, μ) the space $L_1(\mu; B)$ is also $AUMD_1$.*

The proof when $\dim(B) = 1$ is an easy consequence of (6.57) for $p = 1$ and Corollary 5.37.

Then the proof for $B = L_1$ can be completed easily invoking Fubini's Theorem. A similar argument works for $L_1(B)$ with B assumed AUMD_1.

It is easy to show by the same argument as in Remark 5.40 that AUMD implies the ARNP (and actually the super-ARNP).

However, the non-commutative L_1-spaces, such as the trace class, are not AUMD_1 (see [261]).

6.8 UMD operators*

In this section, we briefly review the natural extension of the UMD property for operators $u: B \to C$ between two Banach spaces in place of a single space B. When the operator is the identity on B we recover the UMD property of B. It will be convenient to allow for the space C to be merely a quasi-Banach space (see Remark (1.8) for the definition).

Definition 6.39. Let $1 < p < \infty$. Let B be a Banach and C a quasi-Banach space. A bounded linear operator $u: B \to C$ is called UMD_p if there is a constant C such that for any martingale (f_n) converging in $L_p(B)$ we have for any choice of signs $\varepsilon_n = \pm 1$

$$\sup_n \left\| \sum\nolimits_0^n \varepsilon_k u(df_k) \right\|_{L_p(B)} \leq C \sup_n \|f_n\|_{L_p(B)}. \tag{6.59}$$

We will denote the best C in (6.59) by $C_p(u)$. We will say that u is UMD if this holds for some $1 < p < \infty$.

The basic facts extend with the same proofs to operators. Mainly, the property UMD_p does not depend on $1 < p < \infty$ and the dyadic version of UMD implies UMD also in case of operators. In the case of operators between Banach spaces, we have $C_p(u) = C_{p'}(u^*)$.

The possible lack of convexity in the space C prevents us to invoke convexity to show directly that a UMD operator is such that (6.59) still holds for all predictable sequences (ε_k) in the unit ball of L_∞. In [141] such operators are called MT operators. However, it is proved in [433] that UMD operators are indeed MT for any quasi-normed space C.

Remark 6.40. A similar extension is possible for the Hilbert transform $H^{\mathbb{T}}$: We say that u is an HT_p operator if $H^{\mathbb{T}} \otimes u$ extends to a bounded operator from $L_p(\mathbb{T}, B)$ to $L_p(\mathbb{T}, C)$. Again since this does not depend on $1 < p < \infty$, we say simply that u is an HT operator. The equivalence $UMD \Leftrightarrow HT$ for operators seems to be an open problem (this is of course related to Remark 6.28, Corollary 6.16 and Problem 6.17 since the respective constants would then be automatically equivalent). Indeed, Bourgain's argument apparently only shows that the composition of two HT_p operators is UMD_p, and similarly for the converse direction. We refer the reader to [238–240] for more on this. In particular, in [240] Geiss, Montgomery-Smith and Saksman observe that the operator $v_1^n \to \ell_\infty^n$ (induced by the identity of $\mathbb{R}^n$ or $\mathbb{C}^n$), considered in Example 8.82 and in much of Chapters 9 and 12, would lead to a counter-example for the implication $UMD \Rightarrow HT$ for operators, if it could be proved that its UMD_p constant is $o(\operatorname{Log} n)$.

The most interesting example of UMD operator is probably the following one due to Bourgain and Davis [141].

Theorem 6.41. *For any $0 < r < 1$ and any probability space (S, Σ, μ) the inclusion mapping $u : L_1(\mu) \to L_r(\mu)$ is UMD_p (and hence HT) for all $1 < p < \infty$.*

Sketch of proof. By (5.31) for scalar valued martingales, we know that for some constant C we have

$$\| \sup |\tilde{f}_n| \|_{1,\infty} \le C \sup \|f_n\|_1.$$

A fortiori, for any $0 < r < 1$ there is a constant C_r such that for any n

$$\|\tilde{f}_n\|_r \le C_r \|f_n\|_1.$$

Consider now a martingale (f_n) with values in $B = L_1(\mu)$. Let $u : L_1(\mu) \to L_r(\mu)$ denote the inclusion map. Integrating the last inequality with respect to μ and using Fubini, we find

$$\|\tilde{f}_n\|_{L_r(\mathbb{P}\times\mu)} \le C_r \|f_n\|_{L_1(\mathbb{P}\times\mu)},$$

which can be rewritten as

$$\|u(\tilde{f}_n)\|_{L_r(\mathbb{P};L_r(\mu))} \le C_r \|f_n\|_{L_1(\mathbb{P};L_1(\mu))}. \tag{6.60}$$

We will now use Burkholder's reverse Hólder principle as described at the end of Chapter 5. By multiplying by an independent choice of sign, we may assume without loss of generality that f_n is symmetric (when viewed as $L_1(\mu)$ valued). We may apply the last inequality to $N^{-1}(f_n^{(1)} + \cdots + f_n^{(N)})$ where $f_n^{(1)}, \ldots, f_n^{(N)}$ are i.i.d. copies of f_n. Let us denote by $g_n^{(j)}$ the corresponding transform, so that $g_n^{(j)} = \tilde{f}_n^{(j)}$. Note that (1.43) (after a suitable integration with respect to μ) implies

$$\int \sup_{1\le j\le N} |N^{-1}(g_n^{(1)} + \cdots + g_n^{(j)})|^r d\mathbb{P} d\mu \le 2 \int |N^{-1}(g_n^{(1)} + \cdots + g_n^{(N)})|^r \mathbb{P} d\mu,$$

and a fortiori we have

$$\int \sup_{1\le j\le N} |N^{-1}g_n^{(j)}|^r d\mathbb{P}d\mu \le 4\int |N^{-1}(g_n^{(1)}+\cdots+g_n^{(N)})|^r \mathbb{P}d\mu,$$

and by (6.60) and the triangle inequality in $L_1(\mathbb{P}; L_1(\mu))$ we find

$$\int \sup_{1\le j\le N} |N^{-1}g_n^{(j)}|^r d\mathbb{P}d\mu \le 4C_r^r \|f_n\|^r_{L_1(\mathbb{P};L_1(\mu))}.$$

Now Corollary 5.76 (applied here with $q = r$, $p = 1$) leads us for some constant C' to

$$\|u(\tilde{f}_n)\|_{L_{1,\infty}(\mathbb{P};L_r(\mu))} \le C'\|f_n\|_{L_1(\mathbb{P};L_1(\mu))}.$$

Assume now that (f_n) is a dyadic martingale. Then the extrapolation lemma 5.42 (extended to the present quasi-norm situation) and Doob's inequality imply that for any $1 < q < \infty$ we have

$$\|u(\tilde{f}_n)\|_{L_q(\mathbb{P};L_r(\mu))} \le C'\|f_n\|_{L_q(\mathbb{P};L_1(\mu))}.$$

Thus we conclude that u is UMD_q for all such q. □

Remark 6.42. In particular, taking for μ the uniform probability on $\{1,\ldots,n\}$ we find that the inclusion $L_1^n \to L_r^n$ is a UMD operator with a constant independent of n. More generally (see [141]), this remains valid for the inclusion $E_n \to E_n(r)$ where E_n is any n-dimensional normed space with a symmetric basis (e_k), assumed q-concave for some fixed $q < \infty$, and where $E_n(r)$ denotes the same space but equipped with the quasi-norm defined by $\|x\| = \|\sum_k |x_k|^r e_k\|_{E_n}^{1/r}$ for $x \in E_n$.

Corollary 6.43. *Let $B = L_1(\mu)$ and let $Y \subset B$ be a reflexive subspace. Let $1 \le q < \infty$. Then any $f \in \tilde{H}^q(\mathbb{T}; B/Y)$ can be lifted to an element of $\tilde{H}^q(\mathbb{T}; B)$. Moreover, with the notation in §6.2, any $f \in \tilde{H}^q(G; B/Y)$ can be lifted to an element of $\tilde{H}^q(G; B)$.*

Sketch of proof. Let $Q: B \to B/Y$ denote the quotient map. It suffices to show that for some constant C, for any polynomial f in the open unit ball of $\tilde{H}^q(\mathbb{T}; B/Y)$ there is $g \in \tilde{H}^q(\mathbb{T}; B)$ with $\|g\|_{H^q(\mathbb{T};B)} \le C$ such that $Q(g) = f$. Let $h \in L_q(\mathbb{T}; B)$ with $\|h\|_{L_q(\mathbb{T};B)} \le 1$ such that $Q(h) = f$. We may assume that μ is a probability. By a well know result due to Kadec and Pełczyński [297], if $Y \subset L_1(\mu)$ is reflexive, the inclusion $u: L_1(\mu) \to L_r(\mu)$ is an isomorphism when restricted to Y. Let $h = \sum \hat{h}(n)e^{int}$ be the formal Fourier series of h. Since $Q(h)$ is analytic we know that $\hat{h}(n) \in Y$ for any $n < 0$.

Let $1 < q < \infty$. By the preceding Theorem we know that the Hilbert transform applied to $u(h)$ is controlled in $L_q(L_r)$. It follows that $h_- = \sum_{n<0} \hat{h}(n)e^{int}$

is in $L_q(L_r)$, but since it is also in $L_q(Y)$ it must be in $L_q(B)$. Subtracting the latter to h, we find a function $g = h - h_- \in \tilde{H}^q(\mathbb{T}; B)$ that lifts f, and recollecting the estimates we find $\|g\|_{H^q(\mathbb{T};B)} \le C$. This settles the case $1 < q < \infty$.

By factorization, we will reduce the case $q = 1$ to the case $q > 1$ or say for simplicity $q = 2$. Let f be a polynomial in the open unit ball of $\tilde{H}^1(\mathbb{T}; B/Y)$. By Corollary 4.19 we can write $f = \varphi\psi$ with $\|\varphi\|_{H^2(\mathbb{T})}\|\psi\|_{H^2(\mathbb{T};B/Y)} \le 1$. By Remark 4.23 we may assume $\psi \in \tilde{H}^2(\mathbb{T}; B/Y)$. Then the preceding lifting applied to ψ yields the desired lifting for f, settling the case $q = 1$.

The second assertion follows the same argument but using the fact that Theorem 6.13 and Corollary 6.15 remain clearly valid with essentially the same proof for HT operators. Moreover, the factorization $H^1(G) = H^2(G).H^2(G)$ and its B-valued analogue hold because the theory of outer functions does extend to ordered groups (see [84, §8.4]). □

6.9 Notes and remarks

Pichorides [385] showed that the norm of the Hilbert transform on L_p either on $\mathbb{T}$ or on $\mathbb{R}$ is equal to $\cot(\pi/2p^*)$ for any $1 < p < \infty$ where $p^* = \max\{p, p'\}$. Thus for any B with $\dim(B) \ge 1$ we have

$$C_p^{HT}(B) \ge \cot(\pi/2p^*).$$

Theorem 6.5 is a classical fact due to Elias Stein. Propositions 6.7 and 6.8 are due to de Leeuw [323]. The extension to the Banach space valued case is obvious. Let $\mathcal{F}_G$ denote the Fourier transform on a locally compact Abelian group G. Given a function φ on $\hat{G}$, the (convolution) operator T_φ associated to φ is defined on $L_2(G)$ by $T_\varphi f = \mathcal{F}_G^{-1}(\varphi\mathcal{F}_G(f))$. The same proof shows that if φ is 'regular' (in the sense of [323]) i.e. is such that $\varphi * g_\varepsilon(x) \to \varphi(x)$ when $\varepsilon \to 0$ for all $x \in \mathbb{R}$, then T_φ is bounded on $L_p(\mathbb{R}; B)$ iff it is bounded on $L_p(b\mathbb{R}; B)$ where $b\mathbb{R}$ is the Bohr compactification of $\mathbb{R}$, i.e. the compact group that is dual to $\mathbb{R}$ endowed with its discrete topology. Here $b\mathbb{R}$ is equipped with its normalized Haar measure. The connection with Proposition 6.7 stems from the fact that the mapping $f \mapsto \underline{f}$ defines an isometric embedding $J : L_p(\mathbb{T}, m; B) \to L_p(b\mathbb{R}; B)$ such that, for any 'regular' φ with bounded T_φ, we have $T_\varphi J = JT_{\varphi_{|\mathbb{Z}}}$. In this light, Proposition 6.7 says that

$$\|T_{\varphi_{|\mathbb{Z}}}\|_{B(L_p(\mathbb{T},m;B))} \le \|T_\varphi\|_{B(L_p(\mathbb{R};B))} = \|T_\varphi\|_{B(L_p(b\mathbb{R};B))}.$$

Historically, the implication UMD $\Rightarrow$ HT came first. It was proved by Burkholder with McConnell (see [169]). Their proof is a discretization of a beautiful idea appearing in [176]. The latter paper showed that if one uses the (continuous time) filtration adapted to Brownian motion, then the Hilbert transform itself appears as a martingale transform. So, in retrospect, [176] contained the fundamental idea that the Hilbert transform is somehow 'subordinated' to martingale transforms. The converse turned out to be more delicate but it was proved by Bourgain in [134]. (See also [103].) We include this (i.e. HT $\Rightarrow$ UMD) in §6.1, and the converse implication in §6.3. For the latter (i.e. UMD $\Rightarrow$ HT) we present S. Petermichl's beautiful proof from [378] (see also [379] for a multi-dimensional extension) as well as a more traditional (but more technical) proof based on stochastic integrals (see §6.4).

Note that Figiel already gave in [225, 226] several interesting representation formulae (of a different kind) for general singular integrals in terms of sums of products of some simpler transformations on the Haar system, allowing him to prove that the so-called generalized Calderón-Zygmund operators are bounded on $L_p(B)$ if B is UMD.

In §6.5 we present Bourgain's proof of the Littlewood-Paley inequalities for Fourier series in $L_p(\mathbb{T}; B)$ assuming B UMD, following [137].

In [299] Kalton presents an example (derived from Bourgain's examples of super-reflexive Banach lattices failing UMD) of a Banach space X with a subspace $Y \subset X$ such that both Y and X/Y are isomorphic to ℓ_2 but X fails UMD. Such a space is called a twisted sum of two Hilbert spaces. In the same paper however, he shows that the Kalton-Peck space Z_2 (or Z_p for $1 < p < \infty$) is UMD. He also shows that given a complex interpolation scale B_θ $(0 < \theta < 1)$ of Köthe function spaces, the set of θs such that B_θ is UMD is either open (possibly void) or reduced to a single point.

The reader will find in [403] several versions of the Rubio de Francia–Littlewood-Paley inequalities in UMD Banach lattices. See also [440] for recent work on analytic semi-groups, admitting an H^∞ functional calculus, on $L_p(B)$ when B is a UMD Banach lattice.

In §6.6, we prove the analogue of the equivalence UMD $\Leftrightarrow$ HT for the Walsh system on $\{-1, 1\}^{\mathbb{N}}$. In preparing this part we found Kashin and Saakyan's book [46] very helpful. See also [75] for a parallel discussion of Haar and Fourier analysis.

See [253] for a discussion of dyadic martingale transforms in the matrix valued case, with an estimate of their bound depending logarithmically on the size of the matrix. See also [359] on the same theme.

It is a natural question whether Carleson's theorem about the a.e. convergence of Fourier series extends for functions in $L_2(B)$ when B is a UMD space. The paper [269] (see also [270]) comes very close to solving this. The authors require that, for some $0 < \theta < 1$, B be of the form $(B_0, B_1)_\theta$ with B_1

Hilbertian and B_0 UMD. All known examples of UMD spaces are of this form, in particular the Schatten p-classes ($1 < p < \infty$).

Concerning the AUMD property, the basic facts come from [233]. A characterization of AUMD as a Littlewood-Paley inequality (a kind of analytic version of Theorem 6.30) is given in [126].

As for analytic analogues of the characterization of UMD by ζ-convexity, we refer the reader to [382, 383].

7
Banach space valued H^1 and BMO

7.1 Banach space valued H^1 and BMO: Fefferman's duality theorem

Fefferman's classical Theorem asserts that, in the scalar valued case, the space BMO (defined later) of functions with 'bounded mean oscillation' is the dual of H^1. We will see a martingale analogue of this in §7.3. For functions on the torus $\mathbb{T}$, this duality can be summarized by the identity $H^1(D)^* = BMO_a$, where the space $BMO_a \subset BMO(\mathbb{T})$ is the subspace formed of all functions $f \in BMO(\mathbb{T})$ with Fourier transform vanishing on the negative integers, and the duality is defined by

$$\forall f \in H^1(D)\ \forall \varphi \in BMO_a \qquad \langle\langle \varphi, f\rangle\rangle = \int f(t)\varphi(-t)dm(t).$$

Here, we abuse the notation: a priori we only know $\varphi \in L_1(\mathbb{T})$ so the meaning of the integral $\int f(t)\varphi(-t)dm(t)$ is unclear but it is well defined (and absolutely convergent) when f is a polynomial or when f is (the boundary value of a function) in $H^\infty(D)$, and since $f \mapsto \int f(t)\varphi(-t)dm(t)$ is a continuous linear form on $H^1(D)$ we can extend this meaning by density. Equivalently we may define the duality by setting

$$\langle\langle \varphi, f\rangle\rangle = \lim_{r\uparrow 1} \int f_r(t)\varphi(-t)dm(t),$$

but we must first show the existence of the limit, which is part of the duality Theorem. In fact, since we know that $f_r \to f$ in $H^1(D)$ and $f_r \in H^\infty(D)$, it suffices to show that $f \mapsto \langle\langle \varphi, f\rangle\rangle$ is a continuous linear form on $H^\infty(D) \subset H^1(D)$ equipped with the norm induced on it by $H^1(D)$.

For the definition of $BMO(\mathbb{T})$, we return to our general Banach space framework. We will define $BMO(\mathbb{T}; B)$ (resp. $BMO(\mathbb{R}; B)$) and then, to abbreviate, we set $BMO(\mathbb{T}) = BMO(\mathbb{T}; \mathbb{R})$ (resp. $BMO(\mathbb{R}) = BMO(\mathbb{R}; \mathbb{R})$).

Given a Banach space B, let $\varphi : \mathbb{T} \to B$ (resp. $\varphi : \mathbb{R} \to B$) be a Bochner measurable function. Let us assume that for all subarcs $I \subset \mathbb{T}$ (resp. all bounded intervals $I \subset \mathbb{R}$) the restriction of φ to I is in $L_1(I; B)$. For $\mathbb{T}$ this simply means that $\varphi \in L_1(\mathbb{T}; B)$. To cover both cases together we will say that φ is locally in $L_1(B)$. More generally, we will say that a function φ is locally in a space Z if $1_I\varphi$ is in Z for any bounded interval I as earlier.

Throughout this chapter, for short, we use the term 'bounded interval' I in both cases, so by this we mean an arc in the case of $\mathbb{T}$, and we denote by m_I the normalized Lebesgue measure on I. Note that in the case of $\mathbb{T}$, $m(I)^{-1}dm(t) = |I|^{-1}1_I dt$, where $|I|$ is the length of I, so in both cases we have

$$dm_I(t) = |I|^{-1}1_I dt.$$

The space $BMO(\mathbb{T}; B)$ (resp. $BMO(\mathbb{R}; B)$) is defined as formed of all functions φ, locally in $L_1(B)$, such that for any bounded interval I

$$|I|^{-1}\int_I \|\varphi - \varphi_I\| dt = \int \|\varphi - \varphi_I\| dm_I \le C, \tag{7.1}$$

where we set

$$\varphi_I = |I|^{-1}\int_I \varphi(t)dt = \int \varphi dm_I.$$

It is customary to denote by $\|\varphi\|_*$ the smallest C for which this holds. Note however that this vanishes when φ is a constant function. To take this into account, $BMO(\mathbb{T}; B)$ (resp. $BMO(\mathbb{R}; B)$) is equipped with the norm

$$\|\varphi\|_{BMO(\mathbb{T};B)} = \max\left\{\left\|\int \varphi\right\|, \|\varphi\|_*\right\},$$

$$\left(\text{resp. } \|\varphi\|_{BMO(\mathbb{R};B)} = \max\left\{\left\|\int_0^1 \varphi\right\|, \|\varphi\|_*\right\}\right).$$

We have obvious inclusions

$$L_\infty(\mathbb{T}; B) \subset BMO(\mathbb{T}; B) \text{ and } L_\infty(\mathbb{R}; B) \subset BMO(\mathbb{R}; B).$$

Note that, in several contexts, it is more natural to replace $BMO(\mathbb{R}; B)$ by its quotient by the subspace formed of all the constant B-valued functions. We equip the latter space with the norm $\varphi \mapsto \|\varphi\|_*$ for which it is a Banach space. We will refer losely to this quotient as '$BMO(\mathbb{R}; B)$ modulo constant functions' and we will use this term also for the space $\mathcal{BMO}(\mathbb{R}; B^*)$ defined later.

In the next result, $BMO(B)$ means either $BMO(\mathbb{T}; B)$ or $BMO(\mathbb{R}; B)$. In the sequel, a similar notation will be used for function spaces in several statements enunciated and proved for both cases $\mathbb{R}$ and $\mathbb{T}$.

The next one is the famous John-Nirenberg theorem. The classical proof deduces it from the Calderón-Zygmund decomposition and an iteration trick. Since it appears in many places and is identical in the scalar and Banach space cases, we will skip this argument for now. However, we will indicate later on in this section how a proof can be recovered from further results, as a dualization of the maximal inequalities.

Theorem 7.1 (John-Nirenberg). *There is a constant c_1 such that any φ in the unit ball of $BMO(B)$ satisfies for all I:*

$$\forall\lambda > 0 \quad m_I(\{x \in I \mid \|\varphi(x) - \varphi_I\| > \lambda\}) \leq e\exp[-c_1\lambda]. \tag{7.2}$$

Moreover, there is a constant c_2 such that for any such φ and any $1 \leq p < +\infty$

$$\sup_I \|\varphi - \varphi_I\|_{L^p(m_I;B)} \leq c_2\, p. \tag{7.3}$$

Let B be a complex Banach space. We define $BMO_a(\mathbb{T}; B) \subset BMO(\mathbb{T}; B)$ as formed of all functions $\varphi \in BMO(\mathbb{T}; B)$ with Fourier transform vanishing on the negative integers. We can now state (for the moment without proof) a first Banach space valued version of Fefferman's theorem:

Theorem 7.2. *Let B be a UMD Banach space (over $\mathbb{C}$). Then for any $\varphi \in BMO_a(\mathbb{T}; B^*)$ and any $f \in H^1(D; B)$ the limit*

$$\langle\langle f, \varphi\rangle\rangle = \lim_{r\uparrow 1} \int \langle f_r(t), \varphi(-t)\rangle dm(t),$$

exists. With this duality

$$BMO_a(\mathbb{T}; B^*) = H^1(D; B)^*$$

with equivalent norms.

In particular, when $B = \mathbb{C}$, we have

Corollary 7.3 (Fefferman's Duality Theorem).

$$BMO_a(\mathbb{T}; \mathbb{C}) = H^1(D)^*$$

with equivalent norms.

Remark 7.4. A similar duality holds for $H^1(U)$. This requires to first produce a dense subspace $\mathcal{V} \subset H^1(U)$ such that for any $\varphi \in BMO(\mathbb{R}; \mathbb{C})$ we have $\varphi f \in L_1(\mathbb{R})$ for any $f \in \mathcal{V}$. Let then $\eta_\varphi(f) = \int f(t)\varphi(-t)dt$. In analogy with Corollary 7.3, one can show that η_φ extends by density so that $\eta_\varphi \in H^1(U)^*$ and that any element of $H^1(U)^*$ is of the form η_φ for some $\varphi \in BMO(\mathbb{R}; \mathbb{C})$. But we prefer to skip the technical details (see e.g. [33, p. 245]), since we will later prove a real variable version of essentially the same result (see Corollary 7.8 and Remark 7.18 for clarification).

For $\mathbb{T}$ it is clear that functions in BMO admit a well defined Poisson integral since they are integrable. For $\mathbb{R}$, the analogous fact is a consequence of the next, rather elementary, lemma.

Lemma 7.5. *For any $\varphi \in BMO(\mathbb{R}; B)$ we have $\int_{\mathbb{R}} \|\varphi(t)\| \frac{dt}{1+t^2} < \infty$.*

Proof. Assume $\|\varphi\|_* \leq 1$. Let $I(n) = (-2^n, 2^n)$. Since φ is locally L_1, it suffices to show

$$\sum_{n \geq 0} \int_{I(n) \setminus I(n-1)} \|\varphi(t)\| 2^{-2n} dt < \infty. \tag{7.4}$$

We claim that

$$\|\varphi_{I(n)} - \varphi_{I(0)}\| \leq 2n.$$

From this (7.4) is immediate because

$$\int_{I(n) \setminus I(n-1)} \|\varphi(t)\| dt \leq \int_{I(n) \setminus I(n-1)} \|\varphi - \varphi_{I(n)}\| dt$$
$$+ |I(n) \setminus I(n-1)| \|\varphi_{I(n)} - \varphi_{I(0)}\| + |I(n) \setminus I(n-1)| \|\varphi_{I(0)}\|,$$

and hence

$$\int_{I(n) \setminus I(n-1)} \|\varphi(t)\| dt \leq 2^{n+1}(1 + n + \|\varphi_{I(0)}\|)$$

from which (7.4) follows.

We now prove the claim. When I, J are intervals with $I \subset J$, we have by Jensen

$$\|\varphi_I - \varphi_J\| \leq \tfrac{1}{|I|} \int_I \|\varphi - \varphi_J\| \leq \tfrac{1}{|I|} \int_J \|\varphi - \varphi_J\| \leq \tfrac{|J|}{|I|}.$$

In particular

$$\|\varphi_{I(k+1)} - \varphi_{I(k)}\| \leq 2$$

which implies the claim by the triangle inequality. □

7.2 Atomic B-valued H^1

Actually, there is a formulation of Fefferman's duality Theorem that is valid for an arbitrary (separable) Banach space B, but this involves a different, atomic, version of the space $H^1(D; B)$, that we denote by $h^1_{\text{at}}(\mathbb{T}; B)$, and a slightly extended version of $BMO(\mathbb{T}; B^*)$. To define this we first need the notion of 'atom' in the harmonic analysis context. Let B be a real Banach space. On the torus $\mathbb{T}$ (resp. on the line $\mathbb{R}$) a B-valued atom is a function a in $L_\infty(B)$ supported on an arc (resp. a bounded interval) I such that

$$\int_I a = 0 \text{ and } \|a\|_\infty \leq 1/|I|.$$

For $\mathbb{T}$, we make a special exception: When $\dim(B) = 1$ we consider the constant function a_0 equal to 1 on $\mathbb{T}$ as a real valued atom, even though its mean is equal to $1 \neq 0$; when B is a general Banach space, we include in the collection of B-valued atoms any constant function on $\mathbb{T}$ equal to a unit vector in B.

The space $h^1_{\text{at}}(\mathbb{T}; B)$ (resp. $h^1_{\text{at}}(\mathbb{R}; B)$) is defined as formed of all functions f in $L_1(\mathbb{T}; B)$ (resp. $L_1(\mathbb{R}; B)$) that can be represented as a series

$$f = \sum \lambda_n a_n \tag{7.5}$$

where a_n are B-valued atoms and λ_n are scalars such that $\sum |\lambda_n| < \infty$. Note that since every B-valued atom is in the unit ball of $L_1(B)$ this series is absolutely convergent in $L_1(\mathbb{T}; B)$ (resp. $L_1(\mathbb{R}; B)$), and in particular it converges in any order in $L_1(\mathbb{T}; B)$ (resp. $L_1(\mathbb{R}; B)$).

Then we set

$$\|f\|_{h^1_{\text{at}}(\mathbb{T};B)} = \inf\left\{\sum |\lambda_n|\right\}$$

and similarly for $\|f\|_{h^1_{\text{at}}(\mathbb{R};B)}$ where the inf is over all possible such representations of f. Thus the convex hull of the set of atoms is dense in the unit ball of $h^1_{\text{at}}(\mathbb{T}; B)$ (resp. $h^1_{\text{at}}(\mathbb{R}; B)$). Note that $\int_{\mathbb{R}} f(t)dt = 0$ for any $f \in h^1_{\text{at}}(\mathbb{R}; B)$ (but it is not so on $\mathbb{T}$).

Let $1 \leq p < \infty$. We remind the reader that, assuming B separable, we identified in §2.3 the dual of $L_p(\mathbb{T}; B)$ (resp. $L_p(\mathbb{R}; B)$) as the space $\underline{\Lambda}_{p'}(\mathbb{T}; B^*)$ (resp. $\underline{\Lambda}_{p'}(\mathbb{R}; B^*)$). We define the space $\mathcal{BMO}(\mathbb{T}; B^*)$ (resp. $\mathcal{BMO}(\mathbb{R}; B^*)$) as formed of all functions φ in $\underline{\Lambda}_1(\mathbb{T}; B^*)$ (resp. locally in $\underline{\Lambda}_1(\mathbb{R}; B^*)$) such that (7.1) holds for some C. We then define $\|\varphi\|_*$ and the norm of φ exactly as before. To abbreviate, we will denote the latter simply by $\|\varphi\|_{\mathcal{BMO}}$.

Thus the space $\mathcal{BMO}(\mathbb{T}; B^*)$ (resp. $\mathcal{BMO}(\mathbb{R}; B^*)$) is a larger space than $BMO(\mathbb{T}; B^*)$ (resp. $BMO(\mathbb{R}; B^*)$) since the measurability requirement is relaxed to 'weak* scalarly measurable' as opposed to Bochner measurable, but its norm is defined in the same way. In particular we have isometric embeddings $BMO(\mathbb{T}; B^*) \subset \mathcal{BMO}(\mathbb{T}; B^*)$ (resp. $BMO(\mathbb{R}; B^*) \subset \mathcal{BMO}(\mathbb{R}; B^*)$). By Corollary 2.30 there is equality if B^* has the RNP and hence (since UMD implies the RNP by Remark 5.40) a fortiori if B is UMD.

For any B-valued atom a and any $\varphi \in \mathcal{BMO}(\mathbb{T}; B^*)$ (resp. $\mathcal{BMO}(\mathbb{R}; B^*)$) we set

$$\langle \varphi, a \rangle = \int_{\mathbb{T}} \varphi(t)(a(t))dm(t) \quad (\text{resp.} \quad \langle \varphi, a \rangle = \int_{\mathbb{R}} \varphi(t)(a(t))dt).$$

Note that in both cases

$$|\langle \varphi, a \rangle| \le \|\varphi\|_{\mathcal{BMO}}. \tag{7.6}$$

Indeed, since a has mean zero, assuming a associated to I as earlier, we have

$$\langle \varphi, a \rangle = \langle \varphi - \varphi_I, a \rangle$$

and hence

$$|\langle \varphi, a \rangle| \le \|a\|_\infty \|\varphi - \varphi_I\|_1 \le |I|^{-1} \|\varphi - \varphi_I\|_1 \le \|\varphi\|_{\mathcal{BMO}},$$

and (7.6) follows.

For any $f \in h^1_{\text{at}}(\mathbb{T}; B)$ (resp. $h^1_{\text{at}}(\mathbb{R}; B)$) of the form $f = \sum \lambda_n a_n$ and any $\varphi \in \mathcal{BMO}(\mathbb{T}; B^*)$ (resp. $\mathcal{BMO}(\mathbb{R}; B^*)$) we set

$$\langle \varphi, f \rangle = \sum \lambda_n \langle \varphi, a_n \rangle.$$

Note that the series is absolutely convergent and we will show (this is less obvious than may seem at first glance, see Remark 7.15) that the limit depends only on $f \in L_1(\mathbb{T}; B)$ (resp. $L_1(\mathbb{R}; B)$) and not on the particular representation (7.5).

We have clearly by (7.6) (for either $\mathbb{T}$ or $\mathbb{R}$)

$$|\langle \varphi, f \rangle| \le \|f\|_{h^1_{\text{at}}} \|\varphi\|_{\mathcal{BMO}}. \tag{7.7}$$

This shows that any $\varphi \in \mathcal{BMO}(\mathbb{T}; B^*)$ (resp. $\mathcal{BMO}(\mathbb{R}; B^*)$) defines a continuous linear form ξ_φ on $h^1_{\text{at}}(\mathbb{T}; B)$ (resp. $h^1_{\text{at}}(\mathbb{R}; B)$). We can now formulate a simple duality Theorem valid for a general Banach space B but we postpone the proof till after that of Theorem 7.7.

Theorem 7.6. *Let B be any separable Banach space. Then the correspondence $\varphi \mapsto \xi_\varphi$ is an isomorphism from $\mathcal{BMO}(\mathbb{T}; B^*)$ (resp. $\mathcal{BMO}(\mathbb{R}; B^*)$ modulo constant functions) to $h^1_{\text{at}}(\mathbb{T}; B)^*$ (resp. $h^1_{\text{at}}(\mathbb{R}; B)^*$).*

The essence of Fefferman's Theorem is now transferred to a statement comparing two different kinds of H^1-spaces, namely $h^1_{\text{at}}(B)$ and $H^1(B)$, but we will need a third notion of B-valued h^1, associated to maximal functions. We will denote it by $h^1_{\max}$:

We denote by $h^1_{\max}(D; B)$ (resp. $h^1_{\max}(U; B)$) the space of all harmonic functions $u \in h^1(D; B)$ (resp. $h^1(U; B)$) such that the non-tangential maximal function denoted by u^*, associated to $\alpha = 1$ (recall that $u^*(t)$ is the sup of $\|u(z)\|$ over all z in $\Gamma_\alpha(t)$), is in L_1 and we equip it in both cases with the norm

$$\|u\|_{h^1_{\max}} = \|u^*\|_1.$$

The case $1 < p < \infty$ is excluded because, with a similar definition of $h^p_{\max}$, (3.48) shows that $h^p(D; B) = h^p_{\max}(D; B)$ (and similarly for U), and we have nothing new to add. The case $0 < p < 1$ is of some interest, but we exclude it for simplicity.

By the non-tangential maximal inequality (4.5), we know that $H^1 \subset h^1_{\max}$. Moreover any real valued function u that is the real part of a function in H^1 must be in $h^1_{\max}$. We can summarize this (both for D or U) as

$$\Re(H^1) \subset h^1_{\max}.$$

As we will see, the key ingredient for Fefferman's Theorem is that conversely (both for D or U) in the real valued case

$$h^1_{\max} \subset \Re(H^1).$$

Thus $h^1_{\max} = \Re(H^1)$. Moreover, the corresponding norms are equivalent (where $\Re(H^1)$ is equipped with the norm

$$\|u\|_{\Re(H^1)} = \|F\|_{H^1} \tag{7.8}$$

for the unique $F \in H^1$ such that $\Re F = u$, and also, in the case of D, such that $\Im F(0) = 0$.

Returning to the B-valued case (for either $\mathbb{T}$ or $\mathbb{R}$), we recall that, by the non-tangential maximal inequality (see Corollary 3.11 and (4.35)), there is an absolute constant C such that for any $1 < p \le \infty$ and any $f \in L_p(B)$ with Poisson integral u we have

$$\|u^*\|_p \le Cp'\|f\|_{L_p(B)}. \tag{7.9}$$

A fortiori, we have $L_p(\mathbb{T}; B) \subset h^1_{\max}(\mathbb{T}; B)$ and we may write

$$\|u^*\|_{L_1(\mathbb{T})} = \|f\|_{h^1_{\max}(\mathbb{T};B)} \le Cp'\|f\|_{L_p(\mathbb{T};B)}. \tag{7.10}$$

In the B-valued case, let us denote by

$$\widetilde{h}^1_{\max}(D; B) \quad \text{(resp. } \widetilde{h}^1_{\max}(U; B))$$

the subspace of $h^1_{\max}(D; B)$ (resp. $h^1_{\max}(U; B)$) formed of all the harmonic functions u that are the Poisson integral of a function $f \in L_1(\mathbb{T}; B)$ (resp. $f \in L_1(\mathbb{R}; B)$).

Theorem 7.7. *Let B be any Banach space. Then the Poisson integral $f \mapsto u$ defines a surjective isomorphism (with equivalent norms)*

$$h^1_{\mathrm{at}}(\mathbb{T}; B) \simeq \widetilde{h}^1_{\max}(D; B). \tag{7.11}$$

Moreover, the same result holds with $\mathbb{R}, U$ in place of $\mathbb{T}, D$.

Taking $B = \mathbb{R}$ and recalling (v) in Theorem 3.1, we find

Corollary 7.8. *In the real valued case, either for D or U we have*

$$\Re(H^1) = h^1_{at} = h^1_{\max} (= \widetilde{h}^1_{\max}),$$

with equivalent norms.

The more delicate part of Theorem 7.7 is the inclusion $\widetilde{h}^1_{\max}(D; B) \subset h^1_{\mathrm{at}}(\mathbb{T}; B)$, or equivalently the following lemma:

Lemma 7.9. *There is a constant c_4 such that for any $u \in \widetilde{h}^1_{\max}(D; B)$, that is the Poisson integral of a function $f \in L_1(\mathbb{T}; B)$, we have $\|f\|_{h^1_{\mathrm{at}}(\mathbb{T};B)} \leq c_4 \|u^*\|_1$. Moreover, the same result holds with $\mathbb{R}, U$ in place of $\mathbb{T}, D$.*

We will need the following classical geometric fact.

Lemma 7.10. *Let u be a B-valued harmonic function on an open domain of the complex plane containing a square. Then the average of u over the perimeter of the square is equal to its average over the union of its two diagonals.*

More generally, let $u \in h^1_{\max}(U; B)$ be the Poisson integral on U of a function $f \in L_1(\mathbb{R}; B)$. Then, viewing u as extended by f to $\bar{U}$, this also holds for any square in $\bar{U}$ with one of its sides on $\mathbb{R} = \partial U$.

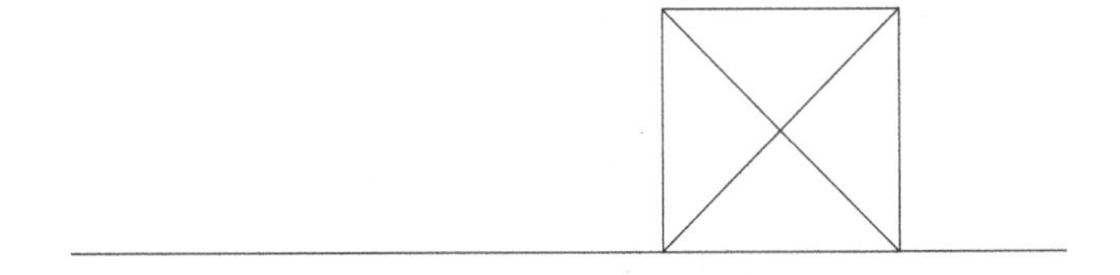

Figure 7.1. Square in $\overline{U}$.

Proof. By 'scalarization' it is easy to reduce the B-valued case to the real valued one. Moreover, we may assume that the square has its sides parallel to the axes, as in the figure. The idea is then to decompose the square into the four triangles determined by the diagonals and to apply Cauchy's theorem to the analytic function $u + i\tilde{u}$ for each of the triangles. The rest is just a computational checking left to the reader. For the second assertion, since the a.e.

non-tangential convergence towards f is dominated, it is easy to deduce this case from the one of a square slightly moved vertically inside U. □

Proof of Lemma 7.9. For simplicity we give the proof only in the case of U, for which Lemma 7.10 yields a spectacularly simple proof, taken from [187]. Assume $\int u^* dt = 1$.

Claim: For each $c > 0$ there is a decomposition $f = g_c + b_c$ with

$$|g_c| \leq 7c \quad a.e. \quad \text{and} \quad b_c = \sum_{I \in \mathcal{C}_c} (f - f_I) 1_I$$

where $\mathcal{C}_c$ is the collection of the disjoint intervals consisting of the connected components of the open set $\{t \in \mathbb{R} \mid u^*(t) > c\}$. Moreover, we have

$$\|g_c\|_1 \leq \int_{\{u^* \leq c\}} \|f\| + 7cm(\{u^* > c\}) \tag{7.12}$$

and

$$\|b_c\|_1 \leq 2 \int_{\{u^* > c\}} \|f\| \tag{7.13}$$

Proof of the Claim:

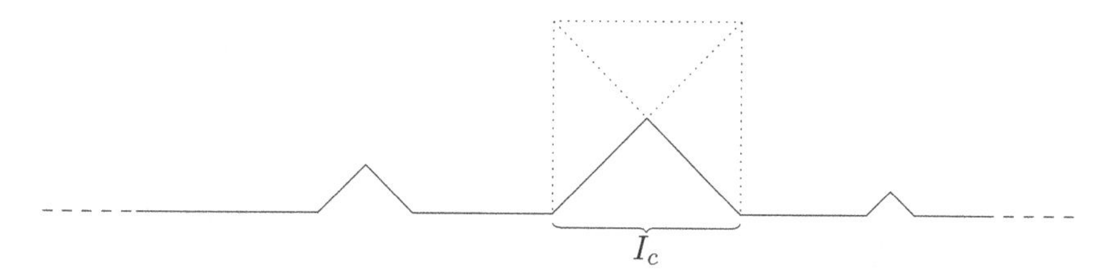

Figure 7.2. $\{u^* > c\}$.

Since $\{u^* > c\} = \cup_{I \in \mathcal{C}_c} I$, the decomposition is necessarily as follows:

$$g_c = 1_{\{u^* \leq c\}} f + \sum_{I \in \mathcal{C}_c} f_I 1_I.$$

All we need to check is the (crucial) fact that

$$\forall I \in \mathcal{C}_c \quad \|f_I\| \leq 7c.$$

Indeed, (7.12) and (7.13) are immediate consequences. This fact follows from Lemma 7.10. Indeed, let S^I be the open square

$$S^I = \{z = x + iy \mid x \in I, 0 < y < |I|\}.$$

We will view u as extended to $\partial U = \mathbb{R} \subset \mathbb{C}$ by its non-tangential limits, i.e. we set $u(t) = f(t)$ for $t \in \mathbb{R}$. Then, by Lemma 7.10, the average of u over the two diagonals of $\overline{S^I}$ is equal to its average over its four sides. But, going back to the very meaning of $I \in \mathcal{C}_c$ shows that $\|u(z)\| \leq c$ for any z on each of the two diagonals and on each of the 3 sides complementing I to form $\overline{S^I}$. Thus, if α denotes the average of f over I and if β, γ, δ denote the averages of u with respect to the three other sides of the square with base I then

$$\left|\frac{1}{4}(\alpha + \beta + \gamma + \delta)\right| \leq c$$

hence

$$|\alpha| \leq 4c + |\beta| + |\gamma| + |\delta| \leq 7c.$$

This proves the claim.

Now we choose $c = 2^k$ ($k \in \mathbb{Z}$). By the claim we have

$$f = g^k + b^k \tag{7.14}$$

$$|g^k| \leq 7 \cdot 2^k \quad \text{and} \quad b^k = \sum_{I \in \mathcal{C}_k} (f - f_I) 1_I$$

where the set $\{u^* > 2^k\}$ is the disjoint union of all the intervals I for $I \in \mathcal{C}_k$. We will define the atoms as follows. For each $k \in \mathbb{Z}$ and each $I \in \mathcal{C}_k$ let

$$a_I = \frac{1}{21 \cdot 2^k |I|} \left(b^k - b^{k+1}\right) 1_I.$$

Note that by (7.13), since u^* is in L^1, b^k must tend to zero in L_1 when $k \to \infty$. On the other hand by (7.12), if $k \to -\infty$ then g_k must tend to zero in L_1, hence $b^k \to f$ in L_1. Therefore, we have (the convergence being in $L_1(B)$)

$$f = \lim_{N \to +\infty} \sum_{-N}^{N} \left(b^k - b^{k+1}\right).$$

Actually, the reader can check easily that this series also converges a.e. Since $b^k - b^{k+1}$ is supported in $\{f^* > 2^k\} = \cup_{I \in \mathcal{C}_k} I$, we have

$$b^k - b^{k+1} = \sum_{I \in \mathcal{C}_k} (b^k - b^{k+1}) 1_I.$$

Hence, letting $\lambda_{k,I} = 21 \cdot 2^k |I|$ and $a_I = (21 \cdot 2^k |I|)^{-1} (b^k - b^{k+1}) 1_I$ we find

$$f = \sum_{k=-\infty}^{k=+\infty} \sum_{I \in \mathcal{C}_k} \lambda_{k,I} a_I$$

and

$$\sum_k \sum_{I \in \mathcal{C}_k} \lambda_{k,I} = \sum_{k=-\infty}^{\infty} 21 \cdot 2^k m(u^* > 2^k)$$
$$\le \sum_{k=-\infty}^{\infty} 21 \cdot 2 \int_{2^{k-1}}^{2^k} m(u^* > t)dt = 42 \int u^* dt.$$

Moreover, the a_I are *atoms*. Indeed, recalling (7.14) we have $g^{k+1} - g^k = b^k - b^{k+1}$ hence

$$\left\| b^k - b^{k+1} \right\|_{L^\infty(I;B)} = \left\| g^{k+1} - g^k \right\|_{L^\infty(I;B)} \le 7.2^{k+1} + 7.2^k = 21 \cdot 2^k.$$

Therefore, $\|a_I\|_\infty \le \frac{1}{|I|}$, moreover

$$a_I = \frac{1}{21 \cdot 2^k |I|} \left((f - f_I)1_I - \sum\nolimits_{J \subset I, J \in \mathcal{C}_{k+1}} (f - f_J)1_J \right),$$

which shows that $\int a_I dt = 0$. Thus, we have proved that

$$\forall f \in h^1_{\max}(B) \quad \|f\|_{h^1_{\mathrm{at}}(B)} \le 42 \|f\|_{h^1_{\max}(B)},$$

which concludes the proof. □

Remark 7.11. It is important to observe that if, in the preceding proof, we assume f in $L_\infty(B)$ and supported in a compact interval K then, in the representation of f obtained previously as $f = \sum_{k=-\infty}^{k=+\infty} \sum_{I \in \mathcal{C}_k} \lambda_{k,I} a_I$, the sum over k is actually a finite sum. Indeed, on the one hand we find $b_c = 0$ for any $c > \|f\|_{L_\infty(B)}$ since the set $\{u^* > c\}$ is void, and hence $b^k - b^{k+1} = 0$ for $2^k > c$, and on the other hand since u^* never vanishes (except in the trivial case when $f \equiv 0$), there is $\varepsilon > 0$ small enough so that $\{u^* > \varepsilon\} \supset K$, from which it follows that $b^k - b^{k+1} = 0$ for all k such that $2^k < \varepsilon$.

Now consider $\varphi \in \mathcal{BMO}(B^*)$, we claim that

$$\left| \int \langle \varphi, f \rangle \right| \le 42 \|\varphi\|_{\mathcal{BMO}} \|f\|_{h^1_{\max}(B)}. \tag{7.15}$$

Indeed, since the sum over k is finite and $b_k - b_{k+1} = -(g_k - g_{k+1}) \in L_\infty(B)$, we may write (note all the integrals are now absolutely convergent and the intervals in $\mathcal{C}_k$ are disjoint)

$$\int \langle \varphi, f \rangle = \sum_k \sum_{I \in \mathcal{C}_k} \lambda_{k,I} \int_I \langle \varphi, a_I \rangle = \sum_k \sum_{I \in \mathcal{C}_k} \lambda_{k,I} \int_I \langle \varphi - \varphi_I, a_I \rangle$$

and hence

$$\left| \int \langle \varphi, f \rangle \right| \le \sum_k \sum_{I \in \mathcal{C}_k} |\lambda_{k,I}| |I|^{-1} \int_I \|\varphi - \varphi_I\| \le 42 \|\varphi\|_{\mathcal{BMO}} \|f\|_{h^1_{\max}(B)}.$$

Remark 7.12. One can also obtain an 'atomic' decomposition of the whole space $h^1_{\max}(B)$ (and not only of $\widetilde{h}^1_{\max}(B)$), but the notion of atom has to be generalized: we say that a bounded B-valued vector measure μ on $\mathbb{R}$ (or $\mathbb{T}$) is a virtual atom if it is supported by a bounded interval (or an arc) I such that $\mu(I) = 0$ and $|\mu| \le |I|^{-1} 1_I m$ (where m denotes the Lebesgue measure on $\mathbb{R}$ (or $\mathbb{T}$) and where $|I| = m(I)$). If we use'virtual atoms' instead of 'atoms', then any $f \in h^1_{\max}(B)$ admits an atomic decomposition as in the proof of Lemma 7.9.

For the converse to Lemma 7.9, it suffices to prove the following (much easier) estimate:

Lemma 7.13. *There is a constant $c' > 0$ such that for any atom a with Poisson integral u we have (on either $\mathbb{R}$ or $\mathbb{T}$) $\|u^*\|_1 \le c'$. Therefore, for any $f \in h^1_{at}(B)$ we have*

$$\|f\|_{h^1_{\max}(B)} \le c' \|f\|_{h^1_{at}(B)}. \tag{7.16}$$

Proof. For simplicity we will prove this only for $\mathbb{R}$, the case of $\mathbb{T}$ is entirely similar. By translation we can reduce for simplicity to

$$\int a\, dx = 0,\ \operatorname{supp}(a) \subset [-\delta, \delta],\ \|a\|_{L_\infty(B)} \le \frac{1}{2\delta}.$$

Then let u be the Poisson integral of a

$$\begin{aligned} u(x+iy) = P_y * a(x) &= \frac{1}{2\delta}\int_{-\delta}^{\delta} P_y(x-s)a(s)ds \\ &= \frac{1}{2\delta}\int_{-\delta}^{\delta} \left[P_y(x-s) - P_y(x)\right] a(s)ds. \end{aligned}$$

Then recall $u^*(t) = \sup_{|x-t|<y} \|u(x+iy)\|$ (see §4.6). Since obviously

$$u^*(t) \le \frac{1}{2\delta}, \tag{7.17}$$

it suffices to majorize conveniently $\int_{t \ge 2\delta} u^*(t)dt$. For that purpose, we note

$$P_y(s) = \frac{1}{\pi}\frac{y}{s^2+y^2}, \quad P_y'(s) = -\frac{1}{\pi}\frac{2sy}{(s^2+y^2)^2} \quad \text{and} \quad |P_y'(s)| \le \frac{1}{\pi}\frac{1}{s^2+y^2}.$$

Moreover

$$\forall t \in [-\delta, \delta] \quad |P_y(x-t) - P_y(x)| \le \sup_{x-\delta \le s \le x+\delta} |P_y'(s)|\delta.$$

Let us observe that if $z = x + iy \in \Gamma_\alpha(t)$ we have $|x - t| < y$, so that if $|x - s| < \delta$, we have

$$|t| < |x - t| + |x| \le y + s + \delta \le 2^{1/2}(y^2 + s^2)^{1/2} + \delta,$$

$$\text{hence } |t| - \delta \le 2^{1/2}(y^2 + s^2)^{1/2}.$$

Therefore, we can write if $|t| > \delta$,

$$|u^*(t)| \le \sup_{|x-t|<y} \frac{1}{2\delta} \int_{-\delta}^{\delta} |P_y(x - s) - P_y(x)| ds \le \sup_{|x-t|<y} \sup_{x-\delta\le s\le x+\delta} \frac{1}{\pi} \frac{1}{s^2 + y^2} \delta$$
$$\le \frac{2\delta}{\pi(|t| - \delta)^2}.$$

So that

$$u^*(t) \le \frac{2}{\pi} \frac{\delta}{(|t| - \delta)^2} \text{ if } |t| > \delta$$

and hence

$$\int_{|t|>2\delta} u^*(t)\, dt \le 2\frac{2\delta}{\pi} \int_{2\delta}^{\infty} \frac{1}{(t - \delta)^2} dt \le \frac{4\delta}{\pi} \left(-\frac{1}{t - \delta} \right)_{2\delta}^{\infty} = \frac{4}{\pi}. \tag{7.18}$$

Recalling (7.17), for any atom we find

$$\int_{-\infty}^{\infty} u^*(t)\, dt \le \int_{|t|\le 2\delta} + \int_{|t|>2\delta} \le 4\delta \cdot \frac{1}{2\delta} + \frac{4}{\pi} = 2 + 4/\pi.$$

We conclude, with a convexity argument, that for all $f \in h^1_{at}(B)$, we have $\|u^*\|_1 \le (2 + 4/\pi)\|f\|_{at}$. □

Remark 7.14. Let $1 < p \le \infty$. A function $a \in L_p(B)$ will be called a (B-valued) p-atom if it is supported in a bounded interval I such that $a_I = 0$ and $\|a\|_{L_p(B)} \le |I|^{-1/p'}$. If $p = \infty$ we recover the previous notion of atom. It is easy to modify the preceding argument to show using (7.9) instead of (7.17) that there is an absolute constant c'' such that any p-atom a satisfies

$$\|a\|_{h^1_{\max}(B)} \le c'' p'. \tag{7.19}$$

Proof of Theorem 7.7. By Lemma 7.9 we have a bounded inclusion $\widetilde{h}^1_{\max}(B) \subset h^1_{at}(B)$. Then Lemma 7.13 shows that the converse inclusion is also bounded. □

Proof of the John-Nirenberg Theorem 7.1. It suffices to prove this for a finite-dimensional Banach space B, with constants independent of the dimension. (To check this reduction write $\|x\| = \sup_{\xi\in J} |\xi(x)|$ where J is the unit ball of B^*. Then it suffices to prove the statement when J is an arbitrary finite subset.) The proof that follows works both for $\mathbb{R}$ or $\mathbb{T}$. Let $1 < p < \infty$. Assume $\|\varphi\|_{BMO(B)} \leq 1$. Let I be a bounded interval (or an arc) and consider $b \in L_{p'}(I; B^*)$ such that $\|b\|_{L_{p'}(I;B^*)} \leq 1$. Note that $a = 1_I((b - b_I)/2)|I|^{-1/p}$ is a B^* valued p'-atom. We also assume $b \in L_\infty(B^*)$. Then by (7.15) and (7.10) (reversing the roles of B and B^*) we have

$$\left|\int \langle \varphi, a\rangle\right| \leq 42\|\varphi\|_{BMO(B)}\|a\|_{h^1_{\max}(B^*)} \leq 42Cp\|a\|_{L_{p'}(B)} \leq 42Cp,$$

but also

$$\int \langle \varphi, a\rangle = (2|I|)^{-1/p} \int_I \langle \varphi, b - b_I\rangle = 2|I|^{-1/p} \int_I \langle \varphi - \varphi_I, b\rangle$$

so we obtain for any b as earlier

$$(2|I|)^{-1/p} \left|\int_I \langle \varphi - \varphi_I, b\rangle\right| \leq 42Cp$$

and taking the supremum over all b as earlier yields

$$(2|I|)^{-1/p}\|\varphi - \varphi_I\|_{L_p(I;B)} \leq 42Cp,$$

so we obtain (7.3) with $c_2 = 84C$. Then by Lemma 5.80, (7.2) follows. □

Remark 7.15. Recall that we defined the duality between $f = \sum \lambda_n a_n \in h^1_{\text{at}}(B)$ (where the a_n's are all atoms) and $\varphi \in \mathcal{BMO}(B^*)$ by setting

$$\langle \varphi, f\rangle = \sum \lambda_n \langle \varphi, a_n\rangle. \tag{7.20}$$

Let us now verify that this depends only on f and not on the particular 'atomic' representation of f. Equivalently, assuming $f = 0$ and $\sum |\lambda_n| < \infty$ we must check that $\sum \lambda_n \langle \varphi, a_n\rangle = 0$. For any $N \geq 1$, let $\varepsilon_N = \sum_{n>N} |\lambda_n|$. Let $f_N = \sum_{n\leq N} \lambda_n a_n$. Note that $f_N \in L_\infty(B)$ and since $f_N = -\sum_{n>N} \lambda_n a_n$ we have $\|f_N\|_{h^1_{\text{at}}(B)} \leq \varepsilon_N$. By (7.15) and (7.16) we have

$$\left|\int \langle \varphi, f_N\rangle\right| \leq 42\|\varphi\|_{\mathcal{BMO}}\|f_N\|_{h^1_{\max}(B)} \leq 42c'\varepsilon_N\|\varphi\|_{\mathcal{BMO}},$$

and hence

$$\sum \lambda_n \langle \varphi, a_n\rangle = \lim_{N\to\infty} \int \langle \varphi, f_N\rangle = 0.$$

Another approach to the same duality, is to consider the linear subspace $V \subset h^1_{\text{at}}(B)$ formed of all $f \in h^1_{\text{at}}(B) \cap L_\infty(B)$ with compact support. Then one can define, for $\varphi \in \mathcal{BMO}(B^*)$ and $f \in V$, $\langle \varphi, f\rangle$ by the following absolutely convergent integral (for $\mathbb{T}$ replace dt by $dm(t)$)

$$\langle \varphi, f\rangle = \int \langle \varphi(t), f(t)\rangle dt.$$

By (7.15) and (7.16), this defines a continuous linear form on $V \subset h^1_{\text{at}}(B)$ for the norm induced on V by $h^1_{\text{at}}(B)$ or equivalently by $h^1_{\max}(B)$. Since V is obviously dense in $h^1_{\text{at}}(B)$, we can extend this linear form to $h^1_{\text{at}}(B)$. It is easy to check that the resulting extension coincides with our earlier definition 7.20.

Proof of Theorem 7.6. Let $\xi \in h^1_{\text{at}}(B)^*$. We first assume that there is a function φ (on $\mathbb{T}$ or $\mathbb{R}$) with values in B^* that is in $\underline{\Lambda}_1(I; B^*)$ for any bounded interval I (by this we mean an arc in the case of $\mathbb{T}$), and such that $\xi(f) = \int \varphi(f)$ for any $f \in V$ (where V is as in the preceding proof). Fix $\varepsilon > 0$. Since B is separable, we can find $a \in L_\infty(B)$ with support in I with $\|a\|_{L_\infty(B)} \le 1/|I|$ such that

$$\begin{aligned}|I|^{-1}\int_I \|\varphi - \varphi_I\| \le \int_I \langle \varphi - \varphi_I, a\rangle + \varepsilon = \int_I \langle \varphi, a - a_I\rangle + \varepsilon \\ = \xi(1_I(a - a_I)) + \varepsilon\end{aligned}$$

but since $1_I(a - a_I)/2$ is clearly an atom, this implies

$$|I|^{-1}\int_I \|\varphi - \varphi_I\| \le 2\|\xi\|_{h^1_{\text{at}}(B)^*} + \varepsilon$$

and hence we conclude $\|\varphi\|_{\mathcal{BMO}} \le 2\|\xi\|_{h^1_{\text{at}}(B)^*}$.

It remains to justify the existence of the function φ. Let $1 < p < \infty$. Note that, in the case of $\mathbb{T}$, by (7.15) we have $L_p(\mathbb{T}; B) \subset h^1_{\text{at}}(\mathbb{T}; B)$ so that ξ defines a fortiori an element of $L_p(\mathbb{T}; B)^* = \underline{\Lambda}_{p'}(\mathbb{T}; B^*) \subset \underline{\Lambda}_1(\mathbb{T}; B^*)$, so this case is clear.

In the case of $\mathbb{R}$, by Remark 7.14 it is easy to check that, for any bounded interval $I \subset \mathbb{R}$, any function $f \in L_p(\mathbb{R}; B)$ supported in I and such that $f_I = 0$ is in $h^1_{\text{at}}(\mathbb{R}; B)$. Let $V_I \subset L_p(\mathbb{R}; B)$ be the subspace formed by such functions. We have a continuous 'inclusion' $V_I \subset h^1_{\text{at}}(\mathbb{R}; B)$ and the same reasoning produces a function $\varphi \in \underline{\Lambda}_1(I; B^*)$ representing ξ when restricted to V_I. One can then easily patch together the functions obtained for the various I's. Note that since the integral of any function in $h^1_{\text{at}}(\mathbb{R}; B)$ vanishes, the resulting φ is only determined modulo constant functions. We skip the remaining details. □

Lemma 7.16. *Assume that B is HT_p $(1 < p < \infty)$ on $\mathbb{R}$ with constant C. Let a be a B-valued atom. We have then*

$$\|\tilde{a}\|_{L_1(B)} \le 2\pi^{-1} + 2^{1/p'}C. \tag{7.21}$$

A similar result holds for $\mathbb{T}$.

Proof. Assume B HT_p on $\mathbb{R}$ (the case of $\mathbb{T}$ is similar). By translation again, it suffices to prove this for a supported in $I = [-\delta, \delta]$ with $\int a\,dx = 0$ and $\|a\|_\infty \le \frac{1}{2\delta}$. Note that if $|t| > 2\delta$ and $s \in I = [-\delta, \delta]$ we have $|t - s| \ge |t| - \delta > |t|/2$. Moreover since $\int a = 0$,

$$\tilde{a}(t) = \pi^{-1} \int a(s)((t-s)^{-1} - t^{-1})ds = \pi^{-1} \int_I a(s)s(t(t-s))^{-1}ds.$$

Therefore, for any $|t| > 2\delta$

$$\|\tilde{a}(t)\| \le \pi^{-1}\|a\|_{L_1(B)}2\delta/t^2 \le 2\pi^{-1}\delta/t^2, \tag{7.22}$$

and hence

$$\int_{\{|t|>2\delta\}} \|\tilde{a}(t)\| \le 2\pi^{-1}.$$

We only now use the property HT_p. We have

$$\|\tilde{a}(t)\|_{L_p(B)} \le C\|a(t)\|_{L_p(B)} \le C|I|^{-1/p'}$$

and hence by Hölder

$$\int_{\{|t|\le 2\delta\}} \|\tilde{a}(t)\| \le (4\delta)^{1/p'}\|\tilde{a}(t)\|_{L_p(B)} \le (4\delta)^{1/p'}C(2\delta)^{-1/p'} = 2^{1/p'}C,$$

and (7.21) follows. □

In the next statement, given $f \in L_1(B)$ on $\mathbb{T}$ (resp. $\mathbb{R}$) we say, for short, that $\tilde{f} \in L_1(B)$ if there is a function $g \in L_1(B)$ such that $\widehat{g}(x) = -i\,\mathrm{sign}(x)\widehat{f}(x)$ for any $x \in \mathbb{Z}$ (resp. $\mathbb{R}$) as in (6.8).

Lemma 7.17. *If f and $\tilde{f}$ are both in $L_1(B)$ on $\mathbb{T}$ or $\mathbb{R}$, then u (the Poisson integral of f) is in $h^1_{\max}(B)$. Moreover, there is a constant c_5 such that for any B and any such f we have*

$$\|u^*\|_1 \le c_5 \max\{\|f\|_{L_1(\mathbb{T};B)}, \|\tilde{f}\|_{L_1(\mathbb{T};B)}\}. \tag{7.23}$$

Proof. We view B as a *real* Banach space (even if it happens to be complex). Then we consider a complexification $B^{\mathbb{C}} = B + iB$ containing B isometrically as a real linear subspace in such a way that $\|a\| \le \|a + ib\|$ for any $a, b \in B$ (for instance the space of $\mathbb{R}$-linear maps from $\mathbb{R}^2$ (Euclidean) to B equipped with its usual operator norm, or equivalently $\mathbb{R}^2 \otimes B$). Let then $F = u + i\tilde{u}$. Clearly (e.g. by scalarization using (6.6)) the Poisson integral of F is a $B^{\mathbb{C}}$ valued analytic function such that $\|u(z)\| \le \|F(z)\|$ for all z in D or U. Therefore, by the maximal inequality for $H^1(B^{\mathbb{C}})$ on D or U (see (4.5) with (3.13) and §4.6) we have for some constant c

$$\|u^*\|_1 \le \|F^*\|_1 \le c\|F\|_{L_1(\mathbb{T};B)} = c\|f + i\tilde{f}\|_{L_1(\mathbb{T};B)} \le c(\|f\|_{L_1(B)} + \|\tilde{f}\|_{L_1(B)}).$$ □

The proof that follows is meant as a recapitulation.

Proof of Corollary 7.8. The equality $h^1_{at} = \widetilde{h}^1_{\max}$ was included in Theorem 7.7. The inclusion $\Re(H^1) \subset \widetilde{h}^1_{\max}$ essentially follows from the non-tangential maximal inequality, as we saw in the proof of Lemma 7.17. To conclude it suffices to show $h^1_{at} \subset \Re(H^1)$. To see this remaining inclusion, observe that a real valued function f is in $\Re(H^1)$ (i.e. is the real part of a function in H^1) iff both f and $\tilde{f}$ are in L_1. Thus (7.21) implies $h^1_{at} \subset \Re(H^1)$. These arguments also yield the equivalence of the 3 norms. □

Remark 7.18. We will now justify the assertion made in Remark 7.4. First we note that for any real valued atom a, the product $\varphi(a + i\tilde{a})$ is locally in L_1, because both factors are locally in L_2 (for φ this follows from Theorem 7.1, and for the other factor because $\|\tilde{a}\|_2 = \|a\|_2 < \infty$). Thus, by (7.22) and by Lemma 7.5 we can take for $\mathcal{V} \subset H^1(U)$ the subspace of functions with real parts in the real linear span of real valued atoms in $\Re(H^1(U))$. In other words, $\mathcal{V}$ is spanned by the functions $f \in H^1(U)$ with boundary values of the form $a + i\tilde{a}$ for some real valued atom a. Let $\xi \in H^1(U)^*$. By (7.8) we know that $\Re(H^1(U))$ and $H^1(U)$ are $\mathbb{R}$-linearly isometrically isomorphic. Therefore $F \mapsto \Re(\xi(F))$ and $F \mapsto \Im(\xi(F))$ can be viewed as bounded $\mathbb{R}$-linear forms on $\Re(H^1(U)) = h^1_{at}$. Let us write $F = f + i\tilde{f}$, so that $\Re(F) = f$ and $\Re(iF) = -\tilde{f}$ on $\mathbb{R} = \partial U$. By Theorem 7.6 (applied to $B = \mathbb{R}$), there are real valued functions $\varphi_1, \varphi_2 \in BMO(\mathbb{R})$ such that $\xi(F) = \xi_{\varphi_1}(f) + i\xi_{\varphi_2}(f)$. Since

$$\xi(iF) = \xi_{\varphi_1}(\Re(iF)) + i\xi_{\varphi_2}(\Re(iF)) = \xi_{\varphi_1}(-\tilde{f}) + i\xi_{\varphi_2}(-\tilde{f}) = -\int (\varphi_1 + i\varphi_2)\tilde{f}$$

we find

$$\xi(F) = (1/2)(\xi(F) - i\xi(iF)) = (1/2)\int (\varphi_1 + i\varphi_2)(F),$$

and hence

$$\xi = \eta_\varphi$$

with $\varphi \in BMO(\mathbb{R}; \mathbb{C})$ defined by

$$\forall t \in \mathbb{R} \quad \varphi(t) = (\varphi_1(-t) + i\varphi_2(-t))/2.$$

The same chain of arguments shows that conversely any such η_φ is in $H^1(U)^*$.

In the Banach space valued setting, the ideas revolving around the Fefferman-Stein Theory [222], together with the atomic viewpoint, lead to the following general statement.

Theorem 7.19. *Let B be a any Banach space. The following are equivalent:*

(i) The space B is HT on $\mathbb{R}$ (or equivalently UMD).
(ii) There is a constant C such that for any (B-valued) atom a we have

$$\|(H^{\mathbb{R}}a)\|_{L_1(B)} \leq C.$$

(iii) The Hilbert transform $f \mapsto \tilde{f}$ is bounded on $h^1_{\text{at}}(\mathbb{R}; B)$.
(iv) The Hilbert transform defines a bounded linear map on $BMO(\mathbb{R}; B)$.
(v) The Hilbert transform defines a bounded linear map from $L_\infty(\mathbb{T}; B)$ to $L_1(\mathbb{T}; B)$.

Moreover, a similar result is valid with $\mathbb{T}$ in place of $\mathbb{R}$ in (i)–(iv).

Sketch of proof. (i) $\Rightarrow$ (ii) $\Rightarrow$ (iii) follows easily from (7.21) and a convexity argument. In the case of $\mathbb{T}$, (iii) $\Rightarrow$ (v) and (iv) $\Rightarrow$ (v) are obvious because we have $L_\infty(\mathbb{T}; B) \subset X \subset L_1(\mathbb{T}; B)$ when X is either $h^1_{\text{at}}(\mathbb{T}; B)$ or $BMO(\mathbb{T}; B)$. In the case of $\mathbb{R}$, let us show that either (iii) or (iv) imply (v). Let $f \in L_\infty(\mathbb{R}; B)$ be a function supported in $[-\pi, \pi]$ with mean 0. Assuming either (iii) or (iv) we know $\tilde{f}$ is locally in $L_1(\mathbb{R}; B)$ so $\tilde{f}_{|[-\pi,\pi]} \in L_1([-\pi, \pi]; B)$. Then arguing as in Remark 6.9, we conclude that (v) holds.

The proof of (v) $\Rightarrow$ (i) is based on extrapolation and type/cotype arguments. We skip the details. Lastly, to show (i) $\Rightarrow$ (iv) we may assume B reflexive (since UMD implies reflexive, see Remark 5.40) and then (iii) and the duality in Theorem 7.6 imply that (iv) holds for B^*. Since HT or UMD are self-dual this completes the proof. □

Corollary 7.20. *Assume B HT (or UMD). Let f be a B-valued Bochner measurable function on $\mathbb{T}$ or $\mathbb{R}$. Then $f \in h^1_{\text{at}}(B)$ iff both f and $\tilde{f}$ are in $L_1(B)$ (in the sense of Lemma 7.17).*

Proof. If f and $\tilde{f}$ are in $L_1(B)$, (and if B is arbitrary) then Lemma 7.17 implies that its Poisson integral u is in $\widetilde{h}^1_{\max}(B)$ ($= h^1_{\text{at}}(B)$ by (7.11)). For the converse we use the HT or UMD assumption: Indeed, obviously $h^1_{\text{at}}(B) \subset L_1(B)$ and (i) $\Rightarrow$ (ii) in Theorem 7.19 shows that if $f \in h^1_{\text{at}}(B)$ then $\tilde{f} \in L_1(B)$. □

Corollary 7.21. *Assume B HT (or UMD). Let $\varphi \in L_1(\mathbb{T}; B)$. Then $\varphi \in BMO(\mathbb{T}; B)$ iff it can be written as a sum*

$$\varphi = g + \tilde{h}$$

with $g, h \in L_\infty(\mathbb{T}; B)$. Moreover, the BMO norm of φ is equivalent to the following one

$$N(\varphi) = \inf\{\|g\|_{L_\infty(\mathbb{T};B)} + \|h\|_{L_\infty(\mathbb{T};B)}\}$$

where the inf runs over all possible decompositions. Moreover, a similar result holds on $\mathbb{R}$ for any φ locally in $L_1(\mathbb{R}; B)$.

Proof. We may assume B reflexive (see Remark 5.40). Let $\varphi \in BMO(\mathbb{T}; B)$. For simplicity, we prove this only on $\mathbb{T}$. By duality, to prove the announced decomposition, it suffices to show that for some c we have $|\langle \varphi, f \rangle| \leq c$ for any $f \in L_1(\mathbb{T}; B^*)$ with finitely supported Fourier transform such that

$$\max\{\|f\|_{L_1(\mathbb{T};B^*)}, \|\tilde{f}\|_{L_1(\mathbb{T};B^*)}\} \leq 1.$$

But then, by Lemma 7.17, the Poisson integral of f denoted by u satisfies $\|u^*\|_1 \leq c_5$ and, by Lemma 7.9, $\|f\|_{h^1_{\mathrm{at}}(\mathbb{T};B)} \leq c_4 c_5$. Recall that in the reflexive case the distinction between $\mathcal{BMO}$ and BMO is irrelevant. Then by (7.7) (applied with B^* in place of B) we conclude that $|\langle \varphi, f \rangle| \leq c_4 c_5 \|\varphi\|_{BMO}$, and hence $N(\varphi) \leq c_4 c_5 \|\varphi\|_{BMO}$. This part used only the reflexivity of B.

To prove the converse, we will use the HT property. By (iv) in Theorem 7.19, since $L_\infty \subset BMO$, the Hilbert transform is a fortiori bounded from $L_\infty(\mathbb{T}; B)$ to $BMO(\mathbb{T}; B)$, say with norm $= c$. Then we have

$$\|f\|_{BMO(\mathbb{T};B)} = \|g + \tilde{h}\|_{BMO(\mathbb{T};B)} \leq \|g\|_{L_\infty(\mathbb{T};B)} + c\|h\|_{L_\infty(\mathbb{T};B)} \leq (1 + c)N(\varphi).$$

This completes the proof. □

By the same argument as in Remark 6.2, we deduce the following consequence of (iv) in Theorem 7.7.

Corollary 7.22. *If B is a complex UMD space, the natural projection from $BMO(\mathbb{T}; B)$ onto the subspace $BMO_a(\mathbb{T}; B)$ is bounded.*

Proof of Theorem 7.2. To connect Theorem 7.19 to Theorem 7.2, note that for any B we have by Theorem 4.14 a natural isomorphic embedding

$$H^1(D; B) \subset h^1_{\max}(D; B),$$

and if B is ARNP we have by Corollary 4.31 and Theorem 7.7 a natural isomorphic embedding

$$H^1(D; B) \subset \widetilde{h}^1_{\max}(D; B) \simeq h^1_{\mathrm{at}}(\mathbb{T}; B),$$

taking $f = \sum a_n z^n$ to the boundary value (say radial limit) $t \mapsto \sum a_n e^{int}$.

From this follows that $H^1(D; B)^*$ is a quotient of $h^1_{\mathrm{at}}(\mathbb{T}; B)^* \simeq \mathcal{BMO}(D; B^*)$, by the annihilator of $\{e^{int} b \mid n \geq 0,\ b \in B\}$.

Lastly, if B is UMD, its dual B^* is also UMD, and hence by Remark 5.40 B and B^* have the RNP, in particular $\mathcal{BMO}(D; B^*) = BMO(D; B^*)$. Then Corollary 7.22 allows us to identify the preceding quotient of $\mathcal{BMO}(D; B^*)$

with the subspace $BMO_a(D; B^*)$, and hence Theorem 7.2 now appears as a consequence. □

7.3 H^1, BMO and atoms for martingales

Let $(\mathcal{A}_n)_{n\geq 0}$ be a fixed filtration on a probability space $(\Omega, \mathcal{A}, \mathbb{P})$. We assume (without loss of generality) that $\mathcal{A} = \mathcal{A}_\infty$.

The Hardy space H^1 has several analogues in martingale theory. The main one is probably as follows: we define the space 'martingale-H^1' relative to $(\mathcal{A}_n)_{n\geq 0}$ to be the space of scalar valued martingales $(f_n)_{n\geq 0}$ such that the maximal function $f^* = \sup_{n\geq 0} |f_n|$ is in L^1.

We denote this space by $h^1_{\max}(\{\mathcal{A}_n\})$ and we set

$$\|(f_n)_{n\geq 0}\|_{h^1_{\max}(\{\mathcal{A}_n\})} = \mathbb{E} \sup_{n\geq 0} |f_n|.$$

The B. Davis inequality (5.67) shows that this norm is equivalent to $\mathbb{E}S$ where $S = (\sum |d_n|^2)^{1/2}$ is the square function, or to

$$\sup_{\xi_n=\pm 1} \sup_N \| \sum\nolimits_{n\leq N} \xi_n df_n \|_1.$$

Equivalently, $(f_n)_{n\geq 0}$ is in $h^1_{\max}(\{\mathcal{A}_n\})$ iff all its martingale transforms by choices of signs are uniformly bounded in L_1 (incidentally, if they are all in L_1, then a routine argument shows that they are automatically uniformly bounded in L_1). Since martingale transforms are analogous to the Hilbert transform, this explains the analogy with the identity $h^1_{\max} = \Re H^1$ from Corollary 7.8. Another explanation lies of course in the Brownian analogue of non-tangential convergence as in §4.5.

The space 'martingale-BMO', denoted by $BMO(\{\mathcal{A}_n\})$, is then defined as the space of all martingales $f_n = \mathbb{E}^{\mathcal{A}_n} f$ with $f \in L_1$ such that

$$\sup_{n\geq 1} \|\mathbb{E}_n |f - f_{n-1}|\|_\infty < \infty.$$

With the convention $f_{-1} = 0$, we equip this space with the norm

$$\|f\|_{BMO(\{\mathcal{A}_n\})} = \sup_{n\geq 0} \|\mathbb{E}_n |f - f_{n-1}|\|_\infty < \infty.$$

We will often abbreviate this to $\|f\|_{BMO}$.

We will identify BMO with the space of all $\mathcal{A}_\infty$-measurable f's for which this is finite.

Remark. Clearly $f \in L_\infty$ suffices for $f \in BMO$. More precisely, it suffices to have the square function $S = (\sum |d_n|^2)^{1/2}$ in L_∞, and

$$\|f\|_{BMO(\{\mathcal{A}_n\})} \leq \|S\|_\infty. \tag{7.24}$$

This is very easy to check using the identity $\mathbb{E}_n|f - f_{n-1}|^2 = \mathbb{E}_n \sum_{k\geq n} |d_k|^2$.

The martingale version of the Fefferman duality theorem then says that the space BMO can be identified with the dual of $h^1_{\max}(\{\mathcal{A}_n\})$, the duality being:

$$\langle g, f\rangle = \lim_{n\to\infty} \mathbb{E}(g_n f_n).$$

All this can be extended rather easily to the Banach space valued case, in analogy with what we just saw for the classical Hardy spaces.

Let B be a Banach space. We will denote by $h^1_{\max}(\{\mathcal{A}_n\}; B)$ the space of all B-valued martingales $(f_n)_{n\geq 0}$ such that

$$\mathbb{E} \sup_{n\geq 0} \|f_n\|_B < \infty,$$

equipped with the norm $\|f\|_{h^1_{\max}(\{\mathcal{A}_n\};B)} = \mathbb{E} \sup_{n\geq 0} \|f_n\|_B$. We will again denote by f^* the maximal function, i.e. we set

$$f^*(\cdot) = \sup_{n\geq 0} \|f_n(\cdot)\|_B.$$

We will denote by $\widetilde{h}^1_{\max}(\{\mathcal{A}_n\}; B)$ the subspace formed of all the martingales $(f_n)_{n\geq 0}$ that, in addition, converge in $L_1(B)$ (or equivalently converge a.s. since the convergence is dominated). Note that the latter subspace coincides with the closure in $h^1_{\max}(\{\mathcal{A}_n\}; B)$ of the subspace formed by the finite martingales, which is perhaps a more natural definition.

Note that by Theorem 2.9 when B has the RNP (in particular when $B = \mathbb{C}$) we have $h^1_{\max}(\{\mathcal{A}_n\}; B) = \widetilde{h}^1_{\max}(\{\mathcal{A}_n\}; B)$, so we only need to consider $h^1_{\max}(\{\mathcal{A}_n\}; B)$. When $\dim(B) = 1$, as before we denote this space simply by $h^1_{\max}(\{\mathcal{A}_n\})$.

Again, the duality between H^1 and BMO can be reformulated nicely using 'atoms', but with a specific new definition as follows. For simplicity we use again the term (B-valued) 'atom' in the martingale context but the risk of confusion is minimal. A function $a\colon\ \Omega \to B$ in $L^1(\Omega, \mathbb{P}; B)$ is called a B-valued atom (relative to our fixed filtration $(\mathcal{A}_n)_{n\geq 0}$) if there is an integer $n \geq 0$ and a set $A \in \mathcal{A}_n$ such that

$$\{a \neq 0\} \subset A,\ \ \mathbb{E}_n(a) = 0 \text{ and } \|a\|_{L_\infty(B)} \leq 1/\mathbb{P}(A). \tag{7.25}$$

By the pointwise inequality $\|a(\cdot)\|_B \le 1_A(\cdot)/\mathbb{P}(A)$, we must have

$$\sup_m \|\mathbb{E}_m(a)\|_B = \sup_{m>n} \|\mathbb{E}_m(a)\|_B \le \sup_{m>n} \mathbb{E}_m(1_A(\cdot)/\mathbb{P}(A)) = 1_A(\cdot)/\mathbb{P}(A)$$

and hence for any B-valued atom a we have

$$\|a\|_{h^1_{\max}(\{\mathcal{A}_n\};B)} \le 1. \tag{7.26}$$

The space $h^1_{\mathrm{at}}(\{\mathcal{A}_n\}; B)$ is then defined as the space of all functions f in $L_1(B)$ that can be written as an absolutely convergent series of the form

$$f = \mathbb{E}_0(f) + \sum\nolimits_{n=1}^{\infty} \lambda_n a_n \tag{7.27}$$

where a_n are B-valued atoms and $\sum |\lambda_n| < \infty$. We define

$$\|f\|_{h^1_{\mathrm{at}}(\{\mathcal{A}_n\};B)} = \|\mathbb{E}_0(f)\|_{L_1(B)} + \inf\left\{\sum |\lambda_n|\right\}$$

where the infimum runs over all possible such representations of f. By (7.26), we have $h^1_{\mathrm{at}}(\{\mathcal{A}_n\}; B) \subset \widetilde{h}^1_{\max}(\{\mathcal{A}_n\}; B)$ and for any $f \in h^1_{\mathrm{at}}(\{\mathcal{A}_n\}; B)$

$$\|f\|_{h^1_{\max}(\{\mathcal{A}_n\};B)} \le \|f\|_{h^1_{\mathrm{at}}(\{\mathcal{A}_n\};B)}. \tag{7.28}$$

We will denote by $BMO(\{\mathcal{A}_n\}; B)$ the space of all martingales $\varphi = (\varphi_n)_{n\ge 0}$ in $L_1(B)$ such that $\sup_{n\ge 0} \sup_{m\ge n} \|\mathbb{E}_n \|\varphi_m - \varphi_{n-1}\|_B\|_\infty < \infty$, with the convention $\varphi_{-1} = 0$ (that we maintain throughout this section) and we equip it with the norm

$$\|(\varphi_n)\|_{BMO(\{\mathcal{A}_n\};B)} = \sup_{n\ge 0} \sup_{m\ge n} \|\mathbb{E}_n \|\varphi_m - \varphi_{n-1}\|_B\|_\infty.$$

We will sometimes abbreviate this by $\|(\varphi_n)\|_{BMO}$.

When φ_n converges to a limit φ in $L_1(B)$, we have

$$\|(\varphi_n)\|_{BMO} = \sup_n \|\mathbb{E}_n \|\varphi - \varphi_{n-1}\|_B\|_\infty.$$

Remark 7.23. Note that obviously $\|\,\|\varphi_n - \varphi_{n-1}\|_B\|_\infty \le \|(\varphi_n)\|_{BMO}$ and hence if we define

$$\|(\varphi_n)\|_{\sim} = \sup_{n\ge 0} \sup_{m\ge n} \|\mathbb{E}_n \|\varphi_m - \varphi_n\|_B\|_\infty, \tag{7.29}$$

and

$$\|(\varphi_n)\|_{\approx} = \max\{\|(\varphi_n)\|_{\sim},\ \sup_{n\ge 0} \|\varphi_n - \varphi_{n-1}\|_{L_\infty(B)}\},$$

we have $\|(\varphi_n)\|_{\sim} \le 2\|(\varphi_n)\|_{BMO}$ and a fortiori

$$(1/2)\|(\varphi_n)\|_{BMO} \le \|(\varphi_n)\|_{\approx} \le 2\|(\varphi_n)\|_{BMO}. \tag{7.30}$$

The norm (7.29) is a 'competing' definition for the BMO-norm of φ (assuming $\varphi_0 = 0$), but although it is equivalent to it for filtrations that are regular in the sense of §7.4 (see (7.51)), in general it is not equivalent to it, which is the source of certain complications.

Remark 7.24. Assume that the σ-algebra $\mathcal{A}_n$ is atomic, which means it is generated by a partition of the probability space Ω into 'atoms' (here this term refers to the sets that are minimal elements of $\mathcal{A}_n$ up to negligible sets). Then, assuming $\varphi_n \to \varphi$ in $L_1(B)$, the condition

$$\|\mathbb{E}_n\|\varphi - \varphi_n\|_B\|_\infty \leq 1$$

is equivalent to the assertion that for any atom I of $\mathcal{A}_n$ we have

$$\frac{1}{\mathbb{P}(I)}\int_I \|\varphi - \varphi_I\|_B d\mathbb{P} \leq 1,$$

where $\varphi_I = \frac{1}{\mathbb{P}(I)}\int_I \varphi d\mathbb{P}$. To check this, just observe that $\mathbb{E}_n\varphi = \varphi_I$ on I. This uncovers the link with (7.1).

Note that we try to avoid the clash in terminology by systematically calling B-valued atoms the functions that are customarily called simply 'atoms'.

Now let a be a B-valued atom related to $A \in \mathcal{A}_n$ as in (7.25). Let $A = \cup_k I_k$ be the decomposition of A into atoms of $\mathcal{A}_n$. We have then

$$a = \sum \lambda_k a_k$$

where each a_k is an atom related to I_k and where $\sum |\lambda_k| \leq 2$. Indeed, we just set

$$a_k = 1_{I_k}(a - a_{I_k})/\lambda_k \quad \text{with} \quad \lambda_k = 2\mathbb{P}(I_k)/\mathbb{P}(A).$$

Thus we could restrict our B-valued atoms to be associated to atoms of $\mathcal{A}_n$!

With a view to duality, let us now define a more general B^* valued analogue of BMO. For simplicity, we assume B separable. Recall that the space $\underline{\Lambda}_1(\Omega, \mathcal{A}, \mathbb{P}; B^*)$ (often abbreviated in what follows to $\underline{\Lambda}_1(B^*)$) is defined as formed of all the weak* scalarly measurable functions $\varphi : \Omega \to B^*$ such that the function $\omega \mapsto \|\varphi(\omega)\|$, which is measurable since B is separable, is in L_1. See §2.4. For any $\varphi \in \underline{\Lambda}_1(\Omega, \mathcal{A}, \mathbb{P}; B^*)$ we can define $\mathbb{E}\varphi$ as the unique linear form $\varphi_0 \in B^*$ such that $\varphi_0(b) = \mathbb{E}\varphi(b)$ for any $b \in B$. More generally, for any σ-subalgebra $\mathcal{A}_n \subset \mathcal{A}$ we define $\mathbb{E}_n\varphi$ as the unique $\varphi_n \in \underline{\Lambda}_1(\Omega, \mathcal{A}_n, \mathbb{P}; B^*)$ such that

$$\forall b \in B \quad \varphi_n(b) = \mathbb{E}_n(\varphi(b)).$$

The existence and uniqueness of this function (up to a negligible set) is easy to check (combine Theorem 2.29 with Remark 2.25).

Note that for any $n \geq 0$ we have

$$\sup_{m \geq n} \|\mathbb{E}_n \|\varphi_m\|_{B^*}\|_\infty = \|\mathbb{E}_n \|\varphi\|_{B^*}\|_\infty. \tag{7.31}$$

Indeed, this is clear when we replace the norm in B^* by $y \mapsto \sup_{\xi \in I} |\xi(y)|$ for any finite subset I of the unit ball of B^*. Note that $\|\mathbb{E}_n \|\varphi\|_{B^*}\|_\infty \leq 1$ iff $\mathbb{E}(1_A \|\varphi\|_{B^*}) \leq \mathbb{P}(A)$ for any $A \in \mathcal{A}_n$. Then (7.31) follows by taking the sup of both sides over all such I's.

We will denote by $\mathcal{BMO}(\{\mathcal{A}_n\}; B^*)$ the space of all functions $\varphi \in \underline{\Lambda}_1(B^*)$ with associated martingale $(\varphi_n)_{n \geq 0}$ such that $\sup_{n \geq 0} \|\mathbb{E}_n \|\varphi - \varphi_{n-1}\|_{B^*}\|_\infty < \infty$ and we equip it with the norm

$$\|\varphi\|_{\mathcal{BMO}(\{\mathcal{A}_n\}; B^*)} = \sup_{n \geq 0} \|\mathbb{E}_n \|\varphi - \varphi_{n-1}\|_{B^*}\|_\infty.$$

By (7.31) this is the same as $\sup_{n \geq 0} \sup_{m \geq n} \|\mathbb{E}_n \|\varphi_m - \varphi_{n-1}\|_{B^*}\|_\infty$. We will sometimes abbreviate this by $\|\varphi\|_{\mathcal{BMO}}$.

Note that for any $\varphi \in BMO(\{\mathcal{A}_n\}; B^*)$ (resp. $\mathcal{BMO}(\{\mathcal{A}_n\}; B^*)$), arguing as in Remark 7.23 we have $\varphi_n - \varphi_{n-1} \in L_\infty(B^*)$ (resp. $\varphi_n - \varphi_{n-1} \in \underline{\Lambda}_\infty(B^*)$) and in any case

$$\|\varphi_n - \varphi_{n-1}\|_{\underline{\Lambda}_\infty(B^*)} \leq \|\varphi\|_{\mathcal{BMO}(\{\mathcal{A}_n\}; B^*)}. \tag{7.32}$$

Note that, by our convention that $\varphi_{-1} = 0$, this implies that for any $m \geq 0$ and any $\varphi \in BMO(\{\mathcal{A}_n\}; B^*)$ (resp. $\mathcal{BMO}(\{\mathcal{A}_n\}; B^*)$), we have

$$\varphi_m \in L_\infty(B^*) \quad (\text{resp. } \varphi_m \in \underline{\Lambda}_\infty(B^*)). \tag{7.33}$$

Let $\varphi \in \mathcal{BMO}(\{\mathcal{A}_n\}; B^*)$. Note that for any B-valued atom a associated to a set $A \in \mathcal{A}_n$ as earlier we have a fortiori $\mathbb{E}_{n-1} a = \mathbb{E}_{n-1}\mathbb{E}_n a = 0$ and hence for any $m \geq n$

$$\mathbb{E}\langle \varphi_m, a\rangle = \mathbb{E}\langle \varphi_m, a - \mathbb{E}_{n-1}\, a\rangle = \mathbb{E}\langle \varphi_m - \varphi_{n-1}, a\rangle$$

and hence since $\|a\|_B \leq \mathbb{P}(A)^{-1} 1_A$ pointwise and $A \in \mathcal{A}_n$

$$|\mathbb{E}\langle \varphi_m, a\rangle| \leq \mathbb{P}(A)^{-1}\mathbb{E}(1_A \|\varphi_m - \varphi_{n-1}\|_{B^*}) \leq \|\varphi\|_{\mathcal{BMO}}. \tag{7.34}$$

Consequently, any $g \in h^1_{\text{at}}(\{\mathcal{A}_n\}; B)$ satisfies

$$|\mathbb{E}\langle \varphi_m, g\rangle| \leq \|\varphi\|_{\mathcal{BMO}} \|g\|_{h^1_{\text{at}}(\{\mathcal{A}_n\}; B)}. \tag{7.35}$$

Note that $BMO(\{\mathcal{A}_n\}; B^*) \subset \mathcal{BMO}(\{\mathcal{A}_n\}; B^*)$ can be identified (isometrically) to the subspace formed of all the martingales $(\varphi_n)_{n \geq 0}$ for which each φ_n is Bochner measurable. As in the preceding section, the two spaces coincide if B^* has the RNP by Corollary 2.30.

The most basic version of the atomic decomposition for martingales is given in the following statement analogous to Lemma 7.9.

Lemma 7.25. *Let B be any Banach space and let (g_n) be a martingale with respect to $(\mathcal{A}_n)$ converging in $L_1(B)$ to g with $g_0 = 0$. Assume that there is a sequence (λ_n) adapted to $(\mathcal{A}_n)$ such that $\|g_n\| \le \lambda_{n-1}$ a.s. for any $n \ge 1$ and satisfying $\mathbb{E}\sup_n \lambda_n < \infty$. Then $g \in h^1_{at}(\{\mathcal{A}_n\}; B)$ and*

$$\|g\|_{h^1_{at}(\{\mathcal{A}_n\};B)} \le 9\mathbb{E}\sup_n \lambda_n. \tag{7.36}$$

Proof. Let us assume that

$$\mathbb{E}\sup_n \lambda_n \le 1. \tag{7.37}$$

By homogeneity it suffices to prove

$$\|g\|_{h^1_{at}(\{\mathcal{A}_n\};B)} \le 9. \tag{7.38}$$

As usual we let $d_n = g_n - g_{n-1}$. Then for any $m \ge 0$ we introduce the stopping time

$$T_m = \inf\{n \ge 0 \mid \lambda_n > 2^m\}.$$

We observe that

$$\|g_{T_m}\| \le 2^m. \tag{7.39}$$

Indeed, if $T_m = n > 0$ we have $\|g_n\| \le \lambda_{n-1} \le 2^m$, and otherwise $g_{T_m} = g_0 = 0$.

We can now conclude: clearly $T_m \uparrow \infty$, thus we know that $g_{T_m} \to g_\infty$ a.e. and in $L_1(B)$ (by dominated convergence), so that we can rewrite g as the sum of the following series convergent in $L_1(B)$

$$g = g_{T_0} + \sum\nolimits_{m\ge 1} (g_{T_m} - g_{T_{m-1}}). \tag{7.40}$$

Let $a_0 = g_{T_0}$. Then $\|a_0\|_\infty \le 1$, $\mathbb{E}_0(a_0) = \mathbb{E}g = 0$, therefore a_0 is a B-valued atom relative to $\mathcal{A}_0$ (with support included in Ω). For any $m \ge 1$ and $n > 0$ we set

$$a_{m,n} = (g_{T_m} - g_{T_{m-1}}) \cdot 1_{\{T_{m-1}=n\}} (2^{m+1}\mathbb{P}\{T_{m-1} = n\})^{-1}.$$

Then $a_{m,n}$ is a B-valued atom: indeed it is supported on $\{T_{m-1} = n\}$ and (7.39) implies $\|a_{m,n}\|_\infty \le \mathbb{P}\{T_{m-1} = n\}^{-1}$. Here we assume $\mathbb{P}\{T_{m-1} = n\} > 0$ otherwise we set for notational convenience $a_{m,n} = 0$. Finally $\mathbb{E}_n(a_{m,n}) = 0$. Indeed, we have $T_{m-1} \le T_m$, so that $T_m \wedge n = T_{m-1} \wedge n$ when $T_{m-1} = n$, and since $\{T_{m-1} = n\} \in \mathcal{A}_n$, by (1.11) and (1.25) we may write

$$\mathbb{E}_n(1_{\{T_{m-1}=n\}}(g_{T_m} - g_{T_{m-1}})) = 1_{\{T_{m-1}=n\}}(g_{T_m\wedge n} - g_{T_{m-1}\wedge n}) = 0.$$

We can now complete the proof of (7.38). We have

$$g = a_0 + \sum\nolimits_{m,n>0} 2^{m+1}\mathbb{P}\{T_{m-1} = n\}a_{m,n}$$

therefore

$$\|g\|_{h^1_{\mathrm{at}}(\{\mathcal{A}_n\};B)} \le 1 + \sum_{m,n>0} 2^{m+1}\mathbb{P}\{T_{m-1} = n\} = 1 + \sum_{m>0} 2^{m+1}\mathbb{P}\{T_{m-1} < \infty\}.$$

Note that

$$\mathbb{P}(T_{m-1} < \infty) \le \mathbb{P}\{\sup\nolimits_n \lambda_n > 2^{m-1}\}$$

and we obtain as announced

$$\|g\|_{h^1_{\mathrm{at}}(\{\mathcal{A}_n\};B)} \le 1 + 8\mathbb{E}\sup\nolimits_n \lambda_n \le 9.$$

□

Remark 7.26. We will now use the Davis decomposition. We need to first observe that for any $h \in L_1(B)$ such that $h_0 = 0$ and $\sum \|dh_n\|_{L_1(B)} < \infty$ we have

$$|\mathbb{E}\langle\varphi_m, h\rangle| = |\mathbb{E}\langle\varphi_m, h_m\rangle| \le \sum\nolimits_n \|dh_n\|_{L_1(B)}\|\varphi\|_{\mathcal{BMO}(\{\mathcal{A}_n\};B^*)}. \quad (7.41)$$

Indeed, we have

$$|\mathbb{E}\langle\varphi_m, h_m\rangle| = \left|\sum\nolimits_{n\le m} \mathbb{E}\langle d\varphi_n, dh_n\rangle\right| \le \sum\nolimits_{n\le m} \|d\varphi_n\|_{\underline{\Delta}_\infty(B^*)}\|dh_n\|_{L_1(B)}$$

and hence (7.32) implies (7.41).

We now invoke the Davis decomposition $(f_n) = (g_n) + (h_n)$ from Corollary 5.33. Note $\|g_n\| \le \|g_{n-1}\| + \|dg_n\| \le \lambda_{n-1}$ with $\lambda_{n-1} = \|g_{n-1}\| + 8f^*_{n-1}$. Using (5.62), (7.35), (7.36) and (7.41) we find a numerical constant c_6 such that for any $m \ge 1$

$$|\mathbb{E}\langle\varphi_m, f_m\rangle| \le c_6\|\varphi\|_{\mathcal{BMO}(\{\mathcal{A}_n\};B^*)}\mathbb{E}\sup\nolimits_n \|f_n\|. \quad (7.42)$$

Remark 7.27. Note that one can also use a suitable notion of 'virtual atom' in the martingale setting to formulate a remark similar to the earlier Remark 7.12.

We can now define the duality pairing between any $\varphi \in \mathcal{BMO}(\{\mathcal{A}_n\}; B^*)$ and any $f \in \widetilde{h}^1_{\max}(\{\mathcal{A}_n\}; B)$.

Note that by (7.33) $\omega \mapsto \langle\varphi_m(\omega), f_m(\omega)\rangle$ is in L_1 for any m.

We set for any $f \in \widetilde{h}^1_{\max}(\{\mathcal{A}_n\}; B)$

$$\langle\langle\varphi, f\rangle\rangle = \lim_{m\to\infty} \mathbb{E}\langle\varphi_m, f_m\rangle, \quad (7.43)$$

but we must first check that this limit exists. Note that for any $f \in \widetilde{h}^1_{\max}(\{\mathcal{A}_n\}; B)$ we have (denoting $f_n = \mathbb{E}^{\mathcal{A}_n} f$) for any $m \geq n$

$$(f - f_n)^* = \sup_{m \geq n} \|f_m - f_n\|.$$

Moreover, by martingale convergence, $f_n \to f$ a.s. and hence by dominated convergence, $(f - f_n)^* \to 0$ a.s. and in $L_1(B)$. This shows that $\|f - f_n\|_{h^1_{\max}(\{\mathcal{A}_n\}; B)} \to 0$ when $n \to \infty$. Using (7.42) (with $f_m - f_n$ in place of f), it follows that, since $\|\varphi\|_{\mathcal{BMO}} \leq 1$, we have

$$\begin{aligned}\sup_{m \geq n} |\mathbb{E}\langle \varphi_m, f_m \rangle - \mathbb{E}\langle \varphi_n, f_n \rangle| &= \sup_{m \geq n} |\mathbb{E}\langle \varphi_m, f_m \rangle - \mathbb{E}\langle \varphi_m, f_n \rangle| \\ &\leq c_6 \sup_{m \geq n} \|f_m - f_n\|_{h^1_{\max}(\{\mathcal{A}_n\}; B)} \to 0,\end{aligned}$$

which shows, as promised, that the limit (7.43) exists. Furthermore, (7.42) implies

$$|\langle\langle \varphi, f \rangle\rangle| \leq c_6 \|\varphi\|_{\mathcal{BMO}} \|f\|_{h^1_{\max}(\{\mathcal{A}_n\}; B)}. \tag{7.44}$$

Thus, any $\varphi \in \mathcal{BMO}(\{\mathcal{A}_n\}; B^*)$ defines a continuous linear form $\xi_\varphi \in \widetilde{h}^1_{\max}(\{\mathcal{A}_n\}; B)^*$ by setting $\xi_\varphi(f) = \langle\langle \varphi, f \rangle\rangle$. We have then the following easy result:

Theorem 7.28. *With respect to the preceding duality, we have*

$$\widetilde{h}^1_{\max}(\{\mathcal{A}_n\}; B)^* = \mathcal{BMO}(\{\mathcal{A}_n\}; B^*).$$

More precisely, we have

$$(c_6)^{-1} \|\xi_\varphi\|_{\widetilde{h}^1_{\max}(\{\mathcal{A}_n\}; B)^*} \leq \|\varphi\|_{\mathcal{BMO}(\{\mathcal{A}_n\}; B^*)} \leq 2\|\xi_\varphi\|_{\widetilde{h}^1_{\max}(\{\mathcal{A}_n\}; B)^*}. \tag{7.45}$$

Proof. The first half of (7.45) follows from (7.44). Conversely, we claim that for any $\varphi \in \mathcal{BMO}(\{\mathcal{A}_n\}; B^*)$

$$\|\varphi\|_{\mathcal{BMO}(\{\mathcal{A}_n\}; B^*)} \leq \sup\{|\langle\langle \varphi, f \rangle\rangle| \mid f \in \mathcal{C}\}, \tag{7.46}$$

where $\mathcal{C}$ is the class formed of the functions f for which there are $m \geq n \geq 0$, a set $A \in \mathcal{A}_n$ and a function b supported on A with $\|b\|_{L_\infty(\mathcal{A}_m; B)} \leq 1$, such that $f = (b - b_{n-1})/\mathbb{P}(A)$ (with $b_{-1} = 0$). Indeed, if $|\langle\langle \varphi, f \rangle\rangle| \leq 1$ for all such f, then we have

$$|\langle\langle \varphi, f \rangle\rangle| = \mathbb{P}(A)^{-1} |\langle \varphi_m, (b - b_{n-1}) \rangle| \leq 1$$

or equivalently

$$\mathbb{P}(A)^{-1} \left| \int_A \langle \varphi_m - \varphi_{n-1}, b \rangle \right| \leq 1$$

which implies (taking the sup over all b's) that

$$\mathbb{P}(A)^{-1}\int_A \|\varphi_m - \varphi_{n-1}\|_{B^*} \le 1,$$

and hence recalling (7.31)

$$\mathbb{E}_n\|\varphi - \varphi_{n-1}\|_{B^*} \le 1$$

completing the proof of (7.46) (actually this shows that equality holds in (7.46)).

But for any $f \in \mathcal{C}$, since $\mathbb{E}_{n-1}(f) = 0$ we have $f^* = \sup_{k\ge n}\|f_k\|$ and since $\|b\| \le 1_A$ and $A \in \mathcal{A}_n$, we have $\|b_k\| \le \mathbb{E}_k\|b\| \le 1_A$ for all $k \ge n$ and

$$\|f_k\| \le \mathbb{P}(A)^{-1}(\|b_k\| + \|b_{n-1}\|) \le \mathbb{P}(A)^{-1}(1_A + \mathbb{E}_{n-1}1_A).$$

Thus $f^* \le \mathbb{P}(A)^{-1}(1_A + \mathbb{E}_{n-1}1_A)$ and we obtain

$$\|f\|_{\widetilde{h}^1_{\max}(\{\mathcal{A}_n\};B)} = \mathbb{E}f^* \le 2.$$

Then the right-hand part of (7.45) follows. Lastly, we should check that any $\xi \in \widetilde{h}^1_{\max}(\{\mathcal{A}_n\};B)^*$ is of the form $\xi(f) = \langle\langle\varphi, f\rangle\rangle$ for some $\varphi \in \mathcal{BMO}(\{\mathcal{A}_n\};B^*)$. To verify this, just note that, for any $1 < p' < \infty$ we have by Doob's maximal inequality (1.38) $L_{p'}(B) \subset \widetilde{h}^1_{\max}(\{\mathcal{A}_n\};B)$ so that ξ restricted to $L_{p'}(\mathcal{A}_m;B)$ is associated to a function $\varphi_m \in \underline{\Lambda}_p(\mathcal{A}_m;B^*) \subset \underline{\Lambda}_1(B^*)$. Clearly $(\varphi_m)_{m\ge 0}$ is a martingale, which defines an element of $\varphi \in \underline{\Lambda}_p(\mathcal{A}_m;B^*)$ (see the proof of Theorem 2.29). By the preceding remarks, $\varphi \in \mathcal{BMO}(\{\mathcal{A}_n\};B^*)$ with associated linear form ξ. □

Remark. There is also an analogue of the John-Nirenberg Theorem: there is c_1 such that for any φ in the unit ball of either $BMO(\{\mathcal{A}_n\};B)$ or $\mathcal{BMO}(\{\mathcal{A}_n\};B^*)$, we have

$$\forall n \ge 0\ \forall\lambda > 0 \quad \mathbb{E}_n\{\|\varphi - \varphi_{n-1}\| > \lambda\} \le e\exp[-c_1\lambda]. \tag{7.47}$$

Therefore, there is c_2 such that for any such φ and any $1 \le p < +\infty$ we have

$$\forall n \ge 0 \quad (\mathbb{E}_n\|\varphi - \varphi_{n-1}\|^p)^{1/p} \le c_2 p. \tag{7.48}$$

As in the preceding section, we may assume without loss of generality that B is finite-dimensional, so that the roles of B and B^* become interchangeable. The proof we gave in §7.1 can be adapted to the martingale setting, using Doob's maximal inequality (1.38) in place of (7.9). More precisely, returning to the end of the preceding proof, since $\|\sup\|f_n\|\|_{p'} \le p\sup\|f_n\|_{p'}$ we find $\|L_{p'}(B) \subset h^1_{\max}(\{\mathcal{A}_n\};B)\| \le p$ and hence, with the same notation, taking the sup over all b in the unit ball of $L_{p'}(B)$ we find

$$(\mathbb{P}(A)^{-1}\int_A \|\varphi_m - \varphi_n\|_{B^*}^p)^{1/p} \le 2p,$$

and (7.48) follows (with B^* in place of B). Then (7.47) follows again from Lemma 5.80.

Corollary 7.29. *In the same situation, if B^* has the RNP, then*

$$BMO(\{\mathcal{A}_n\}; B^*) = \widetilde{h}^1_{\max}(\{\mathcal{A}_n\}; B)^*.$$

Moreover, for any $\varphi \in BMO(\{\mathcal{A}_n\}; B^*)$ the associated martingale converges in $L_p(B^*)$ for all $p < \infty$. If in addition B has the RNP, then $\widetilde{h}^1_{\max}(\{\mathcal{A}_n\}; B) = h^1_{\max}(\{\mathcal{A}_n\}; B)$.

Proof. If B^* has the RNP, $\mathcal{BMO}(B^*) = BMO(B^*)$, so the identity is clear, and since we have a bounded inclusion $BMO(B^*) \subset L_p(B^*)$ the second assertion, as well as the last one, follows from Theorem 2.9. □

Remark 7.30. Let us denote by $h^1_{\mathrm{d}}(\{\mathcal{A}_n\}; B)$ the set of $h \in L_1(B)$ such that $h_0 = 0$ and $\sum_{n\ge1} \|h_n - h_{n-1}\|_{L_1(B)} < \infty$ equipped with the norm

$$\|h\|_{h^1_{\mathrm{d}}} = \sum\nolimits_{n\ge1} \|h_n - h_{n-1}\|_{L_1(B)}.$$

Then (7.30) (suitably adapted to $\mathcal{BMO}$) shows that $h^1_{\mathrm{at}}(B) + h^1_{\mathrm{d}}(B)$ (equipped with the norm $\inf\{\|g\|_{h^1_{\mathrm{at}}(B)} + \|h\|_{h^1_{\mathrm{d}}(B)} \mid f = g + h\}$ can be obviously viewed as a predual of $\mathcal{BMO}(B^*)$, but this is rather superficial.

The Davis decomposition allows us to go much deeper: we have an identity

$$h^1_{\mathrm{at}}(B) + h^1_{\mathrm{d}}(B) = \widetilde{h}^1_{\max}(B), \tag{7.49}$$

and the corresponding norms are equivalent.

Indeed, using the Davis decomposition as in Remark 7.26, Lemma 7.25 implies $\widetilde{h}^1_{\max}(B) \subset h^1_{\mathrm{at}}(B) + h^1_{\mathrm{d}}(B)$. Conversely $h^1_{\mathrm{d}}(B) \subset \widetilde{h}^1_{\max}(B)$ is obvious and $h^1_{\mathrm{at}}(B) \subset \widetilde{h}^1_{\max}(B)$ follows from (7.28).

7.4 Regular filtrations

Several of the preceding statements become a bit simpler when restricted to 'regular' filtrations.

Definition 7.31. A filtration $(\mathcal{A}_n)_{n\ge0}$ is called regular if there is a constant $C \ge 1$ such that, for all $n \ge 1$ and for all $w \ge 0$ in $L_1(\Omega, \mathcal{A}, \mathbb{P})$, we have

$$\mathbb{E}_n(w) \le C\mathbb{E}_{n-1}(w). \tag{7.50}$$

For example it is easy to see that the dyadic filtration is regular. More generally, if $\mathcal{A}_n$ is finite for all n and there is $\delta > 0$ such that, for all $n \geq 1$, for all atoms α of $\mathcal{A}_{n-1}$ and all atoms $\alpha' \subset \alpha$ of $\mathcal{A}_n$, we have $\mathbb{P}(\alpha')\mathbb{P}(\alpha)^{-1} \geq \delta$, then the filtration is regular (the dyadic case corresponds to $\delta = 1/2$).

If the filtration is regular and (7.50) holds, then

$$C^{-1}\|f\|_{BMO} \leq \sup_{n\geq 0} \|\mathbb{E}_n|f - f_n|\|_\infty \leq 2\|f\|_{BMO}. \tag{7.51}$$

Indeed, we have

$$\mathbb{E}_n|f - f_{n-1}| \leq C\mathbb{E}_{n-1}|f - f_{n-1}| \leq C \sup_{n\geq 0} \|\mathbb{E}_n|f - f_n|\|_\infty,$$

whence the first inequality. Also $|f_n - f_{n-1}| \leq \mathbb{E}_n|f - f_{n-1}| \leq \|f\|_{BMO}$ and since $|f - f_n| \leq |f - f_{n-1}| + |f_n - f_{n-1}|$, we obtain the other side.

Theorem 7.32. *In the regular case (for example in the dyadic case), the mapping $f \mapsto (f_n)_{n\geq 0}$ (here $f_n = \mathbb{E}_n(f)$) is an isomorphism from $h^1_{\rm at}(\{\mathcal{A}_n\}; B)$ to $\widetilde{h}^1_{\max}(\{\mathcal{A}_n\}; B)$ and the corresponding norms are equivalent, with equivalence constants independent of B.*

Proof. By (7.28) we already know that $h^1_{\rm at}(\{\mathcal{A}_n\}; B) \subset \widetilde{h}^1_{\max}(\{\mathcal{A}_n\}; B)$. To prove the converse direction, by (7.49) it suffices to check the following:

Claim: We have $h^1_{\rm d}(B) \subset h^1_{\rm at}(B)$ and $\|f\|_{h^1_{\rm at}(B)} \leq 2C\|f\|_{h^1_{\rm d}(B)}$ for any $f \in h^1_{\rm d}(B)$. To verify this, we may clearly assume that, for a *fixed* $n \geq 1$, f is $\mathcal{A}_n$-measurable and $\mathbb{E}_{n-1}f = 0$. Let $w = \|f\|$ and $w' = C\mathbb{E}_{n-1}w$. We then write

$$f = \sum\nolimits_{k\in\mathbb{Z}} f 1_{\{2^{k-1} < w' \leq 2^k\}} = \sum \lambda_k a_k$$

where $\lambda_k = 2^k\mathbb{P}\{2^{k-1} < w' \leq 2^k\}$ and $a_k = (f 1_{\{2^{k-1} < w' \leq 2^k\}})/\lambda_k$.

By (7.50), $w = \|f\| \leq w'$ and hence $\|a_k\| \leq \mathbb{P}\{2^{k-1} < w' \leq 2^k\}^{-1}$. Note that $\mathbb{E}_{n-1}a_k = 0$ (since w' is $\mathcal{A}_{n-1}$-measurable and $\mathbb{E}_{n-1}f = 0$). Moreover, $\{2^{k-1} < w' \leq 2^k\} \in \mathcal{A}_{n-1}$. Thus each a_k is an atom, and hence $\|f\|_{h^1_{\rm at}(B)} \leq \sum_{k\in\mathbb{Z}} \lambda_k \leq 2\mathbb{E}w' = 2C\mathbb{E}\|f\|$. This proves the claim. □

We do not state the analogues for martingales of Theorem 7.19 since these are rather easy to deduce and are already (essentially) contained in Theorem 5.34 and Remark 5.35.

7.5 From dyadic BMO to classical BMO

In this section, we follow [353].

Consider the dyadic filtration $(\mathcal{A}^{\mathbb{T}}_n)$ on $\mathbb{T}$. More precisely, the σ-algebra $\mathcal{A}^{\mathbb{T}}_n$ is generated by the partition of $\mathbb{T}$ obtained as the image by $t \mapsto e^{2\pi it}$ of the partition of $[0, 1)$ formed by the atoms of $\mathcal{A}_n$, namely the collection

$$\left[\frac{k-1}{2^n}, \frac{k}{2^n}\right) \quad 1 \le k \le 2^n.$$

Let $A_n(k) = \{e^{2\pi i t} \mid \frac{k-1}{2^n} \le t < \frac{k}{2^n}\}$ $(1 \le k \le 2^n)$ be the partition of $\mathbb{T}$, formed by the atoms of $\mathcal{A}_n^{\mathbb{T}}$. We will refer to the whole collection of subarcs $\{A_n(k) \mid n \ge 0, 1 \le k \le 2^n\}$ of $\mathbb{T}$ as dyadic intervals in $\mathbb{T}$.

We then set

$$BMO_d(\mathbb{T}; B) = BMO(\{\mathcal{A}_n^{\mathbb{T}}\}; B).$$

It is easy to check that for any $\varphi \in L_1(\mathbb{T}; B)$ we have $\|\varphi\|_{BMO_d(\mathbb{T};B)} \le C$ iff φ satisfies (7.1) for any dyadic interval $I \subset \mathbb{T}$. Indeed, it suffices to check $\mathbb{E}_n\|\varphi - \varphi_n\|_B \le C$ on the atoms of $\mathcal{A}_n^{\mathbb{T}}$.

Thus we have $BMO \subset BMO_d$ but it is well known that this inclusion is strict. For instance, consider a real valued function that is bounded when $|t| > 1/2$ and such that $\varphi(e^{2\pi i t}) = 1_{t>0} \operatorname{Log} t$ when $|t| \le 1/2$. Then $\varphi \in BMO_d \setminus BMO$.

However, it turns out that the space BMO coincides with the intersection of BMO_d with one of its (suitably chosen) translates. For any $\alpha \in \mathbb{R}$, we define

$$d(\alpha) = \inf\{|2^n\alpha - k| \mid n \ge 0, k \in \mathbb{Z}\}.$$

Note that $d(\alpha) \le 1$.

We will select any number α such that $d(\alpha) > 0$. For instance $\alpha = \frac{1}{3}$ clearly works. Given $\varphi \in L_1(\mathbb{T}; B)$ we denote by φ_α the translated function defined by

$$\varphi_\alpha(e^{2\pi i t}) = \varphi(e^{2\pi i(t-\alpha)}).$$

Theorem 7.33. *Let α be such that $d(\alpha) > 0$. Let B be any Banach space. Consider a B-valued function $\varphi \in L_1(\mathbb{T}; B)$. Then $\varphi \in BMO(\mathbb{T}; B)$ iff both φ and φ_α are in $BMO_d(\mathbb{T}; B)$. Moreover, assuming $\int \varphi dm = 0$, we have*

$$\|\varphi\|_{BMO(\mathbb{T};B)} \le \frac{4}{d(\alpha)} \max\left\{ \|\varphi\|_{BMO_d}, \|\varphi_\alpha\|_{BMO_d} \right\}.$$

For the proof, the following simple lemma will be essential.

Lemma 7.34. *Let $\alpha \in \mathbb{R}$ be such that $d(\alpha) > 0$. Let $I \subset \mathbb{R}$ be an interval such that for some $n \ge 0$ we have*

$$d(\alpha)2^{-n-1} \le |I| < d(\alpha)2^{-n}. \tag{7.52}$$

Then there is $k \in \mathbb{Z}$ such that either $I \subset [k2^{-n}, (k+1)2^{-n})$ or $I \subset \alpha + [k2^{-n}, (k+1)2^{-n})$. In particular, for any interval $I \subset \mathbb{R}$ with $|I| < d(\alpha)$ there is a dyadic interval I' with $|I'| \le \frac{2}{d(\alpha)}|I|$ such that either $I \subset I'$ or $I \subset \alpha + I'$.

Proof. Since $d(\alpha) \le 1$, we have $|I| < 2^{-n}$. It follows that there is at most one $k \in \mathbb{Z}$ such that $k2^{-n} \in I$, and also at most one $l \in \mathbb{Z}$ for which $\alpha + l2^{-n} \in I$. We claim that we cannot have both $k2^{-n} \in I$ and $\alpha + l2^{-n} \in I$. Indeed, this would imply $|\alpha + l2^{-n} - k2^{-n}| \le |I|$ and hence a fortiori $2^{-n}d(\alpha) \le |I|$, in contradiction with (7.52). It follows that either $\{k \in \mathbb{Z} | k2^{-n} \in I\} = \phi$ or $\{k \in \mathbb{Z} | \alpha + k2^{-n} \in I\} = \phi$. In either case, we obtain the desired conclusion. □

Proof of Theorem 7.33. By Remark 7.24, it suffices to show that if both φ and φ_α satisfy (7.1) for all dyadic intervals, then φ satisfies (7.1) for all intervals I in $\mathbb{T}$ up to a factor $4/d(\alpha)$. Assume $\max\{\|\varphi\|_{BMO_d}, \|\varphi_\alpha\|_{BMO_d}\} \le 1$. Let $f(t) = \varphi(e^{2\pi it})$. Let $I' \subset \mathbb{R}$ be any dyadic interval. Then we have

$$\frac{1}{|J|}\int_J \|f - f_J\| dt \le 1$$

either if $J = I'$ or if $J = \alpha + I'$.

Let $I \subset \mathbb{R}$ be an arbitrary interval with $|I| < d(\alpha)$. By Lemma 7.34, there is an interval $J \supset I$ with $|J| \le \frac{2}{d(\alpha)}|I|$ such that

$$\frac{1}{|J|}\int_J \|f - f_J\| dt \le 1,$$

and a fortiori

$$\frac{1}{|I|}\int_I \|f - f_J\| dt \le \frac{|J|}{|I|} \le \frac{2}{d(\alpha)}.$$

By Jensen, this implies $\|f_I - f_J\| \le \frac{2}{d(\alpha)}$, and hence by the triangle inequality

$$\frac{1}{|I|}\int_I \|f - f_I\| dt \le \frac{4}{d(\alpha)}.$$

This settles the case when $|I| < d(\alpha)$. But if $d(\alpha) \le |I| \le 1$, there is an interval $J \supset I$ with $|J| = 1$ and since f is 1-periodic, we have again

$$\frac{1}{|I|}\int_I \|f - f_J\| dt \le \frac{|J|}{|I|} \le \frac{1}{d(\alpha)},$$

and we conclude by the same argument that

$$\frac{1}{|I|}\int_I \|f - f_I\| \le \frac{2}{d(\alpha)} \le \frac{4}{d(\alpha)}.$$

□

We will now tackle the same question for $BMO(\mathbb{R}; B)$. The result will be similar, except that we cannot get the second dyadic filtration to be a translate of the other one. Moreover, we must use conditional expectations on $\mathbb{R}$ equipped with its Lebesgue measure, which is an infinite measure space, but this makes little difference.

Consider a filtration $(\mathcal{A}_n)_{n\in\mathbb{Z}}$ of σ-algebras on $\mathbb{R}$. We say that it is a dyadic filtration if each $\mathcal{A}_n$ is generated by a partition of $\mathbb{R}$ into intervals of length 2^{-n} of the form

$$[\alpha_n + k2^{-n},\ \alpha_n + (k+1)2^{-n}) \qquad k \in \mathbb{Z} \tag{7.53}$$

in such a way that any such interval is the disjoint union of two intervals from the partition defining $\mathcal{A}_{n+1}$. (The presence or the absence of the end points in all these intervals is irrelevant since singletons are negligible.) Thus the intervals in (7.53) are the atoms of the σ-algebra $\mathcal{A}_n$. Any interval of the form (7.53) for some $n \in \mathbb{Z}$ will be called $\{\mathcal{A}_n \mid n \in \mathbb{Z}\}$-dyadic. Let us denote by $\mathcal{E}_n$ the endpoints of these intervals, so that

$$\mathcal{E}_n = \alpha_n + \mathbb{Z}2^{-n}.$$

Note that we may assume that $0 \le \alpha_n < 2^{-n}$.

The 'dyadic' nature of $(\mathcal{A}_n)_{n\in\mathbb{Z}}$ is then characterized by the condition

$$\forall n \in \mathbb{Z} \qquad \mathcal{E}_n \subset \mathcal{E}_{n+1}. \tag{7.54}$$

Note that if we are given $\mathcal{A}_{n(0)}$ (or equivalently $\alpha_{n(0)}$), for some $n(0) \in \mathbb{Z}$, then the algebras $\mathcal{A}_n$ for $n > n(0)$ are all determined by $\mathcal{A}_{n(0)}$. However, there are two possible choices for $\mathcal{A}_{n(0)-1}$ and 2^k possibilities for $\mathcal{A}_{n(0)-k}$ for $k > 0$.

Consider for simplicity the case $n(0) = 0$, and assume that $\mathcal{A}_0$ is given to us. Recall $\mathcal{E}_0 = \alpha_0 + \mathbb{Z}$. Then $\mathcal{E}_{-1} \subset \mathcal{E}_0$ has two solutions: either $\mathcal{E}_{-1} = \alpha_0 + 2\mathbb{Z}$ or $\mathcal{E}_{-1} = \alpha_0 + 2\mathbb{Z} + 1$. Thus we have $\mathcal{E}_{-1} = \alpha_0 + 2\mathbb{Z} + \xi_1$ for some $\xi_1 \in \{0, 1\}$. Iterating this, we see that for any given sequence $(\xi_n)_{n>0}$ in $\{0, 1\}$ there is a dyadic filtration $(\mathcal{A}_n)_{n\in\mathbb{Z}}$, coinciding with our given $\mathcal{A}_0$ for $n = 0$, such that

$$\forall n > 0 \quad \mathcal{E}_{-n} = \alpha_0 + 2^n\mathbb{Z} + 2^{n-1}\xi_n + \cdots + 2\xi_2 + \xi_1.$$

Equivalently

$$\alpha_{-n} = \alpha_0 + 2^{n-1}\xi_n + \cdots + \xi_1 \ (\text{ modulo } 2^n).$$

Consider now the following specific choice

$$\forall j > 0 \qquad \xi_{2j} = 0 \text{ and } \xi_{2j+1} = 1. \tag{7.55}$$

A simple calculation then shows that, modulo 2^n we have $\alpha_{-n} = \alpha_0 + \frac{1}{3}(2^n - 1)$ if n is even and $\alpha_{-n} = \alpha_0 - \frac{1}{3}(2^n + 1)$ if n is odd. In any case, since $|\alpha_0| < 1$, for all n large enough, say for all $n > N > 0$, we have

$$\operatorname{dist}(\alpha_{-n}, 2^n\mathbb{Z}) \ge 2^n/4. \tag{7.56}$$

Choosing among these possibilities, one can produce two dyadic filtrations such that the property appearing in Lemma 7.34 remains valid for large $|I|$.

Lemma 7.35. *There is a constant $c > 0$ and two dyadic filtrations $(\mathcal{A}'_n)_{n\in\mathbb{Z}}$ and $(\mathcal{A}''_n)_{n\in\mathbb{Z}}$ in $\mathbb{R}$ such that for any interval $I \subset \mathbb{R}$ there is an interval $J \supset I$ that is either $\{\mathcal{A}'_n \mid n \in \mathbb{Z}\}$-dyadic or $\{\mathcal{A}''_n \mid n \in \mathbb{Z}\}$-dyadic such that $|J| \le c|I|$.*

Proof. Let us denote by $(\alpha'_n, \mathcal{E}'_n)$ (resp. $(\alpha''_n, \mathcal{E}''_n)$) the parameters of $\{\mathcal{A}'_n\}$ (resp. $\{\mathcal{A}''_n\}$). For $\{\mathcal{A}'_n\}$, the choice is simple: We simply choose $\alpha'_n = 0$ and thus $\mathcal{E}'_n = 2^{-n}\mathbb{Z}$ for all $n \in \mathbb{Z}$. For $\{\mathcal{A}''_n\}$ we make the choice that guarantees (7.56). Then, since $\text{dis}(\alpha''_{-n} - \alpha'_{-n}, 2^n\mathbb{Z}) \ge 2^n/4$ for all $n > N$, we have

$$\forall n > N \quad \inf\{|s-t| \mid s \in \mathcal{E}'_{-n},\ t \in \mathcal{E}''_{-n}\} \ge 2^n/4. \tag{7.57}$$

Thus, it is clear that any interval with $|I| < 2^n/4$ is contained in some $J \supset I$ with $|J| = 2^n$ that is either $\{\mathcal{A}'_n \mid n \in \mathbb{Z}\}$-dyadic or $\{\mathcal{A}''_n \mid n \in \mathbb{Z}\}$-dyadic. Now assume, say, $|I| \ge 2^N/4$. For the minimal $n > N$ for which $|I| < 2^n/4$, we have $2^{n-1}/4 \le |I|$ and hence $|J| = 2^n \le 8|I|$. This takes case of I when $|I|$ is large. To take case of small $|I|$, we choose $\alpha_0 = \alpha$ for α such that $d(\alpha) > 0$ as in Lemma 7.34. There only remains the case when $d(\alpha) \le |I| < 2^N/4$, which is now essentially trivial: by (7.57) we may choose $J \supset I$ with $|J| = 2^{N+1}$ that is dyadic with respect to either $\{\mathcal{A}'_n\}$ or $\{\mathcal{A}''_n\}$ and we have $|J| = 2^{N+1} \le c|I|$ with $c = 2^{N+1}/d(\alpha)$. □

Remark 7.36. Let Mf be the Hardy-Littlewood maximal function from Theorem 3.7. By the preceding lemma, for some constant c, we have $Mf \le c\max\{M'f, M''f\}$ where $M'f, M''f$ are the maximal functions associated to the filtrations $(\mathcal{A}'_n)_{n\in\mathbb{Z}}$ and $(\mathcal{A}''_n)_{n\in\mathbb{Z}}$. Using this, one can deduce the Hardy-Littlewood maximal inequalities from Doob's.

For any dyadic filtration $\{\mathcal{A}_n \mid n \in \mathbb{Z}\}$ on $\mathbb{R}$, we denote by

$$BMO(\{\mathcal{A}_n \mid n \in \mathbb{Z}\}; B)$$

the space of functions locally in $L_1(\mathbb{R}; B)$ such that

$$\|\varphi\|_{*,\{\mathcal{A}_n|n\in\mathbb{Z}\}} = \sup_{n\in\mathbb{Z}} \|\mathbb{E}_n\|\varphi - \mathbb{E}_n\varphi\|\|_\infty < \infty.$$

When $B = \mathbb{R}$, we denote this space simply by $BMO(\{\mathcal{A}_n \mid n \in \mathbb{Z}\})$.

Equivalently, $\|\varphi\|_{*,\{\mathcal{A}_n|n\in\mathbb{Z}\}} \le C$ iff (7.1) holds for any $\{\mathcal{A}_n \mid n \in \mathbb{Z}\}$-dyadic interval I.

Using Lemma 7.35, the same reasoning as earlier shows

Theorem 7.37. *For any φ locally in $L_1(\mathbb{R}; B)$, $\|\varphi\|_*$ is equivalent to*

$$\max\{\|\varphi\|_{*,\{\mathcal{A}'_n|n\in\mathbb{Z}\}}, \|\varphi\|_{*,\{\mathcal{A}''_n|n\in\mathbb{Z}\}}\}.$$

In particular we have

$$BMO(\mathbb{R}) = BMO(\{\mathcal{A}'_n \mid n \in \mathbb{Z}\}) \cap BMO(\{\mathcal{A}''_n \mid n \in \mathbb{Z}\}.$$

Assuming B separable, Theorem 7.33 and Theorem 7.37 clearly hold for the spaces $\mathcal{BMO}(\mathbb{T}; B^*)$ and $\mathcal{BMO}(\mathbb{R}; B^*)$ with the same proof. Then, the duality Theorem 7.6 immediately implies the following decomposition of h^1_{at} as a sum of two dyadic h^1_{at}s.

Corollary 7.38. *With the preceding notation, there are two dyadic filtrations* $\{\mathcal{A}'_n \mid n \geq 0\}$ *and* $\{\mathcal{A}''_n \mid n \geq 0\}$ *on* $\mathbb{T}$, *one being a translate of the other, such that*

$$h^1_{\text{at}}(\mathbb{T}; B) = h^1_{\text{at}}(\{\mathcal{A}'_n\}; B) + h^1_{\text{at}}(\{\mathcal{A}''_n\}; B).$$

There are two dyadic filtrations $\{\mathcal{A}'_n \mid n \in \mathbb{Z}\}$ *and* $\{\mathcal{A}''_n \mid n \in \mathbb{Z}\}$ *on* $\mathbb{R}$, *such that*

$$h^1_{\text{at}}(\mathbb{R}; B) = h^1_{\text{at}}(\{\mathcal{A}'_n \mid n \in \mathbb{Z}\}; B) + h^1_{\text{at}}(\{\mathcal{A}''_n \mid n \in \mathbb{Z}\}; B).$$

Moreover, the corresponding norms are equivalent.

Note that by Remark 7.24 this means that every $f \in h^1_{\text{at}}(\mathbb{T}; B)$ can be written as a series of the form (7.27) with B-valued atoms a_n associated to intervals that are either $\{\mathcal{A}'_n\}$-dyadic or $\{\mathcal{A}''_n\}$-dyadic, and similarly for $f \in h^1_{\text{at}}(\mathbb{R}; B)$.

7.6 Notes and remarks

Concerning the very important space BMO, there is an extensive literature, see e.g. [32, 33, 86]. The main classical results on the H^1-BMO duality for scalar valued functions are due to Fefferman and to Fefferman and Stein [222].

The UMD valued case, as stated in Theorem 7.2, combines Bourgain's results in [137], with those of García-Cuerva and O. Blasco [123].

The atomic decomposition of $\Re(H^1)$ was observed by Fefferman as a reformulation of his duality theorem (see [183]). A direct real variable proof of the atomic decomposition of $\Re(H^1)$ was then given by R. Coifman [183]. All this applies to functions on $\mathbb{R}^n$ or $\mathbb{T}^n$ but we chose to restrict our presentation to $n = 1$.

According to Coifman in [183], the atomic decomposition in the martingale case was known to Herz [266, 267]. We refer to [121] for the atomic decomposition of martingale H^1 in continuous time.

What we call a (B-valued) atom is called a simple atom by Weisz in [97], which contains much more information and references on martingale atoms. See also Long's book [58].

The extension to the general Banach valued case of the atomic decomposition of functions in $H^1(\mathbb{R})$ or $H^1(\mathbb{T})$ (related to §7.3) is due independently to García-Cuerva and Bourgain ([137]). Pursuing on this theme, O. Blasco [123] extended the H^1-BMO duality to the B-valued case.

Theorem 7.7 in the scalar case can be found e.g. in [7, p. 371] or [32, p. 254]. The latter proofs extend to the B-valued case with minor changes. Our proof follows the one in Bennett and Sharpley's book [7, p. 371].

The landmark Theorem 7.1 is due to John and Nirenberg. Its proof is identical in the B-valued case. Garnett and Jones ([236, 237]) proved several important refinements. Notably in [236] they describe an equivalent norm to the BMO-distance of a function $\varphi \in BMO$ to L_∞, that also remains valid in the Banach valued case.

In [237] they use an averaging argument, similar in spirit but different from the one in §6.3, to pass from the dyadic case to the general one.

The last section, §7.5, is due to Tao Mei [353]. In the case of $\mathbb{T}^n$ (resp. $\mathbb{R}^n$) for $n > 1$, it is proved in the same note (resp. in [188] and [271, pp. 42–43]) that BMO is the intersection of $n+1$ copies of dyadic-BMO. This result was previously known (somewhat implicitly) notably to Garnett, Jones and Christ, with 3^n copies. The number $n+1$ is optimal (see [188]).

See Petersen's lecture notes [76] for the connection between BMO and Brownian motion, which we decided to skip altogether.

See Blasco and Pott's [125] for a sample of the current active study of operator valued BMO spaces.

See Pereyra's [380] for a presentation of the dyadic side of many results from Fourier analysis, related to BMO and the theory of A_p-weights.

We omitted many important topics related to BMO, such as Carleson measures, Poisson integrals of BMO functions, VMO, BLO, paraproducts and A_p-weights, some of which seem relevant for the Banach space valued case. We refer the reader to Garnett's book [33] or to Elias Stein's [86] to fill these unfortunate gaps, without which this book would probably never have been completed…

8

Interpolation methods (complex and real)

In this chapter, we describe methods to construct a family of 'interpolated' Banach spaces $(B_\theta)_{\theta\in[0,1]}$ starting from a pair (B_0, B_1). We will need to assume that the initial pair (B_0, B_1) is 'compatible'. This means that we are given a (Hausdorff) topological vector space V and continuous injections

$$j_0\colon B_0 \to V \quad \text{and} \quad j_1\colon B_1 \to V.$$

This very rudimentary structure is just what is needed to define the intersection $B_0 \cap B_1$ and the sum $B_0 + B_1$.

The space $B_0 \cap B_1$ is defined as $j_0(B_0) \cap j_1(B_1)$ equipped with the norm

$$\|x\| = \max\{\|j_0^{-1}(x)\|_{B_0}, \|j_1^{-1}(x)\|_{B_1}\}.$$

The space $B_0 + B_1$ is defined as the setwise sum $j_0(B_0) + j_1(B_1)$ equipped with the norm

$$\|x\|_{B_0+B_1} = \inf\{\|x_0\|_{B_0} + \|x_1\|_{B_1} \mid x = j_0(x_0) + j_1(x_1)\}.$$

It is an easy exercise to check that $B_0 \cap B_1$ and $B_0 + B_1$ are Banach spaces. Following a well established tradition, we will identify B_0 and B_1 with $j_0(B_0)$ and $j_1(B_1)$, so that j_0 and j_1 become the inclusion mappings $B_0 \subset V$ and $B_1 \subset V$. We then have $\forall i = 0, 1$

$$B_0 \cap B_1 \subset B_i \subset B_0 + B_1$$

and these inclusions have norm ≤ 1. Note that if we wish we may now replace V by $B_0 + B_1$, so that we may as well assume that V is a Banach space.

Remark 8.1. Let $A_j \subset B_j$ denote the closure of $B_0 \cap B_1$ in B_j $(j = 0, 1)$ with the norm induced by B_j. Consider $x \in A_0 + A_1$, say $x = a_0 + a_1$ with $a_j \in A_j$.

Then, for any decomposition $x = b_0 + b_1$ with $b_j \in B_j$, we have $a_0 - b_0 = b_1 - a_1$, and hence $a_0 - b_0$ and $b_1 - a_1$ both belong to $B_0 \cap B_1$, showing that automatically $b_0 \in A_0$ and $b_1 \in A_1$. In particular, this observation implies that the inclusion $A_0 + A_1 \subset B_0 + B_1$ is isometric when both spaces are equipped with their intrinsic norms as sum spaces. Moreover, it is obvious that $A_0 \cap A_1 = B_0 \cap B_1$ and easy to check that $A_0 + A_1$ is the closure of $B_0 \cap B_1$ in $B_0 + B_1$.

8.1 The unit strip

We will need the following notation.

Let

$$S = \{z \in \mathbb{C} \mid 0 < \mathrm{Re}(z) < 1\}$$
$$\partial_0 = \{z \in \mathbb{C} \mid \mathrm{Re}(z) = 0\}$$
$$\partial_1 = \{z \in \mathbb{C} \mid \mathrm{Re}(z) = 1\}.$$

Note that $\partial S = \partial_0 \cup \partial_1$. Given a subset $X \subset \mathbb{C}$ and a Banach space B, we denote by $C_b(X; B)$ the space of bounded continuous functions $f\colon X \to B$ equipped with the norm

$$\|f\|_{C_b(X;B)} = \sup\nolimits_{z \in X} \|f(z)\|_B.$$

Remark 8.2. Let $0 < \theta < 1$. The function $\chi_\theta\colon S \to D$ defined by

$$\chi_\theta(z) = \frac{e^{i\pi z} - e^{i\pi\theta}}{e^{i\pi z} - e^{-i\pi\theta}} \tag{8.1}$$

is a conformal (bijective biholorphic) mapping from S onto D, such that $\chi_\theta(\theta) = 0$. This mapping is a continuous injection from ∂S to the unit circle. More precisely, it takes ∂_1 to the interior of the arc I_1 of length $2\pi\theta$ joining 1 and $e^{2i\pi\theta}$, and takes ∂_0 to the interior of the complementary arc I_0 of length $2\pi - 2\pi\theta$. Note that $m(I_0) = 1 - \theta$ and $m(I_1) = \theta$.

The inverse map $\chi_\theta^{-1} : D \to S$ also takes the part of ∂D where it is defined continuously to ∂S but there are two exceptional points, namely $e^{2i\theta}$ and 1, corresponding to

$$\lim_{t \to \pm\infty} \chi_\theta(z + it)$$

where it is not defined, and near which it is unbounded. These two points seem worrisome, but since they form a negligible set, we will be able to transfer to S the boundary value analysis for $L_\infty(\partial D)$ from Chapter 3. Indeed, by Remark 3.20, we know that both χ_θ and its inverse preserve non-tangential convergence at almost every boundary point, and this is enough for our purposes.

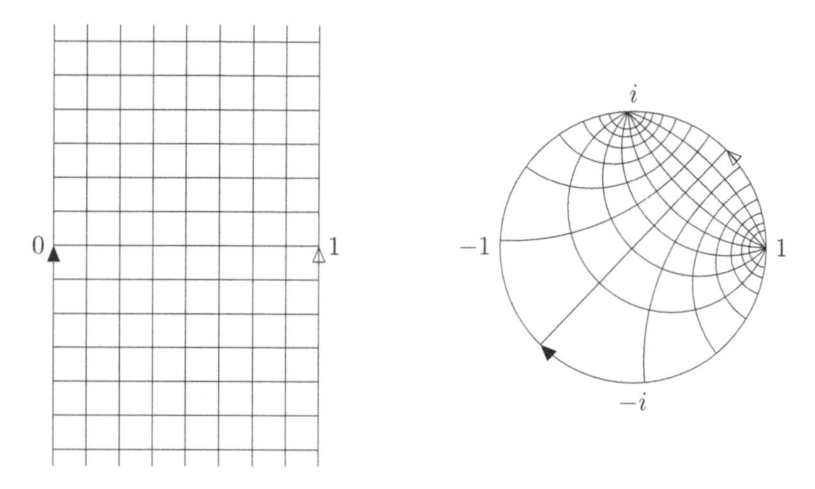

Figure 8.1. The conformal map $\chi_\theta : z \mapsto \frac{e^{i\pi z} - e^{i\pi\theta}}{e^{i\pi z} - e^{-i\pi\theta}}$ for $\theta = 1/4$.

For instance (see e.g. the proof of Theorem 8.3) we will show that any bounded harmonic function on S admits non-tangential limits at a.e. point of ∂S. Here a.e. refers to the Lebesgue measure on ∂S, but actually $\chi_\theta^{-1} : D \to S$ takes Lebesgue measure on ∂D (or P_D^z for any $z \in D$) to a measure equivalent to the Lebesgue measure on ∂S, and hence the negligible sets are the same for all these measures.

Before we explicitly do this, it is preferable to place the discussion in a broader framework:

Given a connected domain $G \subset \mathbb{C}$ with smooth boundary, it is a classical fact that there exists, for any z in G, a unique probability measure P_G^z on ∂G such that

$$f(z) = \int_{\partial G} f \, dP_G^z$$

for any bounded harmonic function on G that extends continuously on $\overline{G}$. The unicity of P_S^z is due to the fact that we can solve the Dirichlet problem for G (i.e. extend harmonically inside G any bounded continuous function on ∂G).

If we are given a conformal equivalence (i.e. an analytic homeomorphism) $\varphi\colon\ G_1 \to G_2$ that extends continuously to a homeomorphism, up to a few negligible exceptional points, from ∂G_1 to ∂G_2, then, by the unicity of P_G^z, the image measure of $P_{G_1}^z$ under φ is $P_{G_2}^{\varphi(z)}$, i.e. we have

$$P_{G_2}^{\varphi(z)} = \varphi(P_{G_1}^z). \tag{8.2}$$

For any θ in S we denote by μ_θ the harmonic measure of θ relative to ∂S, i.e. μ_θ is the unique probability on ∂S such that for any continuous bounded $f\colon \overline{S} \to \mathbb{C}$ that is analytic (or merely harmonic) inside S, we have

$$f(\theta) = \int_{\partial S} f(\xi)\mu_\theta(d\xi). \tag{8.3}$$

If $0 < \theta < 1$, we have $\mu_\theta(\partial_0) = 1 - \theta$ and $\mu_\theta(\partial_1) = \theta$, so that there are probabilities μ_θ^0 and μ_θ^1 supported respectively by ∂_0 and ∂_1 such that

$$\mu_\theta = (1-\theta)\mu_\theta^0 + \theta\mu_\theta^1.$$

Then for any Banach space valued bounded continuous function $f\colon \overline{S} \to B$ that is analytic (or merely harmonic) on S we may rewrite (8.3) as

$$f(\theta) = (1-\theta)\int_{\partial_0} f(\xi)\mu_\theta^0(d\xi) + \theta\int_{\partial_1} f(\xi)\mu_\theta^1(d\xi). \tag{8.4}$$

Using the conformal equivalence between S and the unit disc D, given e.g. by the preceding mapping $\chi_\theta\colon S \to D$ (defined in (8.1)), we have more generally:

Theorem 8.3. *Let $f\colon S \to B$ be a bounded harmonic function admitting non-tangential limits $f(\xi)$ at a.e. point ξ on ∂S. Then f is the 'Poisson integral' of $\xi \mapsto f(\xi)$ i.e. we have*

$$\forall z \in S \qquad f(z) = \int_{\partial S} f(\xi)\, dP_S^z(\xi). \tag{8.5}$$

Moreover, if f is analytic on S we have

$$\mathrm{Log}\|f(z)\| \le \int_{\partial S} \mathrm{Log}\|f\|\, dP_S^z. \tag{8.6}$$

In particular, since $P_S^\theta \equiv \mu_\theta = (1-\theta)\mu_\theta^0 + \theta\mu_\theta^1$ we have for any $0 < \theta < 1$:

$$\mathrm{Log}\|f(\theta)\| \le (1-\theta)\int_{\partial_0} \mathrm{Log}\|f\| d\mu_\theta^0 + \theta\int_{\partial_1} \mathrm{Log}\|f\| d\mu_\theta^1, \tag{8.7}$$

$$\|f(\theta)\| \le \left(\int_{\partial_0} \|f(\xi)\|^p \mu_\theta^0(d\xi)\right)^{\frac{1-\theta}{p}} \left(\int_{\partial_1} \|f(\xi)\|^p \mu_\theta^1(d\xi)\right)^{\frac{\theta}{p}}. \tag{8.8}$$

Proof. The analogue of this for the disc is part of Theorem 3.13. To deduce the case of S, we will apply this to the composition $g = f \circ \chi_\theta{}^{-1}\colon D \to B$, and note that χ_θ preserves the non-tangential character of the limits on the boundary (see Remark 3.20). Fix $z \in S$. Let $w = \chi_\theta(z) \in D$. By Theorem 3.13, g admits non-tangential limits $g(\eta)$ for a.e. $\eta \in \partial D$, and we have $g(w) = \int g(\eta)P_D^w(d\eta)$.

Therefore, $f = g \circ \chi_\theta$ admits non-tangential limits $f(\xi)$ for a.e. $\xi \in \partial S$ and we have $f(z) = g(w) = \int g(\eta)P_D^w(d\eta)$. But by (8.2) we have $P_D^w = \chi_\theta(P_S^z)$, and hence $f(z) = \int g(\eta)P_D^w(d\eta) = \int g(\chi_\theta(\xi))P_S^z(d\xi) = \int f(\xi)P_S^z(d\xi)$. This proves (8.5). If f is analytic, we use Proposition 4.7 and (4.1). Then (8.6) (and hence (8.7)) follows similarly from $\mathrm{Log}\|g(w)\| \le \int_{\partial D} \mathrm{Log}\|g\|\, dP_D^w$. Since the function $t \mapsto \exp pt$ is convex and increasing on $\mathbb{R}$ for any $p > 0$, applying this function to (8.7) yields (8.8). □

The following celebrated and classical three-line lemma is the crucial tool to develop complex interpolation (the extension to the B-valued case is straightforward).

Lemma 8.4 (Three line-lemma). *Let $f\colon \overline{S} \to B$ be a bounded analytic function on S admitting non-tangential limits a.e. on ∂S. Then for any $0 < \theta < 1$*

$$\|f(\theta)\| \le (\sup\nolimits_{\partial_0} \|f\|)^{1-\theta}(\sup\nolimits_{\partial_1} \|f\|)^{\theta}. \tag{8.9}$$

Proof. Note that (8.8) implies (8.9), which corresponds to the case $p = \infty$. □

Remark 8.5. Clearly, for any bounded harmonic function $f\colon S \to B$ as earlier we have for any $t \in \mathbb{R}$ and $z \in S$

$$f(z - it) = \int_{\partial S} f(w)\, dP_S^{z-it}(w) = \int_{\partial S} f(w - it)\, dP_S^z(w) = \int_{\partial S} f(w)\, dP_S^z(w + it)$$

and hence by the unicity of P_S^z

$$dP_S^{z-it}(.) = dP_S^z(. + it).$$

Remark. The exact value of the densities of the probability measures μ_θ^0 and μ_θ^1 on ∂_0 and ∂_1 respectively are not relevant for anything that follows. Nevertheless, we feel we should mention them here, for the interested reader. We have

$$\mu_\theta^0(d\xi) = (1-\theta)^{-1}\frac{\sin(\pi\theta)}{\cosh(\pi\xi) - \cos(\pi\theta)}d\xi/2$$

$$\mu_\theta^1(d\xi) = (\theta)^{-1}\frac{\sin(\pi\theta)}{\cosh(\pi\xi) + \cos(\pi\theta)}d\xi/2.$$

8.2 The complex interpolation method

In this section all Banach spaces will be over the complex field of scalars $\mathbb{C}$.

We will denote by $\mathcal{F}(B_0, B_1)$ (or often simply by $\mathcal{F}$) the space of all functions f in $C_b(\overline{S}; B_0 + B_1)$ such that $f_{|S}\colon S \to B_0 + B_1$ is analytic and

$f_{|\partial_0} \in C_b(\partial_0, B_0)$, $f_{|\partial_1} \in C_b(\partial_1, B_1)$. We set

$$\|f\|_{\mathcal{F}} = \max_{j=0,1}\{\|f_{|\partial_j}\|_{C_b(\partial_j;B_j)}\}.$$

Let $0 < \theta < 1$. The complex interpolation space $(B_0, B_1)_\theta$ is defined as follows:

$$(B_0, B_1)_\theta = \{x \in B_0 + B_1 \mid \exists f \in \mathcal{F}\ f(\theta) = x\}.$$

It is equipped with the norm

$$\|x\|_{(B_0,B_1)_\theta} = \inf\{\|f\|_{\mathcal{F}} \mid f \in \mathcal{F}\ f(\theta) = x\}. \tag{8.10}$$

It is easy to check that it can be identified isometrically with the quotient of $\mathcal{F}$ by the closed subspace $\{f \in \mathcal{F} \mid f(\theta) = 0\}$, and hence it is a Banach space.

Remark 8.6. Let us denote by $\mathcal{F}_0(B_0, B_1)$ (or simply by $\mathcal{F}_0$) the subspace of those f in $\mathcal{F}$ such that $\lim_{|y|\to\infty} \|f(j+iy)\|_{B_j} = 0$ for $j = 0, 1$. For any f in $\mathcal{F}$, let $f_\delta(z) = e^{\delta z^2} f(z)$. Then it is easy to check that $f_\delta \in \mathcal{F}_0$, $\|f_\delta\|_{\mathcal{F}} \le e^\delta \|f\|_{\mathcal{F}}$ and $f_\delta(z) \to f(z)$ uniformly over all compact subsets of S when $\delta \to 0$. Since $f(\theta) = e^{-\delta\theta^2} f_\delta(\theta)$, it is easy to deduce from this that the definition of $(B_0, B_1)_\theta$ is unchanged if we replace $\mathcal{F}$ by $\mathcal{F}_0$. Moreover, if f already lies in $\mathcal{F}_0$, it is easy to check that $\|f - f_\delta\|_{\mathcal{F}} \to 0$.

Lemma 8.7. *We have*

$$B_0 \cap B_1 \subset (B_0, B_1)_\theta \subset B_0 + B_1.$$

Moreover, for any x in $B_0 \cap B_1$

$$\|x\|_{B_0+B_1} \le \|x\|_{(B_0,B_1)_\theta} \le \|x\|_{B_0}^{1-\theta} \|x\|_{B_1}^{\theta} \le \|x\|_{B_0\cap B_1}. \tag{8.11}$$

Proof. Assume $x = f(\theta)$. By (8.4) we have $x = f(\theta) = (1-\theta)x_0 + \theta x_1$ with $x_0 \in B_0$, $x_1 \in B_1$ and $\|x_j\|_{B_j} \le \|f\|_{\mathcal{F}}$ $(j = 0, 1)$, so that

$$\|x\|_{B_0+B_1} \le (1-\theta)\|f\|_{\mathcal{F}} + \theta\|f\|_{\mathcal{F}} \le \|f\|_{\mathcal{F}}.$$

This shows $(B_0, B_1)_\theta \subset B_0 + B_1$ with $\|x\|_{B_0+B_1} \le \|x\|_{(B_0,B_1)_\theta}$. The inclusion $B_0 \cap B_1 \subset (B_0, B_1)_\theta$ is obvious (just take a constant function f). More precisely if we set for $0 \ne x \in B_0 \cap B_1$

$$f(z) = x(\|x\|_{B_0}^{1-z} \|x\|_{B_1}^{z})^{-1}$$

then $f \in \mathcal{F}$ and $\|f\|_{\mathcal{F}} \le 1$, so that $\|f(\theta)\|_{(B_0,B_1)_\theta} \le 1$. Therefore

$$\|x\|_{(B_0,B_1)_\theta} \le \|x\|_{B_0}^{1-\theta} \|x\|_{B_1}^{\theta}.$$ □

Remarks.

(i) If $B_0 = B_1$ then $(B_0, B_1)_\theta = B_0 = B_1$.
(ii) More generally if $B_1 \subset B_0$ with $\|\cdot\|_{B_0} \le \|\cdot\|_{B_1}$, then $B_1 \subset (B_0, B_1)_\theta \subset B_0$, and $\|\cdot\|_{B_0} \le \|\cdot\|_{(B_0,B_1)_\theta} \le \|\cdot\|_{B_1}$. The last two facts are obvious consequences of the preceding lemma.
(iii) Note that $(B_1, B_0)_\theta = (B_0, B_1)_{1-\theta}$ with identical norms. This is immediate since $f(z) \to f(1-z)$ defines an isometric isomorphism $\mathcal{F}(B_0, B_1) \to \mathcal{F}(B_1, B_0)$.

The fundamental 'interpolation theorem' is the following basic property of $(B_0, B_1)_\theta$. Before stating it, we introduce some terminology. Let $T_0\colon\ B_0 \to C_0$ and $T_1\colon\ B_1 \to C_1$ be bounded linear operators. We say that T_0 and T_1 are 'essentially the same' if $T_0(x) = T_1(x)$ for any x in $B_0 \cap B_1$ (the equality being meant of course in $C_0 + C_1$). Then, there is clearly an operator

$$T\colon\ B_0 + B_1 \to C_0 + C_1$$

extending both T_0 and T_1. Note that such a T cannot exist unless T_0 and T_1 are 'essentially the same'.

Theorem 8.8. *Fix $0 < \theta < 1$. Let (B_0, B_1) be as earlier. Let (C_0, C_1) be another compatible pair of Banach spaces. Let $T_0\colon\ B_0 \to C_0$ and $T_1\colon\ B_1 \to C_1$ be bounded linear operators that are 'essentially the same'. Then, the resulting operator $T\colon\ B_0 + B_1 \to C_0 + C_1$ takes $(B_0, B_1)_\theta$ to $(C_0, C_1)_\theta$ and its restriction $T_\theta\colon\ (B_0, B_1)_\theta \to (C_0, C_1)_\theta$ satisfies*

$$\|T_\theta\| \le \|T_0\|^{1-\theta} \|T_1\|^\theta. \tag{8.12}$$

Proof. Assume first $\|T_0\| \le 1$ and $\|T_1\| \le 1$. We can clearly define a bounded operator $T_\Sigma\colon\ B_0 + B_1 \to C_0 + C_1$ that is essentially the same as T_0 or T_1, and $\|T_\Sigma\| \le 1$. Fix $x \in (B_0, B_1)_\theta$. Consider $f \in \mathcal{F}(B_0, B_1)$ such that $f(\theta) = x$. Clearly $g\colon\ z \mapsto T_\Sigma f(z) \in \mathcal{F}(C_0, C_1)$ and $\|g\|_\mathcal{F} \le \|f\|_\mathcal{F}$. Therefore, since $x = T_\Sigma f(\theta) = g(\theta)$ we find

$$\|T_\Sigma x\|_{(C_0,C_1)_\theta} \le \|x\|_{(B_0,B_1)_\theta}.$$

This shows that T_θ is well defined as the restriction of T_Σ to $(B_0, B_1)_\theta$. Assume $\|x\|_{(B_0,B_1)_\theta} < 1$. Let $f \in \mathcal{F}$ be such that $\|f\|_\mathcal{F} < 1$ and $f(\theta) = x$. To verify (8.12), we replace f by the function

$$h\colon\ z \mapsto (\|T_0\|^{1-z} \|T_1\|^z)^{-1} f(z).$$

It is easy to check that $z \mapsto T_\Sigma h(z)$ is in the unit ball of $\mathcal{F}(C_0, C_1)$. Hence $\|T_\Sigma h(\theta)\|_{(C_0,C_1)_\theta} \le 1$, or equivalently

$$\|T_\theta(x)\|_{(C_0,C_1)_\theta} \le \|T_0\|^{1-\theta}\|T_1\|^{\theta}.$$

□

A well-known drawback of all interpolation methods is that interpolation between subspaces of B_0, B_1 does not yield in general a subspace of $(B_0, B_1)_\theta$ (see Remark 8.17 for an example). However, if the subspaces are 'simultaneously complemented' as in the following statement, then all goes well.

Lemma 8.9. *Let $\mathcal{S} \subset B_0 + B_1$ be a closed subspace. Let $A_0 = B_0 \cap \mathcal{S} \subset B_0$ and $A_1 = B_1 \cap \mathcal{S} \subset B_1$ with the norms induced respectively by B_0 and B_1. Assume that there is a bounded linear projection $P\colon\ B_0 + B_1 \to \mathcal{S}$ such that $P(B_0) \subset A_0$, $P(B_1) \subset A_1$ with $\|P_{|B_0}\colon\ B_0 \to A_0\| = c_0$, $\|P_{|B_1}\colon\ B_1 \to A_1\| = c_1$. Then*

$$(A_0, A_1)_\theta = (B_0, B_1)_\theta \cap \mathcal{S} \tag{8.13}$$

and we have

$$\forall x \in (B_0, B_1)_\theta \cap \mathcal{S} \quad \|x\|_{(B_0,B_1)_\theta} \le \|x\|_{(A_0,A_1)_\theta} \le c_0^{1-\theta}c_1^{\theta}\|x\|_{(B_0,B_1)_\theta}. \tag{8.14}$$

In particular, if $c_0 = c_1 = 1$ then (8.13) is isometric.

Proof. The inclusions $(A_0, A_1)_\theta \subset \mathcal{S}$ and $(A_0, A_1)_\theta \subset (B_0, B_1)_\theta$ as well as the left-hand side of (8.14) are obvious. To prove the converse, let $x \in (B_0, B_1)_\theta \cap \mathcal{S}$. Then, by Theorem 8.8, $Px \in (A_0, A_1)_\theta$ with

$$\|Px\|_{(A_0,A_1)_\theta} \le c_0^{1-\theta}c_1^{\theta}\|x\|_{(B_0,B_1)_\theta}.$$

But, since $x \in \mathcal{S}$, $Px = x$. □

Remark 8.10. It will be useful to observe that the conclusion of the preceding lemma holds just as well (with the same proof) if for any $x \in \mathcal{S} \cap B_0 \cap B_1$ there is a linear mapping $P = P_x\colon\ B_0 + B_1 \to \mathcal{S}$ (allowed to depend on x) such that $P_x(x) = x$, $P(B_0) \subset B_0 \cap \mathcal{S}$, and $P(B_1) \subseteq B_1 \cap \mathcal{S}$ with $\|P_{|B_0}\colon\ B_0 \to B_0 \cap \mathcal{S}\| = c_0$, $\|P_{|B_1}\colon\ B_1 \to B_1 \cap \mathcal{S}\| = c_1$.

The following fact is very useful. It shows that, for $x \in B_0 \cap B_1$ we can compute the norm of x in $(B_0, B_1)_\theta$ using very simple functions among those in $\mathcal{F}_0$, namely the functions f in $\mathcal{G}$, which form a dense subset, but it is crucial in (8.15) to have $f(\theta) = x$ and not just $f(\theta)$ close to x.

Lemma 8.11. *$B_0 \cap B_1$ is dense in $(B_0, B_1)_\theta$ $(0 < \theta < 1)$. Moreover, for any x in $B_0 \cap B_1$ we have*

$$\|x\|_{(B_0,B_1)_\theta} = \inf\{\|f\|_{\mathcal{F}} \mid f \in \mathcal{G}\ f(\theta) = x\} \tag{8.15}$$

where $\mathcal{G} \subset \mathcal{F}$ *is the collection of all functions* $f\colon\ \overline{S} \to B_0 \cap B_1$ *of the form* $f(z) = \sum_1^N f_k(z)b_k$ *where* $b_k \in B_0 \cap B_1$ *and* $f_k \in \mathcal{F}(\mathbb{C}, \mathbb{C})$ *are such that* $\lim_{z\in\bar{S}, \Im(z)\to\pm\infty} |f_k(z)| = 0$*. Furthermore,* $\mathcal{G}$ *is dense in* $\mathcal{F}_0(B_0, B_1)$*.*

To prove this, we will need the following.

Lemma 8.12. *Let* $g \in \mathcal{F}$ *and* $T > 0$ *be such that* $g(z) = g(z + iT)$ *for all* z *in* $\overline{S}$*. Then for any* $\varepsilon > 0$ *there are elements* b_n *in* $B_0 \cap B_1$ *and* N *such that*

$$\left\| g - \sum\nolimits_{-N}^{N} b_n e^{2\pi n z/T} \right\|_{\mathcal{F}} < \varepsilon.$$

In particular, there is b *in* $B_0 \cap B_1$ *such that*

$$\|g(\theta) - b\|_{(B_0,B_1)_\theta} < \varepsilon.$$

Proof. A priori, since $y \to g(z + iy)$ is T-periodic we can write (only as a formal Fourier series for the moment)

$$g(x + iy) = \sum\nolimits_{n\in\mathbb{Z}} a_n(x) e^{2\pi i n y/T}$$

where $a_n(x) = \frac{1}{T}\int_0^T g(x + iy)e^{-2\pi i n y/T}\, dy$ or equivalently by periodicity for any integer $N > 0$

$$a_n(x) = \frac{1}{2NT}\int_{-NT}^{NT} g(x + iy)e^{-2\pi i n y/T}\, dy.$$

Let $b_n(x) = a_n(x)e^{-2\pi n x/T}$. Then we have

$$\forall z = x + iy \qquad\qquad g(z) = \sum\nolimits_{n\in\mathbb{Z}} b_n(x)e^{2\pi n z/T},$$

and hence we may write also

$$b_n(x) = \frac{1}{2iNT}\int_{-iNT}^{iNT} g(z)e^{-2\pi n z/T}\, dz.$$

We claim that actually $(b_n(x))_{n\in\mathbb{Z}}$ does not depend on $x \in [0, 1]$.

Indeed, consider the rectangle with vertical sides $[-iNT, iNT]$ and $[x - iNT, x + iNT]$. Let γ be the boundary path of that rectangle. By analyticity, we have

$$\frac{1}{2NiT}\int_\gamma g(z)e^{-2\pi n z/T}\, dz = 0$$

but the vertical sides contribute to this

$$b_n(x) - b_n(0)$$

while the horizontal sides γ_+ and γ_- contribute:

$$\frac{1}{2NT}\left\| \int_{\gamma_+} + \int_{\gamma_-} \right\|_{B_0+B_1} \to 0 \quad \text{when} \quad N \to \infty.$$

Thus we conclude $b_n(x) = b_n(0)$, and analogously $b_n(x) = b_n(1)$. Let $b_n = b_n(x) = b_n(0) = b_n(1)$. Since $b_n(0) \in B_0$ and $b_n(1) \in B_1$, we have $b_n \in B_0 \cap B_1$. Let $g_N(z) = \sum_{|n|\le N} b_n \left(1 - \frac{|n|}{N}\right) e^{2\pi nz/T}$. By a well-known property of the Fejer kernel,

$$\sup_{y\in\mathbb{R}} \|(g - g_N)(j + iy)\|_{B_j} \to 0,$$

hence we conclude $\|g - g_N\|_{\mathcal{F}} \to 0$. □

Proof of Lemma 8.11. We will show that $\mathcal{G}$ is dense in $\mathcal{F}_0$. We start with f in $\mathcal{F}_0$, with $\|f\|_{\mathcal{F}} \le 1$, and again let $f_\delta(z) = e^{\delta z^2} f(z)$. It is easy to see that $\|f_\delta - f\|_{\mathcal{F}} \to 0$ when $\delta \to 0$, so we may replace f by f_δ if we wish. We introduce a periodic function g_T in $\mathcal{F}$ as follows (here $T \ge 1$):

$$g_T(z) = \sum_{k\in\mathbb{Z}} f_\delta(z + ikT).$$

Note that $g_T \in \mathcal{F}$. Indeed, since $\|f\|_{\mathcal{F}} \le 1$, we have for any $z = x + iy$ in $\overline{S}$

$$\max\{\|g_T(iy)\|_{B_0}, \|g_T(1 + iy)\|_{B_1}, \|g_T(x + iy)\|_{B_0+B_1}\} \le e^{\delta} \sum_{k\in\mathbb{Z}} e^{-\delta(y+kT)^2}$$

and the series on the right-hand side obviously converges uniformly over the set $\{y \mid 0 \le y \le T\}$; but since it is periodic of period T, it converges for any y and is bounded by

$$C(\delta, T) = e^{\delta} \left(\sum_{k\in\mathbb{Z}} e^{-\delta||k|-1|^2 T^2} + 2\right). \tag{8.16}$$

This shows that $g_T \in \mathcal{F}$ and $\|g_T\|_{\mathcal{F}} \le C(\delta, T)$. Moreover, we have clearly

$$g_T(z) - f_\delta(z) = \sum_{k\neq 0} f_\delta(z + ikT)$$

and hence we have

$$\lim \|g_T(z) - f_\delta(z)\|_{B_0+B_1} = 0$$

uniformly on compact subsets of $\overline{S}$ when $T \to \infty$. But we will need a more precise estimate. For $0 < s < 1$, we let

$$F_{s,T}(z) = e^{sz^2}(g_T(z) - f_\delta(z)).$$

We now claim that

$$\lim_{(s,T)\to(0,\infty)} \|F_{s,T}\|_{\mathcal{F}} = 0.$$

Note that if we set $s = 1/R$ then

$$|y| = |\mathrm{Im}(z)| \ge R \Rightarrow |e^{sz^2}| \le e^{1-R}.$$

Hence we may concentrate on y's in $[-R, R]$. We have for all $T > 2R$

$$\max_{j=0,1} \sup\nolimits_{|y|\le R}\{\|(g_T - f_\delta)(j+iy)\|_{B_j}\} \le C_0(\delta, T)$$

where $C_0(\delta, T) = e^\delta \sum_{k\neq 0} e^{-\delta(|k|T-R)^2}$. Note that $C_0(\delta, T) \to 0$ when $T \to \infty$ and that $|e^{sz^2}| \le e^s \le e$ for all z in $\overline{S}$. Therefore, we obtain taking $s = 1/R$

$$\|F_{s,T}\|_{\mathcal{F}} \le eC_0(\delta, T) + e^{1-R}(\|g_T\|_{\mathcal{F}} + \|f_\delta\|_{\mathcal{F}})$$

and this proves our claim since $\|g_T\|_{\mathcal{F}} \le C(\delta, T)$ remains bounded when $T \to \infty$. Note that obviously (since $f_\delta \in \mathcal{F}_0$)

$$e^{sz^2} f_\delta(z) \to f_\delta(z) \quad \text{in} \quad \mathcal{F}$$

when $s \to 0$. Therefore

$$\lim_{(s,T)\to(0,\infty)} \|e^{sz^2} g_T - f_\delta\|_{\mathcal{F}} = 0. \tag{8.17}$$

A fortiori, we have

$$\lim_{(s,T)\to(0,\infty)} \|e^{s\theta^2} g_T(\theta) - e^{\delta\theta^2} x\|_{(B_0,B_1)_\theta} = 0.$$

By the preceding lemma, this shows that $B_0 \cap B_1$ is dense in $(B_0, B_1)_\theta$.

Applying the preceding lemma again to g_T, we find that the function $z \mapsto e^{sz^2} g_T(z)$ can be approximated by functions in $\mathcal{G}$ in the $\mathcal{F}$-norm. So the same is true for f_δ. Since f_δ tends to f in $\mathcal{F}$ when $\delta \to 0$, (8.17) shows that $\mathcal{G}$ is dense in $\mathcal{F}_0(B_0, B_1)$.

We can now complete the proof: Let $x \in B_0 \cap B_1$. By Remark 8.6, given $\varepsilon > 0$ we can find f in $\mathcal{F}_0$ such that $f(\theta) = x$ and $\|f\|_{\mathcal{F}} < \|x\|_{(B_0,B_1)_\theta} + \varepsilon$. By the first part of the proof, we can find f_0 in $\mathcal{G}$ such that $\|f - f_0\|_{\mathcal{F}} < \varepsilon$. However, we want the function f_0 to be *equal* to f at the point θ, and this requires a more refined procedure. We will produce instead a function $f_\bullet$ with the desired property. Let χ_θ be the function appearing in (8.1). Note that χ_θ is an inner function (a 'Blaschke factor' for S) in $\mathcal{F}(\mathbb{C}, \mathbb{C})$ such that $\chi_\theta(\theta) = 0$, $\chi_\theta'(\theta) \neq 0$ and $|\chi_\theta(z)| = 1\ \forall z \in \partial S$. Since $z \to F(z) = f(z) - e^{z^2-\theta^2}x$ vanishes when $z = \theta$, and belongs to $\mathcal{F}_0(B_0, B_1)$ (because $x \in B_0 \cap B_1$), we have $F(z) = \chi_\theta h$ with $h \in \mathcal{F}_0(B_0, B_1)$. This holds because, in addition to having θ as a simple zero, χ_θ is such that $|\chi_\theta|^{-1}$ is bounded outside any compact neighbourhood of θ, and has modulus 1 on ∂S. One can also argue by conformal equivalence that on D the composition $z \mapsto F\chi_\theta^{-1}$ vanishes at the origin and hence is divisible by z. Choose $g \in \mathcal{G}$ such that $\|g - h\|_{\mathcal{F}} < \varepsilon$. Note that

$\|\chi_\theta(g-h)\|_{\mathcal{F}} = \|g-h\|_{\mathcal{F}} < \varepsilon$. Then let $f_\bullet(z) = e^{z^2-\theta^2}x + \chi_\theta g$. Clearly $f_\bullet \in \mathcal{G}$, $f_\bullet(\theta) = x$ and finally

$$\begin{aligned}\|f_\bullet\|_{\mathcal{F}} &\le \|e^{z^2-\theta^2}x + \chi_\theta h\|_{\mathcal{F}} + \|g-h\|_{\mathcal{F}}\\ &\le \|f\|_{\mathcal{F}} + \varepsilon\\ &\le \|x\|_{(B_0,B_1)_\theta} + 2\varepsilon.\end{aligned}$$

□

Remark 8.13. Let B_0, B_1 be a compatible couple. Let $A_j \subset B_j$ be the closure of $B_0 \cap B_1$ in B_j $(j = 0, 1)$. Then the following isometric identities hold

$$\mathcal{F}_0(B_0, B_1) = \mathcal{F}_0(A_0, A_1), \tag{8.18}$$

$$(B_0, B_1)_\theta = (A_0, A_1)_\theta. \tag{8.19}$$

Indeed, we have obviously $A_0 \cap A_1 = B_0 \cap B_1$ and hence $\mathcal{G}(B_0, B_1) = \mathcal{G}(A_0, A_1)$. Thus both identities follow from Lemma 8.11 and Remark 8.6.

Remark 8.14. For future reference, we record here an elementary observation: Let β denote the open unit ball of a Banach space B. Let $\Xi \subset \beta$ be a dense subset. Then for any $0 < \varepsilon < 1$ and any $x \in \beta$ there are $s_0, s_1, \ldots$ in Ξ such that $x = \sum_0^\infty \varepsilon^n s_n$. Indeed, there is $s_0 \in \Xi$ such that $\|x - s_0\| < \varepsilon$, then $s_1 \in \Xi$ such that $\|\varepsilon^{-1}(x - s_0) - s_1\| < \varepsilon$, and so on...

The main results on the complex interpolation of L_p-spaces are the next two statements. For brevity we denote here simply by L_p the space $L_p(\Omega, \mathcal{A}, m)$ relative to an arbitrary measure space.

Theorem 8.15. *Let $1 \le p_0, p_1 \le \infty$. Then for any $0 < \theta < 1$ we have*

$$(L_{p_0}, L_{p_1})_\theta = L_{p_\theta} \tag{8.20}$$

with identical norms, where p_θ is defined by the equality $p_\theta^{-1} = (1-\theta)p_0^{-1} + \theta p_1^{-1}$.

Proof. By Lemma 8.11, it suffices to show that $\|x\|_{(L_{p_0}, L_{p_1})_\theta} = \|x\|_{p_\theta}$ for any x in $L_{p_0} \cap L_{p_1}$ or even for any simple function x in $L_{p_0} \cap L_{p_1}$ (since simple functions are obviously dense in $L_{p_0} \cap L_{p_1}$ and in L_{p_θ}). Furthermore, we clearly may replace Ω by the support of $x \in L_{p_0} \cap L_{p_1}$ (which has finite measure), so we may as well assume that $m(\Omega) < \infty$ and that x is a simple function such that $|x| \ge \varepsilon$ for some $\varepsilon > 0$. Let

$$\alpha(z) = p_\theta[(1-z)p_0^{-1} + zp_1^{-1}].$$

Note $\alpha(\theta) = 1$, $\Re\alpha(it) = p_\theta/p_0$ and $\Re\alpha(1+it) = p_\theta/p_1$ for all $t \in \mathbb{R}$. Then the function f defined by $f(z) = \arg(x)|x|^{\alpha(z)}$ (where $\arg(x) = x|x|^{-1}$) is in

$\mathcal{F}(L_{p_0}, L_{p_1})$. Note that $f(\theta) = x$. Assume $\|x\|_{p_\theta} \le 1$. Then $\|f\|_{\mathcal{F}} \le 1$, and hence $\|x\|_{(L_{p_0}, L_{p_1})_\theta} \le 1$. This shows

$$\|x\|_{(L_{p_0}, L_{p_1})_\theta} \le \|x\|_{p_\theta}. \tag{8.21}$$

For the converse we will invoke Theorem 8.11. By density, it suffices to show that $\|x\|_{p_\theta} \le 1$ for any $x \in L_{p_0} \cap L_{p_1}$ in the open unit ball of $(L_{p_0}, L_{p_1})_\theta$. Let $F \in \mathcal{G} = \mathcal{G}(L_{p_0}, L_{p_1})$ be such that $F(\theta) = x$ and $\|F\|_{\mathcal{F}} < 1$. By the simple form of the functions in $\mathcal{G}$, for fixed $\omega \in \Omega$ we may apply (8.8) (with say $p = 1$) to the function $F(z)(\omega)$. This yields $|F(\theta)(\omega)| \le |f_0(\omega)|^{1-\theta} |f_1(\omega)|^\theta$ with $f_j(\omega) = \int_{\partial_j} |F(\xi)(\omega)| \mu_\theta^j(d\xi)$, and since μ_θ^j is a probability and $\|F\|_{\mathcal{F}} < 1$, f_j is in the unit ball of L_{p_j} by Jensen's inequality. By Hölder's inequality, this shows $\|x\|_{p_\theta} = \|F(\theta)\|_{p_\theta} \le 1$. □

We can now derive the famous and classical Riesz-Thorin interpolation theorem, which is at the root of interpolation theory.

Corollary 8.16. *Let $1 \le p_0, p_1, q_0, q_1 \le \infty$. Let T be an operator that is simultaneously of norm $\le c_j$ from L_{p_j} to L_{q_j}, for $j = 0, 1$, or more precisely such that*

$$\forall f \in L_{p_0} \cap L_{p_1} \qquad \|T(f)\|_{q_0} \le c_0 \|f\|_{p_0} \quad \text{and} \quad \|T(f)\|_{q_1} \le c_1 \|f\|_{p_1}.$$

Then T extends to a bounded operator from L_{p_θ} to L_{q_θ} with

$$\|T : L_{p_\theta} \to L_{q_\theta}\| \le c_0{}^{1-\theta} c_1{}^\theta,$$

where

$$\frac{1}{p_\theta} = \frac{1-\theta}{p_0} + \frac{\theta}{p_1} \quad \text{and} \quad \frac{1}{q_\theta} = \frac{1-\theta}{q_0} + \frac{\theta}{q_1}.$$

Proof. Since $L_{p_0} \cap L_{p_1}$ is dense in both L_{p_0} and L_{p_1}, T unambiguously extends to an operator that is simultaneously of norm ≤ 1 on L_{p_0} and on L_{p_1}, to which we may apply Theorem 8.8. The latter shows that T must have norm $\le c_0{}^{1-\theta} c_1{}^\theta$ from $(L_{p_0}, L_{p_1})_\theta$ to $(L_{q_0}, L_{q_1})_\theta$ for any $0 < \theta < 1$. By Theorem 8.15, we obtain the announced statement. □

Remark 8.17. Consider the pair $B_0 = L_1, B_1 = L_\infty$ over (Δ, ν) (recall (Δ, ν) is defined in (5.13)). Then, $(B_0, B_1)_\theta = L_{p_\theta}$ $(0 < \theta < 1)$. Let $\mathcal{S}$ be the subspace spanned in $L_1 + L_\infty$ by the sequence of coordinates (ε_n) on Δ (i.e. essentially the 'Rademacher functions'). Let $A_j = B_j \cap \mathcal{S}$ $(j = 0, 1)$. Then, on one hand $A_0 = B_0 \cap \mathcal{S} \simeq \ell_2$, and also (since $p_\theta < \infty$) $(B_0, B_1)_\theta \cap \mathcal{S} \simeq \ell_2$ by the Khintchin inequality (5.7), but on the other hand $A_1 = B_1 \cap \mathcal{S} \simeq \ell_1$.Therefore, in sharp contrast with Lemma 8.9, we have in this case:

$$(\ell_2, \ell_1)_\theta = (A_0, A_1)_\theta \neq (B_0, B_1)_\theta \cap \mathcal{S} = \ell_2.$$

Remark 8.18. Calderón [177] proved a far reaching generalization of Theorem 8.15 to pairs of Banach lattices over a σ-finite measure space $(\Omega, \mathcal{A}, m)$. We will now briefly describe this mostly without proof. We refer the reader to [51, p. 240] for complete details, and to [66] for a full presentation of Banach lattice Theory.

As usual, we denote by $L_0(\Omega, \mathcal{A}, m)$, or for short $L_0(m)$ the topological vector space of classes of measurable functions on $(\Omega, \mathcal{A}, m)$ (equipped with the topology of convergence in measure). In this remark all inequalities or equalities are meant to hold in $L_0(m)$ or equivalently a.e. For short, we will say that a Banach space B is a Banach lattice on $(\Omega, \mathcal{A}, m)$ if B is continuously included in $L_0(m)$ and if the following holds: If $f, g \in L_0(m)$ are such that $|g| \leq |f|$, then $f \in B$ implies $g \in B$ and $\|g\|_B \leq \|f\|_B$. (These are called ideal Banach lattices in [51].) We will restrict consideration to Banach lattices formed of complex valued functions. This is somehow irrelevant, because it is easy to pass by complexification from a real Banach lattice to a complex one (and conversely).

We will say that B has the domination property if any sequence $g_n \in B$ tending to 0 a.e. and such that $0 \leq g_n \leq f$ for some $f \in B$ must satisfy $\|g_n\|_B \to 0$. It is easy to see that this is equivalent to the following simpler reformulation: For any $f \in B$ and any decreasing sequence of measurable sets A_n with $m(A_n) \to 0$ we have $\||f|1_{A_n}\|_B \to 0$. (Such lattices are called 'regular' in [51, p. 244] and σ-order continuous in [57].)

Let (B_0, B_1) be a pair of Banach lattices over a common $(\Omega, \mathcal{A}, m)$. The inclusions in $L_0(m)$ allow us to view them as a compatible pair, and it is easy to show that $(B_0, B_1)_\theta$ (as well as $B_0 \cap B_1$ or $B_0 + B_1$) is also a Banach lattice over $(\Omega, \mathcal{A}, m)$. However, it turns out that there is an alternate very accurate description of $(B_0, B_1)_\theta$. Fix $0 < \theta < 1$. We define the space $B_0^{1-\theta} B_1^\theta$ as formed of the functions $f \in L_0(m)$ such that there are $f_0 \in B_0$ and $f_1 \in B_1$ such that $|f| = |f_0|^{1-\theta}|f_1|^\theta$, and we equip it with the norm

$$\|f\|_{B_0^{1-\theta}B_1^\theta} = \inf\{\|f_0\|_{B_0}^{1-\theta}\|f_1\|_{B_1}^\theta\} \tag{8.22}$$

where the infimum is over all possible $f_j \in B_j$ such that $|f| = |f_0|^{1-\theta}|f_1|^\theta$. It is easy to see that this definition remains unchanged if we replace $|f| = |f_0|^{1-\theta}|f_1|^\theta$ by $|f| \leq |f_0|^{1-\theta}|f_1|^\theta$. Then the concavity of $(x_0, x_1) \mapsto x_0^{1-\theta}x_1^\theta$ on $\mathbb{R}_+^2$ implies that (8.22) is subadditive. The resulting space is a Banach lattice over $(\Omega, \mathcal{A}, m)$, continuously included in $B_0 + B_1$ (indeed, setting $B = B_0 + B_1$ we have ${B_0}^{1-\theta}{B_1}^\theta \subset B^{1-\theta}B^\theta = B$).

For instance if $B_j = L_{p_j}$ we easily check that $B_0^{1-\theta}B_1^\theta = L_{p_\theta}$, with p_θ as before.

Calderón showed that this is a general phenomenon:

Theorem 8.19. *In the situation of the preceding remark, if the space $B_0^{1-\theta}B_1^\theta$ has the domination property, then we have an isometric identity*

$$(B_0, B_1)_\theta = B_0^{1-\theta}B_1^\theta. \tag{8.23}$$

Proof. Let f be in the open unit ball of $B_0^{1-\theta}B_1^\theta$, so that $|f| = |f_0|^{1-\theta}|f_1|^\theta$ with f_j in the the open unit ball of B_j. Let Ξ be the subset formed of those f for which in addition both $|f_0|$ and $|f_1|$ are bounded above and below by positive numbers on the support of f. The domination property implies that Ξ is norm dense in the unit ball of $B_0^{1-\theta}B_1^\theta$. Now consider $f \in \Xi$ with $|f_0|$ and $|f_1|$ as earlier. Since we may replace Ω by the support of f we may assume that $f > 0$ everywhere and that $|f_0|$ and $|f_1|$ are bounded above and below by positive numbers on Ω, and hence $z \mapsto (|f_1|/|f_0|)^z$ is an entire function with values in $L_\infty(m)$. Then, since the function $F(z) = |f_0|^{1-z}|f_1|^z$ lies in $\mathcal{F}(B_0, B_1)$, f lies in the unit ball of $(B_0, B_1)_\theta$. This shows that $\|f\|_{(B_0,B_1)_\theta} \le 1$ for any f in a dense subset (namely Ξ) of the unit ball of $B_0^{1-\theta}B_1^\theta$. By a standard argument (see Remark 8.14), it follows that we have a norm 1 inclusion $B_0^{1-\theta}B_1^\theta \subset (B_0, B_1)_\theta$.

For the converse we may argue exactly as in the second part of the proof of Theorem 8.15. We leave the easy details to the reader. □

It is an easy but useful observation that if one of the spaces B_0, B_1 has the domination property, then $B_0^{1-\theta}B_1^\theta$ also does and (8.23) holds. This observation follows from $\||f_0|^{1-\theta}|f_1|^\theta\|_{B_0^{1-\theta}B_1^\theta} \le \|f_0\|_{B_0}^{1-\theta}\|f_1\|_{B_1}^\theta$.

Note that if we denote $|U(B)| = \{|f| \mid f \in B, \|f\|_B < 1\}$, then we have a setwise equality

$$|U(B_0^{1-\theta}B_1^\theta)| = |U(B_0)|^{1-\theta}|U(B_1)|^\theta.$$

Thus we may think of $(B_0, B_1)_\theta = B_0^{1-\theta}B_1^\theta$ as a sort of 'geometric mean' of B_0 and B_1 (the arithmetic mean analogue is the basic tool in real interpolation).

We now turn to a generalization of (8.7).

Lemma 8.20. *For any f in $\mathcal{F}(B_0, B_1)$ we have*

$$\operatorname{Log}\|f(\theta)\|_{(B_0,B_1)_\theta} \le (1-\theta)\int_{\partial_0} \operatorname{Log}\|f(\xi)\|_{B_0}\mu_\theta^0(d\xi) + \theta\int_{\partial_1} \operatorname{Log}\|f(\xi)\|_{B_1}\mu_\theta^1(d\xi), \tag{8.24}$$

where the right-hand side may be equal to $-\infty$ (and then, of course, $f(\theta) = 0$). Moreover, for any $p_0 > 0$, $p_1 > 0$ we have

$$\|f(\theta)\|_{(B_0,B_1)_\theta} \le \|f\|_{L_{p_0}(\partial_0,\mu_\theta^0;B_0)}^{1-\theta}\|f\|_{L_{p_1}(\partial_1,\mu_\theta^1;B_1)}^\theta. \tag{8.25}$$

Consequently, for any x in $B_0 + B_1$, $\|x\|_{(B_0,B_1)_\theta}$ is equal to the infimum of the right-hand side of (8.25) *over all f in $\mathcal{F}(B_0, B_1)$ such that $f(\theta) = x$.*

Proof. The proof uses 'outer functions'. Fix $\varepsilon > 0$. Consider the function $\psi\colon\ \partial S \to \mathbb{R}$ equal to $\|f(\cdot)\|_{B_0} + \varepsilon$ on ∂_0 and equal to $\|f(\cdot)\|_{B_1} + \varepsilon$ on ∂_1. We will use an (outer) analytic function F such that $|F| = \psi$ on the boundary of S. In our situation, this is easy to produce because $\mathrm{Log}(\psi)$ is bounded and continuous, therefore there is a harmonic function $u\colon\ S \to \mathbb{R}$ (continuous on $\bar{S}$) extending $\mathrm{Log}(\psi)$ (see (ii) in Theorem 3.1). Being harmonic, u is the real part of an analytic function g, say $g(z) = u(z) + iv(z)$. Let $F(z) = \exp(g(z)) = \exp(u + iv)$. Note that

$$\forall \xi \in \partial_j \qquad |F(\xi)| = e^{u(\xi)} = \|f(\xi)\|_{B_j} + \varepsilon.$$

Now let $h(z) = f(z)e^{-g(z)} = f(z)F(z)^{-1}$. Clearly $h \in \mathcal{F}(B_0, B_1)$ but now

$$\forall \xi \in \partial_j \qquad \|h(\xi)\|_{B_j} = \|f(\xi)\|_{B_j} e^{-u(\xi)} \le 1$$

and hence

$$\|h(\theta)\|_{(B_0,B_1)_\theta} \le 1$$

which implies $\|f(\theta)\|_{(B_0,B_1)_\theta} \le |F(\theta)|$ and therefore by (8.7) (applied to F) we obtain

$$\mathrm{Log}\|f(\theta)\|_{(B_0,B_1)_\theta} \le (1-\theta)\int \mathrm{Log}(\|f(\xi)\|_{B_0} + \varepsilon)\mu_\theta^0(d\xi)$$
$$+\,\theta\int \mathrm{Log}(\|f(\xi)\|_{B_1} + \varepsilon)\mu_\theta^1(d\xi).$$

Letting $\varepsilon \to 0$, we obtain the announced result (8.24). The second inequality follows from the first one: Indeed, by Jensen's inequality and the convexity of the exponential function we have for $j = 0, 1$

$$\exp\left(\int \mathrm{Log}\|f(\xi)\|_{B_j}^{p_j}\mu_\theta^j(d\xi)\right) \le \int \|f(\xi)\|_{B_j}^{p_j}\mu_\theta^j(d\xi) = \|f\|_{L_{p_j}(\mu_\theta^j;B_j)}^{p_j}$$

and trivially

$$\exp\left(\int \mathrm{Log}\|f(\xi)\|_{B_j}\mu_\theta^j(d\xi)\right) = \left(\exp\int \mathrm{Log}\|f(\xi)\|_{B_j}^{p_j}\mu_\theta^j(d\xi)\right)^{1/p_j}$$

therefore (8.24) implies (8.25). □

In the next result, given a Banach space B and a measure space $(\Omega, \mathcal{A}, m)$, we denote simply by $L_p(B)$ the space $L_p(\Omega, \mathcal{A}, m; B)$. It will be convenient to introduce the space $L_\infty^0(B)$ that is the closure in $L_\infty(B)$ of the set of B-valued simple functions that are integrable (i.e. in $L_1(B)$). Let (B_0, B_1) be a compatible pair of

Banach spaces. Let $1 \le p_0, p_1 \le \infty$. We can view $(L_{p_0}(m; B_0), L_{p_1}(m; B_1))$ as compatible; indeed, both spaces are continuously injected into the topological vector space of Bochner measurable functions with values in $B_0 + B_1$, equipped with the topology of convergence in measure. Equivalently, as explained in the beginning of this chapter, this allows us to consider them as included in the space $L_{p_0}(m; B_0) + L_{p_1}(m; B_1)$.

Theorem 8.21. *Let* $1 \le p_0, p_1 \le \infty$. *Assume* $\min\{p_0, p_1\} < \infty$. *Then, for any* $0 < \theta < 1$, *we have*

$$(L_{p_0}(B_0), L_{p_1}(B_1))_\theta = L_{p_\theta}((B_0, B_1)_\theta) \tag{8.26}$$

with identical norms, where $p_\theta^{-1} = (1-\theta)p_0^{-1} + \theta p_1^{-1}$. *Moreover, in case* $p_0 = p_1 = \infty$ *we have*

$$(L^0_\infty(B_0), L^0_\infty(B_1))_\theta = L^0_\infty((B_0, B_1)_\theta) \tag{8.27}$$

with equal norms.

Proof. By routine arguments, we can reduce this to the case of a finite measure space, so we may as well assume that m is a probability.

Note that $L_{p_0}(B_0) \cap L_{p_1}(B_1)$ is dense both in $(L_{p_0}(B_0), L_{p_1}(B_1))_\theta$ (see Lemma 8.11) and in $L_{p_\theta}((B_0, B_1)_\theta)$ (because a $(B_0, B_1)_\theta$ valued simple function is approximable by a $B_0 \cap B_1$ valued one). Therefore, it suffices to show that

$$\|F\|_{(L_{p_0}(B_0), L_{p_1}(B_1))_\theta} = \|F\|_{L_{p_\theta}((B_0, B_1)_\theta)} \tag{8.28}$$

for any F in $L_{p_0}(B_0) \cap L_{p_1}(B_1)$. We claim that it suffices to check (8.28) when F is a simple function or equivalently when $\mathcal{A}$ is a finite σ-algebra.

To prove this claim, consider F in $L_{p_0}(B_0) \cap L_{p_1}(B_1)$. We may assume $p_0 < \infty$. Fix $\varepsilon > 0$. Since the simple functions are dense, we know that there is a *finite* σ-subalgebra $\mathcal{C} \subset \mathcal{A}$ such that

$$\|F - \mathbb{E}^{\mathcal{B}} F\|_{L_{p_0}(B_0)} < \varepsilon \text{ and } \|F - \mathbb{E}^{\mathcal{B}} F\|_{L_{p_1}(B_1)} \le 2\|F\|_{L_{p_1}(B_1)}.$$

For the same reason, there is a finite σ-subalgebra $\mathcal{C} \subset \mathcal{A}$ such that

$$\|F - \mathbb{E}^{\mathcal{C}} F\|_{L_{p_\theta}((B_0, B_1)_\theta)} < \varepsilon. \tag{8.29}$$

Replacing them by $\mathcal{B} \vee \mathcal{C}$, we may assume $\mathcal{B} = \mathcal{C}$. By (8.11), we have then both (8.29) and

$$\|F - \mathbb{E}^{\mathcal{C}} F\|_{(L_{p_0}(B_0), L_{p_1}(B_1))_\theta} < (\varepsilon)^{1-\theta} (2\|F\|_{L_{p_1}(B_1)})^\theta. \tag{8.30}$$

Therefore, we are reduced to prove (8.28) for $\mathbb{E}^{\mathcal{C}} F$ and $\mathbb{E}^{\mathcal{C}} F$ is a simple function with values in $B_0 \cap B_1$. This justifies the preceding claim.

It remains to check (8.28) when F is $\mathcal{C}$-measurable, with $\mathcal{C}$ finite. By Lemma 8.9 (applied with P equal to $\mathbb{E}^{\mathcal{C}}$), we may assume that $\mathcal{A} = \mathcal{C}$. Then, we may as well assume that Ω is a finite set, say $\Omega = [1, 2, \ldots, n]$. In that case, the isometric identity

$$\mathcal{F}(L_\infty(B_0), L_\infty(B_1)) = L_\infty(\mathcal{F}(B_0, B_1)),$$

obviously holds and immediately implies that

$$(L_\infty(B_0), L_\infty(B_1))_\theta = L_\infty((B_0, B_1)_\theta) \tag{8.31}$$

with equal norms.

From this we can derive

$$\|F\|_{(L_{p_0}(B_0), L_{p_1}(B_1))_\theta} \le \|F\|_{L_{p_\theta}((B_0,B_1)_\theta)}. \tag{8.32}$$

Indeed, if $\|F\|_{L_{p_\theta}((B_0,B_1)_\theta)} < 1$ we can factorize F trivially as $F = wG$ with $w \in L_{p_\theta}$, $\|w\|_{p_\theta} < 1$ and with

$$\|G\|_{L_\infty((B_0,B_1)_\theta)} < 1.$$

By (8.21), we can find $f \in \mathcal{F}(L_{p_0}, L_{p_1})$ with $\|f\|_{\mathcal{F}} < 1$ such that $f(\theta) = w$ and by (8.31) we can find $g \in \mathcal{F}(L_\infty(B_0), L_\infty(B_1))$ with $\|g\|_{\mathcal{F}} < 1$ such that $g(\theta) = G$. Then the function $z \to f(z)g(z)$ is in $\mathcal{F}(L_{p_0}(B_0), L_{p_1}(B_1))$ with norm < 1 and $f(\theta)g(\theta) = wG = F$, so that we obtain $\|F\|_{(L_{p_0}(B_0), L_{p_1}(B_1))_\theta} < 1$. By homogeneity this proves (8.32).

We now check the converse to (8.32). Consider f in $\mathcal{F}(L_{p_0}(B_0), L_{p_1}(B_1))$ with norm < 1. Note that for any fixed ω (recall Ω is a finite set) $z \to f(z)(\omega)$ is in $\mathcal{F}(B_0, B_1)$ and by (8.25) we can write

$$\|f(\theta)(\omega)\|_{(B_0,B_1)_\theta} \le \|f(\cdot)(\omega)\|^{1-\theta}_{L_{p_0}(\mu^0_\theta;B_0)} \|f(\cdot)(\omega)\|^{\theta}_{L_{p_1}(\mu^1_\theta;B_1)}$$

hence by Hölder's inequality

$$\begin{aligned}\|f(\theta)\|_{L_{p_\theta}((B_0,B_1)_\theta)} &\le \|f\|^{1-\theta}_{L_{p_0}(\mu^0_\theta \times m;B_0)} \|f\|^{\theta}_{L_{p_1}(\mu^1_\theta \times m;B_1)} \\ &\le \sup_{\partial_0} \|f\|^{1-\theta}_{L_{p_0}(B_0)} \sup_{\partial_1} \|f\|^{\theta}_{L_{p_1}(B_1)} \\ &< 1.\end{aligned}$$

By homogeneity, this yields the converse to (8.32).

The proof of (8.27) being entirely similar, we will skip it. □

To illustrate the preceding result, here is an immediate consequence.

Corollary 8.22. *If B_0 and B_1 are UMD, then so is $(B_0, B_1)_\theta$ for any $0 < \theta < 1$.*

Proof. More precisely, by Theorems 8.21 and 8.8, if B_j is UMD$_{p_j}$ ($j = 0, 1$), then $(B_0, B_1)_\theta$ is UMD$_{p_\theta}$. □

Remark 8.23. Let $\mathbb{B}$ be a Banach lattice over $(\Omega, \mathcal{A}, m)$ (see Remark 8.18) and let B be a Banach space. We denote by $\mathbb{B}(B)$ the space of B-valued Bochner measurable functions $f : \Omega \to B$ such that the function $\|f(.)\|_B$ belongs to $\mathbb{B}$, and we equip it with the norm

$$\|f\|_{\mathbb{B}(B)} = \| \|f(.)\|_B \|_{\mathbb{B}},$$

with which it becomes a Banach space.

Let $(\mathbb{B}_0, \mathbb{B}_1)$ be a pair of Banach lattices over a common $(\Omega, \mathcal{A}, m)$ as in Remark 8.18. Fix $0 < \theta < 1$. Let us denote $\mathbb{B}_\theta = \mathbb{B}_0^{1-\theta}\mathbb{B}_1^\theta$. Calderón [177] proved the following generalization of Theorem 8.21: Assuming that $\mathbb{B}_\theta$ has the domination property, we have an isometric identity

$$(\mathbb{B}_0(B_0), \mathbb{B}_1(B_1))_\theta = \mathbb{B}_\theta((B_0, B_1)_\theta). \tag{8.33}$$

We now return to the situation of Lemma 8.20, but we no longer assume the function f continuous on ∂S.

Theorem 8.24. *Let $1 \le p_0, p_1 \le \infty$. Let $0 < \theta < 1$. Let $f\colon \partial S \to B_0 + B_1$ be such that $f_{|\partial_j} \in L_{p_j}(\partial_j, \mu_\theta^j; B_j)$ ($j = 0, 1$). Then f can be extended inside S to a harmonic function $u\colon S \to B_0 + B_1$ admitting f as its a.e. non-tangential limits, defined by*

$$\forall z \in S \quad u(z) = \int f(x)dP_S^z(x).$$

If u is analytic, then $u(\theta) \in (B_0, B_1)_\theta$ and for any $p_0, p_1 > 0$

$$\|u(\theta)\|_{(B_0,B_1)_\theta} \le \|f\|_{L_{p_0}(\partial_0,\mu_\theta^0;B_0)}^{1-\theta} \|f\|_{L_{p_1}(\partial_1,\mu_\theta^1;B_1)}^{\theta}. \tag{8.34}$$

Moreover, for any x in $(B_0, B_1)_\theta$, $\|x\|_{(B_0,B_1)_\theta}$ is equal to the infimum of the right-hand side of (8.34) *over all f as earlier for which u is analytic and $u(\theta) = x$.*

Proof. Let $p = \min(p_0, p_1)$. Since $f \in L_p(\partial S, \mu_\theta; B_0 + B_1)$, by conformal equivalence with the disc, $f \circ \chi_\theta{}^{-1} \in L_p(\partial D, m; B_0 + B_1)$, and hence it admits a 'Poisson integral' that is harmonic inside D with $f \circ \chi_\theta{}^{-1}$ as non-tangential boundary value on ∂D (see Theorem 3.1 and Remark 3.20). Therefore, f itself admits a 'Poisson integral' that is harmonic inside S with f as its a.e. non-tangential limits.

Note in passing that for any θ there are positive constants a, b (depending on θ) such that $a\mu_{1/2} \le \mu_\theta \le b\mu_{1/2}$; so $f \in L_p(\mu_\theta)$ iff $f \in L_p(\mu_{1/2})$ (again going back to the disc, this reduces to the boundedness from above and below of the Poisson kernel for any fixed $z \in D$).

We now consider the case when u is analytic. We assume first that $p_0 = p_1 = p = \infty$ so that $u \in H^\infty(S; B_0 + B_1)$. We will use the net

$$\varphi_\alpha(x) = (2\alpha)^{-1} 1_{[-\alpha,\alpha]}$$

with $\alpha > 0$ tending to 0.

Consider then the function

$$u_\alpha(z) = \int u(z - it)\varphi_\alpha(t)\, dt. \tag{8.35}$$

Clearly, by Remark 8.5, u_α is the Poisson integral of the function f_α defined for $z \in \partial S$ by $f_\alpha(z) = \int f(z - it)\varphi_\alpha(t)\, dt$. It is easy to check that f_α is the non-tangential limit of u_α (in the norm of $B_0 + B_1$) and also that $u_\alpha \in \mathcal{F}$. By the Banach valued case of the classical Lebesgue differentiation theorem (see Remark 3.9) we have $f_{\alpha|\partial_j} \to f_{|\partial_j}$, $(j = 0, 1)$ a.e. on $\partial S = \partial_0 \cup \partial_1$. It follows that (by dominated convergence)

$$\lim_\alpha \|f_\alpha - f\|_{L_1(\partial_j, \mu_\theta^j)} = 0.$$

By (8.25), this implies that $(u_\alpha(\theta))_\alpha$ is Cauchy in $(B_0, B_1)_\theta$, i.e.

$$\lim_{\alpha,\beta} \|u_\alpha(\theta) - u_\beta(\theta)\|_{(B_0,B_1)_\theta} = 0,$$

so that $u_\alpha(\theta)$ converges in $(B_0, B_1)_\theta$ to a limit $x \in (B_0, B_1)_\theta$, and since

$$\begin{aligned}\|u_\alpha(\theta)\|_{(B_0,B_1)_\theta} &\le \|f_\alpha\|^{1-\theta}_{L_\infty(\partial_0,\mu_\theta^0;B_0)} \|f_\alpha\|^{\theta}_{L_\infty(\partial_1,\mu_\theta^1;B_1)} \\ &\le \|f\|^{1-\theta}_{L_\infty(\partial_0,\mu_\theta^0;B_0)} \|f\|^{\theta}_{L_\infty(\partial_1,\mu_\theta^1;B_1)},\end{aligned}$$

we have

$$\|x\|_{(B_0,B_1)_\theta} = \lim_\alpha \|u_\alpha(\theta)\|_{(B_0,B_1)_\theta} \le \|f\|^{1-\theta}_{L_\infty(\partial_0,\mu_\theta^0;B_0)} \|f\|^{\theta}_{L_\infty(\partial_1,\mu_\theta^1;B_1)}.$$

Since obviously $u_\alpha(\theta) \to u(\theta)$ in $B_0 + B_1$, we must have $u(\theta) = x$ and we obtain the announced result (8.34) for $p_0 = p_1 = \infty$. Once (8.34) has been proved, it is clear that the infimum of its right-hand side over all f such that $u(\theta) = x$ is equal to $\|x\|_{(B_0,B_1)_\theta}$.

Now to treat the general case, we use outer functions exactly as we did previously to prove Lemma 8.20. This yields (8.34) for arbitrary values of $0 < p_0, p_1 \le \infty$. □

Theorem 8.25. *If one of the spaces B_0, B_1 is reflexive, then $(B_0, B_1)_\theta$ is reflexive.*

Proof. Let $B_\theta = (B_0, B_1)_\theta$. Let $x^{**} \in B_\theta^{**}$. Let (x_n) be a bounded net (or a 'generalized sequence' indexed by a directed set) in B_θ tending to x^{**} in the sense of the $\sigma(B_\theta^{**}, B_\theta^*)$-topology. Let (f_n) be bounded in $\mathcal{F}$ such that $f_n(\theta) = x_n$. We will show that $x^{**} \in B_\theta$ if either B_0 or B_1 is reflexive. Let $X_j = L_2(\partial_j, \mu^j_\theta; B_j)$. Assume that (say) B_1 is reflexive. By Corollary 2.24, X_1 is reflexive. Hence the net $\{f_{n|\partial_1}\}$ (which is bounded in X_1 since of $\|f_n\|_{X_1} \le \|f_n\|_{\mathcal{F}}$) admits a subnet converging weakly in X_1. Recall that the weak and strong closures of a convex set coincide ('Mazur's theorem'). This implies that there are convex combinations of $\{f_{n|\partial_1}\}$ of the form

$$g_m = \sum\nolimits_{n\ge m} \lambda_n^{(m)} f_n, \quad \lambda_n^{(m)} \ge 0 \quad \sum \lambda_n^{(m)} = 1$$

such that $g_{m|\partial_1}$ converges in norm in X_1. By (8.25), the corresponding convex combinations of $\{x_n\}$, $\sum_{n\ge m} \lambda_n^{(m)} x_n$ satisfy the Cauchy criterion in B_θ and hence converge in norm to a limit $x \in B_\theta$. Thus we conclude $x^{**} = x \in B_\theta$, i.e. B_θ is reflexive. □

We now wish to enlarge the scope of the fundamental interpolation Theorem 8.8. Firstly, we show that complex interpolation (an advantage over the real method) allows to interpolate multi-linear maps.

Theorem 8.26. *Let $(B_0^{(1)}, B_1^{(1)}), \ldots, (B_0^{(K)}, B_1^{(K)})$ be K compatible couples and let (C_0, C_1) be another compatible couple. Let $T_0\colon\ B_0^{(1)} \times \cdots \times B_0^{(K)} \to C_0$ and $T_1\colon\ B_1^{(1)} \times \cdots \times B_1^{(K)} \to C_1$ be bounded K-linear mappings that are 'essentially the same', i.e. coincide on $(B_0^{(1)} \cap B_1^{(1)}) \times \cdots \times (B_0^{(K)} \cap B_1^{(K)})$. Then the resulting K-linear map*

$$T\colon\ (B_0^{(1)} + B_1^{(1)}) \times \cdots \times (B_0^{(K)} + B_1^{(K)})$$

takes $(B_0^{(1)},\ B_1^{(1)})_\theta \times \cdots \times (B_0^{(K)},\ B_1^{(K)})_\theta$ to $(C_0,\ C_1)_\theta$ with norm $\le \|T_0\|^{1-\theta}\|T_1\|^\theta$.

Sketch. The proof is essentially the same as for Theorem 8.8. For simplicity we restrict ourselves to the case $K = 2$. Let x^1 (resp. x^2) be in the open unit ball of $(B_0^{(1)}, B_1^{(1)})_\theta$ (resp. $(B_0^{(2)}, B_1^{(2)})_\theta$). Let $f^{(1)}$ (resp. $f^{(2)}$) be in the unit ball of $\mathcal{F}(B_0^{(1)}, B_1^{(1)})$ (resp. $\mathcal{F}(B_0^{(2)}, B_1^{(2)})$) such that $f^{(1)}(\theta) = x^1$ (resp. $f^{(2)}(\theta) = x^2$), then $z \mapsto T(f^{(1)}(z), f^{(2)}(z))$ is in the unit ball of $\mathcal{F}(C_0, C_1)$, and hence $\|T(f^{(1)}(\theta), f^{(2)}(\theta))\|_{(C_0,C_1)_\theta} \le 1$.

We leave the remaining details to the reader. □

Remark 8.27. Note that the same result remains true if the map T is anti-linear with respect to some of its coordinates. Indeed, take for instance $K = 2$ and assume T linear in the first variable and anti-linear in the second. Then, if

$$f^{(1)} \in \mathcal{F}(B_0^{(1)}, B_1^{(1)}) \quad \text{and} \quad f^{(2)} \in \mathcal{F}(B_0^{(2)}, B_1^{(2)})$$

the function $z \longmapsto T(f^{(1)}(z), f^2(\bar{z}))$ belongs to $\mathcal{F}(C_0, C_1)$.

The next result was first considered by E. Stein for pairs of L_p-spaces in 1956, before the introduction of 'the complex interpolation method' (which it partially motivated) and is now called the 'Stein interpolation principle'. It is a very useful generalization of the basic interpolation property in Theorem 8.8.

Theorem 8.28. *Consider two compatible pairs (B_0, B_1) and (C_0, C_1). Let $z \mapsto T(z)$ be an analytic function from S to $B(B_0 \cap B_1, C_0 + C_1)$ such that for any x in $B_0 \cap B_1$, the function $z \mapsto T(z)x$ extends continuously to a function in $\mathcal{F}(C_0, C_1)$. Assume that we have constants M_0, M_1 such that*

$$\forall j = 0, 1 \quad \forall x \in B_0 \cap B_1 \quad \forall z \in \partial_j \quad \|T(z)x\|_{C_j} \le M_j \|x\|_{B_j}.$$

Then, for any $0 < \theta < 1$, $T(\theta)$ extends to a bounded operator from $(B_0, B_1)_\theta$ to $(C_0, C_1)_\theta$ such that

$$\|T(\theta)\| \le M_0^{1-\theta} M_1^{\theta}.$$

Proof. Let $x \in B_0 \cap B_1$ and let $f \in \mathcal{G}(B_0, B_1)$ such that $f(\theta) = x$. Let $g(z) = (M_0^{1-z} M_1^{z})^{-1} T(z) f(z)$. Our assumptions imply that $g \in \mathcal{F}(C_0, C_1)$, $g(\theta) = (M_0^{1-\theta} M_1^{\theta})^{-1} T(\theta)x$ and $\|g\|_{\mathcal{F}(C_0,C_1)} \le \|f\|_{\mathcal{F}(B_0,B_1)}$. Thus by Lemma 8.11 we obtain

$$\|T(\theta)x\|_{(C_0,C_1)_\theta} \le M_0^{1-\theta} M_1^{\theta} \|x\|_{(B_0,B_1)_\theta}. \qquad \square$$

See [196] for a more refined statement.

Remark 8.29. Let $1 \le p_0, p_1 < \infty$. Let $B_j = L_{p_j}(w_j.m)$ where w_j is a density (meaning it is measurable and positive a.e.). Let $\varphi_j = w_j^{1/p_j}$, so that $B_j = \{x \mid x\varphi_j \in L_{p_j}(m)\}$ with norm $\|x\|_{B_j} = \|x\varphi_j\|_{L_{p_j}(m)}$. Let $\varphi_\theta = \varphi_0^{1-\theta} \varphi_1^{\theta}$. Then

$$(B_0, B_1)_\theta = \{x \mid x\varphi_\theta \in L_{p_\theta}(m)\}$$

and $\|x\|_{(B_0,B_1)_\theta} = \|x\varphi_\theta\|_{L_{p_\theta}(m)}$. Equivalently, we have

$$(L_{p_0}(w_0.m), L_{p_1}(w_1.m))_\theta = L_{p_\theta}(w.m) \tag{8.36}$$

for w such that $w^{1/p_\theta} = \varphi_\theta$. This follows from Theorem 8.28 (and Theorem 8.15) using $C_j = L_{p_j}(m)$ and the mapping $T(z)x = \varphi_0^{1-z}\varphi_1^{z} x$, and its inverse, which is an isometric isomorphism from B_j to C_j when $z \in \partial_j$. The analyticity of $z \mapsto T(z)x$ can be reduced to the easier case when φ_0 and φ_1 are bounded above and below by positive constants on the support of x. We skip the details. Note that (8.36) also follows from (8.23).

Alternatively, assuming $p_0 \ne p_1$, let $\varphi = (w_1/w_0)^{\frac{1}{p_0 - p_1}}$ so that $\varphi^{p_0} w_0 = \varphi^{p_1} w_1$. Then $\|x\|_{B_j} = \|\varphi^{-1}x\|_{L_{p_j}(m')}$ where $m' = (\varphi^{p_0} w_0).m = (\varphi^{p_1} w_1).m$, from which it is immediate that $\|x\|_{(B_0,B_1)_\theta} = \|\varphi^{-1}x\|_{(L_{p_0}(m'),L_{p_1}(m'))_\theta}$, and hence (note $\varphi^{-p_\theta} m' = w.m$) we obtain (8.36) as a consequence of (8.20). This useful trick (due to Stein and Weiss) reduces interpolation (for any method) of L_p-spaces with weights to the usual pair $(L_{p_0}(m'), L_{p_1}(m'))$. It is also applicable to pairs of Banach space valued L_p-spaces.

8.3 Duality for the complex method

We will now describe the dual of the space $(B_0, B_1)_\theta$. We assume that $B_0 \cap B_1$ is dense in both B_0 and B_1. Then the duals B_0^* and B_1^* are continuously injected into $(B_0 \cap B_1)^*$, so that we may view (B_0^*, B_1^*) as 'compatible'. Let $(\Omega, \mathcal{A}, m)$ be a finite measure space and let B be a Banach space. Let $1 \le p < \infty$. We will use the description of the dual of $L_p(\Omega, \mathcal{A}, m; B)$ given in §2.4. We will need to work with a generalization of $(B_0, B_1)_\theta$, which will be denoted by $(B_0, B_1)^\theta$, where the function f on ∂S is replaced by a vector measure, in analogy with boundary values of bounded B-valued analytic functions when B fails the ARNP.

Recall that we denote by $\Lambda_p(\Omega, \mathcal{A}, m; B)$, or simply by $\Lambda_p(m; B)$, or even simply $\Lambda_p(B)$ the space of all vector measures $\mu\colon \mathcal{A} \to B$ such that $|\mu| \ll m$ and such that the density $\frac{d|\mu|}{dm}$ is in $L_p(\Omega, \mathcal{A}, m)$, equipped with the norm

$$\|\mu\|_{\Lambda_p(m;B)} = \left\| \frac{d|\mu|}{dm} \right\|_p .$$

See §2.4 for details, including the duality between $L_p(m; B)$ and $\Lambda_{p'}(m; B^*)$.

We now consider $\Omega = \partial S$ equipped with the measure $m^S = \mu_{1/2}$. We will use the conformal mapping $\chi = \chi_{1/2}$ introduced in (8.1), that takes S onto D and sends $1/2 \in S$ to $0 \in D$.

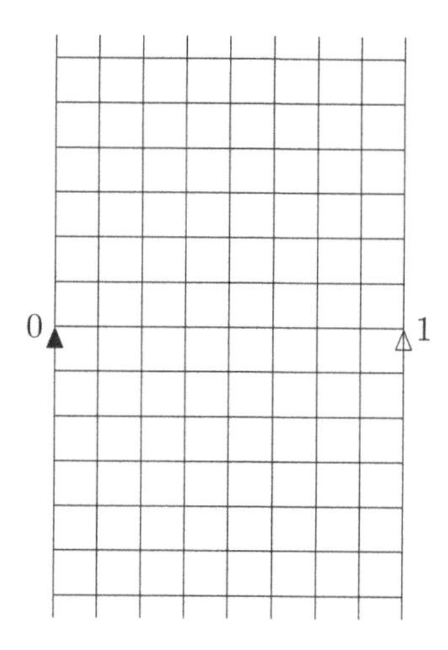

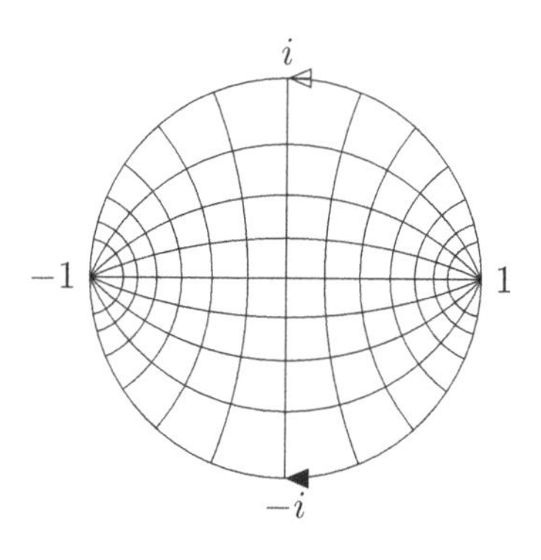

Figure 8.2. The conformal map $\chi = \chi_{1/2} : z \mapsto \frac{e^{i\pi z} - i}{e^{i\pi z} + i}$.

Recall that $\chi(\partial S) \subset \partial D$. We set $m_0 = (1/2)\mu^0_{1/2}$ and $m_1 = (1/2)m^1_{1/2}$ so that $m^S = m_0 + m_1$, and χ takes m^S to the normalized Lebesgue measure m on ∂D, i.e. for any Borel subset $A \subset \partial D$ we have

$$m(A) = m^S(\chi^{-1}(A)).$$

For any z in S, the harmonic probability measure P^z_S on ∂S is absolutely continuous with respect to m^S. Let us denote its density on ∂S by W^z so that

$$P^z_S = W^z.(m_0 + m_1) = W^z.m^S.$$

Note that $W^z \in L_1(m^S) \cap L_\infty(m^S)$. The conformal equivalence χ takes P^z_S to $P^{\chi(z)}_D$ and W^z to the density relative to m of the classical Poisson kernel of D with respect to $\chi(z)$.

Let B be any Banach space. Let $\mu \in \Lambda_p(\partial S, m^S; B)$ $(1 \le p \le \infty)$. Then the function $z \mapsto u(z) = \int_{\partial S} W^z \, d\mu$ is well defined by (2.14) since $W^z \in L_1 \cap L_\infty \subset L_{p'}$, and is harmonic inside S (by the scalarization principle). We will say (abusively) that u is the Poisson integral or the Poisson transform of μ in S.

For any ξ in B^*, the function $z \to \xi(u(z))$ is the harmonic extension of the Radon-Nikodým derivative $D\xi(\mu) = \frac{d\xi(\mu)}{dm^S}$. The latter is clearly in $L_p(m^S)$ by definition of $\Lambda_p(m^S; B)$. Therefore we have non-tangential convergence of $\xi(u(z))$ to $D\xi(\mu)$ at a.e. point of ∂S. Thus, in some 'weak' sense, we recover μ as boundary limit of u. In particular, two such vector measures with the same 'Poisson transform' must coincide.

Note that if $\mu \in \Lambda_\infty(\partial S, m^S; B)$ with $\|\mu\|_{\Lambda_\infty(\partial S, m^S; B)} \le 1$ we have by (2.4)

$$\|u(z)\|_B \le \int W^z d|\mu| \le \int W^z dm^S = 1. \tag{8.37}$$

Let us now assume that $B = B_0 + B_1$, with $\mu \in \Lambda_\infty(m^S; B_0 + B_1)$ and assume that $f(z) = \int_{\partial S} W^z \, d\mu$ is analytic in S with values in $B_0 + B_1$, but also that the restrictions of μ to ∂_0 and ∂_1 define vector measures $\mu_0 \in \Lambda_\infty(m_0; B_0)$ and $\mu_1 \in \Lambda_\infty(m_1; B_1)$ (i.e. we assume $\forall A \subset \partial_j$ Borel subset $\mu_j(A) \in B_j$ $\forall j = 0, 1$). We denote by $\widetilde{\mathcal{F}}$ the space of all such functions f. We equip it with the norm

$$\|f\|_{\widetilde{\mathcal{F}}} = \max\{\|\mu_0\|_{\Lambda_\infty(m_0;B_0)}, \|\mu_1\|_{\Lambda_\infty(m_1;B_1)}\}.$$

Let $0 < \theta < 1$. The space $(B_0, B_1)^\theta$ is then defined by

$$(B_0, B_1)^\theta = \{x \in B_0 + B_1 \mid \exists f \in \widetilde{\mathcal{F}} \quad f(\theta) = x\}.$$

We equip it with the norm

$$\|x\|_{(B_0,B_1)^\theta} = \inf\{\|f\|_{\widetilde{\mathcal{F}}} \mid f \in \widetilde{\mathcal{F}} \quad f(\theta) = x\}.$$

It is easy to see that

$$(B_0, B_1)_\theta \subset (B_0, B_1)^\theta$$

and this inclusion has norm ≤ 1. Indeed, for any $f \in \mathcal{F} = \mathcal{F}(B_0, B_1)$, we have

$$f(z) = \int_{\partial S} f dP_S^z = \int_{\partial S} W^z f dm^S.$$

In other words if we let $\mu = f.m^S$ then f is the Poisson integral of μ and $|\mu| \leq \|f\|_{\mathcal{F}} m^S$, therefore $f \in \widetilde{\mathcal{F}}$ and $\|f\|_{\widetilde{\mathcal{F}}} \leq \|f\|_{\mathcal{F}}$.

Moreover, $(B_0, B_1)^\theta$ satisfies the fundamental interpolation property: given $T_0\colon B_0 \to C_0, T_1 : B_1 \to C_1$ as in Theorem 8.8, the operator T is bounded from $(B_0, B_1)^\theta$ to $(C_0, C_1)^\theta$ with norm $\leq \|T_0\|^{1-\theta}\|T_1\|^\theta$.

The traditional definition of $(B_0, B_1)^\theta$ by Calderón and later authors is different but we will show that it is equivalent to ours. One usually introduces a space, that we will denote by $\widetilde{\mathcal{C}}(B_0, B_1)$ or simply by $\widetilde{\mathcal{C}}$, formed of all analytic functions $f : S \to B_0 + B_1$ admitting a primitive $F : \bar{S} \to B_0 + B_1$ (i.e. $F' = f$ on S) that is analytic inside S, continuous on $\bar{S}$, such that for some constant C we have $\|F(z)\|_{B_0+B_1} \leq C(1 + |z|)$ for all $z \in \bar{S}$, and such that for $j = 0, 1$ we have

$$\forall s, t \in \mathbb{R} \quad F(j + is) - F(j + it) \in B_j \quad \text{and}$$

$$\sup_{t \neq s} \frac{\|F(j + is) - F(j + it)\|_{B_j}}{|s - t|} < \infty.$$

One equips $\widetilde{\mathcal{C}}$ with the norm

$$\|f\|_{\widetilde{\mathcal{C}}} = \max\left\{\sup_{t \neq s} \frac{\|F(is) - F(it)\|_{B_0}}{|s - t|}, \sup_{t \neq s} \frac{\|F(1 + is) - F(1 + it)\|_{B_1}}{|s - t|}\right\}.$$

The following elementary lemma will help to clarify the picture.

Lemma 8.30. *Let $f \in \widetilde{\mathcal{C}}(\mathbb{C}, \mathbb{C})$ with primitive F on S. Then $f = F' \in H^\infty(S)$. Moreover, the functions $F_0 : s \mapsto F(is)$ and $F_1 : s \mapsto F(1 + is)$ are differentiable a.e. and such that $F_0' \in L_\infty(\partial_0)$ and $F_1' \in L_\infty(\partial_1)$. Let $\mathcal{D}F : \partial S \to \mathbb{C}$ be the function defined a.e. on ∂S by*

$$\mathcal{D}F(is) = (1/i)F_0'(is) \text{ and } \mathcal{D}F(1 + is) = (1/i)F_1'(1 + is).$$

Let $\mu = (1/i)(F_0' m_0 + F_1' m_1) = (\mathcal{D}F)m^S$. We have then

$$\forall z \in S \quad F'(z) = \int_{\partial S} \mathcal{D}F dP_S^z = \int_{\partial S} W^z d\mu. \tag{8.38}$$

Proof. Assume $\|f\|_{\widetilde{C}(\mathbb{C},\mathbb{C})} \le 1$. Let F be a primitive of f. We first claim that F satisfies the following Poisson formula

$$F(z) = \int_{\partial S} F(\xi) P_S^z(d\xi). \tag{8.39}$$

To verify this, let $\varepsilon > 0$ and let $F_\varepsilon(z) = e^{\varepsilon z^2} F(z)$. By the assumption that $|F(z)| \in O(|z|)$ when $z \to \infty$, it follows that F_ε is continuous and bounded on S, so we have clearly (e.g. by conformal equivalence with D)

$$F_\varepsilon(z) = \int_{\partial S} F_\varepsilon(\xi) P_S^z(d\xi).$$

Moreover, since $\int |\xi| P_S^z(d\xi) < \infty$ (a calculation left to the reader e.g. using (8.1)), letting ε tend to zero, we find (8.39) by dominated convergence. Incidentally, $\int |\xi| P_S^z(d\xi) < \infty$ implies that $F(\chi^{-1})$ is in $H^1(D)$.

For $\delta > 0$ we set $f_\delta(z) = \frac{F(z+i\delta)-F(z)}{i\delta}$. Then $f_\delta \in H^\infty(S)$ and $\|f_\delta\|_{H^\infty(S)} \le 1$, and, since F is assumed continuous on $\bar{S}$, so is f_δ. Fix $z \in S$. By (8.39), we have

$$f_\delta(z) = \int_{\partial S} (i\delta)^{-1}(F(\xi + i\delta) - F(\xi)) P_S^z(d\xi). \tag{8.40}$$

Since F is Lipschitzian with norm 1 on ∂S, the integrand of the right-hand side of (8.40) is of modulus ≤ 1, the limit

$$\lim_{\delta \to 0} (i\delta)^{-1}(F(\xi + i\delta) - F(\xi))$$

exists, is equal to $\mathcal{D}F(\xi)$ for a.e. $\xi \in \partial S$ (this is the RNP for $\mathbb{C}$; see Remark 2.17), and hence by dominated convergence the right-hand side of (8.40) tends to $\int_{\partial S} \mathcal{D}F(\xi) P_S^z(d\xi)$. Letting $\delta \to 0$ in (8.40) we find

$$F'(z) = \int_{\partial S} \mathcal{D}F(\xi) P_S^z(d\xi).$$

Since $|\mathcal{D}F(\xi)| \le 1$ a.e. we also have $|F'(z)| \le 1$ and hence $F' \in H^\infty(S)$. □

Theorem 8.31. *For any compatible couple (B_0, B_1) we have $\widetilde{C}(B_0, B_1) = \widetilde{\mathcal{F}}(B_0, B_1)$ isometrically. Moreover, if A_j denotes the closure of $B_0 \cap B_1$ in B_j $(j = 0, 1)$ with the induced norm, then $\widetilde{C}(B_0, B_1) = \widetilde{C}(A_0, A_1)$, and hence $(B_0, B_1)^\theta = (A_0, A_1)^\theta$ for any $0 < \theta < 1$.*

Proof. Let λ_j denote the Lebesgue measure on ∂_j. Consider f in the unit ball of $\widetilde{C}$ and let F be its primitive as previously. Let ν_0 be the B_0 valued vector measure on ∂_0 such that for any $s > t$ $\nu_0([it, is[) = F(is) - F(it)$, defined as in Remark 2.17. We have then $|\nu_0| \le \lambda_0$. Let $\mu_0 = \psi_0 \nu_0$ where ψ_0 is the density of m_0 with respect to Lebesgue measure on ∂_0. Clearly, μ_0 (as a B_0 valued vector measure) satisfies $|\mu_0| \le m_0$. Similarly, we define a B_1 valued vector measure μ_1 on ∂_1, satisfying (as a B_1 valued vector measure) $|\mu_1| \le m_1$. Thus we obtain $\max\{\|\mu_0\|_{\Lambda_\infty(m_0;B_0)}, \|\mu_1\|_{\Lambda_\infty(m_1;B_1)}\} \le 1$. Let $\mu = \mu_0 + \mu_1$ be the associated vector measure on ∂S. We will show that $f = F'$ is the Poisson integral of μ. It suffices to show that, for any $\xi \in (B_0 + B_1)^*$, $\xi(F')$ is the Poisson integral of $\xi(\mu)$ and this follows from Lemma 8.30 applied to the function $\xi(F)$. Thus $f = F' \in \widetilde{\mathcal{F}}$ and $\|f\|_{\widetilde{\mathcal{F}}} \le 1$.

Conversely, let f be in the unit ball of $\widetilde{\mathcal{F}}$, let F be a primitive of f, and let μ be the $(B_0 + B_1)$ valued vector measure of which f is the Poisson integral. Since by (8.37) $\sup\{\|f(z)\|_{B_0+B_1} \mid z \in S\} \le 1$, we have $\|F(z) - F(z')\|_{B_0+B_1} \le |z - z'|$ for all $z, z' \in S$, so that F extends by continuity to a continuous $(B_0 + B_1)$ valued function on $\bar{S}$, that is $O(|z|)$ when $|z| \to \infty$. For any $\xi \in (B_0 + B_1)^*$, the function $\xi(F)$ is in $\widetilde{\mathcal{C}}(\mathbb{C}, \mathbb{C})$ so by Lemma 8.30 we have $\forall z \in S \quad \xi(F'(z)) = \int_{\partial S} \mathcal{D}\xi(F) dP_S^z = \int_{\partial S} \mathcal{D}\xi(F) W^z dm^S$, and hence $\xi(F'(z))$ is the Poisson integral of the measure $\mathcal{D}\xi(F)m^S$. But it is also that of $\xi(\mu)$. Therefore $\xi(\mu) = \mathcal{D}\xi(F)m^S$. Let ν_0 (resp. ν_1) be, as earlier, the vector measure such that for any $s > t$ $\nu_0([it, is[) = F(is) - F(it)$ (resp. $\nu_1([1 + it, 1 + is[) = F(1 + is) - F(1 + it)$), and let ψ_j be the density of m_j with respect to Lebesgue measure on ∂_j. (Recall $m^S = m_0 + m_1$.) Recall that λ_j denotes Lebesgue measure on ∂_j.

Since, for any $\xi \in (B_0 + B_1)^*$, $\xi(\mu) = \mathcal{D}\xi(F)m^S$, we have

$$\xi(\mu_j) = \mathcal{D}\xi(F)m_j = \mathcal{D}\xi(F)\psi_j\lambda_j.$$

Therefore, we have

$$\int_{[j+it, j+is[} \psi_j^{-1} d\xi(\mu_j) = \int_{[j+it, j+is[} \mathcal{D}\xi(F) d\lambda_j = \xi(F(j + it) - F(j + is))$$

and hence

$$\int_{[j+it, j+is[} \psi_j^{-1} d\mu_j = F(j + it) - F(j + is).$$

This shows that $F(j + it) - F(j + is) \in B_j$ and moreover

$$\|F(j + it) - F(j + is)\|_{B_j} = \left\| \int_{[j+it, j+is[} (\psi_j)^{-1} d\mu_j \right\|_{B_j} \le \int_{[j+it, j+is[} \psi_j^{-1} d|\mu_j|$$

and hence since $|\mu_j| \le m_j = \psi_j\lambda_j$

$$\le \int_{[j+it, j+is[} d\lambda_j = s - t.$$

In other words we conclude that $f \in \widetilde{\mathcal{C}}$ with norm ≤ 1.

By Remark 8.1, to prove the second assertion, it suffices to show the following claim: For any $f \in \widetilde{\mathcal{C}}(B_0, B_1)$, its primitive F is such that $F(j + is) - F(j + it) \in A_j$ for any $s, t \in \mathbb{R}$ and $z \mapsto F(z) - F(0) \in A_0 + A_1$ for any $z \in \bar{S}$. Indeed, replacing F by $F - F(0)$, the claim implies (by Remark 8.1) that $f \in \widetilde{\mathcal{C}}(A_0, A_1)$, and the converse is trivial.

We will now prove this claim. Since $F(1) - F(0) \in B_0 + B_1$, there are $x_j \in B_j$ such that $F(1) - F(0) = x_0 - x_1$. Consider then the function

$$G(z) = F(z) - F(0) - x_0 = F(z) - F(1) - x_1. \tag{8.41}$$

Note $G(z) \in B_j$ for any $z \in \partial_j$. Moreover, by the Lipschitz conditions imposed on F, G is continuous from ∂_j to B_j. Fix $\varepsilon > 0$ and let $G_\varepsilon(z) = e^{\varepsilon z^2} G(z)$. Then $G_\varepsilon \in \mathcal{F}_0(B_0, B_1)$. Therefore, by (8.18) $G_\varepsilon \in \mathcal{F}_0(A_0, A_1)$. Since $G = e^{-\varepsilon z^2} G_\varepsilon$, we have $G(j + it) \in A_j$ for any $t \in \mathbb{R}$ and also $G(z) \in A_0 + A_1$ for any $z \in S$, and hence, by (8.41), $F(j + it) - F(j) \in x_j + A_j$ for any $t \in \mathbb{R}$. Applying this for $t = 0$, we find $0 \in x_j + A_j$, which means that $x_j \in A_j$ automatically holds ($j = 0, 1$). Now, since $x_0 \in A_0$, $F(z) - F(0) = G(z) + x_0 \in A_0 + A_1$ for any $z \in S$. This proves the claim. □

Lemma 8.32. *The open unit ball of $(B_0, B_1)^\theta$ is included in the closure in $B_0 + B_1$ of the open unit ball of $(B_0, B_1)_\theta$.*

Proof. Let $x \in (B_0, B_1)^\theta$ with norm < 1. Let $f \in \widetilde{\mathcal{C}}$ with $\|f\|_{\widetilde{\mathcal{C}}} < 1$ such that $f(\theta) = x$ and let F be a primitive of f as in the definition of $\widetilde{\mathcal{C}}$. For any $z \in S$ and $\delta > 0$, let $F_\delta(z) = (i\delta)^{-1}(F(z + i\delta) - F(z))$. Clearly $F_\delta(\theta) \to F'(\theta) = f(\theta)$ in $B_0 + B_1$ when $\delta \to 0$. Moreover, $F_\delta \in \mathcal{F}$ with $\|F_\delta\|_{\mathcal{F}} < 1$. Therefore, $\|F_\delta(\theta)\|_{(B_0,B_1)_\theta} < 1$, and $F_\delta(\theta) \to x$. □

Remark 8.33. In the situation of Theorem 8.26, there is a refinement: we are allowed to replace B_θ by B^θ but only for a single one of the factors, say the first one. So the K-linear map T takes $(B_0^{(1)}, B_1^{(1)})^\theta \times (B_0^{(2)}, B_1^{(2)})_\theta \times \cdots \times (B_0^{(K)}, B_1^{(K)})_\theta$ to $(C_0, C_1)^\theta$ also with norm $\leq \|T_0\|^{1-\theta} \|T_1\|^\theta$. The proof is similar to that for Theorem 8.26, but observing that when $f^{(1)}$ is only in the unit ball of $\widetilde{\mathcal{F}}(B_0^{(1)}, B_1^{(1)})$, while the other factors are still in the unit ball of $\mathcal{F}(B_0^{(m)}, B_1^{(m)})$ $(2 \leq m \leq K)$, the map $z \mapsto T(f^{(1)}(z), \ldots, f^{(K)}(z))$ is in the unit ball of $\widetilde{\mathcal{F}}(C_0, C_1)$.

Theorem 8.34. *If one of the spaces B_0, B_1 has the RNP then*

$$(B_0, B_1)_\theta = (B_0, B_1)^\theta$$

isometrically for any $0 < \theta < 1$. Moreover, the same equality holds if $B_0 \subset B_1$ and B_1 has the ARNP.

Proof. Consider $x \in (B_0, B_1)^\theta$ with norm < 1. It suffices to show that x belongs to the unit ball of $(B_0, B_1)_\theta$. Consider $f \in \widetilde{\mathcal{F}}$ with $\|f\|_{\widetilde{\mathcal{F}}} < 1$ and $f(\theta) = x$. Let (φ_α) and f_α be as in the proof of Theorem 8.24. It is easy to see that $f_\alpha \in \mathcal{F}$ with $\|f_\alpha\|_{\mathcal{F}} < 1$. Indeed, if F is the primitive of f, we have $f_\alpha(z) = (2\alpha)^{-1}(F(z+i\alpha) - F(z-i\alpha))$, so $\|f_\alpha\|_{\mathcal{F}} \le \|f\|_{\widetilde{\mathcal{C}}} = \|f\|_{\widetilde{\mathcal{F}}} < 1$ by Theorem 8.31. Now, if we assume that for instance B_1 has the RNP, then $\mu_1 = \psi.m_1$ for some $\psi \in L_\infty(m_1; B_1)$ and

$$\forall z \in \partial_1 \qquad f_\alpha(z) = \int \psi(z - it)\varphi_\alpha(t)\, dt,$$

therefore, by (3.35), $\lim_\alpha f_\alpha(z) = \psi(z)$ for a.e. z in ∂_1 and hence

$$\lim_\alpha \|f_{\alpha|\partial_1} - \psi\|_{L_1(\partial_1, m_1; B_1)} = 0.$$

By (8.34), $f_\alpha(\theta)$ is Cauchy in $(B_0, B_1)_\theta$ and hence

$$f(\theta) = \lim_\alpha f_\alpha(\theta) \quad \text{in} \quad (B_0, B_1)_\theta$$

so we conclude $x \in (B_0, B_1)_\theta$ and

$$\|x\|_{(B_0,B_1)_\theta} \le \lim_\alpha \|f_\alpha(\theta)\|_{(B_0,B_1)_\theta} \le 1.$$

Assuming $B_0 \subset B_1$, we automatically have $f \in H^\infty(S, B_1)$, and hence the ARNP of B_1 ensures, using the conformal equivalence of S and D and Theorem 4.30, that f admits non-tangential B_1-limits a.e. on ∂S, of which f is the Poisson integral as a B_1 valued function. In particular, restricting this to ∂_1, we find again $\mu_1 = \psi.m_1$ for some $\psi \in L_\infty(m_1; B_1)$ and we conclude as in the first part of the proof. □

Remark 8.35. The preceding proof actually shows a bit more: if $f \in \widetilde{\mathcal{F}}$ is the Poisson integral of a vector measure μ with restrictions $\mu_j \in \Lambda_\infty(\partial_j, m_j; B_j)$ $(j = 0, 1)$, and if either μ_0 or μ_1 is differentiable with respect to λ_j (or equivalently with respect to m_j) *on a set of positive Lebesgue measure* then $f(\theta) \in (B_0, B_1)_\theta$. (Hint: To check this, use (8.24) instead of (8.34).)

Remark 8.36. There is an analogue of (8.25) and (8.34) for the Poisson integral u of a vector measure μ such that $\mu_j \in \Lambda_{p_j}(\partial_j, \mu_\theta^j; B_j)$, $j = 0, 1$, with $0 < p_0, p_1 < \infty$. Namely, if u is analytic on S, then

$$\|u(\theta)\|_{(B_0,B_1)^\theta} \le \|\mu_0\|^{1-\theta}_{\Lambda_{p_0}(\partial_0, \mu_\theta^0; B_0)} \|f\|^\theta_{\Lambda_{p_1}(\partial_1, \mu_\theta^1; B_1)}. \tag{8.42}$$

Fix $\varepsilon > 0$. Let $w_j = \frac{d|\mu_j|}{d\mu_\theta^j}$. Let F be the outer function such that

$$\forall \xi \in \partial_j \quad |F(\xi)| = w_j + \varepsilon.$$

Then we can derive (8.42) by the same reasoning as for (8.25). Thus, for any $x \in B_0 + B_1$, the norm of x in $(B_0, B_1)^\theta$ is equal to the infimum of the right-hand side of (8.42) over all $\mu = \mu_0 + \mu_1$ such that $\mu_j \in \Lambda_{p_j}(\partial_j, \mu_\theta^j; B_j)$, $j = 0, 1$ admitting an analytic Poisson integral u such that $u(\theta) = x$.

We will now describe the dual of $(B_0, B_1)_\theta$. The proof is much clearer if one reformulates the definition of $(B_0, B_1)_\theta$ on the unit disc, using the conformal equivalence of S and D. This will require some preparation, summarized in the following remarks.

Note that the conformal mapping $\chi_\theta\colon\ S \to D$ (defined in (8.1)) maps ∂_0 (resp. ∂_1) to an open arc $I_0 \subset \partial D$ ($I_1 \subset \partial D$) with $m(I_0) = 1 - \theta$ ($m(I_1) = \theta$) (here m denotes the normalized Lebesgue measure on the unit circle). Moreover ∂D is the union of I_0 and I_1 and the two extremities of these arcs. Since $\chi_\theta(\theta) = 0$, χ_θ takes μ_θ to m and hence transforms $(1 - \theta)\mu_\theta^0$ (resp. $\theta\mu_\theta^1$) into $m_{|I_0}$ (resp. $m_{|I_1}$). Let $\nu_j = m(I_j)^{-1} m_{|I_j}$ ($j = 0, 1$). Then (8.34) can be rewritten like this: given $f_j \in L_{p_j}(I_j, \nu_j; B_j)$, denote by $f = f_0 \oplus f_1$ the $(B_0 + B_1)$ valued measurable function on ∂D such that $f_{|I_j} = f_j$. Now assume that f admits an analytic Poisson integral, still denoted by $z \mapsto f(z)$ inside D. Then (8.34) implies

$$\|f(0)\|_{(B_0,B_1)_\theta} \le \|f_0\|^{1-\theta}_{L_{p_0}(I_0,\nu_0;B_0)} \|f_1\|^{\theta}_{L_{p_1}(I_1,\nu_1;B_1)}$$

and a fortiori taking $p_0 = p_1 = 2$ by the arithmetic/geometric mean inequality

$$\|f(0)\|_{(B_0,B_1)_\theta} \le ((1-\theta)\|f_0\|^2_{L_2(I_0,\nu_0;B_0)} + \theta\|f_1\|^2_{L_2(I_1,\nu_1;B_1)})^{1/2}$$

or equivalently

$$\|f(0)\|_{(B_0,B_1)_\theta} \le (\|f_0\|^2_{L_2(I_0,m_{|I_0};B_0)} + \|f_1\|^2_{L_2(I_1,m_{|I_1};B_1)})^{1/2}.$$

By Theorem 8.24, using the conformal mapping χ_θ and the preceding inequality, recalling the notation for $\oplus_2$ in (1.9), we may write

$$\|x\|_{(B_0,B_1)_\theta} = \inf\{\|(f_0, f_1)\|_{L_2(I_0,m_{|I_0};B_0)\oplus_2 L_2(I_1,m_{|I_1};B_1)}\} \tag{8.43}$$

where the infimum is over all pairs

$$(f_0, f_1) \quad \text{in} \quad L_2(I_0, m_{|I_0}; B_0) \oplus_2 L_2(I_1, m_{|I_1}; B_1)$$

such that the Poisson integral f of $f_0 \oplus f_1$ is analytic on D and $f(0) = x$.

By a similar argument one can show

$$\|x\|_{(B_0,B_1)^\theta} = \inf\{(\|\mu_0\|^2_{\Lambda_2(I_0,m_{|I_0};B_0)} + \|\mu_1\|^2_{\Lambda_2(I_1,m_{|I_1};B_1)})^{1/2}\} \tag{8.44}$$

where the infimum is over all pairs

$$(\mu_0, \mu_1) \quad \text{in} \quad \Lambda_2(I_0, m_{|I_0}; B_0) \oplus_2 \Lambda_2(I_1, m_{|I_1}; B_1)$$

such that the Poisson integral f of $\mu_0 + \mu_1 \in \Lambda_2(\partial D, m; B_0 + B_1)$ is analytic and $f(0) = x$.

Theorem 8.37. *Assume that $B_0 \cap B_1$ is dense both in B_0 and B_1 (note that by Remark 8.13, we can easily reduce to this case). We have then an isometric identification*

$$(B_0, B_1)_\theta^* \simeq (B_0^*, B_1^*)^\theta.$$

If one of the spaces B_0^*, B_1^* is reflexive or has the RNP, or if $B_0 \subset B_1$ and B_0^* has the ARNP then

$$(B_0, B_1)_\theta^* \simeq (B_0^*, B_1^*)_\theta.$$

Proof (finite-dimensional case). To convince the reader that the proof is actually very simple and natural, we first prove the case when B_0, B_1 are both finite-dimensional to avoid all technicalities. We may as well assume $B_j = \mathbb{C}^n$ equipped with a norm $\|\cdot\|_{B_j}$. Assuming θ fixed, let us denote

$$\Lambda(B_0, B_1) = L_2(I_0, m_{|I_0}; B_0) \oplus_2 L_2(I_1, m_{|I_1}; B_1).$$

We have obviously (isometrically)

$$\Lambda(B_0, B_1)^* = \Lambda(B_0^*, B_1^*).$$

We will use a general elementary fact: Let $E_0 \subset E \subset \Lambda$ be (closed) subspaces of a Banach space Λ. We have then $E^\perp \subset E_0^\perp \subset \Lambda^*$ and isometrically

$$(E/E_0)^* = E_0^\perp / E^\perp. \tag{8.45}$$

Now let $E(B_0, B_1) \subset \Lambda(B_0, B_1)$ be the subset formed by the boundary values of analytic functions f in $H^2(D; \mathbb{C}^n)$, and let $E_0(B_0, B_1) \subset E(B_0, B_1)$ be the subspace of those f such that $f(0) = 0$. Note that $E_0(B_0, B_1) = zE(B_0, B_1)$. We have then isometrically by (8.43)

$$(B_0, B_1)_\theta = E(B_0, B_1)/E_0(B_0, B_1).$$

To describe the dual, simply observe that

$$E(B_0, B_1)^\perp = E_0(B_0^*, B_1^*) \text{ and } E_0(B_0, B_1)^\perp = E(B_0^*, B_1^*).$$

Thus (8.45) gives us the isometric identity $(B_0, B_1)_\theta^* \simeq (B_0^*, B_1^*)_\theta$. □

Proof (general case). Let $\Lambda = L_2(I_0, m_{|I_0}; B_0) \oplus_2 L_2(I_1, m_{|I_1}; B_1)$. Let $E \subset \Lambda$ be the subspace formed by the elements admitting an analytic Poisson integral f and let $E_0 \subset E$ be the subspace of those f such that $f(0) = 0$. Note that $E_0 = zE$. We claim that the set of all analytic polynomials $\sum z^k b_k$ with $b_k \in B_0 \cap B_1$

is dense in E. Consider $f \in E$. Let $F = f \circ \chi_\theta$. As in the proof of Theorem 8.24, let

$$F_\alpha(z) = \int F(z - it)\varphi_\alpha(t)\,dt.$$

By dominated convergence, $F_\alpha \circ \chi_\theta^{-1}$ tends to $F \circ \chi_\theta^{-1} = f$ in $E \subset \Lambda$. Replacing F_α by $e^{\delta z^2} F_\alpha$, and letting δ tends to 0, we find by Lemma 8.11, that the set of all $g \circ \chi_\theta^{-1}$ with $g \in \mathcal{G}$ is dense in E. From this it is easy to deduce the preceding claim (since this reduces the approximation problem to the case of a scalar valued $f \in \mathcal{F}_0$). We now return to (8.43). This just says that

$$(B_0, B_1)_\theta = E/E_0$$

isometrically. Consequently $(B_0, B_1)_\theta^* = (E/E_0)^* = E_0^\perp/E^\perp$ where $E^\perp \subset E_0^\perp \subset \Lambda^*$. Note that by Proposition 2.28, $\Lambda^* = \Lambda_2(I_0, m_{|I_0}; B_0^*) \oplus_2 \Lambda_2(I_1, m_{|I_1}; B_1^*)$. By what precedes, the set of $f = f_0 \oplus f_1$ of the form $f = \sum_0^N z^n b_n$ with $b_n \in B_0 \cap B_1$ is dense in E. Therefore $E_0^\perp$ consists in the set of μs in Λ^* such that $\langle \mu, z^n b\rangle = 0\ \forall n > 0\ \forall b \in B_0 \cap B_1$, or equivalently such that $\hat{\mu}(-n) = 0$ in $B_0^* + B_1^*\ \forall n > 0$. In other words, $E_0^\perp$ is precisely the subspace of those μ admitting an analytic Poisson integral. Similarly $E^\perp$ is the subset of $E_0^\perp$ of those μ for which the analytic extension vanishes at 0. Consequently, by (8.44) we have an isometric identity

$$E_0^\perp/E^\perp = (B_0^*, B_1^*)^\theta.$$

The last assertion follows from Theorem 8.34. □

Remark. The interest of the preceding statement is that it includes as special case the pair L_∞, L_1 over a finite measure space since L_∞^*, being an 'abstract L_1-space' has the ARNP (see Theorem 4.32).

Remark (A digression about twisted sums)**.** With the notation in the preceding proof, let $X = E/z^2E$. This is sometimes viewed as the 'derivative' of the scale, because of the following equivalent description. Let $x_0, x_1 \in B_0 + B_1$. Then $x_0 + zx_1$ defines an element of X iff there is $f \in E$ such that $f(0) = x_0$ and $f'(0) = x_1$. Moreover, the norm of X is the infimum of $\|f\|_E$ over all possible such f's.

Let $Y \subset X$ be defined by $Y = zE/z^2E$. Then both Y and X/Y are isometric to the space $(B_0, B_1)_\theta$. Indeed, since the shift is isometric, Y is isometric to E/zE, and similarly for X/Y. Thus X is what is called a 'twisted sum' of two copies of $(B_0, B_1)_\theta$. In the particular case $(B_0, B_1) = (\ell_\infty, \ell_1)$ with $\theta = 1/2$, we find a twisted sum of two copies of ℓ_2. However, it can be shown that the latter space is *not isomorphic to* ℓ_2. Actually, the resulting space X is isomorphic to the Kalton-Peck space Z_2. See [299, 303].

Lemma 8.38. *Assuming $B_0 \cap B_1$ is dense in both B_0 and B_1, for any $x \in (B_0, B_1)^\theta$ and $\xi \in B_0^* \cap B_1^*$ we have*

$$|\langle x, \xi \rangle| \le \|x\|_{(B_0,B_1)^\theta} \|\xi\|_{(B_0^*,B_1^*)_\theta}.$$

Proof. This is just a particular case of the mult-ilinear interpolation, in the variant indicated in Remark 8.33, applied to $T(x, \xi) = \langle x, \xi \rangle$ (and $C_0 = C_1 = \mathbb{C}$), which has norm 1 both from $B_0 \times B_0^*$ to $\mathbb{C}$ and from $B_1 \times B_1^*$ to $\mathbb{C}$, and hence from $(B_0, B_1)^\theta \times (B_0^*, B_1^*)_\theta$ to $(\mathbb{C}, \mathbb{C})^\theta = \mathbb{C}$. Alternately, we can argue using Lemma 8.32: we have $x_n \in (B_0, B_1)_\theta$ with $\|x_n\|_{(B_0,B_1)_\theta} \to \|x\|_{(B_0,B_1)^\theta}$, such that $x_n \to x$ in $B_0 + B_1$ and the inequality

$$|\langle x_n, \xi \rangle| \le \|x_n\|_{(B_0,B_1)_\theta} \|\xi\|_{(B_0^*,B_1^*)_\theta}$$

is immediate, so since $\xi \in B_0^* \cap B_1^* = (B_0 + B_1)^*$ implies $\langle x_n, \xi \rangle \to \langle x, \xi \rangle$, we obtain another proof of the announced inequality. □

Theorem 8.39. *The inclusion*

$$(B_0, B_1)_\theta \subset (B_0, B_1)^\theta$$

is isometric. Therefore, $(B_0, B_1)_\theta$ is the closure of $B_0 \cap B_1$ in $(B_0, B_1)^\theta$.

Proof. By (8.19) and Theorem 8.31 it suffices to prove this assuming that $B_0 \cap B_1$ is dense in both B_0 and B_1. Moreover, by Lemma 8.11, $B_0 \cap B_1$ (resp. $B_0^* \cap B_1^*$) is dense in $(B_0, B_1)_\theta$ (resp. $(B_0^*, B_1^*)_\theta$). Let x be in the unit ball of $(B_0, B_1)^\theta$ and assume $x \in B_0 \cap B_1$. By the preceding lemma and by the density of $B_0^* \cap B_1^*$ in $(B_0^*, B_1^*)_\theta$, for any ξ in the unit ball of $(B_0^*, B_1^*)_\theta$ we have $|\langle x, \xi \rangle| \le 1$. By Lemma 8.32 applied to the dual pair, we still have $|\langle x, \xi \rangle| \le 1$ for any ξ in the unit ball of $(B_0^*, B_1^*)^\theta$, but, by Theorem 8.37, the latter is the unit ball of $(B_0, B_1)_\theta^*$, so taking the supremum over it, we find $\|x\|_{(B_0,B_1)_\theta} \le 1$. Thus we have proved that $\|x\|_{(B_0,B_1)_\theta} \le \|x\|_{(B_0,B_1)^\theta}$ for any $x \in B_0 \cap B_1$. Since the reverse inequality is obvious, this means that the norms of $(B_0, B_1)_\theta$ and $(B_0, B_1)^\theta$ coincide on $B_0 \cap B_1$, and since $B_0 \cap B_1$ is dense in $(B_0, B_1)_\theta$, the Theorem follows. □

The following important result is called the 'reiteration theorem'.

Theorem 8.40. *Let (B_0, B_1) be a compatible couple. Let $0 \le \theta_0 < \theta_1 \le 1$. Let $B_\theta = (B_0, B_1)_\theta$. Then for any $0 < s < 1$ we have an isometric identity*

$$(B_{\theta_0}, B_{\theta_1})_s = B_\theta$$

where θ is determined by $\theta = (1 - s)\theta_0 + s\theta_1$.

Sketch of Proof. By Remark 8.13 and Lemma 8.11, we may assume that $B_0 \cap B_1$ is dense in both B_0 and B_1 and also in B_{θ_0} and B_{θ_1}. We will prove the statement under the supplementary assumption that $B_0 \cap B_1$ is dense in $(B_{\theta_0}, B_{\theta_1})_s$. Let $x \in B_\theta$ with $\|x\|_{B_\theta} < 1$. Let $f \in \mathcal{F}_0(B_0, B_1)$ be such that $f(\theta) = x$ and $\|f\|_{\mathcal{F}(B_0,B_1)} < 1$. Then, for any $0 \le \theta \le 1$, the function $t \mapsto f(\theta + it)$ is continuous from $\mathbb{R}$ to the unit ball of B_θ. The continuity is easy to deduce from (8.34) and the fact that $f \in \mathcal{F}_0$. In particular, $\|f(\theta_j + it)\|_{B_{\theta_j}} < 1$ for any $t \in \mathbb{R}$ and $j = 0, 1$. There is obviously a conformal (actually affine) mapping w taking the strip S to the strip $\{z \mid \theta_0 < \Re(z) < \theta_1\}$, such that $w(s) = \theta$ and $w(\partial_j) \subset \{\Re(z) = \theta_j\}$ for $j = 0, 1$. Namely $w(z) = \theta_0 + z(\theta_1 - \theta_0)$. Thus, considering $z \mapsto f(w(z))$, we obtain $\|f(\theta)\|_{(B_{\theta_0}, B_{\theta_1})_s} \le 1$. In other words, we have an inclusion $B_\theta \subset (B_{\theta_0}, B_{\theta_1})_s$ of norm ≤ 1. A similar reasoning shows that if $0 < \theta_0 < \theta_1 < 1$ we have an inclusion $(B_0, B_1)^\theta \subset (B^{\theta_0}, B^{\theta_1})^s$ of norm ≤ 1. Applying this to the dual couple (B_0^*, B_1^*), and using the duality Theorem 8.37, we find an inclusion $B_\theta^* \subset (B_{\theta_0}, B_{\theta_1})_s^*$ of norm ≤ 1. This implies that the inclusion $B_\theta \subset (B_{\theta_0}, B_{\theta_1})_s$ is isometric. By our supplementary assumption it has dense range, and hence is an isometric equality. The preceding restrictive assumption was removed in [194], to which we refer the reader. □

8.4 The real interpolation method

Let (B_0, B_1) be a compatible couple. For any $t > 0$ and any x in $B_0 + B_1$, we start by the definition of the so-called K-functional.

$$K_t(x; B_0, B_1) = \inf\{\|b_0\|_{B_0} + t\|b_1\|_{B_1} \mid x = b_0 + b_1\}.$$

We will often abbreviate and write simply $K_t(x)$ instead of $K_t(x; B_0, B_1)$ when the context leaves no room for ambiguity. Let $0 < \theta < 1$ and $1 \le q \le \infty$. We define

$$(B_0, B_1)_{\theta,q} = \left\{ x \in B_0 + B_1 \;\middle|\; \int_0^\infty (t^{-\theta} K_t(x))^q \frac{dt}{t} < \infty \right\} \tag{8.46}$$

and we equip it with the norm

$$\|x\|_{(B_0,B_1)_{\theta,q}} = \left(\int_0^\infty (t^{-\theta} K_t(x))^q \frac{dt}{t} \right)^{1/q}.$$

Of course, when $q = \infty$, this should be understood as meaning $\sup_t t^{-\theta} K_t(x)$.

Note

$$K_t(x; B_0, B_1) = t K_{t^{-1}}(x; B_1, B_0)$$

and hence

$$(B_0, B_1)_{\theta,q} = (B_1, B_0)_{1-\theta,q} \quad \text{isometrically.} \tag{8.47}$$

When $B_0 = B_1$ are identical say when $B_0 = B_1 = C$ we have obviously $K_t(x; C, C) = (1 \wedge t)\|x\|_C$ for any $t > 0$ and hence

$$\|x\|_{(C,C)_{\theta,q}} = C(\theta, q)\|x\|_C, \tag{8.48}$$

where we have set

$$\psi_\theta(t) = t^{-\theta} \wedge t^{1-\theta} \text{ and } C(\theta, q) = \|\psi_\theta\|_{L_q(dt/t)}. \tag{8.49}$$

Note that $t \to K_t(x)$ is by definition the infimum of a family of affine functions, hence it is concave on $\mathbb{R}_+$, non-negative and non-decreasing. For the same reason, $t \mapsto K_t(x)/t$ is non-increasing.

Since K_t is non-decreasing, we may write

$$(\theta q)^{-1} t^{-\theta q} K_t(x)^q = K_t(x)^q \int_t^\infty s^{-\theta q} ds/s \le \int_t^\infty (s^{-\theta} K_s(x))^q ds/s \le \|x\|^q_{(B_0,B_1)_{\theta,q}},$$

and hence

$$\|x\|_{(B_0,B_1)_{\theta,\infty}} \le (\theta q)^{1/q} \|x\|_{(B_0,B_1)_{\theta,q}}. \tag{8.50}$$

This shows

$$(B_0, B_1)_{\theta,q} \subset (B_0, B_1)_{\theta,\infty}. \tag{8.51}$$

More generally, for any $q_0 < q_1$ we have similarly

$$(B_0, B_1)_{\theta,q_0} \subset (B_0, B_1)_{\theta,q_1}. \tag{8.52}$$

Indeed, let $0 < \tau < 1$ be determined by $\frac{1}{q_1} = \frac{1-\tau}{q_0} + \frac{\tau}{\infty}$. By Hölder's inequality applied to $t \mapsto t^{-\theta} K_t(x)$, we have $\|x\|_{\theta,q_1} \le \|x\|^{1-\tau}_{\theta,q_0} \|x\|^{\tau}_{\theta,\infty}$, and hence by (8.50)

$$\|x\|_{\theta,q_1} \le ((\theta q_0)^{1/q_0})^\tau \|x\|_{\theta,q_0}.$$

Remark 8.41. Obviously we have inclusions (with norms bounded by constants independent of (B_0, B_1))

$$B_0 \cap B_1 \subset (B_0, B_1)_{\theta,q} \subset B_0 + B_1.$$

Indeed, by (8.50) $\|x\|_{B_0+B_1} = K_1(x) \le \|x\|_{\theta,\infty} \le (\theta q)^{1/q} \|x\|_{(B_0,B_1)_{\theta,q}}$, and also $K_t(x) \le (1 \wedge t)\|x\|_{B_0 \cap B_1}$, and hence $\|x\|_{(B_0,B_1)_{\theta,q}} \le C(\theta, q)\|x\|_{B_0 \cap B_1}$.

Remark 8.42. If we assume $B_0 \subset B_1$, then it is easy to check that, when $0 < \theta_0 < \theta_1 < 1$, for *arbitrary* $1 \le q_0, q_1 \le \infty$, we have bounded inclusions

$$B_0 \subset (B_0, B_1)_{\theta_0,q_0} \subset (B_0, B_1)_{\theta_1,q_1} \subset B_1. \tag{8.53}$$

Indeed, now $B_0 = B_0 \cap B_1$ and $B_1 = B_0 + B_1$. Let $x \in (B_0, B_1)_{\theta_0, q_0}$. By what precedes we have $x \in (B_0, B_1)_{\theta_0,\infty} \subset B_1$, and hence $K_t(x) \le t\|x\|_{B_1}$ for $0 < t < 1$, and $K_t(x) \le t^{\theta_0}\|x\|_{(B_0,B_1)_{\theta_0,\infty}}$ for $t \ge 1$, from which we deduce $\|x\|_{(B_0,B_1)_{\theta_1,q_1}} \le c\|x\|_{(B_0,B_1)_{\theta_0,q_0}}$ for some constant c. We skip the details.

Remark 8.43. Let $A_j \subset B_j$ be as before the closure of $B_0 \cap B_1$ in B_j $(j = 0, 1)$ with the norm induced by B_j. Clearly, for any $x \in B_0 \cap B_1$ we have $K_t(x; A_0, A_1) = K_t(x; B_0, B_1)$ for any $t > 0$. Thus $\|x\|_{(B_0,B_1)_{\theta,q}} = \|x\|_{(A_0,A_1)_{\theta,q}}$ for any such x. We will show in Remark 8.71 that $B_0 \cap B_1$ is dense in $(B_0, B_1)_{\theta,q}$ for any $q < \infty$; this implies that $(B_0, B_1)_{\theta,q} = (A_0, A_1)_{\theta,q}$ isometrically.

Remark 8.44. With the same notation, if $A_0 = B_0$ then for any $x \in B_0 + B_1$ we have

$$\lim_{t\to 0} K_t(x; B_0, B_1) = 0.$$

Indeed, assume $x = b_0 + b_1$ with $b_j \in B_j$, then for any $\varepsilon > 0$ there is $b_0' \in B_0 \cap B_1$ such that $\|b_0 - b_0'\|_{B_0} < \varepsilon$, and hence $x = (b_0 - b_0') + (b_0' + b_1)$ and $K_t(x; B_0, B_1) \le \|b_0 - b_0'\|_{B_0} + t\|b_0' + b_1\|_{B_1}$. Therefore, $\lim_{t\to 0} K_t(x; B_0, B_1) \le \|b_0 - b_0'\|_{B_0} < \varepsilon$, and the announced result follows.

Remark 8.45. Reviewing the proof of (8.50) shows that if $x \in (B_0, B_1)_{\theta,q}$ then $\lim_{t\to\infty} t^{-\theta} K_t(x; B_0, B_1) = 0$. We claim that also $\lim_{t\to 0} t^{-\theta} K_t(x; B_0, B_1) = 0$. Indeed, using the fact that $t \mapsto t^{-1}K_t(x)$ is non-increasing, we have for any $\varepsilon > 0$ and any $t < \varepsilon$

$$t^{-\theta}K_t(x) = t^{1-\theta}t^{-1}K_t(x) \le (q(1-\theta))^{1/q}\left(\int_0^t (s^{1-\theta}s^{-1}K_s(x))^q ds/s\right)^{1/q}$$

and hence $\sup_{t<\varepsilon} t^{-\theta}K_t(x) \le (q(1-\theta))^{1/q}(\int_0^\varepsilon (s^{-\theta}K_s(x))^q ds/s)^{1/q}$, which implies our claim.

Just as for the complex case, the fundamental interpolation property holds:

Theorem 8.46. *Let (B_0, B_1) and (C_0, C_1) be two compatible couples. Let $T_0\colon C_0 \to B_0$ and $T_1\colon C_1 \to B_1$ be bounded operators that are 'essentially the same'. Then the resulting operator $T\colon C_0 + C_1 \to B_0 + B_1$ maps $(C_0, C_1)_{\theta,q}$ to $(B_0, B_1)_{\theta,q}$ for any $0 < \theta < 1$ and $1 \le q \le \infty$, and moreover, if we denote its restriction by $T_{\theta,q}\colon (C_0, C_1)_{\theta,q} \to (B_0, B_1)_{\theta,q}$, have*

$$\|T_{\theta,q}\| \le \|T_0\|^{1-\theta}\|T_1\|^{\theta}.$$

Proof. We obviously can write for any x in $C_0 + C_1$, say $x = x_0 + x_1$ with $x_j \in C_j$

$$Tx = T_0x_0 + T_1x_1$$

and hence

$$\|T_0x_0\|_{B_0} + t\|T_0\|\|T_1\|^{-1}\|T_1x_1\|_{B_1} \le \|T_0\|(\|x_0\|_{C_0} + t\|x_1\|_{C_1})$$

so that

$$K_{t\|T_0\|\|T_1\|^{-1}}(Tx; B_0, B_1) \le \|T_0\|K_t(x; C_0, C_1). \tag{8.54}$$

Let $\lambda = \|T_0\|\|T_1\|^{-1}$. Since $\frac{dt}{t}$ is a Haar measure over the multiplicative group $(0, \infty)$, we have by (8.54)

$$\begin{aligned}\|t^{-\theta}K_t(Tx)\|_{L^q(\frac{dt}{t})} &= \|(t\lambda)^{-\theta}K_{t\lambda}(Tx)\|_{L^q(\frac{dt}{t})}\\ &\le \lambda^{-\theta}\|t^{-\theta}\|T_0\|K_t(x)\|_{L^q(\frac{dt}{t})}\\ &\le \|T_0\|\lambda^{-\theta}\|t^{-\theta}K_t(x)\|_{L^q(\frac{dt}{t})},\end{aligned}$$

and hence

$$\|T_{\theta,q}\| \le \|T_0\|\lambda^{-\theta} = \|T_0\|^{1-\theta}\|T_1\|^{\theta}. \qquad \square$$

Although the next inequality (8.55) can be proved directly, it is somewhat instructive to derive it from the preceding Theorem:

Corollary 8.47. *For any $x \in (B_0, B_1)_{\theta,q}$ $(0 < \theta < 1, 1 \le q \le \infty)$ we have*

$$\|x\|_{(B_0,B_1)_{\theta,q}} \le C(\theta, q)\|x\|_{B_0}^{1-\theta}\|x\|_{B_1}^{\theta}. \tag{8.55}$$

Proof. Fix $x \in (B_0, B_1)_{\theta,q}$. We apply the theorem with $C_0 = C_1 = \mathbb{R}$ (or $C_0 = C_1 = \mathbb{C}$ in the complex case) and with $T(\lambda) = \lambda x$. We have then by (8.48)

$$\|x\|_{(B_0,B_1)_{\theta,q}} = \|T(1)\|_{(B_0,B_1)_{\theta,q}} \le \|1\|_{(C_0,C_1)_{\theta,q}} = C(\theta, q),$$

where $C(\theta, q)$ is as in (8.49). $\square$

Remark 8.48. Just like in the complex case, considered in Lemma 8.9, real interpolation 'commutes' with the operation of passing to subspaces provided we make the rather strong assumption that the subspaces are simultaneously complemented. More precisely, in the situation of Lemma 8.9, we have obviously for any $x \in A_0 + A_1$

$$K_t(x; B_0, B_1) \le K_t(x; A_0, A_1) \le c_0K_{tc_1/c_0}(x; B_0, B_1).$$

This implies that for any $0 < \theta < 1$ and any $1 \le q \le \infty$, the inclusion $(A_0, A_1)_{\theta,q} \subset (B_0, B_1)_{\theta,q}$ is an isomorphic embedding, and the norm of $(A_0, A_1)_{\theta,q}$ is equivalent to the norm induced on it by $(B_0, B_1)_{\theta,q}$. If $c_0 = c_1 = 1$, the embedding is isometric and the latter norms coincide.

Remark. It is proved in [328] that except for 'trivial couples' the space $(A_0, A_1)_{\theta,q}$ contains as a subspace an isomorphic copy of ℓ_q when $1 \leq q < \infty$ and of c_0 when $q = \infty$. Here by 'trivial couple' we mean one such that the norms of A_0 and A_1 are equivalent on $A_0 \cap A_1$.

The efficiency of the K-functional is partly due to the following classical Hardy inequalities, comparing the integrability properties of a non-negative function $t \mapsto f(t)$ to that of the function $t \mapsto \frac{1}{t}\int_0^t f(s)ds$.

Lemma 8.49. *Let $f : (0, \infty) \to \mathbb{R}_+$ be any measurable function. Then for any $-\infty < \lambda < 1$ and any $1 \leq q \leq \infty$ we have*

$$\left(\int_0^\infty [t^\lambda \frac{1}{t}\int_0^t f(s)ds]^q dt/t\right)^{1/q} \leq (1-\lambda)^{-1}\left(\int_0^\infty [t^\lambda f(t)]^q dt/t\right)^{1/q},$$

and

$$\left(\int_0^\infty [t^{1-\lambda}\int_t^\infty f(s)ds/s]^q dt/t\right)^{1/q} \leq (1-\lambda)^{-1}\left(\int_0^\infty [t^{1-\lambda} f(t)]^q dt/t\right)^{1/q}.$$

with the usual convention if $q = \infty$.

Proof. We apply Hölder's inequality to the product $f(s) = (s^{-\lambda/q'}) \times (s^{\lambda/q'} f(s))$. This yields

$$\frac{1}{t}\int_0^t f(s)ds \leq \left(\frac{1}{t}\int_0^t s^{-\lambda}ds\right)^{1/q'}\left(\frac{1}{t}\int_0^t s^{\lambda q/q'} f(s)^q ds\right)^{1/q}.$$

Since $q/q' = q - 1$ and $t^{\lambda q}(\frac{1}{t}\int_0^t s^{-\lambda}ds)^{q/q'}t^{-2} = (1-\lambda)^{1-q}t^{\lambda-2}$ we find

$$\int_0^\infty [t^\lambda \frac{1}{t}\int_0^t f(s)ds]^q dt/t \leq (1-\lambda)^{1-q}\int_0^\infty s^{\lambda(q-1)} f(s)^q \left(\int_s^\infty t^{\lambda-2}dt\right) ds,$$

which leads to the first Hardy inequality. The proof of the other one is similar and left to the reader. □

Remark. The reader will check easily, with cosmetic changes, that all the results of the preceding section remain valid for a pair (B_0, B_1) of quasi-Banach spaces in the sense of Remark 1.8, assumed compatible in the usual sense. Of course, in that case, the resulting space $(B_0, B_1)_{\theta,q}$ also is a quasi-Banach space.

8.5 Real interpolation between L_p-spaces

The fundamental example is the case of L_p-spaces: Let $(\Omega, \mathcal{A}, m)$ be a measure space, and let $f\colon\ \Omega \to \mathbb{R}$ be a measurable function, such that $m(\{|f| > c\}) < \infty$ for some $c > 0$. We will need to consider the decreasing rearrangement of f.

Traditionally, this is denoted by f^*, but since this notation is also used for maximal functions and since $x \mapsto x^*$ is also the involution on a non-commutative L_p-space, we choose, to avoid confusion, to denote it instead by $f^\dagger$.

We define the decreasing rearrangement $f^\dagger\colon (0,\infty) \to \mathbb{R}_+$ of f by setting

$$f^\dagger(t) = \inf\{c > 0 \mid m(\{|f| > c\}) \le t\}.$$

Then $f^\dagger \ge 0$ is non-increasing and right continuous.

Moreover $m(\{|f| > f^\dagger(t)\}) \le t$ and $m(\{|f| \ge f^\dagger(t)\}) \ge t$. If the distribution function of f is continuous, the value $c = f^\dagger(t)$ is characterized by $m(\{|f| > c\}) = t$. Then, $f^\dagger$ is equal to the inverse of $c \mapsto m(\{|f| > c\})$. In any case:

$$\forall c > 0 \quad |\{f^\dagger > c\}| = m(\{|f| > c\}). \tag{8.56}$$

The latter equality shows that $f^\dagger$ and $|f|$ have the same distribution relative respectively to Lebesgue measure on $(0,\infty)$ and m. Recall that

$$\int |f|^p dm = \int_0^\infty pc^{p-1} m(\{|f| > c\})\, dc.$$

As an immediate consequence of (8.56) we have in particular

$$\forall p > 0 \qquad \int_0^\infty f^\dagger(t)^p dt = \int |f|^p\, dm.$$

More generally, the Lorentz spaces $L_{p,q}(\Omega, m)$ (or simply $L_{p,q}$) are defined ($0 < p, q < \infty$) as formed of the functions f such that

$$\int_0^\infty (t^{1/p} f^\dagger(t))^q \frac{dt}{t} < \infty$$

equipped with the quasi-norm

$$\|f\|_{p,q} = \left(\int_0^\infty (t^{1/p} f^\dagger(t))^q \frac{dt}{t} \right)^{1/q}. \tag{8.57}$$

Note that $L_{p,p} = L_p$ isometrically. When $q = \infty$, the preceding should be understood as

$$\|f\|_{p,\infty} = \sup_{t>0} t^{1/p} f^\dagger(t) = (\sup_{c>0} c^p m(\{|f| > c\}))^{1/p}.$$

The space $L_{p,\infty}$ is usually called 'weak L_p'; we already gave a brief introduction to it in §5.14. A priori, $f \mapsto \|f\|_{p,q}$ is only a quasi-norm. However, as the next statement shows, when $p > 1$ (and only then), it is equivalent to a norm.

Theorem 8.50. *Let $L_p = L_p(\Omega, \mathcal{A}, m)$ on an arbitrary measure space. Consider $f \in L_1 + L_\infty$. Then*

$$K_t(f; L_1, L_\infty) = \int_0^t f^\dagger(s) ds. \tag{8.58}$$

Consequently, for any $0 < \theta < 1$, *and* $1 \le q \le \infty$

$$(L_1, L_\infty)_{\theta,q} = L_{p,q}$$

where $1 < p < \infty$ *is determined by* $\frac{1}{p} = \frac{1-\theta}{1} + \frac{\theta}{\infty}$, *and the norm of* $(L_1, L_\infty)_{\theta,q}$ *is equivalent to the quasi-norm* $f \mapsto \|f\|_{p,q}$. *In particular* $(L_1, L_\infty)_{\theta,p} = L_p$.

Proof. Let w be the 'sign' (or the argument in the complex case) of f, i.e. we set $w(x) = f(x)|f(x)|^{-1}$ if $f(x) \neq 0$, and, say, $w(x) = 1$ otherwise (this value is irrelevant). Fix $t > 0$. Let

$$\begin{aligned} f_0 &= 1_{\{|f|>f^\dagger(t)\}}(f - f^\dagger(t)w) = w1_{\{|f|>f^\dagger(t)\}}(|f| - f^\dagger(t)) \\ f_1 &= f - f_0. \end{aligned}$$

Note $|f_1| = |f| \wedge f^\dagger(t)$. Then let $\Omega_t = \{|f| > f^\dagger(t)\}$. We have

$$\begin{aligned} K_t(f; L_1, L_\infty) &\le \|f_0\|_1 + t\|f_1\|_\infty \\ &\le \int_{\Omega_t} [|f| - f^\dagger(t)]\, dm + tf^\dagger(t) \\ &= \int_0^{m(\Omega_t)} (f^\dagger(s) - f^\dagger(t))\, ds + tf^\dagger(t) \end{aligned}$$

and hence since $m(\Omega_t) \le t$ and $f^\dagger(s) = f^\dagger(t)$ on $[m(\Omega_t), t]$, we obtain

$$K_t(f) \le \int_0^t f^\dagger(s)\, ds.$$

Conversely, assume that $f = f_0 + f_1$, $f_0 \in L_1$, $f_1 \in L_\infty$. Clearly $m(\{|f| > c_0 + c_1\}) \le m(\{|f_0| > c_0\}) + m(\{|f_1| > c_1\})$, and hence for any $0 < \varepsilon < 1$

$$f^\dagger(s) \le f_0^\dagger((1-\varepsilon)s) + f_1^\dagger(\varepsilon s),$$

so that

$$\begin{aligned} \int_0^t f^\dagger(s)ds &\le \int_0^t f_0^\dagger((1-\varepsilon)s)\, ds + \int_0^t f_1^\dagger(\varepsilon s)ds \\ &\le \int_0^t f_0^\dagger((1-\varepsilon)s)ds + tf_1^\dagger(0) \\ &\le (1-\varepsilon)^{-1}\|f_0^\dagger\|_1 + t\|f_1^\dagger\|_\infty = (1-\varepsilon)^{-1}\|f_0\|_1 + t\|f_1\|_\infty. \end{aligned}$$

Taking the limit when $\varepsilon \to 0$ and the infimum over f_0, f_1 yields

$$\int_0^t f^\dagger(s)\, ds \le K_t(f; L_1, L_\infty).$$

To complete the proof it suffices to prove

$$\|f\|_{p,q} \le \|f\|_{(L_1,L_\infty)_{\theta,q}} \le \theta^{-1}\|f\|_{p,q}, \tag{8.59}$$

where $1-\theta = p^{-1}$ (and hence $\theta^{-1} = p'$).

Let $f^{\dagger\dagger}(t) = t^{-1}\int_0^t f^\dagger(s)ds$. Note that $f^\dagger(t) \le f^{\dagger\dagger}(t)$, and hence

$$\|f\|_{p,q} = \|t^{1-\theta} f^\dagger\|_{L_q(\frac{dt}{t})} \le \|t^{1-\theta} f^{\dagger\dagger}\|_{L_q(t^{-1}dt)}.$$

For the converse direction, we write $\int_0^t f^\dagger(s)ds = \int_0^1 f^\dagger(st)t\ ds$, so that $f^{\dagger\dagger}(t) = \int_0^1 f^\dagger(st)ds$. Then by Jensen's inequality (since $q \ge 1$) we have

$$\|t^{1-\theta} f^{\dagger\dagger}\|_{L_q(t^{-1}dt)} \le \int_0^1 \|t^{1-\theta} f^\dagger(st)\|_{L_q(t^{-1}dt)}ds$$
$$= \|t^{1-\theta} f^\dagger(t)\|_{L_q(t^{-1}dt)} \int_0^1 s^{\theta-1}ds = \theta^{-1}\|f\|_{p,q},$$

which proves (8.59). □

We define $L_{p,q}(B)$ as formed of those $f \in L_0(B)$ such that $\omega \mapsto \|f(\omega)\|_B$ is in $L_{p,q}$ and we equip this space with the quasi-norm

$$\|f\|_{L_{p,q}(B)} = \|\,\|f(\cdot)\|_B\|_{L_{p,q}}. \tag{8.60}$$

Then the following fact is immediate.

Corollary 8.51. *For any $x \in L_1(B) + L_\infty(B)$ and any $t > 0$ we have*

$$K_t(x; L_1(B), L_\infty(B)) = K_t(\|x(\cdot)\|_B; L_1, L_\infty).$$

Therefore

$$(L_1(B), L_\infty(B))_{\theta,q} = L_{p,q}(B)$$

and the quasi-norm of $L_{p,q}(B)$ is equivalent to the norm of $(L_1(B), L_\infty(B))_{\theta,q}$.

We leave the proof as an exercise.

Remark 8.52. Let (Ω, m) be $\mathbb{R}_+$ equipped with Lebesgue's measure. Let $f : \mathbb{R}_+ \to \mathbb{R}_+$ be a non-increasing continuous (or merely Borel) function on $\mathbb{R}_+$ Then, since $\{f > f(t)\} \subset [0,t)$ we have $|\{f > f(t)\}| \le t$ and hence $f^\dagger(t) \le f(t)$ for any $t > 0$. But since f and $f^\dagger$ have the same distribution we must have $f = f^\dagger$ almost everywhere on $\mathbb{R}_+$.

This applies in particular to the function $f^{\dagger\dagger}$ (note that since $t \mapsto K_t(f)/t$ is always non-increasing, so is $f^{\dagger\dagger}$ by (8.58)) and of course also to $f^\dagger$. Thus we have $(f^{\dagger\dagger})^{\dagger} = f^{\dagger\dagger}$ a.e. and consequently (8.58) implies

$$\|f\|_{(L_1,L_\infty)_{\theta,q}} = \|f^{\dagger\dagger}\|_{p,q} = \left(\int_0^\infty (t^{1/p} f^{\dagger\dagger}(t))^q \frac{dt}{t}\right)^{1/q}. \tag{8.61}$$

Thus we may rewrite (8.59) as

$$\|f^{\dagger}\|_{p,q} \le \|f^{\dagger\dagger}\|_{p,q} \le \theta^{-1}\|f^{\dagger}\|_{p,q}. \tag{8.62}$$

Lemma 8.53. *Let $f \in L_1 + L_\infty$ and $t > 0$. Then*

$$\int_0^t f^{\dagger}(s)ds = \sup\left\{\int w|f|dm \mid w \ge 0, \int wdm = t\right\}.$$

If m is non-atomic, we have

$$\int_0^t f^{\dagger}(s)ds = \sup\left\{\int_E |f|dm \mid m(E) = t\right\}.$$

Proof. The first equality follows from the second one applied to the function $(\omega, t) \mapsto f(\omega)$ defined on $(\Omega, m) \times ([0, 1], dt)$, which is non-atomic.

We now turn to the second one. It is easy to check that $(1_E f)^{\dagger}(s) = 0$ for all $s > m(E)$ and also that $(1_E f)^{\dagger}(s) \le f^{\dagger}(s)$ for all $s > 0$. Therefore,

$$\int_E |f|dm = \int_0^\infty (1_E f)^{\dagger}(s)ds \le \int_0^t f^{\dagger}(s)ds.$$

This yields

$$\sup_{m(E)=t} \int_E |f|dm \le \int_0^t f^{\dagger}(s)ds. \tag{8.63}$$

If $m\{|f| = f^{\dagger}(t)\} = 0$, the converse inequality is easy: we have $|\{f^{\dagger} > f^{\dagger}(t)\}| = m(\{|f| > f^{\dagger}(t)\}) = t$ and hence the choice of $E = \{|f| > f^{\dagger}(t)\}$ shows that (8.63) is an equality. Indeed, since $|f|$ and $f^{\dagger}$ have the same distribution, we have

$$\int_{\{|f|>f^{\dagger}(t)\}} |f|dm = \int_{\{f^{\dagger}>f^{\dagger}(t)\}} f^{\dagger}(s)ds = \int_0^t f^{\dagger}(s)ds.$$

If $m\{|f| = f^{\dagger}(t)\} > 0$, a little more care is needed. We have $m(\{|f| > f^{\dagger}(t)\}) \le t \le m(\{|f| \ge f^{\dagger}(t)\})$. We will use the assumption that m is non-atomic to select a set E such that $\{|f| > f^{\dagger}(t)\} \subset E \subset \{|f| \ge f^{\dagger}(t)\}$, with $m(E) = t$. Let $t' = m\{|f| > f^{\dagger}(t)\}$. Since $|f|$ and $f^{\dagger}$ have the same distribution and $\{f^{\dagger} > f^{\dagger}(t)\} = [0, t')$, we have then

$$\int_{\{|f|>f^{\dagger}(t)\}} |f|dm = \int_0^{t'} f^{\dagger}ds$$

and hence

$$\int_E |f|dm \le \int_{\{|f|>f^\dagger(t)\}} |f|dm + (t-t')f^\dagger(t) = \int_0^{t'} f^\dagger ds + (t-t')f^\dagger(t)$$
$$= \int_0^t f^\dagger(s)ds.$$

□

Remark 8.54. Let $0 < r < 1$. We return to the situation and notation of Theorem 8.50 but we replace L_1 by the r-normed space L_r. Let $f \in L_r + L_\infty$. It is easy to check that there is a constant $c > 0$ (depending on r) such that for all $t > 0$

$$c^{-1}\left(\int_0^{t^r} f^\dagger(s)^r ds\right)^{1/r} \le K_t(f; L_r, L_\infty) \le c\left(\int_0^{t^r} f^\dagger(s)^r ds\right)^{1/r}. \quad (8.64)$$

Consequently, for any $0 < \theta < 1$ and $r \le q \le \infty$

$$(L_r, L_\infty)_{\theta,q} = L_{p,q}$$

where $\frac{1}{p} = \frac{1-\theta}{r} + \frac{\theta}{\infty}$, and the quasi-norm of $(L_r, L_\infty)_{\theta,q}$ is equivalent to the quasi-norm $f \mapsto \|f\|_{p,q}$. Moreover, the latter is an r-norm. In particular $(L_r, L_\infty)_{\theta,p} = L_p$. The role of $f^{\dagger\dagger}(t)$ is now played by $(\int_0^{t^r} f^\dagger(s)^r ds)^{1/r}$. This generalization of Theorem 8.50 is easy to check by a simple modification of its proof.

More generally, using a suitable version of the reiteration theorem for the real method (described in Theorem 8.72) we will identify the interpolation spaces between the Lorentz spaces.

Theorem 8.55. *Consider* $0 < p_0, q_0, p_1, q_1 \le \infty$. *Assume* $p_0 \ne p_1$. *Then, for any* $0 < \theta < 1$ *and* $0 < q \le \infty$,

$$(L_{p_0,q_0}, L_{p_1,q_1})_{\theta,q} = L_{p_\theta,q}$$

with equivalent norms, where

$$1/p_\theta = (1-\theta)/p_0 + \theta/p_1.$$

In particular, $(L_{p_0}, L_{p_1})_{\theta,q} = L_{p_\theta,q}$, *and the latter space coincides with* L_{p_θ} *if* $q = p_\theta$. *Moreover, if* $p_0 = p_1 = p$ *then we have*

$$(L_{p,q_0}, L_{p,q_1})_\theta = L_{p,q_\theta}$$

with equivalent norms where $1/q_\theta = (1-\theta)/q_0 + \theta/q_1$.

Proof. See Remark 8.73. □

Remark. Historically, the preceding result was inspired by, and appears as an abstract version of the Marcinkiewicz interpolation Theorem (see Theorem 5.78). It implies it as an easy corollary: if an operator is bounded both from L_{p_0} to $L_{p_0,\infty}$ and from L_{p_1} to $L_{p_1,\infty}$ ($p_0 \neq p_1$), then it is bounded from $(L_{p_0}, L_{p_1})_{\theta,p}$ to $(L_{p_0,\infty}, L_{p_1,\infty})_{\theta,p}$, and hence, choosing $p = p_\theta$, we conclude, by the preceding Theorem that it is bounded from L_{p_θ} to itself.

Remark 8.56. Let (B_0, B_1) be a compatible pair. Let $0 < \theta < 1$, $1 \le q \le \infty$ and let p be determined by $p^{-1} = 1 - \theta$. Consider $x \in B_0 + B_1$ and let $f_x(t) = t^{-1}K_t(x; B_0, B_1)$. By Remark 8.52, since f_x is non-increasing on $\mathbb{R}_+$, we have $(f_x)^\dagger = f_x$ a.e. and hence

$$\|x\|_{(B_0,B_1)_{\theta,q}} = \|f_x\|_{p,q} = \left(\int_0^\infty (t^{1/p}f_x(t))^q \frac{dt}{t}\right)^{1/q}.$$

8.6 The K-functional for $(L_1(B_0), L_\infty(B_1))$

Throughout the rest of this section, (B_0, B_1) will be an arbitrary compatible couple of Banach spaces. Then the pair $(\ell_1(B_0), \ell_\infty(B_1))$ can be viewed as 'compatible' in an obvious way. More generally, for any measure space (Ω, m), we can view the couple $(L_1(m; B_0), L_\infty(m; B_1))$ as compatible as already explained earlier before Theorem 8.21.

The easy case when $B_0 = B_1 = B$ was settled by Corollary 8.51.

We will first consider the case of discrete measure spaces, or equivalently vector valued sequence spaces. We start by a lemma conveniently reducing the sequence space case to that of finite sequences.

Lemma 8.57. *Let P_n denote the projection from $\ell_1(B_0) + \ell_\infty(B_1)$ onto $\ell_1(B_0) + \ell_\infty(B_1)$ that preserves the first n coordinates and annihilates the other ones. Then*

$$\forall x \in \ell_1(B_0) + \ell_\infty(B_1) \quad K_t(x; \ell_1(B_0), \ell_\infty(B_1)) = \sup_n K_t(P_n(x); \ell_1(B_0), \ell_\infty(B_1)). \tag{8.65}$$

Proof. To clarify the notation, if x is a sequence of elements in a Banach space, just for the present proof, we denote by $x(k)$ the k–th coordinate of x. Fix $t > 0$. Clearly the right-hand side of (8.65) is not more than its left-hand side. Conversely, assume that the right-hand side of (8.65) is < 1. We will show that the left side also is at most 1.

By our assumption, for all x as in (8.65) and for all n, there is a decomposition $P_n(x) = x_0^n + x_1^n$ such that

$$\|x_0^n\|_{\ell_1(B_0)} + t\|x_1^n\|_{\ell_\infty(B_1)} < 1. \tag{8.66}$$

Let $\mathcal{U}$ be a non-trivial ultrafilter on the positive integers. We let n tend to infinity along $\mathcal{U}$ and we denote simply by $\lim_{\mathcal{U}}$ the various resulting limits. Let

$$R = \lim_{\mathcal{U}} \|x_1^n\|_{\ell_\infty(B_1)} \quad \text{and} \quad a_k = \lim_{\mathcal{U}} \|x_0^n(k)\|_{B_0}.$$

Observe that (8.66) implies

$$\forall K \in \mathbb{N} \quad \left(\sum\nolimits_{k<K} a_k\right) + tR \le 1. \tag{8.67}$$

Now fix $\varepsilon > 0$. For each integer k we can find an integer $n_k > k$ large enough so that

$$\|x_0^{n_k}(k)\|_{B_0} < a_k + \varepsilon 2^{-k} \quad \text{and} \quad \|x_1^{n_k}\|_{\ell_\infty(B_1)} < R + \varepsilon.$$

Then we can define

$$x_0(k) = x_0^{n_k}(k) \quad \text{and} \quad x_1(k) = x_1^{n_k}(k).$$

Clearly $x(k) = x_0(k) + x_1(k)$ for all k, and moreover

$$\forall K \quad \sum\nolimits_{k<K} \|x_0(k)\|_{B_0} + t \sup_{k<K} \|x_1(k)\|_{B_1} < \sum\nolimits_{k<K} a_k + \varepsilon 2^{-k} + t(R+\varepsilon)$$

hence by (8.67)

$$\le 1 + \varepsilon(2+t).$$

Since this holds for all K, we conclude that $x_0 \in \ell_1(B_0)$, $x_1 \in \ell_\infty(B_1)$ and $\|x_0\|_{\ell_1(B_0)} + t\|x_1\|_{\ell_\infty(B_1)} \le 1 + \varepsilon(2+t)$, and since $\varepsilon > 0$ is arbitrary we indeed finally obtain

$$K_t(x; \ell_1(B_0), \ell_\infty(B_1)) \le 1. \qquad \square$$

Theorem 8.58. *Consider the pair $(\ell_1(B_0), \ell_\infty(B_1))$. Then, $\forall x = (x_i) \in \ell_1(B_0) + \ell_\infty(B_1)$, we have*

$$K_t(x; \ell_1(B_0), \ell_\infty(B_1)) = \sup\left\{\sum\nolimits_i K_{t_i}(x_i; B_0, B_1),\ t_i \ge 0,\ \sum t_i \le t\right\}. \tag{8.68}$$

Proof. By the preceding lemma it suffices to prove this assuming that $i \mapsto x_i$ is finitely supported. Let us denote by C_t the right-hand side of the preceding identity (8.68). It is easy to check that $C_t \le K_t(x; \ell_1(B_0), \ell_\infty(B_1))$. Let us check the converse. Let x be such that $C_t < 1$. This means

$$\sup\nolimits_{\sum t_i \le t} \left\{ \inf_{x_i = a_i + b_i} \left(\sum \|a_i\|_{B_0} + t_i \|b_i\|_{B_1} \right) \right\} < 1. \tag{8.69}$$

We want to deduce from this the same inequality but with the inf and the sup interchanged. This can be viewed as a variant of the minimax lemma. We prefer to deduce it, slightly more directly, from the Hahn-Banach theorem, as follows. The inequality (8.69) clearly implies (choosing $t_i = t\xi_i$) that for any non-negative sequence $\xi = (\xi_i)$ such that $\sum \xi_i < 1$ there is, for each index i a decomposition $x_i = \alpha_i + \beta_i$ in $B_0 + B_1$ such that

$$\sum_i \xi_i \left[\left(\sum_k \|\alpha_k\|_{B_0} \right) + t\|\beta_i\|_{B_1} \right] < 1. \tag{8.70}$$

We will show that the left side of (8.68) is ≤ 1. We assume that, for some n, we have $x_i = 0$ for all indices $i \geq n$. Let $C \subset \mathbf{R}^n$ be the set of all points $y = (y_i)$ of the form

$$y_i = \left(\sum_{k \geq 0} \|a_k\|_{B_0} \right) + t\|b_i\|_{B_1} \text{ where } x_i = a_i + b_i,\ a_i \in B_0,\ b_i \in B_1.$$

We claim that the convex hull of C, denoted by $\mathrm{conv}(C)$ intersects $(-\infty, 1)^n$. Otherwise, by Hahn-Banach (we separate a convex set from an open convex one) we would find a separating functional ξ and a real number r such that $\xi < r$ on $(-\infty, 1)^n$ and $\xi > r$ on C. But, since we obviously can assume $r = 1$ and hence $\xi_i \geq 0$ for all i and $\sum \xi_i < 1$, this would contradict (8.70). This shows that $\mathrm{conv}(C)$ intersects $(-\infty, 1)^n$, hence we can find decompositions $x_i = a_i^m + b_i^m$, $1 \leq m \leq M$ and positive scalars $\lambda_1, ..., \lambda_m, ..., \lambda_M$ with $\sum_m \lambda_m = 1$, such that we have for every index i

$$\sum_m \lambda_m \left[\left(\sum_{k \geq 0} \|a_k^m\|_{B_0} \right) + t\|b_i^m\|_{B_1} \right] < 1. \tag{8.71}$$

We can then set

$$a_i = \sum_m \lambda_m a_i^m,\ b_i = \sum_m \lambda_m b_i^m.$$

Note that $x_i = a_i + b_i$. Moreover, by (8.71) and the triangle inequality, for every index i

$$\sum_{k \geq 0} \|a_k\|_{B_0} + t\|b_i\|_{B_1} \leq 1,$$

which clearly implies $K_t(x; \ell_1(B_0), \ell_\infty(B_1)) \leq 1$. By homogeneity, this completes the proof of (8.68). □

Remark 8.59. The formula (8.68) remains valid with the same proof as earlier if the spaces B_0 and B_1 are replaced by families of Banach spaces respectively (B_0^n) and (B_1^n). Let us denote by $\ell_1(\{B_0^n\})$ and $\ell_\infty(\{B_1^n\})$ the corresponding spaces (these are sometimes called the direct sum of the families (B_0^n) and (B_1^n) respectively in the sense of ℓ_1 and ℓ_∞). This gives us the following generalized version of (8.68) : for all x in $\ell_1(\{B_0^n\}) + \ell_\infty(\{B_1^n\})$

$$K_t(x; \ell_1(\{B_0^n\}), \ell_\infty(\{B_1^n\})) = \sup\left\{\sum\nolimits_i K_{t_i}(x_i; B_0^i, B_1^i),\ t_i \geq 0,\ \sum t_i \leq t\right\}. \tag{8.72}$$

In particular, given a couple (B_0, B_1), and weights $w_i > 0$, if we denote by $\ell_1(w; B_0)$ the set of sequences (x_i)s in B_0 such that $\|(x_i)\|_{\ell_1(w;B_0)} = \sum w_i \|x_i\|_{B_0}$ is finite, we find for any $x = (x_i) \in \ell_1(w; B_0) + \ell_\infty(B_1)$

$$K_t(x; \ell_1(w; B_0), \ell_\infty(B_1)) = \sup\left\{\sum\nolimits_i w_i K_{t_i}(x_i; B_0, B_1),\ t_i \geq 0,\ \sum w_i t_i \leq t\right\}. \tag{8.73}$$

We can now deduce the function space case.

Theorem 8.60. *Let $f \in L_1(\Omega, \mathcal{A}, m; B_0) + L_\infty(\Omega, \mathcal{A}, m; B_1)$. For all $t > 0$*

$$K_t(f; L_1(\Omega, \mathcal{A}, m; B_0), L_\infty(\Omega, \mathcal{A}, m; B_1)) = \sup_{\int \phi dm \leq t} \int K_{\phi(\omega)}(f(\omega); B_0, B_1) dm(\omega), \tag{8.74}$$

where the sup runs over all non-negative measurable functions ϕ defined on $(\Omega, \mathcal{A})$ with integral at most t.

Proof. We may clearly assume that the measure space is σ-finite. We need a preliminary observation: Let f be any Bochner measurable function with values in a Banach space B. Then for any $\varepsilon > 0$ there is a countable measurable partition of the measure space such that the oscillation of f for the norm of B is $\leq \varepsilon$. Indeed, since f takes values in a separable subspace, we may assume that B is separable. Then B itself is the union of countably many balls of diameter $\leq \varepsilon$. From this we may form a countable Borel partition of B of sets $\{B_n\}$ each of diameter $\leq \varepsilon$. If we then partition Ω by the collection $\{f^{-1}(B_n)\}$, we obtain the observation.

We now claim that for any $f \in L_1(\Omega, \mathcal{A}, m; B_0) + L_\infty(\Omega, \mathcal{A}, m; B_1)$ and any $\varepsilon > 0$, we can find a σ-subalgebra $\mathcal{B} \subset \mathcal{A}$ generated by a countable measurable partition of Ω into sets $\Omega_i \subset \Omega$ of finite measure such that, if we set

$$f^{\mathcal{B}} = \sum\nolimits_i 1_{\Omega_i} \frac{1}{m(\Omega_i)} \int_{\Omega_i} f dm$$

(this is nothing but the conditional expectation of f with respect to $\mathcal{B}$), then $f^{\mathcal{B}} \in L_1(\Omega, \mathcal{B}, m; B_0) + L_\infty(\Omega, \mathcal{B}, m; B_1)$ and we have

$$K_t(f - f^{\mathcal{B}}; L_1(\Omega, \mathcal{A}, m; B_0), L_\infty(\Omega, \mathcal{A}, m; B_1)) < \varepsilon.$$

Indeed, given a function $f_0 \in L_1(\Omega, \mathcal{A}, m; B_0)$ (resp. $f_1 \in L_\infty(\Omega, \mathcal{A}, m; B_1)$) there is a countable measurable partition of Ω into pieces on each of which the oscillation of f_0 for the norm of B_0 (resp. B_1) is small. On the other hand, since the measure space is σ-finite, it admits a countable measurable partition into sets of finite measure, so that, by refining the partitions, we can always assume that the sets have finite measure and that the same partition works for both f_0 and f_1. From this the claim becomes clear.

This claim reduces the proof of (8.74) to the case when $\mathcal{A}$ is generated by a countable measurable partition of Ω into sets Ω_i of finite measure $w_i > 0$. In that case, (8.74) reduces to (8.73). □

Remark 8.61. Let $1 \le p < \infty$. With exactly the same proof, we can give a similar formula for a suitable modification of the K-functional for the couple $(L_p(\Omega, \mathcal{A}, m; B_0), L_\infty(\Omega, \mathcal{A}, m; B_1))$. Namely, if we set for any $x \in B_0 + B_1$

$$K_t^{[p]}(x; B_0, B_1) = \inf\{(\|b_0\|_{B_0}^p + t^p\|b_1\|_{B_1}^p)^{1/p} \mid x = b_0 + b_1\},$$

then for any $f \in L_p(m; B_0) + L_\infty(m; B_1)$, we have: For all $t > 0$

$$K_t^{[p]}(f; L_p(m; B_0), L_\infty(m; B_1))^p = \sup_{\int \phi^p dm \le t^p} \int K_{\phi(\omega)}^{[p]}(f(\omega); B_0, B_1)^p dm(\omega), \tag{8.75}$$

where the sup runs over all non-negative measurable functions ϕ defined on $(\Omega, \mathcal{A})$ such that $\int \phi^p dm \le t^p$. Note that

$$K_t^{[p]}(x; B_0, B_1) \le K_t(x; B_0, B_1) \le 2^{1/p'} K_t^{[p]}(x; B_0, B_1).$$

Therefore, replacing K_t by $K_t^{[p]}$ in their definition does not affect the interpolation spaces $(B_0, B_1)_{\theta,q}$, only the norm is changed to an equivalent one.

8.7 Real interpolation between vector valued L_p-spaces

In the situation of Theorem 8.60, let us assume (for simplicity) that the intersection $B_0 \cap B_1$ is dense in B_0. Then, following [7, p. 303], we can write for all $x \in B_0 + B_1$

$$K_t(x; B_0, B_1) = \int_0^t k(x, s; B_0, B_1) ds, \tag{8.76}$$

where the k-functional $k(x, s; B_0, B_1)$ is a uniquely defined non-negative, non-increasing, right-continuous function of $s > 0$. In the case of the (scalar valued) couple (L_1, L_∞) over a σ-finite measure space, we recover

$$k(x, s; L_1, L_\infty) = x^\dagger(s)$$

where $x^\dagger$ is the decreasing rearrangement of $|x|$.

Indeed, since $t \mapsto K_t(x)$ is concave and non-decreasing we know that it admits left and right derivatives at all points, the latter coincide except at countably many points and the derived function is non-increasing. Thus, we can simply define $k(x, s; B_0, B_1)$ as the right-hand side derivative of $t \mapsto K_t(x)$ at the point s. Elementary arguments then lead to

$$K_t(x; B_0, B_1) = \lim_{s \to 0^+} K_s(x; B_0, B_1) + \int_0^t k(x, s; B_0, B_1) ds,$$

but by Remark 8.44, since we assume $B_0 \cap B_1$ dense in B_0, (8.76) holds.

Recall the notation $x^{\dagger\dagger}(t) = t^{-1} \int_0^t x^{\dagger}(s) ds$, so that $K_t(x; L_1, L_\infty) = t x^{\dagger\dagger}(t)$. If $0 < p \leq \infty$, $1 \leq q \leq \infty$ we also recall the definition of the quasi-norm $\|x\|_{p,q}$ in the Lorentz space $L_{p,q}$ over a σ-finite measure space as follows:

$$\|x\|_{p,q} = \left(\int_0^\infty [t^{1/p} x^{\dagger}(t)]^q \frac{dt}{t} \right)^{1/q}$$

with the usual convention when $q = \infty$.

If $1 < p \leq \infty$, $1 \leq q \leq \infty$, then either (8.62) or Hardy's classical inequality (see Lemma 8.49) show that this is equivalent to the following norm

$$\|x\|_{(p,q)} = \left(\int_0^\infty [t^{1/p} x^{\dagger\dagger}(t)]^q \frac{dt}{t} \right)^{1/q}$$

with the usual convention when $q = \infty$. In particular $L_{p,p}$ is the same as L_p with an equivalent norm.

With this notation, we can state

Corollary 8.62. *In the same situation as Theorem 8.60, assuming (for simplicity) that the intersection $B_0 \cap B_1$ is dense in B_0, we denote for all f in $L_1(\Omega, \mathcal{A}, m; B_0) + L_\infty(\Omega, \mathcal{A}, m; B_1)$,*

$$\forall s > 0 \, \forall \omega \in \Omega \quad \Psi_f(s, \omega) = k(f(\omega), s; B_0, B_1).$$

Let us abbreviate $(L_1(\Omega, m; B_0), L_\infty(\Omega, m; B_1))$ by $(L_1(B_0), L_\infty(B_1))$. Then

$$\begin{aligned} &K_t(f; L_1(B_0), L_\infty(B_1)) \\ &\quad = K_t(\Psi_f; L_1(\Omega \times]0, \infty[, dmds), L_\infty(\Omega \times]0, \infty[, dmds)). \end{aligned} \tag{8.77}$$

Moreover, for $1 < p \leq \infty$, $1 \leq q \leq \infty$ and $1/p = 1 - \theta$, we have

$$\|f\|_{(L_1(\Omega,m;B_0), L_\infty(\Omega,m;B_1))_{\theta,q}} = \|\Psi_f\|_{(p,q)} \tag{8.78}$$

where the Lorentz space norm is relative to the product space $(\Omega \times]0, \infty[, dmds)$.

Proof. By (8.74) we have

$$K_t(\Psi_f; L_1(\Omega\times]0,\infty[, dmds), L_\infty(\Omega\times]0,\infty[, dmds))$$
$$= \sup_{\int \phi dm \le t} \int K_{\phi(\omega)}(\Psi_f(.,\omega); L_1(]0,\infty[, ds), L_\infty(]0,\infty[, ds))dm(\omega)$$

using (8.74) again this yields (8.77) since we have obviously

$$\forall t>0, \forall \omega \in \Omega \quad K_t(\Psi_f(.,\omega); L_1(]0,\infty[, ds), L_\infty(]0,\infty[, ds))$$
$$= \int_0^t \Psi_f(s,\omega)ds = K_t(f(\omega); B_0, B_1).$$

Clearly (8.78) is an immediate consequence of (8.77) by applying $K_t(x; L_1, L_\infty) = tx^{\dagger\dagger}(t)$ on the product space with $x = \Psi_f$. □

The following very useful result can be viewed as the analogue for real interpolation of Theorem 8.21.

Theorem 8.63. *Let $1 \le p < \infty$ and $0 < \theta < 1$. Let (B_0, B_1) be a compatible pair and let (Ω, m) be any measure space.*

(i) Then

$$(L_p(m; B_0), L_p(m; B_1))_{\theta,p} = L_p(m; (B_0, B_1)_{\theta,p}) \tag{8.79}$$

with equivalent norms.

(ii) More generally, if $1 \le p_0 \ne p_1 \le \infty$ are such that $\frac{1-\theta}{p_0} + \frac{\theta}{p_1} = \frac{1}{p}$, then

$$(L_{p_0}(m; B_0), L_{p_1}(m; B_1))_{\theta,p} = L_p(m; (B_0, B_1)_{\theta,p}) \tag{8.80}$$

with equivalent norms.

First part of the proof. The proof of (i) is rather easy. Although it is a special case of (ii), we prefer to outline a quick proof. For simplicity we write $L_p(B)$ instead of $L_p(m; B)$. Let $f \in L_p(B_0) + L_p(B_1)$. We will show

$$2^{-1/p'} K_t(f; L_p(B_0), L_p(B_1)) \le \left(\int K_t(f(\omega); B_0, B_1)^p dm(\omega)\right)^{1/p} \tag{8.81}$$
$$\le K_t(f; L_p(B_0), L_p(B_1)).$$

Indeed, if $f = f_0 + f_1$ with $f_j \in L_p(B_j)$ $(j = 0, 1)$ then

$$K_t(f(\omega); B_0, B_1) \le \|f_0(\omega)\|_{B_0} + t\|f_1(\omega)\|_{B_1}$$

from which the second inequality in (8.81) is immediate. To prove the first inequality, fix $\varepsilon > 0$, and let $f(\omega) = f_0(\omega) + f_1(\omega)$ be such that f_0, f_1 are

Bochner measurable and such that $\|f_0(\omega)\|_{B_0} + t\|f_1(\omega)\|_{B_1} \le (1+\varepsilon)K_t(f(\omega); B_0, B_1)$. We have then

$$\big\| \|f_0(\cdot)\|_{B_0} + t\|f_1(\cdot)\|_{B_1} \big\|_p \le (1+\varepsilon)\|K_t(f(\cdot); B_0, B_1)\|_p,$$

and then using

$$\begin{aligned}\|f_0\|_{L_p(B_0)} + t\|f_1\|_{L_p(B_1)} &\le 2^{1/p'}\|(\|f_0(\cdot)\|^p_{B_0} + \|tf_1(\cdot)\|^p_{B_1})^{1/p}\|_p \\ &\le 2^{1/p'}\big\|(\|f_0(\cdot)\|_{B_0} + t\|f_1(\cdot)\|_{B_1})\big\|_p\end{aligned}$$

the first inequality in (8.81) follows immediately. Clearly, (8.81) implies (8.79) by integration.

We will now derive (ii) from Corollary 8.62. Let $p_0 = 1$ and $p_1 = \infty$. In the situation of Corollary 8.62, if $q = p$ and $1/p = 1 - \theta$, we have

$$(L_1(\Omega, \mathcal{A}, m; B_0), L_\infty(\Omega, \mathcal{A}, m; B_1))_{\theta,p} = L_p(\Omega, \mathcal{A}, m; (B_0, B_1)_{\theta,p}).$$

Indeed, when $p = q > 1$ Hardy's classical inequality (see Lemma 8.49) shows that for all x in $B_0 + B_1$, $\|k(x, s; B_0, B_1)\|_{L_p(ds)}$ is equivalent to the norm of x in $(B_0, B_1)_{\theta,p}$. Therefore, by (8.78), since $\|\Psi_f\|_{(p,p)}$ is equivalent to $\|\Psi_f\|_{L_p(dmds)}$, it is equivalent to the norm of f in $L_p(\Omega, \mathcal{A}, m; (B_0, B_1)_{\theta,p})$. This proves (ii).

A similar argument works, using Remark 8.61, when $1 < p_0 \le p_1 = \infty$. For the remaining case $1 < p_0, p_1 < \infty$ we will use a special case of a result called the power theorem in [6], for which we postpone the proof to the end of this §. □

Notation. We will use the notation $a \sim b$ if there is a positive constant c (depending only on on p_0, p_1, p) such that $c^{-1}a \le b \le ca$.

Lemma 8.64. *In the situation of the preceding Theorem, assume* $1 < p_0$, $p_1 < \infty$ *and for* $x \in B_0 + B_1$ *let*

$$\kappa_t(x; B_0, B_1) = \inf\{\|x_0\|^{p_0} + t\|x_1\|^{p_1} \mid x = x_0 + x_1, x_0 \in B_0, x_1 \in B_1\}. \quad (8.82)$$

Then for any $x \in B_0 + B_1$ *such that* $t^{-\theta}K_t(x) \to 0$ *when either* $t \to 0$ *or* $t \to \infty$ *we have*

$$\|t^{-\theta p/p_1}\kappa_t(x; B_0, B_1)\|_{L_1(dt/t)} \sim \|t^{-\theta}K_t(x; B_0, B_1)\|^p_{L_p(dt/t)}. \quad (8.83)$$

End of the proof of Theorem 8.63. Let f be (Bochner) measurable with values in $B_0 + B_1$. Using Lemma 8.64 and Remark 8.45 the proof boils down to the observation (left to the reader) that

$$\iint t^{-\frac{\theta p}{p_1}}\kappa_t(f(\omega); B_0, B_1)dm(\omega)dt/t = \int t^{-\frac{\theta p}{p_1}}\kappa_t(f(\omega); L_{p_0}(B_0), L_{p_1}(B_1))dt/t.$$

Indeed, by (8.83) and Fubini, this implies

$$\int \|t^{-\theta}K_t(f(\omega); B_0, B_1)\|^p_{L_p(dt/t)}dm(\omega) \sim \int \|t^{-\theta}K_t(f; L_{p_0}(B_0), L_{p_1}(B_1))\|^p_{L_p(dt/t)},$$

or equivalently after taking the p-th root

$$\|f\|_{L_p((B_0,B_1)_{\theta,p})} \sim \|f\|_{(L_{p_0}(B_0),L_{p_1}(B_1))_{\theta,p}},$$

which completes the proof. □

Remark 8.65. More generally, the proof of (i) in Theorem 8.63 yields that for any $q \geq p$ (resp. $q \leq p$) we have a bounded inclusion

$$(L_p(B_0), L_p(B_1))_{\theta,q} \supset L_p((B_0, B_1)_{\theta,q})$$
$$(\text{resp. } (L_p(B_0), L_p(B_1))_{\theta,q} \subset L_p((B_0, B_1)_{\theta,q})).$$

This follows again by integration but using the fact ('Hölder-Minkowski' see §5.13) that $L_q(\frac{dt}{t}, L_p) \supset L_p(L_q(\frac{dt}{t}))$ (resp. $L_q(\frac{dt}{t}, L_p) \subset L_p(L_q(\frac{dt}{t}))$). See [193] for counter-examples to the other inclusions.

Remark 8.66. When $p_1 = \infty$, (8.80) becomes

$$(L_{p_0}(B_0), L_\infty(B_1))_{\theta,p} = L_p((B_0, B_1)_{\theta,p}). \tag{8.84}$$

Recall however that $L_\infty(B_1)$ is defined as the space of essentially bounded *Bochner measurable* B_1 valued functions. This is rather restrictive in certain 'concrete' situations. To extend the scope of (8.84) we record here a simple observation: Assume $L_\infty(B_1)$ isometrically embedded in an priori larger space $\mathcal{L}$ of B_1 valued functions (or classes of functions), for instance $\mathcal{L} = \Lambda_\infty(B_1)$. Intuitively, $\mathcal{L}$ is formed of bounded B_1 valued functions but measurable in a broader sense, and we assume that $L_\infty(B_1) \subset \mathcal{L}$ is formed of those elements in $\mathcal{L}$ that are Bochner measurable. Assume $B_0 \subset B_1$. Then, for any x that is a Bochner measurable B_1 valued function, we have

$$\forall t > 0 \qquad K_t(x; L_{p_0}(B_0), L_\infty(B_1)) = K_t(x; L_{p_0}(B_0), \mathcal{L}).$$

Indeed, if $x = x_0 + x_1$ with $x_0 \in L_{p_0}(B_0)$ and $x_1 \in \mathcal{L}$, then a fortiori $x_0 \in L_{p_0}(B_1)$, so that $x_1 = x - x_0$ is Bochner-measurable as a B_1 valued function and hence automatically in $L_\infty(B_1)$. Consequently, the norms of such an x in the (θ, q) interpolated spaces is the same for the two pairs $(L_{p_0}(B_0), L_\infty(B_1))$ and $(L_{p_0}(B_0), \mathcal{L})$. We will use this for the following example: $B_1 = \ell_\infty(B)$, so that $L_\infty(B_1) = L_\infty(\ell_\infty(B))$ and $\mathcal{L} = \ell_\infty(L_\infty(B))$. Note that with our (Bochner sense) definition of $L_\infty(B)$, when we take as measure space $\mathbb{N}$ equipped with the counting measure, the space $L_\infty(B)$ is in general smaller than $\ell_\infty(B)$, but the latter coincides in that case with $\Lambda_\infty(B)$.

Proof of Lemma 8.64. Let $K_t(x) = K_t(x; B_0, B_1))$ and $\kappa_t(x) = \kappa_t(x; B_0, B_1))$. We set

$$h_t(x) = \inf \max\{\|x_0\|_{B_0}^{p_0}, t\|x_1\|_{B_1}^{p_1}\} \quad \text{and} \quad g_t(x) = \inf \max\{\|x_0\|_{B_0}, t\|x_1\|_{B_1}\}$$

where each infimum is as in (8.82). Obviously $h_t(x) \sim \kappa_t(x)$ and $g_t(x) \sim K_t(x)$. We claim that if $s = t^{p_1}(g_t(x))^{p_0-p_1}$ then $h_s(x) = (g_t(x))^{p_0}$. Taking this claim for granted, let us complete the proof.

To abbreviate we set $g_t = g_t(x)$ and $h_s = h_s(x)$. We will compute our integrals using the change of variable by $t \mapsto s(t) = t^{p_1}(g_t)^{p_0-p_1}$, which varies continuously from 0 to ∞ (see the subsequent justification). Integrating by parts we find

$$(\theta p/p_1)\int_0^\infty s^{-\theta p/p_1} h_s ds/s = \int_0^\infty s^{-\theta p/p_1} dh_s - [s^{-\theta p/p_1} h_s]_0^\infty.$$

Note $p_0 - (\theta p/p_1)(p_0 - p_1) = p$, so that, if we replace s by $s(t)$, $s^{-\theta p/p_1} h_s$ becomes $t^{-\theta p} g_t^p$. Thus, using our claim, this becomes after our change of variable

$$(\theta p/p_1)\int_0^\infty s^{-\theta p/p_1} h_s ds/s = \int_0^\infty (t^{p_1} g_t^{p_0-p_1})^{-\theta p/p_1} dg_t^{p_0} - [t^{-\theta p} g_t^p]_0^\infty.$$

By our assumption on x, the last term vanishes, so we find

$$(\theta p/p_1)\int_0^\infty s^{-\theta p/p_1} h_s ds/s = \int_0^\infty (t^{p_1} g_t^{p_0-p_1})^{-\theta p/p_1} dg_t^{p_0} = p^{-1}\int t^{-\theta p} dg_t^p$$

which, by integration by parts again, is

$$= \theta \int t^{-\theta p} g_t(x)^p dt/t.$$

Replacing g_t and h_s by the equivalent terms K_t and κ_s, we obtain the lemma. We now justify the legitimacy of the change of variable by $s(t)$. We first show that $t \mapsto g_t$ and $s \mapsto h_s$ are continuous on $(0, \infty)$. Note that for any fixed $1 < q < \infty$ we have $\ell_t \le g_t \le 2^{1/q}\ell_t$ where

$$\ell_t = \inf \max\{(\|x_0\|_{B_0}^q + (t\|x_1\|_{B_1})^q)^{1/q} \mid x = x_0 + x_1\}.$$

But, since $t \mapsto (\ell_{t^{1/q}})^q$ is concave (as infimum of a family of affine functions), it is continuous, and hence so is $t \mapsto \ell_t$, and since we can choose q arbitrarily large, $t \mapsto g_t$ is also continuous. A similar argument shows that $s \mapsto h_s$ is continuous. Note that since $t \mapsto t/g_t$ and $t \mapsto g_t$ are both non-decreasing, the same is true for $t \mapsto s(t) = (t/g_t)^{p_1} g_t^{p_0}$. Moreover, since $g_t \sim K_t \ge t\|x\|_{B_0+B_1}$ for $t \le 1$, $t \mapsto t/g_t$ remains bounded when $t \to 0$, and hence $s(t) \to 0$ when $t \to 0$. Similarly, since $g_t \sim K_t \le t\|x\|_{B_0+B_1}$ for $t \ge 1$, $t \mapsto t/g_t$ remains bounded below when $t \to \infty$, so if $g_t \to \infty$ when $t \to \infty$, then $s(t) \to \infty$, and if g_t remains bounded $p_1 > 0$ implies the same.

We now check the claim. Let $\varepsilon > 0$. Let $x_j \in B_j$ be such that $x = x_0 + x_1$ and $g_t(x) \le \max\{\|x_0\|_{B_0}, t\|x_1\|_{B_1} \le (1+\varepsilon)g_t(x)$. Then obviously

$$1 \le \max\{(\|x_0\|_{B_0}/g_t(x))^{p_0}, (t\|x_1\|_{B_1}/g_t(x))^{p_1} \le \varphi(\varepsilon)$$

where $\varphi(\varepsilon) = \max\{(1+\varepsilon)^{p_0}, (1+\varepsilon)^{p_1}\}$ is such that $\varphi(\varepsilon) \to 1$ when $\varepsilon \to 0$. Therefore, we have

$$\inf_{x_0+x_1=x} \max\{(\|x_0\|_{B_0}/g_t(x))^{p_0}, (t\|x_1\|_{B_1}/g_t(x))^{p_1} = 1$$

and if we multiply this by $(g_t(x))^{p_0}$ we obtain the claim $h_s(x) = (g_t(x))^{p_0}$. □

8.8 Duality for the real method

The duality for the real method is given by the following.

Theorem 8.67. *Let (B_0, B_1) be a compatible couple of Banach spaces. Assume that $B_0 \cap B_1$ is dense in both B_0 and B_1, so that the pair (B_0^*, B_1^*) is naturally compatible. Then, for any $0 < \theta < 1$ and $1 \le q < \infty$, setting as usual $q' = q/(q-1)$, we have*

$$(B_0, B_1)^*_{\theta,q} = (B_0^*, B_1^*)_{\theta,q'},$$

with equivalent norms.

The clearest way to prove the preceding statement is to introduce a companion of the K-functional, the so-called J-functional, which provides an equivalent (but at the same time somewhat dual) description of the norm of $(B_0, B_1)_{\theta,q}$.

Let (B_0, B_1) be a compatible couple. For any $x \in B_0 \cap B_1$ we define

$$J_t(x; B_0, B_1) = \max\{\|x\|_{B_0}, t\|x\|_{B_1}\}.$$

It is best to immediately observe that if $B_0 \cap B_1$ is dense in both spaces (B_0, B_1), so that (B_0^*, B_1^*) is compatible, then for any $\xi \in B_0^* \cap B_1^*$ we have

$$J_{1/t}(\xi; B_0^*, B_1^*) = \text{the dual norm to } K_t(\cdot; B_0, B_1). \tag{8.85}$$

The proof is immediate since the polar of the convex hull of two convex symmetric bodies C_0 and C_1 is the intersection of their polars, so we may apply this with C_0 (resp. C_1) equal to the unit ball of B_0 (resp. t^{-1} times the unit ball of B_1), both viewed as bodies sitting in $B_0 + B_1$.

We will often abbreviate and write simply $J_t(x)$ instead of $J_t(x; B_0, B_1)$ when there is no risk of confusion.

Let $0 < \theta < 1$ and $1 \le q \le \infty$. We then define the Banach space $(B_0, B_1)_{\theta,q,J}$ as formed of all $x \in B_0 + B_1$ for which there is a function $f \in L_1(dt/t; B_0 + B_1)$ such that

$$x = \int f(t)dt/t$$

and the function $t \mapsto t^{-\theta} J_t(f(t))$ is in $L_q(dt/t)$. We define (with the usual convention for the case $q = \infty$)

$$\|x\|_{(B_0,B_1)_{\theta,q,J}} = \inf\{\|t^{-\theta} J_t(f(t))\|_{L_q(dt/t)}\}$$

where the infimum runs over all f such that $x = \int f(t)dt/t$.

Note that the integral $\int f(t)dt/t$ is well defined. Indeed, since $\|x\|_{B_0+B_1} \le (1 \wedge t^{-1})J_t(x)$ we have $\|f(t)\|_{B_0+B_1} \le (t^{\theta} \wedge t^{-(1-\theta)})[t^{-\theta} J_t(f(t))]$, and hence, since

$$\forall q' \in [1, \infty) \quad t^{\theta} \wedge t^{-(1-\theta)} \in L^{q'}(dt/t), \tag{8.86}$$

the condition $\|t^{-\theta} J_t(f(t))\|_{L_q(dt/t)} < \infty$ ensures by itself that $f \in L_1(dt/t; B_0 + B_1)$ as soon as f is Bochner measurable into $B_0 + B_1$.

We note for future reference that, since the measure dt/t is left invariant by the map $t \mapsto 1/t$, we have for any $f \in L_1(dt/t; B_0 + B_1)$ $\int f(t)dt/t = \int f(1/t)dt/t$ and also $\|t^{-\theta} J_t(f(t))\|_{L_q(dt/t)} = \|t^{\theta} J_{1/t}(f(1/t))\|_{L_q(dt/t)}$. Thus we can define equivalently

$$\|x\|_{(B_0,B_1)_{\theta,q,J}} = \inf\left\{\|t^{\theta} J_{1/t}(f(t))\|_{L_q(dt/t)} \mid x = \int f(t)dt/t\right\}. \tag{8.87}$$

For emphasis, we will sometimes denote by $(B_0, B_1)_{\theta,q,K}$ the space $(B_0, B_1)_{\theta,q}$ introduced in (8.46).

Remark 8.68. There is a useful way to discretize the integrals appearing in the definition of the norms of the spaces $(B_0, B_1)_{\theta,q,K}$ and $(B_0, B_1)_{\theta,q,J}$:

Let $r > 1$ be a fixed number. We simply consider the Riemann sums associated to the partition of $(0, \infty)$ formed by the intervals $[r^n, r^{n+1})$ $(n \in \mathbb{Z})$. More precisely, we define

$$\|x\|_{\theta,q,K,r} = \left(\sum\nolimits_{n\in\mathbb{Z}} (\text{Log } r)((r^n)^{-\theta} K_{r^n}(x))^q\right)^{1/q}.$$

Then since $t \mapsto K_t(x)$ is non-decreasing and $\int_{r^n}^{r^{n+1}} dt/t = \text{Log } r$ we have

$$r^{-\theta} \|x\|_{\theta,q,K,r} \le \|x\|_{\theta,q,K} \le r^{\theta} \|x\|_{\theta,q,K,r}. \tag{8.88}$$

Similarly, assume that we have a representation $x = \sum_{n\in\mathbb{Z}} d_n$ as a series converging in $B_0 + B_1$. We have then

$$\|x\|_{\theta,q,J} \le (r/\mathrm{Log}\, r)\left(\sum\nolimits_{n\in\mathbb{Z}} (\mathrm{Log}\, r)((r^n)^{-\theta} J_{r^n}(d_n))^q\right)^{1/q}. \tag{8.89}$$

To verify this, one just defines a function f taking constant values on the intervals $[r^n, r^{n+1})$ $(n \in \mathbb{Z})$: we set $f(t) = d_n/(\mathrm{Log}\, r)$ on $[r^n, r^{n+1})$, then $\int f(t)dt/t = \sum d_n$ and $\|t^{-\theta} J_t(f(t))\|_{L_q(dt/t)} \le (\mathrm{Log}\, r)^{-1} r(\sum_{n\in\mathbb{Z}} (\mathrm{Log}\, r)((r^n)^{-\theta} J_{r^n}(d_n))^q)^{1/q}$.

The main result about the J-functional is the following equivalence theorem that says that both the J- and K-methods lead to the same spaces with equivalent norms:

Theorem 8.69. *For any $0 < \theta < 1$ and $1 \le q \le \infty$, we have*

$$(B_0, B_1)_{\theta,q,J} = (B_0, B_1)_{\theta,q}$$

with equivalent norms.

Proof. For any $t, s \in (0, \infty)$ we have obviously

$$K_t(x) \le (1 \wedge (t/s))J_s(x).$$

Therefore, if $x = \int f(s)ds/s$ we have by Jensen

$$t^{-\theta} K_t(x) \le t^{-\theta} \int K_t(f(s))ds/s \le \int [(t/s)^{-\theta} \wedge (t/s)^{1-\theta}][s^{-\theta} J_s(f(s))]ds/s$$

and hence if we let again

$$\psi_\theta(t) = t^{-\theta} \wedge t^{1-\theta}$$

we have (the convolution being in the multiplicative group $(0, \infty)$)

$$t^{-\theta} K_t(x) \le \psi_\theta * \varphi$$

with $\varphi(s) = s^{-\theta} J_s(f(s))$, and hence

$$\|t^{-\theta} K_t(x)\|_{L_q(dt/t)} \le \|\varphi\|_{L_q(dt/t)} \|\psi_\theta\|_{L_1(dt/t)}$$

and since, by (8.86), we have $C(\theta, 1) = \|\psi_\theta\|_{L_1(dt/t)} < \infty$, we find

$$\|x\|_{(B_0,B_1)_{\theta,q}} \le C(\theta, 1)\|x\|_{(B_0,B_1)_{\theta,q,J}},$$

and $(B_0, B_1)_{\theta,q,J} \subset (B_0, B_1)_{\theta,q}$.

To prove the converse, we will need to prove the following important:

Claim. Let $x \in B_0 + B_1$. Assume that $(1 \wedge 1/t)K_t(x) \to 0$ when $t \to 0$ and when $t \to \infty$. Then for any $\varepsilon > 0$ and any $n \in \mathbb{Z}$ there is $d_n \in B_0 \cap B_1$ such that

$$J_{2^n}(d_n) \leq (3 + \varepsilon)K_{2^n}(x), \tag{8.90}$$

and $x = \lim_{N\to\infty} \sum_{-N}^{N} d_n$ in $B_0 + B_1$.

Using this it is easy to conclude. We first note that, by (8.50) the assumption of the claim holds if $x \in (B_0, B_1)_{\theta,q}$ and $0 < \theta < 1$. Then we have by (8.88) (with $r = 2$)

$$\|((2^n)^{-\theta} J_{2^n}(d_n))\|_{\ell_q(\mathbb{Z})} \leq (3+\varepsilon)\|((2^n)^{-\theta} K_{2^n}(x))\|_{\ell_q(\mathbb{Z})} \leq \frac{(3+\varepsilon)2^\theta}{(\mathrm{Log}\, 2)^{1/q}} \|x\|_{(B_0,B_1)_{\theta,q}}.$$

Thus we conclude by (8.89) that $x \in (B_0, B_1)_{\theta,q,J}$ with

$$\|x\|_{(B_0,B_1)_{\theta,q,J}} \leq \frac{2(3+\varepsilon)2^\theta}{\mathrm{Log}\, 2} \|x\|_{(B_0,B_1)_{\theta,q}}.$$

To prove the claim we choose $x_n^0 \in B_0$, $x_n^1 \in B_1$ such that $x_n = x_n^0 + x_n^1$ such that $\|x_n^0\|_{B_0} + 2^n\|x_n^1\|_{B_1} \leq K_{2^n}(x)(1 + \varepsilon/3)$. By our assumption we have $\lim_{N\to\infty} x - x_N^0 = 0$ in B_1 and $\lim_{N\to\infty} x_{-N}^0 = 0$ in B_0. A fortiori, $x = \lim_{N\to\infty} x_N^0 - x_{-N}^0$ in $B_0 + B_1$. Thus if we set $d_n = x_n^0 - x_{n-1}^0$ we have $x = \lim_{N\to\infty} \sum_{-N}^{N} d_n$ in $B_0 + B_1$. Moreover, since $x_n^0 + x_n^1 = x_{n-1}^0 + x_{n-1}^1$ we have $d_n = -(x_n^1 - x_{n-1}^1)$ and hence $J_{2^n}(d_n) \leq (\|x_n^0\|_{B_0} + \|x_{n-1}^0\|_{B_0}) \vee 2^n(\|x_n^1\|_{B_1} + \|x_{n-1}^1\|_{B_1})$, from which (8.90) follows immediately. □

As we will see later on, with respect to reiteration, the extremal endspaces $(B_0, B_1)_{\theta,1}$ and $(B_0, B_1)_{\theta,\infty}$ (recall (8.52)) play a special role in real interpolation. The next lemma helps to recognize when a space is intermediate between them.

Lemma 8.70. *Let B be any Banach space. Consider an operator $T\colon B_0 \cap B_1 \to B$. Fix $0 < \theta < 1$. If, for any $x \in B_0 \cap B_1$, we have*

$$\|Tx\|_B \leq \|x\|_{B_0}^{1-\theta} \|x\|_{B_1}^{\theta},$$

then T extends to a bounded operator from $(B_0, B_1)_{\theta,1}$ to B with

$$\|T\colon (B_0, B_1)_{\theta,1} \to B\| \leq C,$$

where C is a constant depending only on θ.

Conversely, if $\|T\colon (B_0, B_1)_{\theta,1} \to B\| \leq 1$, then $\|Tx\|_B \leq C(\theta, 1)\|x\|_{B_0}^{1-\theta} \|x\|_{B_1}^{\theta}$, where $C(\theta, 1)$ is as in (8.49).

Proof. By Theorem 8.69, it suffices to show that

$$\|Tx\|_B \le \|x\|_{(B_0,B_1)_{\theta,1,J}}. \tag{8.91}$$

By our assumption, we have for any $t > 0$ $\|Tx\|_B \le t^{-\theta} J_t(x)$. Thus, if $x = \int f(t)dt/t$ we have by Jensen

$$\|Tx\|_B \le \int \|T(f(t))\| dt/t \le \int t^{-\theta} J_t(f(t))dt/t.$$

Taking the infimum over all f's, we get the announced inequality (8.91).

The converse part is obvious by (8.55) applied with $q = 1$. □

Remark 8.71. As announced in Remark 8.43 we will now show, using the equivalence in Theorem 8.69 that $B_0 \cap B_1$ is dense in $(B_0, B_1)_{\theta,q}$ for any $q < \infty$. Let $x \in (B_0, B_1)_{\theta,q,J}$. Let f be such that $x = \int f(t)dt/t$ and $\|t^{-\theta} J_t(f(t))\|_{L_q(dt/t)} < \infty$. Fix $\varepsilon > 0$ (small) and $R > 0$ (large) and let $x_{\varepsilon,R} = \int_\varepsilon^R f(t)dt/t$. Note that $\varepsilon\|f(t)\|_{B_0\cap B_1} \le J_t(f(t)) \le R\|f(t)\|_{B_0\cap B_1}$ if $t \in [\varepsilon, R]$, and hence $x_{\varepsilon,R} \in B_0 \cap B_1$. But then

$$\|x - x_{\varepsilon,R}\|_{(B_0,B_1)_{\theta,q,J}} \le \|1_{[0,\varepsilon]\cup[R,\infty]} t^{-\theta} J_t(f(t))\|_{L_q(dt/t)}$$

and, since we assume $q < \infty$, the latter tends to 0 when $\varepsilon \to 0$ and $R \to \infty$.

Proof of the duality Theorem 8.67. For proper emphasis, let us denote $(B_0, B_1)_{\theta,q}$ by $(B_0, B_1)_{\theta,q,K}$. By the equivalence Theorem 8.69, it suffices to prove that we have isometrically

$$(B_0, B_1)^*_{\theta,q,K} = (B_0^*, B_1^*)_{\theta,q',J}.$$

By (8.85), any $f \in L_1(dt/t; B_0^* + B_1^*)$ such that $t \mapsto t^\theta J_{1/t}(f(t); B_0^*, B_1^*)$ is in the unit ball of $L_{q'}(dt/t)$ defines a linear form

$$B_0 \cap B_1 \ni x \mapsto \int \langle f(t), x\rangle dt/t \tag{8.92}$$

in the unit ball of $(B_0, B_1)^*_{\theta,q,K}$ (since $B_0 \cap B_1$ is dense in $(B_0, B_1)_{\theta,q,K}$). Thus $\xi = \int f(t)dt/t$ is in the unit ball of $(B_0, B_1)^*_{\theta,q,K}$. Therefore, by (8.87), any ξ in the unit ball of $(B_0^*, B_1^*)_{\theta,q',J}$ defines an element in the unit ball of $(B_0, B_1)^*_{\theta,q,K}$. Conversely, let ξ be in the unit ball of $(B_0, B_1)^*_{\theta,q,K}$. Using (8.88) and again (8.85), we can find a function $f \in L_1(dt/t; B_0^* + B_1^*)$ that is *constant on each interval* $[r^n, r^{n+1})$ $(n \in \mathbb{Z})$, such that both

$$\xi(x) = \int \langle f(t), x\rangle dt/t = \sum\nolimits_{n\in\mathbb{Z}} (\operatorname{Log} r)\langle f(r^n), x\rangle$$

and

$$\left(\sum_{n\in\mathbb{Z}}(\operatorname{Log} r)((r^n)^\theta J_{1/r^n}(f(r^n); B_0^*, B_1^*))^{q'}\right)^{1/q'} \leq r^\theta.$$

A fortiori, since $t \in [r^n, r^{n+1})$ implies $J_{1/t} \leq J_{1/r^n}$ and $\int_{r^n}^{r^{n+1}} dt/t = \operatorname{Log} r$ we have

$$\left(\int (t^\theta J_{1/t}(f(t); B_0^*, B_1^*))^{q'} dt/t\right)^{1/q'} \leq r^{2\theta},$$

and hence the linear form ξ is associated via (8.92) to a function f such that $t \mapsto r^{-2\theta} t^\theta J_{1/t}(f(t); B_0^*, B_1^*)$ is in the unit ball of $L_{q'}(dt/t)$. In other words, by (8.87), we have

$$\|\xi\|_{(B_0^*, B_1^*)_{\theta,q',J}} \leq r^{2\theta}.$$

Letting $r \to 1$, we conclude that the correspondence just defined between $(B_0^*, B_1^*)_{\theta,q',J}$ and $(B_0, B_1)^*_{\theta,q,K}$ is bijective and isometric. □

8.9 Reiteration for the real method

We will use later (especially in Chapter 9) the real interpolation analogue of the reiteration theorem, as follows (cf. [6, p. 50]):

Theorem 8.72. *Let (B_0, B_1) be a compatible couple of Banach spaces. Let $0 \leq \theta_0 \neq \theta_1 \leq 1$ and $1 \leq q_0, q_1 \leq \infty$. Consider the couple X_0, X_1 where we set:*

$$X_j = (B_0, B_1)_{\theta_j, q_j} \text{ if } 0 < \theta_j < 1 \ (j = 0, 1),$$
$$X_j = B_0 \text{ if } \theta_j = 0, \quad \text{and } X_j = B_1 \text{ if } \theta_j = 1.$$

Then for any $0 < \theta < 1$ and $1 \leq q \leq \infty$ we have

$$(X_0, X_1)_{\theta,q} = (B_0, B_1)_{\tau,q}$$

(with equivalent norms) where $\tau = (1-\theta)\theta_0 + \theta\theta_1$.

We will prove this using an estimate of the K-functional for the couple (X_0, X_1) due to Holmstedt.

Remark 8.73. The reiteration theorem can be viewed as an 'abstract' version of the Marcinkiewicz theorem, and in fact part of Theorem 8.55 (which, as remarked earlier, implies the Marcinkiewicz theorem) is a particular case of Theorem 8.72. Indeed, if we let $(B_0, B_1) = (L_1, L_\infty)$ and $X_j = L_{p_j,q_j} = (B_0, B_1)_{\theta_j,q_j}$ (here $1 - \theta_j = 1/p_j$) and also $1 - \tau = 1/p$, we obtain Theorem 8.55, at least when $1 < p_0, p_1 < \infty$.

Notation. We will often denote $B_{\theta,q} = (B_0, B_1)_{\theta,q}$.

Remark 8.74. We should emphasize a very remarkable feature of Theorem 8.72, namely the fact that the space $(X_0, X_1)_{\theta,q}$ *does not depend on* the values of q_0 and q_1!

In particular the resulting space is the same for $q_0 = q_1 = 1$ as for $q_0 = q_1 = \infty$. This implies that if (X_0, X_1) are any spaces such that (for each $j = 0, 1$)

$$B_{\theta_j,1} \subset X_j \subset B_{\theta_j,\infty}$$

(a space X_j satisfying this is called 'of exponent θ_j'), then $(X_0, X_1)_{\theta,q}$ is still the same space $B_{\tau,q}$ since it is sandwiched between two copies of that same space, obtained with each of the two choices $(X_0, X_1) = (B_{\theta_0,1}, B_{\theta_1,1})$ and $(X_0, X_1) = (B_{\theta_0,\infty}, B_{\theta_1,\infty})$.

Note that it is crucial for this phenomenon that $\theta_0 \neq \theta_1$. In case $\theta_0 = \theta_1$, the result is as follows. Since we do not use it in the sequel, we skip its proof (see [6, p. 51 and p. 112]).

Theorem 8.75. *With the same notation as in Theorem 8.72, assume now* $0 < \theta_0 = \theta_1 < 1$. *In that case, we have*

$$(X_0, X_1)_{\theta,q} = B_{\tau,q}$$

(with equivalent norms) where $\tau = \theta_0 = \theta_1$, *but where* q *is now restricted to satisfy* $\frac{1}{q} = \frac{1-\theta}{q_0} + \frac{\theta}{q_1}$.

The reiteration theorem will be derived from the following beautiful formula due to Holmstedt. Here, given two functions $f, g : (0, \infty) \to \mathbb{R}$ we make the convention to write $f(t) \approx g(t)$ if there is a constant $c > 0$ such that $c^{-1}f(t) \leq g(t) \leq cf(t)$ for all $t > 0$.

Theorem 8.76 (Holmstedt's formula)**.** *With the same notation as in Theorem 8.72, let* $\delta = \theta_1 - \theta_0$. *Assume first* $0 < \theta_0 < \theta_1 < 1$. *Then for any* $x \in X_0 + X_1$

$$K_t(x; X_0, X_1) \approx a_{t^{1/\delta}}(x) + t b_{t^{1/\delta}}(x)$$

where (with the usual convention when either $q_0 = \infty$ *or* $q_1 = \infty$*)*

$$a_t(x) = \left(\int_0^t [s^{-\theta_0} K_s(x; B_0, B_1)]^{q_0} ds/s \right)^{\frac{1}{q_0}}$$

$$b_t(x) = \left(\int_t^\infty [s^{-\theta_1} K_s(x; B_0, B_1)]^{q_1} ds/s \right)^{\frac{1}{q_1}}.$$

Moreover, if $\theta_0 = 0 < \theta_1 < 1$ *we have*

$$K_t(x; B_0, X_1) \approx t b_{t^{1/\delta}}(x)$$

and if $0 < \theta_0 < \theta_1 = 1$ *we have*

$$K_t(x; X_0, B_1) \approx a_{t^{1/\delta}}(x).$$

Proof. Assume $0 < \theta_0 < \theta_1 < 1$. We will prove the announced result in the following clearly equivalent form:

$$K_{t^\delta}(x; X_0, X_1) \approx a_t(x) + t^\delta b_t(x).$$

We first claim that there is a constant γ such that

$$a_t(x) + t^\delta b_t(x) \le \gamma K_{t^\delta}(x; X_0, X_1). \tag{8.93}$$

Let $x = x_0 + x_1$ with $x_j \in X_j$ $(j = 0, 1)$. We have $a_t(x) \le a_t(x_0) + a_t(x_1)$. Obviously $a_t(x_0) \le \|x_0\|_{X_0}$. Let $\alpha = (1/\delta q_0)^{1/q_0}$ and $\beta = \alpha(\theta_1 q_1)^{1/q_1}$. Recall that $K_s(x_1; B_0, B_1) \le s^{\theta_1}\|x_1\|_{\theta_1,\infty}$ for any $s > 0$. Thus, by (8.50) we have

$$a_t(x_1) \le \left(\int_0^t [s^{-\theta_0} s^{\theta_1}]^{q_0} ds/s\right)^{\frac{1}{q_0}} \|x_1\|_{\theta_1,\infty} = \alpha t^\delta \|x_1\|_{\theta_1,\infty} \le \beta t^\delta \|x_1\|_{\theta_1,q_1}.$$

Recollecting, we find $a_t(x) \le \|x_0\|_{X_0} + \beta t^\delta \|x_1\|_{X_1}$. Arguing similarly, we find a constant β' so that $t^\delta b_t(x) \le \beta' \|x_0\|_{X_0} + t^\delta \|x_1\|_{X_1}$. Thus we obtain (8.93) with $\gamma = \max\{1 + \beta, 1 + \beta'\}$, which is one side of the promised equivalence, assuming $0 < \theta_0 < \theta_1 < 1$, but the cases $\theta_0 = 0$ and $\theta_1 = 1$ are also settled by the same argument.

We now turn to the converse direction. Fix $t > 0$. Let $x \in B_0 + B_1$ be such that $a_t(x) + t^\delta b_t(x) < \infty$. We will show that x belongs to $X_0 + X_1$ and satisfies the converse to (8.93). More precisely, let $x = x_0 + x_1$ be any decomposition such that

$$\|x_0\|_{B_0} + t\|x_1\|_{B_1} \le 2K_t(x; B_0, B_1). \tag{8.94}$$

We will show that $x_j \in X_j$ for $j = 0, 1$ and that there is a constant c such that

$$K_{t^\delta}(x; X_0, X_1) \le \|x_0\|_{X_0} + t^\delta \|x_1\|_{X_1} \le c(a_t(x) + t^\delta b_t(x)). \tag{8.95}$$

To prepare for this let us record here two elementary bounds: Let $c_1, c_2, \ldots$ denote positive constants that may depend on θ_j, q_j but do not depend on t.

Since $s \mapsto K_s(x; B_0, B_1)/s$ is non-increasing, $K_t(x; B_0, B_1)/t \le K_s(x; B_0, B_1)/s$ for any $s \le t$, so that

$$t^{-\theta_0} K_t(x; B_0, B_1) \le c_1 a_t(x). \tag{8.96}$$

Similarly, since $K_t \le K_s$ when $t \le s$, we have

$$t^{-\theta_1} K_t(x; B_0, B_1) \le c_2 b_t(x). \tag{8.97}$$

Now let us set

$$a_{>t}(x) = \left(\int_t^\infty [s^{-\theta_0} K_s(x; B_0, B_1)]^{q_0} ds/s \right)^{\frac{1}{q_0}},$$

$$b_{<t}(x) = \left(\int_0^t [s^{-\theta_1} K_s(x; B_0, B_1)]^{q_1} ds/s \right)^{\frac{1}{q_1}}.$$

so that $\|x_0\|_{X_0} \le a_t(x_0) + a_{>t}(x_0)$ and $\|x_1\|_{X_1} \le b_{<t}(x_1) + b_t(x_1)$.

We claim that $\|x_0\|_{X_0} \le c a_t(x)$ and $\|x_1\|_{X_1} \le c b_t(x)$ for some constant c. From this (8.95) will follow. We first prove $\|x_0\|_{X_0} \le c a_t(x)$. Since $x_0 = x - x_1$, by the triangle inequality, we may write $a_t(x_0) \le a_t(x) + a_t(x_1)$, and hence

$$\|x_0\|_{X_0} \le a_t(x_0) + a_{>t}(x_0) \le a_t(x) + a_t(x_1) + a_{>t}(x_0),$$

so it now suffices to show two estimates of the form $a_t(x_1) \le c' a_t(x)$ and $a_{>t}(x_0) \le c' a_t(x)$. We have obviously by (8.94)

$$K_s(x_1; B_0, B_1) \le s\|x_1\|_{B_1} \le 2(s/t) K_t(x; B_0, B_1) \tag{8.98}$$

for any $s > 0$, therefore

$$a_t(x_1) \le (2/t) \left(\int_0^t [s^{1-\theta_0}]^{q_0} ds/s \right)^{\frac{1}{q_0}} K_t(x; B_0, B_1) \le c_3 t^{-\theta_0} K_t(x; B_0, B_1),$$

and by (8.96) we obtain $a_t(x_1) \le c_3 c_1 a_t(x)$.

We shall now majorize $a_{>t}(x_0)$. We have obviously again by (8.94) for all $s > 0$

$$K_s(x_0; B_0, B_1) \le \|x_0\|_{B_0} \le 2K_t(x; B_0, B_1), \tag{8.99}$$

and hence

$$a_{>t}(x_0) \le 2 \left(\int_t^\infty [s^{-\theta_0}]^{q_0} ds/s \right)^{\frac{1}{q_0}} K_t(x; B_0, B_1) = c_4 t^{-\theta_0} K_t(x; B_0, B_1)$$

and hence by (8.96) $a_{>t}(x_0) \le c_4 c_1 a_t(x)$. This shows that $\|x_0\|_{X_0} \le c a_t(x)$.

By an entirely analogous argument, we will now prove that for some constant c we have $\|x_1\|_{X_1} \le c b_t(x)$. We write $\|x_1\|_{X_1} \le b_{<t}(x_1) + b_t(x_1) \le b_{<t}(x_1) + b_t(x_0) + b_t(x)$. By (8.99), we then have $b_t(x_0) \le c_5 t^{-\theta_1} K_t(x; B_0, B_1)$ and by (8.97) we find $b_t(x_0) \le c_5 c_2 b_t(x)$. Similarly, by (8.98) and again (8.97) we have $b_{<t}(x_1) \le c_6 t^{-\theta_1} K_t(x; B_0, B_1) \le c_6 c_2 b_t(x)$, completing the proof of $\|x_1\|_{X_1} \le c b_t(x)$ and of our claim.

Thus we obtain (8.95) and the proof is complete in the case $0 < \theta_0 < \theta_1 < 1$.

Now assume $\theta_0 = 0$ and $X_0 = B_0$. We claim that, for some c, we have $\|x_0\|_{B_0} \le ct^\delta b_t(x)$. Indeed, since $K_t(x; B_0, B_1) \le K_s(x; B_0, B_1)$ for any $s \ge t$ and $\delta = \theta_1$ we have

$$\|x_0\|_{B_0} \le 2K_t(x; B_0, B_1) \le c_7 t^\delta b_t(x),$$

and since the preceding bound for $\|x_1\|_{X_1}$ remains valid, we conclude that $K_{t^\delta}(x; B_0, X_1) \le \|x_0\|_{B_0} + t^\delta\|x_1\|_{X_1} \le c_8 t^\delta b_t(x)$. This settles the case $\theta_0 = 0$. The case $\theta_1 = 1$ can be checked similarly. □

Proof of Theorem 8.72. We will use only that

$$B_{\theta_j,1} \subset X_j \subset B_{\theta_j,\infty}, \tag{8.100}$$

where when $\theta_j \in \{0, 1\}$ we make the convention that $B_{\theta_j,1} = B_{\theta_j,\infty} = B_{\theta_j}$. We will first show that

$$(X_0, X_1)_{\theta,q} \subset B_{\tau,q}. \tag{8.101}$$

By (8.100) it suffices to show this inclusion for $X_j = B_{\theta_j,\infty}$. But then the Holmstedt formula with $q_0 = q_1 = \infty$ (and with $X_0 = B_0$ or $X_1 = B_1$ if either $\theta_0 = 0$ or $\theta_1 = 1$) and (8.96) show that for any $x \in X_0 + X_1$ for some c

$$\forall t > 0 \quad t^{-\theta_0} K_t(x; B_0, B_1) \le cK_{t^\delta}(x; X_0, X_1),$$

which implies since $\tau = \theta\delta + \theta_0$

$$t^{-\tau} K_t(x; B_0, B_1) \le c(t^\delta)^{-\theta} K_{t^\delta}(x; X_0, X_1)$$

and using the change of variable $s = t^\delta$ (note $dt/t = \delta^{-1} ds/s$) we find

$$\|x\|_{B_{\tau,q}} \le c\delta^{-1/q}\|x\|_{X_{\theta,q}},$$

proving (8.101).

To prove the converse, using (8.100) it suffices to show

$$B_{\tau,q} \subset (X_0, X_1)_{\theta,q} \tag{8.102}$$

for $X_j = B_{\theta_j,1}$. In that case, the Holmstedt formula with $q_0 = q_1 = 1$ and $0 < \theta_0 < \theta_1 < 1$ gives us for some c'

$$(t^\delta)^{-\theta} K_{t^\delta}(x; X_0, X_1) \le c' t^{-\tau} \left(t^{\theta_0} \int_0^t s^{-\theta_0} f(s) ds/s + t^{\theta_1} \int_t^\infty s^{-\theta_1} f(s) ds/s \right)$$

where $f(s) = K_s(x; B_0, B_1)$. Then Hardy's inequalities (see Lemma 8.49), and the same change of variable $s = t^\delta$, imply (8.102). If either $\theta_0 = 0$ or $\theta_1 = 1$, the Holmstedt formula reduces to a single term and the same argument remains valid. □

8.10 Comparing the real and complex methods

The next result gives the main general known connection between the real and complex methods.

Theorem 8.77. *Let (B_0, B_1) be a compatible couple of complex Banach spaces. Then, for any $0 < \theta < 1$, the following bounded inclusions hold*

$$(B_0, B_1)_{\theta,1} \subset (B_0, B_1)_\theta \subset (B_0, B_1)_{\theta,\infty}. \tag{8.103}$$

Proof. By (8.11) we may apply Lemma 8.70 with $T\colon\ B_0 \cap B_1 \to (B_0, B_1)_\theta$ equal to the natural inclusion. This implies $(B_0, B_1)_{\theta,1} \subset (B_0, B_1)_\theta$. For the converse, note that if $f \in \mathcal{F}$ and $f(\theta) = x$ we may apply (8.4) not only to f but also, for any fixed $t > 0$, to the function $f_t(z) = t^{\theta - z} f(z)$, this gives us

$$x = f_t(\theta) = (1-\theta)\int_{\partial_0} f_t(\xi)\mu_\theta^0(d\xi) + \theta \int_{\partial_1} f_t(\xi)\mu_\theta^1(d\xi)$$

and hence $x = (1-\theta)x_0 + \theta x_1$ with $\|x_0\|_{B_0} \le t^\theta \|f\|_{\mathcal{F}}$ and $\|x_1\|_{B_1} \le t^{\theta-1}\|f\|_{\mathcal{F}}$. Thus we obtain $K_t(x) \le (1-\theta)\|x_0\|_{B_0} + t\theta\|x_1\|_{B_1} \le t^\theta \|f\|_{\mathcal{F}}$, and hence $\|x\|_{(B_0,B_1)_{\theta,\infty}} \le \|f\|_{\mathcal{F}}$. Taking the infimum over f, we obtain the second inclusion $(B_0, B_1)_\theta \subset (B_0, B_1)_{\theta,\infty}$ with norm at most 1. □

By Remark 8.74 this implies

Corollary 8.78. *Let $0 \le \theta_0 \ne \theta_1 \le 1$, $0 < \theta < 1$ and $1 \le q \le \infty$. Let $B_\theta = (B_0, B_1)_\theta$ for $0 < \theta < 1$. Then*

$$(B_{\theta_0}, B_{\theta_1})_{\theta,q} = (B_0, B_1)_{\tau,q}$$

(with equivalent norms) where $0 < \tau < 1$ is determined by $(1-\theta)\theta_0 + \theta\theta_1 = \tau$.

Remark 8.79. We quote without proof a useful analogous result for the reverse situation, when we apply the complex method to a pair of real interpolation spaces: If $1 \le q_j \le \infty$ $(j = 0, 1)$ are such that $1/q = (1-\theta)/q_0 + \theta/q_1$ and the rest of the notation is as in the preceding corollary, we have

$$(B_{\theta_0,q_0}, B_{\theta_1,q_1})_\theta = (B_0, B_1)_{\tau,q}.$$

See [6, p. 103] for a proof.

Remark 8.80. Let G be a locally compact Abelian group. A Banach space B is called of G-Fourier type p $(1 \le p \le 2)$ if the Fourier transform on G is bounded from $L_p(G; B)$ to $L_{p'}(\hat{G}; B)$. It is proved in [376] that, if both B_0, B_1 are of $\mathbb{R}$-Fourier type p, then (8.103) can be improved: we have

$$(B_0, B_1)_{\theta,p} \subset (B_0, B_1)_\theta \subset (B_0, B_1)_{\theta,p'}.$$

More generally, this holds if B_j is of $\mathbb{R}$-Fourier type p_j with $1/p = (1-\theta)/p_0 + \theta/p_1$. In particular if B_0, B_1 are both Hilbertian, then $(B_0, B_1)_\theta = (B_0, B_1)_{\theta,2}$.

In [353, 354] this notion is called weak $\mathbb{R}$-Fourier type p, reserving the term Fourier type p for the case when the Fourier transform has norm 1.

In this context, Bourgain [132, 133] proved the following important Theorem: If a Banach space B is of type p for some $p > 1$ in the sense of (10.38), then for some (possibly different) $p > 1$ the space B is of $\mathbb{Z}$-Fourier type p, and hence by [310] of $\mathbb{R}$-Fourier type p. See [101, 310] for comparisons between G-Fourier type p for various groups G, and [230] for a survey of this whole direction.

8.11 Symmetric and self-dual interpolation pairs

In this section, we show that the identities $(L_\infty, L_1)_{1/2} = L_2$ and $(L_\infty, L_1)_{1/2,2} = L_2$ can be derived as a special case of an elegant general principle, based on the self-duality of the interpolation method we use. We will see that the same idea applies to any 'abstract' self-dual pair.

We first discuss the very simple case of 'symmetric pairs' consisting of a space B and its dual B^*. To avoid technicalities, we restrict first to pairs of reflexive spaces.

Let (B_0, B_1) be an interpolation pair of reflexive spaces, such that $B_0 \cap B_1$ is dense both B_0, B_1 so that the dual pair B_0^*, B_1^* is viewed as compatible in the usual way. Assume given an isometric linear map

$$T\colon\ B_1 \to B_0^*$$

that is symmetric (or self-dual), i.e. such that

$$\forall x, y \in B_0 \cap B_1 \qquad T(x)(y) = T(y)(x). \tag{8.104}$$

Theorem 8.81. *In the preceding situation, let $0 < \theta < 1$ and $1 \le q < \infty$.*

(i) For the complex method, we have isometrically (via T)

$$(B_0, B_1)_\theta^* \simeq (B_0, B_1)_{1-\theta}.$$

In particular, $(B_0, B_1)_{1/2}$ is isometric to its dual.

(ii) For the real one, we have isomorphically (via T)

$$(B_0, B_1)_{\theta,q}^* \simeq (B_0, B_1)_{1-\theta,q'}.$$

In particular $(B_0, B_1)_{1/2,2}$ is isomorphic to its dual.

Proof. The key is simply to observe that the pair (B_0^*, B_1^*) can be identified with (B_1, B_0) via T. Indeed, since $T\colon\ B_1 \to B_0^*$ is an isometric isomorphism, its adjoint $T^*\colon\ B_0 \to B_1^*$ is also one, but the symmetry assumption shows that T and T^* are essentially the same in the sense of Theorem 8.8. Applying the latter Theorem to T and its inverse, we find that T defines an isometric isomophism from $(B_0, B_1)_{1-\theta}$ to $(B_1^*, B_0^*)_{1-\theta} = (B_0^*, B_1^*)_\theta = (B_0, B_1)_\theta^*$. The last identity because of the duality Theorem 8.37. This completes the proof of (i). In the real case, the proof is the same but now we use Theorem 8.67. □

The following instructive example will be used in Chapter 12:

Example 8.82. Let $\mathbb{K}$ denote either $\mathbb{R}$ or $\mathbb{C}$.

Let v_1^n denote $\mathbb{K}^n$ equipped with the norm

$$\|x\|_{v_1^n} = |x_1| + |x_2 - x_1| + \cdots + |x_n - x_{n-1}|.$$

We consider the interpolation spaces $(v_1^n, \ell_\infty^n)_{\theta,q}$ and $(v_1^n, \ell_\infty^n)_\theta (0 < \theta < 1, 1 \le q \le \infty)$. Let $p = (1-\theta)^{-1}$. We denote

$$\mathcal{W}_{p,q}^n = (v_1^n, \ell_\infty^n)_{p,q} \quad \text{and} \quad \mathcal{W}_p^n = \mathcal{W}_{p,p}^n.$$

Note that $\|x\|_{v_1^n} = \|Tx\|_{\ell_1^n}$ where

$$Tx = (x_n - x_{n-1}, x_{n-1} - x_{n-2}, \ldots, x_2 - x_1, x_1). \tag{8.105}$$

Note that T satisfies (8.104) with respect to the canonical duality on $\mathbb{K}^n$ (equivalently the matrix of T is symmetric). Note that by (8.47) we have isometrically

$$\mathcal{W}_{p,q}^n = (\ell_\infty^n, v_1^n)_{1-\theta,q}.$$

Therefore (exchanging the roles of v_1^n and ℓ_∞^n for convenience) Theorem 8.81 yields:

Corollary 8.83. *In the complex case we have isometrically* $(v_1^n, \ell_\infty^n)_\theta^* = (v_1^n, \ell_\infty^n)_{1-\theta}$ *and in particular* $(v_1^n, \ell_\infty^n)_{1/2}$ *is isometric to its dual, via the mapping* $T\colon\ (v_1^n, \ell_\infty^n)_{1-\theta} \to (v_1^n, \ell_\infty^n)_\theta^*$ *defined in* (8.105).

Let $(e_1, \ldots, e_n)$ denote the canonical basis in $\mathbb{K}^n$ and let $(e_1^*, \ldots, e_n^*)$ be the biorthogonal functionals in $(\mathbb{K}^n)^*$.

Corollary 8.84. *For all* $1 < p < \infty$ *and* $1 \le q \le \infty$ *there is a constant* C *(independent of* n*) such that*

$$\|T\colon\ \mathcal{W}_{p',q'}^n \to (\mathcal{W}_{p,q}^n)^*\| \le C \quad \text{and} \quad \|T^{-1}\colon\ (\mathcal{W}_{p,q}^n)^* \to \mathcal{W}_{p',q'}^n\| \le C.$$

Moreover, if we let $\sigma_j = \sum_1^j e_k$ $(1 \le j \le n)$ *then for all* x *in* $\mathbb{K}^n$ *we have*

$$\frac{1}{2C}\left\|\sum\nolimits_1^n x_j e_j^*\right\|_{(\mathcal{W}^n_{p,q})^*} \le \left\|\sum\nolimits_1^n x_j \sigma_j\right\|_{\mathcal{W}^n_{p',q'}} \le 2C\left\|\sum\nolimits_1^n x_j e_j^*\right\|_{(\mathcal{W}^n_{p,q})^*} \quad (8.106)$$

Proof. The first part is but a particular case of Theorem 8.81. Let $x = Ty$, $y \in \mathbb{K}^n$. We have

$$\frac{1}{C}\left\|\sum y_j e_j\right\|_{\mathcal{W}^n_{p',q'}} \le \left\|\sum x_j e_j^*\right\|_{(\mathcal{W}^n_{p,q})^*} \le C\left\|\sum y_j e_j\right\|_{\mathcal{W}^n_{p',q'}}. \quad (8.107)$$

Let $V\colon\ \mathbb{K}^n \to \mathbb{K}^n$ be defined by $V(z_1, \dots, z_n) = (z_n, \dots, z_1)$. Note that $\|Vz\|_{v_1^n} \le 2\|z\|_{v_1^n}$, V is an isometry on ℓ_∞^n and $V = V^{-1}$, therefore we have

$$\forall z \in \mathbb{K}^n \qquad 2^{-1}\|z\|_{\mathcal{W}^n_{p',q'}} \le \|Vz\|_{\mathcal{W}^n_{p',q'}} \le 2\|z\|_{\mathcal{W}^n_{p',q'}}.$$

But then we have

$$y = T^{-1}x = (x_n, x_n + x_{n-1}, \dots, x_n + \cdots + x_1) = V\sum\nolimits_1^n x_j\sigma_j.$$

Therefore

$$2^{-1}\left\|\sum\nolimits_1^n x_j\sigma_j\right\|_{\mathcal{W}^n_{p',q'}} \le \|y\|_{\mathcal{W}^n_{p',q'}} \le 2\left\|\sum\nolimits_1^n x_j\sigma_j\right\|_{\mathcal{W}^n_{p',q'}}$$

and (8.106) follows from (8.107). □

Let B be a Banach space. Throughout the sequel, for any $x \in B^*$, viewing $b \in B$ as an element of B^{**}, we denote $b(x) = x(b)$.

Suppose we have an inclusion $i\colon\ B^* \subset B$ with which we view (B^*, B) as compatible, and we set $(B_0, B_1) = (B^*, B)$. Assume for a moment that B is reflexive. Then we may take for $T\colon\ B_1 \to B_0^*$ simply the identity map on B. However, the requirement that T be 'symmetric' in the preceding sense imposes a restriction of symmetry on i itself, namely that $i(x)(y) = i(y)(x)$ for all $x, y \in B^*$. Thus, for any inclusion mapping $i\colon\ B^* \subset B$ satisfying the latter symmetry condition, the associated interpolation scale inherits the symmetry described in Theorem 8.81.

We will now see (without any reflexivity assumption on B) that if the inclusion $B^* \subset B$ is 'positive definite', then Hilbert space appears as a central point in the scale. The typical illustration of this situation is of course the inclusion $L_\infty \subset L_1$ over a finite measure space.

It is worthwhile to first clarify how the interpolation scale depends from the 'compatibility' convention.

Proposition 8.85. *Let* (B_0, B_1) *be a compatible pair, so that* $B_0 + B_1$ *is well defined. For each* $j = 0, 1$, *let* $v_j\colon\ C_j \to B_j$ *be an isometric isomorphism from another Banach space* C_j. *We make the pair* (C_0, C_1) *compatible using the* *embeddings* $v_j : C_j \to B_j \subset B_0 + B_1$. *Let* $T : C_0 + C_1 \to B_0 + B_1$ *be the isometric isomorphism induced by* v_0, v_1. *Then, for any* $0 < \theta < 1$ *and* $1 \leq q \leq \infty$, T *defines an isometric isomorphism from* $(C_0, C_1)_\theta$ *(resp.* $(C_0, C_1)_{\theta,q}$*) to* $(B_0, B_1)_\theta$ *(resp.* $(B_0, B_1)_{\theta,q}$*).*

Proof. The proof of this is obvious going back to the definitions. □

In the next statement we denote by $\bar{B}$ the complex conjugate space, i.e. the same as B but with the complex multiplication changed to $\alpha \cdot x = \bar{\alpha} x$. Clearly $\bar{B}$ is anti-linearly isometric to B. Note that any Hilbert space H is canonically linearly isometric to its anti-dual $\overline{H^*}$. Thus for any linear map $t : H \to B$, its adjoint $t^* : B^* \to H^*$ defines a $\mathbb{C}$-linear map $\overline{t^*} : \overline{B^*} \to \overline{H^*}$, so that since $\overline{H^*} \simeq H$ we can form the composition $i = t\overline{t^*} : \overline{B^*} \to B$.

Proposition 8.86. *Let B be any Banach space and let $H \subset B$ be a Hilbert space included as a dense linear subspace of B. In the preceding situation, Let $t : H \to B$ denote the inclusion mapping. With the preceding notation, we view the pair $(\overline{B^*}, B)$ as compatible using the inclusion $i = t\overline{t^*} : \overline{B^*} \to B$. Let $B_\theta = (\overline{B^*}, B)_\theta$ and $B_{\theta,q} = (\overline{B^*}, B)_{\theta,q}$. Then $B_{1/2} = B_{1/2,2} = H$ and*

$$\forall x \in \overline{B^*} \qquad \|x\|_{B_{1/2}} = \|x\|_H.$$

Proof. First observe that since t is injective with dense range, $\overline{t^*}$ and hence the composition i has the same properties. This allows us to consider the dual pair $(\overline{\overline{B^*}}^*, B^*)$ as compatible. We may obviously identify $\overline{\overline{B^*}}^*$ with $\overline{B^{**}}$ or with $\overline{B}^{**}$. Let $B_0 = \overline{B^*}$ and $B_1 = B$. Although it is the identity map let us denote by $x \mapsto \bar{x}$ the anti-linear isometry from B^* to $\overline{B^*}$. We have a sesquilinear map

$$\Phi\colon\ (\bar{x}, \bar{y}) \to i(\bar{x})(y) = \overline{i(\bar{y})(x)} = \langle \overline{t^*}(\bar{x}), \overline{t^*}(\bar{y})\rangle_H$$

defined on $B_0 \times B_0$ such that $|\Phi(x, y)| \leq \|\bar{x}\|_{B_0} \|i(\bar{y})\|_{B_1}$ and also $|\Phi(x, y)| \leq \|i(\bar{x})\|_{B_1} \|\bar{y}\|_{B_0}$ for all $\bar{x}, \bar{y} \in B_0 \times B_0$. This means that Φ extends by density to a map of norm ≤ 1 both on $B_0 \times B_1$ and on $B_1 \times B_0$. By the fundamental interpolation property (see Theorem 8.26 and recall Remark 8.27), we have $\|\Phi\colon\ B_{1/2} \times B_{1/2} \to \mathbb{C}\| \leq 1$. Thus

$$\forall x, y \in B_0 \qquad |i(\bar{x})(y)| \leq \|i(\bar{x})\|_{B_{1/2}} \|i(\bar{y})\|_{B_{1/2}}$$

and with $y = x$, taking the square root (note that $\overline{t^*}(\bar{x})$ and $i(\bar{x})$ are identified to the same point in $\overline{H^*} = H$) we find for any $x \in B_0$

$$|i(\bar{x})(x)|^{1/2} = \|i(\bar{x})\|_H \leq \|i(\bar{x})\|_{B_{1/2}}. \tag{8.108}$$

This means that we have mapping of norm ≤ 1 from $(i(B_0), \|\cdot\|_{1/2})$ to $(t(H), \|\cdot\|_H)$. Since $i(B_0)$ is dense in both $B_{1/2}$ and H, to conclude it suffices to show the converse to (8.108). But by duality, (8.108) implies for any $y \in H$

$$\|y\|_{B^*_{1/2}} \le \|y\|_H,$$

and we know from Theorem 8.39 that the norm induced by $B^*_{1/2}$ on $B^*_0 \cap B^*_1 = B^*_1 = B^*$ coincides with the norm of $(B^*_0, B^*_1)_{1/2}$. But $(B^*_0, B^*_1)_{1/2} = (\overline{B^{**}}, B^*)_{1/2}$ and the complex conjugate of the latter space is clearly identical to $(B^{**}, \overline{B^*})_{1/2}$ and by Remark 8.13 this is the same as $(B, \overline{B^*})_{1/2}$ or equivalently $B_{1/2}$. Thus we conclude that for any $\bar{y} \in B_0 = \overline{B^*}$

$$\|i(\bar{y})\|_{B_{1/2}} \le \|i(\bar{y})\|_H$$

and hence that (8.108) is an equality. The proof of the real case is entirely similar using Theorem 8.67 and Remark 8.43. □

In 'concrete' examples, in particular for non-commutative L_p-spaces, it is desirable to eliminate the complex conjugation from the preceding statement. Here is one general way to do that.

Proposition 8.87. *Let B be any Banach space and let $i: B^* \subset B$ be an injective, bounded linear map with dense range, with which we view the pair (B^*, B) as compatible. Let $B_\theta = (B^*, B)_\theta$ and $B_{\theta,q} = (B^*, B)_{\theta,q}$.*

(i) If B is a real space, assume $i(x)(x) \ge 0$ for all x in B^* and $i(x)(y) = i(y)(x)$ for all $x, y \in B^*$. Then $B_{1/2,2}$ is isomorphic to a (real) Hilbert space and, when restricted to B, its norm is equivalent to $x \mapsto i(x)(x)^{1/2}$.

(ii) In the complex case, assume there is an isometric anti-linear involution $J\colon B^* \to B^*$ such that $i(Jx)(x) \ge 0$ for all x in B^*. Then $B_{1/2}$ is isometric to a Hilbert space and

$$\forall x \in B^* \qquad \|x\|_{B_{1/2}} = (i(x)(Jx))^{1/2}.$$

In addition, $B_{1/2,2}$ is isomorphic to a (complex) Hilbert space.

Proof. We will prove only the complex case. The proof of the real case is entirely similar. Consider the inner product defined on B^* by $\langle x, y\rangle = i(Jx)(y)$ (linear in y, anti-linear in x). Then the Cauchy-Schwarz inequality shows that $i(Jx)(x) = 0$ implies $i(Jx) = 0$ and hence $x = 0$. Thus $x \mapsto i(Jx)(x)^{1/2}$ is a norm on B^*. Let K be the Hilbert space obtained after completion of B^* for this norm. We have a natural continuous injection $B^* \to K$, and hence taking adjoints we find a map $v: K^* \to B^{**}$, but since $i(x) \in B$ for any $x \in B^*$ it is easy to check that this map actually takes its values in B, so that we write it as $v: K^* \to B$ and its adjoint $v^*: B^* \to K$ is the natural injection we started from. Let $H \subset B$ be the subset defined by $H = v(K^*)$, and let $u: K^* \to H$ be the same mapping but viewed as acting into H. We equip H with the Hilbert space structure transplanted from K^* using u so that $u: K^* \to H$ is unitary. Let $t: H \to B$ denote the inclusion. Note that $t\overline{t^*} = (tu)\overline{(tu)^*}$ and $v = tu$. Thus, letting $\mathcal{I} = t\overline{t^*} : \overline{B^*} \to B$, we find a composition

$$\mathcal{I}: \overline{B^*} \to \overline{H^*} = H \to B.$$

By the preceding Theorem, using $\mathcal{I}$ for the compatibility of $(\overline{B^*}, B)$, we have $H = (\overline{B^*}, B)_{1/2}$, and for any $\bar{x} \in \overline{B^*}$

$$\|\mathcal{I}(\bar{x})\|^2_{(\overline{B^*},B)_{1/2}} = \|\mathcal{I}(\bar{x})\|^2_H = \mathcal{I}(\bar{x})(x) = \|\overline{t^*}(\bar{x})\|^2_{\bar{H}} = \|\overline{v^*}(\bar{x})\|^2_{\bar{K}} = \|v^*(x)\|^2_K$$

and hence

$$\|\mathcal{I}(\bar{x})\|^2_{(\overline{B^*},B)_{1/2}} = i(Jx)(x). \tag{8.109}$$

We now invoke Proposition 8.85 to pass from the compatibility defined for $(B_0, B_1) = (\overline{B^*}, B)$ by $\mathcal{I}$ to the one defined for $(C_0, C_1) = (B^*, B)$ by $i: B^* \to B$. Let $v_0: B^* \to \overline{B^*}$ be the $\mathbb{C}$-linear isometry defined by $v_0(x) = \overline{J(x)}$ and let v_1 be the identity on B. Then for the resulting map $T: B^* + B \to \overline{B^*} + B$ (as defined in Proposition 8.85) we have $T(x) = \mathcal{I}\overline{J(x)}$ for any $x \in C_0 = B^*$. By Proposition 8.85, it follows that for any $x \in B^*$ we have

$$\|x\|_{(B^*,B)_{1/2}} = \|T(x)\|_{(\overline{B^*},B)_{1/2}} = \|T(x)\|_H = \|\mathcal{I}\overline{J(x)}\|_H$$

so by (8.109) this gives us as announced

$$\|x\|_{(B^*,B)_{1/2}} = i(x)(Jx). \qquad \square$$

Remark 8.88. There are nice illustrations of the preceding statement to non-commutative L_p-spaces. For instance (see §14.1 for unexplained terminology) let M be a von Neumann algebra with predual M_* and let $f \in M_*$ be a normal state, i.e. a positive weak* continuous functional on M such that $f(1) = 1$ (this is a non-commutative analogue of a probability measure), that is 'faithful' i.e. such that $f(x) > 0$ for any $0 \neq x \geq 0$. Then let $fx \in M_*$ be defined by $fx(y) = f(xy)$. We may apply Proposition 8.87 to the continuous injective mapping $i: M \to M_*$ defined by $i(x) = fx$ and to the involution $x \mapsto x^*$ on M. This tells us that $(M, M_*)_{1/2}$ is the completion of M for the norm $\|x\| = (f(xx^*))^{1/2}$. For more on this viewpoint, see Kosaki's [309].

8.12 Notes and remarks

The material in this chapter is mostly classical. The basic classical reference, for both the real and complex cases, is [6], to which we owe a lot. The book

[51] covers the same area, but from a different viewpoint. See also [95]. For more on real interpolation, we refer the reader to [12] and also [7].

The complex interpolation method was introduced independently by A. Calderón and by J. L. Lions around 1960 (apparently the dates of publication do not reflect the order or priority). There are also two notes by S.G. Krein (in Doklady that same year) that describe a method, the theory of 'scales', closely related to Calderón's (see [51] for more details on this).

While Lions wrote nothing but the Comptes Rendus note [334], Calderón published a very detailed, very thorough account of all aspects of his theory. His memoir [177] remains must reading for anyone interested in the subject. The very useful basic formula (8.15), complementing Calderón's definition, appears as Lemma 2.5 in Stafney's [427].

There are some delicate issues that arise if one tries to relax the definition of the space $\mathcal{F}$ in the definition of complex interpolation. See [196] for a counter-example related to this.

A priori, the complex method requires pairs of Banach spaces. Nevertheless, complex interpolation can be satisfactorily developed for couples of quasi-Banach spaces in some special situations, see [120, 198].

In turn, J. L. Lions concentrated his efforts on the real interpolation in collaboration with J. Peetre, in the landmark paper [335]. Later on, Peetre introduced the K- and J-methods that replaced advantageously the Lions-Peetre methods and have been tremendously successful in analysis and approximation theory. In particular the formula (8.58) for $K_t(f; L_1, L_\infty)$ in Theorem 8.50 is due to him. The latter formula was, much later, extended to $K_t(f; L_1(B_0), L_\infty(B_1))$ in [393], which we followed closely in §8.6. Nevertheless, Theorem 8.60 already appears in Lions and Peetre's original work [335]. We essentially follow [6, p. 68] in the proof of the 'power' Lemma 8.64. See [193] for related information. As mentioned in [393], Lemma 8.57 is due to B. Maurey.

There is an extensive literature on interpolation of vector valued L_p and H^p-spaces. See [439] for a contribution on the real interpolation of H^p-spaces with values in UMD Banach lattices.

The self-duality results in §8.11 go back to the early days of interpolation, both real and complex. However, the original versions (as the one we chose to present for simplicity) required extra assumptions such as reflexivity, which were lifted later on. See [434] and the references there for the state of the art in that direction. Although previous versions were somewhat implicit in [177] or [6], Theorem 8.34 was observed in [261] together with the part involving the ARNP.

Theorem 8.39 is due to Bergh [117].

We made a special effort in Theorems 8.31 and 8.34 to clarify the relationship of the dual method $(. \,, .\,)^\theta$ with the differentiability properties of vector measures. See [195] for a detailed presentation of many questions related to Calderón's duality theorem.

In [285], building upon ideas of Aronszajn and Gagliardo and also Jaak Peetre, Svante Janson explores and develops useful (and unified) ways of understanding the real and complex interpolation methods and also a few other more exotic methods, all as special cases of two Aronszajn-Gagliardo constructions. For related developments, see [182] and also Ovchinnikov's method of orbits [372].

Around 1978 a new wave of interest for complex interpolation surged from the discovery by Coifman, Cwikel, Rochberg, Sagher, and Weiss that the complex method worked just as well for families of Banach spaces. To avoid technicalities, to describe this, we restrict to the finite-dimensional case, i.e. we will interpolate norms on $\mathbb{C}^n$. In this new theory it is more natural to replace the strip S by the disc and to use as initial data for the interpolation problem a measurable family $\|\cdot\|_{e^{it}}$ of norms on $\mathbb{C}^n$ indexed by the points of ∂D, instead of just a pair of norms. Then (under some mild non-degeneracy condition) assuming that $t \mapsto \|x\|_{e^{it}}$ is in $L_\infty(\mathbb{T}, m)$ for any $x \in \mathbb{C}^n$ one can extend the initial data to a family $\|\cdot\|_z$ of norms indexed by $z \in D$. More precisely, one defines for any $z \in D$

$$\|x\|_z = \inf\{\text{ess sup}_{\xi\in\partial D} \|f(\xi)\|_\xi \mid f(z) = x\},$$

where the infimum runs over all bounded analytic functions $f: D \to \mathbb{C}^n$ with non-tangential boundary values denoted by $\xi \mapsto f(\xi)$.

For example, in the case of ℓ_p-norms: if $\|\cdot\|_{e^{it}} = \|\cdot\|_{\ell^n_{p(t)}}$ for some measurable function $p: t \mapsto p(t) \in [1, \infty)$, one finds for $z \in D$ $\|\cdot\|_z = \|\cdot\|_{\ell^n_{p(z)}}$ where $z \mapsto 1/p(z)$ is the Poisson integral of $t \mapsto 1/p(t)$, i.e. we have

$$\frac{1}{p(z)} = \int_{\partial D} \frac{1}{p(t)} P^z(dt).$$

When the family of norms $\|\cdot\|_{e^{it}}$ takes only two values, say $\|\cdot\|_{B_0}$ and $\|\cdot\|_{B_1}$, we recover the usual complex interpolation. More precisely, if $\|\cdot\|_{e^{it}} = \|\cdot\|_{B_0}$ (resp. $\|\cdot\|_{e^{it}} = \|\cdot\|_{B_1}$) on a subset $A_0 \subset \mathbb{T}$ (resp. $A_1 \subset \mathbb{T}$) with $m(A_0) = 1 - \theta$ (resp. $m(A_1) = \theta$) then for the central value $z = 0$, we find $\|\cdot\|_0 = \|\cdot\|_{(B_0,B_1)_\theta}$. This extends the case of a two valued family of norms described in (8.24) (take there $p_0 = p_1 = \infty$).

See [184–187, 196, 197] for more information on all this. See also Yanqi Qiu's note [407] for an interesting recent example.

It may be worthwhile to develop the notions of RNP or ARNP for families of norms, with Theorem 8.34 (see also Remark 8.35) as a starting point.

Corollaries 8.83 and 8.84 go back to some 1974 discussions with Bernard Maurey.

This chapter being already far too long we had to skip many results on real and complex interpolation applied to pairs of 'concrete' function spaces. In particular for pairs of H^p-spaces, much is known in any dimension. See e.g. Janson and Jones's [288] in connection with martingale filtrations and BMO, Kalton's [300] for H^p-analogues of (8.23), and the surveys [303, 306].

Another major omission is Tom Wolff's beautiful work. In [436] he showed that both for the real and complex methods, the following principle holds under minimal assumptions: If we are given four compatible Banach spaces $B_1, \ldots, B_4$ such that $B_2 = (B_1, B_3)_\alpha$ and also $B_3 = (B_2, B_4)_\beta$ with $0 < \alpha, \beta < 1$, then actually $B_2 = (B_1, B_4)_{\theta(2)}$ and $B_3 = (B_1, B_4)_{\theta(3)}$ where $\theta(2)$ and $\theta(3)$ are the same barycentric coefficients that we would obtain, were we to consider four points $x_1, \ldots, x_4$ on the real line with convex combinations given by $x_2 = (1 - \alpha)x_1 + \alpha x_3$ and $x_3 = (1 - \beta)x_2 + \beta x_4$.

9

The strong p-variation of scalar valued martingales

In this chapter, we hope to illustrate how useful the real interpolation method can be to study the variations of sequences of (scalar) random variables. In general, we feel the interaction interpolation/probability, which has been historically rather limited, deserves to be developed further. This partly motivates the detailed treatment of real interpolation in §8.4 and §8.7. In Chapter 12, we will considerably expand the study of the spaces $v_p(B)$ and $\mathcal{W}_{p,q}(B)$ initiated in the present chapter,

Let $0 < p < \infty$ and let $x = (x_n)$ be a sequence in a Banach space B. The strong p-variation of $x = (x_n)$, denoted by $V_p(x)$, is defined as follows:

$$V_p(x) = \sup \left(\|x_0\|^p + \sum\nolimits_{j \geq 1} \|x_{n(j)} - x_{n(j-1)}\|^p \right)^{1/p}$$

where the supremum runs over all increasing sequences of integers $0 = n(0) < n(1) < n(2) < \cdots$. We denote by $v_p(B)$ the space of all sequences $x = (x_n)$ such that $V_p(x) < \infty$.

We set $v_p = v_p(\mathbb{K})$ where $\mathbb{K} = \mathbb{R}$ or $\mathbb{C}$.

Note that for all $0 < p < q < \infty$ we have

$$V_q(x) \leq V_p(x). \tag{9.1}$$

Clearly, when $p \geq 1$, the spaces $v_p(B)$ and v_p are Banach spaces. The extreme cases $p = \infty$ and $p = 1$ are especially simple. Indeed, the analogue of $V_p(x)$ for $p = \infty$ is equivalent to $\sup_{n \geq 0} \|x_n\|$, so it is natural to set $v_\infty(B) = \ell_\infty(B)$. As for $p = 1$, the triangle inequality shows that

$$V_1(x) = \|x_0\| + \|x_1 - x_0\| + \|x_2 - x_1\| + \cdots$$

so that $v_1(B)$ is just the space of sequences in B with bounded variation.

We will make crucial use of real interpolation. Consider a measure space $(\Omega, \mathcal{A}, m)$ and a Banach space B. As usual, we set $L_p(B) = L_p(\Omega, \mathcal{A}, m; B)$. Let (B_0, B_1) be an interpolation pair of Banach spaces. Consider the interpolation pair $(L_1(B_0), L_\infty(B_1))$. Let $0 < \theta < 1$ and $\frac{1}{p} = \frac{1-\theta}{1} + \frac{\theta}{\infty}$ (so that $1/p = 1 - \theta$). We will use repeatedly a special case of (8.80), namely:

$$L_p((B_0, B_1)_{\theta,p}) = (L_1(B_0), L_\infty(B_1))_{\theta,p}, \tag{9.2}$$

with equivalent norms.

In this chapter the couple $(v_1(B), \ell_\infty(B))$ plays the central role. Let $1 < p < \infty$ and $\theta = 1 - 1/p$. We denote

$$\mathcal{W}_{p,q}(B) = (v_1(B), \ell_\infty(B))_{\theta,q} \quad (0 < \theta < 1,\ 1 \le q \le \infty).$$

We also set

$$\mathcal{W}_p(B) = \mathcal{W}_{p,p}(B).$$

We now apply (9.2) to the couple $(v_1(B), \ell_\infty(B))$. This gives us assuming $p = (1-\theta)^{-1}$ (i.e. $\frac{1}{p} = \frac{1-\theta}{1} + \frac{\theta}{\infty}$):

$$L_p(\mathcal{W}_p(B)) = (L_1(v_1(B)), L_\infty(\ell_\infty(B)))_{\theta,p} \tag{9.3}$$

with equivalent norms.

The connection between $\mathcal{W}_p(B)$ and the strong p-variation lies in the following (in the converse direction, we will show in (12.29) and (12.31) that $v_p(B) \subset \mathcal{W}_{p,\infty}(B) \subset \mathcal{W}_{p+\varepsilon}(B)$ for any $\varepsilon > 0$).

Lemma 9.1. *If $1 < p < \infty$ and $1 - \theta = \frac{1}{p}$, then $\mathcal{W}_p(B) \subset v_p(B)$ and this inclusion has norm bounded by a constant $K(p)$ depending only on p.*

Proof. This is easy to prove. Indeed for any fixed sequence $0 = n(0) < n(1) < \dots$ we introduce the operator $T\colon\ v_1(B) \to \ell_1(B)$ defined by

$$T(x) = (x_0, x_{n(1)} - x_0, \dots, x_{n(k)} - x_{n(k-1)}, \dots).$$

This has clearly norm ≤ 1. On the other hand, considered as operator from $\ell_\infty(B)$ into $\ell_\infty(B)$, T has norm ≤ 2. Therefore it follows from the interpolation theorem (cf. Theorem 8.46), that T has norm ≤ 2 as an operator from $\mathcal{W}_p(B)$ into $(\ell_1(B), \ell_\infty(B))_{\theta,p}$. By Theorem 8.63, this space can be identified with $\ell_p(B)$ with an equivalent norm. This yields (for some constant $K(p)$)

$$\left(\|x_0\|^p + \sum \|x_{n(k)} - x_{n(k-1)}\|^p\right)^{1/p} \le K(p)\|x\|_{\mathcal{W}_p(B)},$$

and the announced result clearly follows from this. □

In this chapter we study the strong p-variation of scalar martingales. We will return to the B-valued case in a later chapter. Our main result is the following:

Theorem 9.2. *Assume* $1 \le p < 2$.

(i) There is a constant C_p such that every martingale $M = (M_n)_{n\ge 0}$ in L_p satisfies (with the convention $M_{-1} \equiv 0$)

$$\mathbb{E}V_p(M)^p \le (C_p)^p \sum\nolimits_{n\ge 0} \mathbb{E}|M_n - M_{n-1}|^p.$$

(ii) More generally, for any $1 \le r < \infty$, there is a constant C_{pr} such that every martingale $M = (M_n)_{n\ge 0}$ in L_r satisfies

$$\|V_p(M)\|_r \le C_{pr} \left\| \left(\sum\nolimits_{n\ge 0} |M_n - M_{n-1}|^p \right)^{1/p} \right\|_r.$$

Throughout the sequel, we will set by convention $M_{-1} = 0$ whenever $M = (M_n)_{n\ge 0}$ is a martingale. All the r.v.'s are assumed to be defined on a given probability space $(\Omega, \mathcal{A}, \mathbb{P})$. We will need the following key lemma.

Lemma 9.3. *For any martingale M in L_2, we have*

$$\|M\|_{(L_1(v_1),L_\infty(\ell_\infty))_{\frac{1}{2}\infty}} \le 2\left(\sum\nolimits_{n\ge 0} \mathbb{E}|M_n - M_{n-1}|^2 \right)^{1/2}.$$

Note that by orthogonality we have

$$\sum\nolimits_{n\ge 0} \mathbb{E}|M_n - M_{n-1}|^2 = \sup_{n\ge 0} \mathbb{E}|M_n|^2. \tag{9.4}$$

Proof of Lemma 9.3. Given a sequence of r.v.'s $X = (X_n)_{n\ge 0}$, we denote simply by $K_t(X)$ the K_t-norm of X with respect to the couple $(L_1(v_1), L_\infty(\ell_\infty))$. Explicitly, assuming that (X_n) converges a.s., we have

$$K_t(X) = \inf\left\{ \|X_0^0\|_1 + \sum\nolimits_{n\ge 1} \|X_n^0 - X_{n-1}^0\|_1 + t \sup_n \|X_n^1\|_\infty \right\} \tag{9.5}$$

where the infimum runs over sequences of r.v.'s X^0 and X^1 such that $X_n = X_n^0 + X_n^1$ for all $n \ge 0$. Note that the assumed a.s. convergence allows us to invoke Remark 8.66 with $\mathcal{L} = \ell_\infty(L_\infty)$.

Let (M_n) be a martingale, relative to an increasing sequence of σ-algebras $(\mathcal{A}_n)_{n\ge 0}$, and let $0 \le T_0 \le T_1 \le \cdots$ be a sequence of stopping times (relative to $(\mathcal{A}_n)_{n\ge 0}$) with values in $\mathbb{N} \cup \{\infty\}$. We assume that (M_n) is bounded in L_2, hence M_n converges a.s. (and in L_2) to a limit, denoted by M_∞, which is in L_2. Moreover, we have $M_n = \mathbb{E}(M_\infty|\mathcal{A}_n)$ and $M_T = \mathbb{E}(M_\infty|\mathcal{A}_T)$ for any stopping time T with values in $\mathbb{N} \cup \{\infty\}$ (see (1.24)). Therefore, the sequence $(M_{T_k})_{k\ge 0}$

is a martingale, and (9.4) implies

$$\mathbb{E}\left(|M_{T_0}|^2 + \sum\nolimits_{k\geq 1} |M_{T_k} - M_{T_{k-1}}|^2\right) \leq \sup \mathbb{E}|M_{T_k}|^2 \tag{9.6}$$

$$\leq \mathbb{E}|M_\infty|^2 = \sum\nolimits_{n\geq 0} \mathbb{E}|M_n - M_{n-1}|^2.$$

To prove Lemma 9.3, we may assume for simplicity that $\|M_\infty\|_2 \leq 1$. Then we define by induction starting with $T_0 = \inf\{n \geq 0, |M_n| > t^{-1/2}\}$,

$$T_1 = \inf\{n > T_0, |M_n - M_{T_0}| > t^{-1/2}\}$$

$$\vdots$$

$$T_k = \inf\{n > T_{k-1}, |M_n - M_{T_{k-1}}| > t^{-1/2}\}$$

and so on. As usual, we make the convention $\inf \emptyset = +\infty$, i.e., we set $T_k = +\infty$ on the set where

$$\sup_{n>T_{k-1}} |M_n - M_{T_{k-1}}| \leq t^{-1/2}.$$

Clearly $\{T_k\}$ is an increasing sequence of stopping times so that (9.6) holds. We note that if $T_0(\omega) < \infty$ then $|M_{T_0(\omega)}(\omega)| \geq t^{-1/2}$ and

$$\text{if} \quad T_k(\omega) < \infty \quad \text{and} \quad k \geq 1 \quad \text{then} \quad (M_{T_k} - M_{T_{k-1}})(\omega) \geq t^{-1/2}. \tag{9.7}$$

Moreover, we have for all $k \geq 0$

$$\sup_{n<T_0} |M_n| \leq t^{-1/2} \text{ a.s.} \quad \text{and} \quad \sup_{T_k\leq n<T_{k+1}} |M_n - M_{T_k}| \leq t^{-1/2} \text{ a.s.} \tag{9.8}$$

Hence, we can write $M_n = X_n^0 + X_n^1$, with X^0, X^1 defined as follows:

$$X_n^0 = \sum\nolimits_{k\geq 0} 1_{\{T_k\leq n<T_{k+1}\}} M_{T_k}$$

$$X_n^1 = 1_{\{n<T_0\}} M_n + \sum\nolimits_{k\geq 0} 1_{\{T_k\leq n<T_{k+1}\}}(M_n - M_{T_k}).$$

By (9.8), on one hand we have

$$\| \sup |X_n^1| \|_\infty \leq t^{-1/2}. \tag{9.9}$$

On the other hand, let $\Delta_0 = |M_{T_0}|$ and $\Delta_k = |M_{T_k} - M_{T_{k-1}}|$ for $k \geq 1$.

We have

$$|X_0^0| + \sum\nolimits_{n\geq 1} |X_n^0 - X_{n-1}^0| = 1_{\{T_0<\infty\}}\Delta_0 + \sum\nolimits_{k\geq 1} \Delta_k 1_{\{T_k<\infty\}}. \tag{9.10}$$

This can be estimated as follows. We have by (9.7)

$$t^{-1/2}\left(1_{\{T_0<\infty\}} + \sum\nolimits_{k\geq 1} 1_{\{T_k<\infty\}}\right) \leq \Delta_0 1_{\{T_0<\infty\}} + \sum\nolimits_{k\geq 1} \Delta_k 1_{\{T_k<\infty\}}. \tag{9.11}$$

Let $N = 1_{\{T_0<\infty\}} + \sum_{k\geq 1} 1_{\{T_k<\infty\}}$. By Cauchy-Schwarz, (9.11) implies

$$Nt^{-1/2} \leq N^{1/2}\left(|\Delta_0|^2 + \sum\nolimits_{k\geq 1} |\Delta_k|^2\right)^{1/2}. \tag{9.12}$$

Clearly N is finite a.s. (since M_n converges a.s.), thus (9.12) implies

$$N^{1/2} \leq t^{1/2}\left(|\Delta_0|^2 + \sum |\Delta_k|^2\right)^{1/2}$$

and hence by (9.6)

$$(\mathbb{E}N)^{1/2} \leq t^{1/2}\|M_\infty\|_2 \leq t^{1/2}. \tag{9.13}$$

Now going back to (9.10) we find again by Cauchy-Schwarz and (9.13)

$$\mathbb{E}\left(|X_0^0| + \sum\nolimits_{n\geq 1} |X_n^0 - X_{n-1}^0|\right) \leq (\mathbb{E}N)^{1/2}\left\|\left(|\Delta_0|^2 + \sum |\Delta_k|^2\right)^{1/2}\right\|_2$$
$$\leq t^{1/2}.$$

By (9.9) and (9.5), this yields $K_t(M) \leq 2t^{1/2}$ so that

$$\|M\|_{(L_1(v_1), L_\infty(\ell_\infty))_{1/2,\infty}} \leq 2.$$

By homogeneity, this completes the proof of Lemma 9.3. □

Proof of Theorem 9.2. Let $(\mathcal{A}_n)_{n\geq 0}$ be a fixed increasing sequence of σ-subalgebras of $\mathcal{A}$. All martingales that follow will be with respect to $(\mathcal{A}_n)_{n\geq 0}$. For $1 \leq p \leq \infty$, we will denote by D_p the subspace of $\ell_p(L_p)$ formed of all the sequences $\varphi = (\varphi_n)_{n\geq 0}$ such that φ_n is $\mathcal{A}_n$-measurable for all $n \geq 0$ and $\mathbb{E}(\varphi_n|\mathcal{A}_{n-1}) = 0$ for all $n \geq 1$. We first claim that if $1 \leq p_0, p_1 \leq \infty$ and if $\frac{1}{p} = \frac{1-\theta}{p_0} + \frac{\theta}{p_1}$ then

$$D_p = (D_{p_0}, D_{p_1})_{\theta,p}. \tag{9.14}$$

This follows from an argument well known to interpolation theorists, described in Remark 8.48. Indeed, to check this, we first note that by (9.2) we have

$$(L_{p_0}(\ell_{p_0}), L_{p_1}(\ell_{p_1}))_{\theta,p} = L_p(\ell_p), \tag{9.15}$$

with equivalent norms.

We may clearly identify isometrically $L_p(\ell_p)$ and $\ell_p(L_p)$. There is a projection $P\colon\ L_p(\ell_p) \to D_p$ defined by

$$\forall X = (X_n)_{n\geq 0} \in L_p(\ell_p) \quad P(X) = (\varphi_n)_{n\geq 0}$$

with

$$\varphi_0 = \mathbb{E}(X_0|\mathcal{A}_0) \quad \text{and} \quad \varphi_n = \mathbb{E}(X_n|\mathcal{A}_n) - \mathbb{E}(X_n|\mathcal{A}_{n-1}).$$

Clearly, P is a bounded projection onto D_p satisfying

$$\|P(X)\|_{D_p} \leq 2\|X\|_{L_p(\ell_p)}.$$

Since this holds in particular for both $p = p_0$ and $p = p_1$, we have the simultaneous complementation considered in Remark 8.48, and consequently, for any X in D_p,

$$\|X\|_{(L_{p_0}(\ell_{p_0}),L_{p_1}(\ell_{p_1}))_{\theta,p}} \leq \|X\|_{(D_{p_0},D_{p_1})_{\theta,p}} \leq 2\|X\|_{(L_{p_0}(\ell_{p_0}),L_{p_1}(\ell_{p_1}))_{\theta,p}}.$$

By (9.15), this implies that for some constant $C = C(p_0, p_1, \theta)$

$$C^{-1}\|X\|_{D_p} = C^{-1}\|X\|_{L_p(\ell_p)} \leq \|X\|_{(D_{p_0},D_{p_1})_{\theta,p}} \leq C\|X\|_{L_p(\ell_p)} = C\|X\|_{D_p},$$

which proves Claim 9.14.

We can now complete the proof of Theorem 9.2 (i). Let us denote by T the operator that associates to any φ in D_1 the martingale $(M_n)_{n\geq 0}$ defined by $M_n = \sum_{i\leq n}\varphi_i$. Clearly $\|T(\varphi)\|_{L_1(v_1)} \leq \|\varphi\|_{D_1}$. On the other hand, Lemma 9.3 implies that T is bounded from D_2 into $B_1 = (L_1(v_1), L_\infty(\ell_\infty))_{\frac{1}{2}\infty}$, with norm ≤ 2. Therefore if $1 < p < 2$ the interpolation Theorem 8.46 implies that T is bounded from $(D_1, D_2)_{\theta,p}$ into $(L_1(v_1), B_1)_{\theta,p}$. By the reiteration Theorem 8.72 we have $(L_1(v_1), B_1)_{\theta,p} = (L_1(v_1), L_\infty(\ell_\infty))_{\delta,p}$ with $\delta = \theta/2$. Now if θ is chosen so that $\frac{1}{p} = 1 - \delta$, we have by (9.3) and Lemma 9.1

$$(L_1(v_1), L_\infty(\ell_\infty))_{\delta,p} = L_p(\mathcal{W}_p) \subset L_p(v_p).$$

On the other hand, by (9.14) we have (since $\frac{1-\theta}{1} + \frac{\theta}{2} = \frac{1}{p}$) $(D_1, D_2)_{\theta,p} = D_p$. Recapitulating, we find a constant $C = C(p)$ depending only on $1 \leq p < 2$ such that for all φ in D_p we have

$$\|T(\varphi)\|_{L_p(v_p)} \leq C\|\varphi\|_{D_p}.$$

This establishes the first part of Theorem 9.2. Note in passing that we obtained an a priori stronger inequality, namely

$$\|M\|_{L_p(\mathcal{W}_p)} \leq C(\sum\nolimits_{n\geq 0} \mathbb{E}|M_n - M_{n-1}|^p)^{1/p}.$$

The second part follows from the standard arguments used to prove the Burkholder-Davis-Gundy inequalities. We use the general method described in Lemma 5.23. Let $g(\omega) = (g_n(\omega))_{n\geq 0}$ be a martingale in L_r. We set $v_\infty(\omega) = V_p(g(\omega))$ and for any $N \geq 1$ we denote by $v_N(\omega)$ the strong p-variation of $\{g_0(\omega), \ldots, g_N(\omega)\}$ i.e. the strong p-variation of the *restriction* of our martingale to $[0, 1, \ldots, N]$. Equivalently, $v_N(\omega)$ is the strong p-variation of $(g_{n\wedge N}(\omega))_{n\geq 0}$. We set $w_N = (\sum_0^N |dg_k|^p)^{1/p}$. Applying (i) to the martingale $(1_{\{T>0\}}g_{n\wedge T})_{n\geq 0}$, we find that (5.42) holds for any stopping times $R \leq T$. If we assume $|dg_{n+1}| \leq \psi_n$ for all $n \geq 0$ with (ψ_n) adapted, then Lemma 5.26 yields that for any $0 < r < \infty$ we have for some constant $C_1 = C_1(p, r)$

$$\|v_\infty\|_r \leq C_1(\|w_\infty\|_r + \|\psi^*\|_r). \tag{9.16}$$

Let us denote in the rest of this proof

$$\psi_n = 4 \sup_{0 \leq k \leq n} |dM_k|.$$

We now invoke Theorem 5.32 (B. Davis decomposition) with r replacing p. This gives us a decomposition $M_n = h_n + g_n$ with $h_0 = 0$, $|dg_n| \leq \psi_{n-1}$ and

$$\|\sum |dh_n|\|_r \leq 4r\|\psi^*\|_r. \tag{9.17}$$

Since $V_p(M) \leq V_p(g) + \sum |dh_n|$, $(\sum |dg_n|^p)^{1/p} \leq (\sum |dM_n|^p)^{1/p} + \sum |dh_n|$, (9.16) implies that for some constant $C_2 = C_2(p, r)$ we have

$$\|V_p(M)\|_r \leq C_2 \left(\left\| \left(\sum |dM_n|^p \right)^{1/p} \right\|_r + \|\psi^*\|_r \right),$$

but since obviously

$$\|\psi^*\|_r \leq \left\| \left(\sum |dM_n|^p \right)^{1/p} \right\|_r$$

we obtain (ii). □

The next result is an immediate consequence of Theorem 9.2.

Corollary 9.4. *Let $1 \leq p < 2$. Let $M = (M_t)_{t \geq 0}$ be a martingale in L_p. Assume that the paths of M are right continuous and admit left limits and that the continuous part of M is 0. Let*

$$V_p(M) = \sup_{0=t_0 \leq t_1 \leq \ldots} \left(|M_o|^p + \sum\nolimits_{i \geq 1} |M_{t_i} - M_{t_{i-1}}|^p \right)^{1/p}$$

and

$$S_p(M) = \left(\sum\nolimits_{t \in [0,\infty[} |M_t - M_{t^-}|^p \right)^{1/p}.$$

Then, for all $1 \leq r < \infty$, we have for any martingale M in L_r

$$\|V_p(M)\|_r \leq C_{pr} \|S_p(M)\|_r. \tag{9.18}$$

Remark. There are also inequalities similar to Theorem 9.2 (ii) or (9.18) with a 'moderate' Orlicz function space instead of L_r, cf. [165, 175].

Our method gives (with almost no extra effort) a new proof of the following results of Lépingle [326].

Theorem 9.5. *Assume $2 < p < \infty$ and $1 \le r < \infty$. Then there is a constant C_{pr} such that every martingale $M = (M_n)_{n \ge 0}$ in L_r satisfies*

$$\|V_p(M)\|_r \le C_{pr} \| \sup_n |M_n| \|_r.$$

Moreover, there is a constant C'_p such that every martingale $M = (M_n)_{n \ge 0}$ in L_1 satisfies

$$\|V_p(M)\|_{1,\infty} \le C'_p \sup_n \|M_n\|_1. \tag{9.19}$$

Proof. We first consider the particular case $r = p$. With the preceding notation, consider the operator $S\colon\ L_\infty \to L_\infty(\ell_\infty)$ defined for φ in L_∞ by

$$S(\varphi) = (\mathbb{E}(\varphi|\mathcal{A}_n))_{n \ge 0}.$$

Clearly $\|S\| \le 1$. Let $B_0 = (L_1(v_1), L_\infty(\ell_\infty))_{1/2,\infty}$. By Lemma 9.3, S is bounded from L_2 into B_0. By the interpolation Theorem S must be bounded from $(L_2, L_\infty)_{\theta,p}$ into $(B_0, L_\infty(\ell_\infty))_{\theta,p}$ $(0 < \theta < 1, 1 \le p \le \infty)$. Now assume that $\frac{1}{p} = \frac{1-\theta}{2} + \frac{\theta}{\infty}$. Then, by (9.2), $(L_2, L_\infty)_{\theta,p} = L_p$. Moreover, by the reiteration Theorem 8.72

$$(B_0, L_\infty(\ell_\infty))_{\theta,p} = (L_1(v_1), L_\infty(\ell_\infty))_{\omega p}$$

for $\omega = \frac{1-\theta}{2} + \theta = \frac{1+\theta}{2}$. Note that $\frac{1-\omega}{1} + \frac{\omega}{\infty} = \frac{1}{p}$, hence by (9.2), the last equality implies $(B_0, L_\infty(\ell_\infty))_{\theta,p} = L_p(\mathcal{W}_p)$. Recapitulating, we find that S is bounded from L_p into $L_p(\mathcal{W}_p)$ with norm $\le C_1(p)$ for some constant $C_1(p)$ depending only on p. Let $\varphi \in L_p$ and let $M_n = \mathbb{E}(\varphi|\mathcal{A}_n)$. Applying Lemma 9.1 again we conclude that

$$\|M\|_{L_p(v_p)} \le K(p)\|M\|_{L_p(\mathcal{W}_p)} \le K(p)C_1(p)\|\varphi\|_p.$$

This proves Theorem 9.5 in the case $r = p$.

We now turn to the case $1 \le r < \infty$. We can argue as earlier for the second part of Theorem 9.2. Consider a martingale (M_n) in L_r. We apply the B. Davis decomposition (Lemma 5.33) in L_r, i.e. we have $M = g + h$ with $h_0 = 0$, $|dg_n| \le 8M^*_{n-1}$ and $\| \sum |dh_n| \|_r \le 6r\|M^*\|_r$. We define v_n and v_∞ as in the proof of Theorem 9.2, but we set

$$w_n = \sup_{k \le n} |g_n| \quad \text{and} \quad w_\infty = \sup_n |g_n|.$$

Then we invoke Remark 5.29. We leave the remaining details as an exercise. The weak-type 1-1 inequality (9.19) is a direct application of Theorem 5.15 (Gundy's decomposition). □

Remark. Of course, there is also a version of Theorem 9.5 in the case of a continuous parameter martingale $(M_t)_{t>0}$.

One can easily derive from Theorems 9.2 and 9.5 (by a classical stopping time argument) the following analogous 'almost sure' statements.

Proposition 9.6. *Let $M = (M_n)$ be a martingale with $\mathbb{E}\sup_{n\geq 1}|M_n - M_{n-1}| < \infty$. If $1 \leq p < 2$, then $\{V_p(M) < \infty\} \overset{a.s.}{=} \{\sum_{n\geq 1} |M_n - M_{n-1}|^p < \infty\}$.*

Moreover if $2 < p < \infty$, then $\{V_p(M) < \infty\} \overset{a.s.}{=} \{\sup|M_n| < \infty\}$.

Proof. Let $B_p = \{V_p(M) < \infty\}$, $A_p = \{\sum_0^\infty |dM_n|^p < \infty\}$, and $A_p(t) = \{\sum_0^\infty |dM_n|^p \leq t\}$. To prove the first assertion it suffices to show that $A_p(t) \subset B_p$ for all $0 < t < \infty$. Fix $0 < t < \infty$. Let $T = \inf\{n \mid \sum_0^n |dM_n|^p > t\}$. We may assume $M_0 = 0$ and hence $T > 0$ and $A_p(t) = \{T = \infty\}$. Let $f_n = M_{n\wedge T}$. Then $\sum |df_n|^p = \sum_{n\leq T} |dM_n|^p \leq t + \sup|dM_n|^p$. Therefore $(\sum |df_n|^p)^{1/p} \in L_1$. By (ii) (case $r = 1$) in Theorem 9.2 $V_p((f_n)) \in L_1$ and hence $V_p((f_n)) < \infty$ a.s., but on $\{T = \infty\}$ we have $(f_n) = (M_n)$ so $A_p(t) \subset B_p$. To prove the second assertion in case $2 < p < \infty$, we set $T = \inf\{n \mid |M_n| > t\}$ and again $f_n = M_{n\wedge T}$. Then $\{T = \infty\} = \{\sup|f_n| \leq t\}$, $\sup|f_n| \leq t + \sup|dM_n|$ and hence $\sup|f_n| \in L_1$. By Theorem 9.5 (case $r = 1$), $V_p((f_n)) \in L_1$ and hence $V_p((f_n)) < \infty$ a.s. and since $(f_n) = (M_n)$ on $\{T = \infty\}$ we conclude that $V_p(M) < \infty$ a.s. on the set $\{\sup|f_n| \leq t\}$. This proves the second assertion. □

9.1 Notes and remarks

This chapter closely follows Xu and the author's paper [400]. The K-functional for the pair (v_1, ℓ_∞) was identified by Bergh and Peetre in [118]. This is a crucial ingredient in [400].

Theorem 9.5 was obtained first by Lépingle [327] using the Skorokhod embedding of martingales into Brownian motion. Our proof is very different. Indeed we prove both Theorem 9.5 and Theorem 9.2 using the same idea, combining Lemma 9.3 and reiteration for real interpolation.

Note that while the Skorokhod embedding is unavailable in the Banach space valued case, our method allows to extend Theorem 9.2 (resp. 9.5) to the B-valued case, assuming B p-uniformly smooth (resp. p-uniformly convex), see §10.6.

The random variable $N = 1_{\{T_0<\infty\}} + \sum_{k\geq 1} 1_{\{T_k<\infty\}}$ appearing in the proof of Lemma 9.3 is the same as the 'number of escapes' defined by Burkholder in [173], which itself is similar to notions long considered in probability by Doob ('upcrossings') L. Dubins ('method of rises') and B. Davis ('ε-escursions'), see [173, p. 23].

There is an extensive literature on the strong p-variation both in probability theory, function theory and harmonic analysis. We will only give in the following a few sample references.

Prior to [400], analogous questions had been considered mainly for sequences or processes with independent increments (cf. e.g. [146, 352, 355]). For a more recent approach to Corollary 9.4, see [384]. See [340] for a study of the strong p-variation of (strong) Markov processes.

See [426] for a more recent result on the strong p-variation of α-stable processes for $0 < \alpha \leq 2$ and $p > \alpha$.

See [23] for more information on the relations between p-variation, differentiability and empirical processes.

See [139, 209, 291] for inequalities analogous to those of this chapter in ergodic theory and [324] in the semi-group context.

Our subsequent Chapter 12 contains a detailed study of the interpolation spaces $(v_1, \ell_\infty)_{\theta,q}$, which is quite useful to understand the spaces of functions with strong p-variation finite.

10

Uniformly convex Banach space valued martingales

10.1 Uniform convexity

This chapter is based mainly on [387]. The main result is:

Theorem 10.1. *Any uniformly convex Banach space B admits an equivalent norm $|\cdot|$ satisfying for some constant $\delta > 0$ and some $2 \le q < \infty$*

$$\forall x, y \in B \qquad \left|\frac{x+y}{2}\right|^q + \delta\left|\frac{x-y}{2}\right|^q \le \frac{|x|^q + |y|^q}{2}, \tag{10.1}$$

or equivalently

$$\forall x, y \in B \qquad |x|^q + \delta|y|^q \le \frac{|x+y|^q + |x-y|^q}{2}. \tag{10.2}$$

In other words, B with its new norm is at least as uniformly convex as L^q (for some $2 \le q < \infty$). The argument crucially uses martingale inequalities, but the relevant inequalities (see Corollary 10.7) are 'weaker' than those expressing the UMD property.

We recall:

Definition 10.2. A Banach space B is called uniformly convex if for any $0 < \varepsilon \le 2$ there is a $\delta > 0$ such that for any pair x, y in B the following implication holds

$$(\|x\| \le 1, \|y\| \le 1, \|x-y\| \ge \varepsilon) \Rightarrow \left\|\frac{x+y}{2}\right\| \le 1 - \delta.$$

The modulus of uniform convexity $\delta_B(\varepsilon)$ is defined as the 'best possible' δ i.e.

$$\delta_B(\varepsilon) = \inf\left\{1 - \left\|\frac{x+y}{2}\right\| \;\middle|\; \|x\| \le 1, \|y\| \le 1, \|x-y\| \ge \varepsilon\right\}. \tag{10.3}$$

Note that uniform convexity obviously passes from B to any subspace (resp. any quotient B/S) $S \subset B$ with $\delta_S(\varepsilon) \geq \delta_B(\varepsilon)$ (resp. $\delta_{B/S}(\varepsilon) \geq \delta_B(\varepsilon)$) for all $0 < \varepsilon \leq 2$.

It is easy to see that if $B = \mathbb{C}$ or if B is a Hilbert space of $\mathbb{R}$-dimension ≥ 2, we have $\delta_B(\varepsilon) = 1 - (1 - \varepsilon^2/4)^{1/2}$. Indeed, the parallelogram identity can be equivalently written as

$$\left\| \frac{x+y}{2} \right\|^2 + \left\| \frac{x-y}{2} \right\|^2 = \frac{\|x\|^2 + \|y\|^2}{2}$$

from which $\delta_B(\varepsilon) \geq 1 - (1 - \varepsilon^2/4)^{1/2}$ $(\geq \varepsilon^2/8)$ can be deduced and this is obviously optimal.

Since by Dvoretzky's theorem (see Theorem 10.43) any infinite-dimensional Banach space B contains ℓ_2^ns almost isometrically (in particular for $n = 2$), we must have $\delta_B(\varepsilon) \leq (1 - \varepsilon^2/4)^{1/2}$ hence $\delta_B(\varepsilon) \in O(\varepsilon^2)$ when $\varepsilon \to 0$. Actually, by [367], this already holds for any B with $\dim(B) > 1$. We will show in §10.3 that

$$\delta_{L_p}(\varepsilon) \sim \begin{cases} C_p \varepsilon^2 & \text{if } 1 < p \leq 2 \\ C_p \varepsilon^p & \text{if } 2 \leq p < \infty. \end{cases}$$

Moreover, it is easy to see that L_1 and ℓ_1 are not uniformly convex. Also (note that ℓ_1 isometrically embeds in L_∞ or ℓ_∞) L_∞ and ℓ_∞ are not uniformly convex.

The following result (often attributed to David P. Milman, and independently Pettis [380]) is classical.

Theorem 10.3. *Any uniformly convex Banach space is reflexive.*

Proof. Let U_B denote the unit ball of B. Fix $x^{**} \in B^{**}$ with $\|x^{**}\| = 1$. Let (x_i) be a generalized sequence in the unit ball of B converging to $x^{**} \in B^{**}$ for the topology $\sigma(B^{**}, B^*)$, i.e. such that $\langle x_i, \xi \rangle \to \langle x^{**}, \xi \rangle$ for any ξ in U_{B^*}. Clearly this implies $\|x_i\| \to \|x^{**}\|$ (indeed, for $\varepsilon > 0$ choose ξ such that $\langle x^{**}, \xi \rangle > 1 - \varepsilon$ and note $|\langle x_i, \xi \rangle| \leq \|x_i\|$), and similarly $\|2^{-1}(x_i + x_j)\| \to \|x^{**}\| = 1$ when $i, j \to \infty$. If B is assumed uniformly convex, this forces $\|x_i - x_j\| \to 0$ when $i, j \to \infty$ and hence by Cauchy's criterion x_i converges in norm to $x \in B$. Obviously we must have $x^{**} = x$ so we conclude $B^{**} = B$. □

We will use the following results due to Figiel ([223, 224]), completing earlier ones due to Day [208]. Although their proofs are elementary, the details are tedious so we skip them here.

Lemma 10.4. *Let B be uniformly convex.*

(i) *The function $\varepsilon \to \delta_B(\varepsilon)/\varepsilon$ is non-decreasing on $[0, 2]$.*
(ii) *For any measure space (Ω, m), and any $1 < r < \infty$ the space $L_r(m; B)$ (in particular $L_2(m; B)$) is uniformly convex.*

Remark 10.5. With our definition of $\delta_B(\varepsilon)$, it is obvious that $\varepsilon \to \delta_B(\varepsilon)$ is non-decreasing. This is less obvious (but nevertheless true) for the function $\varepsilon \to \hat{\delta}_B(\varepsilon)$ defined by

$$\hat{\delta}_B(\varepsilon) = \inf\left\{1 - \left\|\frac{x+y}{2}\right\| \;\middle|\; \|x\| = \|y\| = 1,\ \|x-y\| = \varepsilon\right\}.$$

Indeed, it turns out that $\hat{\delta}_B(\varepsilon) = \delta_B(\varepsilon)$ if the (real) dimension of B is at least 2 (see e.g. [223]).

To illustrate the next statement by a concrete example, let us anticipate the forthcoming §10.3 and consider the case of $B = L_q$. As we will show in §10.3: If $2 \le q < \infty$ we have

$$\forall x, y \in L_q, \qquad \left\|\frac{x+y}{2}\right\|_q^q + \left\|\frac{x-y}{2}\right\|_q^q \le \frac{\|x\|_q^q + \|y\|_q^q}{2},$$

which implies $\delta_{L_q}(\varepsilon) \ge 1 - (1 - (\varepsilon/2)^q)^{1/q} \sim q^{-1}(\varepsilon/2)^q$.

If $1 < q \le 2$, we will show

$$\forall x, y \in L_q \qquad \left(\left\|\frac{x+y}{2}\right\|_q^2 + (q-1)\left\|\frac{x-y}{2}\right\|_q^2\right)^{1/2} \le \left(\frac{\|x\|_q^q + \|y\|_q^q}{2}\right)^{1/q},$$

from which we deduce $\delta_{L_q}(\varepsilon) \ge 1 - (1 - (q-1)(\varepsilon/2)^2)^{1/2} \ge (q-1)\varepsilon^2/8$.

Theorem 10.6. *Let $2 \le q < \infty$ and let $\alpha > 0$ and C be fixed positive constants. The following two properties of a Banach space B are equivalent:*

(i) *There is a norm $|\cdot|$ on B such that for all x, y in B we have $\alpha\|x\| \le |x| \le \|x\|$ and*

$$\left|\frac{x+y}{2}\right|^q + \left\|\frac{x-y}{2C}\right\|^q \le \frac{|x|^q + |y|^q}{2}. \tag{10.4}$$

(ii) *For all B-valued dyadic martingales $(M_n)_{n\ge 0}$ in $L_q(B)$ we have*

$$\alpha^q \mathbb{E}\|M_0\|^q + C^{-q}\sum\nolimits_1^\infty \mathbb{E}\|dM_n\|^q \le \sup\nolimits_{n\ge 0} \mathbb{E}\|M_n\|^q. \tag{10.5}$$

Moreover, this implies:

(iii) All B-valued martingales $(M_n)_{n\geq 0}$ in $L_q(B)$ satisfy

$$\alpha^q \mathbb{E}\|M_0\|^q + 2(2C)^{-q}\sum\nolimits_1^\infty \mathbb{E}\|dM_n\|^q \leq \sup\nolimits_{n\geq 0}\mathbb{E}\|M_n\|^q.$$

Proof. (i) ⇒ (ii) Consider a dyadic martingale on $\Omega = \{-1, 1\}^{\mathbb{N}_*}$ associated to $\mathcal{A}_n = \sigma(\varepsilon_1, \ldots, \varepsilon_n)$, where $\varepsilon_n\colon\ \Omega \to \{-1, 1\}$ denotes the n-th coordinate. Then $\forall n \geq 1$ $dM_n = \varepsilon_n\delta_{n-1}$ with δ_{n-1} $(n-1)$-measurable. Let $x = M_{n-1}(\omega)$, $y = \delta_{n-1}(\omega)$. Then (i) implies for any fixed ω

$$|M_{n-1}(\omega)|^q + C^{-q}\|dM_n(\omega)\|^q \leq \int |M_{n-1}(\omega) + \varepsilon_n(\omega')\delta_{n-1}(\omega)|^q dP(\omega').$$

Integrating this with respect to ω, we find (since ε_n and $\mathcal{A}_{n-1}$ are independent)

$$\mathbb{E}|M_{n-1}|^q + C^{-q}\mathbb{E}\|dM_n\|^q \leq \mathbb{E}|M_n|^q,$$

which yields after a summation over $n \geq 1$

$$\mathbb{E}|M_0|^q + C^{-q}\sum\nolimits_{n\geq 1}\mathbb{E}\|dM_n\|^q \leq \sup \mathbb{E}|M_n|^q.$$

Finally, replacing $|\ \ |$ by the equivalent norm $\|\ \ \|$, we obtain (ii).
(i) ⇒ (iii) The proof is similar to the preceding. We take $x = M_{n-1}, y = M_n$. This yields after integration of (10.4)

$$\mathbb{E}|M_{n-1} + 2^{-1}dM_n|^q + C^{-q}\mathbb{E}\|2^{-1}dM_n\|^q \leq 2^{-1}(\mathbb{E}|M_{n-1}|^q + \mathbb{E}|M_n|^q) \quad (10.6)$$

but also we have trivially (Jensen)

$$\mathbb{E}|M_{n-1}|^q \leq \mathbb{E}|M_{n-1} + 2^{-1}dM_n|^q$$

hence plugging this into (10.6) we may subtract $2^{-1}\mathbb{E}|M_{n-1}|^q$ to both sides of the resulting inequality and after multiplication by 2 we find

$$\mathbb{E}|M_{n-1}|^q + 2C^{-q}\mathbb{E}\|2^{-1}dM_n\|^q \leq \mathbb{E}|M_n|^q$$

then the proof of (i) ⇒ (iii) is completed exactly as earlier for (i) ⇒ (ii).

(ii) ⇒ (i) Assume (ii). We define the norm $|\ \ |$ as follows: for any x in B we set

$$|x|^q = \inf\left\{\mathbb{E}\|M_N\|^q - C^{-q}\sum\nolimits_{n=1}^N \mathbb{E}\|dM_n\|^q\right\}$$

where the infimum runs over all N and all (finite) dyadic martingales $(M_0, M_1, \ldots, M_N)$ that start at x, i.e. such that $M_0 = x$.

By (10.5), we have for any x in B

$$\alpha^q\|x\|^q \leq |x|^q$$

and consideration of the trivial martingale $M_n \equiv x$ yields

$$|x|^q \leq \|x\|^q, \quad (10.7)$$

so that $|\ \ |$ is indeed equivalent to the original norm on B. Now consider x, y in B and fix $\varepsilon > 0$. Let M', M'' be finite martingales with $x = M'_0$ and $y = M''_0$ such that (note that we may clearly increase N, by adding null increments, in order to use the same N for both martingales)

$$\mathbb{E}\|M'_N\|^q - C^{-q}\sum\nolimits_1^N \mathbb{E}\|dM'_n\|^q < |x|^q + \varepsilon$$
$$\mathbb{E}\|M''_N\|^q - C^{-q}\sum\nolimits_1^N \mathbb{E}\|dM''_N\|^q < |y|^q + \varepsilon.$$

Then, let (M_n) be the martingale that starts at $(x+y)/2$, i.e. $M_0 \equiv (x+y)/2$, jumps with M_1 either to x or to y with equal probability 1/2 and then continues along the paths of M' or M'' depending on $M_1 = x$ or $M_1 = y$. More precisely, we can write M_n as follows: (Since M'_k and M''_k depend only on $\varepsilon_1, \ldots, \varepsilon_k$ we may denote them as $M'_k(\varepsilon_1, \ldots, \varepsilon_k)$ and $M''_k(\varepsilon_1, \ldots, \varepsilon_k)$.)

$$M_0 \equiv (x+y)/2,$$
$$M_1 = (x+y)/2 + \varepsilon_1(x-y)/2, \ldots,$$
$$M_n = ((1+\varepsilon_1)/2)M'_{n-1}(\varepsilon_2, \ldots, \varepsilon_n) + ((1-\varepsilon_1)/2)M''_{n-1}(\varepsilon_2, \ldots, \varepsilon_n).$$

Finally, we clearly have

$$\mathbb{E}\|M_{N+1}\|^q = (\mathbb{E}\|M'_N\|^q + \mathbb{E}\|M''_N\|^q)/2$$

and

$$\sum_1^{N+1} \mathbb{E}\|dM_n\|^q = \|(x-y)/2\|^q + \left(\sum\nolimits_1^N \mathbb{E}\|dM'_n\|^q + \sum\nolimits_1^N \mathbb{E}\|dM''_N\|^q\right)/2,$$

thus we find (recalling the original choice of M' and M'')

$$|(x+y)/2|^q \le \mathbb{E}\|M_{N+1}\|^q - C^{-q}\sum\nolimits_1^{N+1} \mathbb{E}\|dM_n\|^q$$
$$\le (|x|^q + |y|^q)/2 - C^{-q}\|(x-y)/2\|^q + \varepsilon$$

so we obtain

$$|(x+y)/2|^q + C^{-q}\|(x-y)/2\|^q \le (|x|^q + |y|^q)/2, \tag{10.8}$$

and hence by (10.7)

$$|(x+y)/2|^q + C^{-q}|(x-y)/2|^q \le (|x|^q + |y|^q)/2.$$

It is not entirely evident that $|\cdot|$ is a norm, but (10.8) guarantees that for any pair x, y in B the function $f\colon\ \mathbb{R} \to \mathbb{R}$ defined by $f(t) = |x+ty|^q$ satisfies $f((t_1+t_2)/2) \le (f(t_1)+f(t_2))/2$ for any $t_1, t_2 \in \mathbb{R}$ ('midpoint convexity'), and the latter implies (see e.g. [82]) that f is a convex (and hence continuous) function on $\mathbb{R}$. Knowing this, it becomes obvious that $\{x \mid |x| \le 1\}$ is a convex set, so that $|\cdot|$ is indeed a norm on B.

This completes the proof of (ii) $\Rightarrow$ (i). $\square$

Note that when $\alpha = 1$ we have $\|x\| = |x|$ for all x, so that the original norm coincides with the 'new' one and hence is uniformly convex. The next result corresponds to the case $\alpha < 1$.

Corollary 10.7. *Fix $2 \le q < \infty$. The following properties of a Banach space B are equivalent.*

(i) There is an equivalent norm $|\cdot|$ on B such that (10.1) *holds for some $\delta > 0$.*

(ii) There is a constant C such that all B-valued martingales $(M_n)_{n\ge 0}$ in $L_q(B)$ satisfy (recall the convention $dM_0 = M_0$)

$$\sum\nolimits_{n\ge 0} \mathbb{E}\|dM_n\|^q \le C^q \sup\nolimits_{n\ge 0} \mathbb{E}\|M_n\|^q. \tag{10.9}$$

(iii) Same as (ii) for all dyadic martingales, i.e. all martingales based on the dyadic filtration of [0,1], or the corresponding one on $\{-1, 1\}^{\mathbb{N}}$.

We now turn to the main point, i.e. the proof that any uniformly convex B satisfies (10.9) for some q and C. We first place ourselves in a more 'abstract' setting, replacing martingales by monotone basic sequences, defined as follows:

Definition 10.8. A finite sequence $\{x_1, \dots, x_N\}$ of elements in a Banach space is called a monotone basic sequence if for any sequence of scalars $\lambda_1, \dots, \lambda_N$ we have

$$\sup\nolimits_{1\le n\le N} \left\|\sum\nolimits_1^n \lambda_k x_k\right\| \le \left\|\sum\nolimits_1^N \lambda_k x_k\right\|.$$

An infinite sequence $(x_n)_{n\ge 1}$ is called a monotone basic sequence if $(x_1, \dots, x_N)$ is one for any $1 \le N < \infty$.

Independently of James's work on basic sequences in super-reflexive spaces analogous results (such as (10.10)) were proved in the USSR by the Gurarii brothers [259] for uniformly convex spaces.

Theorem 10.9. *Let B be a uniformly convex Banach space. Then for any monotone basic sequence $(x_1, \dots, x_N)$ in B, the following implication holds*

$$\left\|\sum\nolimits_1^N x_k\right\| \le 1 \Rightarrow \|x_1\| + \sum\nolimits_2^N \delta_B(\|x_k\|) \le 1. \tag{10.10}$$

Consequently, there are $1 \le q < \infty$ and a constant C such that any monotone basic sequence $(x_1, \ldots, x_N)$ satisfies

$$\left(\sum \|x_k\|^q\right)^{1/q} \le C \left\|\sum_1^N x_k\right\|. \tag{10.11}$$

Proof. Let $S_n = \sum_1^n x_k$ and set $a_k = \|S_k\|$. Assume $\|S_N\| \le 1$. Fix $2 \le k \le N$. Using $x = a_k^{-1} S_k$ and $y = a_k^{-1} S_{k-1}$ we find

$$a_k^{-1} \|S_{k-1} + x_k/2\| \le 1 - \delta_B(a_k^{-1}\|x_k\|)$$

and since (by monotone basicity) $a_{k-1} \le \|S_{k-1} + x_k/2\|$, we find

$$a_{k-1} \le a_k(1 - \delta_B(a_k^{-1}\|x_k\|))$$

or equivalently for all $k \ge 2$

$$a_k \delta_B(a_k^{-1}\|x_k\|) \le a_k - a_{k-1}.$$

But then, since $a_k^{-1} \ge 1$, by Lemma 10.4 (i)

$$\delta_B(\|x_k\|) \le a_k - a_{k-1}$$

from which (10.10) follows immediately. We deduce from (10.10) that for all $N \ge 2$

$$\left\|\sum_1^N x_k\right\| \le 1 \Rightarrow \inf_{1 \le k \le N} \delta_B(\|x_k\|) \le (N-1)^{-1}.$$

Let $\varepsilon(N)$ be the largest $\varepsilon > 0$ such that $\delta_B(\varepsilon) \le (N-1)^{-1}$. Note that $\varepsilon(N) \to 0$ since $\delta_B(\varepsilon) > 0$ for all $\varepsilon > 0$ and δ_B is non-decreasing. Hence

$$\left\|\sum_1^N x_k\right\| \le 1 \Rightarrow \inf \|x_k\| \le \varepsilon(N),$$

which we may rewrite by homogeneity

$$\inf_{1 \le k \le 1} \|x_k\| \le \varepsilon(N) \left\|\sum_1^N x_k\right\|. \tag{10.12}$$

We will now show that (10.12) implies the second assertion of Theorem 10.9. This follows by a very general principle based on the fact that (10.12) automatically holds for any sequence of N ‘blocks’ built out of a longer monotone basic sequence. More precisely, let us denote by $b(N)$ the best constant b such that for any monotone basic sequence $(x_1, \ldots, x_N)$ we have

$$\inf_{1 \le k \le N} \|x_k\| \le b \left\|\sum_1^N x_k\right\|.$$

It is easy to see that $b(N) \ge b(N+1)$ for all $N \ge 1$. Moreover, a moment of thought shows that b is ‘sub-multiplicative’ i.e. for all integers N, K we have

$$b(NK) \le b(N)b(K).$$

(Hint: Given $y_1, \ldots, y_{NK}$ consider $x_1 = y_1 + \cdots + y_K$, $x_2 = y_{K+1} + \cdots + y_{2K}$, $x_N = y_{(K-1)N+1} + \cdots + y_{NK}$.)

But now (10.12) ensures that $b(N) \le \varepsilon(N)$ and hence that $b(N) \to 0$ when $N \to \infty$. Let us choose an integer m such that $b(m) < 1$ and let $0 < r < \infty$ be determined by $b(m) = m^{-1/r}$. Then, by sub-multiplicativity, we have $b(m^k) \le (m^k)^{-1/r}$ for any $k \ge 1$. If n is arbitrary we choose k so that $m^k \le n < m^{k+1}$ and hence, since $b(\cdot)$ is non-increasing we find finally $b(n) \le m^{1/r}n^{-1/r}$ for all $n \ge 1$.

Let $x_1, \ldots, x_N$ be a monotone basic sequence with $\left\| \sum_1^N x_k \right\| \le 1$. Let $(x_{\sigma(1)}, \ldots, x_{\sigma(N)})$ be a permutation chosen so that $\|x_{\sigma(1)}\| \ge \cdots \ge \|x_{\sigma(N)}\|$. Note that of course this is a priori no longer a monotone basic sequence. Fix j. Let $1 \le m(1) < m(2) < \cdots < m(j) \le N$ be the places corresponding to $\{\sigma(1), \ldots, \sigma(j)\}$ in $[1, \ldots, N]$. Let

$$y_1 = \sum\nolimits_1^{m(1)} x_k, \quad y_2 = \sum\nolimits_{m(1)<k\le m(2)} x_k, \; \ldots, \; y_j = \sum\nolimits_{m(j-1)<k\le N} x_k.$$

We have then

$$\inf_{1\le t\le j} \|y_t\| \le b(j)$$

and moreover by the triangle inequality and the 'monotony'

$$\|x_{m(1)}\| \le \|y_1\| + \left\| \sum\nolimits_1^{m(1)-1} x_k \right\| \le 2\|y_1\|$$

and similarly

$$\|x_{m(2)}\| \le 2\|y_2\|, \ldots, \|x_{m(j)}\| \le 2\|y_j\|$$

so that we find

$$\|x_{\sigma(j)}\| = \inf_{1\le t\le j} \|x_{m(t)}\| \le 2 \inf_{t\le j} \|y_t\| \le 2b(j).$$

We conclude $\|x_{\sigma(j)}\| \le 2m^{1/r} j^{-1/r}$ and hence for any $q > r$

$$\sum \|x_j\|^q = \sum \|x_{\sigma(j)}\|^q \le (2m^{1/r})^q \cdot \sum j^{-q/r}.$$

Thus, for any $q > r$, setting $C = 2m^{1/r}(\sum_1^\infty j^{-q/r})^{1/q}$, we obtain the announced result (10.11). □

Corollary 10.10. *Let B be isomorphic to a uniformly convex Banach space. Fix $1 < s < \infty$. Then there is a number $2 \le q < \infty$ and a constant C such that any B-valued martingale (f_n) in $L_q(B)$ satisfies*

$$\left(\sum\nolimits_0^\infty \|f_n - f_{n-1}\|_{L_s(B)}^q \right)^{1/q} \le C \sup \|f_n\|_{L_s(B)}. \tag{10.13}$$

Proof. If B is uniformly convex, so is $L_s(B)$ by Lemma 10.4. So this follows from the preceding Theorem. □

We will need a very simple 'dualization' of the preceding inequality:

Proposition 10.11. *Let $(\mathcal{A}_n)_{n\geq 0}$ be a filtration on a probability space $(\Omega, \mathcal{A}, \mathbb{P})$ with $\mathcal{A} = \mathcal{A}_\infty$. Let $1 < s < \infty$ and $1 \leq q' \leq 2 \leq q \leq \infty$ with $\frac{1}{q} + \frac{1}{q'} = 1$. The following properties of a Banach space B are equivalent.*

(i) *There is a constant C such that for all B-valued martingales $(f_n)_{n\geq 0}$ adapted to $(\mathcal{A})_{n\geq 0}$ we have (recall $f_{-1} \equiv 0$ by convention)*

$$\left(\sum\nolimits_0^\infty \|f_n - f_{n-1}\|_{L_s(B)}^q\right)^{1/q} \leq C \sup \|f_n\|_{L_s(B)}.$$

(ii) *There is a constant C' such that for all B^*-valued martingales $(g_n)_{n\geq 0}$ adapted to $(\mathcal{A}_n)_{n\geq 0}$ we have*

$$\sup \|g_n\|_{L_{s'}(B^*)} \leq C' \left(\sum\nolimits_0^\infty \|g_n - g_{n-1}\|_{L_{s'}(B^*)}^{q'}\right)^{1/q'}.$$

Moreover the best constants C and C' satisfy $C/2 \leq C' \leq C$.

Proof. Assume (i). Fix n. Let $g_n \in L_{s'}(B^*)$. For any $\varepsilon > 0$ there is f_n with $\|f_n\|_{L_s(B)} = 1$ such that

$$\|g_n\|_{L_{s'}(B^*)} \leq (1+\varepsilon)|\langle f_n, g_n\rangle|,$$

but $\langle f_n, g_n\rangle = \sum_0^n \langle df_k, dg_k\rangle$ and hence

$$\begin{aligned}\|g_n\|_{L_{s'}(B^*)} &\leq (1+\varepsilon)\left|\sum\nolimits_0^n \langle df_k, dg_k\rangle\right| \\ &\leq (1+\varepsilon)\left(\sum\nolimits_0^n \|df_k\|_{L_s(B)}^q\right)^{1/q}\left(\sum\nolimits_0^n \|dg_k\|_{L_{s'}(B^*)}^{q'}\right)^{1/q'},\end{aligned}$$

so that by (i) we find

$$\|g_n\|_{L_{s'}(B^*)} \leq (1+\varepsilon)C\left(\sum\nolimits_0^n \|dg_k\|_{L_{s'}(B^*)}^{q'}\right)^{1/q'}$$

and (ii) follows immediately with $C' \leq C$. Conversely, assume (ii). Fix n and let $f_n \in L_s(B)$. For any $\varepsilon > 0$ there are $\varphi_0, \ldots, \varphi_n$ in $L_{s'}(\Omega, \mathcal{A}, \mathbb{P}; B^*)$ with $(\sum_0^n \|\varphi_k\|_{L_{s'}(B^*)}^{q'})^{1/q'} \leq 1+\varepsilon$ such that

$$\left|\sum\nolimits_0^n \langle df_k, \varphi_k\rangle\right| = \left(\sum\nolimits_0^n \|df_k\|_{L_s(B)}^q\right)^{1/q}.$$

Note that (with the convention $\mathbb{E}_{-1} = 0$)

$$\sum\nolimits_0^n \langle df_k, \varphi_k\rangle = \sum\nolimits_0^n \langle df_k, (\mathbb{E}_k - \mathbb{E}_{k-1})\varphi_k\rangle = \langle f_n, g_n\rangle$$

where $g_n = \sum_0^n (\mathbb{E}_k - \mathbb{E}_{k-1})(\varphi_k)$. In addition since $dg_k = \mathbb{E}_k\varphi_k - \mathbb{E}_{k-1}\varphi_k$ we have $\|dg_k\|_{L_{s'}(B^*)} \leq 2\|\varphi_k\|_{L_{s'}(B^*)}$ and hence $(\sum \|dg_k\|_{L_{s'}(B^*)}^{q'})^{1/q'} \leq 2$.

Thus we obtain by (ii)

$$\left(\sum_0^n \|df_k\|_{L_s(B)}^q\right)^{1/q} = |\langle f_n, g_n\rangle| \le \|f_n\|_{L_s(B)}\|g_n\|_{L_{s'}(B^*)} \le 2C'\|f_n\|_{L_s(B)}.$$

This shows that (ii) $\Rightarrow$ (i) with $C \le 2C'$. □

To prove Theorem 10.1 we apply Corollary 10.10 with B replaced by $L_2(B)$. We can use just as well $L_s(B)$ for any $1 < s < \infty$, but the reader should note that we have a priori no control over how q depends on s so we cannot just set $q = s$! Thus the main difficulty is to pass from (10.13) to (10.9). This is precisely what the next crucial result achieves, with a slight loss on the exponent (of course the case $s = 2$ being the obvious one there, we assume $s > 2$).

Lemma 10.12. *Let $2 < s < \infty$. Let B be a Banach space. Assume that for some constant $\chi \ge 1$, all B-valued martingales $(f_n)_{n\ge 0}$ satisfy*

$$\forall N \ge 0 \qquad \left(\|f_0\|_{L_2(B)}^s + \sum_{n\ge 1} \|df_n\|_{L_2(B)}^s\right)^{1/s} \le \chi \sup_n \|f_n\|_{L_2(B)}. \quad (10.14)$$

Then for each $q > s$ there is a constant $C = C(q, s)$ such that all dyadic B-valued martingales, on $\{1, 1\}^{\mathbb{N}_*}$ with the usual filtration $\mathcal{D}_n = \sigma(\varepsilon_j, 1 \le j \le n)$, satisfy (10.9).

By the dualization given by Proposition 10.11, Lemma 10.12 is equivalent to the next one.

Lemma 10.13. *Let $1 < r < 2$. Let B be a Banach space. Assume that for some constant $\chi \ge 1$, all B-valued martingales $(f_n)_{n\ge 0}$ satisfy*

$$\sup_{n\ge 0} \|f_n\|_{L_2(B)} \le \chi \left(\|f_0\|_{L_2(B)}^r + \sum_{n\ge 1} \|df_n\|_{L_2(B)}^r\right)^{1/r}. \quad (10.15)$$

Then for each $1 < p < r$ there is a constant $C_1 = C_1(p, r)$ such that all dyadic B-valued martingales satisfy

$$\sup_{n\ge 0} \mathbb{E}\|f_n\|^p \le C_1^p \left(\mathbb{E}\|f_0\|^p + \sum_{n\ge 1} \mathbb{E}\|df_n\|^p\right). \quad (10.16)$$

The next statement, possibly of independent interest, is the main new ingredient to prove Lemma 10.12. For the reader's benefit, we emphasize that the inclusions $\ell_s(L_2) \subset L_2(\ell_q)$ or $\ell_s(L_2(B)) \subset L_2(\ell_q(B))$ *fail* when $q > s > 2$. Nevertheless, this lemma shows that something like that becomes true in a specific situation involving martingales.

Lemma 10.14. *Let $(f_n)_{n\ge 0}$ be a dyadic (or merely calibrated) martingale with values in an arbitrary Banach space B, and converging in $L_2(B)$. Let $\mathcal{A}_n = \sigma(f_1, \ldots, f_n)$ be the associated filtration with $\mathcal{A}_0$ trivial on $(\Omega, \mathbb{P})$.*

Let $2 \le s < \infty$. Assume that for any sequence of stopping times $T(1) \le T(2) \le \cdots$ we have

$$\left(\|f_{T(1)}\|^s_{L_2(B)} + \sum\nolimits_{k\ge 2} \|f_{T(k)} - f_{T(k-1)}\|^s_{L_2(B)}\right)^{1/s} \le 1. \tag{10.17}$$

As usual let $d_k = f_k - f_{k-1}$ for $k \ge 1$ and $d_0 = f_0$. Let

$$A_t(n) = \left\{\omega \in \Omega \mid \sum\nolimits_{k\ge 0} 1_{\{\|d_k(\omega)\|>t\}} \ge n\right\}.$$

Then for any $t > 0$ and any $n \ge 1$ we have

$$t^2\mathbb{P}(A_t(n)) \le 4n^{-2/s}. \tag{10.18}$$

Moreover, for each $q > s$ there is a constant $C'(q,s) \ge 1$ such that

$$\left\|\left(\sum\nolimits_{k\ge 0} \|d_k\|^q\right)^{1/q}\right\|_{2,\infty} \le C'(q,s). \tag{10.19}$$

Proof. The reader who wishes to can easily reduce this to the case of a finite martingale, but this is not really needed. Fix $t > 0$. Let $T(1) = \inf\{k \ge 0 \mid \|d_k\| > t\}$, $T(2) = \inf\{k > T(1) \mid \|d_k\| > t\}$, and

$$T(m+1) = \inf\{k > T(m) \mid \|d_k\| > t\},$$

with the usual convention $\inf \phi = \infty$. We have

$$A_t(n) = \{T(n) < \infty\}.$$

By our assumption

$$\left(\|f_{T(1)}\|^2_{L_2(B)} + \sum\nolimits_{2\le k\le n} \|f_{T(k)} - f_{T(k-1)}\|^2_{L_2(B)}\right)^{1/2} \le n^{1/2-1/s}.$$

We first claim that $\|d_{T(k)}\|_{L_2(B)} \le 2\|f_{T(k)} - f_{T(k-1)}\|_{L_2(B)}$ for any $k > 1$. Indeed, since $\|d_{n+1}\|$ is $\mathcal{A}_n$-measurable for any $n \ge 0$, $T(k) - 1$ is a stopping time for any $k > 1$ (with the convention $\infty - 1 = \infty$). Therefore by (1.26) we have

$$f_{T(k)-1} - f_{T(k-1)} = \mathbb{E}^{\mathcal{A}_{T(k)-1}}(f_{T(k)} - f_{T(k-1)})$$

so that $\|f_{T(k)-1} - f_{T(k-1)}\|_{L_2(B)} \le \|f_{T(k)} - f_{T(k-1)}\|_{L_2(B)}$ for any $k > 1$, and hence

$$\|d_{T(k)}\|_{L_2(B)} \le 2\|f_{T(k)} - f_{T(k-1)}\|_{L_2(B)}.$$

Similarly, we claim that $\|d_{T(1)}\|_{L_2(B)} \le 2\|f_{T(1)}\|_{L_2(B)}$. If $T(1) = 0$ a.s. then $d_{T(1)} = f_{T(1)}$ so this is obvious. Otherwise, since the set $\{T(1) = 0\}$ is in $\mathcal{A}_0 = \{\phi, \Omega\}$ we may assume $T(1) > 0$ a.s. and then again $T(1) - 1$ is a stopping time and we have

$$\|d_{T(1)}\|_{L_2(B)} \le 2\|f_{T(1)}\|_{L_2(B)}.$$

In any case we find

$$\mathbb{E}\sum\nolimits_{1\le k\le n}\|d_{T(k)}\|^2 \le 4\mathbb{E}(\|f_{T(1)}\|^2+\sum\nolimits_{k\ge 2}\|f_{T(k)}-f_{T(k-1)}\|^2)\le 4n^{1-2/s}.$$

But on $A_t(n)$ we have $\sum_{1\le k\le n}\|d_{T(k)}\|^2 > nt^2$ and hence

$$t^2\mathbb{P}(A_t(n))\le 4n^{-2/s}.$$

Let $b_1\ge b_2\ge\cdots b_n\ge\cdots$ be a non-increasing rearrangement of the sequence $(\|d_k\|)$. Then

$$A_t(n)=\{b_n>t\}.$$

Choose r such that $s<r<q$. We will use the elementary fact (left to the reader) that there is a constant $\gamma(r,q)$ such that

$$\left(\sum\nolimits_{k\ge 0}\|d_k\|^q\right)^{1/q}=\left(\sum\nolimits_{k\ge 0}b_k^q\right)^{1/q}\le\gamma(r,q)\sup\nolimits_{m\ge 0}2^{m/r}b_{2^m}.$$

Changing t to $tn^{-1/r}$ we find

$$\mathbb{P}\{n^{1/r}b_n>t\}\le 4t^{-2}n^{2/r-2/s}$$

and hence, setting $\varepsilon=2/s-2/r>0$, we find for any integer $m\ge 0$

$$\mathbb{P}\{2^{m/r}b_{2^m}>t\}\le 4t^{-2}2^{-\varepsilon m}$$

and hence

$$\mathbb{P}\{\sup\nolimits_m 2^{m/r}b_{2^m}>t\}\le 4t^{-2}\sum\nolimits_m 2^{-\varepsilon m}=c_\varepsilon t^{-2},$$

or equivalently

$$\|\sup\nolimits_m 2^{m/r}b_{2^m}\|_{2,\infty}\le c_\varepsilon^{1/2}.$$

A fortiori this yields

$$\left\|\left(\sum\nolimits_{k\ge 0}\|d_k\|^q\right)^{1/q}\right\|_{2,\infty}=\left\|\left(\sum\nolimits_{k\ge 0}b_k^q\right)^{1/q}\right\|_{2,\infty}\le\gamma(r,q)c_\varepsilon^{1/2}.\qquad\square$$

Proof of Lemma 10.12. Let $C'=C'(r,s)$ be the constant appearing in (10.19). We will apply Lemma 5.26 with (v_n) scalar valued and (w_n) B-valued. Explicitly, we let $v_n=(\sum_0^n\|d_k\|^q)^{1/q}$ and $w_n=C'\chi f_n$. Let $\psi_n=C'\chi\|d_{n+1}\|$. Recall $C'\chi\ge 1$. Then, by the triangle inequality, (5.40), (5.41) and (5.52) obviously hold. By homogeneity, we may assume that $\sup_n\|f_n\|_{L_2(B)}\le 1$. Note that since we assume (10.14) for all martingales we may apply it to the martingale $(f_{T(k)})$. This shows that the martingale $\chi^{-1}(f_n)$ satisfies (10.17). Then, by homogeneity, (10.19) implies $\|v_n\|_{2,\infty}\le\|w_n\|_{L_2(B)}$. Now for any pair $R\le T$ of stopping times, we may apply (10.19) to the martingale $M_n = f_{T\wedge n} - f_{R\wedge n}$. This gives us

$$\left\| \left(\sum_{R<k\le T} \|d_k\|^q \right)^{1/q} \right\|_{2,\infty} \le \|w_T - w_R\|_{L_2(B)}.$$

Note $v_T - v_R \le (\sum_{R<k\le T} \|d_k\|^q)^{1/q}$. Therefore the assumption (5.42) in Lemma 5.26 holds with $p = 2$. By (5.44) we have

$$\left(\sum\nolimits_k \mathbb{E}\|d_k\|^q\right)^{1/q} = \|v_\infty\|_q \le \chi(2,q)C'\chi \| \sup_n \|f_n\| \|_q$$

and since by Doob's inequality (1.38) $\| \sup_n \|f_n\| \|_q \le q' \sup_n \|f_n\|_{L_q(B)}$ we obtain (10.9). □

Remark. Starting from (10.19), there is another route using the softer extrapolation Lemma 5.23 (and Remark 5.25) instead of Lemma 5.26: one uses it to show $\|v_\infty\|_p \lesssim \|w\|_p$ for any $p \in (1,2)$ then one dualizes and uses Lemma 5.23 again to replace the resulting p', p' estimate by an r', r' one for any $0 < r' < p'$, in particular for any $r' < 2$. Then one more dualization brings us back to $\|v_\infty\|_r \lesssim \|w\|_r$ for any $r > 2$ and in particular for $r = q$. The dualizations are justified by the same argument as for Proposition 10.11.

Remark 10.15. Let us denote by $\{\|d_k\|^*\}$ the random sequence obtained by rearranging $\{\|d_k\|\}$ in non-increasing order. Changing t to $tn^{-1/s}$ in (10.18) shows that (10.18) is equivalent to $\sup_{n\ge1} \sup_{t>0} t^2\mathbb{P}\{n^{1/s}\|d_n\|^* > t\} \le 4$, or

$$\sup_{n\ge1} \|n^{1/s}\|d_n\|^*\|_{2,\infty} \le 2.$$

A similar result holds if we replace $L_2(B)$ by $L_p(B)$ in (10.17) for some $1 \le p \le s < \infty$, we then obtain $\sup_{n\ge1} \|n^{1/s}\|d_n\|^*\|_{p,\infty} \le 2$. This specific kind of estimate seems new in the martingale context.

Proof of Theorem 10.1. If B is uniformly convex, Corollary 10.10 (with $s = 2$) shows that B satisfies the assumption of Lemma 10.12. Therefore we conclude by Corollary 10.7. □

Theorem 10.1 admits the following refinement:

Theorem 10.16. *Let $2 < q_0 < \infty$. If a uniformly convex Banach space B satisfies*

$$\delta_B(\varepsilon)\varepsilon^{-q_0} \to \infty \quad \textit{when} \quad \varepsilon \to 0$$

then there is an equivalent norm on B for which the associated modulus of convexity δ satisfies for some $q < q_0$

$$\inf_{0<\varepsilon\le2} \delta(\varepsilon)\varepsilon^{-q} > 0.$$

Proof. By (ii) in Lemma 10.12, we may replace B by $L_2(B)$. We can then argue exactly as in the preceding proof of Theorem 10.1. Here is a slightly more direct argument: Let $a(N)$ be the smallest constant C such that for all N-tuples of B-valued martingale differences $d_1, \ldots, d_N$ we have

$$N^{-1} \sum_{1 \le n \le N} \|d_n\|_{L_2(B)} \le C\|d_1 + \cdots + d_N\|_{L_2(B)}.$$

Then it is easy to check that $a(NK) \le a(N)a(K)$ for all $N, K \ge 1$. Applying (10.10) in $L_2(B)$ shows that $a(N)N^{-1/q_0} \to 0$ when $N \to \infty$. Then the sub-multiplicativity implies that there is $q_1 < q_0$ such that $a(N)N^{-1/q_1}$ is bounded. Arguing as for (10.11), we find that (10.14) holds whenever $s > q_1$, so the conclusion follows again, with $q_1 < s < q < q_0$, from Lemma 10.12 and Corollary 10.7. □

Definition 10.17. We will say that a Banach space B is q-uniformly convex if there is a constant $c > 0$ such that $\delta_B(\varepsilon) \ge c\varepsilon^q$ for all $0 < \varepsilon \le 2$.

As we will show in Proposition 10.31, B is q-uniformly convex iff there is $C > 0$ such that $\forall x, y \in B$

$$\left\|\frac{x+y}{2}\right\|^q + C^{-q}\left\|\frac{x-y}{2}\right\|^q \le \frac{\|x\|^q + \|y\|^q}{2}.$$

We discuss complex interpolation for this property in Proposition 11.44.

With this terminology, let us recapitulate:

B is q-uniformly convex iff (10.5) holds with $\alpha = 1$.

Moreover, B is isomorphic to a q-uniformly convex space iff it satisfies (10.9) for some constant C.

Lastly, any uniformly convex space is isomorphic to a q-uniformly convex one for some $q < \infty$ (see Theorem 10.1).

10.2 Uniform smoothness

Uniform smoothness is dual to uniform convexity: B is uniformly smooth (resp. uniformly convex) iff B^* is uniformly convex (resp. uniformly smooth). Therefore many of its properties can be deduced from the corresponding properties of uniform convexity. Nevertheless, the intrinsic geometric significance of uniform smoothness is of considerable interest in many questions involving e.g. differentiability of functions on B.

Definition 10.18. A Banach space B is called uniformly smooth if there is a function $t \to \rho(t)$ on $\mathbb{R}_+$ that is $o(t)$ when $t \to 0$ such that for any x, y in the unit sphere of B we have

$$\frac{\|x+ty\|+\|x-ty\|}{2} \le 1+\rho(t).$$

The modulus of (uniform) smoothness $\rho_B(t)$ is defined as the 'best possible' ρ, i.e.

$$\rho_B(t) = \sup\{2^{-1}(\|x+ty\|+\|x-ty\|)-1 \mid x, y \in B,\ \|x\| = \|y\| = 1\}.$$

With this notation, B is uniformly smooth iff

$$\lim_{t\to 0} \rho_B(t)/t = 0.$$

For example, for a Hilbert space H, we have $\rho_H(t) = (1+t^2)^{1/2} - 1 \simeq t^2/2$. By Dvoretzky's theorem (see Theorem 10.43), for any infinite-dimensional space B, we must have $\rho_B \ge \rho_H$ for $H = \ell_2^n$ and hence $\rho_B(t) \ge (1+t^2)^{1/2} - 1$.

The following formula due to Lindenstrauss [330] (see also [207]) illustrates the dual relationship between δ_B and ρ_{B^*}.

Lemma 10.19. *For any (real or complex) Banach space B*

$$\rho_{B^*}(t) = \sup\{t\varepsilon/2 - \delta_B(\varepsilon) \mid 0 < \varepsilon \le 2\}. \tag{10.20}$$

Proof. Let $U_B = \{x \in B \mid \|x\| \le 1\}$ and $S_B = \{x \in B \mid \|x\| = 1\}$. By definition we have in the real case

$$\begin{aligned}
\rho_{B^*}(t) &= \sup\{2^{-1}(\|\xi+t\eta\|+\|\xi-t\eta\|)-1 \mid \xi,\eta \in S_{B^*}\}\\
&= \sup\{2^{-1}(\langle \xi+t\eta, x\rangle + \langle \xi-t\eta, y\rangle)-1 \mid \xi,\eta \in S_{B^*}, x, y \in U_B\}\\
&= \sup\left\{\left\|\frac{x+y}{2}\right\| + t\left\|\frac{x-y}{2}\right\| - 1 \,\middle|\, x, y \in U_B\right\}\\
&= \sup_{0<\varepsilon\le 2}\sup\left\{\left\|\frac{x+y}{2}\right\| + t\varepsilon/2 - 1 \,\middle|\, x, y \in U_B,\ \|x-y\| \ge \varepsilon\right\}\\
&= \sup_{0<\varepsilon\le 2}\{t\varepsilon/2 - \delta_B(\varepsilon)\}.
\end{aligned}$$

In the complex case, just use $\|\xi \pm t\eta\| = \sup\{\Re(\langle \xi \pm t\eta, x\rangle) \mid x \in U_B\}$. □

It is natural to wonder whether conversely δ_{B^*} is in duality with ρ_B. Unfortunately, this is not true because, unlike ρ_B, the function δ_{B^*} is in general not convex (see [333]). Nevertheless, if we denote by $\widetilde{\delta}_B$ the largest convex function dominated by δ_B, we have a nice duality, and moreover, $\widetilde{\delta}_B$ and δ_B are essentially equivalent. We refer to [223, 228] for more on this.

Lemma 10.20. *For any (real or complex) Banach space B*

$$\widetilde{\delta}_{B^*}(\varepsilon) = \sup\{t\varepsilon/2 - \rho_B(t) \mid 0 < t < \infty\}. \tag{10.21}$$

Moreover for any $0 < \gamma < 1$ and $\varepsilon > 0$ we have

$$(\gamma^{-1} - 1)\delta_{B^*}(\gamma\varepsilon) \le \widetilde{\delta}_{B^*}(\varepsilon) \le \delta_{B^*}(\varepsilon).$$

Proof. The first formula is proved just like (10.20), and we find $\sup\{t\varepsilon/2 - \rho_B(t) \mid 0 < t < \infty\} = \sup_{t>0}(\inf_{0<s\le 2}\{\delta_B(s) + t(\varepsilon - s)/2\}$ but note that being the supremum of affine functions the right-hand side of (10.21) is a convex function, that majorizes any affine function f (say $f(\varepsilon) = a\varepsilon + b$) such that $f \le \delta_{B^*}$ because it is easy to see that

$$\sup_{t>0}(\inf_{0<s\le 2}\{f(s) + t(\varepsilon - s)/2\} = f(\varepsilon).$$

This establishes (10.21). The second assertion is more delicate, we refer the reader to [223]. □

As a consequence we have

Proposition 10.21. *A Banach space B is uniformly convex (resp. uniformly smooth) iff its dual B^* is uniformly smooth (resp. uniformly convex).*

Proof. If B is uniformly convex, the formula (10.20) clearly implies by elementary calculus that B^* is uniformly smooth (note that ρ_{B^*} is essentially the Legendre conjugate of δ_B). Conversely, if B is uniformly smooth, then Lemma 10.20 implies that B^* is uniformly convex. Note that by Theorem 10.3, B is reflexive if either B or B^* is uniformly convex. From this it is easy to complete the proof. □

In view of the preceding (almost perfect) duality, it is not surprising that the results of §6.1 have analogues for uniform smoothness, so we will content ourselves with a brief outline with mere indications of proofs.

Theorem 10.22. *Let $1 < p \le 2$ and let $\alpha > 0$ and $C > 0$ be fixed constants. The following two properties of a Banach space B are equivalent:*

(i) *There is a equivalent norm $|\ |$ on B such that for all x, y in B we have $\|x\| \le |x| \le \alpha^{-1}\|x\|$ and*

$$2^{-1}(|x+y|^p + |x-y|^p) \le |x|^p + C^p\|y\|^p. \qquad (10.22)$$

(ii) *For any dyadic B-valued martingale $(M_n)_{n\ge 0}$ in $L_p(B)$ we have*

$$\sup \mathbb{E}\|M_n\|^p \le \alpha^{-p}\mathbb{E}\|M_0\|^p + C^p \sum\nolimits_1^\infty \mathbb{E}\|dM_n\|^p. \qquad (10.23)$$

Moreover, this implies:

(iii) *All B-valued martingales in $L_p(B)$ satisfy*

$$\sup \mathbb{E}\|M_n\|^p \le \alpha^{-p}\mathbb{E}\|M_0\|^p + 2C^p \sum\nolimits_1^\infty \mathbb{E}\|dM_n\|^p.$$

Proof. (ii) $\Rightarrow$ (i). Assume (ii). We define

$$|x|^p = \sup \left\{ \mathbb{E}\|M_N\|^p - C^p \sum\nolimits_1^N \mathbb{E}\|dM_n\|^p \right\}$$

where the supremum runs over all $N \geq 1$ and all dyadic martingales M_0, $M_1, \ldots, M_N$ such that $M_0 = x$. Note that we trivially have $|x| \geq \|x\|$ (by choosing $M_N = x$), and by (ii) we have $|x| \leq \alpha^{-1}\|x\|$, so that $|\cdot|$ and $\|\cdot\|$ are equivalent. The same idea as in the previous section shows that

$$\forall x, y \in B \qquad 2^{-1}(|x|^p + |y|^p) \leq |2^{-1}(x+y)|^p + \|2^{-1}(x-y)\|^p$$

or equivalently (replace (x, y) by $(x+y, x-y)$)

$$2^{-1}(|x+y|^p + |x-y|^p) \leq |x|^p + C^p\|y\|^p \leq |x|^p + C^p|y|^p$$

so we obtain (i). Lastly, in case $x \to |x|$ is not a norm we define

$$|x|_1 = \inf \left\{ \sum |x_k| \right\}$$

over all the decompositions $x = \sum x_k$ as a finite sum of elements of B. Note that for any $t > |x|_1$ we can write $x = \sum \lambda_k x_k$ with $\lambda_k \geq 0$, $\sum \lambda_k = 1$ and $|x_k| \leq t$. Using this, it is then easy to check that (10.22) remains true when $|\ \ |_1$ replaces $|\ \ |$, completing the proof that (ii) $\Rightarrow$ (i).

(i) $\Rightarrow$ (iii) and (i) $\Rightarrow$ (ii). For any $n \geq 1$ and ω we have

$$\begin{aligned} 2^{-1}\ &(|M_{n-1}(\omega) + dM_n(\omega)|^p + |M_{n-1}(\omega) - dM_n(\omega)|^p) \\ &\leq |M_{n-1}(\omega)\ |^p + C^p\|dM_n(\omega)\|^p \end{aligned}$$

and hence after integration

$$2^{-1}(\mathbb{E}|M_n|^p + \mathbb{E}|M_{n-1} - dM_n|^p) \leq \mathbb{E}|M_{n-1}|^p + C^p\mathbb{E}\|dM_n\|^p$$

but since $\mathbb{E}|M_{n-1}|^p \leq \mathbb{E}|M_{n-1} - dM_n|^p$ we deduce

$$\mathbb{E}|M_n|^p \leq \mathbb{E}|M_{n-1}|^p + 2C^p\mathbb{E}\|dM_n\|^p$$

and hence (note the telescoping sum)

$$\sup \mathbb{E}|M_n|^p \leq \mathbb{E}|M_0|^p + 2C^p \sum\nolimits_1^\infty \mathbb{E}\|dM_n\|^p.$$

Since $|\cdot|$ is an equivalent norm, (iii) follows. To check (i) $\Rightarrow$ (ii) just observe that in the dyadic case $\mathbb{E}|M_{n-1} - dM_n|^p = \mathbb{E}|M_{n-1} + dM_n|^p = \mathbb{E}|M_n|^p$ (so the factor 2 disappears in the preceding argument). □

Corollary 10.23. *Fix $1 < p \le 2$. The following properties of a Banach space B are equivalent:*

(i) There is an equivalent norm $|\cdot|$ on B such that for some constant C we have

$$\forall x, y \in B \qquad 2^{-1}(|x+y|^p + |x-y|^p) \le |x|^p + C^p|y|^p. \quad (10.24)$$

(ii) There is a constant C such that all B-valued martingales $(M_n)_{n\ge 0}$ in $L_p(B)$ satisfy (recall $dM_0 = M_0$ by convention)

$$\sup \mathbb{E}\|M_n\|^p \le C^p \sum\nolimits_0^\infty \mathbb{E}\|dM_n\|^p. \quad (10.25)$$

(iii) Same as (ii) for all dyadic martingales.

The next result is the dual analogue of Theorem 10.9, and although we prefer to give a direct argument, it can be proved by duality.

Theorem 10.24. *Assume B uniformly smooth. Then for any (finite or infinite) monotone basic sequence (x_n) in B, we have*

$$\sup\nolimits_n \|x_n\| + 2\sum\nolimits_1^\infty \rho_B(\|x_n\|) \le 1 \Rightarrow \sup\nolimits_n \|x_1 + \cdots + x_n\| \le 2.$$

Consequently, there is a constant C and $1 < p \le 2$ such that for all N and all monotone basic sequences (x_n) we have

$$\left\|\sum\nolimits_1^N x_n\right\| \le C\left(\sum \|x_n\|^p\right)^{1/p}.$$

Proof. Let $S_n = x_1 + \cdots + x_n$. We have

$$\|S_{n-1}\|^{-1}(\|S_{n-1} + x_n\| + \|S_{n-1} - x_n\|) - 2 \le 2\rho_B(\|x_n\|\,\|S_{n-1}\|^{-1})$$

and also $1 \le \|S_{n-1}\|^{-1}\|S_{n-1} - x_n\|$ by monotony. Put together, this yields

$$\|S_n\| \le \|S_{n-1}\| + 2\|S_{n-1}\|\rho_B(\|x_n\|\,\|S_{n-1}\|^{-1}).$$

Assume $\|S_{n-1}\| \ge 1$. Since $t \to \rho_B(t)/t$ is non-decreasing (since ρ_B is convex) we have $\|S_{n-1}\|\rho_B(\|x_n\|\|S_{n-1}\|^{-1}) \le \rho_B(\|x_n\|)$ and we find

$$\|S_n\| \le \|S_{n-1}\| + 2\rho_B(\|x_n\|).$$

This yields (telescoping sum) that if $\|S_N\| \ge 1$ we have for all $n \ge N$

$$\|S_n\| \le \|S_N\| + 2\sum\nolimits_{n>N} \rho_B(\|x_n\|).$$

Let N be the first integer (if any) such that $\|S_N\| \ge 1$. Then $\|S_N\| \le 1 + \|x_N\|$ so we obtain $\sup \|S_n\| \le 1 + \|x_N\| + 2\sum_{n>N} \rho_B(\|x_n\|) \le 2$. $\square$

The analogue of Theorem 10.1 for smoothness is now immediate:

Theorem 10.25. *Any uniformly smooth Banach space B admits an equivalent norm $|\cdot|$ satisfying for some constant $C > 0$ and some $1 < p \le 2$*

$$\forall x, y \in B \qquad \frac{|x+y|^p + |x-y|^p}{2} \le |x|^p + C|y|^p. \tag{10.26}$$

Proof. Using Proposition 10.21, this can be easily deduced from Theorem 10.1 by duality. Alternatively, a direct proof can be obtained by combining Theorem 10.9 with Lemma 10.13 and Corollary 10.23. □

Theorem 10.25 admits the following refinement:

Theorem 10.26. *Let $1 < r < 2$. If a uniformly smooth Banach space B satisfies*

$$\rho_B(t)t^{-r} \to 0 \quad \text{when} \quad t \to 0 \tag{10.27}$$

then there is an equivalent norm on B for which the associated modulus of smoothness ρ satisfies for some $p > r$

$$\sup_{t>0} \rho(t)t^{-p} < \infty.$$

Proof. By the Lindenstrauss duality formula (see (10.20) and Lemma 10.20), this can be immediately deduced from Theorem 10.16 by duality. □

Definition. We will say that B (its unit sphere or its norm) is smooth if for any x, y in B with $x \neq 0$ the function $t \to \|x + ty\|$ is differentiable at $t = 0$.

Remark. Fix $x, y \in B$. Let $f(t) = \|x + ty\|$ $(t \in \mathbb{R})$. Assume that

$$(f(t) + f(-t))/2 - 1 \to 0$$

when $t \to 0$. Then f is differentiable at 0. Indeed, since f is a convex function, it admits left and right derivatives everywhere, in particular at $t = 0$ where we denote them by $f'_-(0)$ and $f'_+(0)$ respectively, but our assumption implies $f'_-(0) = f'_+(0)$ so $f'(0)$ exists (and the converse is obvious). Let us denote $\xi_x(y) = f'(0)$. We now assume that B is smooth, i.e. $\xi_x(y)$ exists for any y in B. We will show that, if B is a *real* Banach space

$$\xi_x \in B^*, \quad \|\xi_x\|_{B^*} = 1 \quad \text{and} \quad \xi_x(x) = \|x\|. \tag{10.28}$$

Taking $y = x$, we immediately find $\xi_x(x) = \|x\|$. Note that $\xi_x(sy) = s\xi_x(y)$ for any $s \in \mathbb{R}$. Moreover, from $\|x + t(y_1 + y_2)/2\| \le (\|x + ty_1\| + \|x + ty_2\|)/2$ we deduce easily that if $\xi_x(y)$ exists for any y, then we must have $\xi_x(y_1 + y_2) = \xi_x(y_1) + \xi_x(y_2)$, so that $y \to \xi_x(y)$ is a linear form on B. Moreover, from $|f(t) - \|x\|| \le |t|\|y\|$ we deduce $|\xi_x(y)| \le \|y\|$ so that (since $\xi_x(x) = \|x\|$) $\|\xi_x\|_{B^*} = 1$.

In addition, ξ_x is the unique $\xi \in B^*$ satisfying (10.28). Indeed, for any such ξ we have (when $|t| \to 0$)

$$\|x\| + t\xi(y) \leq \|x + ty\| = \|x\| + t\xi_x(y) + o(|t|)$$

and hence $\xi = \xi_x$. □

By the preceding remark, if B is uniformly smooth, a fortiori its unit sphere S_B is 'smooth,' and for any $x \neq 0$ in B there is a *unique* $\xi_x \in S_{B^*}$ satisfying (10.28). It is useful to observe that when B is uniformly smooth the map

$$x \mapsto \xi_x \colon\ B - \{0\} \to S_{B^*}$$

is uniformly continuous when restricted to closed bounded subsets of $B - \{0\}$. More precisely we have (here we reproduce a proof in [5]):

Proposition 10.27. *Let B be a uniformly smooth Banach space. Then*

$$\forall x, y \in B \qquad \|\xi_x - \xi_y\| \leq 2\rho_B(2\big\|x\|x\|^{-1} - y\|y\|^{-1}\big\|)/\big\|x\|x\|^{-1} - y\|y\|^{-1}\big\|. \tag{10.29}$$

In particular, if $\|x\| = \|y\| = 1$

$$\|\xi_x - \xi_y\| \leq 2\rho_B(2\|x - y\|)/\|x - y\|.$$

Proof. Recall that, by definition of ξ_a, for all a, b in B with $a \neq 0$

$$\lim_{|t| \to 0} t^{-1}(\|a + tb\| - \|a\|) = \langle \xi_a, b\rangle.$$

By convexity of $t \mapsto \psi(t) = \|a + tb\| - \|a\|$, the function $t \mapsto \psi(t)/t$ must be non-decreasing on $\mathbb{R}_+$, and hence

$$\langle \xi_a, b\rangle \leq \|a + b\| - \|a\|. \tag{10.30}$$

Since $\xi_x = \xi_{x\|x\|^{-1}}$ it suffices to prove (10.29) when $\|x\| = \|y\| = 1$. In that case, (10.29) becomes

$$\|x - y\|\,\|\xi_x - \xi_y\| \leq 2\rho_B(2\|x - y\|). \tag{10.31}$$

Let $z \in B$ be such that $\|z\| = \|x - y\|$. Assuming $\|x\| = \|y\| = 1$, we have by repeated use of (10.30) (note also $\langle \xi_x, x - y\rangle = 1 - \langle \xi_x, y\rangle \geq 0$)

$$\begin{aligned}
\langle \xi_y, z\rangle - \langle \xi_x, z\rangle &\leq \|y + z\| - 1 - \langle \xi_x, z\rangle \\
&\leq \|y + z\| - 1 + \langle \xi_x, x - y - z\rangle \\
&\leq \|y + z\| - 1 + \|2x - y - z\| - 1 \\
&= \|x + (y - x + z)\| + \|x - (y - x + z)\| - 2 \\
&\leq 2\rho_B(\|y - x + z\|) \leq 2\rho_B(2\|y - x\|).
\end{aligned}$$

The last step because $\|z\| = \|y - x\|$. Taking the supremum of the preceding over all z with $\|z\| = \|x - y\|$, we obtain (10.31). □

Corollary 10.28. *For $0 < \delta < R < \infty$, let $B(\delta, R) = \{x \in B \mid \delta \leq \|x\| \leq R\}$. The following properties of a Banach space B are equivalent.*

(i) B is uniformly smooth.
(ii) B is smooth and $x \to \xi_x$ is uniformly continuous on the unit sphere S_B.
(iii) B is smooth and $x \to \xi_x$ is uniformly continuous on $B(\delta, R)$ for any $0 < \delta < 1 < R < \infty$.

Proof. (i) ⇒ (ii) follows from Proposition 10.27 and (ii) ⇔ (iii) is easy using $\xi_x = \xi_{x\|x\|^{-1}}$. If (iii) holds, then assuming $\|x\| = \|y\| = 1$ and $|t| < \min(1 - \delta, R - 1)$ we have by the 'fundamental formula of calculus'

$$\|x + ty\| - \|x\| = \int_0^t \langle \xi_{x+sy}, y \rangle ds$$

and hence

$$2^{-1}(\|x + ty\| + \|x - ty\| - 2\|x\|) = \int_0^t \langle \xi_{x+sy} - \xi_{x-sy}, y \rangle ds/2.$$

Therefore, we find

$$\rho_B(t) \leq |t| \sup\{\|\xi_x - \xi_{x'}\| \mid x, x' \in B(\delta, R), \|x - x'\| \leq t\}/2,$$

from which (iii) ⇒ (i) is immediate. □

Corollary 10.29. *Let $1 < p \leq 2$. Assume that $\rho_B(t) \in O(t^p)$ when $t \to 0$. Then, for any $0 < \delta < R < \infty$, there is a constant $C = C_{\delta,R}$ such that*

$$\forall x, y \in B(\delta, R) \qquad \|\xi_x - \xi_y\| \leq C\|x - y\|^{p-1}.$$

In particular, if $p = 2$, the map $x \mapsto \xi_x$ is Lipschitzian on $B(\delta, R)$.

Proof. This is an immediate consequence of (10.29) by elementary calculus. □

We refer the reader to [19, 38] for supplementary information and more references.

The property appearing in Corollary 10.29 was already considered in early pioneering work by Fortet and Mourier on the strong law of large numbers for Banach space valued random variables, cf. [229]. As we will show in the next chapter (see Theorem 11.43), the validity of the strong law of large numbers for B-valued martingales is equivalent to the super-reflexivity of B.

Definition 10.30. We will say that a Banach space B is p-uniformly smooth if there is a constant $c > 0$ such that $\rho_B(t) \le ct^p$ for all $t > 0$.

We discuss complex interpolation for this property in Proposition 11.44.

Proposition 10.31. *Let $1 < p \le 2 \le q < \infty$. Let $p' = p/(p-1)$ as usual. A Banach space B is p-uniformly smooth iff its dual B^* is p'-uniformly convex. Moreover:*

(i) The space B is p-uniformly smooth iff there is $C > 0$ such that

$$\forall x, y \in B \quad \frac{\|x+y\|^p + \|x-y\|^p}{2} \le \|x\|^p + C^p\|y\|^p.$$

(ii) The space B is q-uniformly convex iff there is $C > 0$ such that

$$\forall x, y \in B \quad \|x\|^q + C^{-q}\|y\|^q \le \frac{\|x+y\|^q + \|x-y\|^q}{2},$$

or equivalently

$$\forall x, y \in B \quad \|(x+y)/2\|^q + C^{-q}\|(x-y)/2\|^q \le \frac{\|x\|^q + \|y\|^q}{2}.$$

Proof. By the duality formulae (10.20) and (10.21), the first assertion follows (as implicitly observed in the proof of Proposition 10.21). Thus, by an easy dualization, it suffices to prove (i). The if part of (i) (or of (ii)) is obvious. Let us check the converse. Assume that B is p-uniformly smooth. Then there is a constant $c > 0$ such that if $\|x\| = 1$ we have for any $y \in B$

$$\frac{\|x+y\| + \|x-y\|}{2} \le 1 + c\|y\|^p. \tag{10.32}$$

Our goal is to replace the left-hand side of (10.32) by $\frac{\|x+y\|^p+\|x-y\|^p}{2}$. The result will then follow by homogeneity. Let $f(p) = (\frac{\|x+y\|^p+\|x-y\|^p}{2})^{1/p}$. Note that $1 \le f(1) \le f(p)$. We have

$$f(p) - f(1) = f(1)\left[\left(\frac{(1+\beta)^p + (1-\beta)^p}{2}\right)^{1/p} - 1\right],$$

where $\beta = \frac{\|x+y\|-\|x-y\|}{\|x+y\|+\|x-y\|}$. Note that $1 - \beta \ge 0$ and $|\beta| \le \|y\|/f(1) \le \|y\|$. Moreover, we have $(\frac{(1+\beta)^p+(1-\beta)^p}{2})^{1/p} \le (\frac{(1+\beta)^2+(1-\beta)^2}{2})^{1/2} = (1+\beta^2)^{1/2}$. Thus if $\|y\| \le 1 = \|x\|$ we have for some $c' > 0$ $f(p) - f(1) \le f(1)c'\|y\|^2$ and hence by (10.32) we find for some $C > 0$ that $f(p)^p \le (1 + c\|y\|^p)^p(1 + c'\|y\|^2)^p \le 1 + C^p\|y\|^p$.

This is restricted to $\|y\| \le 1$, but we may choose $C \ge 1$ large enough so that $(1+t)^p \le C^pt^p$ for any $t > 1$. Then we have by the triangle inequality

$f(p)^p \le (1+\|y\|)^p \le C^p\|y\|^p$ whenever $\|y\| > 1$. Thus we conclude, as announced that $f(p)^p \le 1 + C^p\|y\|^p$ for any $y \in B$, and hence by homogeneity that $f(p)^p \le \|x\|^p + C^p\|y\|^p$ for any $x, y \in B$. This establishes the only if part in (i). □

With this terminology, let us recapitulate:
B is p-uniformly smooth iff there is $C > 0$ such that (10.23) holds with $\alpha = 1$.

Moreover, B is isomorphic to a p-uniformly smooth space iff it satisfies (10.25) for some constant C.

Lastly, any uniformly smooth space is isomorphic to a p-uniformly smooth one for some $p > 1$ (see Theorem 10.25).

10.3 Uniform convexity and smoothness of L_p

We first recall that any Hilbert space H is both uniformly convex and uniformly smooth, by the 'parallelogram identity'

$$\forall x, y \in H \qquad 2^{-1}(\|x+y\|^2 + \|x-y\|^2) = \|x\|^2 + \|y\|^2.$$

The latter implies

$$\delta_H(\varepsilon) = (1-(1-\varepsilon^2/4)^{1/2} \simeq \varepsilon^2/8 \quad \text{and} \quad \rho_H(t) = (1+t^2)^{1/2} - 1 \simeq t^2/2.$$

In this section, we denote simply by L_p the space $L_p(\Omega, \mathcal{A}, m)$ where $(\Omega, \mathcal{A}, m)$ is an arbitrary measure space. Our goal is to prove

Theorem 10.32.

(i) If $1 < p \le 2$, we have: $\forall t > 0\ \forall \varepsilon \in [0, 2]$

$$\rho_{L_p}(t) \le t^p/p \quad \textit{and} \quad \delta_{L_p}(\varepsilon) \ge (p-1)\varepsilon^2/8$$

(ii) If $2 \le p' < \infty$, we have $\forall t > 0\ \forall \varepsilon \in [0, 2]$

$$\rho_{L_{p'}}(t) \le (p'-1)t^2/2 \qquad \delta_{L_{p'}}(\varepsilon) \ge (\varepsilon/2)^{p'}/p'.$$

Remark. The constants in the preceding estimates are sharp, i.e. they give the right order of magnitude when t or ε are small. For instance, if $1 < p \le 2$, we have $\rho_{L_p}(t) = t^p/p + o(t^p)$ when $t \to 0$, and similarly for the other estimates.

Part of the preceding statement is very easy to prove by interpolation:

Lemma 10.33. *Let $1 < p \le 2 \le p' < \infty$. We have then:*

$$\forall x, y \in L_p \qquad \left(\frac{\|x+y\|_p^p + \|x-y\|_p^p}{2}\right)^{1/p} \le (\|x\|_p^p + \|y\|_p^p)^{1/p}, \tag{10.33}$$

$$\forall x, y \in L_{p'} \qquad (\|x\|_{p'}^{p'} + \|y\|_{p'}^{p'})^{1/p'} \le \left(\frac{\|x+y\|_{p'}^{p'} + \|x-y\|_{p'}^{p'}}{2}\right)^{1/p'}. \tag{10.34}$$

Proof. Assume $L_p = L_p(\Omega, m)$. Let $\Delta_1 = \{-1, 1\}$ equipped with $\nu_1 = (\delta_1 + \delta_{-1})/2$. For (10.33), we consider the operator

$$T\colon\ \ell_p^{(2)}(L_p) \to L_p(\Delta_1, \nu_1; L_p)$$

defined by $T(x, y) = x + \varepsilon_1 y$ (here $\varepsilon_1(\omega) = \omega\ \forall \omega \in \Delta_1$). Note that $\ell_p^{(2)}(L_p)$ (resp. $L_p(\Delta_1, \nu_1; L_p)$) can obviously be identified with the L_p-space associated to the disjoint union of two copies of (Ω, m) (resp. with $L_p(\Delta_1 \times \Omega, \nu_1 \times m)$). Clearly T is a contraction both when $p = 1$ (triangle inequality) and when $p = 2$ (parallelogram inequality). Therefore by interpolation (cf. Corollary 8.16) (10.33) is valid for any intermediate value: $1 < p < 2$. The proof of (10.34) can be done in a similar fashion by considering the operator

$$T^*\colon\ L_{p'}(\Delta_1, \nu; L_{p'}) \to \ell_{p'}^{(2)}(L_{p'})$$

interpolating between $p' = 2$ and $p' = \infty$. Alternatively, one can simply observe that (10.34) follows from $\|T\| = \|T^*\|$ when $p^{-1} + p'^{-1} = 1$. □

The other estimates follow from:

Lemma 10.34. *If $1 < p < 2$, then for all x, y in L_p*

$$(\|x\|_p^2 + (p-1)\|y\|_p^2)^{1/2} \le \left(\frac{\|x+y\|_p^p + \|x-y\|_p^p}{2}\right)^{1/p} \tag{10.35}$$

or equivalently

$$\left(\left\|\frac{x+y}{2}\right\|_p^2 + (p-1)\left\|\frac{x-y}{2}\right\|_p^2\right)^{1/2} \le \left(\frac{\|x\|_p^p + \|y\|_p^p}{2}\right)^{1/p}. \tag{10.36}$$

Proof. By the 2-point hypercontractive inequality (see Theorem 5.4 with $q = 2$) we know that for any fixed ω

$$(|x(\omega)|^2 + (p-1)|y(\omega)|^2)^{1/2} \le \left(\frac{|(x+y)(\omega)|^p + \|(x-y)(\omega)|^p}{2}\right)^{1/p}.$$

Then, taking the L_p-norm of both sides (and using the Hölder-Minkowski contractive inclusion $L_p(\ell_2) \subset \ell_2(L_p)$) we find (10.35). □

Proof of Theorem 10.32.

(i) Assume $1 < p \le 2$. By (10.33) we have

$$2^{-1}(\|x + ty\|_p + \|x - ty\|_p) \le (1 + t^p)^{1/p} \le 1 + t^p/p$$

and hence $\rho_{L_p}(t) \le t^p/p$. By (10.36), if $\|x\|_p, \|y\|_p \le 1$ and $\|x - y\|_p \ge \varepsilon$ then $\left\|\frac{x+y}{2}\right\|_p \le (1 - (p-1)\varepsilon^2/4)^{1/2} \le 1 - (p-1)\varepsilon^2/8$, and hence $\delta_{L_p}(\varepsilon) \ge (p-1)\varepsilon^2/8$.

(ii) Now assume $2 \leq p' < \infty$. Replacing $(x+y, x-y)$ by (x, y) in (10.34) we obtain

$$\left\|\frac{x+y}{2}\right\|_{p'}^{p'} + \left\|\frac{x-y}{2}\right\|_{p'}^{p'} \leq \frac{\|x\|_{p'}^{p'} + \|y\|_{p'}^{p'}}{2}$$

and hence we find

$$\begin{aligned}\delta_{L_{p'}}(\varepsilon) &\geq 1 - (1 - (\varepsilon/2)^{p'})^{1/p'} \\ &\geq (\varepsilon/2)^{p'}/p'.\end{aligned}$$

By duality, (10.35) implies

$$\left(\frac{\|x+y\|_{p'}^{p'} + \|x-y\|_{p'}^{p'}}{2}\right)^{1/p'} \leq (\|x\|_{p'}^2 + (p-1)^{-1}\|y\|_{p'}^2)^{1/2} \qquad (10.37)$$

and hence

$$\rho_{L_{p'}}(t) \leq (1 + (p-1)^{-1}t^2)^{1/2} - 1 \leq t^2/(2(p-1)) = (p'-1)t^2/2.$$

□

Remark 10.35. Let (B_0, B_1) be a compatible pair of complex Banach spaces, such that B_1 is isometric to a Hilbert space (but we make no assumption on B_0), and let $B = (B_0, B_1)_\theta$. In [388] such spaces are called θ-Hilbertian. By interpolation it is immediate that the proof of Lemma 10.33 is valid in such a space with p determined by $1/p = (1-\theta)/1 + \theta/2$ or equivalently $\theta = 2/p'$. More precisely, the same argument implies that for any $x, y \in B$ we have

$$\left(\frac{\|x+y\|_p^p + \|x-y\|_p^p}{2}\right)^{1/p} \leq (\|x\|_p^p + \|y\|_p^p)^{1/p}$$

and

$$(\|x\|_{p'}^{p'} + \|y\|_{p'}^{p'})^{1/p'} \leq \left(\frac{\|x+y\|_{p'}^{p'} + \|x-y\|_{p'}^{p'}}{2}\right)^{1/p'}.$$

A fortiori, θ-Hilbertian spaces are p-uniformly convex and p'-uniformly smooth.

10.4 Type, cotype and UMD

The notions of type/cotype provide a classification of Banach spaces that parallels in many ways the one given by uniform smoothness/uniform convexity. To give a more complete picture for the reader, we feel the need to describe the basic results of that theory, but since it is only loosely related to martingales, we limit ourselves to a survey without proofs (for more detailed information see [68, 347, 348, 391] and also [54, Chapter 9] and [56, Chapter 4]).

Let (Δ, ν) and ε_n be as defined in (5.13), so that the sequence (ϵ_n) is an i.i.d. sequence of symmetric $\{-1, 1\}$ valued random variables on (Δ, ν). Let B be a Banach space. We will denote simply by $\| \ \|_{L_p(B)}$ the 'norm' in the space $L_p(\Delta, \nu; B)$, for $0 < p \leq \infty$.

Definitions.

(i) Let $1 \leq p \leq 2$. A Banach space B is called of type p if there is a constant C such that, for all finite sequences (x_j) in B

$$\left\| \sum \epsilon_j x_j \right\|_{L_2(B)} \leq C \left(\sum \|x_j\|^p \right)^{1/p}. \tag{10.38}$$

We denote by $T_p(B)$ the smallest constant C for which (10.38) holds.

(ii) Let $2 \leq q \leq \infty$. A Banach space B is called of cotype q if there is a constant C such that for all finite sequences (x_j) in B

$$\left(\sum \|x_j\|^q \right)^{1/q} \leq C \left\| \sum \epsilon_j x_j \right\|_{L_2(B)}. \tag{10.39}$$

We denote by $C_q(B)$ the smallest constant C for which (10.39) holds. Clearly, if $p_1 \leq p_2$ then type $p_2 \Rightarrow$ type p_1 while cotype $p_1 \Rightarrow$ cotype p_2. Let us immediately observe that every Banach space is of type 1 and of cotype ∞ with constants equal to 1. In some cases this cannot be improved, for instance if $B = \ell_1$ it is easy to see that (10.38) holds for no $p > 1$. Similarly, if $B = \ell_\infty$ or c_0, then (10.39) holds for no $q < \infty$. We make this more precise in Remark 10.46. At the other end of the classification, if B is a Hilbert space then

$$\forall x_1, \ldots, x_n \in B \qquad \left\| \sum \epsilon_j x_j \right\|_{L_2(B)} = \left(\sum \|x_j\|^2 \right)^{1/2}.$$

Therefore a Hilbert space is of type 2 and cotype 2 (with constants 1). More generally, any space B that is isomorphic to a Hilbert space is of type 2 and cotype 2. It is a striking result of Kwapień [315] that the converse is true: if B is of type 2 and cotype 2, then B must be isomorphic to a Hilbert space.

Actually, by Kahane's inequality (Theorem 5.2), the choice of the norm in $L_2(\Delta, \nu; B)$ plays an inessential role in the preceding definitions. In the case $B = \mathbb{R}$, Kahane's inequality reduces to Khintchin's (5.7). These inequalities make it very easy to analyse the type and cotype of the L_p-spaces:

Proposition 10.36. *If $1 \leq p \leq 2$, every L_p-space is of type p and of cotype 2. If $2 \leq p < \infty$, any L_p-space is of type 2 and of cotype p.*

These are essentially best possible. The space L_∞ contains isometrically any separable Banach space, in particular the already mentioned ℓ_1 and c_0. Therefore, L_∞ is of type 1 and cotype ∞ and nothing more.

Using Kahane's inequality, one can easily generalize the preceding observation.

Proposition 10.37. *Let B be a Banach space of type p and of cotype q. Let (Ω, m) be any measure space. Then $L_r(\Omega, m; B)$ is of type $r \wedge p$ and of cotype $r \vee q$.*

Similar ideas lead to the following result that shows how to use type and cotype to study sums of independent random variables.

Proposition 10.38. *Let $(\Omega, \mathcal{A}, \mathbb{P})$ be a probability space. Let (Y_n) be a sequence of independent mean zero random variables with values in a Banach space B. Assume that B is of type p and cotype q, and that the series $\sum Y_n$ is a.s. convergent. Then for $0 < r < \infty$, we have*

$$\alpha \mathbb{E}\left(\sum \|Y_n\|^q\right)^{r/q} \leq \mathbb{E}\left\|\sum Y_n\right\|^r \leq \beta \mathbb{E}\left(\sum \|Y_n\|^p\right)^{r/p}$$

where α and β are positive constants depending only on r, q and B.

Proof. Assume first that each Y_n is symmetric. Consider the sequence $(\epsilon_n Y_n)_{n\geq 1}$ defined on $(\Delta \times \Omega, \nu \times \mathbb{P})$. This sequence has the same distribution as $(Y_n)_{n\geq 1}$. It is therefore easy to deduce Proposition 10.38 in that case from (10.38), (10.39) and Kahane's inequality. The general case follows by an easy symmetrization argument. □

In particular, taking $r = p$ (resp. $r = q$) in Proposition 10.38 we find

$$\mathbb{E}\left\|\sum Y_n\right\|^p \leq \beta \sum \mathbb{E}\|Y_n\|^p \tag{10.40}$$

$$\left[\text{resp.} \quad \alpha \sum \mathbb{E}\|Y_n\|^q \leq \mathbb{E}\left\|\sum Y_n\right\|^q\right]. \tag{10.41}$$

We now compare type and cotype with the notions introduced in Definitions 10.17 and 10.30.

Proposition 10.39. *Let B be a Banach space. If B is isomorphic to a p-uniformly smooth (resp. q-uniformly convex) space then B is of type p (resp. cotype q).*

Proof. This is an immediate consequence of (10.25) (resp. (10.9)) applied to the martingale $M_n = \sum_1^n \varepsilon_j x_j$. □

The converse to the preceding proposition is not true in general. This is obvious if we consider only cotype: Indeed L_1 or ℓ_1 is of cotype 2 but being non-reflexive has no equivalent uniformly convex (or smooth) norm.

For type, this is much less obvious, but we will present in Chapter 12 examples of non-reflexive spaces of type 2 and cotype $q > 2$ (see Corollary 12.20). Again, being non-reflexive, these necessarily admit no equivalent uniformly convex (or smooth) norm.

The situation changes dramatically for the class of UMD spaces. In the latter class, the notions we are comparing actually coincide:

Proposition 10.40. *Assume B is UMD. Then B is isomorphic to a p-uniformly smooth (resp. q-uniformly convex) space iff B is of type p (resp. cotype q).*

Proof. Assume B UMD$_p$ so that (5.23) holds. Then if B is of type p we find that (5.23) implies (10.25). The proof for cotype is similar. □

Definition 10.41. We will say that a Banach space B is of M-type p (resp. M-cotype q) if it satisfies (10.25) (resp. (10.9)) for some constant C.

This terminology provides us with a convenient way to summarize rather simply the preceding discussion like this:

M-type p (resp. M-cotype q) implies type p (resp. cotype q) but not conversely. However, for UMD spaces, the converse does hold.

The next statement shows that type interpolates well. Note that for *cotype*, the similar interpolation property *fails*, see [235]. See Remark 10.66 (or Proposition 11.44) for the analogous results for M-type and M-cotype.

Proposition 10.42. *Let (B_0, B_1) be an interpolation pair of Banach spaces, such that B_j is of type p_j for $j = 0, 1$. Let $0 < \theta < 1$ and let p_θ be such that $1/p_\theta = (1-\theta)/p_0 + (\theta)/p_1$. Let $1 < q < \infty$. Then, in the complex (resp. real) case, $(B_0, B_1)_\theta$ (resp. $(B_0, B_1)_{\theta,q}$) is of type p_θ (resp. $p_\theta \wedge q$).*

Proof. The complex case follows by applying the interpolation Theorem 8.21 to the operator $T : \ell_{p_j}(B_j) \to L_{p_j}(B_j)$ defined by $T((x_j)) = \sum \varepsilon_j x_j$. For the real case, by Kahane's inequality, we may consider the same operator T but acting from $\ell_{p_j}(B_j)$ to $L_q(B_j)$. Then, using Remark 8.65 we obtain the case $q \geq p_\theta$. Now if $q < p_\theta$, we may find $q_j < p_j$ such that the associated q_θ is equal to p_θ. Since B_j is a fortiori of type q_j, the result follows. □

In particular, this shows that, if $p \neq 2$, the Lorentz space $L_{p,q}$ is of type $p \wedge q \wedge 2$ and, using (i) in Proposition 10.55, of cotype $p \vee q \vee 2$. Moreover, $L_{2,q}$ is of type $q \wedge 2 - \varepsilon$ and cotype $q \vee 2 + \varepsilon$ for any $\varepsilon > 0$ (but not for $\varepsilon = 0$, see [57, p. 97]).

The notions of type and cotype have appeared in various problems involving the analysis of vector valued functions or random variables. One of the great advantages of the classification of Banach spaces in terms of type and cotype is the existence of a rather satisfactory 'geometric' characterization of these notions. We first explain the characterization of spaces that have a non-trivial type or a non-trivial cotype. The reader should compare this with the characterizations of super-reflexivity in the next chapter, for instance (i) in Theorem 10.45 is reminiscent of the equivalence between J-convexity and the existence of $p > 1$ such that (10.25) holds.

Definition. Let $1 \le p \le \infty$. Fix $\lambda > 1$. We say that B contains ℓ_p^n's λ-uniformly if, for all n, there exist $x_1, \dots, x_n$ in B such that

$$\forall (\alpha_j) \in \mathbb{R}^n \quad \left(\sum |\alpha_j|^p\right)^{1/p} \le \left\|\sum_1^n \alpha_j x_j\right\| \le \lambda \left(\sum |\alpha_j|^p\right)^{1/p}. \quad (10.42)$$

For future reference we recall here a fundamental result (see [227]):

Theorem 10.43 (Dvoretzky's Theorem). *For any $\varepsilon > 0$, any infinite-dimensional Banach space contains ℓ_2^ns $(1+\varepsilon)$-uniformly.*

Remark 10.44. Krivine proved [312] that if a Banach space B contains ℓ_p^ns $(1+\varepsilon)$-uniformly for some $\varepsilon > 0$ then it also contains them $(1+\varepsilon)$-uniformly for all $\varepsilon > 0$. The cases $p = 1$ and $p = \infty$ (see Theorem 11.5 for that one), go back to James [275]. Therefore, from now on we simply say in that case that B contains ℓ_p^ns uniformly.

Theorem 10.45 ([348]). *Let B be a Banach space.*

(i) B is of type p for some $p > 1$ iff B does not contain ℓ_1^ns uniformly.
(ii) B is of cotype q for some $q < \infty$ iff B does not contain ℓ_∞^ns uniformly.

Remark 10.46. In such results, the 'only if' part is trivial. Indeed, assume (10.42). Then we have

$$n^{1/p} \le \left\|\sum \epsilon_j x_j\right\|_{L_2(B)} \le \lambda n^{1/p},$$

and

$$n^{1/r} \le \left(\sum \|x_j\|^r\right)^{1/r} \le \lambda n^{1/r}.$$

Therefore B cannot be of type $r > p$ or of cotype $r < p$. In particular if $p = 1$ (resp. $r = \infty$) B cannot have a non-trivial type (resp. cotype).

Corollary 10.47. *Any UMD space is isomorphic to a p-uniformly smooth (resp. q-uniformly convex) space for some $p > 1$ (resp. $q < \infty$).*

Proof. By (5.70) (recall also (5.22)) a UMD space B cannot contain ℓ_p^ns uniformly for either $p = 1$ or $p = \infty$. Therefore, B must be of type p and cotype q for some $p > 1$ and $q < \infty$. By Proposition 10.40, the corollary follows. □

Remark 10.48. It is natural to wonder whether conversely uniform convexity or uniform smoothness implies UMD. We will see in the next chapter that this is the same as asking whether super-reflexivity implies UMD. A first counter-example was given by the author in [386]. Later on, Bourgain [134] produced a counter-example among Banach lattices, and, more precisely, he constructed, for any $2 < q < \infty$, a *Banach lattice* that is 2-uniformly smooth and q-uniformly convex but fails the UMD property (see also [234] for an example failing a 'one sided' version of UMD).

Recently, a considerably simpler example was produced by Yanqi Qiu in [405]. He showed that for any $1 < p \neq q < \infty$ the spaces

$$B_N(p,q) = L_p(m; L_q(m; L_p(m; L_q(m; \cdots))))$$

iterated N-times have unbounded UMD_2-constant when $N \to \infty$. Here the measure m must be assumed non-degenerate. For simplicity, let m be equal to the probability on the 2-point space giving measure $1/2$ to each point. It follows that if one considers, for example, the space $B(p,q)$ (resp. $B_\infty(p,q)$) defined as the direct sum over N in the ℓ_2-sense (resp. the inductive limit in a natural sense) of the spaces $\{B_N(p,q)\}$, then this space fails UMD if $p \neq q$. Let us now take $1 < p < 2 < p' < \infty$, $q = p'$ and $\theta = 2/p'$. With the latter choice, the space $B(p,p')$ is clearly θ-Hilbertian (because by suitably iterating Theorem 8.21 we find $B_N(p,p') = (B_N(1,\infty), B_N(2,2))_\theta$) and hence it is p-uniformly smooth and p'-uniformly convex by Remark 10.35. We refer the reader to [405] for many interesting complements on UMD spaces.

Actually, Theorem 10.45 can be extended as follows: Let $1 \le p_0 < 2 < q_0 < \infty$. A space B is of type p for some $p > p_0$ iff B does not contain $\ell_{p_0}^n$s uniformly. The type and cotype indices are defined as follows:

$$p(B) = \sup\{p \mid B \text{ is of type } p\} \tag{10.43}$$

$$q(B) = \inf\{q \mid B \text{ is of cotype } q\}. \tag{10.44}$$

Corollary 10.49. *If $p(B) > 1$ then $q(B) < \infty$.*
Moreover, $p(B) > 1$ iff $p(B^*) > 1$.

Proof. These statements follow easily from Theorem 10.45. Indeed, if we note that ℓ_1^n embeds isometrically (in the real case) into $\ell_\infty^{2^n}$, we immediately see that B contains ℓ_1^ns uniformly as soon as it contains ℓ_∞^ns uniformly. This shows that $p(B) > 1$ implies $q(B) < \infty$. Similarly, it is easy to see that B contains ℓ_1^ns uniformly iff its dual B^* also does. We leave this as an exercise to the reader (use the fact that it is the same to embed ℓ_1^n in a quotient of B^* or in B^* itself). □

Remark. In addition, it is rather easy to show that B is of type p (resp. cotype q) iff its bidual B^{**} has the same property (actually this follows from Theorem 11.3).

The main theorem relating the type and cotype of B to the geometry of B is

Theorem 10.50 ([312, 348]). *Let B be an infinite-dimensional Banach space. Then for each $\epsilon > 0$, B contains ℓ_p^ns $(1+\epsilon)$-uniformly both for $p = p(B)$ and $p = q(B)$.*

By Theorem 10.50 and Remark 10.46, we have

$$p(B) = \inf\{p \mid B \text{ contains } \ell_p^n\text{s uniformly}\} \tag{10.45}$$

$$q(B) = \sup\{p \mid B \text{ contains } \ell_p^n\text{s uniformly}\}. \tag{10.46}$$

For classical concrete spaces, the type and cotype has been completely elucidated. For instance, the case of Banach lattices is completely clear, cf. [345]. Here are the main results in that case (which includes Orlicz spaces, Lorentz spaces, etc.). Let us consider a Banach lattice B that is a sublattice of the lattice of all measurable functions on a measure space (Ω, m). Then if $x_1, \ldots, x_n$ are elements of B and if $0 < p < \infty$, the function $(\sum |x_j|^p)^{1/p}$ is well defined as a measurable function and is also in B (by the lattice property).

Maurey proved a Banach lattice generalization of Khintchin's inequality, which reduces the study of type and cotype for lattices to some very simple 'deterministic' inequalities:

Theorem 10.51 ([345]). *Let B be a Banach lattice as previously. Assume $q(B) < \infty$. Then there is a constant β depending only on B such that for all $x_1, \ldots, x_n$ in B we have*

$$\frac{1}{\sqrt{2}} \left\| \left(\sum |x_j|^2\right)^{1/2} \right\| \le \left\| \sum \epsilon_j x_j \right\|_{L_2(B)} \le \beta \left\| \left(\sum |x_j|^2\right)^{1/2} \right\|. \tag{10.47}$$

Note: The left-hand side of (10.47) holds in any Banach lattice B; it follows from Khintchin's inequality, see (5.122).

It follows immediately that B (as earlier) is of type p (resp. cotype q) iff there is a constant C such that any finite sequence (x_j) in B satisfies

$$\left\| \left(\sum |x_j|^2\right)^{1/2} \right\| \le C \left(\sum \|x_j\|^p\right)^{1/p}$$

$$\left(\text{resp.} \quad \left(\sum \|x_j\|^q\right)^{1/q} \le C \left\| \left(\sum |x_j|^2\right)^{1/2} \right\| \right).$$

In the case $p < 2$ (or $q > 2$), one can even obtain a much simpler result as shown by the following:

Theorem 10.52 ([345]). *Let B be a Banach lattice as previously.*

(i) *Let $2 < q < \infty$. Then B is of cotype q iff there is a constant C such that any sequence (x_j) of* disjointly supported *elements of B satisfies*

$$\left(\sum \|x_j\|^q\right)^{1/q} \le C \left\|\sum x_j\right\|.$$

(ii) *Assume $q(B) < \infty$. Let $1 < p < 2$. Then B is of type p iff there is a constant C such that any sequence (x_j) of disjointly supported elements satisfies*

$$\left\|\sum x_j\right\| \le C \left(\sum \|x_j\|^p\right)^{1/p}.$$

Remark. For $q = 2$ (or $p = 2$) the preceding statement is false, the Lorentz spaces $L^{2,1}$ (or $L^{2,q}$ for $2 < q < \infty$) provide counter-examples.

Note that for a disjointly supported sequence (x_j) we have

$$\left\|\sum |x_j|\right\| = \|\sup |x_j|\,\| = \left\|\sum x_j\right\|.$$

Remark. In the particular case of Banach lattices, type and cotype are closely connected to the moduli of uniform smoothness or uniform convexity. This is investigated in great detail in the paper [224].

Remark 10.53. Theorem 10.51 has the following useful consequence: For B as in Theorem 10.51, then, for some β', for any finite collection $(x_{i,j})$ in B we have

$$\frac{1}{2}\left\|\left(\sum\nolimits_{i,j} |x_{i,j}|^2\right)^{1/2}\right\|_B \le \left\|\sum\nolimits_{i,j} \varepsilon_i\varepsilon'_j x_{i,j}\right\|_{L_2(\Delta\times\Delta;B)} \le \beta' \left\|\left(\sum\nolimits_{i,j} |x_{i,j}|^2\right)^{1/2}\right\|_B.$$

One proves this by iterating (10.47). This shows that the series $\sum_{i,j} \varepsilon_i\varepsilon'_j x_{i,j}$ is unconditionally convergent when it converges. The unconditionality of such series is sometimes called property α. If one leaves the class of Banach lattices, there are UMD spaces that fail this property (e.g. the Schatten p-classes for $1 < p \neq 2 < \infty$).

We should mention that there are several relatively natural spaces for which the type or cotype is not well understood. For instance, by [432] the projective tensor product $\ell_2\widehat{\otimes}\ell_2$ is of cotype 2, but it remains unknown whether $\ell_2\widehat{\otimes}\ell_2\widehat{\otimes}\ell_2$ is also of cotype 2. However, by [147], the space $\ell_p\widehat{\otimes}\ell_q\widehat{\otimes}\ell_r$ is of no finite cotype if $p^{-1} + q^{-1} + r^{-1} \le 1$, in particular $\ell_3\widehat{\otimes}\ell_3\widehat{\otimes}\ell_3$ contains ℓ_∞^ns uniformly.

In the rest of this section we briefly review the notion of K-convexity that is the key to the duality between type and cotype. More precisely, let B be a Banach space. We will see (Proposition 10.54) that if B is of type p, then B^* is of cotype p' with $\frac{1}{p} + \frac{1}{p'} = 1$, the converse fails in general, but it is true if B is a K-convex space. The real meaning of K-convexity was elucidated in [389], where it is proved that a Banach space B is K-convex if (and only if) B does not contain ℓ_1^ns uniformly. The spaces that do not contain ℓ_1^ns uniformly are sometimes called B-convex; so that with this terminology B- and K-convexity are equivalent properties.

We now define K-convexity. We need some notation. We denote by I_B the identity operator on a Banach space B. Let us denote by R_1 the orthogonal projection from $L_2(\Delta, \nu)$ onto the closed span of the sequence $\{\epsilon_n \mid n \in \mathbb{N}\}$. A Banach space B is called K-convex if the operator $R_1 \otimes I_B$ (defined a priori only on $L_2(\Delta, \nu) \otimes B$) extends to a bounded operator from $L_2(\Delta, \nu; B)$ into itself. We will denote by $K(B)$ the norm on $R_1 \otimes I_B$ considered as an operator acting on $L_2(\Delta, \nu; B)$. Clearly $R_1 \otimes I_B$ is bounded on $L_2(B)$ iff $R_1 \otimes I_{B^*}$ is bounded on $L_2(B^*)$. Let us first treat a simple example, the case when $B = \ell_1^k$ with $k = 2^n$. Then, we may isometrically identify B with $L_1(\Delta_n)$ where $\Delta_n = \{-1, +1\}^n$, equipped with its normalized Haar measure. Let us denote by b_j the j-th coordinate on $\{-1, +1\}^n$ considered as an element of $L_1(\Delta_n)$. Consider then the B-valued function $F\colon\ \Delta \to B$ defined by $F(\omega) = \prod_{j=1}^n (1 + \epsilon_j(\omega) b_j)$. We have $\|F(\omega)\|_B = 1$ hence $\|F\|_{L_2(B)} = 1$. But on the other hand, we have clearly

$$((R_1 \otimes I_B)F)(\omega) = \sum\nolimits_{j=1}^n \epsilon_j(\omega) b_j, \tag{10.48}$$

so that by (5.7)

$$\|(R_1 \otimes I_B)(F)\|_{L_2(B)} = \mathbb{E}\left|\sum\nolimits_1^n \epsilon_j\right| \geq A_1 n^{1/2} \tag{10.49}$$

for some positive numerical constant A_1. Returning the definition of $K(B)$, we find

$$K(\ell_1^{2^n}) \geq A_1 n^{1/2}.$$

In particular, $K(\ell_1^n)$ is unbounded when $n \to \infty$. From this (and the observation that if S is a closed subspace of B then $K(S) \leq K(B)$) we deduce immediately:

Proposition 10.54. *A K-convex Banach space cannot contain ℓ_1^ns uniformly.*

We now turn to the duality between type and cotype. We first state some simple observations.

Proposition 10.55. *Let B be a Banach space. Let $1 \leq p \leq 2 \leq p' \leq \infty$ be such that $\frac{1}{p} + \frac{1}{p'} = 1$.*

(i) If B is of type p, then B^* is of cotype p'.
(ii) If B is K-convex, and if B^* is of cotype p' then B is of type p.

To clarify the proof we state the following:

Lemma 10.56. *Consider $x_1, \ldots, x_n$ in an arbitrary Banach space B. Define*

$$|||(x_j)||| = \sup\left\{ |\sum\nolimits_1^n \langle x_j, x_j^*\rangle| \mid x_j^* \in B^* \ \left\|\sum\nolimits_1^n \epsilon_j x_j^*\right\|_{L_2(B^*)} \leq 1 \right\}. \quad (10.50)$$

Then

$$|||(x_j)||| = \inf\left\{ \left\|\sum\nolimits_1^n \epsilon_j x_j + \Phi\right\|_{L_2(B)} \right\} \quad (10.51)$$

where the infimum is over all Φ in $L_2(B)$ such that $\mathbb{E}(\epsilon_j \Phi) = 0$ for all $j = 1, 2, \ldots, n$ (or equivalently over all Φ in $L_2 \otimes B$ such that $(R_1 \otimes I_B)(\Phi) = 0$).

Proof of Lemma 10.56. We consider the natural duality between $L_2(B)$ and $L_2(B^*)$. Let $S \subset L_2(B^*)$ be the subspace

$$S = \left\{ \sum\nolimits_1^n \epsilon_j x_j^* \mid x_j^* \in B^* \right\}.$$

The norm that appears on the right-hand side of (10.51) is the norm of the space $\mathcal{X} = L_2(B)/S^{\perp}$. Clearly $\mathcal{X}^* = S^{\perp\perp} = S$. Therefore, the identity (10.51) is nothing but the familiar equality

$$\forall z \in \mathcal{X} \qquad \sup\{|\langle z, z^*\rangle| \mid z^* \in \mathcal{X}^*, \ \|z^*\| \leq 1\} = \|x\|.$$ □

Proof of Proposition 10.55. We leave part (i) as an exercise for the reader. Let us prove (ii). Assume B^* of cotype p' so that $\exists C \ \forall n \ \forall x_j^* \in B^*$

$$\left(\sum \|x_j^*\|^{p'}\right)^{1/p'} \leq C \left\|\sum \epsilon_j x_j^*\right\|_{L_2(B^*)}.$$

This implies for all x_j in B

$$|||(x_j)||| \leq C\left(\sum \|x_j\|^p\right)^{1/p}.$$

Assume $\sum \|x_j\|^p < 1$. By (10.51) there is a Φ in $L_2(B)$ such that $\mathbb{E}(\epsilon_j \Phi) = 0$ for all j and such that

$$\left\|\sum \epsilon_j x_j + \Phi\right\|_{L_2(B)} < C.$$

We have

$$\sum \epsilon_j x_j = (R_1 \otimes I_B)\left(\sum \epsilon_j x_j + \Phi\right)$$

hence

$$\left\|\sum \epsilon_j x_j\right\|_{L_2(B)} \leq K(B)C.$$

By homogeneity, this proves that B is of type p with constant not more than $K(B)C$. □

We come now to the main result on the type/cotype duality, the converse to Proposition 10.54.

Theorem 10.57 ([389]). *A Banach space B is K-convex if (and only if) it does not contain ℓ_1^n's uniformly.*

The projection R_1 can be replaced by all kinds of projections that behave similarly in the preceding statement. For instance, let (g_n) be an i.i.d. sequence of normal Gaussian r.v.'s on some probability space $(\Omega, \mathcal{A}, \mathbb{P})$, and let G_1 be the orthogonal projection from $L_2(\Omega, \mathcal{A}, \mathbb{P})$ onto the closed span of $\{g_n \mid n \in \mathbb{N}\}$. Then a space B is K-convex iff $G_1 \otimes I_B$ is a bounded operator from $L_2(\Omega, \mathcal{A}, \mathbb{P}; B)$ into itself.

We can proceed similarly in the context of Proposition 10.38, by introducing a projection Q_1 as follows. Let $(\Omega, \mathcal{A}, \mathbb{P})$ be a probability space. We write simply L_2 for $L_2(\Omega, \mathcal{A}, \mathbb{P})$. Let $(\mathcal{C}_n)_{n\geq 1}$ be a sequence of *independent* σ-subalgebras of $\mathcal{A}$. Let S_0 be the (one-dimensional) subspace of L_2 formed by the constant functions. Let S_1 be the subspace formed by all the functions of the form

$$\sum\nolimits_{n=1}^{\infty} y_n$$

with $y_n \in L_2(\mathcal{C}_n)$ for all n, $\mathbb{E}y_n = 0$ and $\sum \mathbb{E}|y_n|^2 < \infty$. We denote by Q_1 the orthogonal projection from L_2 onto S_1. One can then show (see Theorem 10.58) that if B is K-convex then $Q_1 \otimes I_B$ is bounded on $L_2(B)$. Note that, in the case $(\Omega, \mathbb{P}) = (\Delta, \nu)$, if we take for $\mathcal{C}_n$ the σ-algebra generated by ϵ_n then Q_1 coincides with R_1.

Let us return to our probability space $(\Omega, \mathcal{A}, \mathbb{P})$. We may as well assume that $\bigcup_n \mathcal{C}_n$ generates the σ-algebra $\mathcal{A}$. Actually, we can define a sequence of projections $(Q_k)_{k\geq 0}$ as follows. Let us denote by F_k the closed subspace of L_2 spanned by all the functions f for which there are $n_1 < n_2 < \cdots < n_k$ such that f is measurable with respect to the σ-algebra generated by $\mathcal{C}_{n_1} \cup \cdots \cup \mathcal{C}_{n_k}$.

{Consider the following special case: let (θ_n) be a sequence of independent r.v.'s and let $\mathcal{C}_n$ be the σ-algebra generated by θ_n. Then F_k is the subspace of all the functions in L_2 that depend on at most k of the functions $\{\theta_n \mid n \geq 1\}$.}

Note that $F_k \subset F_{k+1}$ and $\cup F_k$ is dense in L_2. Let then $S_k = F_k \cap F_{k-1}^{\perp}$, and let Q_k be the orthogonal projection from L_2 onto S_k. When $k = 0$, we denote by Q_0 the orthogonal projection onto the subspace of constant functions.

{Note: In the special case considered earlier, let us denote by λ_n the law of θ_n. Then S_k is the subspace spanned by all the functions of the form $F(\theta_{n_1}, \ldots, \theta_{n_k})$ such that

$$\int F(x_1, \ldots, x_j, \ldots, x_k) d\lambda_{n_j}(x_j) = 0$$

for all $j = 1, 2, \ldots, k$.}

We can now formulate a strengthening of Theorem 10.57.

Theorem 10.58. *Let $(Q_k)_{k\geq 0}$ be as earlier. If a Banach space B does not contain ℓ_1^ns uniformly then $Q_k \otimes I_B$ defines a bounded operator on $L_p(\Omega, \mathcal{A}, \mathbb{P}; B)$ for $1 < p < \infty$ and any $k \geq 0$. Moreover there is a constant $C = C(p, B)$ such that the norm of $Q_k \otimes I_B$ on $L_p(B)$ satisfies*

$$\|Q_k \otimes I_B \colon L_p(B) \to L_p(B)\| \leq C^k \quad \textit{for all} \quad k \geq 0.$$

Clearly Theorem 10.57 is a consequence of Theorem 10.58. The proofs of these results are intimately connected with the theory of *holomorphic semi-groups*. Since this would take us too far from our main theme, we refer the reader to [389, 390] (or to [347]) for complete proofs and details.

10.5 Square function inequalities in q-uniformly convex and p-uniformly smooth spaces

Let $(f_n)_{n\geq 0}$ be a B-valued martingale in $L_1(B)$. We denote (with the convention $df_0 = f_0$)

$$S_p(f)(\omega) = \left(\sum\nolimits_0^\infty \|df_n(\omega)\|_B^p\right)^{1/p} \quad \text{and} \quad f^*(\omega) = \sup\nolimits_{n\geq 0} \|f_n(\omega)\|_B.$$

Recall that by Doob's inequality (1.38) we have for any $1 < r < \infty$

$$\sup\nolimits_n \|f_n\|_{L_r(B)} \leq \|f^*\|_r \leq r' \sup\nolimits_n \|f_n\|_{L_r(B)}. \tag{10.52}$$

As for $S_p(f)$, when $p = 2$ and B is either $\mathbb{R}$, $\mathbb{C}$ or a Hilbert space, we recover the classical square function, see §5.1. In that case, we already know that, for any $1 \leq r < \infty$, $\|S_2(f)\|_r$ and $\|f^*\|_r$ are equivalent, see (5.35) and (5.63). Our main result in this section is an analogue of this for $S_p(f)$ (resp. $S_q(f)$) in case B is p-uniformly smooth (resp. q-uniformly convex). Unfortunately however, we cannot take $p = q$ in general (unless B is a Hilbert space) and hence the analogous inequalities are only one sided, as in the next two statements.

Theorem 10.59. *Let B be a Banach space. Fix $2 \leq q < \infty$. The properties in Corollary 10.7 (in other words M-cotype q) are equivalent to:*

(iv) For any $1 \leq r < \infty$, there is a constant $C = C(q, r)$ such that all B-valued martingales $(f_n)_{n\geq 0}$ in $L_r(B)$ satisfy

$$\|S_q(f)\|_r \leq C\|f^*\|_r. \tag{10.53}$$

Theorem 10.60. *Let B be a Banach space. Fix $1 < p \le 2$. The properties in Corollary 10.23 (in other words M-type p) are equivalent to:*

(iv) For any $1 \le r < \infty$ there is a constant $C' = C'(p, r)$ such that all B-valued martingales $(f_n)_{n\ge 0}$ in $L_r(B)$ satisfy

$$\|f^*\|_r \le C'\|S_p(f)\|_r. \tag{10.54}$$

To clarify the duality between (10.53) and (10.54) the following lemma will be used. We already proved this (with a different constant) in Corollary 5.61. Nevertheless we include a quick alternate proof illustrating the use of interpolation for the spaces $L_r(\ell_p)$.

Lemma 10.61. *Let $1 < r, p < \infty$, let $(\mathcal{A}_n)_{n\ge 0}$ be any filtration and set as usual $\mathbb{E}_n = \mathbb{E}^{\mathcal{A}_n}$. Then for any sequence $(\varphi_n)_{n\ge 0}$ in $L_r(\Omega, \mathcal{A}, \mathbb{P})$ we have for any $1 < p < \infty$*

$$\left\|\left(\sum |\mathbb{E}_n\varphi_n|^p\right)^{1/p}\right\|_r \le c_{r,p}\left\|\left(\sum |\varphi_n|^p\right)^{1/p}\right\|_r, \tag{10.55}$$

where $c_{r,p} = r(r-1)^{\frac{1}{p}-1}$.

Proof. By Doob's inequality (1.32) (resp. the dual Doob inequality (1.33)) (10.55) holds for $p = \infty$ (resp. $p = 1$) with $C(0) = r' = r(r-1)^{-1}$ (resp. $C(1) = r$). Therefore by the complex interpolation of 'mixed normed spaces' (see Theorem 8.21), (10.55) holds for a general $1 < p < \infty$ with $c_{r,p} = C(0)^{1-\theta}C(1)^{\theta}$ where $\theta = 1/p$. This yields $c_{r,p} = (r')^{1/p'}(r)^{1/p} = r(r-1)^{-1/p'}$. □

Proposition 10.62. *Fix $1 < r, r' < \infty$ and $1 < p, p' < \infty$ with $\frac{1}{r} + \frac{1}{r'} = 1$ and $\frac{1}{p} + \frac{1}{p'} = 1$. For a Banach space B and a given filtration $(\mathcal{A}_n)_{n\ge 0}$ on $(\Omega, \mathcal{A}, \mathbb{P})$ the following are equivalent:*

(i) There is a constant C such that all B-valued martingales $(f_n)_{n\ge 0}$ in $L_r(B)$, adapted to $(\mathcal{A}_n)_{n\ge 0}$, satisfy

$$\sup \|f_n\|_r \le C\|S_p(f)\|_r. \tag{10.56}$$

(ii) There is a constant C' such that all B^*-valued martingales $(g_n)_{n\ge 0}$ in $L_{r'}(B^*)$, adapted to $(\mathcal{A}_n)_{n\ge 0}$, satisfy

$$\|S_{p'}(g)\|_{r'} \le C' \sup \|g_n\|_{r'}. \tag{10.57}$$

Moreover, we may exchange the roles of B and B^* if we wish.

Proof. The proof that (ii) $\Rightarrow$ (i) is very easy: assuming (ii), for $\varepsilon > 0$, choose g in the unit ball of $L_{r'}(B^*)$ so that $\|f_n\|_{L_r(B)} \le (1+\varepsilon)\langle g, f_n\rangle$. Let $g_n = \mathbb{E}_n g$. Then note $\langle g, f_n\rangle = \langle g_n, f_n\rangle = \sum_0^n \langle dg_k, df_k\rangle$, it follows $\|f_n\|_{L_r(B)} < (1+\varepsilon)\mathbb{E}|\sum_0^n \langle dg_k, df_k\rangle| \le \ (1+\varepsilon)\mathbb{E}(S_{p'}(g)S_p(f)) \le (1+\varepsilon)\|S_{p'}(g)\|_{r'}\|S_p(f)\|_r \le (1+\varepsilon)C'\|S_p(f)\|_r$ so we obtain (i) with $C \le C'$. Conversely, assume (i). To prove (10.57) we may assume $(g_n)_{n\ge 0}$ is a finite martingale so that $g_k = g_n$ for all $k \ge n$. Fix $\varepsilon > 0$. Let $\varphi_0, \ldots, \varphi_n \in L_r(B)$ be such that

$$\left\| \left(\sum\nolimits_0^n \|\varphi_k\|^p\right)^{1/p} \right\|_r \le 1 \quad \text{and} \quad \mathbb{E}\sum\nolimits_0^n \langle \varphi_k, dg_k\rangle \ge (1+\varepsilon)\|S_{p'}(g)\|_{r'}. \tag{10.58}$$

Note that

$$\mathbb{E}\sum\nolimits_0^n \langle \varphi_k, dg_k\rangle = \mathbb{E}\sum\nolimits_0^n \langle df_k, dg_k\rangle = \mathbb{E}\langle f_n, g_n\rangle \le \|f_n\|_{L_r(B)}\|g_n\|_{L_{r'}(B^*)}, \tag{10.59}$$

where $df_k = (\mathbb{E}_k - \mathbb{E}_{k-1})\varphi_k$ and $f_n = \sum_0^n df_k$. Moreover, by the triangle inequality and (10.55), we have

$$\|S_p(f)\|_r \le 2c_{r,p} \left\| \left(\sum \|\varphi_k\|^p\right)^{1/p} \right\|_r \le 2c_{r,p}. \tag{10.60}$$

Thus we obtain by (10.58), (10.59), (10.56) and (10.60)

$$(1+\varepsilon)\|S_{p'}(g)\|_{r'} \le \|f_n\|_{L_r(B)}\|g_n\|_{L_{r'}(B^*)} \le 2Cc_{r,p}\|g_n\|_{L_{r'}(B^*)},$$

so we obtain (ii) with $C' \le 2Cc_{r,p}$. □

It will be convenient to break the proofs of Theorems 10.59 and 10.60 in two. The first parts are formulated in the next two lemmas.

Lemma 10.63. *Let us denote by* (iv)$_r$ *the assertion* (iv) *in Theorem 10.59 for a fixed value* $1 \le r < \infty$. *Then* (iv)$_q \Rightarrow$ (iv)$_r$ *for any* $1 \le r < q$.

Proof. We will use the 'extrapolation method' described in Lemma 5.23 and the B. Davis decomposition in Corollary 5.33. By Corollary 5.33, we have $f_n = g_n + h_n$ with $h_0 = 0$, $\|dg_n\|_B \le 8f^*_{n-1}$ for all $n \ge 1$ and $\left\| \sum_0^\infty \|dh_n\| \right\|_r \le 8r\|f^*\|_r$ for any $1 \le r < \infty$. We set $v_n(\omega) = (\sum_0^n \|dg_k(\omega)\|_B^q)^{1/q}$ and $w_n(\omega) = \|g_n(\omega)\|_B$. Applying (iv)$_q$ to the martingale $(1_{\{T>0\}}g_{n\wedge T})_{n\ge 0}$, we find

$$\|1_{\{T>0\}}v_T\|_q \le C(q,q)\|1_{\{T>0\}}w_T\|_q. \tag{10.61}$$

Fix r such that $1 \le r < q$. By Lemma 5.23

$$C(q,q)^{-1}\|S_q(g)\|_r \le (q^{1/r}(q-r)^{-1/r}+1)\|g^*\|_r + 8\|f^*\|_r. \tag{10.62}$$

But since g is essentially a 'perturbation of f by h' we have $S_q(f) \le S_q(g) + \sum \|dh_n\|$ and $g^* \le f^* + \sum \|dh_n\|$, and hence $\|S_q(f)\|_r \le \|S_q(g)\|_r + 6r\|f^*\|_r$

and $\|g^*\|_r \le (1+6r)\|f^*\|_r$, so that (10.62) yields (10.53) with $C(q,r) \le C(q,q)((q^{1/r}(q-r)^{-1/r}+1)(1+6r)+8+6r)$. □

Lemma 10.64. *Assume B p-uniformly smooth (actually we use only type p). Then there is a constant t_p such that for any $1 \le r < \infty$ and any martingale $(f_n)_{n\ge 0}$ in $L_r(B)$, there is a choice of sign $\xi_n \pm 1$ such that the transformed martingale $\tilde{f}_n = \sum_0^n \xi_k df_k$ satisfies*

$$\|\tilde{f}^*\|_r \le t_p \|S_p(f)\|_r. \tag{10.63}$$

Proof. Since B is p-uniformly smooth, a fortiori by Proposition 10.39, it is of type p, i.e. there is a constant C such that for any finite sequence (x_j) in B we have

$$\left\|\sum \varepsilon_j x_j\right\|_{L_r(\nu;B)} \le C\left(\sum \|x_j\|^p\right)^{1/p}.$$

By (1.42) we have

$$\left\|\sup_n \left\|\sum\nolimits_0^n \varepsilon_j x_j\right\|\right\|_r^r \le 2C^r \left(\sum \|x_j\|^p\right)^{r/p}.$$

Replacing x_j by $df_j(\omega)$ and integrating in ω we find $I \le 2^{1/r}C\|S_p(f)\|_r$, where

$$I = \left(\iint \sup_n \left\|\sum\nolimits_0^n \varepsilon_j(\omega')df_j(\omega)\right\|^r \, d\nu(\omega')d\mathbb{P}(\omega)\right)^{1/r}.$$

Thus to conclude it suffices to choose $\xi_j = \varepsilon_j(\omega')$ so that $\|\tilde{f}^*\|_r \le I$ (the latter because the infimum over ω' is at most the average). □

Lemma 10.65. *Let us denote by* (iv)^r *the assertion* (iv) *in Theorem 10.60 for a fixed value of* $1 \le r < \infty$. *Then* (iv)$^p \Rightarrow$ (iv)^r *for all* $1 \le r < p$.

Proof. The idea is the same as for Lemma 10.63 but there is an extra difficulty that is overcome by using Lemma 10.64. As earlier, we use the decomposition in Corollary 5.33: we have $f = g + h$ with $h_0 = 0$, $\|dg_n\| \le 8f^*_{n-1}$ for all $n \ge 1$ and $\|\sum \|dh_n\|_B\|_r \le 8r\|f^*\|_r$ for all $1 \le r < \infty$. Let $\xi_n = \pm 1$ be an arbitrary choice of signs. Again we denote

$$\tilde{g}_n = \sum\nolimits_0^n \xi_k dg_k \quad \text{and} \quad \tilde{f}_n = \sum\nolimits_0^n \xi_k df_k.$$

We set $v_n(\omega) = \tilde{g}^*_n(\omega)$ and $w_n(\omega) = (\sum_0^n \|dg_n(\omega)\|^p)^{1/p}$. Assuming (iv)^p, we find for any stopping time T

$$\|1_{\{T>0\}} v_T\|_p \le C'(p,p)\|1_{\{T>0\}} w_T\|_p.$$

Fix r such that $1 \le r \le p$. By (5.38) (applied with $\psi_n = 8f_n^*$) there is a constant C (depending on r and p) such that

$$\|\tilde{g}^*\|_r \le C\|S_p(g)\|_r + 8\|f^*\|_r. \tag{10.64}$$

Since $|\tilde{f}^*|_r \le \|\tilde{g}^*\|_r + \left\| \sum \|dh_n\| \right\|_r$ and $\|S_p(g)\|_r \le \|S_p(f)\|_r + \left\| \sum \|dh_n\| \right\|_r$ we deduce from (10.64)

$$\|\tilde{f}^*\|_r \le C\|S_p(f)\|_r + 8(r(C+1)+1)\|f^*\|_r.$$

Since this holds for any choice of signs $\xi_n = \pm 1$ we may exchange the roles of f and $\tilde{f}$ (note that $\tilde{\tilde{f}} = f$!) and we find

$$\|f^*\|_r \le C\|S_p(f)\|_r + 8(r(C+1)+1)\|\tilde{f}^*\|_r.$$

If we now choose ξ_n according to Lemma 10.64 we obtain (10.54) with $C' \le C + 8(r(C+1)+1)t_p$. □

Proof of Theorem 10.59. The case $1 \le r \le q$ is covered by Lemma 10.63. Recall $\|f^*\|_r \le r' \sup_n \|f_n\|_r$ by Doob's inequality. Let $p = q'$ so that $p^{-1} + q^{-1} = 1$. If $q \le r' < \infty$ then $1 < r \le p$, and, by Proposition 10.62, (10.56) holds at least with $r = p$. By Lemma 10.65, (10.56) holds for all $1 < r < p$, therefore by Proposition 10.62 again, (10.57) (and a fortiori also (10.53)) holds for all r' with $q = p' < r' < \infty$. □

Proof of Theorem 10.60. The argument is the same as for Theorem 10.59: The case $1 \le r \le p$ is covered by Lemma 10.65 and the case $p < r < \infty$ (i.e. $1 < r' < p'$) can be deduced from Lemma 10.63 (applied to $q = p'$) by duality using Proposition 10.62. □

Remark. The detour through duality in the preceding proofs can be avoided by using Lemma 5.26 instead of Lemma 5.23.

Remark 10.66. Let (B_0, B_1) be an interpolation pair of Banach spaces, such that B_j is of M-type p_j for $j = 0, 1$. Let $0 < \theta < 1$ and let p_θ be such that $1/p_\theta = (1-\theta)/p_0 + (\theta)/p_1$. Let $1 < q < \infty$. Then, in the complex (resp. real) case, $(B_0, B_1)_\theta$ (resp. $(B_0, B_1)_{\theta,q}$) is of M-type p_θ (resp. $p_\theta \wedge q$). The complex case is immediate just as in Proposition 10.42. For the real case, combining Theorem 10.60 with (10.55) (choosing $r = q$), one can show that B_j is of M-type p_j iff the mapping $T : L_q(\ell_{p_j}(B_j)) \to L_q(B_j)$ defined by $T((f_j)) = \sum(\mathbb{E}_j - \mathbb{E}_{j-1})(f_j)$ is bounded. Then if $q \ge p_\theta$ the announced result follows from (i) in Theorem 8.63 and Remark 8.65, and if $q < p_\theta$ one can replace p_0, p_1 by smaller numbers (as in the proof of Proposition 10.42) to achieve $q = p_\theta$. By duality, the analogous result for M-cotype also holds. The present remark can also be deduced from the subsequent Proposition 11.44.

When $r = 1$, Theorem 10.59 (iv) tells us that $f \in h^1_{\max}(B)$ implies $S_q(f) \in L_1$. It is natural to expect a dual result involving BMO, in analogy with (7.24). This turns out to be very easy to prove directly:

Proposition 10.67. *If B is isomorphic to a p-uniformly smooth space ($1 \le p \le 2$), then there is a constant C such that all B-valued martingales (f_n) with respect to any filtration $(\mathcal{A}_n)$ such that $S_p(f) \in L_\infty$ satisfy*

$$\|(f_n)\|_{BMO(\{\mathcal{A}_n\};B)} \le C\|S_p(f)\|_\infty. \tag{10.65}$$

Proof. Assume $\|S_p(f)\|_\infty \le 1$. By (10.25) we know that f_n converges in $L_p(B)$ (and a.s.) to a limit f_∞. It suffices to show that for some C we have for any $n \ge 1$

$$\mathbb{E}_n\|f_\infty - f_{n-1}\| \le C.$$

Equivalently, it suffices to check that for any non-negligible $A \in \mathcal{A}_n$

$$\int_A \|f_\infty - f_{n-1}\| d\mathbb{P}/\mathbb{P}(A) \le C.$$

Fix such a set $A \in \mathcal{A}_n$. For any k such that $n \le k \le \infty$, let $\mathcal{A}'_k$ be the trace of $\mathcal{A}_k$ on A. Let $\mathbb{P}_A$ be the conditional probability on A such that $\mathbb{P}_A(\alpha) = \mathbb{P}(\alpha)/\mathbb{P}(A)$ for any $\alpha \in \mathcal{A}'_\infty$ ($\alpha \subset A$), and for any $k \ge n$ let

$$M_k = (f_k - f_{n-1})_{|A}.$$

Then it is easy to check that $(M_k)_{k\ge n}$ is a martingale on $(A, \mathbb{P}_A)$ with respect to $(\mathcal{A}'_k)_{k\ge n}$ satisfying

$$\left(\sum\nolimits_{k\ge n} \|dM_k\|^p_{L_p(\mathbb{P}_A;B)}\right)^{1/p} \le \|S_p(f)\|_\infty \le 1.$$

By (10.25), we have $\sup_{k\ge n} \|M_k\|_{L_p(\mathbb{P}_A;B)} \le C$ and a fortiori

$$\frac{1}{\mathbb{P}(A)} \int_A \|f_\infty - f_{n-1}\| = \lim_{k\to\infty} \int \|M_k\| d\mathbb{P}_A \le C,$$

which is the desired result. □

Remark. Conversely, (10.65) implies that B is isomorphic to a p-uniformly smooth space. See [387, p.344] for a detailed argument.

10.6 Strong p-variation, uniform convexity and smoothness

We will now extend the method presented in Chapter 9 to the Banach space valued case. The extension to the Hilbert space valued case is straightforward, but the martingale inequalities (10.9) (resp. (10.25)) satisfied by q-uniformly convex (resp. p-uniformly smooth) spaces allow us to go much further:

Theorem 10.68. *Let $1 < p_1 \le 2 \le q_0 < \infty$.*

(i) Assume that B is isomorphic to a p_1-uniformly smooth space. Then for all $1 < p < p_1$ there is a constant $C = C(p, p_1)$ such that all B-valued martingales $f = (f_n)_{n\ge 0}$ in $L_p(B)$ satisfy

$$\mathbb{E}V_p(f)^p \le C\mathbb{E}\sum\nolimits_0^\infty \|df_n\|_B^p.$$

(ii) Assume that B is isomorphic to a q_0-uniformly convex space. Then for all $q > q_0$ there is a constant $C = C(q, q_0)$ such that all B-valued martingales in $L_q(B)$ satisfy

$$\mathbb{E}V_q(f)^q \le C\sup\nolimits_n \mathbb{E}\|f_n\|^q.$$

The proof is based on the following key fact:

Lemma 10.69. *Let $1 < r < \infty$ and let $0 < \theta < 1$ be such that $1 - \theta = \frac{1}{r}$. Let $(f_n)_{n\ge 0}$ be a B-valued martingale converging in $L_r(B)$. Assume that for all increasing sequences of stopping times $0 \le T_0 \le T_1 \le T_2 \le \cdots$ we have*

$$\mathbb{E}\|f_{T_0}\|^r + \sum\nolimits_{k\ge 1} \mathbb{E}\|f_{T_k} - f_{T_{k-1}}\|^r \le 1.$$

Then

$$\|\{f_n\}\|_{(L_1(v_1(B)),L_\infty(\ell_\infty(B)))_{\theta,\infty}} \le 2.$$

Proof. This can be proved by an obvious adaptation of the argument for Lemma 9.3. One just chooses $T_k = \inf\{n > T_{k-1} \mid \|f_n - f_{T_{k-1}}\| > t^{\theta-1}\}$. □

We will use repeatedly the identity (see Theorem 8.63)

$$L_p((B_0, B_1)_{\theta,p}) = (L_1(B_0), L_\infty(B_1))_{\theta,p} \tag{10.66}$$

valid for any $0 < \theta < 1$ provided p is linked to θ by $1 - \theta = 1/p$.

Proof of Theorem 10.68. Here again we can adapt the proof of Theorem 9.2. (i) Assume B p_1-uniformly smooth. Then by (10.9) applied (with p replaced by p_1) to the martingale $M_n = f_{T_k\wedge n} - f_{T_{k-1}\wedge n}$ (here k is fixed) we have

(for some constant C_1)

$$\mathbb{E}\|f_{T_k} - f_{T_{k-1}}\|^{p_1} \le C_1 \sum\nolimits_{T_{k-1}<n\le T_k} \mathbb{E}\|f_n - f_{n-1}\|^{p_1}$$

and hence

$$\sum\nolimits_0^\infty \mathbb{E}\|f_{T_k} - f_{T_{k-1}}\|^{p_1} \le C_0 \sum\nolimits_0^\infty \mathbb{E}\|df_n\|^{p_1}.$$

Let $\|f\|_{D(p_1)} = \left(\sum_0^\infty \mathbb{E}\|df_n\|^{p_1}\right)^{1/p_1}$. Let θ_1 be such that $1-\theta_1 = 1/p_1$. On one hand, by the preceding lemma we have a bounded inclusion

$$D(p_1) \subset (L_1(v_1(B)), L_\infty(\ell_\infty(B)))_{\theta_1,\infty} \tag{10.67}$$

and on the other hand we have trivially (actually this is an equality)

$$D(1) \subset L_1(v_1(B)). \tag{10.68}$$

By the same argument as in Chapter 9 we know (see (9.14)) that $D(p_1) = (D(1), D(\infty))_{\theta_1,p_1}$ where we set $D(\infty) = \ell_\infty(L_\infty(B))$. Therefore, by the reiteration Theorem, (10.67) and (10.68) imply that, for any θ with $0 < \theta < \theta_1$ and any $1 \le p \le \infty$, we have

$$(D(1), D(\infty))_{\theta,p} \subset (L_1(v_1(B)), \ell_\infty(L_\infty(B)))_{\theta,p}.$$

We now choose p so that $1-\theta = 1/p$. This gives us $(D(1), D(\infty))_{\theta,p} = D(p)$ and also by (10.66) (see (9.3))

$$(L_1(v_1(B)), \ell_\infty(L_\infty(B)))_{\theta,p} = L_p((v_1(B), \ell_\infty(B))_{\theta,p})$$

but by Lemma 9.1

$$(v_1(B), \ell_\infty(B))_{\theta,p} \subset v_p(B).$$

Thus we obtain that the inclusion

$$D(p) \subset L_p(v_p(B))$$

is bounded, and this is precisely (i).

To prove (ii) assume B q_0-uniformly convex. Let θ_0 be such that $1 - \theta_0 = 1/q_0$. By (10.25) applied to the martingale $(f_{T_k})_{k\ge 0}$ we have for some constant C_0

$$\sum \mathbb{E}\|f_{T_k} - f_{T_{k-1}}\|^{q_0} \le C_0 \sup \mathbb{E}\|f_n\|^{q_0}.$$

Let $T : L_{q_0}(B) \to \ell_\infty(L_{q_0}(B))$ be defined by $T(M) = (\mathbb{E}_n M - \mathbb{E}_{n-1}M)_{n\geq 0}$. By the preceding lemma we have (boundedly)

$$T(L_{q_0}(B)) \subset (L_1(v_1), L_\infty(\ell_\infty(B)))_{\theta_0,\infty} \tag{10.69}$$

and trivially

$$T(L_\infty(B)) \subset \ell_\infty(L_\infty(B)). \tag{10.70}$$

Observe that $L_{q_0}(B) = (L_1(B), L_\infty(B))_{\theta_0,q_0}$ (see (10.66)). Therefore, by the reiteration Theorem, (10.69) and (10.70) imply

$$T((L_1(B), L_\infty(B))_{\theta,q}) \subset (L_1(v_1), \ell_\infty(L_\infty(B)))_{\theta,q}$$

for any θ with $\theta_0 < \theta < 1$ and any $1 \leq q \leq \infty$. If we choose q so that $1 - \theta = 1/q$ we find by (10.66)

$$T(L_q(B)) \subset L_q((v_1(B), \ell_\infty(B))_{\theta,q})$$

and again $(v_1(B), \ell_\infty(B))_{\theta,q} \subset v_q(B)$ so we obtain

$$T(L_q(B)) \subset L_q(v_q(B)),$$

which is exactly (ii). □

Remark. In the situation of Theorem 10.68, fix $1 < p < p_1$ (resp. $q > q_0$). Then for each $1 \leq r \leq p$ (resp. $1 \leq r \leq q$) there is a constant C such that

$$\|V_p(f)\|_r \leq C \left\| \left(\sum \|df_n\|_B^p \right)^{1/p} \right\|_r$$

(resp. $\|V_q(f)\|_r \leq C \| \sup \|f_n\|_B \|_r$ and $\|V_q(f)\|_{1,\infty} \leq C \sup \|f_n\|_{L_1(B)}$). Indeed, this can be proved exactly as in the proofs of part (ii) in Theorems 9.2 and 9.5.

10.7 Notes and remarks

The source of this chapter is mainly [387], but the latter paper was inspired by Enflo's fundamental results on super-reflexivity that are described in detail in the next chapter. Enflo's main result from [220] was that 'super-reflexive' implies 'isomorphic to uniformly convex', thus completing a program initiated by R. C. James ([275, 278]), that we describe in the notes and remarks of the next chapter. While Enflo and James work with 'trees', in our paper [387] the relevance of martingales was recognized and a new proof was given of Enflo's theorem with an improvement: the modulus of convexity can always be found of power type, or equivalently we can always find a renorming satisfying (10.1). We follow the original proof of this from [387], but with a significant shortcut, namely Lemma 10.14 which seems new, and possibly of independent interest (see Remark 10.15). See [231] for complementary results. Another proof of the renorming due to Maurey appears in [4]. Maurey's proof has the advantage that it requires only to assume (10.14) for dyadic martingales. See also G. Lancien's paper [317]. Very recently, Raja [409] proposed a new approach to Theorem 10.1 based on the idea that there exists a super-multiplicative function that is the supremum, in a rather natural ordering, of the set of all the moduli of convexity of equivalent norms. Then the super-multiplicativity forces the existence of a non-trivial exponent, in analogy with what happens e.g. for type and cotype.

In our presentation, we find it preferable to separate the two steps: in this chapter we show that any uniformly convex space is isomorphic to a space with a modulus 'of power type' (i.e. satisfying (10.1)) and only in the next one do we show Enflo's result that 'super-reflexive' implies 'isomorphic to uniformly convex'.

In both chapters, we replace the Banach space B by $L_q(B)$ with $1 < q < \infty$ and we treat martingale difference sequences simply as monotone basic sequences in $L_q(B)$. The corresponding inequalities for basic sequences in uniformly convex (resp. smooth) spaces are due to the Gurarii brothers [259] (resp. to Lindenstrauss [330]).

Proposition 10.39 is an early result proved by Figiel and the author in [228]. By Kwapień's theorem [315] it implies that if B is isomorphic to a 2-uniformly smooth space and to a (possibly different) 2-uniformly convex one, then B is isomorphic to a Hilbert space. Note however that the naive isometric analogue of this fails: There are two-dimensional spaces B such that for some C we have $\rho_B(t) \leq Ct^2$ and $\delta_B(\varepsilon) \geq \varepsilon^2/C$ but B is *not isometric* to a Hilbert space. Such an example can be produced easily, see e.g. [19, pp. 159–160].

We learnt about the work of Fortet-Mourier through unpublished work by J. Hoffmann-Jørgensen. The presentation in §10.2 was strongly influenced by [5], to which we refer the reader interested in non-linear aspects of Banach space theory. The estimates for the modulus of convexity (and of smoothness) of L_p in §10.3 are due to O. Hanner [262]. See also [109] for more recent results including the non-commutative case. The proof of (i) in Proposition 10.31 comes from [109].

The main results on type and cotype are due to B. Maurey and the author, see [347, 391] and the references given in the text for more details. Proposition 10.42 was first observed by Beauzamy.

The results of §10.6 come essentially from [387], while those of §10.5 come from [400].

11

Super-reflexivity

11.1 Finite representability and super-properties

The notion of 'finite representability' is the basis for that of 'super-property'.

Definition. A Banach space X is said to be finitely representable (f.r. in short) in another Banach space Y if for any finite-dimensional subspace $E \subset X$ and for any $\varepsilon > 0$ there is a subspace $\widetilde{E} \subset Y$ that is $(1+\varepsilon)$-isomorphic to E (i.e. there is an isomorphism $u\colon\ E \to \widetilde{E}$ with $\|u\|\,\|u^{-1}\| \le 1+\varepsilon$).

In other words, X f.r. Y means that, although Y may not contain an isomorphic copy of the whole of X, it contains an almost isometric copy of any finite-dimensional subspace of X. In the appendix to this chapter, devoted to background on ultraproducts, we show that X is f.r. Y iff X embeds isometrically in an ultraproduct of Y.

The following simple perturbation argument will be used repeatedly.

Lemma 11.1. *Let X, Y be Banach spaces. Let $E_0 \subset \cdots \subset E_n \subset E_{n+1} \subset \cdots$ be a sequence (or a family directed by inclusion) of finite-dimensional subspaces of X such that $\overline{\cup E_n} = X$. Then for X to be f.r. in Y it suffices that for any $\varepsilon > 0$ and any n there is a subspace $\widetilde{E}_n \subset Y$ that is $(1+\varepsilon)$-isomorphic to E_n.*

Proof. Consider $E \subset X$ with $\dim(E) < \infty$. It suffices to show that for any fixed $\varepsilon > 0$ there is n and $\widehat{E} \subset E_n$ such that E is $(1+\varepsilon)$ isomorphic to $\widehat{E}$. Let $\delta > 0$ to be specified later. Let $x_1, \ldots, x_d$ be a linear basis of E. Choose n and $\hat{x}_1, \ldots, \hat{x}_d$ in E_n such that $\|x_j - \hat{x}_j\| < \delta$ for all $j = 1, \ldots, d$. Let $v\colon\ E \to \widehat{E}$ be the linear map determined by $v(x_j) = \hat{x}_j$. Let $\mathbb{K} = \mathbb{R}$ or $\mathbb{C}$ as usual. For any $(\alpha_j) \in \mathbb{K}^d$, we have by the triangle inequality

$$\left|\left\|v\left(\sum \alpha_j x_j\right)\right\| - \left\|\sum \alpha_j x_j\right\|\right| \le \delta \sum |\alpha_j|, \tag{11.1}$$

but since all norms are equivalent on $\mathbb{K}^d$ there is a constant C_E such that

$$\sum |\alpha_j| \leq C_E \left\| \sum \alpha_j x_j \right\|.$$

Thus, assuming $\delta C_E < 1$, (11.1) implies

$$(1 - \delta C_E) \left\| \sum \alpha_j x_j \right\| \leq \left\| \sum \alpha_j \hat{x}_j \right\| \leq (1 + \delta C_E) \left\| \sum \alpha_j x_j \right\|,$$

and hence $\widehat{E}$ is λ-isomorphic to E with $\lambda = (1 + \delta C_E)(1 - \delta C_E)^{-1}$. To conclude we simply choose δ small enough so that $\lambda < 1 + \varepsilon$. □

Remark. The preceding lemma shows in particular that L_p is f.r. in ℓ_p for any $1 \leq p < \infty$, and that L_∞ is f.r. in c_0.

Definition. Consider a property P for Banach spaces. We say that a Banach space Y has 'super-P' if every Banach space X that is f.r. in Y has P.

Remark. In particular Y is super-reflexive (resp. has the super-RNP) if every X f.r. in Y is reflexive (resp. has the RNP). The passage from P to super-P is a fruitful way to associate to an *infinite-dimensional* property (such as e.g. reflexivity) its finite-dimensional counterpart. If the property P is already stable by finite representability, then P and super-P are the same. Such properties are usually called 'local'. The 'local theory' of Banach spaces designates the part of the theory that studies infinite-dimensional spaces through the collection of their finite-dimensional subspaces.

Remark 11.2. Let B be a complex Banach space. If B is super-reflexive as a real Banach space then it is also super-reflexive as a complex space. Indeed, any complex space X that is f.r. in B must be reflexive as a real space, but this is the same as reflexive as a complex space. Conversely, if B is super-reflexive as a complex space, it is also as a real space, but this is a bit less obvious. It follows e.g. from (i) $\Leftrightarrow$ (iii) in Theorem 11.22, since the notion of separated tree is the same in the real or complex cases. It also follows from Proposition 11.8.

The following result called the 'local reflexivity principle' is classical.

Theorem 11.3 ([332]). *The bidual B^{**} of an arbitrary Banach space B is f.r. in B.*

To study super-reflexivity, we will need the following elementary fact.

Lemma 11.4. *Let B be a Banach space. Then for any b^{**} in B^{**} any $\varepsilon > 0$ and any finite subset $\xi_1, \ldots, \xi_n$ in B^* there is b in B with $\|b\| \leq (1 + \varepsilon)\|b^{**}\|$ such that*

$$\langle \xi_i, b^{**} \rangle = \langle \xi_i, b \rangle \qquad \forall i = 1, \ldots, n.$$

Proof. Let $\mathbb{K} = \mathbb{R}$ or $\mathbb{C}$ be the scalar field. We may clearly assume ξ_i linearly independent. Assume $\|b^{**}\| = 1$ for simplicity. Let $C \subset \mathbb{K}^n$ be the convex set $\{(\langle \xi_i, b\rangle)_{i\le n} \mid b \in B\ \|b\| \le 1\}$. Clearly, since b^{**} is in the $\sigma(B^{**}, B^*)$ closure of the unit ball of B, we know that $(\langle \xi_i, b^{**}\rangle)_{i\le n} \in \overline{C}$. But (since we assumed the ξ_i's independent) C has non-empty interior hence $\overline{C} \subset (1+\varepsilon)C$ for any $\varepsilon > 0$. Thus we conclude that $(\langle \xi_i, b^{**}\rangle)_{i\le n} \in (1+\varepsilon)C$. □

The following result is classical. It combines several known facts, notably (iv) ⇒ (iii) goes back to R. C. James [275].

Theorem 11.5. *The following properties of a Banach space B are equivalent:*

(i) Every Banach space is f.r. in B.
(ii) c_0 is f.r. in B.
(iii) For any $\lambda > 1$ and any $n \ge 1$ there are $x_1^n, \ldots, x_n^n$ in B satisfying

$$\forall (\alpha_j) \in \mathbb{K}^n \qquad \sup|\alpha_j| \le \left\| \sum\nolimits_1^n \alpha_j x_j^n \right\| \le \lambda \sup|\alpha_j|. \tag{11.2}$$

(iv) For some $\lambda > 1$, for any $n \ge 1$ there are $x_1^n, \ldots, x_n^n$ in B satisfying (11.2).
(v) For some $\lambda > 1$, for any $n \ge 1$ there are $x_1^n, \ldots, x_n^n$ in B with norm ≥ 1 and such that

$$\sup\left\{ \left\| \sum\nolimits_1^n \varepsilon_j x_j^n \right\| \mid \varepsilon_j = \pm 1 \right\} \le \lambda.$$

Proof. (i) ⇒ (ii) ⇒ (iii) ⇒ (iv)⇒ (v) are trivial. We will show that (iii) ⇒ (i) and (v) ⇒ (iii). Assume (iii). Let $E \subset B_1$ be a finite-dimensional subspace in an arbitrary Banach space B_1. Let $S = \{s_1, \ldots, s_n\}$ be an ε-net in the unit sphere of E. For each $s_j \in S$, choose $\xi_j \in E^*$ such that $\langle \xi_j, s_j\rangle = 1 = \|\xi_j\|$. We define $u\colon\ E \to \ell_\infty^n$ by setting $u(x) = (\xi_j(x))_{j\le n}$. We have $\|u\| = 1$ and

$$\forall s \in S \qquad \|u(s)\|_{\ell_\infty^n} = 1.$$

Therefore by Lemma 11.65 from the appendix

$$\forall x \in E \qquad (1-\varepsilon)\|x\| \le \|u(s)\|_{\ell_\infty^n} \le \|x\|.$$

This shows that E embeds $(1-\varepsilon)^{-1}$-isomorphically into ℓ_∞^n. Thus (iii) implies that B_1 is f.r. in B, or equivalently (iii) ⇒ (i).

The proof that (v) ⇒ (iii) is a well-known 'blocking trick'. Assume (v). Let $C(n)$ be the smallest constant C such that for any $x_1, \ldots, x_n$ in B we have

$$\inf\nolimits_{j\le n} \|x_j\| \le C \sup_{\varepsilon_j = \pm 1} \left\| \sum\nolimits_1^n \varepsilon_j x_j \right\|.$$

A simple blocking argument shows that $C(nk) \leq C(n)C(k)$ for all n, k. Since we assume (v), we have $\inf_n C(n) \geq \lambda^{-1}$, but by the sub-multiplicativity of $C(n)$ this implies $C(n) \geq 1$ for all n. Therefore, for any n and any $\lambda > 1$ we can find $x_1, \ldots, x_n$ in B such that $\sup_{\varepsilon_j=\pm 1} \|\sum \varepsilon_j x_j\| \leq \lambda$ and $\inf_{j\leq n} \|x_j\| \geq 1$. For each k, choose $\xi_k \in B^*$ such that $\|\xi_k\| = 1$ and $\xi_k(x_k) \geq 1$. Note that if ε_j is the sign of $\xi_k(x_j)$ we have

$$\sum |\xi_k(x_j)| = \left\langle \xi_k, \sum \varepsilon_j x_j \right\rangle \leq \left\| \sum \varepsilon_j x_j \right\| \leq \lambda.$$

Consequently

$$\sum_{j\neq k} |\xi_k(x_j)| \leq \lambda - 1. \tag{11.3}$$

Let C be the set of real scalars $\alpha_1, \ldots, \alpha_n$ with $\sup |\alpha_j| \leq 1$. Note that the maximum value on C of $\|\sum \alpha_j x_j\|$ is attained on an extreme point of C (of the form $\alpha_j = \pm 1$), so we have $\|\sum \alpha_j x_j\| \leq \lambda \sup |\alpha_j|$ for any $(\alpha_1, \ldots, \alpha_n)$ in $\mathbb{R}^n$. Let $x = \sum \alpha_j x_j$. Choose k so that $|\alpha_k| = \sup_j |\alpha_j|$. By (11.3) we have

$$\sup |\alpha_j| = |\alpha_k| = \left| \xi_k \left(\sum \alpha_j x_j \right) - \sum_{j\neq k} \alpha_j \xi_k(x_j) \right| \leq \|x\| + (\lambda - 1) \sup |\alpha_j|$$

and hence we find $(2 - \lambda) \sup |\alpha_j| \leq \|x\|$. Thus we conclude

$$\sup |\alpha_j| \leq (2 - \lambda)^{-1} \|x\| \leq (2 - \lambda)^{-1} \lambda \sup |\alpha_j|,$$

and since $(2 - \lambda)^{-1}\lambda$ is arbitrarily close to 1 this shows that (v) $\Rightarrow$ (iii), at least in the real case. To check the complex case, note that

$$\sup_{z_j \in \mathbb{C}\ |z_j|=1} \left\| \sum z_j x_j \right\| \leq 2 \sup_{\varepsilon_j=\pm 1} \left\| \sum \varepsilon_j x_j \right\|.$$

From this one sees that in (v) we may replace the signs ± 1 by unimodular complex numbers and complete the proof of (v) $\Rightarrow$ (iii) exactly as in the real case. □

Recall from §10.4:

Definition. We say that B contains ℓ_∞^ns uniformly if it satisfies (iii) in Theorem 11.5. We sometimes say λ-uniformly if we wish to keep track of the constant.

A property P (of Banach spaces) is called a super-property if super-$P \Leftrightarrow P$.

Corollary 11.6. *Let P be a non-universal super-property, meaning that there is at least one Banach space failing it. Then a Banach space with property P cannot contain ℓ_∞^ns uniformly.*

The reader will find background on ultrafilters, ultraproducts and ultrapowers in the appendix to this chapter.

Proposition 11.7. *Let X, Y be Banach spaces. Then X is finitely representable in Y (in short X f.r. Y) iff X embeds isometrically into an ultrapower of Y.*

Proof. Assume that X embeds isometrically into an ultrapower $Y^I/\mathcal{U}$ of Y. By Lemma 11.66, for any Y, $Y^I/\mathcal{U}$, and hence a fortiori X, is f.r. in Y, proving the 'if' part. Conversely assume X f.r. in Y. Let I be the set of pairs (E, ε) where $E \subset Y$ is a finite-dimensional subspace and $\varepsilon > 0$. We equip I with the order defined by $i = (E_1, \varepsilon_1) \le j = (E_2, \varepsilon_2)$ if $E_1 \subset E_2$ and $\varepsilon_2 < \varepsilon_1$. Note that obviously for any x in X there is $i = (E, \varepsilon)$ in I such that $x \in E$. Since X f.r. Y, for any $i = (E, \varepsilon)$ there is a linear map $u_i\colon\ E \to Y$ such that

$$\forall x \in E \qquad \|x\| \le \|u_i(x)\| \le (1+\varepsilon)\|x\|. \tag{11.4}$$

Let $\mathcal{U}$ be an ultrafilter adapted to I (meaning refining the net just defined, see §11.8). Then we define $u\colon\ X \to Y^I/\mathcal{U}$ as follows: for any x in X we set $u(x) = \overbrace{u_i(x_i))_{i\in I}}^{\cdot}$ where $x_{(E,\varepsilon)} = x$ whenever $x \in E$ and (say) $x_{(E,\varepsilon)} = 0$ if $x \notin E$. By the observation after (11.57), this indeed defines a *linear* map $u\colon\ X \to Y^I/\mathcal{U}$. Let ε_i denote the second coordinate of i so that $i = (E, \varepsilon_i)$. Note that $\lim \varepsilon_i = 0$ and hence $\lim_{\mathcal{U}} \varepsilon_i = 0$. Therefore, by (11.57) and (11.4) for any x in X

$$\|ux\| = \lim\nolimits_{\mathcal{U}} \|u_i(x)\| = \|x\|.$$

This shows that u is an isometric embedding of X into $Y^I/\mathcal{U}$. □

The following is an immediate consequence of Proposition 11.7:

Proposition 11.8. *Let P be a Banach space property. A Banach space B has super-P iff any space isometric to a subspace of an ultrapower of B has P.*

Proposition 11.9. *Let P be a Banach space property that is stable under isomorphism (for example reflexivity). Then super-P is also stable under isomorphisms.*

Proof. Indeed, if $B_1 \simeq B$ (isomorphically) then, for any $(I, \mathcal{U})$, we have obviously $B_1^I/\mathcal{U} \simeq B^I/\mathcal{U}$ (isomorphically). By Proposition 11.8, if B has super-P then any subspace of $B^I/\mathcal{U}$ has P, and hence (by the stability under isomorphism) any subspace of $B_1^I/\mathcal{U}$ has P, so that B_1 has super-P. □

11.2 Super-reflexivity and inequalities for basic sequences

We will make crucial use of the following beautiful theorem due to V. Ptak [404]. This was later rediscovered by several authors, among which R. C. James who made an extremely deep contribution ([274–277]) to the subject of reflexivity and weak compactness.

Theorem 11.10. *The following properties of a Banach space B are equivalent:*

(i) B is not reflexive.
(ii) For any $0 < \theta < 1$, there is a sequence $(x_n, \xi_n)_{n\geq 1}$ in $B \times B^*$ with $\|x_n\| \leq 1$, $\|\xi_n\| \leq 1$ for all n such that

$$\xi_j(x_i) = 0 \qquad \forall i < j \tag{11.5}$$

$$\xi_j(x_i) = \theta \qquad \forall i \geq j. \tag{11.6}$$

(ii)′ For some $0 < \theta < 1$, there is $(x_n, \xi_n)_{n\geq 1}$ as in (ii).
(iii) For any $0 < \theta < 1$, there is a sequence (x_n) in B such that for any finitely supported scalar sequence (α_n) we have

$$\theta \sup_j \left| \sum_{i\geq j} \alpha_i \right| \leq \left\| \sum \alpha_i x_i \right\| \leq \sum |\alpha_i|. \tag{11.7}$$

(iii)′ For some $0 < \theta < 1$, the same as (iii) holds.
(iv) For any $0 < \theta < 1$, there is a sequence (y_n) in B such that for any finitely supported scalar sequence (β_n) we have

$$\theta \sup_n |\beta_n| \leq \left\| \sum \beta_n y_n \right\| \leq \sum_{n\geq 0} |\beta_n - \beta_{n+1}|. \tag{11.8}$$

(iv)′ For some $0 < \theta < 1$, the same as (iv) holds.
(v) The inclusion mapping $v_1 \to \ell_\infty$ (where v_1 denotes the space of scalar sequences (β_n) with $\sum_{n\geq 0} |\beta_n - \beta_{n+1}| < \infty$) factors through B.

Proof. (ii) $\Rightarrow$ (ii)′ is trivial and (ii)′ $\Rightarrow$ (i) is easy. Indeed, if (ii)′ holds and if x^{**} is a $\sigma(B^{**}, B^*)$ cluster point of (x_n), we must have $\xi_j(x^{**}) = \theta$ by (11.6). Let $\xi \in B^*$ be a $\sigma(B^*, B)$ cluster point of (ξ_n). Then by (11.5) we must have $\xi(x_i) = 0$. If $x^{**} \in B$, on one hand this implies $\xi(x^{**}) = 0$ but on the other hand $\xi_j(x^{**}) = \theta$ implies $\xi(x^{**}) = \theta$. This contradiction shows that $x^{**} \notin B$ and hence that B is not reflexive.

The main point is to show (i) $\Rightarrow$ (ii). Assume (i). Fix $0 < \theta < 1$ and $\varepsilon > 0$. Pick $x^{**} \in B^{**}$ with $\|x^{**}\| = 1$ such that $\text{dist}(x^{**}, B) > \theta$. (Obviously, such an x^{**} must exist, otherwise a simple iteration argument would show that $B^{**} = B$.)

Since $\|x^{**}\| = 1$, there is ξ_1 in B^* with $\|\xi_1\| \le 1$ such that $x^{**}(\xi_1) = \theta$. Hence (see Lemma 11.4), for any $\varepsilon > 0$, there is x_1 in B with $\|x_1\| \le 1 + \varepsilon$ such that $x_1(\xi_1) = \theta$. We will now prove by induction the existence of a sequence as in (ii) except that we will find $\|x_n\| \le 1 + \varepsilon$, but a posteriori we may renormalize (x_n), so this is unimportant.

Let E_1 be the subspace spanned by $\{x_1\}$. Since $\text{dist}(x^{**}, E_1) > \theta$ (and since $B^{**}/E_1 = (B/E_1)^{**} = (E_1^{\perp})^*$), there is ξ_2 in B^* with $\|\xi_2\| \le 1$ such that $\xi_2 \in E_1^{\perp}$ and $x^{**}(\xi_2) = \theta$. Then, by Lemma 11.4, there is x_2 in B with $\|x_2\| \le 1 + \varepsilon$ such that $x_2(\xi_1) = \theta$ and $x_2(\xi_2) = \theta$ and so on. To check the induction step, assume we have constructed $(x_1, \ldots, x_n)$, $(\xi_1, \ldots, \xi_n)$ satisfying (11.5) and (11.6). Let $E_n = \text{span}\{x_1, \ldots, x_n\}$, we find $\xi_{n+1} \in E_n^{\perp}$ with $\|\xi_{n+1}\| \le 1$ such that $x^{**}(\xi_{n+1}) = \theta$, and (using Lemma 11.4) we find x_{n+1} in B with $\|x_{n+1}\| \le 1 + \varepsilon$ such that $x_{n+1}(\xi_i) = \theta \ \forall i \le n + 1$. This completes the induction step and also the proof that (i) implies (ii).

It is an easy exercise to see that (ii) $\Leftrightarrow$ (iii) and (ii)$'$ $\Leftrightarrow$ (iii)$'$.

The equivalences (iii)$\Leftrightarrow$ (iv) and (iii)$'$ $\Leftrightarrow$ (iv)$'$ are obvious: just note the identity ('Abel summation') $\sum \alpha_i x_i = \sum \beta_n y_n$ where we set $x_{-1} = 0$, $y_0 = x_0$ and $y_n = x_n - x_{n-1}$ (or equivalently $x_n = y_0 + \cdots + y_n$), $\alpha_n = \beta_n - \beta_{n+1}$ (or equivalently $\beta_n = \sum_{i \ge n} \alpha_i$).

Lastly, (iv)$\Leftrightarrow$ (v) is easy: (iv) can be interpreted as a factorization $v_1 \to Y \to \ell_\infty$ of the inclusion $v_1 \to \ell_\infty$ through the closed span Y of (y_n) but using Hahn-Banach extensions of the functionals $\sum \beta_n y_n \mapsto \beta_n$ we can extend the second map $Y \to \ell_\infty$ to one from B to ℓ_∞, and this gives the factorization in (v). Conversely if (v) holds i.e. we have a factorization $v_1 \to B \to \ell_\infty$ (with bounded maps) then (iv)$'$ is immediate. □

Theorem 11.11. *The super-RNP is equivalent to super-reflexivity.*

Proof. From reflexive $\Rightarrow$ RNP (see Corollary 2.15), we deduce trivially super-reflexive $\Rightarrow$ super-RNP. To show the converse, it suffices obviously to prove that super-RNP $\Rightarrow$ reflexive. Equivalently it suffices to show that if B is a non-reflexive space then there is a space X that is f.r. in B failing the RNP. Assume B non-reflexive. Then by the preceding theorem there is a sequence (x_n) in B such that for any finitely supported scalar sequence (α_n) we have $\xi_j(\sum \alpha_i x_i) = \theta \sum_{i \ge j} \alpha_i$, hence

$$\theta \sup_j \left| \sum_{i \ge j} \alpha_i \right| \le \left\| \sum \alpha_i x_i \right\| \le \sum |\alpha_i|. \tag{11.9}$$

We will now construct a space X that will be f.r. in B and will contain a $\theta/2$-separated dyadic tree, and hence will fail the RNP. The space X will be defined as the completion of L_1 with respect to the norm $|||\cdot|||$ defined in what follows.

The underlying model for the construction is this: When (x_i) is the canonical basis of ℓ_1 (which satisfies (11.9) with $\theta = 1$) then the construction produces L_1 as the space X.

Let $(\mathcal{A}_n)$ be the dyadic filtration in $L_1 = L_1([0, 1])$. For any f in L_1 we introduce the semi-norm

$$\|f\|_{(n)} = \left\| \sum_{0\le k<2^n} \int_{k2^{-n}}^{(k+1)2^{-n}} f(t)dt \cdot x_k \right\|.$$

Let $\mathcal{U}$ be a non-trivial ultrafilter on $\mathbb{N}$, i.e. an ultrafilter adapted to $\mathbb{N}$ (see our appendix of §11.8). We set

$$|||f||| = \lim_{n,\mathcal{U}} \|f\|_{(n)}.$$

We have by (11.9) for all f in L_1

$$\theta \sup_{0\le s\le 1} \left| \int_s^1 f(t)dt \right| \le |||f||| \le \int |f(t)|\, dt. \tag{11.10}$$

Indeed, (11.9) implies this on the left-hand side with the supremum over s of the form $s = k2^{-n}$, hence (11.10) follows by continuity of $s \to \int_s^1 f(t)dt$.

Let X be the completion of $(L_1, |||\cdot|||)$. By a routine argument one can check that this space X embeds in an ultraproduct of copies of B and hence is f.r. in B. By Lemma 11.12, the unit ball of X contains an infinite $\theta/2$-separated dyadic tree and hence fails the RNP. □

Lemma 11.12. *Let X be a Banach space. Assume that there is a linear map $J\colon L_1([0, 1], dt) \to X$ such that for some $\theta > 0$ we have for all f in L_1*

$$\theta \sup_{0\le s\le 1} \left| \int_s^1 f(t)dt \right| \le \|J(f)\| \le \int_0^1 |f(t)|dt.$$

Then the unit ball of X contains a $\theta/2$-separated dyadic tree and hence X fails the RNP.

Proof. Fix n. To any $(\varepsilon_1, \dots, \varepsilon_n)$ in $\{-1, 1\}^n$ we associate the interval $I(\varepsilon_1, \dots, \varepsilon_n)$ defined by induction as follows: we set $I(1) = [0, \frac{1}{2}[$, $I(-1) = [\frac{1}{2}, 1]$ and if $I(\varepsilon_1, \dots, \varepsilon_n)$ is given we define $I(\varepsilon_1, \dots, \varepsilon_n, +1)$ as the left half of $I(\varepsilon_1, \dots, \varepsilon_n)$ and $I(\varepsilon_1, \dots, \varepsilon_n, -1)$ as its right half.

Note that $|I(\varepsilon_1, \dots, \varepsilon_n)| = 2^{-n}$. Let then $\Omega = \{-1, 1\}^{\mathbb{N}}$. Let $(M_n)_{n \geq 0}$ be the L_1 valued martingale defined for $\varepsilon = (\varepsilon_n)_n \in \Omega$ by $M_0 \equiv 1$ and

$$M_n(\varepsilon) = 2^n \cdot 1_{I(\varepsilon_1,\dots,\varepsilon_n)}.$$

Note that $\|M_n(\varepsilon)\|_{L^1} = 1$ for all ε in Ω and since

$$M_n(\varepsilon) - M_{n-1}(\varepsilon) = 2^{n-1}\varepsilon_n(1_{I(\varepsilon_1,\dots,\varepsilon_{n-1},1)} - 1_{I(\varepsilon_1,\dots,\varepsilon_{n-1},-1)})$$

for all $n \geq 1$ we have

$$\sup_s \left| \int_s^1 (M_n - M_{n-1})(t)\, dt \right| \geq 1/2.$$

Hence the martingale $(J(M_n(\cdot)))$ is a B-valued $\theta/2$-separated dyadic martingale with range in the unit ball of B. □

Remark 11.13. The proof of Theorem 11.11 shows that, if B is not reflexive, then, for any $\theta < 1$ there is a space X f.r. in B satisfying the condition in Lemma 11.12. In the case of real valued scalars, this will be refined in (11.36), but the proof of this improvement is much more delicate.

Remark 11.14. By [143] (see also [130]), there are Banach spaces without RNP that do *not* contain any δ-separated infinite *dyadic* tree, whatever $\delta > 0$ may be. This gives an example of a space X failing the RNP but also failing the assumption of Lemma 11.12.

Definition. Fix a number $\lambda \geq 1$. A finite sequence $\{x_1, \dots, x_N\}$ in a Banach space B is called λ-basic if for any N-tuple of scalars $(\alpha_1, \dots, \alpha_N)$ we have

$$\sup_{1 \leq n \leq N} \left\| \sum_{j=1}^{n} \alpha_j x_j \right\| \leq \lambda \left\| \sum\nolimits_1^N \alpha_j x_j \right\|. \tag{11.11}$$

An infinite sequence (x_n) is called λ-basic if $\{x_1, \dots, x_N\}$ is λ-basic for all $N \geq 1$. The case $\lambda = 1$ has already been distinguished in the preceding chapter: 1-basic sequences are called 'monotone basic' sequences.

Note that (11.11) trivially implies by the triangle inequality

$$\sup |\alpha_j| \|x_j\| \leq 2\lambda \left\| \sum \alpha_j x_j \right\| \tag{11.12}$$

If $\overline{\text{span}}[x_n] = B$, the sequence $\{x_n\}$ is said to be a basis (sometimes called a Schauder basis) of B. Then any x in B has a *unique* representation as the sum of a convergent series $\sum_1^\infty \alpha_j x_j$ with uniquely determined scalar coefficients. Conversely any sequence (x_n) with this property must be λ-basic for some $\lambda \geq 1$ by the classical Banach-Steinhaus principle. Indeed, this property implies that there are biorthogonal functionals x_n^* in B^* such that any b in B can be written as $b = \sum_1^\infty x_n^*(b)x_n$. Let $P_N(b) = \sum_1^N x_n^*(b)x_n$ so that, for any b in B, $P_N(b) \to b$ and hence $\sup_N \|P_N(b)\| < \infty$. By the Banach-Steinhaus principle, we must have $\sup_N \|P_N\| < \infty$, so that (x_n) is λ-basic with $\lambda = \sup_N \|P_N\|$.

Obviously, a λ-basic sequence is a basis for the closed subspace it spans. This justifies the term 'basic'.

The natural basis of ℓ_p $(1 \le p < \infty)$ or c_0 is of course a basis in the preceding sense. Let B be any Banach space. In the sequel we will use repeatedly the observation that a sequence of martingale differences (df_n) in $L_p(B)$ is a monotone (i.e. λ-basic with $\lambda = 1$) basic sequence in $L_p(B)$.

Definition 11.15. A basis (x_n) is called boundedly complete if for any scalar sequence (α_n) such that $\sup_N \| \sum_1^N \alpha_n x_n \| < \infty$ the sum $S_N = \sum_1^N \alpha_n x_n$ converges in B.

Note that if $S_N \to b$ we have automatically $x_n^*(S_N) \to x_n^*(b)$ for each n and hence $\alpha_n = x_n^*(b)$ for all n. Let $P_N\colon\ B \to B$ be, as previously, the projection defined by $P_N(b) = \sum_1^N x_n^*(b)x_n$.

Definition 11.16. A basis (x_n) is called shrinking if for any x^* in B^* we have $\|x^* - P_N^* x^*\| \to 0$. Equivalently, this means that the biorthogonal functionals (x_n^*) form a basis in B^*.

The following classical theorem due to R. C. James characterizes reflexive Banach spaces with a basis.

Theorem 11.17. *Let B be a Banach space with a basis (x_n). Then B is reflexive iff (x_n) is both boundedly complete and shrinking.*

Proof. We may assume that (x_n) is λ-basic for some $\lambda \ge 1$. Assume that B is reflexive. Let $S_N = \sum_1^N \alpha_n x_n$. If $\{S_N\}$ is bounded, by weak compactness of the closed balls, there is a subsequence weakly converging to a limit b in B. Then, for any fixed n, $x_n^*(S_N) \to x_n^*(b)$ (along a subsequence), but $\alpha_n = x_n^*(S_N)$ for all $N > n$, therefore $\alpha_n = x_n^*(b)$ for any n and hence (see the remarks preceding Definition 11.15) $S_N = P_N(b)$ tends to b when $N \to \infty$. This shows that (x_n) is boundedly complete. Since $P_N(b) \to b$ for any b in B, we have $x^*(P_N(b)) \to x^*(b)$ for any x^* in B^* and hence $P_N^* x^* \to x^*$ with respect to $\sigma(B^*, B)$. If B is reflexive, $\sigma(B^*, B) = \sigma(B^*, B^{**})$ is the weak topology on B^*, and hence by Mazur's theorem x^* lies in the norm closure of $\text{conv}\{P_N^* x^* \mid N \ge 1\}$. This yields: $\forall \varepsilon > 0\ \exists m\ \exists \xi \in \text{conv}\{P_N^* x^* \mid 1 \le N \le m\}$ with $\|\xi - x^*\| < \varepsilon$. Since $P_m^* \xi = \xi$ (because $P_N P_m = P_m P_N = P_N\ \forall N \le m$) we have

$$\|(1 - P_m^*)(x^*)\| \le \|(1 - P_m^*)(x^* - \xi)\| \le (1 + \lambda)\varepsilon$$

and hence we conclude that (x_n) is shrinking. Conversely, assume that (x_n) is boundedly complete and shrinking. Consider x^{**} in B^{**}. We can write $P_N^{**}(x^{**}) = \sum_1^N x^{**}(x_n^*)x_n$. We have

$$\sup \left\| \sum\nolimits_1^N x^{**}(x_n^*)x_n \right\| \le \sup \|P_N\| \le \lambda.$$

Since (x_n) is assumed boundedly complete, $\sum_1^N x^{**}(x_n^*)x_n$ converges to an element b in B. But now, for any fixed $n \le N$, we have

$$x^{**}(x_n^*) = x_n^* \left(\sum\nolimits_1^N x^{**}(x_n^*)x_n \right) \to x_n^*(b) \text{ when } N \to \infty,$$

and hence $x^{**}(x_n^*) = b(x_n^*)$ for any n. Finally, if (x_n) is assumed shrinking, $\{x_n^*\}$ is norm total in B^*, so this last equality implies $x^{**}(x^*) = b(x^*)$ for any x^* in B^*, which means $x^{**} = b$. Thus we conclude that B is reflexive. □

Remark 11.18. Let $p > 1$ (resp. $q < \infty$). Let (e_n) be a basic sequence in a Banach space B. We say that (e_n) satisfies an upper p-estimate (resp. a lower q-estimate) if there is a constant C such that for any finite sequence $x_1, \ldots, x_N$ of disjoint consecutive (finite) blocks on (e_n) we have

$$\left\| \sum x_j \right\| \le C \left(\sum \|x_j\|^p \right)^{1/p} \left(\text{resp. } \left(\sum \|x_j\|^q \right)^{1/q} \le C \left\| \sum x_j \right\| \right).$$

If this holds, then (e_n) is shrinking (resp. is boundedly complete).

Indeed, let P_N denote the projection onto span$[e_0, \ldots, e_N]$. Consider $\xi \in B^*$. Dualizing our hypothesis we find that for any increasing sequence $0 = n(0) < n(1) < \cdots$ we have

$$\left(\sum \|(P_{n(k)} - P_{n(k-1)})^* \xi\|^{p'} \right)^{1/p'} \le C \|\xi\|.$$

This implies $(P_{n(k)} - P_{n(k-1)})^* \xi \to 0$ when $k \to \infty$. But we may choose the sequence $n(k)$ inductively so that (say) $\|(P_{n(k+1)} - P_{n(k)})^* \xi\| > (1/2)\|(I - P_{n(k)})^* \xi\|$ so we conclude that $\|\xi - P_N^* \xi\| \to 0$ when $N \to \infty$. The boundedly complete case is similar. We leave the details to the reader. □

Remark 11.19. Fix $1 < p < \infty$. By Theorem 2.9, B has the RNP iff any martingale difference sequence in $L_p(B)$ is boundedly complete when viewed as a monotone basic sequence.

We will use two variants of Theorem 11.10 as follows:

Remark 11.20. If B is not reflexive then for any $\lambda > 1$ there is a sequence (x_n, ξ_n) satisfying (ii) in Theorem 11.10 but moreover such that the sequence (ξ_n) is λ-basic.

Indeed, choose numbers $1 < \lambda_n < \lambda$ such that $\prod \lambda_n < \lambda$. It is easy to modify the induction step to obtain this: at each step where we have produced $(x_j, \xi_j)_{j \le n}$ we can find a finite subset F_n of the unit ball of B such that for any ξ in $\text{span}(\xi_1, \ldots, \xi_n)$ we have $\|\xi\| \le \lambda_n \sup\{|\xi(x)| \mid x \in F_n\}$. Suppose we have produced $x_1, \ldots, x_n, \xi_1, \ldots, \xi_n$. We then replace E_n by $\text{span}(E_n, F_n)$ to find ξ_{n+1} in $E_n^\perp \cap F_n^\perp$ with otherwise the same properties, so we may continue and find x_{n+1} with $\|x_{n+1}\| \le 1 + \varepsilon$ such that $x_{n+1}(\xi_j) = \theta$ for all $j \le n + 1$. The fact that $\xi_{n+1} \in F_n^\perp$ guarantees that for any $\xi \in \text{span}[\xi_1, \ldots, \xi_n]$ and any scalar α, we have

$$\forall x \in F_n \qquad \xi(x) = (\xi + \alpha\xi_{n+1})(x)$$

and hence

$$\|\xi\| \le \lambda_n \sup_{x \in F_n} |\xi(x)| \le \lambda_n \|\xi + \alpha\xi_{n+1}\|. \tag{11.13}$$

Now if we choose our sequence $\lambda_1, \ldots, \lambda_n, \ldots$ as announced so that $\prod \lambda_n < \lambda$, we clearly deduce from (11.13) that (ξ_n) is λ-basic.

Remark 11.21. By an analogous refinement, if B is not reflexive then for any $\lambda > 4$ there is a sequence (x_n, ξ_n) satisfying (ii) in Theorem 11.10 but moreover such that the two sequences (x_n) and $(x_1, x_2 - x_1, x_3 - x_2, \ldots)$ are λ-basic. Let x^{**} be as in the proof of Theorem 11.10. Suppose given $x_1, \ldots, x_n$ and $E_n = \text{span}(x_1, \ldots, x_n)$. Since $d(x^{**}, E_n) > \theta$, we have for any x in E_n and any scalar α

$$\theta|\alpha| \le \|x + \alpha x^{**}\|$$

and hence by the triangle inequality

$$\|x\| \le (1 + \theta^{-1})\|x + \alpha x^{**}\|. \tag{11.14}$$

Let $\varepsilon > 0$. Let G_n be a finite subset of $B_{E_n^*}$ such that

$$\forall x \in E_n \qquad \|x\| \le (1 + \varepsilon) \sup\{|\xi(x)| \mid \xi \in G_n\}.$$

By (11.14), each ξ in $B_{E_n^*}$ (in particular each ξ in G_n) extends to a linear form $\hat{\xi}$ of norm $\le 1 + \theta^{-1}$ on the span of x^{**} and E_n that vanishes on x^{**}. Then we claim that $\hat{\xi}$ extends to $\tilde{\xi} \in B^*$ with $\|\tilde{\xi}\| \le (1 + \theta^{-1})(1 + \varepsilon)$. Indeed, since $\text{span}[E_n, x^{**}] \subset B^{**}$, the Hahn-Banach theorem a priori gives us $\tilde{\xi}$ in B^{***}, extending $\hat{\xi}$ to the whole of B^{**}, but we can use Lemma 11.4, applied to B^* instead of B, to find $\tilde{\xi}$ in B^*. In any case, note that $\tilde{\xi}(x^{**}) = 0$.

Let $\widetilde{G}_n = \{\tilde{\xi} \mid \xi \in G_n\}$. Then in the induction step, we may select x_{n+1} so that $\xi(x_{n+1}) = 0$ for all ξ in $\widetilde{G}_1 \cup \cdots \cup \widetilde{G}_n$. This guarantees that, for any k, all the x_i's for $i > k$ vanish on $\widetilde{G}_k$, so for any x in E_k, say $x = \sum_1^k \alpha_j x_j$ we have

$$\begin{aligned}\|x\| &\le (1+\varepsilon)\sup\{|\xi(x)| \mid \xi \in \widetilde{G}_k\}\\ &\le (1+\varepsilon)\sup\left\{\left|\xi\left(x+\sum_{i>k}\alpha_i x_i\right)\right| \;\middle|\; \xi \in \widetilde{G}_k\right\}\\ &\le (1+\theta^{-1})(1+\varepsilon)^2\left\|x+\sum_{i>k}\alpha_i x_i\right\|.\end{aligned}$$

Thus, if we choose θ and ε so that $(1+\theta^{-1})(1+\varepsilon)^2 \le \lambda'$, we obtain (x_n) λ'-basic (so we can obtain it λ'-basic for any $\lambda' > 2$). In addition, we will show that the sequence (z_i) defined by $z_1 = x_1$ and $z_i = x_i - x_{i-1}$ is $\lambda'(1+\theta^{-1}(1+\varepsilon))$-basic (so we can obtain it λ-basic for any $\lambda > 4$). Indeed, consider scalars (β_i) and let

$$x = \sum\nolimits_1^j \beta_i z_i, \qquad y = \sum\nolimits_{j+1}^n \beta_i z_i.$$

Note that $x \in E_j$ but our problem is that $z_{j+1} = x_{j+1} - x_j$ involves x_j. We must show

$$\|x\| \le \lambda'(1+\theta^{-1}(1+\varepsilon))\|x+y\|.$$

By (11.6) we have $\xi_{j+1}(x+y) = \theta\beta_{j+1}$, and hence

$$\theta|\beta_{j+1}| \le \|x+y\|, \tag{11.15}$$

and then by the triangle inequality

$$\|x+(y+\beta_{j+1}x_j)\| \le \|x+y\| + |\beta_{j+1}|(1+\varepsilon) \le (1+\theta^{-1}(1+\varepsilon))\|x+y\|.$$

But now since (x_i) is λ'-basic and $y+\beta_{j+1}x_j$ is in the span of $\{x_{j+1}, x_{j+2}, \ldots\}$

$$\|x\| \le \lambda'\|x+(y+\beta_{j+1}x_j)\|$$

and hence the announced result

$$\|x\| \le \lambda'(1+\theta^{-1}(1+\varepsilon))\|x+y\|. \qquad \square$$

To state the next result it will be convenient to introduce two sequences of positive numbers attached to a Banach space B, as follows.

For each $n \ge 1$, we set

$$\mathrm{bio}_n(B) = \inf\left\{\sup_{i\le n}\left\|\sum_{j\le i} y_j\right\|_B \sup_{j\le n}\|\xi_j\|_{B^*}\right\}$$

where the infimum runs over all biorthogonal systems $(y_i, \xi_i)_{i\le n}$ in $B \times B^*$ (biorthogonal means here that $\xi_i(y_j) = 0$ if $i \neq j$ and $= 1$ if $i = j$).

Note that obviously $\text{bio}_n(B) \le \text{bio}_{n+1}(B)$ for all $n \ge 1$. Let $c = \sup \|\xi_i\|$. Replacing ξ_i by $c^{-1}\xi_i$ and y_i by cy_i we may assume $c = 1$. By the Hahn-Banach theorem, $(y_1, \dots, y_n)$ admits a biorthogonal system (ξ_j) with $\sup \|\xi_j\|_{B^*} \le 1$ iff for any scalar n-tuple $(\alpha_1, \dots, \alpha_n)$ we have

$$\sup |\alpha_j| \le \left\| \sum \alpha_j y_j \right\|. \tag{11.16}$$

Thus, we can equivalently define $\text{bio}_n(B)$ as the infimum of $\sup_{i\le n} \|\sum_{j\le i} y_j\|$ over all (y_j) satisfying (11.16).

Equivalently, setting $x_i = \theta \sum_{j\le i} y_j$, we have

$$\text{bio}_n(B) = \inf\{\theta^{-1}\} \tag{11.17}$$

where the infimum runs over all $\theta \le 1$ for which there is an n-tuple $(x_1, \dots, x_n)$ in B satisfying for any scalar n-tuple $(\alpha_1, \dots, \alpha_n)$

$$\theta \sup_j \left| \sum_{i\ge j} \alpha_i \right| \le \left\| \sum \alpha_i x_i \right\| \le \sum |\alpha_i|.$$

Note that (11.17) clearly shows that if a space X is f.r. in B then necessarily

$$\text{bio}_n(B) \le \text{bio}_n(X) \qquad \forall n \ge 1. \tag{11.18}$$

In particular, this shows by Theorem 11.3, that $\text{bio}_n(B) \le \text{bio}_n(B^{**})$. The converse is obvious: since $B \subset B^{**}$ we must have $\text{bio}_n(B^{**}) \le \text{bio}_n(B)$. Thus we obtain

$$\text{bio}_n(B^{**}) = \text{bio}_n(B) \qquad \forall n \ge 1. \tag{11.19}$$

Moreover, (11.17) also shows that for any quotient space, say B/S (with $S \subset B$ a closed subspace), we have

$$\text{bio}_n(B) \le \text{bio}_n(B/S). \tag{11.20}$$

Indeed, one verifies this by a trivial lifting of $x_i \in B/S$ up in B.

Note one more equivalent definition of $\text{bio}_n(B)$:

$$\text{bio}_n(B) = \inf\{\theta^{-1}\} \quad \text{where} \quad \theta \le 1 \quad \text{runs over} \tag{11.21}$$

all the numbers for which there is a n-tuple $(x_j, \xi_j)_{j\le n}$ in $B \times B^*$ with $\|x_j\| \le 1$, $\|\xi_j\| \le 1$ such that

$$\xi_j(x_i) = \theta \quad \forall i \ge j \quad \text{and} \quad \xi_j(x_i) = 0 \quad \forall i < j. \tag{11.22}$$

Indeed, if (y_i, ξ_i) is as in the original definition, if we set $x_i = \theta \sum_{j\le i} y_j$ with $\theta = (\sup_i \|\sum_{j\le i} y_j\|)^{-1}$ and $\sup \|\xi_i\| = 1$, we obtain (11.22). Conversely, given (x_j, ξ_j) as in (11.22), if we set $y_j = \theta^{-1}(x_j - x_{j-1})$, $y_1 = \theta^{-1}x_1$ we find $\sup \|\sum_{j\le i} y_j\| \le \theta^{-1}$.

From (11.21), it is immediate (replacing (y_j, ξ_j) by (ξ_{n+1-j}, y_{n+1-j}), $1 \le j \le n$) that

$$\mathrm{bio}_n(B^*) \le \mathrm{bio}_n(B).$$

Hence also $\mathrm{bio}_n(B^{**}) \le \mathrm{bio}_n(B^*)$, and since we already saw $\mathrm{bio}_n(B^{**}) = \mathrm{bio}_n(B)$, we conclude that $\mathrm{bio}_n(B)$ is self-dual:

$$\mathrm{bio}_n(B) = \mathrm{bio}_n(B^*) \quad \forall n \ge 1. \tag{11.23}$$

We also introduce

$$t_n(B) = \inf\{\sup_\omega \|M_n(\omega)\|_B\}$$

where the infimum runs over all dyadic martingales $(M_k)_{k\ge 0}$ such that $\|M_k(\omega) - M_{k-1}(\omega)\| \ge 1$ for all ω and all $1 \le k \le n$. Again we have obviously $t_n(B) \le t_{n+1}(B)$ for all n. Note that $t = \sup_n t_n(B) < \infty$ iff B contains for some $\delta > 0$ arbitrarily long δ-separated finite dyadic trees in its unit ball.

Again we have $t_n(B) \le t_n(X)$ if X f.r. B. Moreover, by an easy lifting argument, this also holds when X is isometric to a quotient of B.

Theorem 11.22. *The following properties of a Banach space B are equivalent:*

(i) *B is super-reflexive.*
(i)′ *B^* is super-reflexive.*
(ii) *$\mathrm{bio}_n(B) \to \infty$ when $n \to \infty$.*
(iii) *$t_n(B) \to \infty$ when $n \to \infty$.*
(iv) *For any $\lambda > 1$, there is $q < \infty$ and a constant C such that for any N and any λ-basic sequence $(y_1, \ldots, y_N)$ in B or in any quotient of B we have*

$$\left(\sum \|y_j\|^q\right)^{1/q} \le C\left\|\sum y_j\right\|. \tag{11.24}$$

(iv)′ *For some $\lambda > 1$, the same as (iv) holds.*
(v) *For any $\lambda > 1$, there is $p > 1$ and a constant C such that for any N and any λ-basic sequence $(y_1, \ldots, y_N)$ in B we have*

$$\left\|\sum y_j\right\| \le C\left(\sum \|y_j\|^p\right)^{1/p}. \tag{11.25}$$

(v)′ *For some $\lambda > 1$, the same as (v) holds.*

Proof. The proofs that (i) $\Rightarrow$ (ii) or that (i) $\Rightarrow$ (iii) are similar to the proof of Theorem 11.11. Assume that $\text{bio}_n(B)$ (resp. $t_n(B)$) remains bounded when $n \to \infty$. We will show that there is a space X (resp. Y) that is f.r. in B and that is not reflexive (resp. fails the RNP). This will show that (i) $\Rightarrow$ (ii) (resp. (i) $\Rightarrow$ (iii)). Let us outline the argument for (i) $\Rightarrow$ (ii). Assume that (ii) fails i.e. that $\text{bio}_n(B) < C$ for all $n \ge 1$. Then, for each n we have (by homogeneity) a biorthogonal system $(y_i^n, \xi_i^n)_{i \le n}$ such that $\sup_{i \le n} \| \sum_{j \le i} y_j^n \| \le C$ and

$$\|\xi_i^n\| = 1 \quad \text{for} \quad i = 1, 2, \ldots, n.$$

We will define the Banach space X as the completion of $\mathbb{K}^{(\mathbb{N})}$ for the norm $\| \cdot \|_X$ defined as follows. For each n we set for any finitely supported scalar sequence $\alpha = (\alpha_k)$

$$\|\alpha\|_n = \left\| \sum_{1 \le k \le n} \alpha_k \sum_{j \le k} y_j^n \right\|.$$

Then we fix a non-trivial ultrafilter $\mathcal{U}$ on $\mathbb{N}$ and we set:

$$\|\alpha\|_X = \lim_{n,\mathcal{U}} \|\alpha\|_n.$$

Let $x = \sum_{1 \le k \le n} \alpha_k \sum_{j \le k} y_j^n$. We have clearly by biorthogonality $\xi_i^n(x) = \sum_{k \ge i} \alpha_k$ hence

$$\sup_{i \le n} \left| \sum_{k \ge i} \alpha_k \right| \le \|\alpha\|_n \le C \sum |\alpha_i|$$

and hence

$$\sup_i \left| \sum_{k \ge i} \alpha_k \right| \le \|\alpha\|_X \le C \sum |\alpha_i|.$$

By (i) $\Leftrightarrow$ (iii) in Theorem 11.10 we see that X is not reflexive, but since X manifestly embeds in an ultraproduct of subspaces of B, X is f.r. in B. This completes the proof that (i) $\Rightarrow$ (ii).

The proof that (i) $\Rightarrow$ (iii) is similar: if $t_n(B) < C$ for all n we produce Y f.r. in B and containing in its unit ball an infinite δ-separated dyadic tree with $\delta = 1/C$ (see the proof of Theorem 11.11); we leave the details to the reader. Note that (ii) $\Rightarrow$ (i) follows from Theorem 11.10. Indeed, by the latter theorem if B is not reflexive $\text{bio}_n(B)$ is bounded; therefore (ii) implies B reflexive. But by (11.18), if B satisfies (ii) then any X f.r. in B also satisfies (ii) and hence must be reflexive. This shows that (ii) $\Rightarrow$ (i).

Similarly, we have (iii) $\Rightarrow$ (i). Indeed, it suffices to show (iii) implies B reflexive. But if B is not reflexive, Remark 11.13 (and $t_n(B) \le t_n(X)$ if X f.r. B) clearly shows that $t_n(B)$ remains bounded when $n \to \infty$; this shows (iii) $\Rightarrow$ (i). Thus we have proved (i) $\Leftrightarrow$ (ii) $\Leftrightarrow$(iii), and hence by (11.23), (i) $\Leftrightarrow$ (i)$'$.

We will now show that (ii) $\Rightarrow$ (iv). Fix $\lambda > 1$. We will show that if (iv) fails for this λ then (ii) also fails. We will argue as we did in the preceding section for monotone basic sequences. Let $b(N, \lambda)$ be the smallest constant b such that for any λ-basic $(y_1, \ldots, y_N)$ in a quotient of B we have

$$\inf_{1 \le k \le N} \|y_k\| \le b \left\| \sum\nolimits_1^N y_k \right\|.$$

Clearly (see the proof of Theorem 10.9)

$$b(NK, \lambda) \le b(N, \lambda) b(K, \lambda) \quad \text{for all} \quad N, K, \tag{11.26}$$

and also $b(K, \lambda) \le \lambda b(N, \lambda)$ for any $K > N$. Therefore, if $b(N, \lambda) < 1$ for some $N > 1$ we find $r < \infty$ and C such that $b(N, \lambda) \le C N^{-1/r}$ for all N and this leads to (see the proof of (10.11))

$$\left(\sum \|y_j\|^q \right)^{1/q} \le C \left\| \sum y_j \right\| \quad \text{for} \quad q > r$$

and some constant C. This argument shows that if (iv) fails for some $\lambda > 1$ we must have $b(n, \lambda) \ge 1$ for all $n > 1$. Equivalently, for any $\varepsilon > 0$ there is $(y_1, \ldots, y_n)$ λ-basic in a quotient of B, say B/S for some subspace $S \subset B$, such that $\left\| \sum_1^n y_j \right\| \le 1 + \varepsilon$ but $\|y_j\| > 1$ for all $1 \le j \le n$. By (11.12), there are functionals (ξ_i) biorthogonal to y_j with $\|\xi_i\| \le 2\lambda$, and by (11.11) we have

$$\sup_i \left\| \sum_{j \le i} y_j \right\| \le \lambda \left\| \sum\nolimits_1^n y_j \right\| \le \lambda(1 + \varepsilon)$$

hence we obtain $\mathrm{bio}_n(B/S) \le \lambda(1 + \varepsilon) 2\lambda$, but by (11.20) we know that $\mathrm{bio}_n(B) \le \mathrm{bio}_n(B/S)$, therefore (ii) fails. This completes the proof that (ii) $\Rightarrow$ (iv).

We now show (ii) $\Rightarrow$ (v). Assume (ii). Then, as we already mentioned, by (11.23), B^* satisfies (ii) and hence, using the already proved implication (ii) $\Rightarrow$ (iv), B^* satisfies (iv), and actually all quotient spaces of B^* satisfy (iv). Then let $(x_1, \ldots, x_n)$ be λ-basic in B. Let $E = \mathrm{span}\{x_1, \ldots, x_n\}$. We have $E^* = B^*/E^\perp$ and the biorthogonal functionals $(x_1^*, \ldots, x_n^*)$ are λ-basic in E^*. By (iv) applied in E^*, we have for any scalar n-tuple (note that $(\alpha_i x_i^*)_{i \le n}$ is also basic if $\alpha_i \ne 0$)

$$\left(\sum |\alpha_i|^q \|x_i^*\|^q \right)^{1/q} \le C \left\| \sum \alpha_i x_i^* \right\|$$

hence, by duality, if $p > 1$ is conjugate to q we find

$$\begin{aligned}\left\|\sum x_i\right\| &= \sup\left\{\left|\left(\sum x_i\right)(x^*)\right| \;\middle|\; x^* \in E^*,\ \|x^*\| \le 1\right\}\\ &= \sup\left\{\left|\sum \alpha_i\right| \;\middle|\; \left\|\sum \alpha_i x_i^*\right\| \le 1\right\}\\ &\le \sup\left\{\left(\sum (|\alpha_i|\|x_i^*\|)^q\right)^{1/q}\left(\sum \|x_i^*\|^{-p}\right)^{1/p} \;\middle|\; \left\|\sum \alpha_i x_i^*\right\| \le 1\right\}\\ &\le C\left(\sum \|x_i^*\|^{-p}\right)^{1/p}\\ &\le C\left(\sum \|x_i\|^p\right)^{1/p}\end{aligned}$$

where for the last line we used $1 = x_i^*(x_i) \le \|x_i^*\|\|x_i\|$. This completes the proof that (ii) $\Rightarrow$ (v).

Note that (iv) $\Rightarrow$ (iv)$'$ and (v) $\Rightarrow$ (v)$'$ are trivial. Now, we prove (v)$'$ $\Rightarrow$ (i): Since (v)$'$ is clearly a super-property, it suffices to show (v)$'$ implies B reflexive. But if B is not reflexive, by Remark 11.20, for any $\lambda > 1$, we can find a λ-basic sequence (ξ_n) with $\|\xi_n\| \le 1$ satisfying (11.6) for some (x_n) in the unit ball of B. This implies

$$\theta n = \sum_{j\le n} \xi_j(x_n) \le \left\|\sum\nolimits_1^n \xi_j\right\|$$

but now (v)$'$ implies $\left\|\sum_1^n \xi_j\right\| \le Cn^{1/p}$ with $p > 1$, which is impossible when $n \to \infty$. This contradiction shows that (v)$'$ implies the reflexivity of B, concluding the proof of (v)$'$ $\Rightarrow$ (i).

It only remains to show (iv)$'$ $\Rightarrow$ (v)$'$. Since the finite-dimensional subspaces of B^* are the duals of the finite-dimensional quotients of B, by duality (iv)$'$ implies that B^* satisfies (v)$'$. Applying the (just proved) implication (v)$'$ $\Rightarrow$ (i) to the space B^*, we conclude that B^* must be super-reflexive, and hence (recall (11.23)) B itself satisfies (ii), and we already proved (ii) $\Rightarrow$ (v) $\Rightarrow$ (v)$'$. So we conclude (iv)$'$ $\Rightarrow$ (v)$'$. □

Remark. Returning to Remark 11.21, recall that the sequence (z_i) (defined by $z_1 = x_1$ and $z_i = x_i - x_{i-1}$ for $i > 1$) can be found λ-basic with $\lambda > 4$, and also $\|z_i\| \ge \xi_i(z_i) = \theta$. But then $\sum_1^n z_i = x_n$ hence $\|\sum_1^n z_i\| \le 1$, which contradicts any estimate of the form $(\sum \|z_i\|^q)^{1/q} \le C\|\sum z_i\|$. This shows that if B itself (without its quotients) satisfies (iv) then B is super-reflexive.

Corollary 11.23. *If B is super-reflexive then for any $\lambda > 1$ there are $p > 1$ and $q < \infty$ and positive constants C' and C'' such that any λ-basic sequence $(x_1, \ldots, x_N)$ is B satisfies*

$$(C')^{-1}\left(\sum \|x_i\|^q\right)^{1/q} \leq \left\|\sum x_i\right\| \leq C''\left(\sum \|x_i\|^p\right)^{1/p}.$$

11.3 Uniformly non-square and J-convex spaces

We start this section by a remarkable result discovered by R. C. James [275].

Theorem 11.24. *In any non-reflexive Banach space B, there is, for any $\delta > 0$, a pair x, y in the unit sphere of B such that*

$$\|x \pm y\| \geq 2 - \delta.$$

Remark. Banach spaces that fail the conclusion of Theorem 11.24 are called uniformly non-square. More precisely, B is 'uniformly non-square' if there is $\delta > 0$ such that for any x, y in the unit ball we have either $\|(x+y)/2\| \leq 1 - \delta$ or $\|(x-y)/2\| \leq 1 - \delta$. This is a weak form of uniform convexity. In fact, this is the same as saying that the uniform convexity modulus $\delta_B(\varepsilon)$ is > 0 for *some* $0 < \varepsilon < 2$ (while uniform convexity is the same but for *all* $0 < \varepsilon < 2$).

Remark. Let $\alpha, \beta \in \mathbb{R}$ such that $|\alpha| + |\beta| = 1$. Assume $\|x\|, \|y\| \leq 1$ and $\|x \pm y\| \geq 2 - \delta$. Then for some $\varepsilon = \pm 1$ we have

$$\begin{aligned}\|\alpha x + \beta y\| &= \|\,|\alpha|x + \varepsilon|\beta|y\| \geq \|x + \varepsilon y\| - \|(1 - |\alpha|)x + \varepsilon(1 - |\beta|)y\| \\ &\geq 2 - \delta - (1 - |\alpha| + 1 - |\beta|) = 1 - \delta.\end{aligned}$$

Therefore by homogeneity we have

$$\forall \alpha, \beta \in \mathbb{R} \qquad (1 - \delta)(|\alpha| + |\beta|) \leq \|\alpha x + \beta y\| \leq |\alpha| + |\beta|.$$

In particular, any non-reflexive Banach space contains for any $\delta > 0$ a two dimensional subspace $(1 + \delta)$-isometric to $\ell_1^{(2)}$.

In the real case, $\ell_1^{(2)}$ is the same (isometrically) as $\ell_\infty^{(2)}$. Explicitly: Given x, y as earlier, let $a = (x + y)/2$ and $b = (x - y)/2$. Then

$$\forall \alpha, \beta \in \mathbb{R} \qquad (1 - \delta)\max\{|\alpha|, |\beta|\} \leq \|\alpha a + \beta b\| \leq \max\{|\alpha|, |\beta|\}.$$

Note however that this is no longer valid in the complex case.

Thus we have

Corollary 11.25. *The two-dimensional space $\ell_1^{(2)}$ (over the reals) is finitely representable in every non-reflexive real Banach space.*

Corollary 11.26. *Any uniformly non-square Banach space is super-reflexive.*

By Proposition 11.9 we can 'automatically' strengthen the preceding statement:

Corollary 11.27. *Any Banach space isomorphic to a uniformly non-square one is super-reflexive.*

Naturally the question was raised whether $\ell_1^{(2)}$ could be replaced by $\ell_1^{(n)}$ for $n > 2$ in particular for $n = 3$, but, in a 1973 tour de force, James himself gave a counter-example ([281], see also [206, 283]). We will give different and simpler examples of the same kind in Chapter 12. In the positive direction, one can generalize Theorem 11.24 as follows. This is also due to James (see [275, 284]).

Theorem 11.28. *Let B be a non-reflexive space. Then for any $n \geq 1$ and any $\delta > 0$ there are $x_1, \ldots, x_n$ in the unit sphere of B such that for any choice of signs $\varepsilon_j = \pm 1$ where the $+$ signs all precede the $-$ signs (we call these 'admissible' choices of signs) we have*

$$\|\varepsilon_1 x_1 + \cdots + \varepsilon_n x_n\| \geq n - \delta.$$

More explicitly we have for any $j = 1, \ldots, n-1$,

$$\|x_1 + \cdots + x_j - x_{j+1} - \cdots - x_n\| \geq n - \delta \text{ and also } \|x_1 + \cdots + x_n\| \geq n - \delta.$$

Definition. A Banach space B is called J-convex if there is an integer $n > 1$ and a number $\delta > 0$ such that for any $x_1, \ldots, x_n$ in the unit ball of B

$$\inf \|\sum \varepsilon_k x_k\| \leq n(1-\delta)$$

where the infimum runs over all admissible choice of signs i.e. such that $\varepsilon_k = \pm 1$ and all the $+$ signs appear before the $-$ signs (if any).

Note that if B is J-convex then any space f.r. in B is automatically J-convex. Using this, Theorem 11.28 can then be rephrased as follows:

Corollary 11.29. *Any J-convex Banach space is reflexive (and actually super-reflexive).*

The next result will be deduced rather easily from this last one.

Corollary 11.30. *J-convexity and super-reflexivity are equivalent properties.*

Remark 11.31. The girth of the unit ball of a real Banach space B is the infimum of the lengths of centrally symmetric simple closed rectifiable curves on its surface. It is proved in [284] (see also [424]) that a Banach space is super-reflexive if and only if the girth of its unit ball is (strictly) more than 4. In sharp contrast, the girth of ℓ_1, c_0 or ℓ_∞ is equal to 4. This is closely connected to the fact that super-reflexivity is equivalent to J-convexity.

The original proofs of both Theorems 11.24 and 11.28 are rather delicate. We follow a simpler approach due to Brunel and Sucheston [149].

We will need the following notion.

Definition. A sequence $(\hat{x}_n)$ in a Banach space will be called subsymmetric if for any integer N, for any $(\alpha_1, \ldots, \alpha_N)$ in $\mathbb{R}^N$ and for any increasing sequence $n(1) < n(2) < \cdots < n(N)$ we have

$$\left\|\sum\nolimits_1^N \alpha_j \hat{x}_j\right\| = \left\|\sum\nolimits_1^N \alpha_j \hat{x}_{n(j)}\right\|.$$

The sequence $(\hat{x}_n)$ will be called 'additive' if for any finite sequence of real scalars (α_j) and for any $m \geq 1$, the preceding term $\left\|\sum_1^N \alpha_j \hat{x}_j\right\|$ is equal to:

$$\left\|\alpha_1 \sum_{0<j\leq m} \hat{x}_j/m + \alpha_2 \sum_{m<j\leq 2m} \hat{x}_j/m + \cdots + \alpha_N \sum_{(N-1)m<j\leq Nm} \hat{x}_j/m\right\|.$$

We will also need

Lemma 11.32. *Let (x_n) be a subsymmetric sequence such that $x_1 \neq x_2$ in a Banach space B. Then the sequence $d_j = x_{2j-1} - x_{2j}$ $(j \geq 1)$ is an unconditional basic sequence with constant 2. More precisely, for any finitely supported sequence of scalars (α_j) and any subset $\beta \subset \mathbb{N}$ we have*

$$\left\|\sum\nolimits_{j\in\beta} \alpha_j d_j\right\| \leq \left\|\sum \alpha_j d_j\right\| \tag{11.27}$$

and hence

$$\sup_{\pm} \left\|\sum \pm\alpha_j d_j\right\| \leq 2\left\|\sum \alpha_j d_j\right\|. \tag{11.28}$$

Proof. Clearly (11.27) implies (11.28) by considering the index sets β_+ and β_- where the sign is $+$ or $-$. By an elementary iteration, it suffices to prove (11.27) when β is the complement of a singleton $\{j\}$. Equivalently, it suffices to prove

$$\|\alpha_1 d_1 + \cdots + \widehat{\alpha_j d_j} + \cdots + a_N d_N\| \leq \left\|\sum \alpha_j d_j\right\|$$

where the hat marks the absence. But now by subsymmetry for any m and any $0 < p \leq m$

$$\left\|\sum \alpha_j d_j\right\| = \left\|\sum\nolimits_{k=1}^{j-1} \alpha_k d_k + \alpha_j D_{j+p} + \sum\nolimits_{k=j+1}^{\infty} \alpha_k d_{k+m}\right\|, \tag{11.29}$$

where $D_{j+p} = x_{2(j-1)+p} - x_{2(j-1)+p+1}$. Note $m^{-1}(D_{j+1} + \cdots + D_{j+m}) \to 0$ when $m \to \infty$ (telescoping sum). Averaging (11.29) over $0 < p \leq m$ and letting $m \to \infty$ we obtain (11.27). □

Notation: Consider a bounded function $f\colon I \times I \to \mathbb{R}$. For each fixed $k \in I$, we can define $\lim_{\mathcal{U}} f(k, i)$ but also $\lim_{\mathcal{U}} f(i, k)$ and of course these differ in general. To avoid ambiguity we will denote by $\lim_{i\,\mathcal{U}} f(i, j)$ the limit (relative to i) when j is kept fixed, and we denote by $\lim_{j\,\mathcal{U}} f(i, j)$ the limit (relative to j) when i is kept fixed. Similarly, given a function $f\colon I^N \to \mathbb{R}$ we can define the iterated limits

$$\lim_{i(1)\,\mathcal{U}} (\lim_{i(2)\,\mathcal{U}} \ldots (\lim_{i(N)\,\mathcal{U}} f(i(1), \ldots, i(N))\ldots).$$

Lemma 11.33. *If B is non-reflexive then there is a subsymmetric and additive sequence (x_n) satisfying (11.7) for some $\theta > 0$ and such that the closed span of $[x_n]$ is f.r. in B.*

Proof. By Theorem 11.10, B contains a sequence (x_n) satisfying (11.7). Let (e_n) be the canonical basis in the space $\mathbb{K}^{(\mathbb{N})}$ of finitely supported sequences of scalars ($\mathbb{K} = \mathbb{R}$ or $\mathbb{C}$). For any N and any (α_j) in $\mathbb{K}^N$ we define

$$\|\alpha_1 e_1 + \cdots + \alpha_N e_N\| = \lim_{i(1)\,\mathcal{U}} (\lim_{i(2)\,\mathcal{U}} \ldots (\lim_{i(N)\,\mathcal{U}} \|\alpha_1 x_{i(1)} + \cdots + \alpha_N x_{i(N)}\|)\ldots).$$

Let B_1 be the completion of $\mathbb{K}^{(\mathbb{N})}$ equipped with this norm. Since span$[e_1, \ldots, e_N]$ is (by definition) a subspace of an N-times iteration of ultrapowers starting with one of B, it must be f.r. in B (see Lemma 11.66). Therefore B_1 itself is f.r. in B. Clearly, if we replace (x_n) by $(x_{i(1)}, x_{i(2)}, \ldots)$ with $i(1) < i(2) < \cdots$ then (11.7) remains valid, therefore (e_n) itself still satisfies (11.7). Lastly, it takes a moment of thought to check that (e_n) is subsymmetric. We will now modify (e_n) to obtain a sequence that is also additive. Consider again a finitely supported sequence of scalars $(\alpha_1, \ldots, \alpha_N, 0, 0, \ldots)$, we define

$$\|(\alpha_j)\|_{(m)} = \left\|\alpha_1 m^{-1} \sum_{j=1}^{m} e_j + \alpha_2 m^{-1} \sum_{j=m+1}^{2m} e_j + \cdots + \alpha_N m^{-1} \sum_{j=m(N-1)+1}^{Nm} e_j\right\|.$$

We claim that $\|(\alpha_j)\|_{(m)}$ converges when $m \to \infty$. By subsymmetry of (e_n) and the triangle inequality, we have obviously

$$\|(\alpha_j)\|_{(m)} \leq \|(\alpha_j)\|_{(1)} = \left\|\sum \alpha_j e_j\right\|.$$

More generally, for any pair of integer k, m we have

$$\|(\alpha_j)\|_{(mk)} \leq \|(\alpha_j)\|_{(m)}. \tag{11.30}$$

Thus for any $n \geq m$, dividing n by m we can write $n = mk + p$ with $p < m$ and we easily check (again by the triangle inequality) that

$$\|(\alpha_j)\|_{(n)} \leq \frac{mk}{n}\|(\alpha_j)\|_{(mk)} + \frac{p}{n}\|(\alpha_j)\|_{(p)}.$$

This gives us by (11.30)

$$\forall m \geq 1 \qquad \overline{\lim_{n\to\infty}}\, \|(\alpha_j)\|_{(n)} \leq \|(\alpha_j)\|_{(m)},$$

and hence $\overline{\lim}_{n\to\infty} \|(\alpha_j)\|_{(n)} = \inf_m \|(\alpha_j)\|_{(m)}$. This proves the announced claim. We now define a norm $|||\cdot|||$ on $\mathbb{R}^{(\mathbb{N})}$ by setting

$$|||(\alpha_j)||| = \lim_{m\to\infty} \|(\alpha_j)\|_{(m)}. \tag{11.31}$$

Let B_2 be the completion of $(\mathbb{K}^{(\mathbb{N})}, |||\cdot|||)$. Let us denote by $(\hat{x}_n)$ the basis (e_n) viewed as sitting in B_2. Then, an easy verification shows that $(\hat{x}_n)$ is subsymmetric and still satisfies (11.7). Moreover, using (11.30) it is easy to see that $(\hat{x}_n)$ is additive. Lastly, note that (11.31) implies that B_2 is f.r. in B_1 and a fortiori in B. □

Proof of Theorem 11.24. By Lemma 11.33, we may assume that B contains a subsymmetric additive sequence (x_n) satisfying (11.7) for some $\theta > 0$. The idea of the proof (going back to [275]) can be roughly outlined as follows. Consider two long sequences of coefficients equal to ± 1 as follows:

$$\begin{matrix} 1 & 0 & -1 & 0 & 1 & 0 & -1 & 0 & \dots & 1 & 0 & -1 & 0 \\ 0 & 1 & 0 & -1 & 0 & 1 & 0 & -1 & \dots & 0 & 1 & 0 & -1 \end{matrix}$$

where the second sequence is obtained from the first one by a single right-hand shift with 0 added at the front and removed at the end. Then if these represent vectors z_1 and z_2 we have $\|z_1\| = \|z_2\|$ by subsymmetry and $\|z_1 + z_2\| = 2\|z_1\|$ by additivity. But moreover (this needs more care) $z_1 - z_2$ is very similar (up to 2 digits) to the vector obtained from z_1 after repeating each of its coefficients, so, again by additivity, we find $\|z_1 - z_2\| \simeq 2\|z_1\|$. More precisely, let (these depend on m but for the moment we deliberately keep m silent):

$$z_1 = x_1 - x_3 + x_5 - x_7 + \cdots + x_{4m-3} - x_{4m-1} \tag{11.32}$$

$$z_2 = x_2 - x_4 + x_6 - x_8 + \cdots + x_{4m-2} - x_{4m}. \tag{11.33}$$

Let $r(m) = \|z_1\|$. Note that $\|z_1\| = \|z_2\| = r(m)$. Observe that the sequence of signs appearing in $z_1 + z_2$ is $(++--++--\cdots)$. Therefore, by additivity, we have

$$\|z_1 + z_2\| = 2\|z_1\| = 2r(m).$$

As for $z_1 - z_2$ the sequence of signs is

$$(+ - - + + - - \cdots - -+).$$

This is as before except for the first and last sign. From this we easily deduce

$$\|z_1 - z_2\| \geq 2\|z_1\| - \| - e_1 + e_2\| = 2r(m) - \|e_2 - e_1\|.$$

We then distinguish two cases.

Case 1. $r(m)$ is unbounded when $m \to \infty$. Let $x = z_1/\|z_1\|$ and $y = z_2/\|z_2\|$. We have $\|x + y\| = 2$ and $\|x - y\| \geq 2 - \delta(m)$ where $\delta(m) = \|e_2 - e_1\|r(m)^{-1} \to 0$ when $m \to \infty$, so the proof is complete in this case.

Case 2. $\sup_m r(m) < \infty$. By Lemma 11.32, we have

$$\sup_{\pm} \left\|\sum\nolimits_1^m \pm(x_{2j-1} - x_{2j})\right\| \leq 2\left\|\sum\nolimits_1^m x_{2j-1} - x_{2j}\right\| = 2r(m).$$

Thus we find for any (α_j) in $\mathbb{R}^m$

$$\left\|\sum\nolimits_1^m \alpha_j(x_{2j-1} - x_{2j})\right\| \leq 2r(m)\sup|\alpha_j|.$$

Moreover by (11.27) we have for any j

$$|\alpha_j|\|x_1 - x_2\| \leq \left\|\sum \alpha_j(x_{2j-1} - x_{2j})\right\|$$

and hence $\sup|\alpha_j|\|x_1 - x_2\| \leq \|\sum \alpha_j(x_{2j-1} - x_{2j})\|$. Thus we conclude in this case that $\overline{\text{span}}[x_{2j-1} - x_{2j}]$ is isomorphic to c_0, and hence that B contains ℓ_∞^ns uniformly. By Theorem 11.5 any Banach space (in particular ℓ_1) is f.r. in B. Thus we obtain the desired conclusion in this case also. □

Proof of Theorem 11.28. The idea is similar to that of the preceding proof. We keep the same notation. We define for $k = 1, 2, \ldots, n$

$$z_k = x_k - x_{n+k} + x_{2n+k} - x_{3n+k} + \cdots + x_{(2m-2)n+k} - x_{(2m-1)n+k}.$$

Note that again, for any $k = 1, \ldots, n$, we have by subsymmetry

$$\|z_k\| = \|x_1 - x_2 + \cdots + x_{2m-1} - x_{2m}\| = r(m),$$

and by additivity

$$\|z_1 + \cdots + z_n\| = n\|z_1\| = nr(m).$$

Consider now $z_1 + \cdots + z_j - (z_{j+1} + \cdots + z_n)$ with $1 \leq j < n$.

The ordered sequence of non-zero basis coefficients of that vector is

$$\overbrace{+\cdots+}^{j}\ \overbrace{-\cdots-}^{n}\ \overbrace{+\cdots+}^{n}\ \cdots\ \overbrace{-\cdots-}^{n}\ \overbrace{+\cdots+}^{n-j}$$

where in the middle we have $2m-1$ series of n equal signs. This implies by additivity that

$$\|z_1+\cdots+z_j-(z_{j+1}+\cdots+z_n)\|\geq nr(m)-(n-j)\|e_1-e_2\|.$$

We may assume that we are in case 1, i.e. $r(m)\to\infty$. Let then

$$z'_j=z_j/\|z_j\|.$$

We find $\|z'_1+\cdots+z'_n\|=n$ and

$$\|z'_1+\cdots+z'_j-(z'_{j+1}+\cdots+z'_n)\|\geq n-\delta'(m)$$

with $\delta'(m)=(n/m)\|e_1-e_2\|\to 0$ when $m\to\infty$. □

Corollary 11.34. *Let B be a non-reflexive or merely a non-J-convex Banach space. Then there is a Banach space $\widetilde{B}$ f.r. in B that contains a sequence (x_n) such that (here we deliberately insist on real scalars)*

$$\forall(\alpha_j)\in\mathbb{R}^{(\mathbb{N})}\qquad \sup_j\left\{\left|\sum_{i<j}\alpha_i\right|+\left|\sum_{i\geq j}\alpha_i\right|\right\}\leq\left\|\sum\alpha_i x_i\right\|\leq\sum|\alpha_i|. \tag{11.34}$$

Equivalently, there are ξ_j in $\widetilde{B}^*$ with $\|\xi_j\|\leq 1$ such that $\xi_j(x_i)=1$ for all $i<j$ and $\xi_j(x_i)=-1$ for all $i\geq j$.

Proof. Choose a sequence δ_n tending to 0, say $\delta_n=1/n$. By Theorem 11.28, we may assume that B is not J-convex, so that for any $n\geq 1$, there are $x_1^{(n)},\ldots,x_n^{(n)}$ in the unit sphere of B such that

$$\left\|\sum\nolimits_1^n\varepsilon_j x_j^{(n)}\right\|\geq n-\delta_n$$

for all the admissible choices of signs. Note that this implies obviously

$$\forall k\leq n\qquad \left\|\sum\nolimits_1^k\varepsilon_j x_j^{(n)}\right\|\geq k-\delta_n. \tag{11.35}$$

For any $(\alpha_j)\in\mathbb{K}^{(\mathbb{N})}$ (recall this means (α_j) is finitely supported) we define

$$\left|\left|\left|\sum\alpha_j e_j\right|\right|\right|=\lim_{n\mathcal{U}}\left\|\sum\nolimits_1^n\alpha_j x_j^{(n)}\right\|.$$

Let $\widetilde{B}$ be the completion of $(\mathbb{K}^{(\mathbb{N})},|||\cdot|||)$ and let $x_j=e_j$ viewed as an element of $\widetilde{B}$. By Lemma 11.66 we know that $\widetilde{B}$ is f.r. in B. Then by (11.35) we have $|||\sum_1^k\varepsilon_j x_j|||\geq k$ (and hence this is $=k$) for any k and any admissible choice of signs. For each n and $j\leq n$ let $\xi_j^{(n)}$ in the unit sphere of $\widetilde{B}^*$ be such that $\xi_j^{(n)}(x_1+\cdots+x_j-x_{j+1}-\cdots-x_n)=n$. Clearly we must have $\xi_j^{(n)}(x_i)=1$ for all $i\leq j$ and $=-1$ for all i such that $j<i\leq n$ (if any). Let ξ_j be a $\sigma(\widetilde{B}^*,\widetilde{B})$

cluster point of $\{\xi_j^{(n)} \mid n \geq 1\}$ (or let $\xi_j = \lim_{n\,\mathcal{U}} \xi_j^{(n)}$). Clearly, (ξ_j) satisfies the property in Corollary 11.34. Then

$$\xi_n\left(\sum \alpha_j x_j\right) = \sum_{j \leq n} \alpha_j - \sum_{j>n} \alpha_j$$

and also

$$\xi_{N+1}\left(\sum_1^N \alpha_j x_j\right) = \sum_1^N \alpha_j.$$

Thus, if we now restrict to *real* scalars, (11.34) follows since

$$|\sum_{j \leq n} \alpha_j| + |\sum_{j>n} \alpha_j| = \sup_{\pm 1} |\sum_{j \leq n} \alpha_j \pm \sum_{j>n} \alpha_j|. \qquad \square$$

Proof of Corollary 11.30. By Corollary 11.29 we know that J-convexity implies super-reflexivity, and Corollary 11.34 implies the converse. Indeed, the space $\widetilde{B}$ appearing in Corollary 11.34 satisfies (11.7) and hence is not reflexive. $\square$

Remark. We suspect that Corollary 11.34 fails in the complex case. More precisely, there might exist non-reflexive complex Banach spaces that do not contain almost isometric copies of the complex version of 'squares', i.e. do not contain almost isometrically the space $\mathbb{C}^2$ equipped with the norm $\|(x, y)\| = |x| + |y|$.

Corollary 11.35. *Let B be a non-reflexive real Banach space. Then there is a space X f.r. in B admitting a linear map $J\colon\ L_1([0, 1], dt; \mathbb{R}) \to X$ such that for any f (real valued) in L_1*

$$\sup_{0 \leq s \leq 1} \left|\int_0^s f(f)dt\right| + \left|\int_s^1 f(t)dt\right| \leq \|J(f)\| \leq \int_0^1 |f(t)|dt. \qquad (11.36)$$

Proof. Let (x_n) be the sequence in the preceding corollary and let X be the associated space via the construction described in the proof of Theorem 11.11. We clearly have the announced property. $\square$

Corollary 11.36. *Let B be a non-reflexive space. Then there is a space X f.r. in B such that there is a dyadic martingale (f_n) in $L_\infty(X)$ satisfying for all $n \geq 1$ and all $\omega \in \{-1, 1\}^n$*

$$\|f_n(\omega)\| \leq 1, \quad \text{but} \quad \|f_n(\omega) - f_{n-1}(\omega)\| = 1.$$

In addition, for all $n \geq 1$ and all $\omega \neq \omega' \in \{-1, 1\}^n$ we have

$$\|f_n(\omega) - f_n(\omega')\| = 2, \quad \|f_n(\omega) - f_{n-1}(\omega')\| = 2.$$

In particular, the unit ball of X contains a 1-separated infinite dyadic tree.

Proof. We just repeat the argument for Lemma 11.12. Then the stronger property (11.36) yields the announced result. □

See Remark 1.35 for concrete examples of infinite trees as described in the preceding statement.

11.4 Super-reflexivity and uniform convexity

The main result of this section is the following.

Theorem 11.37. *The following properties of a Banach space B are equivalent.*

(i) B is super-reflexive.

(ii) There is an equivalent norm on B for which the associated modulus of uniform convexity δ satisfies for some $2 \le q < \infty$

$$\inf_{0<\varepsilon\le 2} \delta(\varepsilon)/\varepsilon^q > 0.$$

(ii)′ There is an equivalent norm on B for which the associated modulus of uniform smoothness ρ satisfies for some $1 < p \le 2$

$$\sup_{t>0} \rho(t)/t^p < \infty.$$

(iii) B is isomorphic to a uniformly convex space.

(iii)′ B is isomorphic to a uniformly smooth space.

(iv) B is isomorphic to a uniformly non-square space.

The equivalence of (i),(iii),(iii)′ and (iv) is a beautiful result due to Enflo [220]. As in the preceding chapter, we will follow the martingale inequality approach of [387] and prove directly that (i) ⇒ (ii) (or equivalently since super-reflexivity is self-dual (i) ⇒ (ii)′).

The proof will use martingale inequalities in $L_2(B)$. So we first need to replace B by $L_s(B)$. This is the content of the next two statements.

Lemma 11.38. *Let $1 < s < \infty$. Then a Banach space B is J-convex iff there are $n > 1$ and $\alpha < 1$ such that for any $x_1, \ldots, x_n$ in B we have*

$$\left(n^{-1} \sum_{\xi\in A(n)} \left\| \sum \xi_j x_j \right\|^s \right)^{1/s} \le \alpha n^{\frac{1}{s'}} \left(\sum \|x_j\|^s \right)^{1/s}, \tag{11.37}$$

where $A(n) \subset \{-1, 1\}^n$ is the subset formed of the n admissible choices of signs.

Proof. Assume B J-convex, so $\exists n\ \exists\delta > 0$ such that $\forall x_1, \ldots, x_n \in B$

$$\inf_{\xi\in A(n)} \left\|\sum\nolimits_1^n \xi_j x_j\right\| \le n(1-\delta)\max\|x_j\|. \tag{11.38}$$

Fix $1 < s < \infty$. We claim that there is $\delta' > 0$ such that $\forall x_1, \ldots, x_n \in B$

$$\inf_{\xi\in A(n)} \left\|\sum \xi_j x_j\right\| \le n^{1/s'}(1-\delta')\left(\sum \|x_j\|^s\right)^{1/s}. \tag{11.39}$$

Indeed, if not then $\forall\delta' > 0\ \exists x_1, \ldots, x_n$ such that

$$(1-\delta')n^{1/s'}\left(\sum \|x_j\|^s\right)^{1/s} < \inf\left\|\sum \xi_j x_j\right\| \le \sum \|x_j\|. \tag{11.40}$$

Moreover we may assume by homogeneity $\sum \|x_j\| = n$. But (11.40) contains an approximate reverse Hölder inequality, so an elementary reasoning shows that (11.40) implies

$$\max\{|\|x_i\| - \|x_j\|| \mid 1 \le i, j \le n\} \le \varphi_n(\delta')$$

with $\varphi_n(\delta') \to 0$ when $\delta' \to 0$. Since $\sum_1^n \|x_j\| = n$, we obtain

$$\max\|x_j\| \le 1 + \varphi_n(\delta') \quad \text{and} \quad \min\|x_j\| \ge \sup\|x_j\| - \varphi_n(\delta') \ge 1 - \varphi_n(\delta').$$

Note $n = \sum \|x_j\| \le n^{1/s'}(\sum \|x_j\|^s)^{1/s}$. But then (11.38) and (11.40) together imply

$$n(1-\delta') < n(1-\delta)(1+\varphi_n(\delta')),$$

and here $\delta > 0$ is fixed while δ' and $\varphi_n(\delta')$ tend to zero, so this is impossible. This establishes (11.39). Then we note that (11.39) trivially implies (11.37) with $\alpha = (n^{-1}((1-\delta')^s + (n-1))^{1/s}$ and $\delta' > 0$ ensures $\alpha < 1$.
Conversely if (11.37) holds then a fortiori $\inf_{\xi\in A(n)} \|\sum \xi_j x_j\| \le \alpha n \sup\|x_j\|$ and hence B is J-convex. □

Proposition 11.39. *Let $1 < s < \infty$ and let $(\Omega, \mathcal{A}, m)$ be any measure space. If B is super-reflexive, then $L_s(m; B)$ also is.*

Proof. By Corollary 11.30 it suffices to show that B is J-convex iff $L_s(m; B)$ also is. By integration at the s-th power, it is clear that B satisfies (11.37) iff $L_s(\mu; B)$ also does. □

Corollary 11.40. *Fix $1 < s < \infty$. If B is super-reflexive, then there are $1 < p \le 2 \le q < \infty$ (a priori depending on s) and positive constants C and C' such that any B-valued martingale (f_n) satisfies:*

$$C^{-1}\left(\sum\nolimits_0^\infty \|df_n\|_{L_s(B)}^q\right)^{1/q} \le \sup\|f_n\|_{L_s(B)} \le C'\left(\sum\nolimits_0^\infty \|df_n\|_{L_s(B)}^p\right)^{1/p}. \tag{11.41}$$

Proof. By the preceding proposition, we may apply Theorem 11.22 to $L_s(B)$. Note that martingale difference sequences are monotone basic sequences in $L_s(B)$. Thus this follows from Corollary 11.23. □

We can outline the proof of Theorem 11.37 like this: if B is super-reflexive, so is $L_2(B)$, so that all monotone basic sequences in $L_2(B)$ satisfy a lower q-estimate of the form (11.24). Applying this to B-valued martingales we find that there is $q < \infty$ and a constant C such that all B-valued martingales $(f_n)_{n\geq 0}$ satisfy (recall the convention $df_0 = f_0$ and $df_k = f_k - f_{k-1}$ for all $k \geq 1$)

$$\forall N \geq 1 \qquad \left(\sum\nolimits_0^N \|df_n\|_{L_2(B)}^q\right)^{1/q} \leq C \left\|\sum\nolimits_0^N df_n\right\|_{L_2(B)}. \tag{11.42}$$

We wish to apply the renorming Corollary 10.7, which assumes (10.9). By Theorem 10.59 (and Doob's maximal inequality (1.38) for $p = 2$), (10.9) is equivalent to an estimate of the form

$$\forall N \geq 1 \qquad \left\|\left(\sum \|df_n\|^q\right)^{1/q}\right\|_2 \leq C' \left\|\sum\nolimits_0^N df_n\right\|_{L_2(B)}. \tag{11.43}$$

The difficulty here is that when $2 \leq q < \infty$ we have always

$$\left(\sum \|df_n\|_{L_2(B)}^q\right)^{1/q} \leq \left\|\left(\sum \|df_n\|^q\right)^{1/q}\right\|_2 \tag{11.44}$$

but not conversely ! So the inequality we need appears significantly stronger than (11.42). But this technical problem was solved in the preceding chapter: we saw how to pass from (11.42) to an inequality of the form (10.9), at the cost of a slight loss of the exponent q.

Proof of Theorem 11.37. We first prove the equivalence of (i)–(iv). The implications (ii) ⇒ (iii) ⇒ (iv) are trivial and (iv) ⇒ (i) is Corollary 11.27. Thus it suffices to show (i) ⇒ (ii). Assume (i). By Proposition 11.39 $L_2(B)$ is super-reflexive. By Theorem 11.22, there is a constant C and $s < \infty$ such that any finite martingale (f_n) in $L_2(B)$ satisfies (10.9). By Lemma 10.12 and Corollary 10.7 we obtain (ii).

We now turn to (ii)′ and (iii)′. Note that B satisfies (ii)′ (resp. (iii)′) iff B^* satisfies (ii) (resp. (iii)) in Theorem 11.37. Thus since B is super-reflexive iff B^* also is (see Theorem 11.22) we can deduce (i) ⇔ (ii)′ ⇔ (iii)′ from the part of Theorem 11.37 that we just proved. However, the reader will surely observe that a direct argument for the main point (i) ⇒ (ii)′ can alternatively be obtained by combining together (i) ⇒ (v) in Theorem 11.22 applied to $L_2(B)$, Lemma 10.13 and Corollary 10.23. □

11.5 Strong law of large numbers and super-reflexivity

We now return to the strong law of large numbers (already considered in §1.9), this time for (Banach space valued) martingales.

Lemma 11.41. *Fix an integer $n \geq 1$. Let $\Omega = \{-1, 1\}^n$, let $\varepsilon_k\colon\ \Omega \to \{-1, 1\}$ denote as usual the k-th coordinate, let $\mathcal{A}_0 = \{\phi, \Omega\}$ be the trivial σ-algebra and let $\mathcal{A}_k = \sigma(\varepsilon_1, \ldots, \varepsilon_k)$ for $k = 1, 2, \ldots, n$. Fix an integer $n \geq 1$. The following properties of a finite-dimensional Banach space B are equivalent:*

(i) There is a B-valued martingale $(f_0, \ldots, f_n)$ adapted to $(\mathcal{A}_0, \ldots, \mathcal{A}_n)$ such that for all $1 \leq k \leq n$ and all $\omega \in \Omega$

$$\|df_k(\omega)\| = 1 \quad \text{and} \quad \|f_n(\omega)\| = 1.$$

(ii) There is a B^*-valued martingale $(g_0, \ldots, g_n)$ adapted to $(\mathcal{A}_0, \ldots, \mathcal{A}_n)$, with $g_0 = 0$ such that for all $1 \leq k \leq n$ and all $\omega \in \Omega$

$$\|g_n(\omega)\| = n \quad \text{and} \quad \|dg_k(\omega)\| = 1.$$

Proof. We start by observing that for any B-valued dyadic martingale $(f_0, \ldots, f_n)$ on $(\mathcal{A}_0, \ldots, \mathcal{A}_n)$ we have for all $k = 1, \ldots, n$ and $1 \leq p \leq \infty$

$$\|df_k\|_{L_p(B)} \leq \|f_k\|_{L_p(B)} \leq \|f_n\|_{L_p(B)}.$$

Indeed since $df_k = \varepsilon_k \psi_{k-1}$ with ψ_{k-1} being $\mathcal{A}_{k-1}$-measurable we have $\|f_k\|_{L_p(B)} = \|f_{k-1} \pm df_k\|_{L_p(B)}$, so this observation follows from the triangle inequality. Note however that this is special to the dyadic filtration, the general case requires an extra factor 2.

Assume (i). Since $1 \leq \|df_k(.)\|$, there is φ_k in the unit ball of $L_\infty(\mathcal{A}_k, B^*)$ such that $1 \leq \langle df_k(.), \varphi_k(.)\rangle$ and a fortiori $1 \leq \mathbb{E}\langle df_k, \varphi_k\rangle$.
Let $g_n = \sum_1^n (\mathbb{E}_k - \mathbb{E}_{k-1})(\varphi_k)$. We have

$$n \leq \sum\nolimits_1^n \mathbb{E}\langle df_k, \varphi_k\rangle = \mathbb{E}\langle f_n, g_n\rangle \leq \|f_n\|_{L_\infty(B)} \|g_n\|_{L_1(B^*)}$$

and hence $n \leq \mathbb{E}\|g_n\|$, but since (by the preceding observation for $p = \infty$) $\|dg_k\|_{L_\infty(B^*)} = \|(\mathbb{E}_k - \mathbb{E}_{k-1})\varphi_k\|_{L_\infty(B^*)} \leq \|\varphi_k\|_{L_\infty(B^*)} \leq 1$ we have $\|g_n\|_{L_\infty(B^*)} \leq n$ and hence $\mathbb{E}\|g_n\| \geq n$ forces $\|g_n(\omega)\| = n$ for all ω. Similarly, since $1 \leq \mathbb{E}\langle df_k, \varphi_k\rangle = \mathbb{E}\langle df_k, dg_k\rangle \leq \mathbb{E}\|dg_k\|$, the fact that $\|dg_k\|_{L_\infty(B^*)} \leq 1$ forces $\|dg_k(\omega)\| = 1$ for all ω.

Conversely, assume (ii). Since $\mathbb{E}\|g_n\| \geq n$, there is f_n in the unit ball of $L_\infty(\mathcal{A}_n, B)$ such that $\mathbb{E}\langle f_n, g_n\rangle \geq n$, and hence $\sum_1^n \mathbb{E}\langle df_k, dg_k\rangle \geq n$. The latter implies $\sum_1^n \mathbb{E}\|df_k\| \geq n$ but (by the preceding observation again with $p = \infty$) we have

$$\|df_k\|_{L_\infty(B)} \leq 1 \quad \text{and hence} \quad \sum\nolimits_1^n \|df_k(\omega)\| \leq n$$

for all ω. It follows that $\|df_k(\omega)\| = 1$ for all $k = 1, \ldots, n$ and all ω. In addition, since we have $1 = \mathbb{E}\|df_n\| \le \mathbb{E}\|f_n\|$, we also obtain $\|f_n(\omega)\| = 1$ for all ω. □

Lemma 11.42. *If a Banach space B is not super-reflexive then for each $n \ge 1$ and any $0 < \theta < 1$ there is a B-valued martingale $(\tilde{g}_0, \ldots, \tilde{g}_n)$ adapted to $\mathcal{A}_0, \ldots, \mathcal{A}_n$ with $\tilde{g}_0 = 0$ such that*

$$\inf_{\omega\in\Omega} \|\tilde{g}_n(\omega)\| \ge \theta n \quad \textit{and} \quad \sup_{1\le k\le n} \sup_{\omega\in\Omega} \|d\tilde{g}_k(\omega)\| \le 1.$$

Proof. By Theorem 11.22 we may assume that B^* is not super-reflexive. By Corollary 11.36, for each n there is a finite-dimensional space E f.r. in B^* containing an E valued martingale satisfying (i) in Lemma 11.41. Fix $\varepsilon > 0$. Since E is $(1+\varepsilon)$-isomorphic to a subspace of B^*, E^* is $(1+\varepsilon)$-isomorphic to a quotient of B. Thus, E^* contains the range of a martingale (g_n) satisfying (ii) in Lemma 11.41. We have $dg_k = \varepsilon_k \psi_{k-1}$ with ψ_{k-1} in the unit ball of $L_\infty(\mathcal{A}_{k-1}, E^*)$. Fix θ so that $0 < \theta < (1+\varepsilon)^{-1}$. Let U_B denote the unit ball of B. Let $Q\colon\ B^* \to E^*$ be a surjection of norm 1 such that $Q(U_B) \supset \theta U_{E^*}$. Then there is $\widetilde{\psi}_{k-1}$ in $L_\infty(\mathcal{A}_{k-1}, B)$ with $\|\widetilde{\psi}_{k-1}\|_{L_\infty(B)} \le \theta^{-1}$ lifting ψ_{k-1}, i.e. such that $Q(\widetilde{\psi}_{k-1}) = \psi_{k-1}$. Let then $\tilde{g}_n = \theta \sum_1^n \varepsilon_k \widetilde{\psi}_{k-1}$. We have $\|df_k\|_{L_\infty(B)} = \theta\|\widetilde{\psi}_{k-1}\|_{L_\infty(B)} \le 1$ and $Q(\tilde{g}_n) = \theta g_n$ therefore

$$\theta n = \|\theta g_n(\omega)\| \le \|\tilde{g}_n(\omega)\|$$

for all ω in Ω. □

The strong law of large numbers yields one more characterization of super-reflexivity:

Theorem 11.43. *Fix $1 < s \le \infty$. The following properties of a Banach space B are equivalent:*

- *(i) B is super-reflexive.*
- *(ii) For any martingale (f_n) in $L_s(B)$ such that* $\sup_n \|df_n\|_{L_s(B)} < \infty$, *we have $n^{-1} f_n \to 0$ almost surely.*
- *(iii) For any dyadic B-valued martingale such that* $\sup_n \|df_n\|_{L_\infty(B)} < \infty$ *we have $n^{-1} f_n \to 0$ almost surely.*
- *(iv) For any dyadic B-valued martingale such that* $\sup_n \|df_n\|_{L_\infty(B)} \le 1$ *we have* $\limsup_{n\to\infty} n^{-1}\|f_n\| < 1$ *almost surely.*

Proof. Assume (i). By Corollary 11.40 there is $p > 1$ and C such that (11.41) holds. If $\sup \|df_n\|_{L_s(B)} < \infty$, this implies that $\sum n^{-1} df_n$ converges in $L_s(B)$ and hence (cf. Theorem 1.30) almost surely. By a classical (elementary) lemma due to Kronecker any sequence $\{x_n\}$ in B such that $\sum n^{-1} x_n$ converges must satisfy $n^{-1}\sum_1^n x_k \to 0$. Therefore we obtain (ii) and (ii) ⇒ (iii)⇒ (iv) are

trivial. Conversely, assume (iv). If B is not super-reflexive, we will construct a dyadic B-valued martingale $(F_n)_{n\geq 0}$ with $\|dF_n\|_{L_\infty(B)} \leq 1$ for all n and such that $\limsup_{n\to\infty} n^{-1}\|F_n\| = 1$ a.s., thus contradicting (iv). This shows that (iv) $\Rightarrow$ (i). We now turn to the announced construction:
Our basic building block will be this: Let $0 < \theta < 1$. By Lemma 11.42, for any N there is a dyadic martingale $g_1^{(N)}, \ldots, g_N^{(N)}$ with $g_0^{(N)} = 0$ such that $\|dg_k^{(N)}\|_{L_\infty(B)} \leq 1$ and $\inf_\omega \|g_N^{(N)}(\omega)\| \geq \theta N$.

Now let $0 < \theta_n < 1$ and $\xi_n > 0$ be sequences such that

$$\lim_n \theta_n = 1 \quad \text{and} \quad \lim_n \xi_n = 0.$$

Let $N(1) < N(2) < \cdots < N(n) < \cdots$ be increasing sufficiently fast so that

$$\frac{N(1) + \cdots + N(n-1)}{N(n)} < \xi_n \quad \text{for all} \quad n \geq 1. \tag{11.45}$$

Let $S(n) = N(1) + \cdots + N(n)$. Let $g_1^{[n]}, \ldots, g_{N(n)}^{[n]}$ (with $g_0^{[n]} = 0$) be the product of our basic building block, when we take $N = N(n)$ and $\theta = \theta_n$. We define a martingale $(F_{S(n)})_{n\geq 1}$ adapted to $(\mathcal{A}_{S(n)})$ as follows: we set $F_{S(1)} = g_{N(1)}^{[1]}$, then $F_{S(2)} - F_{S(1)} = g_{N(2)}^{[2]}(\varepsilon_{S(1)+1}, \ldots, \varepsilon_{S(1)+N(2)}) \ldots$ and

$$F_{S(n)} - F_{S(n-1)} = g_{N(n)}^{[n]}(\varepsilon_{S(n-1)+1}, \ldots, \varepsilon_{S(n-1)+N(n)}). \tag{11.46}$$

Since $\mathbb{E}g_N^{(N)} = g_0^{(N)} = 0$ for all N, $(F_{S(n)})_{n\geq 0}$ is indeed a martingale adapted to $(\mathcal{A}_{S(n)})_{n\geq 1}$. For any $k \leq S(n)$ we set $F_k = \mathbb{E}^{\mathcal{A}_k}(F_{S(n)})$. Then $(F_k)_{k\geq 1}$ is a (dyadic) martingale adapted to $(\mathcal{A}_k)_{k\geq 1}$ and of course $F_k = F_{S(n)}$ if $k = S(n)$. Note that $\|F_k\|_{L_\infty(B)} \leq \sum_1^k \|dF_j\|_{L_\infty(B)} \leq k$ for all $k \geq 1$. We have by (11.46) for any ω

$$\begin{aligned}\|F_{S(n)}(\omega)\| \geq \theta_n N(n) - \|F_{S(n-1)}\| &\geq \theta_n N(n) - S(n-1) \\ &\geq \theta_n S(n) - (1+\theta_n)S(n-1)\end{aligned}$$

and hence by (11.45)

$$\|F_{S(n)}(\omega)\| \geq S(n)(\theta_n - (1+\theta_n)\xi_n).$$

Thus, since $\theta_n - (1+\theta_n)\xi_n \to 1$, for any ω

$$\limsup_{n\to\infty} \|n^{-1}F_n(\omega)\| \geq 1$$

as announced. □

11.6 Complex interpolation: θ-Hilbertian spaces

Let $1 \leq p \leq 2 \leq q \leq \infty$. Following Proposition 10.31, let us say that a Banach space is 'q-uniformly convex with constant C' (resp. 'p-uniformly smooth with constant C') if for all x, y in B we have

$$\|(x+y)/2\|^q + C^{-q}\|(x-y)/2\|^q \le 2^{-1}(\|x\|^q + \|y\|^q) \tag{11.47}$$

$$(\text{resp. } 2^{-1}(\|x+y\|^p + \|x-y\|^p) \le \|x\|^p + C^p\|y\|^p).$$

Note that any Banach space is trivially 1-uniformly smooth and ∞-uniformly convex with constant 1. The next result describes the stability of these notions under complex interpolation.

Proposition 11.44. *Let (B_0, B_1) be a compatible couple of complex Banach spaces. Let $0 < \theta < 1$ and let $B_\theta = (B_0, B_1)_\theta$.*

(i) Let $2 \le q_0, q_1 \le \infty$. If B_j is q_j-uniformly convex with constant C_j ($j = 0, 1$) then B_θ is q_θ-uniformly convex with constant $C_\theta = C_0^{1-\theta} C_1^\theta$ where $q_\theta^{-1} = (1-\theta)q_0^{-1} + \theta q_1^{-1}$.
(ii) Let $1 \le p_0, p_1 \le 2$. If B_j is p_j-uniformly smooth with constant C_j ($j = 0, 1$) then B_θ is p_θ-uniformly smooth with constant $C_\theta = C_0^{1-\theta} C_1^\theta$ where $p_\theta^{-1} = (1-\theta)p_0^{-1} + \theta p_1^{-1}$.

If one of the spaces B_0, B_1 is super-reflexive, then so is B_θ.

Proof. Let $Y(q_j)$ denote the direct sum $B_j \oplus B_j$ equipped with the norm

$$\|(x, y)\|_{Y(q_j)} = ((\|x\|^{q_j} + \|y\|^{q_j})/2)^{1/q_j}.$$

Let $X(q_j)$ denote $B_j \oplus B_j$ equipped with the norm

$$\|(x, y)\|_{X(q_j)} = (\|x\|^{q_j} + C_j^{-q_j}\|y\|^{q_j})^{\frac{1}{q_j}}.$$

By (8.26) and (8.29), we have both $(Y(q_0), Y(q_1))_\theta = Y(q_\theta)$ and $(X(q_0), X(q_1))_\theta = X(q_\theta)$ isometrically for any $1 \le q_0, q_1 \le \infty$. Consider the operator T defined by

$$T(x, y) = \left(\frac{x+y}{2}, \frac{x-y}{2}\right).$$

Note that by our assumption in (i) we have $\|T\colon\ Y(q_j) \to X(q_j)\| \le 1$ both for $j = 0$ and $j = 1$. Therefore by the interpolation Theorem

$$\|T\colon\ Y(q_\theta) \to X(q_\theta)\| \le 1.$$

This proves (i). The proof of (ii) is similar (or can be deduced by duality). The last assertion now follows from Theorems 11.37 and 10.1. □

Corollary 11.45. *If a Banach space B is super-reflexive, there are $p > 1$ and $q < \infty$ and a single equivalent norm $|\cdot|$ satisfying both* (10.1) *and* (10.24) *for some constants $\delta, C > 0$.*

Proof. Assume first that B is a *complex* Banach space. Then the complex interpolation method applied between the two norms appearing respectively in (ii) and (ii)′ in Theorem 11.37 produces an interpolated norm (of course still equivalent to the original one) that satisfies the desired property. Indeed, if the first norm, say $\| \ \|_0$, is q-uniformly convex and the second one, say $\| \ \|_1$, is p-uniformly smooth, then, by Proposition 11.44, the interpolated norm $\| \ \|_\theta$, is both q_θ-uniformly convex and p_θ-uniformly smooth, where $q_\theta^{-1} = (1-\theta)/q + \theta/\infty = (1-\theta)/q$ and $p_\theta^{-1} = (1-\theta) + \theta/p$. If B is a real space, its complexification (e.g. $B(H, B)$ with $H = \mathbb{C}$ viewed as a two-dimensional real Hilbert space) inherits the super-reflexivity of B. Indeed, by Proposition 11.39), it is isomorphic to a super-reflexive (real) space (namely the ℓ_2-sense direct sum $B \oplus B$) and hence (see Remark 11.2) it is super-reflexive as a complex space. Therefore the real case reduces to the complex one. □

Problem: If B is both isomorphic to a p-uniformly smooth space and isomorphic to a q-uniformly convex one, is B isomorphic to a space that is both q-uniformly convex and p-uniformly smooth?

Note that the interpolation argument in Corollary 11.45 yields a norm that is both q_θ-uniformly convex and p_θ-uniformly smooth but with ‘worse’ values $q_\theta > q$ and $p_\theta < p$ and in such a way that $q_\theta \to \infty$ when $p_\theta \to p$ (and $p_\theta \to 1$ when $q_\theta \to q$).

Definition 11.46. Let $0 < \theta < 1$. A complex Banach space B is called θ-Hilbertian if there is a compatible pair of complex Banach spaces B_0, B_1 with B_1 isometric to a Hilbert space such that B is isometric to $(B_0, B_1)_\theta$.

With this notion in mind, it is natural to view arbitrary Banach spaces as 0-Hilbertian, while Hilbert spaces are 1-Hilbertian.

The preceding definition was proposed in [388]. A slightly more general notion is discussed in [397] in the framework of interpolation of families of spaces in the sense of [184] and the term ‘strictly θ-Hilbertian’ is used there for the spaces we call θ-Hilbertian.

By Proposition 11.44, θ-Hilbertian implies super-reflexive.
Although it seems unlikely, the converse direction is open: Is any super-reflexive space isomorphic to a θ-Hilbertian space (or merely to a subspace of a quotient of such a space) for some $0 < \theta < 1$?

For Banach lattices the answer is positive by the next corollary. Let B be a Banach lattice on $(\Omega, \mathcal{A}, m)$ in the sense of Remark 8.18. We say that B is p-convex (resp. p-concave) if for any $x, y \in B$ we have

$$\|(|x|^p + |y|^p)^{1/p}\| \le (\|x\|^p + \|y\|^p)^{1/p} \text{ (resp. } \|(|x|^p + |y|^p)^{1/p}\| \ge (\|x\|^p + \|y\|^p)^{1/p}).$$

Theorem 11.47. *Let $0 < \theta < 1$ and p such that $\frac{1}{p} = \frac{1-\theta}{1} + \frac{\theta}{2}$. Let B be a Banach lattice on $(\Omega, \mathcal{A}, m)$. The following are equivalent:*

(i) *There is a Banach lattice B_0 on $(\Omega, \mathcal{A}, m)$ such that $B = (B_0, L_2(m))_\theta$.*
(ii) *B is both p-convex and p'-concave.*

Sketch of proof. To avoid any technicality, we make a radical simplification: We assume that Ω is a finite set. In other words, we restrict consideration to atomic Banach lattices with finitely many atoms, or finite-dimensional spaces with a (finite) unconditional basis. In that situation, (8.23) holds for any pair B_0, B_1 on $(\Omega, \mathcal{A}, m)$, since all spaces are reflexive. This should allow us to make the proof crystal clear.

First, the implication (i) $\Rightarrow$ (ii) can be checked directly using both (8.26) and (8.33). The main point is the converse. To prove it we first verify the following.

Claim: If B is p-convex then there is a Banach lattice X on $(\Omega, \mathcal{A}, m)$ such that $B = X^{1/p} L_\infty(m)^{1/p'}$.

This is immediate: We just set $X = \{x \mid |x|^{1/p} \in B\}$ and $\|x\|_X = \| |x|^{1/p} \|_B^p$.

Now if B is p'-concave, it is easy to check that the latter space X is p'/p-concave, and hence its dual X^* is r-convex with $r = (p'/p)'$, or equivalently $1/r = 1 - p/p'$. Applying the claim to it, and setting $\alpha = 1 - 1/r = p/p'$ we find $X^* = Z^{1-\alpha} L_\infty(m)^\alpha$ for some Z, or equivalently $X = B_0^{1-\alpha} L_1(m)^\alpha$ for some Banach lattice B_0 on $(\Omega, \mathcal{A}, m)$. The proof can then be completed schematically like this, by an easily justified 'associativity' argument:

$$B = X^{1/p} L_\infty(m)^{1/p'} = (B_0^{1-\alpha} L_1(m)^\alpha)^{1/p} L_\infty(m)^{1/p'} = B_0^{\frac{1-\alpha}{p}} L_1(m)^{\frac{1}{p'}} L_\infty(m)^{\frac{1}{p'}}$$

$$= B_0^{1-2/p'} (L_1(m)^{1/2} L_\infty(m)^{1/2})^{2/p'} = B_0^{1-\theta} L_2(m)^\theta.$$

By (8.23) this proves (ii) $\Rightarrow$ (i). □

Corollary 11.48. *A Banach lattice B on $(\Omega, \mathcal{A}, m)$ is super-reflexive iff it is isomorphic to a θ-Hilbertian space.*

Proof. By Proposition 10.39 (or by Theorem 10.45) we know that a super-reflexive space must be of type p_1 and cotype q_1 for some $p_1 > 1$ and $q_1 < \infty$. If B is a Banach lattice (see [57, pp. 100–101] for a detailed discussion and also [57, pp. 54–55]) this implies that, for any $p < p_1$ and any $q > q_1$, B has an equivalent Banach lattice norm that is both p-convex and q-concave. Replacing p by $p \wedge q'$, we conclude that B is isomorphic to a p-convex and p'-concave Banach lattice for some $p > 1$. Then, by the preceding Theorem, B is θ-Hilbertian up to isomorphism. This proves the only if part. By Prop. 11.44

again, the converse implication holds in general, since super-reflexive is stable under isomorphism (see Proposition 11.9). □

Remark. We can mimic Definition 11.46 for other properties. Let $\mathcal{P}$ be a property of complex Banach spaces. We will say that B has property $\mathcal{P}_\theta$ if B is isometric to $(B_0, B_1)_\theta$ with B_1 satisfying $\mathcal{P}$. There are some interesting open questions related to this kind of 'smoothing' of Banach space properties. For instance, if $\mathcal{P}$ is the property of being of type 2 and $1/p = (1-\theta)/1 + \theta/2$, then $\mathcal{P}_\theta$ obviously implies type p, but the converse remains an open question. A positive answer would have important consequences (see [388]).

11.7 Complex analogues of uniform convexity*

Let B be a complex Banach space. In analogy with ARNP and AUMD, it is natural to expect that there are analytic versions of uniform convexity, where the average over the signs ± 1 is replaced by an average over the unit circle $\mathbb{T} = \partial D \subset \mathbb{C}$. With this substitution, convexity is replaced by plurisubharmonicity (in the sense of Definition 4.52) for scalar valued functions on B.

It is easy to see that the definition of the modulus of convexity $\delta_B(\varepsilon)$ given in (10.3) is equal to the following one:

$$\delta_B(\varepsilon) = \inf\{1 - \|x\| \mid \sup_\pm \|x \pm y\| \le 1, \|y\| \ge \varepsilon/2\}.$$

It is not difficult to check that this is equivalent to the following alternate modulus

$$\delta_B^\infty(\varepsilon) = \inf\{\sup_\pm \|x \pm y\| - 1 \mid \|x\| = 1, \|y\| \ge \varepsilon\}.$$

More generally, if $1 < q \le \infty$, let

$$\delta_B^q(\varepsilon) = \inf\left\{\left(\frac{\|x+y\|^q + \|x-y\|^q}{2}\right)^{1/q} - 1 \mid \|x\| = 1, \|y\| \ge \varepsilon\right\}.$$

Again one can show that this is equivalent to δ_B in the sense that B is uniformly convex iff $\delta_B^q(\varepsilon) > 0$ for all $\varepsilon > 0$ and moreover δ_B^q and δ_B are equivalent as Orlicz function when $\varepsilon \to 0$ (i.e. there are positive constants a, b, a', b' such that $a'\delta_B(b'\varepsilon) \le \delta_B^q(\varepsilon) \le a\delta_B(b\varepsilon)$). The proof follows elementary calculations as e.g. [57, Lemma 1.e.10] (see also [223]). Full details appear in [202].

Let $1 \le q \le \infty$. Let B be a Banach (or quasi-Banach) space. As before, let m be the normalized Lebesgue measure on the unit circle $\mathbb{T} = \partial D \subset \mathbb{C}$. We define for any $\varepsilon > 0$

$$H_B^q(\varepsilon) = \inf\{\|x + zy\|_{L_q(dm(z);B)} - 1 \mid \|x\| = 1, \|y\| \ge \varepsilon\},$$

with the usual convention for the case $q = \infty$.

Consideration of $x \pm zy$ immediately leads to

$$\delta_B^q(\varepsilon) \le H_B^q(\varepsilon)$$

for any $\varepsilon > 0$. Thus the property $H_B^q(\varepsilon) > 0$ for any $\varepsilon > 0$ is analogous to but formally weaker than uniform convexity. By [204, 210], it is known that this property does not depend on the value of $1 \le q \le \infty$.

Definition 11.49. A Banach space B is called uniformly PL-convex if for some (or equivalently for all) $1 \le q \le \infty$ we have

$$\forall \varepsilon > 0 \quad H_B^q(\varepsilon) > 0.$$

Remark. We will see in Corollary 11.59 that any L_1-space is uniformly PL-convex. Thus uniform PL-convexity is strictly more general than uniform convexity.

Remark. Since we leave the realm of convexity, the p-normed case ($0 < p < 1$) is also rather natural (with the example of $B = L_p$ in mind). Note that the property $H_B^p(0) \ge 0$ makes sense in this case, and corresponds to the fact that $x \mapsto \|x\|^p$ is PSH. However, we will concentrate on the Banach space case.

In analogy with the convex case:

Definition 11.50. We will say that B is q-uniformly PL-convex if there is a constant $C > 0$ such that $H_B^q(\varepsilon) \ge C\varepsilon^q$ for all $\varepsilon > 0$.

Note that $H_B^q(\varepsilon)$ is non-decreasing, so only small values of ε say $0 < \varepsilon < 1$ are relevant in the preceding definition.
Moreover, if B is q-uniformly PL-convex, then for any unit vector x and any y we have $\|x + zy\|_{L_q(m;B)}^q \ge (1 + C\|y\|^q)^q \ge 1 + qC\|y\|^q$ and hence by homogeneity, setting $C' = qC$ we find

$$\forall x, y \in B \quad \|x + zy\|_{L_q(m;B)}^q \ge \|x\|^q + C'\|y\|^q. \tag{11.48}$$

We will now relate this notion to a martingale inequality. See Definition 4.54 for PSH-martingales.

Definition 11.51. A B-valued martingale $(f_n)_{0\le n<\infty}$ in $L_1(B)$, with respect to a filtration $(\mathcal{A}_n)$, will be called semi-analytic (resp. semi-Hardy) if the odd increments df_{2j+1} are analytic (resp. Hardy) in the following sense: there is a random variable z_{2j+1} independent of $\mathcal{A}_{2j}$ such that $\mathcal{A}_{2j+1} = \sigma(\mathcal{A}_{2j}, z_{2j+1})$ and $df_{2j+1} = z_{2j+1}\varphi_{2j}$ with φ_{2j} being $\mathcal{A}_{2j}$-measurable (resp. df_{2j+1} is an analytic function of z_{2j+1}).

We say that (f_n) is a semi-PSH martingale if, for any even integer $n \geq 0$, the pair (f_n, f_{n+1}) is a PSH martingale.

Consider for example $\Omega = \{-1, 1\}^{\mathbb{N}} \times \mathbb{T}^{\mathbb{N}}$ equipped with the product probability $\mathbb{P} = \nu \times m^{\mathbb{N}}$, and let us denote respectively by (ε_n) and (z_n) the coordinates on Ω coming respectively from $\{-1, 1\}^{\mathbb{N}}$ and $\mathbb{T}^{\mathbb{N}}$. Let $\mathcal{A}_{2j+1} = \sigma(z_1, \varepsilon_1, \cdots, z_j, \varepsilon_j, z_{j+1})$ and $\mathcal{A}_{2j} = \sigma(z_1, \varepsilon_1, \cdots, z_j, \varepsilon_j)$, and let $\mathcal{A}_0$ be trivial. In that case a semi-analytic martingale is one such that

$$df_{2j+1} = z_{2j+1}\varphi_{2j}(z_1, \varepsilon_1, \cdots, z_j, \varepsilon_j).$$

Note in passing that such martingales are calibrated.
We will call these 'basic semi-analytic' martingales.

Lemma 11.52. *If B is q-uniformly PL-convex there is a constant $C' > 0$ such that any B-valued semi-analytic martingale $(f_n)_{0 \leq n < \infty}$ satisfies*

$$\mathbb{E}\|f_0\|^q + C' \sum\nolimits_{j \geq 0} \mathbb{E}\|df_{2j+1}\|^q \leq \sup\nolimits_{n \geq 1} \mathbb{E}\|f_n\|^q.$$

A fortiori, any B-valued analytic martingale $(f_n)_{0 \leq n < \infty}$ satisfies

$$\mathbb{E}\|f_0\|^q + C' \sum\nolimits_{j \geq 0} \mathbb{E}\|df_j\|^q \leq \sup\nolimits_{n \geq 1} \mathbb{E}\|f_n\|^q. \tag{11.49}$$

Proof. By (11.48) with the preceding notation, for each j and each ω, we have

$$\|f_{2j}(\omega)\|^q + C'\|\varphi_{2j}(\omega)\|^q \leq \int \|df_{2j}(\omega) + z_{2j+1}\varphi_{2j}(\omega)\|^q dm(z_{2j+1})$$

and hence after integration over ω

$$\mathbb{E}\|f_{2j}\|^q + C'\mathbb{E}\|df_{2j+1}\|^q \leq \mathbb{E}\|f_{2j+1}\|^q$$

but by convexity (Jensen's inequality) we have $\mathbb{E}\|f_{2j-1}\|^q \leq \mathbb{E}\|f_{2j}\|^q$, and hence we obtain

$$\mathbb{E}\|f_{2j-1}\|^q + C'\mathbb{E}\|df_{2j+1}\|^q \leq \mathbb{E}\|f_{2j+1}\|^q$$

and also $\mathbb{E}\|f_0\|^q + C'\mathbb{E}\|df_1\|^q \leq \mathbb{E}\|f_1\|^q$. Then a telescoping argument leads to the announced inequality.
The second part can be proved by the same argument. One can also view it as a special case of the first one by viewing an analytic martingale as a semi-analytic one such that $df_{2j} = 0$ for all j. □

We now come to the analogue of Corollary 10.7:

Theorem 11.53. *Let $2 \leq q < \infty$. The following properties of a Banach space B are equivalent:*

(i) There is an equivalent norm on B for which B is q-uniformly PL-convex.
(ii) There is a constant $C > 0$ such that any B-valued basic semi-analytic martingale $(f_n)_{0 \le n < \infty}$ satisfies

$$\mathbb{E}\|f_0\|^q + \sum\nolimits_{j \ge 0} \mathbb{E}\|df_{2j+1}\|^q \le C^q \sum\nolimits_{n \ge 1} \mathbb{E}\|f_n\|^q.$$

Proof. (i) $\Rightarrow$ (ii) follows from Lemma 11.52. Conversely, assume (ii). Just like for Theorem 10.6, we set

$$|x|^q = \inf\{\mathbb{E}\|f_\infty\|^q - C^{-q} \sum\nolimits_{j \ge 0} \mathbb{E}\|df_{2j+1}\|^q\}$$

where the inf runs over all finite basic semi-analytic martingales (f_n) such that $f_0 = x$. Clearly $|x|^q \le \|x\|^q \le C^q |x|^q$. We claim that for any $y \in B$ we have

$$|x|^q + \|y\|^q \le \int |x + zy|^q dm(z).$$

Since the argument is similar to the proof of Theorem 10.6, we merely sketch it. We have for any z (roughly) a finite basic semi-analytic martingale (f_n^z) such that $f_0^z = x + zy$ and

$$|x + zy|^q \approx \mathbb{E}\|f_\infty^z\|^q - C^{-q} \sum\nolimits_{j \ge 0} \mathbb{E}\|df_{2j+1}^z\|^q.$$

We then shift the variables appearing inside f_∞^z by two slots so that they become $(z_2, \varepsilon_2, \cdots, z_j, \varepsilon_j, \cdots)$ and we define

$$f(z_1, \varepsilon_1, \cdots, z_j, \varepsilon_j) = f_\infty^{z_1}(z_2, \varepsilon_2, \cdots).$$

Then the martingale $f_n = \mathbb{E}^{\mathcal{A}_n} f$ is basic semi-analytic and satisfies $f_0 = x$, $f_1 = f_2 = x + z_1 y$ and this yields

$$|x|^q \le \int |x + z_1 y|^q dm(z_1) - C^{-q} \|y\|^q.$$

A fortiori, we have

$$\forall x, y \in B \quad |x + zy|^q_{L_q(m;B)} \ge |x|^q + C^{-q} |y|^q,$$

and hence $|\ |$ satisfies (11.48). We leave the rigorous justification of the existence of a suitably measurable selection $z \mapsto f_\infty^z$ to the reader.
To check that it is a norm, we will use the 'room available' in the even increments. Let $x, y \in B$. We can find (roughly) two finite basic semi-analytic martingales (f_n) and (g_n) such that $f_0 = x$, $g_0 = y$ and

$$|x|^q \approx \mathbb{E}\|f_\infty\|^q - C^{-q} \sum\nolimits_{j \ge 0} \mathbb{E}\|df_{2j+1}\|^q,$$

$$|y|^q \approx \mathbb{E}\|g_\infty\|^q - C^{-q} \sum\nolimits_{j \ge 0} \mathbb{E}\|dg_{2j+1}\|^q.$$

We again shift the variables so that both depend on $(z_2, \varepsilon_2, \cdots, z_j, \varepsilon_j, \cdots)$. We then define another finite basic semi-analytic martingale (M_n) by

$$M_0 = M_1 = (x+y)/2, \; M_2 = 1_{\varepsilon_1=1}x + 1_{\varepsilon_1=-1}y$$

and

$$M_\infty = 1_{\varepsilon_1=1} f_\infty(z_2, \varepsilon_2, \cdots) + 1_{\varepsilon_1=-1} g_\infty(z_2, \varepsilon_2, \cdots).$$

This yields

$$|(x+y)/2|^q \le (|x|^q + |y|^q)/2.$$

Thus the set $\{x \mid |x| \le 1\}$ is convex and hence $|\;|$ is a norm on B (as in the end of the proof of Theorem 10.6). □

Corollary 11.54. *The equivalent properties in Theorem 11.53 imply that B is of cotype q.*

Proof. Let $f_n = \sum_1^n z_k x_k$ $(x_k \in B)$. By (ii) we have

$$\left(\sum \|x_k\|^q\right)^{1/q} \le C \left\|\sum z_k x_k\right\|_{L_q(B)}$$

but it is easy to check (by the 'contraction principle' (5.20)) that

$$\left\|\sum z_k x_k\right\|_{L_q(B)} \le 2 \left\|\sum \varepsilon_k x_k\right\|_{L_q(B)},$$

thus by Kahane's inequality (Theorem 5.2) B is of cotype q. □

One could wonder whether conversely finite cotype implies uniform *PL*-convexity. Since, by a result of Bourgain in [131] (see also [141]), $B = L_1/H^1$ is of cotype 2, the next statement provides a counter-example.

Corollary 11.55. *The space $B = L_1(\mathbb{T}; m)/H^1(\mathbb{T}; m)$ is not isomorphic to a uniformly PL-convex space.*

Proof. To verify this we will examine a concrete example of B-valued analytic martingale $(f_k)_{0 \le k \le n}$ that contradicts the property (ii) in Theorem 11.53.

Let $q: L_1 \to B$ denote the quotient map. Consider the following so-called Riesz product defined for any fixed $(z_1, \cdots, z_n) \in \bar{D}^n$ by $F_0(z_1, \cdots, z_n) = 1$ and

$$F_n(z_1, \cdots, z_n) = \prod_{k=1}^n \left(1 + \frac{\bar{z}_k e^{i3^k t} + z_k e^{-i3^k t}}{2}\right) \in L_1(dm(t)),$$

and let

$$f_n(z_1, \cdots, z_n) = q(F_n(z_1, \cdots, z_n)) \in B.$$

A simple verification shows that

$$\|F_n(z_1, \cdots, z_n)\|_{L_1(dm(t))} = \int_{\mathbb{T}} F_n(z_1, \cdots, z_n)dm(t) = 1$$

and hence $\|f_n(z_1, \cdots, z_n)\|_B \leq 1$. Moreover, we have

$$F_k - F_{k-1} = \left(\frac{\bar{z}_k e^{i3^k t} + z_k e^{-i3^k t}}{2}\right) F_{k-1} \in \frac{z_k e^{-i3^k t}}{2} F_{k-1} + H^1(\mathbb{T}; m)$$

and hence for any $1 \leq k \leq n$

$$f_k - f_{k-1} = q(F_k - F_{k-1}) = z_k q(e^{-i3^k t} F_{k-1}/2).$$

This shows that $(f_k)_{0 \leq k \leq n}$ can be viewed as a B-valued analytic martingale, and for any $z = (z_1, \cdots, z_n) \in \mathbb{T}^n$

$$\|(f_k - f_{k-1})(z)\|_B = \inf_{h \in H^1} \|e^{-i3^k t} F_{k-1}/2 + h\|_{L_1} = \inf_{h \in H^1} \|F_{k-1}/2 + e^{i3^k t} h\|_{L_1}$$

and hence

$$\|(f_k - f_{k-1})(z)\|_B \geq 1/2$$

because $\int (F_{k-1}/2 + e^{i3^k t} h)dm(t) = \int (F_{k-1}/2)dm(t) = 1/2$. Therefore we find $\|f_n(z)\|_B \leq 1$ but also $\|(f_k - f_{k-1})(z)\|_B \geq 1/2$ for any $z = (z_1, \ldots, z_n) \in \mathbb{T}^n$. By (11.49), B cannot be isomorphic to a q-uniformly PL-convex space for any $q < \infty$. In fact it is easy to see that this is an obstruction to B being isomorphic to any uniformly PL-convex space. We leave this to the reader. □

One can define similarly the notion of p-uniformly PL-smoothness and prove the analogous renorming statement. However, several natural questions remain open. For instance, it is not clear whether q-uniformly PL-convex spaces have the ARNP (or the super-ARNP). The difficulty lies in the fact that a subsequence of an analytic martingale (f_n) is not necessarily an analytic one. So if B is q-uniformly PL-convex, it follows easily that any B-valued analytic martingale (f_n) satisfies $\|df_n\|_{L_q(B)} \to 0$ but convergence requires that the same holds for all subsequences (f_{n_k}) and this is unclear. A priori a similar problem arises for Hardy martingales, but that case can be resolved by invoking Theorem 4.57. This is probably part of the motivation that led Edgar to introduce the more general class of PSH-martingales. In connection with the super-ARNP, the following is rather natural.

Definition 11.56. We say that B is of PM-cotype q if there is a constant C such that, for all finite B-valued Hardy martingales (f_n), we have

$$\left(\sum \|df_n\|_{L_1(B)}^q\right)^{1/q} \leq C\|f_\infty\|_{L_1(B)}. \tag{11.50}$$

By Theorem 4.57, this remains valid for all PSH martingales. Since the latter are obviously stable by taking subsequences, it is not hard to show the following dichotomy:
Either there is $q < \infty$ such that B is of PM-cotype q
Or for any n and $\varepsilon > 0$ there is a B-valued Hardy martingale $(f_0, f_1, \cdots, f_n)$ with $f_0 = 0$ such that

$$\forall k = 1, \cdots, n \ \ \|df_k\|_{L_1(B)} = 1 \text{ and } \|f_n\|_{L_1(B)} \le 1 + \varepsilon. \qquad (11.51)$$

Indeed, let a_n be the smallest number a such that

$$\inf_{1 \le k \le n} \|df_k\|_{L_1(B)} \le a \|f_n\|_{L_1(B)}$$

for any such martingale. We have then $a_{nk} \le a_n a_k$ for all n, k and hence either $a_n = 1$ for all n, or there is $q < \infty$ such that $a_n = O(n^{-1/q})$. Then the announced dichotomy follows as e.g. in the proof of Theorem 10.9.

If B is of PM-cotype $q < \infty$, then B has the super-ARNP. Indeed, it is clear that any Hardy (or PSH) martingale that is bounded in $L_1(B)$ converges in $L_1(B)$. Thus, by Theorem 4.46, B has the ARNP and since PM-cotype is clearly a super-property, it automatically implies the super-ARNP. We conjecture that conversely the super-ARNP implies that B is of PM-cotype q for some $q < \infty$. Indeed, it seems likely that if B fails the super-ARNP then (11.51) yields a space finitely representable in B but failing the ARNP (as in the proof of Theorem 11.11).

In connection with the notion of PM-cotype, it is natural to introduce a variant of uniform PL-convexity as follows:

Definition 11.57. Given a Banach space B and $\varepsilon > 0$ we define

$$\Delta_B(\varepsilon) = \inf\{\|f\|_{H^1(D;B)} - 1 \mid f(0) = x, \|x\| = 1, \|f - f(0)\|_{H^1(D;B)} \ge \varepsilon\},$$

and we say that B is uniformly PM-convex if $\Delta_B(\varepsilon) > 0$ for any $\varepsilon > 0$.

Remark. Note that

$$\text{uniform convexity} \Rightarrow \text{uniform PM} - \text{convexity} \Rightarrow \text{uniform PL} - \text{convexity and,}$$

$$\forall \varepsilon > 0 \quad \Delta_B(\varepsilon) \le H_B^1(\varepsilon).$$

We claim that if there is $c > 0$ such that $\Delta_B(\varepsilon) \ge c\varepsilon^q$ for any $\varepsilon > 0$, then B is of PM-cotype q. Indeed, arguing as for (11.48) and setting $c' = qc$, it is easy to check that we have for any $f \in H^1(D; B)$

$$\|f\|_{H^1(D;B)} \ge (\|f(0)\|^q + c'\|f - f(0)\|^q_{H^1(D;B)})^{1/q}. \qquad (11.52)$$

Therefore for any B-valued Hardy martingale (f_n) we have for any $n \geq 1$

$$\|f_n\|^q_{H^1(D;B)} \geq \|f_{n-1}\|^q_{H^1(D;B)} + c'\|f_n - f_{n-1}\|^q_{H^1(D;B)} \tag{11.53}$$

and hence, adding all these, the usual telescoping yields

$$\sup_n \|f_n\|_{H^1(D;B)} \geq (\|f_0\|^q_{H^1(D;B)} + c' \sum\nolimits_{n\geq 1} \|f_n - f_{n-1}\|^q_{H^1(D;B)})^{1/q}. \tag{11.54}$$

A fortiori, we find that B is of PM-cotype q.

By §10.3 and by (10.53), we already know that if $p > 1$ any L_p space is of PM-cotype q with $q = \max\{p, 2\}$. We will now extend this to $p = 1$.

Proposition 11.58. *For any $f \in H^1(D)$ we have*

$$\|f\|_{H^1(D)} \geq (|f(0)|^2 + \frac{1}{2}\|f - f(0)\|^2_{H^1(D)})^{1/2}. \tag{11.55}$$

Proof. Assume $\|f\|_{H^1(D)} = 1$. We write $f = gh$ with $\|g\|_{H^2(D)} = \|g\|_{H^2(D)} = 1$. Then since $f(0) = g(0)h(0)$, we have $f - f(0) = g(0)(h - h(0)) + (g - g(0))h$ and also $\|g - g(0)\|_2^2 = \|g\|_2^2 - |g(0)|^2$ and $\|h - h(0)\|_2^2 = \|h\|_2^2 - |h(0)|^2$, therefore

$$\begin{aligned}\|f - f(0)\|_1 &\leq \|g(0)(h - h(0))\|_1 + \|(g - g(0))h\|_1 \\ &\leq |g(0)|\|h - h(0)\|_2 + \|g - g(0)\|_2\|h\|_2\end{aligned}$$

and hence

$$\begin{aligned}\|f - f(0)\|_1^2 &\leq 2(|g(0)|^2(\|h\|_2^2 - |h(0)|^2) + (\|g\|_2^2 - |g(0)|^2)\|h\|_2^2) \\ &= 2(\|g\|_2^2\|h\|_2^2 - |g(0)|^2|h(0)|^2) = 2(1 - |f(0)|^2).\end{aligned}$$

Thus we obtain as announced $1 \geq (|f(0)|^2 + \frac{1}{2}\|f - f(0)\|_1^2)^{1/2}$. □

Corollary 11.59. *If $B = L_1(m')$ (for some measure space (Ω, m')), then the B-valued analogue of* (11.55) *holds, we have $\Delta_B(\varepsilon) \geq \frac{1}{2}\varepsilon^2$ for any $\varepsilon > 0$ and hence B is of PM-cotype 2. A fortiori, B is uniformly PL-convex.*

Proof. This is easy by integrating (11.55) with respect to m', and then PM-cotype 2 follows from (11.54). □

Remark 11.60. In [261] the preceding is extended to the case when B is a non-commutative L_1-space. The crucial tool there is an extension to $H^1(D; B)$ of the factorization of any function f in the unit ball of H^1 as a product $f = gh$ of two functions g, h in the unit ball of H^2. For instance, when B is the trace class S_1, this holds for any f in the unit ball of $H^1(D; B)$ with g, h in the unit ball of $H^2(D; S_2)$. Once the extended factorization is established, the proof of Proposition 11.58 works identically.

By Remark 4.38, if B satisfies (11.50) then (recalling Lemma 4.48) we have for any $f \in H^1(D; B)$ and any sequence $0 = r(0) < r(1) < \cdots < r(n) < \cdots < 1$

$$\left(\sum \|f_{r(n)} - f_{r(n-1)}\|^q_{L_1(m;B)}\right)^{1/q} \le C\|f\|_{H^1(D;B)}.$$

Similarly, if (11.52) (and hence (11.53)) holds, then for any $0 < r < 1$ we have

$$\|f\|_{H^1(D;B)} \ge \left(\|f_r\|^q_{L_1(m;B)} + c'\|f - f_r\|^q_{H^1(D;B)}\right)^{1/q}.$$

In the particular case of Banach lattices, there is a satisfactory isomorphic characterization of uniform PL-convexity as follows:

Theorem 11.61. *Let B be a Banach lattice. The following are equivalent:*

(i) B is isomorphic to a uniformly PL-convex space.
(ii) c_0 is not finitely representable in B (equivalently by Theorem 10.45 there is $q < \infty$ such that B is of cotype q).
(iii) There is $q < \infty$ such that B is isomorphic to a q-uniformly PL-convex space.
(iv) B has the super-ARNP.

Proof. Assume (i). Note that uniform PL-convexity is obviously a super-property. Since there is a space (namely the space $\ell_\infty(\mathbb{C})$ or just the two-dimensional complex ℓ_∞-space) failing it, Theorem 11.5 guarantees that (i) $\Rightarrow$ (ii). By a well-known renorming (see [54, Prop. 1.d.8, p. 54]) if (ii) holds, B is isomorphic to a Banach lattice that is q-concave with constant 1. By the next remark this implies (iii). Assume (iii). Then (ii) holds by Corollary 11.54, and we know from Corollary 4.34 that (ii) implies the ARNP. Since any ultrapower of B is again a Banach lattice not containing c_0 isomorphically, Corollary 4.34 implies that any ultrapower of B has the ARNP. Since the ARNP passes to subspaces, Proposition 11.8 shows that (iv) holds. This establishes (iii) $\Rightarrow$ (iv). Conversely, since c_0 fails the ARNP, (iv) $\Rightarrow$ (ii) and we already mentioned that (ii) $\Rightarrow$ (iii). Since (iii) $\Rightarrow$ (i) is trivial this completes the proof. □

Remark 11.62. If a Banach lattice is q-concave with constant 1, i.e. we have

$$(\|x\| + \|y\|^q)^{1/q} \le \|(|x| + |y|^q)^{1/q}\| \tag{11.56}$$

for any $x, y \in B$, then B is q-uniformly PL-convex. Indeed, by convexity of the norm we have for any fixed $\delta > 0$

$$\left\| \int |x + zy| dm(z) \right\| \le \int \|x + zy\| dm(z)$$

and by the q-uniform PL-convexity of $\mathbb{C}$ we have for some $\delta > 0$

$$(|x| + \delta|y|^q)^{1/q} \le \int |x + zy| dm(z).$$

Therefore by q-concavity, we have

$$(\|x\| + \delta\|y\|^q)^{1/q} \le \|(|x| + \delta|y|^q)^{1/q}\| \le \| \int |x + zy| dm(z)\| \le \int \|x + zy\| dm(z),$$

and hence B is q-uniformly PL-convex.
In [204] this is proved assuming only that (11.56) holds for disjointly supported pairs x, y.

We now derive more examples.

Proposition 11.63. *Let $Y \subset L_1(\mu)$ be a reflexive subspace. Then the quotient space $L_1(\mu)/Y$ is AUMD, of PM-cotype 2 and a fortiori has the super-ARNP. Moreover, it is isomorphic to a 2-uniformly PL convex space.*

Proof. This is an immediate consequence of the lifting property in Corollary 6.43 together with the AUMD property and the PM-cotype 2 of $L_1(\mu)$. □

In sharp contrast we recall that L_1/H^1 fails the ARNP and is not isomorphic to a uniformly PL convex space (see Remark 4.35 and Corollary 11.55).

11.8 Appendix: ultrafilters, ultraproducts

Let I be a 'directed set'. By this we mean that I is a partially ordered set such that for any i, j in I there exists k in I such that $k \ge i$ and $k \ge j$.

If $(x_i)_{i \in I}$ is a family in a metric space, we view $(x_i)_{i \in I}$ as a 'generalized sequence' so that $x_i \to x$ means that $\forall \varepsilon > 0 \, \exists j$ such that $\forall i \ge j \; d(x_i, x) < \varepsilon$.

Definition. Consider a linear form $\mathcal{U} \in \ell_\infty(I)^*$ that is also a $*$-homomorphism

$$(\text{i.e.} \quad \forall x, y \in \ell_\infty(I) \qquad \mathcal{U}(xy) = \mathcal{U}(x)\mathcal{U}(y) \quad \text{and} \quad \mathcal{U}(\bar{x}) = \overline{\mathcal{U}(x)}).$$

We will say that $\mathcal{U}$ is an ultrafilter adapted to I if for any (x_i) in $\ell_\infty(I)$ such that $x_i \to x$ we have $\mathcal{U}((x_i)) = x$.

Remark 11.64. The existence of ultrafilters adapted to I is easy to check: let $\delta_i \in \ell_\infty(I)^*$ be the evaluation homomorphism defined by $\delta_i(x) = x_i$. Let F_j be the pointwise closure of the set $\{\delta_i \mid i \ge j\}$. Since I is a directed index set, the intersection of finitely many of the F_js is non-empty. Thus, by the weak-$*$ compactness of the unit ball of $\ell_\infty(I)^*$, the intersection of the whole family of sets $\{F_j\}$ is non-void and it is formed of ultrafilters in the preceding sense.

We will denote by convention

$$\lim_{\mathcal{U}} x_i = \mathcal{U}((x_i)_{i\in I}).$$

Given a family of Banach spaces $(B_i)_{i\in I}$, let $\mathcal{B} = \left(\oplus \sum_{i\in I} B_i\right)_\infty$, i.e. $\mathcal{B}$ is formed of families $b = (b_i)_{i\in I}$ with $b_i \in B_i$ for all i such that $\|b\|_{\mathcal{B}} = \sup_{i\in I} \|b_i\| < \infty$. For any b in $\mathcal{B}$ we set

$$p_{\mathcal{U}}(b) = \lim_{\mathcal{U}} \|b_i\|_{B_i}.$$

Then $p_{\mathcal{U}}$ is a semi-norm on $\mathcal{B}$.

The ultraproduct $\prod_{i\in I} B_i/\mathcal{U}$ is defined as the Banach space quotient $\mathcal{B}/\ker(p_{\mathcal{U}})$. Fix an element x in $\prod_{i\in I} B_i/\mathcal{U}$. It is important to observe that for any representative $(b_i)_{i\in I}$ of the equivalence class of x modulo $\ker(p_{\mathcal{U}})$ we have

$$\|x\|_{\prod_{i\in I} B_i/\mathcal{U}} = \lim_{\mathcal{U}} \|b_i\|_{B_i}. \tag{11.57}$$

We will denote by $\dot{b}$ the element of $\prod_{i\in I} B_i/\mathcal{U}$ determined by $b = (b_i)_{i\in I}$ so we can rewrite (11.57) as $\|\dot{b}\| = \lim_{\mathcal{U}} \|b_i\|_{B_i}$. Another useful observation is that if for some j we have $b_i = b'_i\ \forall i \geq j$ then $\dot{b} = \dot{b}'$. Indeed, this implies $\|b_i - b'_i\| \to 0$ (relative to the directed set I) and hence $\lim_{\mathcal{U}} \|b_i - b'_i\| = 0$.

Remark. Let K be a compact subset of a locally convex space L. Let $(y_i)_{i\in I}$ be a family of elements of K. Clearly there is a unique y in K such that for any linear form $\xi \in L^*$ we have $\xi(y) = \lim_{\mathcal{U}} \xi(y_i)$. In that case also we will denote $y = \lim_{\mathcal{U}} y_i$.

When $B_i = B$ for all $i \in I$, we say that $\prod B_i/\mathcal{U}$ is an ultrapower and we denote it by $B^I/\mathcal{U}$.

The following elementary lemma will be useful

Lemma 11.65. *Let E, Y be Banach spaces. let S be an ε-net in the unit sphere of E and let $u\colon E \to Y$ be a linear operator such that*

$$\forall s \in S \qquad 1 - \delta \leq \|u(s)\| \leq 1 + \delta.$$

Then

$$\forall x \in E \qquad \left(\frac{1-\delta-2\varepsilon}{1-\varepsilon}\right)\|x\| \leq \|u(x)\| \leq \left(\frac{1+\delta}{1-\varepsilon}\right)\|x\|$$

Proof. Assume $\dim(E) < \infty$ (this is the only case we will use). Consider $x \in E$ with $\|x\| = 1$ and $\|u\| = \|ux\|$. Choose $s \in S$ such that $\|x - s\| \leq \varepsilon$.

Then $\|u\| = \|ux\| \le \|us\| + \|u(x-s)\| \le (1+\delta) + \varepsilon\|u\|$ and hence $\|u\| \le (1+\delta)(1-\varepsilon)^{-1}$. In the converse direction, if $\|x\| = 1$ we have

$$\|ux\| \ge \|us\| - \varepsilon\|u\| \ge 1 - \delta - \varepsilon(1+\delta)(1-\varepsilon)^{-1} = (1-\delta-2\varepsilon)(1-\varepsilon)^{-1}.$$

The argument can be easily adapted to the infinite-dimensional case. □

Lemma 11.66. *Assume that each space in the family $(B_i)_{i\in I}$ is f.r. in a Banach space B. Then the ultraproduct $\prod B_i/\mathcal{U}$ is f.r. in B. In particular, any ultrapower $B^I/\mathcal{U}$ of B is f.r. in B.*

Proof. Let $E \subset \prod B_i/\mathcal{U}$ be a finite-dimensional subspace. Note that since its unit sphere is compact it admits a *finite* ε-net S. Let $(\dot{e}_1, \ldots, \dot{e}_n)$ be a linear basis of E with representatives $(e_1(i))_{i\in I}, \ldots, (e_n(i))_{i\in I}$. Any $x \in E$ can be uniquely written as $x = \sum_1^n \alpha_j \dot{e}_j$ ($\alpha_j \in \mathbb{K}$). We define $u_i\colon\ E \to B_i$ by setting $u_i(x) = \sum_1^n \alpha_j e_j(i)$ for each i in I. Note that $\forall x \in E$, $\overbrace{(u_i(x))_{i\in I}}^{\cdot} = x$. Therefore by (11.57) we have

$$\forall x \in E \qquad \lim_{\mathcal{U}} \|u_i(x)\| = \|x\|.$$

Fix $\delta > 0$. Since S is finite there is j such that

$$\forall i \ge j\ \forall s \in S \qquad 1 - \delta < \|u_i(s)\| < 1 + \delta$$

and hence by Lemma 11.65 we have

$$\forall x \in E \qquad (1-\delta-2\varepsilon)(1-\varepsilon)^{-1}\|x\| \le \|u_i(x)\| \le (1+\delta)(1-\varepsilon)^{-1}\|x\|.$$

Thus we conclude that E is $(1 + f(\varepsilon,\delta))$-isometric to $u_i(E) \subset B_i$ for some function $(\varepsilon,\delta) \mapsto f(\varepsilon,\delta)$ tending to 0 when ε and δ tend to 0. Since each B_i is f.r. in B we conclude that $\prod B_i/\mathcal{U}$ is f.r. in B. □

11.9 Notes and remarks

The notion of 'finitely representable' and 'super-property' are due to R. C. James [278]. The local reflexivity principle (Theorem 11.3) goes back to Lindenstrauss and Rosenthal [332]. As we mentioned in the text, Ptak's paper [404] seems to be the earliest reference for Theorem 11.10 but it was independently proved by D. Milman and V. Milman and by James. The reformulations in terms of factorizations such as (v) in Theorem 11.10 were emphasized in Lindenstrauss and Pełczyński's influential paper [331].

Theorem 11.11 was stated in [387]. Theorems 11.17 and 11.22 are due to R. C. James as well as Corollary 11.23 and essentially all the results in §11.3. James first proved in [275] that uniformly non-square implies reflexive. In the same paper, he notes that the extension from pairs to triples of vectors leads to a proof that if $n = 3$ for any $\varepsilon > 0$ any J-(n, ε) convex space is reflexive. Later on in [424], the authors observe that the same proof works for any integer $n \geq 2$, thus showing that J-convex implies reflexive. Since J-convex is a super-property, this shows that J-convex implies super-reflexive. But the converse was an easy consequence of James early ideas on reflexivity. Therefore this yielded the equivalence of 'J-convex' and 'super-reflexive'. In the mean time, in [278], having observed the implications (isomorphic to uniformly non-square) $\Rightarrow$ super-reflexive and (isomorphic to uniformly convex) $\Rightarrow$ super-reflexive, James asked whether the converses hold. In his remarkable paper [220], Enflo proved that indeed the converses are true. In Theorem 11.37, this corresponds to the equivalence of (i), (iii), (iii)$'$ and (iv), which all come from [220]. The equivalence with (ii) and (ii)$'$ (i.e. the existence of moduli of power type) was proved later in [387].

We follow [387] throughout §11.4. The strong law of large numbers for super-reflexive spaces given in Theorem 11.41 (essentially from [387]) is modelled on Beck's strong law of large numbers ([114]) for B-convex Banach spaces, that is restricted to martingales with independent increments. See e.g. [437] for more on this theme.

Concerning θ-Hilbertian spaces, we refer to [388, 397]. For a different framework, see Kalton's work [299, 300].

Concerning §11.7 on complex uniform convexity, the earliest reference seems to be Globevnik's paper [254], where the sup over the signs ± 1 is replaced by one over $\{\pm 1, \pm i\}$ or over the whole unit disc. The complex uniform convexity of L_1 is proved there (and earlier unpublished work by L. A. Harris is also mentioned in a note added to the paper). See [374, 375] for complements to Globevnik's [254]. The main references around PL-convexity or PSH-martingales are due to Davis, Garling and Tomczak-Jaegerman [204, 233], Edgar [217, 218], Bourgain and Davis [141]. Theorem 11.53 is essentially from [204] but was simplified in [141]. Corollary 11.55 is due to the author (see [204]). Concerning Banach lattices, the equivalence of (i), (ii) and (iii) (and implicitly (iv)) in Theorem 11.61 comes from [204]. The work of Shangquan Bu and Walter Schachermayer [162] is also relevant here. In [210] Dilworth compares several notions of modulus of PL-convexity. Haagerup was the first who noticed the importance of the (sharp) inequality

$$\forall x, y \in \mathbb{C} \quad \left(|x|^2 + \frac{|y|^2}{2}\right)^{1/2} \leq \int |x + zy| dm(z),$$

and he proved a non-commutative analogue (see [204]): any non-commutative L_1-space (i.e. the predual of a von Neumann algebra) is 2-uniformly PL-convex. This was later on extended in [261].

While writing §11.7, I benefitted from the notes [50] written by Omran Kouba and Antonio Pallarès from a seminar organized with B. Maurey back in 1986–1987.

See [211] for connections between complex geodesics in the unit ball (for the Carathéodory distance) and complex uniform convexity.

See [127] for alternative proofs of the known facts that the Lebesgue spaces L_p and the Schatten classes S_p are q-uniformly PL-convex for $q = \max\{2, p\}$ using the distributional Laplacian of $u(z) = \|x + zy\|^p$.

See [200] for an application of complex uniform convexity to spectral theory.

12

Interpolation between strong p-variation spaces

Let

$$0 < \theta < 1, \quad p = (1-\theta)^{-1} \quad 1 \le q \le \infty. \tag{12.1}$$

In this chapter we will focus on the real interpolation space defined by

$$\mathcal{W}_{p,q} = (v_1, \ell_\infty)_{\theta,q}, \tag{12.2}$$

where v_1 denotes as before the space of sequences with bounded variation. In words, we are interested in interpolating between the properties bounded variation and boundedness.

In Chapter 9, we already considered this to establish several martingale inequalities involving the space v_p of scalar sequences with finite 'strong p-variation', but we mainly made use of the space $\mathcal{W}_p = \mathcal{W}_{p,p}$.

Remark 12.1. Clearly

$$v_1 \subset \mathcal{W}_{p,q} \subset \ell_\infty,$$

and hence $\mathcal{W}_{p,q}$ is *non-reflexive*. Indeed, by Theorem 11.10 any intermediate Banach space between v_1 and ℓ_∞ is necessarily *non-reflexive*. Actually, the argument for (iv) $\Rightarrow$ (i) in Theorem 11.10 shows that the inclusion map $v_1 \to \ell_\infty$ is not weakly compact.

If we replace v_1 by ℓ_1 in (12.2) we obtain the Lorentz space $\ell_{p,q}$ and in particular if $q = p$ we find the space ℓ_p. We will show that although (12.2) is non-reflexive it behaves in many ways like the spaces ℓ_{pq}, and like ℓ_p when $q = p$. In particular, $\mathcal{W}_p$ satisfies 'almost' the same type and cotype and Hölder-Minkowski inequalities as ℓ_p (see Corollary 12.19 and (12.37)).

12.1 The spaces $v_p(B)$, $\mathcal{W}_p(B)$ and $\mathcal{W}_{p,q}(B)$

By $\ell_{p,q}$ we mean the space $L_{p,q}(\Omega, m)$ when (Ω, m) is $\mathbb{N}$ equipped with the counting measure $m = \sum \delta_n$. Equivalently, we have $\ell_{p,q} = (\ell_1, \ell_\infty)_{\theta,q}$. More generally, for any Banach space B, we denote $\ell_{p,q}(B) = L_{p,q}(\Omega, m; B)$ with $(\Omega, m) = (\mathbb{N}, \sum \delta_n)$, or equivalently

$$\ell_{p,q}(B) = (\ell_1(B), \ell_\infty(B))_{\theta,q}, \tag{12.3}$$

with θ, p, q as in (12.1). Let $x = (x_n)$ be a sequence in $\ell_{p,q}(B)$, and let $(x_n^\dagger)$ be a sequence obtained by rearranging (by a permutation of the integers) the sequence $(\|x_n\|)$ in non-increasing order. Then, if $q < \infty$ we let

$$\|x\|_{\ell_{p,q}(B)} = \left(\sum (n^{1/p} x_n^\dagger)^q\right)^{1/q}, \quad \text{and} \quad \|x\|_{\ell_{p,\infty}(B)} = \sup n^{1/p} x_n^\dagger. \tag{12.4}$$

Unless otherwise specified, we equip $\ell_{p,q}(B)$ with this quasi-norm.

Note that

$$\|x\|_{\ell_{p,p}(B)} = \|x\|_{\ell_p(B)}.$$

A simple elementary verification shows that this quasi-norm is equivalent (with constants depending on p, q) to the one defined earlier in (8.60) and (8.57). Moreover, by Corollary 8.51, it is equivalent to the norm of the space appearing in (12.3).

For any auxiliary Banach space B we define, again with $p = (1 - \theta)^{-1}$

$$\mathcal{W}_{p,q}(B) = (v_1(B), \ell_\infty(B))_{\theta,q}, \tag{12.5}$$

$$\mathcal{W}_p(B) = \mathcal{W}_{p,p}(B) = (v_1(B), \ell_\infty(B))_{\theta,p}. \tag{12.6}$$

Let $x = (x_n)$ be a sequence of elements of B. Let

$$V_{p,q}(x) = \sup\{\|(x_0, x_{n(1)} - x_0, \ldots, x_{n(k)} - x_{n(k-1)}, \ldots)\|_{\ell_{p,q}(B)}\}$$

where the supremum runs over all increasing sequence $0 = n(0) < n(1) < n(2) < \cdots$ of integers. Let

$$v_{p,q}(B) = \{x = (x_n) \mid V_{p,q}(x) < \infty\},$$

equipped with the norm defined by

$$\|x\|_{v_{p,q}(B)} = V_{p,q}(x).$$

A simple application of interpolation yields:

Lemma 12.2. *We have $\mathcal{W}_{p,q}(B) \subset v_{p,q}(B)$ and there is a constant $K(p, q)$ such that*

$$\forall x \in \mathcal{W}_{p,q}(B) \quad \|x\|_{v_{p,q}(B)} \le K(p, q)\|x\|_{\mathcal{W}_{p,q}(B)}. \tag{12.7}$$

Proof. The proof is the same as for Lemma 9.1: We just apply the fundamental interpolation property (Theorem 8.46) to the operator T defined by $T(x) = (x_0, x_{n(1)} - x_0, \ldots, x_{n(k)} - x_{n(k-1)}, \ldots)$. This is clearly bounded simultaneously from $v_1(B)$ to $\ell_1(B)$ (with norm ≤ 1) and from $\ell_\infty(B)$ to $\ell_\infty(B)$ (with norm ≤ 2), and hence from $\mathcal{W}_{p,q}(B) = (v_1(B), \ell_\infty(B))_{\theta,q}$ to $\ell_{p,q}(B) = (\ell_1(B), \ell_\infty(B))_{\theta,q}$ (with norm $\le 2^{1-1/p}$). Then we use the equivalence of (12.3) and (12.4). □

By general interpolation theory (see (8.52) and (8.53)), for all $1 < p < r$ and arbitrary $1 \le q_0, q_1 \le \infty$ we have bounded inclusions

$$\mathcal{W}_{p,q_0}(B) \subset \mathcal{W}_{r,q_1}(B). \tag{12.8}$$

This also holds in case $p = r$, but then only if $q_1 \ge q_0$.

We denote as usual by c_0 (resp. $c_0(B)$) the subspace of ℓ_∞ (resp. $\ell_\infty(B)$) formed of all sequences that tend to zero. Similarly we will denote by v_1^0 (resp. $v_1^0(B)$) the subspace of v_1 (resp. $v_1(B)$) formed of all sequences that tend to zero. Recall that $\mathbb{K}$ denotes the scalars i.e. $\mathbb{K} = \mathbb{R}$ or $\mathbb{K} = \mathbb{C}$. Note that, by subtracting its limit from a sequence in v_1 or in $v_1(B)$ we find

$$v_1 \simeq \mathbb{K} \oplus v_1^0 \quad \text{and} \quad v_1(B) \simeq B \oplus v_1^0(B). \tag{12.9}$$

The pair (v_1, ℓ_∞) has a self-dual character that will be crucial in the sequel. Let us describe this duality. For $x \in v_1$, $y \in \ell_\infty$ we set

$$\langle x, y\rangle = x_0 y_0 + \sum\nolimits_1^\infty (x_n - x_{n-1}) y_n. \tag{12.10}$$

Note that $|\langle x, y\rangle| \le \|x\|_{v_1} \|y\|_\infty$. Moreover, with this duality we have

$$(v_1^0)^* \simeq \ell_\infty \quad \text{and} \quad (c_0)^* = v_1.$$

More generally, we have

$$v_1^0(B)^* \simeq \ell_\infty(B^*) \quad \text{and} \quad c_0(B)^* = v_1(B^*) \tag{12.11}$$

with respect to the duality defined either for $x \in v_1(B^*)$ and $y \in c_0(B)$, or for $x \in \ell_\infty(B^*)$ and $y \in v_1^0(B)$, by

$$\langle x, y\rangle = \langle x_0, y_0\rangle + \sum\nolimits_1^\infty \langle x_n - x_{n-1}, y_n\rangle \tag{12.12}$$

$$= \lim_{n\to\infty} (\langle x_0, y_0 - y_1\rangle + \cdots + \langle x_{n-1}, y_{n-1} - y_n\rangle + \langle x_n, y_n\rangle) \tag{12.13}$$

$$= \sum\nolimits_1^\infty \langle x_{n-1}, y_{n-1} - y_n\rangle. \tag{12.14}$$

More precisely, we have

$$2^{-1}\|x\|_{\ell_\infty(B^*)} \le \|x\|_{(v_1^0(B))^*} \le \|x\|_{\ell_\infty(B^*)} \text{ and } \|x\|_{(c_0(B))^*} = \|x\|_{v_1(B^*)}. \tag{12.15}$$

Remark. For any sequence $x = (x_n)$, let $\hat{x}$ be the shifted sequence defined by $\hat{x}_0 = 0$ and $\hat{x}_n = x_{n-1}$ for all $n \geq 1$. If $x = (x_n)$ and $y = (y_n)$ are both finitely supported, then, taking $B = \mathbb{K}$ for simplicity, Abel summation (or integration by parts) shows $\langle x, y\rangle = -\langle y, \hat{x}\rangle$. A similar identity holds if y (resp. x) is a B- (resp. B^*)-valued sequence (but this requires exchanging the roles of B and B^*).

We will now introduce preduals $u^0_{p'}$ and u^0_p respectively for the spaces v_p or $v_{p'}$ when (we keep this notation throughout)

$$1 < p, p' < \infty \quad \text{and} \quad \frac{1}{p} + \frac{1}{p'} = 1.$$

Note that, by the Cauchy criterion, $V_p(x) < \infty$ implies that x_n converges to a limit $x_\infty \in B$ when $n \to \infty$.

Let

$$v^0_p(B) = v_p(B) \cap c_0(B) \quad \text{and} \quad v^0_p = v_p \cap c_0.$$

Note that

$$\forall x \in v_p(B) \qquad (x_n - x_\infty)_{n\geq 0} \in v^0_p(B). \tag{12.16}$$

Let $B^{(\mathbb{N})}$ denote the space of finitely supported functions $b = (b(n))_{n\in\mathbb{N}}$ with $b(n) \in B$ for all n. It is easy to see that $B^{(\mathbb{N})}$ is dense in $v^0_p(B)$ for any $1 \leq p < \infty$ (see the proof of Lemma 12.6).

For any $b = (b(n)) \in B^{(\mathbb{N})}$ there is a finite partition of $\mathbb{N}$ into disjoint intervals $I_0, I_1, \ldots, I_N$ with $I_0 = [0, n(0)]$, $I_1 = (n(0), n(1)], \ldots, I_N = (n(N-1), n(N)]$ and there are $\xi_0, \ldots, \xi_N \in B$ such that

$$\forall n \in \mathbb{N} \qquad b(n) = \sum\nolimits_0^N \xi_k 1_{I_k}(n). \tag{12.17}$$

We require that $\xi_N \neq 0$ and $\xi_k \neq \xi_{k+1}$ for all $0 \leq k < N$ and we set

$$[b]_{p'} = \left(\sum\nolimits_0^N \|\xi_k\|^{p'}\right)^{1/p'}.$$

Note that the preceding requirement minimizes $\sum \|\xi_k\|^{p'}$.

For any $x \in (B^*)^{\mathbb{N}}$ we have

$$\langle x, b\rangle = \langle x_{n(0)}, \xi_0\rangle + \sum\nolimits_1^N \langle x_{n(k)} - x_{n(k-1)}, \xi_k\rangle$$

and hence we have

$$\sup\{|\langle x, b\rangle| \mid b \in B^{(\mathbb{N})},\ [b]_{p'} \leq 1\} = \widetilde{V}_p(x) \tag{12.18}$$

where

$$\widetilde{V}_p(x) = \sup_{0\leq n(0)<n(1)<\cdots} \{(\|x_{n(0)}\|^p + \|x_{n(1)} - x_{n(0)}\|^p + \cdots)^{1/p}\}.$$

Note that

$$V_p(x) \le \widetilde{V}_p(x) \le 2^{1/p'} V_p(x). \tag{12.19}$$

Note also that if $I_0 = [0, n(0)]$ we have

$$\langle x, \xi_0 1_{I_0} \rangle = \langle x_{n(0)}, \xi_0 \rangle. \tag{12.20}$$

We then set for any $b \in B^{(\mathbb{N})}$

$$\|b\|_{u^0_{p'}(B)} = \inf \left\{ \sum\nolimits_1^m [b_j]_{p'} \right\}$$

where the infimum runs over all decompositions $b = \sum_1^m b_j$ ($b_j \in B^{(\mathbb{N})}$). In other words, $\|\cdot\|_{u^0_{p'}(B)}$ is the gauge of the convex hull of $\{b \mid [b]_{p'} \le 1\}$. Then we define the Banach space $u^0_{p'}(B)$ as the completion of $B^{(\mathbb{N})}$ equipped with this norm. Note that $\sup \|b(n)\| \le [b]_{p'}$ and hence $\lim_{n\to\infty} b(n) = 0$ for any b in $u^0_{p'}(B)$ so that we have a bounded inclusion

$$u^0_{p'}(B) \subset c_0(B).$$

In fact, $b \in u^0_{p'}(B)$ iff b can be written as $b = \sum_1^\infty b_j$ with $b_j \in B^{(\mathbb{N})}$ such that $\sum_1^\infty [b_j]_{p'} < \infty$. Moreover (the inf being over all such decompositions):

$$\|b\|_{u^0_{p'}(B)} = \inf \sum\nolimits_1^\infty [b_j]_{p'}.$$

When $B = \mathbb{K}$, we denote simply

$$u^0_{p'} = u^0_{p'}(\mathbb{K}).$$

From (12.18) it is immediate that, with respect to the duality (12.12) we have

$$v_p(B^*) = u^0_{p'}(B)^* \tag{12.21}$$

with equivalent norms. In particular, $v_p = (u^0_{p'})^*$.

More explicitly, any $x \in v_p(B^*)$ defines a linear form f_x on $u^0_{p'}(B)$ by setting $f_x(b) = \langle x, b \rangle$ for any $b \in B^{(\mathbb{N})}$. By (12.18) and (12.19), the latter form admits a unique bounded extension to an element of $u^0_{p'}(B)^*$ satisfying

$$\|x\|_{v_p(B^*)} = V_p(x) \le \|f_x\|_{u^0_{p'}(B)^*} \le 2^{1/p'} \|x\|_{v_p(B^*)}.$$

Conversely, to any linear form $f \in u^0_{p'}(B)^*$, we associate the sequence $(x_n) \in B^{*\mathbb{N}}$ defined by $f(\xi 1_{[0,n]}) = x_n(\xi)$ (for all $\xi \in B$, $n \ge 0$). Then, recalling (12.20), we have $f(b) = f_x(b)$ for any $b \in B^{(\mathbb{N})}$ and (12.18) again shows that $x \in v_p(B^*)$ and of course $f = f_x$. Thus we conclude that the correspondence $x \mapsto f_x$ is a surjective isomorphism from $v_p(B^*)$ onto $u^0_{p'}(B)^*$. In this way, we avoid discussing the possible types of convergence of the series (12.12).

Remark 12.3. By an abuse of notation we will continue to denote by $\langle x, y\rangle$ the duality just established for $x \in v_p(B^*)$ and $y \in u^0_{p'}(B)$. (We adopt that notation also for $x \in u^0_{p'}(B^*)$ and $y \in v^0_p(B)$.) Note however, that this is really defined only when y is finitely supported and extended by density and continuity to the whole of $u^0_{p'}(B)$.

Thus there is a constant $C = C(p)$ such that for all x in $v_p(B^*)$ we have

$$\frac{1}{C}\|x\|_{v_p(B^*)} \le \sup\{|\langle x, y\rangle| \mid y \in u^0_{p'}(B),\ \|y\|_{u^0_{p'}(B)} \le 1\} \le C\|x\|_{v_p(B^*)}. \tag{12.22}$$

Moreover, since $u^0_{p'}(B)^*$ norms $u^0_{p'}(B)$, there is a constant $C' = C'(p)$ such that for any y in $u^0_{p'}(B)$ we have

$$\frac{1}{C'}\|y\|_{u^0_{p'}(B)} \le \sup\{|\langle x, y\rangle| \mid x \in v^0_p(B^*),\ V_p(x) \le 1\} \le C'\|y\|_{u^0_{p'}(B)}. \tag{12.23}$$

Indeed, the last equivalence is clear if we replace $v^0_p(B^*)$ by $v_p(B^*) = u^0_{p'}(B)^*$, but if y is supported say in $[0, \ldots, N]$ then for any x in $v_p(B^*)$

$$\langle x, y\rangle = \langle P_N(x), y\rangle$$

where $P_N(x) = (x_0, x_1, \ldots, x_N, 0 \ldots)$ and we have obviously

$$V_p(P_N(x)) \le 2V_p(x). \tag{12.24}$$

From this (12.23) follows easily for all y in $u^0_{p'}(B)$.

Let us denote $g_y(x) = \langle x, y\rangle$. Then (12.23) can be rewritten

$$\frac{1}{C'}\|y\|_{u^0_{p'}(B)} \le \|g_y\|_{v^0_p(B^*)^*} \le C'\|y\|_{u^0_{p'}(B)}. \tag{12.25}$$

Remark 12.4. Using the $\ell_{p,q}$ norm in place of the ℓ_p norm in the definition of $[\,.\,]_{p'}$, we can define analogously the space $u^0_{p,q}(B)$ and if $1 \le p, q < \infty$ the same argument leads to $u^0_{p,q}(B)^* = v_{p',q'}(B^*)$.

12.2 Duality and quasi-reflexivity

By Theorem 11.10 we already know that v^0_p (and a fortiori v_p) is non-reflexive, but it is much less obvious that, if $1 < p < \infty$, it is quasi-reflexive, i.e. of finite codimension in its bidual. This phenomenon was discovered by James. For that reason, the space v^0_2 is usually denoted by J and called the James space. In fact we have

Theorem 12.5. *Let $1 < p < \infty$ with $\frac{1}{p} + \frac{1}{p'} = 1$. By Remark 12.3, the duality* (12.10) *(or* (12.12) *in the B-valued case) is well defined for $x \in v_p$, $y \in u^0_{p'}$, and*

also for $x \in u^0_{p'}$, $y \in v^0_p$. With respect to that duality, we have

$$(v^0_p)^* = u^0_{p'} \quad \text{and} \quad (u^0_{p'})^* = v_p$$

with equivalent norms. More explicitly, the mapping $y \mapsto g_y$ (resp. $x \mapsto f_x$) extends to an isomorphism from $u^0_{p'}$ to $(v^0_p)^*$ (resp. from v_p to $(u^0_{p'})^*$). In particular if X is either $u^0_{p'}$, v^0_p or v_p we have $\dim(X^{**}/X) = 1$. *More generally, if* $\dim(B) = n$ *then if $X = v_p(B)$ or if $X = u^0_p(B)$ we have* $\dim(X^{**}/X) = n$.

Lemma 12.6. *The canonical basis (e_n) is a basis of v^0_p satisfying an upper p-estimate in the sense of Remark 11.18. In particular, it is a shrinking basis of v^0_p $(1 < p < \infty)$.*

Proof. By (12.24) we already know that (e_n) is a basic sequence in v^0_p. Let $P_N x = (x_0, x_1, \dots, x_N, 0, \dots)$. Assume $V_p(x) < \infty$. We will show that $V_p(x - P_N x) \to 0$ when $N \to \infty$ for any $x \in v^0_p$. Choose $0 = n(0) < n(1) < \cdots < n(K)$ such that

$$V_p(x)^p - \varepsilon < |x_0|^p + |x_{n(1)} - x_0|^p + \cdots + |x_{n(K)} - x_{n(K-1)}|^p.$$

We have then for any $n(K) < n(K+1) < \cdots$

$$\sum\nolimits_{j>K} |x_{n(j)} - x_{n(j-1)}|^p < \varepsilon$$

and hence if $N = n(K)$

$$V_p(x - P_N x)^p \le \sup_{j \ge N} |x_j|^p + \varepsilon,$$

and since we may assume that $N = n(K)$ is as large as we wish we conclude that $V_p(x - P_N x) \to 0$ when $N \to \infty$ for any $x \in v^0_p$.

Note that by (12.19) we have

$$\sup |x_j| \le 2^{1/p'} V_p(x)$$

and hence

$$V_p(P_N x) \le (V_p(x)^p + \sup |x_j|^p)^{1/p} \le 3V_p(x).$$

Thus we conclude that (e_n) is a basis of v^0_p. Note that by (12.19) the norm in v_p is equivalent to

$$\max\left\{\sup |x_j|, \sup_{n(0)<n(1)<\cdots} \left(\sum |x_{n(k)} - x_{n(k-1)}|^p\right)^{1/p}\right\}.$$

From this it is easy to see that there is a constant C so that for any sum of disjoint consecutive blocks $b_1, \dots, b_N$ on (e_n) we have

$$\|b_1 + \cdots + b_N\| \le C(\|b_1\|^p + \cdots + \|b_N\|^p)^{1/p}.$$

By Remark 11.18 the basis (e_n) must be shrinking. □

Proof of Theorem 12.5. In the duality (12.10) (recall (12.20)), the vectors $\sigma_n = \sum_0^n e_k$ are biorthogonal to e_n, i.e. we have $\langle e_k, \sigma_n \rangle = 0$ for all $k \neq n$ and $= 1$ if $k = n$. By Lemma 12.6, (e_n) is a shrinking basis for v_p^0. Therefore (σ_n) is a basis for $(v_p^0)^*$. Note that $\mathrm{span}(\sigma_0, \dots, \sigma_n) = \mathrm{span}(e_0, \dots, e_n)$. By (12.25) we find $(v_p^0)^* = u_{p'}^0$. We already know by (12.21) that $v_p = (u_{p'}^0)^*$. Thus if $X = v_p^0$ we have $X^{**} = v_p$ and hence by (12.16) $\dim(X^{**}/X) = 1$. If $X = v_p$ (resp. $X = u_{p'}^0$) then $X \simeq v_p^0 \oplus \mathbb{K}$ (resp. $X^* \simeq v_p \simeq v_p^0 \oplus \mathbb{K}$) and hence $X^{**} \simeq (v_p^0)^{**} \oplus \mathbb{K} \simeq X \oplus \mathbb{K}$ (resp. $X^{**} \simeq (v_p^0)^* \oplus \mathbb{K} \simeq X \oplus \mathbb{K}$). The other assertions are proved similarly. □

We will now identify the dual of $\mathcal{W}_{p,q}$. Let $\mathcal{W}_{p,q}^0 = (v_1^0, c_0)_{\theta,q}$ with $1 - \theta = 1/p$. Note that

$$\mathcal{W}_{p,q}^0 = \mathcal{W}_{p,q} \cap c_0. \tag{12.26}$$

Indeed, since (see (12.9))

$$v_1 \simeq v_1^0 \oplus \mathbb{K} \quad \text{and} \quad c \simeq c_0 \oplus \mathbb{K},$$

we obviously have

$$\mathcal{W}_{p,q} \simeq (v_1^0, c_0)_{\theta,q} \oplus \mathbb{K} \tag{12.27}$$

where the second coordinate is $x \mapsto \lim x_n$. Therefore (12.26) follows immediately.

By general interpolation (see Remark 8.41) $v_1^0 \cap c_0$ is dense in $\mathcal{W}_{p,q}^0 = (v_1^0, c_0)_{\theta,q}$ $(0 < \theta < 1, 1 \le q < \infty)$, from which it is easy to see that finitely supported sequences form a dense subspace of $\mathcal{W}_{p,q}^0$. Thus by (12.27) $\mathcal{W}_{p,q}^0$ can be identified with the closure in $\mathcal{W}_{p,q}$ of the space of finitely supported sequences.

Theorem 12.7. *Let (e_n) denote the canonical basis of $\mathbb{K}^{(\mathbb{N})}$, let (e_n^*) be the biorthogonal functionals and let $\sigma_n = \sum_0^n e_j$. Let $1 < p < \infty$ and $1 \le q < \infty$. Then (e_n) and (σ_n) each form a basis in $\mathcal{W}_{p,q}^0$. If moreover $q > 1$, (e_n^*) is a basis of $(\mathcal{W}_{p,q}^0)^*$. The linear mapping T defined on $\mathrm{span}[\sigma_n]$ by $T\sigma_n = e_n^*$ $(n \ge 0)$ extends to an isomorphism from $\mathcal{W}_{p',q'}^0$ onto $(\mathcal{W}_{p,q}^0)^*$. In particular, $\mathcal{W}_{p'}^0$ is isomorphic to $(\mathcal{W}_p^0)^*$.*

Proof. Each of (e_n) and (σ_n) is a basis for both spaces c_0 and v_1^0. By interpolation applied to the partial sum operators, it follows that each is also a basis in $\mathcal{W}^0_{p,q}$ for any $1 < p < \infty$, $1 \le q < \infty$. Recall the notion of upper p-estimate from Remark 11.18. Obviously, (e_n) satisfies an upper r-estimate in v_1^0 for $r = 1$, but it also satisfies one in c_0 for any r (or say for $r = \infty$). Therefore, by an interpolation argument based on Theorem 8.63, it follows that (e_n) satisfies an upper r-estimate in $\mathcal{W}^0_{p,q}$ for $1 < r < \min(p, q)$. It follows (see Remark 11.18) that (e_n) is shrinking in $\mathcal{W}^0_{p,q}$. Equivalently this means that (e_n^*) is a basis in $(\mathcal{W}^0_{p,q})^*$. Define $T\colon \operatorname{span}[\sigma_n] \to (\mathcal{W}^0_{p,q})^*$ by $T(\sigma_n) = e_n^*$. By Corollary 8.84 there is a constant C (independent of n) such that for any n and any x in $\operatorname{span}(\sigma_0, \ldots, \sigma_n)$ we have

$$C^{-1}\|x\|_{\mathcal{W}^0_{p',q'}} \le \|T(x)\|_{(\mathcal{W}^0_{p,q})^*} \le C\|x\|_{\mathcal{W}^0_{p',q'}},$$

but since $\operatorname{span}(\sigma_n)$ and $\operatorname{span}(e_n^*)$ are dense respectively in $\mathcal{W}^0_{p',q'}$ and $(\mathcal{W}^0_{p,q})^*$, T extends to an isomorphism from $\mathcal{W}^0_{p',q'}$ to $(\mathcal{W}^0_{p,q})^*$. □

Remark 12.8. One can check, arguing as for Theorem 12.5, that $\dim(X^{**}/X) = 1$ when X is any of the spaces $\mathcal{W}^0_{p,q}$ or $\mathcal{W}_{p,q}$ with $1 < p < \infty$ and $1 < q < \infty$.

Remark 12.9. Moreover, the B-valued analogue of Theorem 12.7 also holds with the obvious adjustments: the dual space $(\mathcal{W}^0_{p,q}(B))^*$ is isomorphic to $\mathcal{W}^0_{p',q'}(B^*)$.

12.3 The intermediate spaces $u_p(B)$ and $v_p(B)$

We will denote by $c(B) \subset \ell_\infty(B)$ the subspace formed of all convergent sequences, equipped with the norm induced by $\ell_\infty(B)$.

Note that for any $0 < \theta < 1$, $1 \le q \le \infty$

$$(v_1(B), \ell_\infty(B))_{\theta,q} = (v_1(B), c(B))_{\theta,q}, \tag{12.28}$$

with identical norms. To check this, we first observe that

$$(v_1(B), \ell_\infty(B))_{\theta,q} \subset c(B).$$

Indeed, for any $x \in (v_1(B), \ell_\infty(B))_{\theta,q}$ we have $\lim_{t\to\infty} K_t(x, v_1(B), \ell_\infty(B)) = 0$. Therefore x lies in the closure of $v_1(B)$ in $\ell_\infty(B)$ which, since $v_1(B) \subset c(B)$, is included in $c(B)$. Moreover, again since $v_1(B) \subset c(B)$, we have obviously

$$K_t(x; v_1(B), \ell_\infty(B)) = K_t(x; v_1(B), c(B))$$

for any $x \in c(B)$, and (12.28) becomes clear.

So far, we only defined $u_p^0(B)$. We now turn to $u_p(B)$.

Definition. For $b = (b(n)) \in c(B)$, let $b(\infty) = \lim b(n) \in B$. We denote by $u_p(B)$ the subspace of $c(B)$ formed of all $b = (b(n))$ such that

$$(b(n) - b(\infty))_{n\in\mathbb{N}} \in u_p^0(B).$$

We equip $u_p(B)$ with the norm

$$\|b\|_{u_p(B)} = \|b(\infty)\| + \|(b(n) - b(\infty))\|_{u_p^0(B)}.$$

Remark 12.10. Thus $u_p(B) \simeq B \oplus u_p^0(B)$. In the *same* decomposition we have *'simultaneously'* $v_1(B) \simeq B \oplus v_1^0(B)$ and $c(B) \simeq B \oplus c_0(B)$. Therefore, by Remark 8.48, we must have also

$$(v_1(B), c(B))_{\theta,q} \simeq B \oplus (v_1^0(B), c_0(B))_{\theta,q}$$

for any $0 < \theta < 1$ and $1 \le q \le \infty$.

The following result will be crucial in the sequel ([118, 399]).

Lemma 12.11. *Let* $0 < \theta < 1$, $p = (1-\theta)^{-1}$ *and let* B *be an arbitrary Banach space. We have bounded inclusions*

$$(v_1(B), \ell_\infty(B))_{\theta,1} \subset u_p(B) \subset v_p(B) \subset (v_1(B), \ell_\infty(B))_{\theta,\infty} \qquad (12.29)$$

$$(v_1^0(B), c_0(B))_{\theta,1} \subset u_p^0(B) \subset v_p^0(B) \subset (v_1^0(B), c_0(B))_{\theta,\infty}. \qquad (12.30)$$

For any $1 < p < s < \infty$ *we have bounded inclusions*

$$\mathcal{W}_p(B) \subset v_p(B) \subset \mathcal{W}_s(B). \qquad (12.31)$$

Proof. By Remark 12.10, it suffices to prove (12.30). The second inclusion in (12.30), namely $u_p^0(B) \subset v_p^0(B)$ is clear from the definition of $u_p^0(B)$, since for any b as in (12.17) we have obviously $V_p(b) \le 2[b]_p$. Let us show $v_p(B) \subset \mathcal{W}_{\theta,\infty}(B) = (v_1(B), \ell_\infty(B))_{\theta,\infty}$. Let $x \in B^{\mathbb{N}}$ with $V_p(x) \le 1$. Fix $t > 1$. Then let $n(1) = \inf\{n > 0 \mid \|x_n - x_0\| \ge t^{-(1-\theta)}\}$, and let $n(2) < n(3) < \cdots$ be defined similarly by $n(k) = \inf\{n > n(k-1) \mid \|x_n - x_{n(k-1)}\| > t^{-(1-\theta)}\}$. Whenever the preceding infimum runs over the void set we set $n(k) = \infty$ and we stop the process. Since $V_p(x) \le 1$, the process has to stop at a certain stage k (so that $n(k) < \infty$ but $n(k+1) = \infty$). We have then on one hand

$$t^{-(1-\theta)}k^{1/p} \le (\|x_{n(1)} - x_0\|^p + \cdots + \|x_{n(k)} - x_{n(k-1)}\|^p)^{1/p} \le 1$$

and hence $k \le t$. But on the other hand we can decompose x as $x = x^0 + x^1$ with $x' = x - x^0$ and x^0 defined by

$$x_n^0 = x_{n(j)} \quad \text{if} \quad n(j) \le n < n(j+1)$$

where we set by convention $n(0) = 0$ and $n(k+1) = \infty$. By definition of $n(0) < n(1) < \cdots$ we have $\|x^1\|_{\ell_\infty(B)} \le t^{-(1-\theta)}$ and also (recall $k \le t$)

$$\begin{aligned}\|x^0\|_{v_1(B)} &\le \|x_0\| + \|x_{n(1)} - x_0\| + \cdots + \|x_{n(k)} - x_{n(k-1)}\| \\ &\le (k+1)^{1/p'} V_p(x) \le 2k^{1/p'} \le 2t^{1/p'} = 2t^\theta,\end{aligned}$$

so we find $K_t(x; v_1(B), \ell_\infty(B)) \le \|x^0\|_{v_1(B)} + t\|x^1\|_{\ell_\infty(B)} \le 3t^\theta$. Thus we conclude

$$\|x\|_{\mathcal{W}_{\theta,\infty}(B)} \le 3\|x\|_{v_p(B)}.$$

Note that if $x \in c_0(B) \cap v_p(B)$, after subtracting their respective limits, we find x^0 and x^1 in $c_0(B)$ also, so the same argument gives us

$$\|x\|_{(v_1^0(B), c_0(B))_{\theta,\infty}} \le 6\|x\|_{v_p^0(B)}.$$

That yields the third inclusion in (12.30). It only remains to prove the first inclusion in (12.30). But by duality the latter is equivalent to

$$u_p^0(B)^* \subset (v_1^0(B), c_0(B))^*_{\theta,1}$$

and by the duality for real interpolation spaces (Theorem 8.67) and by (12.15) this boils down to

$$u_p^0(B)^* \subset (\ell_\infty(B^*), v_1(B^*))_{\theta,\infty} = (v_1(B^*), \ell_\infty(B^*))_{1-\theta,\infty}.$$

Equivalently, since $u_p^0(B)^* = v_{p'}(B^*)$ this reduces to

$$v_{p'}(B^*) \subset (v_1(B^*), \ell_\infty(B^*))_{1-\theta,\infty}$$

and this is but the third inclusion in (12.29) with $B^*, p', 1-\theta$ in place of B, p, θ, which we have established, since it is equivalent to the third inclusion in (12.30).

The last assertion follows from (12.7) and the general fact (see (8.53)) that for an interpolation pair (B_0, B_1) with $B_0 \subset B_1$, for any $0 < \theta < \beta < 1$ we have a bounded inclusion $(B_0, B_1)_{\theta,p} \subset (B_0, B_1)_{\beta,s}$. □

Lemma 12.12. *Let $1 < r < p < s < \infty$. $0 < \alpha, \beta < 1$ be determined by the equalities $\frac{1}{p} = \frac{1-\alpha}{r} + \frac{\alpha}{\infty}$ and $\frac{1}{p} = \frac{1-\beta}{1} + \frac{\beta}{s}$. Then*

$$\mathcal{W}_p(B) = (v_r(B), \ell_\infty(B))_{\alpha,p} \tag{12.32}$$

$$\mathcal{W}_p(B) = (v_1(B), u_s(B))_{\beta,p} \tag{12.33}$$

with equivalent norms. More generally, for any $1 \le q \le \infty$, we have

$$\mathcal{W}_{p,q}(B) = (v_r(B), \ell_\infty(B))_{\alpha,q} \text{ and } \mathcal{W}_{p,q}(B) = (v_1(B), u_s(B))_{\beta,q}.$$

Proof. The key is to use 'reiteration'. By Lemma 12.11 the reiteration Theorem 8.72 implies (12.32) and (12.33). □

The space $v_{p,\infty}(B)$ corresponds to the sequences with variation in weak-ℓ_p. It corresponds to an 'intersection' between the scales $v_{p,q}$ and $\mathcal{W}_{p,q}$ as formulated in the following lemma:

Lemma 12.13. *For any $1 < p < \infty$ and any B we have*

$$v_{p,\infty}(B) = \mathcal{W}_{p,\infty}(B)$$

with equivalent norms.

Proof. By (12.7) we already observed $\mathcal{W}_{p,\infty}(B) \subset v_{p,\infty}(B)$. Conversely, the proof of (12.29) actually shows $v_{p,\infty}(B) \subset (v_1(B), \ell_\infty(B))_{\theta,\infty}$ if $1-\theta = \frac{1}{p}$. □

Let $0 < p < \infty, 0 < q \le \infty$. Let us denote by $\delta_{p,q}(B)$ the space of sequences $x = (x_n) \in B^{\mathbb{N}}$ such that the sequence $y = (y_n)$, defined by $y_0 = x_0$ and $y_n = x_n - x_{n-1}$ for all $n \ge 1$, is in $\ell_{p,q}(B)$. We equip it with the quasi-norm

$$\|x\|_{\delta_{p,q}(B)} = \|y\|_{\ell_{p,q}(B)}.$$

When $0 < r < 1$, the spaces v_r behave slightly surprisingly with respect to interpolation, as the next statement shows.

Theorem 12.14. *Let B be any Banach space. Fix $0 < r < 1$. Let $0 < \theta < 1$, $1 \le q \le \infty$. Let p be determined by $\frac{1}{p} = \frac{1-\theta}{r}$. Then $(v_r(B), \ell_\infty(B))_{\theta,q}$ can be described as follows:*

(i) If $r < p < 1$ (i.e. $0 < \theta < 1 - r$) we have

$$(v_r(B), \ell_\infty(B))_{\theta,q} = \delta_{p,q}(B)$$

with equivalent norms.

(ii) If $1 < p < \infty$ (i.e. $1 - r < \theta < 1$) we have

$$(v_r(B), \ell_\infty(B))_{\theta,q} = \mathcal{W}_{p,q}(B)$$

with equivalent norms.

Proof. Let $X(\alpha, q) = (v_r(B), \ell_\infty(B))_{\alpha,q}$. Since the operator T taking (x_n) to $(x_0, x_1 - x_0, \ldots, x_n - x_{n-1}, \ldots)$ is bounded simultaneously from $v_r(B)$ to $\ell_r(B)$ and from $\ell_\infty(B)$ to itself, it is also bounded (by Theorem 8.46) from $X(\alpha, 1)$ to $(\ell_r(B), \ell_\infty(B))_{\alpha,1} = \ell_1(B)$. Therefore $X(\alpha, 1) \subset v_1(B)$. Then, by the same argument as for Lemma 12.11 we obtain

$$X(\alpha, 1) \subset v_1(B) \subset X(\alpha, \infty)$$

Therefore, by the reiteration Theorem 8.72 (extended to the quasi-normed case see [6, p. 67]) for any $0 < \gamma < 1$, $0 < \delta < 1$ and $1 \le q \le \infty$, we have

$$(v_r(B), v_1(B))_{\gamma,q} = (v_r(B), \ell_\infty(B))_{\theta,q} \tag{12.34}$$

where $\theta = \gamma\alpha$, and

$$(v_1(B), \ell_\infty(B))_{\delta,q} = (v_r(B), \ell_\infty(B))_{\theta,q} \tag{12.35}$$

where $\theta = (1-\delta)\alpha + \delta \doteq \alpha + \delta(1-\alpha)$. As before, define $y = (y_n)$ by $y_0 = x_0$ and $y_n = x_n - x_{n-1}$ for all $n \ge 1$. Since $\|x\|_{v_r(B)} = \|y\|_{\delta_r(B)}$ for any $0 < r \le 1$, we may identify $v_r(B)$ with $\delta_r(B)$, or equivalently with $\ell_r(B)$, so that by Theorem 8.63 we have

$$(v_r(B), v_1(B))_{\gamma,q} = \delta_{p,q}(B).$$

Thus (12.34) implies (i). By definition of $\mathcal{W}_{p,q}(B)$, (12.35) is the same as (ii). □

12.4 L_q-spaces with values in v_p and $\mathcal{W}_p$

In this section, we will show that the spaces $\mathcal{W}_p$ satisfy an analogue of the Hölder-Minkowski inequality (see the appendix of §5.13 in Chapter 5). The latter refers to the fact that, assuming $1 \le r \le p \le \infty$, for any measure spaces (Ω_1, m_1), (Ω_2, m_2) we have a norm 1 inclusion

$$L_r(m_1; L_p(m_2)) \subset L_p(m_2; L_r(m_1)). \tag{12.36}$$

Note that the reverse inclusion only holds when $p \le r$, and when $p = r$, Fubini's theorem gives us isometrically

$$L_p(m_1; L_p(m_2)) \simeq L_p(m_2; L_p(m_1)) \simeq L_p(m_1 \times m_2).$$

Although this is very special to (and in some sense characteristic of) L_p-spaces, it turns out that the space $\mathcal{W}_p$ satisfies an analogous property: If $r < p$ we have a bounded inclusion $L_r(\mathcal{W}_p) \subset \mathcal{W}_p(L_r)$, while if $p < r$ we have the reverse $\mathcal{W}_p(L_r) \subset L_r(\mathcal{W}_p)$. There is however (necessarily) a singularity when $r = p$ that reflects the non-reflexivity of $\mathcal{W}_p$.

Theorem 12.15. *Let (Ω, m) be any measure space and B any Banach space. For simplicity, we set $L_p(B) = L_p(\Omega, m; B)$. Let $1 < p < \infty$. For any $r < p < s$ we have the following bounded natural inclusions:*

$$L_r(\mathcal{W}_p(B)) \subset \mathcal{W}_p(L_r(B)) \tag{12.37}$$

$$\mathcal{W}_p(L_s(B)) \subset L_s(\mathcal{W}_p(B)). \tag{12.38}$$

Proof. We first observe that this can be easily reduced to the case of an atomic measure space with finitely many atoms, and this allows us to ignore all measurability considerations since we may as well assume $L_r = \ell_r^n$ and $L_s = \ell_s^n$. Let $0 < \alpha, \beta < 1$ be as in Lemma 12.12. Now observe that the following inclusions both hold with norm ≤ 1

$$L_r(v_r(B)) \subset v_r(L_r(B)) \quad \text{and} \quad L_r(\ell_\infty(B)) \subset \ell_\infty(L_r(B)).$$

Therefore by interpolation we have

$$(L_r(v_r(B)), L_r(\ell_\infty(B)))_{\alpha,p} \subset (v_r(L_r(B)), \ell_\infty(L_r(B)))_{\alpha,p}$$

but by (12.32) the last space coincides with $\mathcal{W}_p(L_r(B))$ and by Remark 8.65 since $p > r$ we have

$$L_r(\mathcal{W}_p(B)) \subset (L_r(v_r(B)), L_r(\ell_r(B)))_{\alpha,p}$$

and hence (12.37) follows.

The proof of (12.38) is entirely similar but with the inclusions reversed. By the duality between $v_p(B^*)$ and $u_{p'}^0(B)$ we have $u_s^0(L_s(B)) \subset L_s(u_s^0(B))$, or equivalently (see Remark 12.10)

$$u_s(L_s(B)) \subset L_s(u_s(B)),$$

and obviously also

$$v_1(L_s(B)) \subset L_s(v_1(B)).$$

Therefore $(v_1(L_s(B)), u_s(L_s(B)))_{\beta,p} \subset (L_s(v_1(B)), L_s(u_s(B)))_{\beta,p}$ but, since $s > p$, by Remark 8.65 again $(L_s(v_1(B)), L_s(u_s(B)))_{\beta,p}) \subset L_s((v_1(B), u_s(B))_{\beta,p})$, and hence

$$(v_1(L_s(B)), u_s(L_s(B)))_{\beta,p} \subset L_s((v_1(B), u_s(B))_{\beta,p}) = L_s(\mathcal{W}_p(B)),$$

and by (12.33) we obtain (12.38). □

Remark. By Remark 12.9, it is easy to see that (12.37) and (12.38) are actually equivalent by duality.

The next corollary shows how to apply our study of the spaces $\mathcal{W}_{p,q}$ to the more classical spaces v_p. The main point is the fact that the two scales are intertwined in the form expressed by Lemma 12.11.

Corollary 12.16. *In the situation of Theorem 12.15, let $1 < p_0 < p_1 < \infty$. Then for any r, s such that $1 < r < p_0 < p_1 < s < \infty$ we have the following bounded inclusions:*

$$v_{p_0}(L_s(B)) \subset L_s(v_{p_1}(B))$$
$$L_r(v_{p_0}(B)) \subset v_{p_1}(L_r(B)).$$

Proof. Pick p such that $p_0 < p < p_1$. We have (by (12.8) and Lemma 12.11) bounded inclusions $v_{p_0}(B) \subset \mathcal{W}_p(B) \subset v_1(B)$. Moreover, this holds for any B (and hence also with $L_r(B)$ or $L_s(B)$ in place of B). Therefore the result follows immediately from Theorem 12.15. □

Lemma 12.17. *Let $T\colon B_1 \to B_2$ be a bounded operator between Banach spaces. Then, for any $1 < p < \infty$, T extends 'naturally' to a bounded operator $\widetilde{T}\colon \mathcal{W}_p(B_1) \to \mathcal{W}_p(B_2)$ taking $x = (x_n)_{n\geq 0} \in \mathcal{W}_p(B_1)$ to $(Tx_n)_{n\geq 0}$, and moreover $\|\widetilde{T}\| = \|T\|$.*

Proof. This is a direct application of the fundamental interpolation principle (cf. Theorem 8.46): indeed we have clearly $\|\widetilde{T}\colon v_1(B_1) \to v_1(B_2)\| \leq \|T\|$ and $\|\widetilde{T}\colon \ell_\infty(B_1) \to \ell_\infty(B_2)\| \leq \|T\|$, therefore $\|\widetilde{T}\colon \mathcal{W}_p(B_1) \to \mathcal{W}_p(B_2)\| \leq \|T\|$. The converse is obvious by considering the action of $\widetilde{T}$ on sequences (x_n) such that $x_0 \in B_1$ and $x_n = 0$ for all $n > 0$. □

Corollary 12.18. *Let (Ω_1, m_1), (Ω_2, m_2) be two measure spaces. With the notation of Theorem 12.15, assuming $1 \leq r < p < s \leq \infty$, any bounded linear operator $T\colon L_r(m_1) \to L_s(m_2)$ extends to a bounded operator*

$$\widetilde{T}\colon L_r(m_1; \mathcal{W}_p) \to L_s(m_2; \mathcal{W}_p)$$

such that $T(f \otimes x) = T(f) \otimes x$ $(f \in L_r(m_1), x \in \mathcal{W}_p)$.

Proof. By Theorem 12.15, it suffices to show that $\widetilde{T}$ is bounded from $\mathcal{W}_p(L_r)$ to $\mathcal{W}_p(L_s)$, and this follows from the preceding lemma. □

Corollary 12.19. *Let $1 < r < p < s < \infty$. Then $\mathcal{W}_p$ is of type $r \wedge 2$ and of cotype $s \vee 2$.*

Proof. Let (Δ, ν) and ε_n be as defined in (5.13). Consider the operator

$$T\colon \ell^{r\wedge 2} \to L^{s\vee 2}(\nu)$$

defined by $T((\alpha_n)) = \sum \alpha_n \varepsilon_n$. By the Khintchin inequalities (cf. (5.7)), T is bounded, and hence so is $\widetilde{T}$ by the preceding corollary, and that means $\mathcal{W}_p$ is of type $r \wedge 2$. We may argue similarly with the operator $T\colon L_{r\wedge 2}(\nu) \to \ell_{s\vee 2}$ defined by $T(f) = (\int f\varepsilon_n \, d\nu)_{n\geq 0}$ and this shows that $\mathcal{W}_p$ is of cotype $s \vee 2$. □

Note that, by the classical Kwapień theorem [315], a Banach space is of type 2 and cotype 2 iff it is isomorphic to a Hilbert space. In particular, type 2 and cotype 2 forces reflexivity. However, we now can state:

Corollary 12.20. *For any $\varepsilon > 0$, there are non-reflexive Banach spaces of type 2 and of cotype $2 + \varepsilon$.*

Recall that the Banach-Mazur $d(E, F)$ between two (isomorphic) Banach spaces is defined by

$$d(E, F) = \inf\{\|u\| \|u^{-1}\|\}$$

where the infimum runs over all possible isomorphisms $u\colon\ E \to F$.

Remark 12.21. The space $\mathcal{W}_2$ has several remarkable properties reminiscent of Hilbert space: it is isomorphic to its dual and moreover there is a constant C such that any n-dimensional subspace $(n > 1)$ $E \subset \mathcal{W}_2$ satisfies $d(E, \ell_2^n) \leq C \operatorname{Log} n$. This logarithmic growth is sharp. Indeed $\mathcal{W}_2$ is of course non-reflexive (see Remark 12.1) but any Banach space X for which the function $f(n) = \sup\{d(E, \ell_2^n) \mid E \subset X\}$ is $o(\operatorname{Log}(n))$ must be reflexive! Thus, it is as close to ℓ_2 as can be among non-reflexive spaces! See [399] for details.

12.5 Some applications

We wish to illustrate here the usefulness of results such as Theorem 12.15 in analysis and probability theory.

We will give just a sample result about Gaussian variables; by this we mean a random variable with a Gaussian distribution with mean zero and a certain variance $\sigma \geq 0$. Consider a Gaussian process $(X_n)_{n\geq 0}$, i.e. a sequence of real valued random variables such that all the variables in $\operatorname{span}[X_n]$ (in particular the X_n's themselves) are Gaussian. Let γ_p be the L_p-norm of a standard Gaussian variable with variance 1. Then for any $n \leq m$, we have

$$\forall\, 0 < p < \infty \qquad \|X_m - X_n\|_p = \gamma_p \|X_m - X_n\|_2. \tag{12.39}$$

In the next result we show how our study of $\mathcal{W}_p$ allows to derive information on v_p. Part (ii) is due to Jain and Monrad [273].

Theorem 12.22. *Fix $1 < p < \infty$ and consider a Gaussian process (X_n) on a probability space $(\Omega, \mathcal{A}, \mathbb{P})$. Let $L_2 = L_2(\Omega, \mathcal{A}, \mathbb{P})$.*

(i) The random sequence $X(\omega) = \{X_n(\omega) \mid n \geq 0\}$ belongs a.s. to $\mathcal{W}_p$ iff $n \to X_n$ belongs to $\mathcal{W}_p(L_2)$.

(ii) *For any $r < p$, the condition $X \in v_r(L_2)$ is sufficient for $V_p(X) < \infty$ a.s. and the condition $X \in v_p(L_2)$ is necessary.*

Proof. By Fernique's well-known integrability theorem for norms of Gaussian processes (see e.g. [54]), we know that $X \in \mathcal{W}_p$ a.s. (resp. $X \in v_p$ a.s.) iff $\mathbb{E}\|X\|_{\mathcal{W}_p}^r < \infty$ (resp. $\mathbb{E}\|X\|_{v_p}^r < \infty$) for any $1 < r < \infty$.

We then choose $r < p < s$ and note that for a Gaussian process, (12.39) shows that

$$(\gamma_r)^{-1}\|X\|_{\mathcal{W}_p(L_r)} = \|X\|_{\mathcal{W}_p(L_2)} = (\gamma_s)^{-1}\|X\|_{\mathcal{W}_p(L_s)}.$$

Then Theorem 12.15 shows that

$$X \in L_r(\mathcal{W}_p) \Rightarrow X \in \mathcal{W}_p(L_2) \Rightarrow X \in L_s(\mathcal{W}_p)$$

but by Fernique's theorem $X \in L_r(\mathcal{W}_p) \Leftrightarrow X \in L_s(\mathcal{W}_p) \Leftrightarrow X \in \mathcal{W}_p$ a.s., so we obtain (i).

Then (ii) is immediate: indeed we have $\mathcal{W}_p \subset v_p$ and, for any $r < p$, $v_r(L_2) \subset \mathcal{W}_p(L_2)$ (see (12.31)), so the sufficiency follows. By convexity, we have $L_1(v_p) \subset v_p(L_1)$ and hence

$$\|X\|_{v_p(L_2)} = \gamma_1^{-1}\|X\|_{v_p(L_1)} \leq \gamma_1^{-1}\|X\|_{L_1(v_p)},$$

which takes care of the necessity part. □

The reader surely has observed that, except for Fernique's result, we only use the Gaussian character of (X_n) via the condition (12.39). Thus, for instance, we have more generally:

Theorem 12.23. *Consider any sequence of random variables $(X_n)_{n\geq 0}$ such that $X_0 = 0$, let $1 < p < s < \infty$ and assume there is a constant C such that*

$$\forall n \leq m \qquad \|X_n - X_m\|_s \leq C\|X_n - X_m\|_p. \tag{12.40}$$

Then, $X \in \mathcal{W}_p(L_p) \Leftrightarrow X \in L_p(\mathcal{W}_p)$, and the corresponding norms are equivalent.

Proof. Indeed, by our assumption, we have $X \in \mathcal{W}_p(L_p) \Rightarrow X \in \mathcal{W}_p(L_s)$, and by Theorem 12.15

$$\mathcal{W}_p(L_s) \subset L_s(\mathcal{W}_p) \subset L_p(\mathcal{W}_p).$$

For the converse, we claim that (12.40) implies that, for any $r < p$, there is a constant C' such that

$$\forall n \leq m \qquad \|X_n - X_m\|_p \leq C'\|X_n - X_m\|_r.$$

Indeed, by Hölder's inequality, since $r < p < s$, for some $0 < \theta < 1$ we have $\|X_n - X_m\|_s \le C\|X_n - X_m\|_p \le C\|X_n - X_m\|_r^{\theta}\|X_n - X_m\|_s^{1-\theta}$ and hence $\|X_n - X_m\|_p \le \|X_n - X_m\|_s \le C^{1/\theta}\|X_n - X_m\|_r$. This proves the claim with $C' = C^{1/\theta}$. The latter shows $X \in \mathcal{W}_p(L_r) \Rightarrow X \in \mathcal{W}_p(L_p)$. We then obtain $X \in L_p(\mathcal{W}_p) \Rightarrow X \in L_r(\mathcal{W}_p) \Rightarrow X \in \mathcal{W}_p(L_r) \Rightarrow X \in \mathcal{W}_p(L_p)$, where the second implication holds by Theorem 12.15 (since $r < p$) and the first one simply because $L_p \subset L_r$. □

12.6 K-functional for $(v_r(B), \ell_\infty(B))$

For any $x \in B^{\mathbb{N}}$ we denote

$$V_{p,N}(x) = \sup\{(\|x_0\|^p + \|x_{n(1)} - x_0\|^p + \cdots + \|x_{n(N)} - x_{n(N-1)}\|^p)^{1/p}\}$$

where N is fixed and the supremum runs over all increasing N-tuples of integers $n(1) < n(2) < \cdots < n(N)$. Note that

$$V_{p,N}(x) \le 2(1+N)^{1/p}\|x\|_\infty. \tag{12.41}$$

Lemma 12.24 ([118]). *Let $1 \le r < \infty$. For any x in $v_r(B) + \ell_\infty(B)$ we have for any $N \ge 1$*

$$2^{-1-1/r}V_{r,N}(x) \le K_{N^{1/r}}(x, v_r(B), \ell_\infty(B)) \le 2V_{r,N}(x).$$

Proof. For simplicity we set $K_t(x) = K_t(x; v_r(B), \ell_\infty(B))$. We have obviously $V_{r,N}(x) \le V_r(x)$ and $V_{r,N}(x) \le 2(N+1)^{1/r}\|x\|_\infty$. Therefore if $x = x_0 + x_1$ we can write

$$V_{r,N}(x) \le V_{r,N}(x_0) + V_{r,N}(x_1) \le \|x_0\|_{v_r(B)} + 2(N+1)^{1/r}\|x_1\|_\infty$$

and hence $V_{r,N}(x) \le 2^{1+1/r}K_{N^{1/r}}(x)$.

For the converse inequality, we use the same idea as in the proof of Lemma 12.11. By homogeneity we may assume $V_{r,N}(x) = 1$. We let $n(1) = \inf\{n \mid \|x_n - x_0\| > N^{-1/r}\}$, $n(2) = \inf\{n > n(1) \mid \|x_n - x_{n(1)}\| > N^{-1/r}\}$ and so on. The process will stop at some integer k. Note that $N^{-1/r}(k \wedge N)^{1/r} < V_{r,N}(x) = 1$ and hence $k \wedge N < N$, which forces $k < N$. We then define $x_n^0 = x_0$ on $[0, n(1)]$, $x_n^1 = x_1$ on $(n(1), n(2)]$, $\cdots$, $x_n^0 = x_{n(k-1)}$ if $n \in (n(k-1), n(k)]$ and $x^1 = x - x^0$. Then $\|x^1\|_\infty \le N^{-1/r}$ and, since $k < N$, $\|x^0\|_{v_r(B)} \le V_{r,N}(x) = 1$. Thus we obtain

$$K_{N^{1/r}}(x) \le \|x^0\|_{v_r(B)} + N^{1/r}\|x^1\|_\infty \le 2.$$

□

Remark. Actually, the preceding lemma remains valid for $0 < r < 1$ with possibly different constants, with the same proof. In that case, the space v_r is only a quasi-normed space.

Now that we have a more concrete description of the K-functional, we can give a rather nice one for the interpolation spaces $\mathcal{W}_{p,q}(B)$:

Theorem 12.25. *Assume $1 \le r < p < \infty$, $1 \le q \le \infty$. A sequence $x = (x_n)$ in $B^{\mathbb{N}}$ belongs to $\mathcal{W}_{p,q}(B)$ iff the sequence $(N^{-1/r}V_{r,N}(x))_{N\ge 1}$ is in $\ell_{p,q}$ and the corresponding norms (or quasi-norms) are equivalent.*

Proof. We use Lemma 12.12. Let

$$a_{r,N}(x; B) = N^{-1/r}V_{r,N}(x).$$

Simply observe that if $\frac{1}{p} = \frac{1-\alpha}{r} + \frac{\alpha}{\infty}$ and $1 \le q < \infty$, by 'change of variable' (we replace t by $N^{1/r}$)

$$\begin{aligned}\int_0^\infty (t^{-\alpha}K_t(x; v_r(B), \ell_\infty(B))^q \frac{dt}{t} &\simeq \sum\nolimits_{N\ge 1}(N^{-\alpha/r}K_{N^{1/r}}(x; v_r(B), \ell_\infty(B)))^q N^{-1}\\ &\simeq \sum\nolimits_{N\ge 1}(N^{1/p}a_{r,N}(x; B))^q N^{-1},\end{aligned}$$

and the result follows by Remark 8.56. □

Remark 12.26. The preceding result shows that, for any B and any closed subspace $S \subset B$, $\mathcal{W}_{p,q}(S)$ is a closed subspace of $\mathcal{W}_{p,q}(B)$ and its norm is equivalent to the one induced on it by $\mathcal{W}_{p,q}(B)$. Indeed, if $x \in S^{\mathbb{N}}$ the sequence $\{a_{r,N}(x) \mid N \ge 1\}$ is the same whether we view $x \in S^{\mathbb{N}}$ or $x \in B^{\mathbb{N}}$.

Corollary 12.27. *In the situation of Theorem 12.25. Assume that $1 \le r < p$ is such that*

$$\ell_r(\ell_{p,q}) \subset \ell_{p,q}(\ell_r). \tag{12.42}$$

Then, for any Banach space B

$$\ell_r(\mathcal{W}_{p,q}(B)) \subset \mathcal{W}_{p,q}(\ell_r(B)), \tag{12.43}$$

and the same with L_r in place of ℓ_r.

Proof. If we apply Theorem 12.25 with $\ell_r(B)$ in place of B, for any sequence $x = (x_n)$ with $x_n = (x_n(k))_{k\in\mathbb{N}} \in \ell_r(B)$ for each N, we have

$$\|x\|_{\mathcal{W}_{p,q}(\ell_r(B))} \simeq \|(a_{r,N}(x; \ell_r(B)))_{N\ge 1}\|_{\ell_{p,q}}.$$

For each $N \ge 1$ we have

$$a_{r,N}(x; \ell_r(B)) \le \left(\sum\nolimits_k (a_{r,N}(x(k); B))^r\right)^{1/r}$$

where, for each k, $x(k) \in B^{\mathbb{N}}$ is the sequence $(x_n(k))_{n\in\mathbb{N}}$. By (12.42) we find

$$\|(a_{r,N}(x;\ell_r(B)))_{N\geq 1}\|_{\ell_{p,q}} \lesssim \left(\sum\nolimits_k \|(a_{r,N}(x(k);B))_{N\geq 1}\|_{\ell_{p,q}}^r\right)^{1/r},$$

which, by Theorem 12.25 (applied to both B and $\ell_r(B)$), implies (12.43). □

Remark 12.28. One can derive an alternate proof of (12.37) from Corollary 12.27. Then (12.38) follows by a duality argument.

Remark 12.29. Let B, B_1 be arbitrary Banach spaces. Let $Q\colon\ B \to B_1$ be a bounded surjection onto B_1 so that $B_1 \simeq B/\ker(Q)$. Then for any $1 < p < \infty$ the associated map $I \otimes Q$ is a surjection from $\mathcal{W}_p(B)$ onto $\mathcal{W}_p(B_1)$. Indeed, this lifting (or 'projective') property can be proved by duality using Remarks 12.26 and 12.9. An alternate proof follows from Remark 12.32.

Using Remarks 12.26 and 12.29, one can prove a generalization of Corollary 12.18 for operators T from a subspace S_r of $L_r(m_1)$ to a quotient space Q_s of $L_s(m_2)$. For any Banach space B we denote by $S_r(B)$ the closure of $S_r \otimes B$ in $L_r(m_1;B)$. Moreover, assuming $Q_s = L_s(m_2)/F$ for some $F \subset L_s(m_2)$, we denote $Q_s(B) = L_s(m_2;B)/F(B)$.

Corollary 12.30. *With the notation of Corollary 12.18, let S_r be a subspace of $L_r(m_1)$ and Q_s a quotient space of $L_s(m_2)$. Then any bounded linear operator $T\colon\ S_r \to Q_s$ extends to a bounded operator*

$$\widetilde{T}\colon\ S_r(\mathcal{W}_p) \to Q_s(\mathcal{W}_p).$$

We leave the proof as an exercise.

Remark. The preceding corollary implies that if $X = \mathcal{W}_p^*$ then for any r-absolutely summing operator $T : X \to Y$ (with Y an arbitrary Banach space), the adjoint T^* is s-integral (see [78, 316] for the definitions). It follows that any 1-absolutely summing operator $T : X \to Y$ factors through an L_1-space when viewed as acting into Y^{**}. In other words X has the so-called GL-property. The GL-property was used by Gordon and Lewis in [255] as a necessary condition for a space to have local unconditional structure. However, by [290], it is known that any non-reflexive quasi-reflexive space such as $\mathcal{W}_p$ (or its dual) fails local unconditional structure and is not isomorphic to any complemented subspace of a Banach lattice.

12.7 Strong p-variation in approximation theory

Another useful description of the space $\mathcal{W}_p$ can be given in terms of approximation theory. Actually, it would be more natural (as is done in [118]) to work with functions on [0, 1] and to consider approximation by splines, but we prefer to stick to our 'discrete' setting.

Let $\mathcal{S}_N \subset \ell_\infty(B)$ be the subset formed of all $b = (b(n))_{n\geq 0}$ such that $\mathbb{N}$ can be partitioned into N intervals on each of which b is constant. Then let

$$\forall x \in \ell_\infty(B) \quad S_N(x) = \inf\{\|x - b\|_\infty \mid b \in \mathcal{S}_N\}. \tag{12.44}$$

This is simply the distance of x in $\ell_\infty(B)$ to $\mathcal{S}_N$. Note that $\mathcal{S}_N \subset \mathcal{S}_{N+1}$, and hence

$$S_{N+1}(x) \leq S_N(x).$$

For any $x \in \mathcal{S}_N$ we have obviously

$$V_1(x) = \sup_k V_{1,k}(x) = V_{1,N}(x) \leq (1 + 2N)\|x\|_\infty \tag{12.45}$$

Theorem 12.31. *Let $1 < p < \infty$. The following properties of a sequence $x \in B^{\mathbb{N}}$ are equivalent:*

(i) $x \in \mathcal{W}_p(B)$.
(ii) $\sum_N S_N(x)^p < \infty$.

Moreover the corresponding quasi-norm $x \mapsto (\|x_0\|^p + \sum_{N\geq 1} S_N(x)^p)^{1/p}$ is equivalent to the norm in the space $\mathcal{W}_p(B)$.

Proof. The proof of Lemma 12.24 actually shows that $S_N(x) \leq N^{-1/r} V_{r,N}$ (indeed in that proof $x^0 \in \mathcal{S}_N$). Therefore by Theorem 12.25, applied with $r = 1$, (i) implies (ii).

Conversely, assume (ii). Note that $\mathcal{S}_n + \mathcal{S}_k \subset \mathcal{S}_{n+k}$ for any $n, k \geq 1$, and also

$$\sum S_N(x)^p \simeq \sum 2^n S_{2^n}(x)^p < \infty.$$

We claim that there is a constant c such that

$$\forall N \geq 1 \quad N^{-1} V_{1,N}(x) \leq c \left(N^{-1} \sum\nolimits_{n\leq N} S_n(x) + \sum\nolimits_{n>N} n^{-1} S_n(x) \right). \tag{12.46}$$

To show this, let $x^{(n)} \in \mathcal{S}_{2^n}$ be such that $\|x - x^{(n)}\|_\infty \leq 2S_{2^n}(x)$. Let $\Delta_n = x^{(n+1)} - x^{(n)}$ and $x^{(0)} = 0$. Note that x is the limit of $x^{(n)}$ in $\ell_\infty(B)$ or, equivalently $x = \sum_0^\infty \Delta_n$, and $\Delta_n \in \mathcal{S}_{2^n + 2^{n+1}} \subset \mathcal{S}_{2^{n+2}}$. Therefore, by (12.45), we have $V_{1,2^k}(\Delta_n) \leq 2^{n+3}\|\Delta_n\|_\infty$ that we will use when $k \geq n$, while, for any k, we already saw in (12.41) that $V_{1,2^k}(\Delta_n) \leq (2^{k+1}+1)\|\Delta_n\|_\infty$. This allows us to write

$$\begin{aligned}V_{1,2^k}(x) &\leq V_{1,2^k}\left(\sum\nolimits_{n\leq k}\Delta_n\right) + V_{1,2^k}\left(\sum\nolimits_{n>k}\Delta_n\right),\\ &\leq \sum\nolimits_{n\leq k} V_{1,2^k}(\Delta_n) + \sum\nolimits_{n>k}(2^{k+1}+1)\|\Delta_n\|_\infty\\ &\leq \sum\nolimits_{n\leq k} 2^{n+3}\|\Delta_n\|_\infty + (2^{k+1}+1)\sum\nolimits_{n>k}\|\Delta_n\|_\infty.\end{aligned}$$

But $\|\Delta_n\|_\infty \leq \|x^{(n)}-x\|_\infty + \|x-x^{(n+1)}\|_\infty \leq 4S_{2^n}(x)$ so we find for some constant c an estimate of the form

$$2^{-k}V_{1,2^k}(x) \leq c\left(2^{-k}\sum\nolimits_{n\leq k}2^nS_{2^n}(x) + \sum\nolimits_{n>k}S_{2^n}(x)\right),$$

from which Claim 12.46 becomes an elementary verification (possibly with a different c). From this claim, elementary arguments show that

$$\sum S_N(x)^p < \infty \Rightarrow \sum (N^{-1}V_{1,N}(x))^p < \infty.$$

Indeed, by Hardy's classical inequality, for any $1<p<\infty$ and for any scalar sequence $(a_1,a_2,\ldots)$, we have

$$\left\|\left(N^{-1}\sum\nolimits_{n\leq N}a_n\right)_{N\geq 1}\right\|_{\ell_p} \leq p'\|(a_n)\|_{\ell_p}$$

and therefore by duality also for any $1<p'<\infty$ and any sequence $(b_1,b_2,\ldots)$

$$\left\|\left(\sum\nolimits_{N\geq n}b_N/N\right)_{n\geq 1}\right\|_{\ell_{p'}} \leq p'\|(b_N)\|_{\ell_{p'}},$$

and hence $\sum S_N(x)^p<\infty \Rightarrow x\in\mathcal{W}_p$ by Theorem 12.25. □

Remark 12.32. The preceding proof shows that the properties in Theorem 12.31 are also equivalent to

(iii) For each integer $n\geq 1$, there are $\Delta_n \in \mathcal{S}_{2^n}$ such that $x=\sum_n\Delta_n$ and $\sum 2^n \sup_{k\geq 2^n}\|\Delta_k\|^p_{\ell_\infty(B)} < \infty$.

Throughout this chapter we have collected a wealth of information on the real interpolation spaces $\mathcal{W}_{p,q}$. In sharp contrast, the complex analogue remains a long standing open question:

Problem: Describe the complex interpolation spaces between the complex valued versions of v_1 and ℓ_∞.

If $1-\theta=1/p$, is the space $(v_1,\ell_\infty)_\theta$ identical to the space $\mathcal{W}_p$?

12.8 Notes and remarks

This chapter is mainly based on [399]. A key idea comes from Bergh and Peetre's [118]: There they prove Lemma 12.24 and (12.29) in the scalar case but the Banach valued case is identical.

As mentioned in the text, the classical James space J is the one that we denote by v_2^0. Theorem 12.5 and Lemma 12.6 are due to James. See [392] for a proof that J^* is of cotype 2 and has the GL-property. More generally, if $1 < p < \infty$, v_p^* is of cotype p' and has the GL-property. However, the space v_p itself contains ℓ_∞^ns uniformly and hence its type or cotype is trivial (in sharp contrast with $\mathcal{W}_p$).

Our approach can be applied equally well to the couple of function spaces $(V_1(I; B), \ell_\infty(I; B))$ when $I \subset \mathbb{R}$ is an interval (in particular when $I = \mathbb{R}$). Here the definition of $V_p(I; B)$ $(0 < p < \infty)$ is exactly the same as for sequences, or equivalently a function $f\colon\ I \to B$ is in $V_p(I; B)$ iff for any increasing mapping $T\colon\ \mathbf{N} \to \mathbf{I}$, the composition $f \circ T\colon\ n \mapsto f(T(n))$ is in $v_p(B)$ and $\|f\|_{V_p(I;B)}$ is equivalent to $\sup\{\|f \circ T\|_{v_p(B)}\}$ where the sup runs over all possible such increasing mappings T. In case $I = \mathbb{R}$, it is natural to define $V_p^0(I; B)$ as the closure of the subset of compactly supported infinitely differentiable functions, and to replace (as we did for sequence spaces) $V_p(I; B)$ by $V_p^0(I; B)$. See [438] for more information. See [118] for connections with approximation by splines.

We then define exactly as before $\mathcal{W}_{p,q}(I; B) = (V_1(I; B), \ell_\infty(I; B))_{\theta,q}$. Many results of this chapter remain valid, for instance this is the case for Theorem 12.15, Corollaries 12.16 to 12.19 and those in §12.5. Among the few references we know (besides [392]) that study the Banach spaces of functions with finite strong p-variation, we should mention [305] and also the books [13, 23, 24]. See also Koch's [308] for recent results on the well-posedness of dispersive PDEs, which use the analogues of the spaces $v_p(B)$ and $u_p(B)$ for functions on $\mathbb{R}$.

13

Martingales and metric spaces

13.1 Exponential inequalities

The following beautiful inequality from 1967 is due to Azuma [106]. As we will soon show, it is very useful to establish concentration of measure on a certain type of metric probability spaces.

Theorem 13.1. *Any real valued martingale $(M_n)_{n\geq 0}$ in L_∞ with $M_0 = 0$ satisfies for any $n \geq 1$*

$$\mathbb{E}_{n-1}\exp(M_n) \leq \exp(M_{n-1})\exp(\|dM_n\|_\infty^2/2) \tag{13.1}$$

and consequently

$$\sup_n \mathbb{E}\exp(M_n) \leq \exp\left(\sum\nolimits_1^\infty \|dM_n\|_\infty^2/2\right).$$

Proof. We will use the following elementary bound: for any $\lambda > 0$

$$\forall t \in [-1, 1] \qquad \exp(\lambda t) \leq \cosh(\lambda) + t\sinh(\lambda). \tag{13.2}$$

Indeed, by the convexity of $t \to \exp(\lambda t)$ on $[-1, 1]$, since $t = 2^{-1}(t+1)(1) + 2^{-1}(1-t)(-1)$ we have

$$\exp(\lambda t) \leq 2^{-1}(t+1)\exp(\lambda) + 2^{-1}(1-t)\exp(-\lambda),$$

which proves this bound. We will also use

$$\cosh(\lambda) \leq \exp(\lambda^2/2). \tag{13.3}$$

This just follows from Stirling's formula:

$$\cosh(\lambda) = 1 + \sum\nolimits_1^\infty \lambda^{2n}/(2n)! \leq 1 + \sum\nolimits_1^\infty \lambda^{2n}/(2^n n!).$$

We now claim that

$$\mathbb{E}_{n-1}\exp(M_n) \le \exp(M_{n-1})\exp\left(\frac{1}{2}\|dM_n\|_\infty^2\right).$$

Let $t_n = \|dM_n\|_\infty^{-1} dM_n$ and $\lambda_n = \|dM_n\|_\infty$. We have by (13.2)

$$\begin{aligned}\mathbb{E}_{n-1}\exp M_n &= \exp(M_{n-1})\mathbb{E}_{n-1}\exp \lambda_n t_n\\ [2pt] &\le \exp(M_{n-1})\mathbb{E}_{n-1}[\cosh(\lambda_n) + t_n \sinh(\lambda_n)]\\ [2pt] &= \exp(M_{n-1})\cosh(\lambda_n)\end{aligned}$$

hence by (13.3) we obtain

$$\mathbb{E}\exp(M_n) = \mathbb{E}\mathbb{E}_{n-1}\exp M_n \le \mathbb{E}\exp(M_{n-1})\exp(\|dM_n\|_\infty^2/2),$$

and hence by induction

$$\mathbb{E}\exp(M_n) \le \mathbb{E}\exp(M_0)\exp\left(\sum\nolimits_1^n \|dM_k\|_\infty^2/2\right).$$

Since $M_0 = 0$ we obtain the announced inequality. □

Corollary 13.2. *Assume* $\sum_1^\infty \|dM_n\|_\infty^2 \le 1$ *and* $M_0 = 0$. *Let* $M_\infty = \lim M_n$. *Then, for any* $t > 0$

$$\mathbb{P}\{|M_\infty| > t\} \le \mathbb{P}\{\sup_n |M_n| > t\} \le 2\exp(-t^2/2).$$

More generally, for any martingale transform $\tilde{M}_n = \sum_{1\le k\le n} \varepsilon_k dM_k$ *with* $\varepsilon_k = \pm 1$, *we have*

$$\mathbb{P}\{\sup_n |\tilde{M}_n| > t\} \le 2\exp(-t^2/2).$$

Proof. Fix $s > 0$. By Doob's inequality applied to the submartingale $f_n = \exp(sM_n)$, we have, for any $c > 0$, $c\mathbb{P}\{\sup_n f_n > c\} \le \sup_n \mathbb{E}(f_n) \le \exp(s^2/2)$. Choosing $c = \exp(st)$ we find $\mathbb{P}\{\sup_n M_n > t\} \le \exp(s^2/2 - st)$. The optimal choice $s = t$ then yields

$$\mathbb{P}\{\sup_n M_n > t\} \le \exp(-t^2/2). \tag{13.4}$$

Since the same result holds for $-M_n$, we find

$$\mathbb{P}\{\sup_n |M_n| > t\} \le \mathbb{P}\{\sup_n M_n > t\} + \mathbb{P}\{\sup_n -M_n > t\} \le 2\exp(-t^2/2).$$

Then, since $\|dM_n\|_\infty = \|d\tilde{M}_n\|_\infty$ the last assertion is obvious. □

Corollary 13.3. *For any dyadic (or calibrated) real valued martingale $(M_n)_{n\ge 0}$ with $M_0 = 0$, $Z_n = \exp(M_n - \sum_1^n |dM_k|^2/2)$ is a super-martingale satisfying*

$$\forall n \qquad \mathbb{E}\exp\left(M_n - \sum\nolimits_1^n |dM_k|^2/2\right) \le 1.$$

Proof. We have $Z_n = Z_{n-1}\exp(dM_n - |dM_n|^2/2)$, and hence

$$\mathbb{E}_{n-1}Z_n = Z_{n-1}\mathbb{E}_{n-1}\exp(dM_n - |dM_n|^2/2),$$

but by (13.2) and (13.3) we have

$$\mathbb{E}_{n-1}\exp(dM_n) \le \mathbb{E}_{n-1}\cosh(dM_n) \le \mathbb{E}_{n-1}\exp(|dM_n|^2/2) = \exp(|dM_n|^2/2).$$

Thus we obtain $\mathbb{E}_{n-1}Z_n \le Z_{n-1}$ and $\mathbb{E}Z_n \le \cdots \le \mathbb{E}Z_1 \le 1$. □

Corollary 13.4. *In the dyadic (or calibrated) case, if the square function $S = (\sum_0^\infty |dM_n|^2)^{1/2}$ of a martingale is in L_∞, then $M_\infty = \lim M_n$ satisfies $\mathbb{E}\exp\alpha M_\infty^2 < \infty$ for some $\alpha > 0$. Moreover, there is a constant C such that for any $2 \le p < \infty$*

$$\|M_\infty\|_p \le C\sqrt{p}\,\left\|\left(\sum\nolimits_0^\infty |dM_n|^2\right)^{1/2}\right\|_\infty.$$

Proof. We may as well assume (M_n) real valued, $\|S\|_\infty \le 1$ and $M_0 = 0$. By the preceding corollary we have for any real λ

$$\mathbb{E}\exp(\lambda M_\infty) \le \exp(\lambda^2/2)$$

and hence

$$\mathbb{P}(M_\infty > \lambda) \le \exp(-\lambda^2/2)$$

and also (replace M by $-M$)

$$\mathbb{P}(-M_\infty > \lambda) \le \exp(-\lambda^2/2)$$

so that

$$\mathbb{P}(|M_\infty| > \lambda) \le 2\exp(-\lambda^2/2).$$

From this we find for any $\alpha < 1/2$

$$\mathbb{E}\exp(\alpha M_\infty^2) = 1 + \int_0^\infty \frac{d}{dt}e^{\alpha t^2}\mathbb{P}(|M_\infty| > t)\,dt < \infty.$$

The last assertion follows from well-known Lemma 5.80. □

Remark. An interesting application of Azuma's inequality to Sidon sets appears in the recent paper [142].

13.2 Concentration of measure

Martingale inequalities can be very helpful to exhibit concentration of measure phenomena. We will just give a few samples of possible applications in this direction. See [53] for more on this.

Let B be a Banach space. Consider a sequence of independent and mean zero random variables (Y_j) in $L_1(B)$ on a probability space $(\Omega, \mathcal{A}, \mathbb{P})$. Assume that the series $S = \sum_1^\infty Y_j$ converges in $L_1(B)$. Let $\mathcal{A}_n = \sigma(Y_1, \ldots, Y_n)$. Let

$$f(\omega) = \|S(\omega)\|_B, \text{ and } f_n = \mathbb{E}^{\mathcal{A}_n} f.$$

Let $\mathcal{A}_0$ be the trivial σ-algebra so that $f_0 = \mathbb{E}f$. The key observation is that

$$|(f_n - f_{n-1})(\omega)| \le \int \|Y_n(\omega) - Y_n(\omega')\| d\mathbb{P}(\omega'). \tag{13.5}$$

Indeed, let $S_n = \sum_1^n Y_j$ and $R_n = \sum_{j>n} Y_j$.

We have $f_n(\omega) = \mathbb{E}_{\omega'} \|S_n(\omega) + R_n(\omega')\|$ and hence

$$\begin{aligned} |(f_n - f_{n-1})(\omega)| &\le \mathbb{E}_{\omega'} \Big| \|S_n(\omega) + R_n(\omega')\| - \|S_{n-1}(\omega) + R_{n-1}(\omega')\| \Big| \\ &\le \mathbb{E}_{\omega'} \|(S_n - S_{n-1})(\omega) + (R_n - R_{n-1})(\omega')\| \\ &= \mathbb{E}_{\omega'} \|Y_n(\omega) - Y_n(\omega')\|, \end{aligned}$$

which proves (13.5).

The concentration of measure phenomenon depends very much on the concrete situation. First, let us recapitulate what we just saw (this idea goes back to Yurinsky, see [53]).

Proposition 13.5. *With the preceding notation, we have*

$$\|S\| - \mathbb{E}\|S\| = f \quad \textit{and} \quad |df_n| \le \mathbb{E}_{\omega'} \|Y_n - Y_n(\omega')\| \quad \textit{for all} \quad n \ge 1.$$

Corollary 13.6. *Assume* $\sum \|Y_n\|^2_{L_\infty(B)} \le 1$. *Then for any* $t > 0$

$$\mathbb{P}\Big\{ \Big| \|S\| - \mathbb{E}\|S\| \Big| > t \Big\} \le 2 \exp\left(-\frac{t^2}{8}\right).$$

Proof. We have $\|df_n\|_\infty \le 2\|Y_n\|_{L_\infty(B)}$ so the result follows from Corollary 13.2. □

Corollary 13.7. *For any* $1 < p < \infty$ *we have*

$$\Big\| \|S\| - \mathbb{E}\|S\| \Big\|_p \le b_p \left(\Big\| \Big(\sum \|Y_n\|^2\Big)^{1/2} \Big\|_p + \Big(\sum (\mathbb{E}\|Y_n\|)^2\Big)^{1/2} \right),$$

where b_p *is as in* (5.35).

Proof. Note that by the triangle inequality

$$|df_n(\omega)| \le \mathbb{E}_{\omega'}\|Y_n(\omega) - Y_n(\omega')\| \le \|Y_n(\omega)\| + \mathbb{E}\|Y_n\|,$$

and hence $S(f) \le (\sum \|Y_n\|^2)^{1/2} + (\sum(\mathbb{E}\|Y_n\|)^2)^{1/2}$. Thus the result follows from (5.35). □

In [346], Maurey gave a beautiful application of martingale inequalities on the symmetric group. Here is a brief outline of his ideas.

Theorem 13.8. *Fix $n \ge 1$. Let Ω denote the set of all $n!$ permutations of $[1, \ldots, n]$ and let $\mathbb{P}$ denote the uniform probability measure on Ω. Let $\mathcal{T} \subset \Omega$ denote the subset of transpositions. Then any function $f\colon\ \Omega \to \mathbb{R}$ such that*

$$\forall \omega \in \Omega\ \forall \tau \in \mathcal{T} \quad |f(\omega) - f(\omega\tau)| \le 1 \tag{13.6}$$

satisfies

$$\forall t > 0 \quad \mathbb{P}\{f - \mathbb{E}f > t\} \le \exp -(t^2/2n), \tag{13.7}$$

and hence

$$\mathbb{P}\{|f - \mathbb{E}f| > t\} \le 2\exp -(t^2/2n). \tag{13.8}$$

Proof. Let $\mathcal{A}_k$ be the σ-algebra generated by $\{\omega(1), \ldots, \omega(k)\}$ when $k = 1, \ldots, n$. Let $\mathcal{A}_0 = \{\phi, \Omega\}$ be the trivial algebra. Let $f_k = \mathbb{E}^{\mathcal{A}_k} f$ and $d_k = f_k - f_{k-1}$ $(k = 1, \ldots, n)$. By Corollary 13.2 and (13.4), it clearly suffices to prove the following claim:

$$\forall \omega \in \Omega \qquad |d_k(\omega)| \le 1.$$

The proof consists in rewriting d_k as an average of some of the elements appearing in (13.6). To verify this, fix $\pi \in \Omega$ and $k \in [1, \ldots, n]$. We can write

$$f_k(\pi) = \int f(\pi g) d\mu_k(g)$$

where μ_k is the normalized Haar measure on the subgroup $G_k \subset \Omega$ of all permutations leaving $[1, \ldots, k]$ invariant. Consider g' a permutation leaving $[1, \ldots, k-1]$ invariant. Let $j \in [k, \ldots, n]$ be the integer such that $g'(j) = k$ and let τ be the transposition that exchanges j and k (note that τ depends on g'). We have $\tau k = j$ and hence $g'\tau k = k$ so that now $g'\tau$ leaves $[1, \ldots, k]$ invariant, equivalently $g'\tau \in G_k$. Since μ_k is translation invariant on G_k we have for any fixed g'

$$f_k(\pi) = \int f(\pi g g' \tau) d\mu_k(g).$$

Averaging over g' this yields

$$f_k(\pi) = \int \left(\int f(\pi g g' \tau) d\mu_{k-1}(g') \right) d\mu_k(g).$$

Now we use the translation invariance of μ_{k-1} and the observation that the substitution $g' \to g^{-1}g'$ leaves τ unchanged (since $g^{-1}g'(j) = k$). We find

$$f_k(\pi) = \int f(\pi g' \tau) d\mu_{k-1}(g').$$

Thus we conclude

$$f_k(\pi) - f_{k-1}(\pi) = \int (f(\pi g' \tau) - f(\pi g')) d\mu_{k-1}(g')$$

and our claim follows immediately. □

Remark. The group of all $n!$ permutations of $[1, \ldots, n]$ is *generated* by the subset of all $n^2 - n$ transpositions. The inequality proved in the preceding theorem is analogous to an isoperimetric inequality with respect to this group with this set of generators. A similar deviation inequality can be obtained on the combinatorial cube $\Delta_n = \{-1, 1\}^n$, as follows:

Theorem 13.9. *Fix $n \geq 1$. Let $\Delta_n = \{-1, 1\}^n$. We view Δ_n as a group with respect to the pointwise product. For $1 \leq k \leq n$, we denote $\xi_k = (1, \ldots, -1, \ldots, 1)$ where -1 stands at the k-th place and all other coefficients are equal to 1. Note that $\{\xi_1, \ldots, \xi_n\}$ generate the group Δ_n. Let $f \colon \Delta_n \to \mathbb{R}$ be a function such that*

$$\forall \omega \ \forall k \qquad |f(\omega) - f(\omega \xi_k)| \leq 2.$$

Then

$$\forall t > 0 \qquad \mathbb{P}(|f - \mathbb{E}f| > t) \leq 2 \exp -(t^2/2n).$$

Proof. Let $\mathcal{A}_k$ be the σ-algebra generated by the k first coordinates on Δ_n and let $\mathcal{A}_0$ be the trivial σ-algebra. A simple verification shows that

$$f_{k-1}(\omega) = 2^{-1}(f_k(\omega) + f_k(\omega \xi_k))$$

and hence

$$(f_k - f_{k-1})(\omega) = 2^{-1}(f_k(\omega) - f_k(\omega \xi_k)) = \mathbb{E}_k(\delta_k f)$$

where $\delta_k f(\omega) = (f(\omega) - f(\omega \xi_k))/2$. By our assumption we have $\|\delta_k f\|_\infty \leq 1$. Thus we find $|df_k(\omega)| \leq 1$ and the result follows from (13.4) and Corollary 13.2. □

For the next result, we introduce a distance d on the set of permutations Ω by setting

$$d(\pi, \omega) = \text{card}\{k \in [1, \dots, n] \mid \pi(k) \neq \omega(k)\}.$$

Note that for any transposition τ

$$d(\omega\tau, \omega) = 2. \tag{13.9}$$

Corollary. *Let $A \subset \Omega$ be a set of permutations of $[1, \dots, n]$ with $\mathbb{P}(A) = \alpha$. Let A^k denote the set of permutations that differ with an element of A on at most k points. Then*

$$\frac{\text{card}(A^k)}{n!} > 1 - 2\alpha^{-1}e^{-k^2/32n}.$$

Proof. Let $f(\omega) = d(\omega, A)$. Then $A^k = \{\omega \mid f(\omega) \le k\}$. We have clearly by the triangle inequality for the distance d

$$\forall \omega, \pi \qquad |f(\omega) - f(\pi)| \le d(\omega, \pi)$$

and hence by (13.9), for any transposition τ, $|f(\omega) - f(\omega\tau)| \le 2$.

By (13.7) we have for any $t > 0$

$$\mathbb{P}(f - \mathbb{E}f > 2t) \le e^{-t^2/2n}, \tag{13.10}$$

and similarly with $-f$ in place of f. Note that $f(\omega) - f(\pi) > 4t$ implies that either $f(\omega) - \mathbb{E}f > 2t$ or $\mathbb{E}f - f(\pi) > 2t$. Therefore (13.10) implies

$$(\mathbb{P} \times \mathbb{P})\{(\omega, \pi) \mid f(\omega) - f(\pi) > 4t\} \le 2e^{-t^2/2n}.$$

But $f(\pi) = 0$ for any $\pi \in A$, so that, restricting to $\pi \in A$, we have a fortiori

$$\mathbb{P}\{\omega \mid f(\omega) > 4t\}\mathbb{P}(A) \le 2e^{-t^2/2n}.$$

Since $\mathbb{P}(A) = \alpha$, taking $t = k/4$ the announced result follows immediately. □

A similar result can be derived for Δ_n. We leave this as an exercise.

13.3 Metric characterization of super-reflexivity: trees

This section is based on Bourgain's [138]. By general arguments (see [5]) it was known that super-reflexivity is preserved under Lipschitz isomorphism. Therefore knowing this, one would expect there should be a characterization of super-reflexive Banach spaces using only their structure as *metric spaces*. This is precisely the content of Bourgain's characterization in Theorem 13.10.

Definition. Let (T_1, d_1), (T_2, d_2) be metric spaces. A map $F: T_1 \to T_2$ is called Lipschitz (or Lipschitzian) if there is a constant L such that

$$\forall s, t \in T_1 \qquad d_2(F(s), F(t)) \leq L d_1(s, t).$$

We will then say that F is L-Lipschitz. The smallest such constant L will be denoted by $\|F\|_{\mathrm{Lip}}$, i.e.

$$\|F\|_{\mathrm{Lip}} = \sup_{s \neq t}\{d_2(F(s), F(t))/d_1(s, t)\}.$$

When F is injective, the product $\|F\|_{\mathrm{Lip}} \|F^{-1}_{|F(T_1)}\|_{\mathrm{Lip}}$ is called the distortion of F.

Definition. Let (T_n, d_n) be a sequence of metric spaces. We say that a metric space (T, d) contains $\{T_n\}$ Lipschitz uniformly if, for any n, there are injective Lipschitz mappings $F_n: T_n \to T$ with bounded distortion. In other words, there is $\lambda \geq 1$ and positive constants a_n, b_n with $a_n b_n \leq \lambda$ such that for all n

$$\forall s, t \in T_n \qquad (1/a_n) d(s, t) \leq d(F_n(s), F_n(t)) \leq b_n d(s, t).$$

In the latter case we say that (T, d) (or simply T) contains the sequence $\{T_n\}$ λ-uniformly.

Note that when T is a Banach space, we can always, by scaling, normalize F_n as we wish, for instance we may restrict consideration to the F_n's that are distance non-increasing, i.e. for which $b_n = 1$.

Let $\mathcal{T}_n$ be a finite dyadic tree with $1 + 2 + \cdots + 2^n = 2^{n+1} - 1$ vertices (or nodes). We will label these points as $\tau(\varepsilon_1 \dots \varepsilon_j)$, $1 \leq j \leq n$, $\varepsilon_j \in \{-1, 1\}$ and we denote by τ_ϕ the 'root' of the tree.

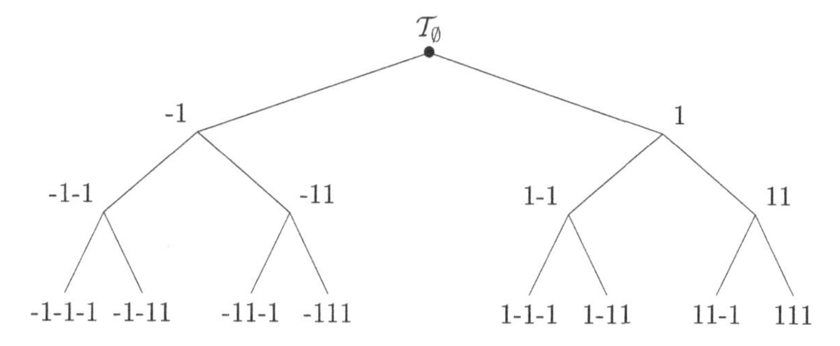

Figure 13.1. $\mathcal{T}_n$ for $n = 3$.

We equip $\mathcal{T}_n$ with its natural 'geodesic' distance as a graph, i.e. we set

$$d(\tau(\varepsilon_1' \dots \varepsilon_j'), \tau(\varepsilon_1'' \dots \varepsilon_k'')) = j + k - 2N$$

where $N = N(\varepsilon', \varepsilon'')$ is the largest N such that $(\varepsilon_1', \dots, \varepsilon_N') = (\varepsilon_1'', \dots, \varepsilon_N'')$.

Theorem 13.10. *A Banach space B is super-reflexive iff B does not contain the sequence $\{\mathcal{T}_n\}$ Lipschitz uniformly.*

The if part follows from:

Lemma 13.11. *If B is not super-reflexive then B contains the sequence $\{\mathcal{T}_n\}$ Lipschitz uniformly.*

Proof. By Theorem 11.10 if B is not super-reflexive, for any $0 < \theta < 1$ and any $n \geq 1$ there are $x_0, \ldots, x_{2^n}$ in B such that for any scalars α_j we have

$$(\theta/2)\sup_j \left(\left|\sum\nolimits_{i<j} \alpha_i\right| + \left|\sum\nolimits_{i\geq j} \alpha_i\right|\right) \leq \left\|\sum \alpha_j x_j\right\| \leq \sum |\alpha_j|. \quad (13.11)$$

There is a natural partial order on $\mathcal{T}_n$: we say that $s < t$ $(s, t \in \mathcal{T}_n)$ if s, t lie on the same branch with s closer to the root. This can also be reformulated by saying s, t are of the form $s = \tau(\varepsilon_1 \ldots \varepsilon_j)$ and $t = \tau(\varepsilon_1 \ldots \varepsilon_k)$ for some $k > j$ and $(\varepsilon_1, \ldots, \varepsilon_k) \in \{-1, 1\}^k$.

We write $s \leq t$ if either $s < t$ or $s = t$. Also we denote $(r, s] = \{w \in \mathcal{T}_n \mid r < w \leq s\}$ for any $r \in \mathcal{T}_n$. Note that there is a bijective mapping $\varphi\colon \mathcal{T}_n \to [1, \ldots, 2^{n+1} - 1]$ such that φ maps any pair of disjoint intervals of the form $(r, s], (r, t]$ into disjoint intervals in $[1, \ldots, 2^{n+1} - 1]$, separated by $\varphi(r)$. More precisely, we have either

$$\forall w \in (r, s]\ \forall w' \in (r, t] \quad \varphi(w) < \varphi(r) < \varphi(w')$$

or

$$\forall w \in (r, s]\ \forall w' \in (r, t] \quad \varphi(w') < \varphi(r) < \varphi(w).$$

The existence of such a φ can be proved either by looking at a picture of a tree or using the expansion of numbers in 'base 2': Just set $\psi(\tau_\phi) = 0$ and $\psi(\tau(\varepsilon_1 \ldots \varepsilon_k)) = \sum_1^k 2^{-j}\varepsilon_j$ and, to obtain φ, just relabel the range of ψ, respecting its order, as $[1, \ldots, 2^{n+1} - 1]$. We can then define an 'embedding' $F_n\colon \mathcal{T}_n \to B$ by setting

$$\forall t \in \mathcal{T}_n \qquad F_n(t) = \sum_{w \leq t} x_{\varphi(w)}.$$

We claim that for all s, t

$$(\theta/2)d(s, t) \leq \|F_n(s) - F_n(t)\| \leq d(s, t). \quad (13.12)$$

Indeed, assume $d(s, t) = j + k$ with

$$s = \tau(\varepsilon_1 \ldots \varepsilon_N \varepsilon'_{N+1} \ldots \varepsilon'_{N+j}) \quad \textit{and} \quad t = \tau(\varepsilon_1 \ldots \varepsilon_N \varepsilon''_{N+1} \ldots \varepsilon''_{N+k}),$$

with $\varepsilon'_{N+1} \neq \varepsilon''_{N+1}$ $(N \geq 0)$. Let $r = \tau(\varepsilon_1 \dots \varepsilon_N)$. Then

$$F_n(s) - F_n(t) = \sum\nolimits_{k \in A'} x_k - \sum\nolimits_{k \in A''} x_k$$

where $A' = \varphi((r, s])$, $A'' = \varphi((r, t])$ are included in disjoint subintervals I', I'' of $[1, \dots, 2^{n+1} - 1]$, separated by $\varphi(r)$, with $|A'| = j$ and $|A''| = k$, and hence (13.11) yields (13.12). □

The only if part of Theorem 13.10 will be deduced from:

Lemma 13.12. *If B is super-reflexive then there is a constant C and $q < \infty$ such that:*

(i) For any $m > 1$ and any family $(x_0, \dots, x_m)$ in B we have

$$\inf_{0 \leq j,\ j+2k \leq m} k^{-1} \|x_j + x_{j+2k} - 2x_{j+k}\| \leq C(\mathrm{Log}\, m)^{-\frac{1}{q}} \sup_{1 \leq j \leq m} \|x_j - x_{j-1}\|. \tag{13.13}$$

(ii) For any $n > 1$ and any $F\colon \mathcal{T}_n \to B$ we have

$$\inf_{2k \leq N+k \leq n} k^{-1} \mathbb{E} \| F(\tau(\varepsilon_1 \dots \varepsilon_N \varepsilon'_{N+1} \dots \varepsilon'_{N+k})) - F(\tau(\varepsilon_1 \dots \varepsilon_N \varepsilon''_{N+1} \dots \varepsilon''_{N+k}))\| \tag{13.14}$$

$$\leq C(\mathrm{Log}(n))^{-1/q} \|F\|_{\mathrm{Lip}}, \tag{13.15}$$

where the expectation sign denotes the (triple) average with respect to $\varepsilon, \varepsilon', \varepsilon''$ in $\{-1, 1\}^{\mathbb{N}_*}$.

Proof. If B (and hence $L_2(B)$) is super-reflexive, we know (see (11.42)) that there is $2 \leq q < \infty$ and C such that for all B-valued dyadic martingales (f_k) we have

$$\left(\sum\nolimits_1^n \|df_k\|^q_{L_2(B)}\right)^{1/q} \leq C\|f_n\|_{L_2(B)}.$$

A fortiori we have

$$\inf\nolimits_{1 \leq \ell \leq n} \|df_\ell\|_{L_1(B)} \leq C n^{-1/q} \|f_n\|_{L_\infty(B)}. \tag{13.16}$$

Let $(\mathcal{A}_\ell)_{\ell \geq 0}$ denote the dyadic filtration on $[0, 1[$ (see §1.4). Recall that $\mathcal{A}_n$ is generated by the 2^n atoms $[(k-1)2^{-n}, k2^{-n}[$ $(1 \leq k \leq 2^n)$. Let $m = 2^n + 1$ and let $f_n\colon [0, 1] \to B$ be the $\mathcal{A}_n$-measurable function defined by

$$f_n = x_k - x_{k-1} \quad \text{on} \quad [(k-1)2^{-n}, k2^{-n}].$$

It is easy to check that all the values of the increments $df_\ell(\omega) = f_\ell(\omega) - f_{\ell-1}(\omega)$ are of the form $(2k)^{-1}(x_j + x_{j+2k} - 2x_{j+k})$ for some $0 \leq j, j + 2k \leq m$. Indeed, when $\ell = n$ this is verified with $k = 1$, when $\ell = n - 1$ with $k = 2$, when $\ell = n-2$ with $k = 2^2$, and so on. Thus, from (13.16), we obtain (13.13) for m of the form $m = 2^n + 1$. For the general case, just choose n such that $2^n + 1 \le m < 2^{n+1} + 1$ and note that $\operatorname{Log} m \simeq \operatorname{Log}(2^n+1) \simeq n$. This completes the proof of (i).

To prove (ii), we apply (i) to $L_2(B)$ in place of B. Set (for $j = 1, \ldots, n$)

$$x_j = F(\tau(\varepsilon_1 \ldots \varepsilon_j)).$$

We view x_j as a B-valued function on $\{-1,1\}^{\mathbb{N}_*}$ that depends only on $\varepsilon_1 \ldots \varepsilon_j$, considered as an element of $L_2(B) = L_2(\{-1,1\}^{\mathbb{N}_*}; B)$. Let $N = j + k$. Let us denote $\xi = (\varepsilon_1 \ldots \varepsilon_N)$, $\eta' = (\varepsilon'_{N+1} \ldots \varepsilon'_{N+k})$, and $\eta'' = (\varepsilon''_{N+1} \ldots \varepsilon''_{N+k})$. Note that x_{j+k} and x_j (and hence $x_j - 2x_{j+k}$) both depend only on $\xi = (\varepsilon_1 \ldots \varepsilon_N)$ so that by the triangle inequality

$$\begin{aligned}
&\|x_{N+k}(\xi,\eta') - x_{N+k}(\xi,\eta'')\|_{L_2(B)} \\
&\quad \le \|x_{N+k}(\xi,\eta') + x_j - 2x_{j+k}\|_{L_2(B)} + \|x_{N+k}(\xi,\eta'') + x_j - 2x_{j+k}\|_{L_2(B)} \\
&\quad = 2\|x_{N+k} + x_j - 2x_{j+k}\|_{L_2(B)}.
\end{aligned}$$

Note that $\|x_j - x_{j-1}\|_{L_2(B)} \le \|x_j - x_{j-1}\|_{L_\infty(B)} \le \|F\|_{\mathrm{Lip}}$, and also that the condition $2k \le N + k \le n$ is equivalent to $0 \le j$, $j + 2k \le n$. Therefore a fortiori we obtain (ii) from (13.13) applied to $L_2(B)$, but this time with $m = n$. □

Remark 13.13. Let $F\colon \mathcal{T}_n \to B$ be an injective map. If B satisfies (ii) in Lemma 13.12, for some constant $C' > 0$ independent of n we have

$$\|F\|_{\mathrm{Lip}} \|F^{-1}_{|F(\mathcal{T}_n)}\|_{\mathrm{Lip}} \ge C'(\operatorname{Log} n)^{1/q}. \tag{13.17}$$

Indeed, we may assume that $\|F^{-1}_{|F(\mathcal{T}_n)}\|_{\mathrm{Lip}} = 1$. Then we have

$$d(F(s), F(t)) \ge d(s,t)$$

for all s, t. From this it is easy to check that $\mathcal{T}_n$ satisfies

$$k^{-1}\mathbb{E} d(\tau(\varepsilon_1 \ldots \varepsilon_N \varepsilon'_{N+1} \ldots \varepsilon'_{N+k}), \tau(\varepsilon_1 \ldots \varepsilon_N \varepsilon''_{N+1} \ldots \varepsilon''_{N+k})) \ge 1$$

(because $\varepsilon'_{N+1} \ne \varepsilon''_{N+1}$ implies that the preceding distance is equal to $2k$ and this event occurs with probability $1/2$). Therefore (13.14) immediately implies (13.17). □

Proof of Theorem 13.10. The if part follows from Lemma 13.11 and the converse from the preceding remark. □

Remark 13.14. Bourgain observed in [138] that already in Hilbert space the estimate of (13.17) is sharp. See also Matoušek's [342] for more on this.

Recently a very simple and pretty proof of Theorem 13.10 was given by Kloeckner [307], as follows. It is based on the following very simple lemma (inspired by [342] and similar to the diamond graph analogue (13.22) from [289], discussed in the next section).

We will use the terminology inspired from genealogy, hopefully transparent for the reader, in which each point $\tau(\varepsilon_1 \dots \varepsilon_j)$ of the tree $\mathcal{T}_n$ is a '(single) parent' with exactly two children namely $\tau(\varepsilon_1 \dots \varepsilon_j - 1)$ and $\tau(\varepsilon_1 \dots \varepsilon_j + 1)$.

Using similar terms for general trees, let Y be the four-vertices tree with one root a_0 that has one child a_1 and two grand-children a_2, a_2'.

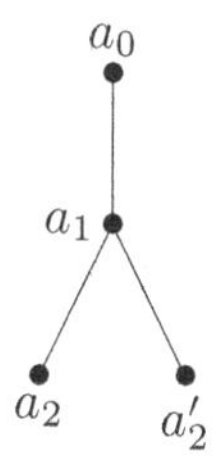

Lemma 13.15. *Let B be a uniformly convex Banach space. For any embedding $\varphi : Y \to B$ that is L-Lipschitz and distance non-decreasing, we have*

$$\min\{\|\varphi(a_0) - \varphi(a_2)\|, \|\varphi(a_0) - \varphi(a_2')\|\} \leqslant 2L(1 - \delta_B(1/L)). \quad (13.18)$$

Proof. Let $x = L^{-1}(\varphi(a_0) - \varphi(a_1))$, $y = L^{-1}(\varphi(a_1) - \varphi(a_2))$ and $y' = L^{-1}(\varphi(a_1) - \varphi(a_2'))$. Since φ is L-Lipschitz, x, y, y' are in the unit ball of B. Assume by contradiction that (13.18) fails. Then both $\|(x + y)/2\| > 1 - \delta_B(1/L)$ and $\|(x + y')/2\| > 1 - \delta_B(1/L)$. By definition of δ_B, this implies both $\|x - y\| < 1/L$ and $\|x - y'\| < 1/L$, and hence $\|y - y'\| < 2/L$. But since φ is distance non-decreasing, we must have $L\|y - y'\| = \|\varphi(a_2) - \varphi(a_2')\| \geq 2$, which is absurd. This proves the lemma. □

Second proof of the 'only if part' of Theorem 13.10. We must show that if B is super-reflexive then B does not contain $\{T_n\}$ Lipschitz uniformly. By Theorem 11.37, we may assume that B is uniformly convex. Let $\varphi : T_n \to B$ a mapping that is L-Lipschitz and distance non-decreasing. We assume that $n = 2^k$. Let $f(L) = L(1 - \delta_B(1/L))$ for $L \geq 1$. Observe that $f(L) \leq L$ for any $L \geq 1$. We claim that necessarily $f^k(L) \geq 0$ and actually we will prove $f^k(L) \geq 1$ (here f^k denotes the composition of f with itself k-times). Assuming this claim, we conclude as follows. Note that $1 \leq f^k(L) \leq f^{k-1}(L) \leq \cdots f(L) \leq L$. Also note (see Lemma 10.4, but actually in the end of the proof we will avoid using this) that $x \mapsto x - f(x) = x\delta_B(1/x)$ is non-increasing, so we know that

$f^j(L) - f^{j+1}(L) \geq L - f(L) = L\delta_B(1/L)$. Thus we find

$$kL\delta_B(1/L) \leq \sum_0^{k-1} f^j(L) - f^{j+1}(L) = L - f^k(L) \leq L,$$

and we obtain $k \leq (\delta_B(1/L))^{-1}$. By Theorem 10.1 (or Theorem 11.37) we may assume $\delta_B(1/L) \geq \delta(1/L)^q$ for some $q < \infty$ and $\delta > 0$. This gives us $k \leq \delta^{-1}L^q$ or equivalently $L \geq \delta^{1/q}k^{1/q}$, proving that L cannot remain bounded when n or k tend to ∞. Note we really used only that $x \mapsto (1/x)^{q-1}$ is non-increasing because we may work with $f(L) = L(1 - \delta(1/L)^q)$.

To prove the claim, we will apply the lemma repeatedly. By the lemma, the root a_0 has at least two grand-children a_2^i ($i = 1, 2$), with different parents, such that

$$2 \leqslant \|\varphi(a_0) - \varphi(a_2^i)\| \leqslant 2f(L)$$

where $f(L) = L(1 - \delta_B(1/L))$. Applying the lemma again, each of a_2^i also has two grand-children with different parents, satisfying similar inequalities, and we can apply the same reasoning for every other generation. Restricting φ to these vertices, we get an embedding of $T_{n/2}$ with Lipschitz-norm at most $f(L)$.

We can iterate these restrictions k-times, and we obtain a restriction of φ that is an embedding of $T_{n/2^k} = T_1$ with Lipschitz-norm $\leq f^k(L)$. Since φ is distance non-decreasing, this must be at least 1 so the claim is proved. □

13.4 Another metric characterization of super-reflexivity: diamonds

This section is based on Johnson and Schechtman's [289], but we use a simplification due to Ostrovskii [368]. Here the sequence $\{\mathcal{T}_n\}$ is replaced by the sequence $\{D_n\}$ of the diamond graphs defined as follows:

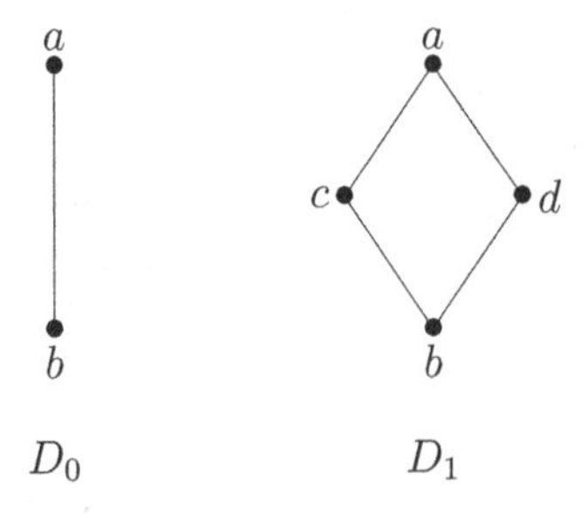

Figure 13.2. Diamond 1.

Let us first describe the relevant picture. We start with a graph D_0 with two vertices (a, b) at distance 1 and one edge. Then, to create D_1 we replace the edge (a, b) by a 'losange' with opposite vertices (a, b) and two more vertices c, d pictured on each side of (a, b). The edges of D_1 are defined by joining (a, c) (c, b) (b, d) and (d, a) (but, in D_1, there is no more a direct edge between (a, b)) all at mutual distance $1/2$. To create D_2 we repeat the same procedure simultaneously for each of the 4 edges (namely (a, c) (c, b) (b, d) and (d, a)) of D_1. This transforms D_1 into a new graph D_2 now with $4 \times 4 = 4^2$ edges, and the distance between the two endpoints of each edge is equal to $1/4$ and so on ... One can see a resemblance with a diamond shape appearing, whence the name. The distance on D_n is such that the distance between the two endpoints of each of its edges is equal to $1/2^n$. Thus it is equal to 2^{-n}-times the usual geodesic distance. In this way $d_{D_0}(a, b) = d_{D_1}(a, b) = 1$ and more generally we have an isometric embedding $D_n \subset D_{n+1}$ for any $n \geq 0$. Moreover, we have

$$\forall s, t \in D_n \quad d_{D_n}(s, t) \leq 1. \tag{13.19}$$

Note that if $V(D_n)$ (resp. $E(D_n)$) denotes the set of vertices (resp. edges) in D_n we have by an elementary induction

$$|E(D_n)| = 4^n \quad \text{and} \quad |V(D_n)| = 2 + 2\sum\nolimits_0^{n-1} 4^j = 2 + 2\frac{4^n - 1}{3}. \tag{13.20}$$

Preliminary observation (from [368]). Assume that a Banach space B contains the sequence $\{D_n\}$ Lipschitz uniformly. Then the unit ball of B contains arbitrary long δ-separated dyadic trees for some $\delta > 0$. Equivalently, with the notation of Theorem 11.22 we have $t_n(B) \leq 1/\delta$ for all n.

Fix $n > 0$. Let $f : D_n \to B$ be such that

$$\forall s, t \in D_n \quad \delta d_{D_n}(s, t) \leq \|f(s) - f(t)\| \leq d_{D_n}(s, t).$$

Restricting f to $D_1 \subset D_n$, we have in particular

$$\forall s, t \in D_1 \quad \delta d_{D_1}(s, t) \leq \|f(s) - f(t)\| \leq d_{D_1}(s, t). \tag{13.21}$$

With the preceding simple minded notation for the vertices of D_1, consider

$$x_0 = f(a) - f(b),$$

and then running around the 4 edges, let $y_1 = 2(f(a) - f(c))$, $y_2 = 2(f(c) - f(b))$, $y_3 = 2(f(a) - f(d))$, $y_4 = 2(f(d) - f(b))$. Note that

$$x_0 = (y_1 + y_2)/2 \quad x_0 = (y_3 + y_4)/2$$

and also by (13.21)

$$\delta \leq \|x_0\| \leq 1 \quad \text{and} \quad \forall j \quad \delta \leq \|y_j\| \leq 1.$$

We claim that either $\|(y_1 - y_2)/2\| \geq \delta$ or $\|(y_3 - y_4)/2\| \geq \delta$. To check this just observe that $(y_1 - y_2) - (y_3 - y_4) = 4(f(d) - f(c))$ and hence

$$\|(y_1 - y_2)\| + \|(y_3 - y_4)\| \geq \|(y_1 - y_2) - (y_3 - y_4)\| \geq 4\delta,$$

from which our claim follows. Thus we may assume e.g. $\|(y_1 - y_2)/2\| \geq \delta$. We then set

$$x_1(1) = y_1 \quad x_1(-1) = y_2.$$

This gives us the beginning of a dyadic tree in the unit ball starting at $x_0 = (x_1(1) + x_1(-1))/2$ and such that both $\|x_0 - x_1(1)\| \geq \delta$ and $\|x_0 - x_1(-1)\| \geq \delta$. But now it is clear how to continue: Indeed, we have selected two new edges (a, c) and (c, b) (corresponding to y_1 and y_2) to which we can repeat exactly the same procedure. We replace (a, b) by (a, c) and repeat the argument (note that the scaling fits this: the function $2f$ restricted to the metric subspace formed of (a, c) and the associated losange based on (a, c) satisfies (13.21) if we identify isometrically the latter losange with D_1). Continuing in this way, we obtain for $k = 1, 2, \ldots, n$ a family $x_k(\varepsilon_1, \ldots, \varepsilon_k)$ in the unit ball such that for all choices of $(\varepsilon_1, \ldots, \varepsilon_k) \in \{1, -1\}^k$

$$\|x_k(\varepsilon_1, \ldots, \varepsilon_k) - x_{k-1}(\varepsilon_1, \ldots, \varepsilon_{k-1})\| \geq \delta.$$

This is precisely the range of a dyadic martingale with δ-separated increments. So we obtain $t_n(B) \leq 1/\delta$ as announced. □

Remark 13.16. Assume that B is uniformly convex. Then for any $f : D_1 \to B$ satisfying (13.21), by definition of δ_B, we have $\|x_0\| \leq 1 - \delta_B(\|x(1) - x(-1)\|)$, or equivalently

$$\|f(a) - f(b)\| \leq 1 - \delta_B(2\delta). \tag{13.22}$$

Moreover, it is easy to deduce from the preceding argument the following quantitative estimates:

Let $M_n = \inf\{\|F\|_{\mathrm{Lip}} \|F^{-1}_{|F(D_n)}\|_{\mathrm{Lip}}\}$ where the infimum runs over all injective $F\colon\ D_n \to B$. If B is q-uniformly convex (i.e. we have a lower bound $\delta_B(\varepsilon) \geq c\varepsilon^q$ for some $c > 0$), then it follows from (10.9) that, for some $c' > 0$, we have $M_n \geq c' n^{1/q}$ for all n.

Theorem 13.17. *A Banach space B is super-reflexive iff B does not contain the sequence $\{D_n\}$ Lipschitz uniformly.*

Remark 13.18. This theorem is 'independent' from Bourgain's characterization (Theorem 13.10) in the following sense. It is easy to check that diamonds

$\{D_n\}$ do not admit embeddings with uniformly bounded distortions into any family of trees. It was shown in [370] that dyadic trees $\{\mathcal{T}_n\}$ do not admit embeddings with uniformly bounded distortions into diamonds.

It will be convenient to view D_n as embedded in the set $\{0, 1\}^{2^n}$ equipped with the so-called Hamming distance

$$\rho(s,t) = \sum_1^{2^n} |s_j - t_j| = |\{j \mid s_j \neq t_j\}|.$$

In Corollary 13.20, we will compare this to the geodesic distance δ_{D_n}, which is the same as before except for the normalization, so we have

$$\forall s,t \in D_n \quad \delta_{D_n}(s,t) = 2^n d_{D_n}(s,t).$$

We will now define what we mean by a labelling of D_n. These will turn out to have good metric embedding properties (see Theorem 13.19). A labelling is an injective mapping $\Phi_n : D_n \to \{0, 1\}^{2^n}$ belonging to a certain class of mappings that will be defined by induction on n. We impose on a labelling the following neighboring condition: For any two neighbors $s, t \in D_n$ the images $\Phi_n(s)$, $\Phi_n(t)$ differ by exactly one coordinate.

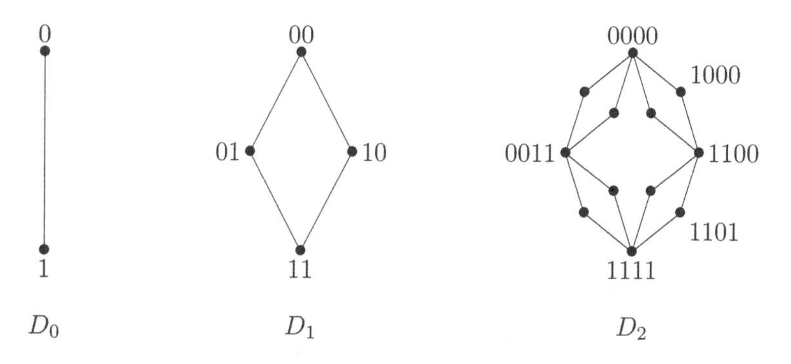

Figure 13.3. Diamond 2.

The inductive construction of labellings is as follows. We first label D_0 by $\Phi_0 : D_0 \to \{0, 1\}$. Note that we have two possible labellings, but we may choose anyone of them (so the points a,b in the preceding discussion are now labelled either as 0, 1 or as 1, 0). Then we consider the set D_0' that is a 'copy' of D_0 formed of the two points $(0, 0)$ and $(1, 1)$, and we label the two 'new' points (c and d in the preceding discussion) of D_1 using $(0, 1)$ and $(1, 0)$. This gives us a labelling of D_1.

Then assuming known a labelling $\Phi_{n-1} : D_{n-1} \to \{0,1\}^{2^{n-1}}$, we define $D'_{n-1} \subset \{0,1\}^{2^n}$ simply by 'doubling' each element $t = (t_1, \ldots, t_{2^{n-1}}) \in \Phi_{n-1}(D_{n-1})$, by repeating each coordinate i.e. we take introduce the doubling correspondence:

$$t = (t_1, \ldots, t_{2^{n-1}}) \mapsto \hat{t} = (t_1, t_1, \ldots, t_{2^{n-1}}, t_{2^{n-1}})$$

and we set $D'_{n-1} = \{\hat{t} \mid t \in \Phi_{n-1}(D_{n-1})\} \subset \{0,1\}^{2^n}$. Then if $(s,t) \in D_{n-1}$ is a pair of neighbours, we know $\Phi_{n-1}(s)$, $\Phi_{n-1}(t)$ differ just by one digit (and hence $(\hat{s}, \hat{t})$ differ by exactly two digits), so there are exactly two points s' and t' in $\{0,1\}^{2^n}$ that are 'midpoints' between $\hat{s}, \hat{t}$, i.e. such that $\rho(s', \hat{s}) = \rho(s', \hat{t}) = 1$ and similarly $\rho(t', \hat{s}) = \rho(t', \hat{t}) = 1$. (More precisely, we obtain s' by changing one of the differing two digits in s, and t' by changing the other one.) We let D''_n denote the collection of all the new points s', t' obtained in this way and we then identify D_n with $D'_{n-1} \cup D''_n \subset \{0,1\}^{2^n}$. This gives us an injective mapping $\Phi_n : D_n \to \{0,1\}^{2^n}$ that still satisfies the neighboring condition. The mappings that can be obtained in this way are what we call labellings.

Notation: Let $t = (t_j) \in \mathbb{R}^{2^n}$. We denote as usual by (e_j) the canonical basis of $\mathbb{R}^{2^n}$. We define

$$\|t\|_0 = \sup_j \left\{ \left|\sum\nolimits_{i<j} t_i\right| + \left|\sum\nolimits_{i \geq j} t_i\right| \right\},$$

and we also set

$$\|t\|_1 = \sum |t_j|.$$

The if part of Theorem 13.17 will follow from:

Theorem 13.19. *Any labelling $\Phi_n : D_n \to \{0,1\}^{2^n}$ satisfies*

$$\forall s, t \in D_n \quad \|\Phi_n(s) - \Phi_n(t)\|_1 \leq \delta_{D_n}(s,t), \tag{13.23}$$

$$\forall s, t \in D_n \quad (1/2)\delta_{D_n}(s,t) \leq \|\Phi_n(s) - \Phi_n(t)\|_0. \tag{13.24}$$

Moreover, equality holds in (13.23) *if either $\Phi_n(s)$ or $\Phi_n(t)$ is equal to* $(1, \ldots, 1)$.

Proof. Consider a pair (s,t) of neighbors in D_n. By construction of the embedding $\Phi_n : D_n \to \{0,1\}^{2^n}$ we know that (s,t) differ on exactly one coordinate, so that $\|\Phi_n(s) - \Phi_n(t)\|_1 = 1$. Then, the triangle inequality shows that an arbitrary pair (s,t) must satisfy (13.23). In addition, the equality case in the last assertion can be checked by an immediate induction argument.

We will prove (13.24) by induction on n. We urge the reader to first carefully check the picture for the case $n = 1$, which is the heart of the matter. Assume we have proved (13.24) for $n-1$ and let us prove it for n. We return to the 'fractal' process generating D_n. The key observation is that once D_1 has been generated the new points will all appear in 4 separate parts corresponding to each of the 4 edges of D_1. The 4 parts are disjoint except for the 4 vertices of D_1, each of which being common to 2 parts. It is easy to check that each of these 4 parts is an isometric copy of D_{n-1}. We can thus picture D_n starting from D_1 but imagining that we have replaced each of its edges by a copy of D_{n-1}.

Observe that the function Φ_n restricted to each of the 4 parts is just a translation of a labelling of D_{n-1} when the respective part is identified with D_{n-1}. Consider, for instance, the part of D_n emanating from the edge linking $(1,0)$ to $(1,1)$. This corresponds (via Φ_n) to the points $(t_j) \in \{0,1\}^{2^n}$ such that all the coordinates in the first half (i.e. the first 2^{n-1} coordinates) are equal to 1. Now we observe that there is a labelling Φ_{n-1} of D_{n-1} such that for any such t there is a unique $t' \in D_{n-1}$ satisfying

$$\Phi_n(t) = (1, \ldots, 1, \Phi_{n-1}(t')).$$

Then, we have for any pair (s,t) of such points

$$\|\Phi_n(t) - \Phi_n(s)\|_0 = \|(1, \ldots, 1, \Phi_{n-1}(t')) - (1, \ldots, 1, \Phi_{n-1}(s'))\|_0,$$

so we find

$$\|\Phi_n(t) - \Phi_n(s)\|_0 = \|\Phi_{n-1}(t') - \Phi_{n-1}(s')\|_0,$$

and since

$$d_{D_n}(t,s) = d_{D_{n-1}}(t',s')$$

the induction hypothesis gives us (13.24) for any pair (s,t) of such points. The other 3 cases are similar.

Thus we only have to worry about a pair (s,t) of points sitting in 2 different parts among the possible 4 ones. If (s,t) sit in two parts that are not disjoint (i.e. have one vertex in common coming from the 'mother' copy of D_1) the verification is rather easy. Assume for instance (all the other 3 cases of this type are similar) that t is as before so that $\Phi_n(t) = (1, \ldots, 1, t_{2^{n-1}+1}, \ldots, t_{2^n})$ and that $\Phi_n(s) = (s_1, \ldots s_{2^{n-1}}, 1, \ldots, 1)$. Let u be the point of D_n that is common to the two parts containing (s,t). Then $\Phi_n(u) = (1, 1, \ldots, 1)$. Consider the path joining (s,t) and passing through u. Then

$$\Phi_n(t) - \Phi_n(s) = \sum_1^{2^{n-1}} (1 - s_j) e_j - \sum_{2^{n-1}+1}^{2^n} (1 - t_j) e_j,$$

has coordinates first ≥ 0 then ≤ 0 with only one change of sign. Therefore,

$$\|\Phi_n(t) - \Phi_n(s)\|_0 = \sum_1^{2^{n-1}} (1 - s_j) + \sum_{2^{n-1}+1}^{2^n} (1 - t_j),$$

or equivalently

$$\|\Phi_n(t) - \Phi_n(s)\|_0 = \|\Phi_n(t) - \Phi_n(u)\|_1 + \|\Phi_n(u) - \Phi_n(s)\|_1.$$

But then by the equality case in (13.23)

$$\delta_{D_n}(t, s) \leq \delta_{D_n}(t, u) + \delta_{D_n}(u, s) = \|\Phi_n(t) - \Phi_n(u)\|_1 + \|\Phi_n(u) - \Phi_n(s)\|_1$$

and hence $\delta_{D_n}(t, s) \leq \|\Phi_n(t) - \Phi_n(s)\|_0$, and a fortiori (13.24) holds.

Lastly consider now the case when (s, t) sit in two disjoint parts. Assume for instance that t is as before such that $\Phi_n(t) = (1, \ldots, 1, t_{2^{n-1}+1}, \ldots, t_{2^n})$ and that $\Phi_n(s) = (0, \ldots, 0, s_{2^{n-1}+1}, \ldots, s_{2^n})$ (the other case of this type is similar). Then $\Phi_n(t) - \Phi_n(s) = \sum_1^{2^{n-1}} e_j + \sum_{2^{n-1}+1}^{2^n} (t_j - s_j) e_j$. Therefore (regardless of the values of s_j, t_j)

$$\|\Phi_n(t) - \Phi_n(s)\|_0 \geq 2^{n-1}$$

and since, by (13.19), we know that $\delta_{D_n}(t, s) \leq 2^n$ for any pair (s, t) we find

$$\|\Phi_n(t) - \Phi_n(s)\|_0 \geq (1/2)\delta_{D_n}(t, s).$$

This completes the induction step, and hence the proof. □

Since $\|.\|_0 \leq \|.\|_1$ we immediately deduce:

Corollary 13.20. *Any labelling Φ_n satisfies*

$$\forall s, t \in D_n \quad (1/2)\delta_{D_n}(s, t) \leq \rho(\Phi_n(s), \Phi_n(t)) \leq \delta_{D_n}(s, t).$$

In particular, there is an embedding Φ_n of (D_n, δ_{D_n}) into the Hamming cube $(\{0, 1\}^{2^n}, \rho)$ with distortion at most 2.

Proof of Theorem 13.17. If B is not super-reflexive then by Corollary 11.34 for any $0 < \theta < 1$ and any n there is a sequence $x_1, \ldots, x_{2^n}$ in B satisfying

$$\forall (\alpha_j) \in \mathbb{R}^{2^n} \qquad \theta\|(\alpha_j)\|_0 \leq \left\|\sum \alpha_i x_i\right\| \leq \|(\alpha_j)\|_1. \tag{13.25}$$

Indeed any such finite sequence in a space finitely representable in B can obviously be 'copied' back into B. Let $f_n : \mathbb{R}^{2^n} \to B$ be defined by $f_n((t_j)) = \sum_1^n t_j x_j$. Then, by Theorem 13.19, the function $F_n \colon D_n \to B$ defined by $F_n(t) = f_n(\Phi_n(t))$ satisfies

$$\forall s, t \in D_n \quad (\theta/2)\delta_{D_n}(s, t) \leq d(F_n(s), F_n(t)) \leq \delta_{D_n}(s, t). \tag{13.26}$$

Thus we obtain the if part. Note that we only need this for *some* $0 < \theta < 1$, so the required ingredient (13.25) can already be derived from (iii) in Theorem 11.10 (changing the θ in (11.7) to $\theta/3$ in (13.25)).

Conversely assume B super-reflexive. Then by (i) $\Rightarrow$ (iii) in Theorem 11.22 (or by Theorem 11.37) B does not contain arbitrarily long δ-separated bounded dyadic trees, so the preliminary observation implies that B also cannot contain $\{D_n\}$ Lipschitz uniformly. □

13.5 Markov type *p* and uniform smoothness

The notion of Markov type p was introduced by K. Ball in [108] using Markov chains on (finite subsets of) the Banach space under consideration.

Let E be an arbitrary finite set and let $X_0, X_1, \dots, X_n$ be a stationary symmetric Markov chain on E with invariant probability measure μ on E. This means that $X_0, \dots, X_n$ are E valued random variables on a probability space $(\Omega, \mathcal{A}, \mathbb{P})$ for which there is a symmetric kernel $P\colon\ E \times E \to \mathbb{R}_+$ ('transition probability') such that for any function $f\colon\ E \to V$ with values in (say) a vector space V we have for any $0 \le k \le n$

$$\mathbb{E}^{\sigma(X_0,\dots,X_k)} f(X_n) = \int P^{n-k}(X_k, \omega) f(\omega) d\mu(\omega). \tag{13.27}$$

Note in particular that, since this last expression depends only on X_k, this encodes the 'Markov property'

$$\mathbb{E}^{\sigma(X_0,\dots,X_k)} f(X_n) = \mathbb{E}^{\sigma(X_k)} f(X_n).$$

The symmetry of the transition probability P implies that the chain is reversible, i.e. that $X_0, X_1, \dots, X_n$ have the same joint distribution as $X_n, X_{n-1}, \dots, X_0$.

Definition 13.21. A Banach space B is called of Markov type p $(1 \le p \le 2)$ if there is a constant C such that, for any n, for any finite set E, any function $f\colon\ E \to B$ and any $(X_0, \dots, X_n)$ as earlier, we have

$$\|f(X_n) - f(X_0)\|_{L_p(B)} \le C n^{1/p} \|f(X_1) - f(X_0)\|_{L_p(B)}.$$

The smallest such C is called the Markov type p constant of B.

The next result from [357] answers a question left open by K. Ball in [108].

Theorem 13.22 ([357]). *Let $1 \le p \le 2$. If a Banach space B is isomorphic to a p-uniformly smooth, then B is of Markov type p.*

Proof. By our assumption on B, we know that all B-valued martingales in $L_p(B)$ satisfy (10.25). The idea of the proof is to show that $f(X_n) - f(X_0)$ can be rewritten as a sum

$$\sum d'_k + \sum d''_k + \delta$$

where (d'_k) and (d''_k) are martingale differences and we have for all $k = 1, \dots, n$

$$\max\{\|d'_k\|_{L_p(B)}, \|d''_k\|_{L_p(B)}, \|\delta\|_{L_p(B)}\} \le 2\|f(X_1) - f(X_0)\|_{L_p(B)}.$$

The desired result will then follow immediately from (10.25).

Let $\mathcal{A}_k = \sigma(X_0, \dots, X_k)$ and $\mathcal{B}_k = \sigma(X_n, \dots, X_k)$. Let $f_n = f(X_n)$ and $\delta_k = f_k - f_{k-1}$. Since δ_k is $\mathcal{A}_k$-measurable, we have obviously

$$f_n - f_0 = \sum\nolimits_1^n \delta_k = \sum\nolimits_1^n d'_k + \sum\nolimits_1^n \mathbb{E}^{\mathcal{A}_{k-1}} \delta_k \tag{13.28}$$

where $d'_k = (\mathbb{E}^{\mathcal{A}_k} - \mathbb{E}^{\mathcal{A}_{k-1}})(\delta_k)$.

We thus obtain $f_n - f_0$ written as a sum of martingale differences $\sum d'_k$ up to another term that we will now estimate. Since δ_{k-1} is $\mathcal{B}_{k-2}$-measurable, we have

$$f_n - f_0 = \sum\nolimits_2^{n+1} \delta_{k-1} = \sum\nolimits_2^{n+1} d''_k + \sum\nolimits_2^{n+1} \mathbb{E}^{\mathcal{B}_{k-1}}(\delta_{k-1}) \tag{13.29}$$

where $d''_k = (\mathbb{E}^{\mathcal{B}_{k-2}} - \mathbb{E}^{\mathcal{B}_{k-1}})(\delta_{k-1})$. Here again (d''_k) are martingale differences.

We now claim that for any $k = 2, \dots, n$

$$\mathbb{E}^{\mathcal{B}_{k-1}}(\delta_{k-1}) = -\mathbb{E}^{\mathcal{A}_{k-1}} \delta_k. \tag{13.30}$$

This is a simple consequence of the reversibility of the chain. Indeed, on one hand we have $\mathbb{E}^{\mathcal{A}_{k-1}} \delta_k = \mathbb{E}^{\mathcal{A}_{k-1}} f_k - f_{k-1}$ and hence by (13.27)

$$\mathbb{E}^{\mathcal{A}_{k-1}} \delta_k = \int P(X_{k-1}, t) f(t) d\mu(t) - f(X_{k-1}).$$

On the other hand, since $(X_n, \dots, X_0)$ and $(X_0, \dots, X_n)$ have the same distribution, we have

$$\mathbb{E}^{\mathcal{B}_{k-1}}(\delta_{k-1}) = \mathbb{E}^{X_n \dots X_{k-1}} f_{k-1} - \mathbb{E}^{X_n \dots X_{k-1}} f_{k-2} = f_{k-1} - \int P(X_{k-1}, t) f(t) d\mu(t),$$

and this proves Claim 13.30. Thus, adding (13.28) and (13.29) yields

$$2(f_n - f_0) = \sum\nolimits_1^n d'_k + \sum\nolimits_2^{n+1} d''_k + \mathbb{E}^{\mathcal{A}_0} \delta_1 + \mathbb{E}^{\mathcal{B}_n} \delta_n. \tag{13.31}$$

We now observe that by the triangle inequality $\|d'_k\|_{L_p(B)} \le 2\|\delta_k\|_{L_p(B)}$ and since (X_k, X_{k-1}) and (X_1, X_0) have the same distribution, we have $\|\delta_k\|_{L_p(B)} = \|f_1 - f_0\|_{L_p(B)}$ for all k, so that

$$\|d'_k\|_{L_p(B)} \le 2\|f_1 - f_0\|_{L_p(B)}$$

and similarly

$$\|d_k''\|_{L_p(B)} \leq 2\|f_1 - f_0\|_{L_p(B)}.$$

Thus we obtain finally from (13.31)

$$\|f_n - f_0\|_{L_p(B)} \leq \left\|\sum\nolimits_1^n d_k'\right\|_{L_p(B)} + \left\|\sum\nolimits_2^{n+1} d_k''\right\|_{L_p(B)} + 2\|f_1 - f_0\|_{L_p(B)}$$

and since by our assumption on B all B-valued martingales satisfy (10.25) we conclude that

$$\|f_n - f_0\|_{L_p(B)} \leq (Cn^{1/p} + Cn^{1/p} + 2)\|f_1 - f_0\|_{L_p(B)}. \qquad \square$$

Remark. The converse to Theorem 13.22 remains an open problem: It is rather easy to show that Markov type $p > 1$ implies type p, but it is unclear whether it implies super-reflexivity.

13.6 Notes and remarks

Concerning concentration inequalities, general references are Milman and Schechtman's [68] and more recently Ledoux's book [53] (see also [54, 92]). See Talagrand's work [431] for an account of many refinements on concentration of measure and isoperimetric inequalities.

In F. Baudier's [111], Bourgain's metric characterization of super-reflexivity is 'patched up' to produce a slightly surprising infinite-dimensional version: if B is not super-reflexive then B contains a bi-Lipschitz copy of the infinite dyadic tree. See also [112, 369], for a more general 'pasting principle' of finite embeddings into infinite ones. See also the more recent paper [113] in connection with property (β), which is somewhat analogous to uniform convexity.

In a similar vein, using Chatterji's Theorem 2.9, Ostrovskii [371] proved the following metric characterization of the RNP: A Banach space X does not have the RNP if and only if there exists a metric space M containing a thick family T of geodesics that admits a bi-Lipschitz embedding into X.

In [322], Lee, Naor and Peres show that an infinite weighted tree admits a bi-Lipschitz embedding into Hilbert space if and only if it does not contain arbitrarily large complete binary trees with uniformly bounded distortion. In that same paper, they introduce the notion of Markov p-convexity. In [351], the authors prove that a Banach space is Markov p-convex iff it is isomorphic to a p-uniformly convex space, or equivalently iff it is of M-cotype p in the sense of Definition 10.41.

In Ostrovskii's [368] the preliminary observation presented after (13.20) was made (this simplifies part of Johnson and Schechtman's original proof from [289]) and various related results were obtained.

See [370] for characterizations of super-reflexivity in terms of bilipschitz embeddability of word hyperbolic groups.

See [321] for information on the growth of the embedding constants of the diamond graphs in L_p.

For more information on the theme of this chapter we refer the reader to M. I. Ostrovskii's book [73].

The main result in §13.5 is due to Naor, Peres, Schramm and Sheffield [357], completing Keith Ball's earlier results in [108]. See [212] for more recent results on Markov type p.

14

An invitation to martingales in non-commutative L_p-spaces*

In this chapter, we turn to non-commutative martingale inequalities. This is a rather recent and still active development, but since we plan to devote an entire second volume to that subject, we merely content ourselves here with a general outline.

14.1 Non-commutative probability space

Let H be a Hilbert space and let $B(H)$ denote the algebra of all bounded operators on H. Recall that for $T \in B(H)$, $T \geq 0$ means $\langle Th, h \rangle \geq 0\ \forall h \in H$. We denote

$$B(H)_+ = \{T \in B(H) \mid T \geq 0\}.$$

In the non-commutative setting, the analogue of the σ-algebra is a von Neumann algebra $M \subset B(H)$ i.e. a self-adjoint unital subalgebra of $B(H)$ that is closed in the weak operator topology (or equivalently in several other inequivalent topologies). A fortiori, M is a C^*-subalgebra, i.e. it is norm-closed. But von Neumann algebras have an additional property (not shared by general C^*-algebras): they are dual spaces. More precisely, there is a specific subspace $M_* \subset M^*$ called the predual such that we have an isometric identification:

$$M \simeq (M_*)^*.$$

We will denote by σ the weak-$*$ topology on M, which is usually denoted by $\sigma(M, M_*)$. The analogue of a finite measure (resp. a probability) is a positive σ-continuous linear functional $\tau\colon\ M \to \mathbb{C}$ (resp. such that $\tau(1) = 1$) that is also a trace (or 'is tracial'). By tracial, we mean

$$\forall x, y \in M \quad \tau(xy) = \tau(yx).$$

Here 'positive' means $\tau(x) \geq 0$ for all x in $M_+ = M \cap B(H)_+$. A standard assumption is that τ is faithful, i.e. if $x \in M_+$ and $\tau(x) = 0$ then necessarily $x = 0$. This is analogous to the removal of negligible sets from the σ-algebra. We will always assume in the sequel that our trace τ is 'standard' by which we mean faithful and σ-continuous (one usually says 'normal' instead of σ-continuous).

The σ-algebra of 'events' is replaced by the set $\mathcal{P}(M)$ of all self-adjoint projections P in M. Equivalently one may choose to identify a projection $P \in \mathcal{P}(M)$ with the closed subspace $E \subset H$ such that $P(H) = E$ (i.e. P is the orthogonal projection onto E). Then the 'measurable sets' are replaced by the closed subspaces of H and 'disjointness' by orthogonality. The fact that τ is normal is essentially equivalent to the σ-additivity of τ: for any family $(P_j)_{j\in J}$ of mutually orthogonal projections in $\mathcal{P}(M)$ we have

$$\tau\left(\sum\nolimits_{j\in J} P_j\right) = \sum\nolimits_{j\in J} \tau(P_j).$$

Note that since we assume τ faithful, the set of j's for which $P_j \neq 0$ is countable if $\sum \tau(P_j) < \infty$.

To summarize, by a 'generalized probability space' we mean a von Neumann algebra M equipped with a 'standard' trace τ.

14.2 Non-commutative L_p-spaces

Given our generalized (possibly non-commutative) probability space (M, τ), we define $L_p(M, \tau)$ as the completion of M for the norm $\|\ \|_p$ defined by

$$\forall x \in M \qquad \|x\|_p = (\tau(|x|^p))^{1/p}. \tag{14.1}$$

Note that when $p = 2$ this is a Hilbert space norm associated to the scalar product defined by:

$$\forall x, y \in M \quad \langle x, y\rangle = \tau(y^*x).$$

When $0 < p < 1$ we obtain only a quasi-Banach space, just like in Remark 1.8.

Here $|x| = (x^*x)^{1/2}$. We recall that any x in $B(H)$ has a unique polar decomposition $x = u|x|$ where u is a partial isometry vanishing on $|x|(H)^\perp$. When $x \in M$, its decomposition $x = u|x|$ respects M, i.e. we have $u \in M$ and $|x| \in M$. Of course we have $|x| \geq 0$. A fortiori $|x|$ is self-adjoint and hence we may apply spectral theory to it. More generally, for any operator $a \in M$ such that $a^*a = aa^*$ (such operators are called 'normal'), we can make sense of $f(a)$ for any continuous function f on the spectrum of a, and we have $f(a) \in M$. Indeed, this is obvious if f is a polynomial in z and $\bar{z}$, and for the general case, since M

is norm-closed, one may approximate f by such a polynomial uniformly over the spectrum of a. This applies in particular to the function $f(t) = t^p$ on any interval containing the spectrum of $|x|$, such as $[0, \|x\|]$. The properties of the spaces $L_p(\tau)$ can be summarized as follows:

Theorem 14.1. *Let $1 \le p < \infty$.*

(i) The space $L_p(\tau) = L_p(M, \tau)$ is a Banach space, and in case $p = 2$ a Hilbert space.

(ii) We have $\|axb\|_p \le \|a\| \|x\|_p \|b\|$ for all a, x, b in M and hence $x \mapsto axb$ extends by density to a bounded linear map on $L_p(\tau)$ such that for any x in $L_p(\tau)$

$$\forall a, b \in M \qquad \|axb\|_p \le \|a\| \|x\|_p \|b\|.$$

(iii) Similarly we have $\|x\|_p = \|x^*\|_p$ for any x in M and hence $x \mapsto x^*$ extends by density to an isometry on $L_p(\tau)$ so that for any x in $L_p(\tau)$

$$\|x^*\|_p = \|x\|_p.$$

(iv) Let $1 \le p, q, r < \infty$ be such that $\frac{1}{p} + \frac{1}{q} = \frac{1}{r}$. We have $\|ab\|_r \le \|a\|_p \|b\|_q$ for any a, b in M and hence the product $(a, b) \mapsto ab$ extends to a bounded bilinear map $L_p(\tau) \times L_q(\tau) \to L_r(\tau)$ so that we have

$$\forall a \in L_p(\tau) \forall b \in L_q(\tau) \qquad \|ab\|_r \le \|a\|_p \|b\|_q.$$

In particular, if $\tau(1) = 1$ and $1 \le r \le p < \infty$, we have $L_p(\tau) \subset L_r(\tau)$ and $\|a\|_r \le \|a\|_p$ for any $a \in L_p(\tau)$.

(v) We have $|\tau(x)| \le \|x\|_1 = \tau(|x|)$ for any x in M and hence the functional $x \mapsto \tau(x)$ extends to a continuous linear form on $L_1(\tau)$ and we have

$$\forall x \in L_1(\tau) \qquad |\tau(x)| \le \|x\|_1.$$

Let $a \in M$ and $x \in L_1(\tau)$. By (ii)*, we have $ax \in L_1(\tau)$. Let us denote*

$$\langle a, x \rangle = \tau(ax), \tag{14.2}$$

so that, by (ii) *again, $|\langle a, x \rangle| \le \|a\| \|x\|_1$. In the duality* (14.2)*, M is isometric to the dual of $L_1(\tau)$.*

(vi) More generally, if $1 < p, p' < \infty$ with $p^{-1} + p'^{-1} = 1$, the duality (14.2) *extends to the case $a \in L_p(\tau)$, $x \in L_{p'}(\tau)$ and we have isometrically*

$$L_p(\tau)^* = L_{p'}(\tau).$$

In addition to the preceding basic properties, the spaces $L_p(\tau)$ behave very nicely with respect to real or complex interpolation. For instance, for any

$1 \le p_0, p_1 < \infty$ the pair $(L_{p_0}(\tau), L_{p_1}(\tau))$ is an interpolation pair and we have isometrically for any $0 < \theta < 1$

$$(L_{p_0}(\tau), L_{p_1}(\tau))_\theta = L_{p_\theta}(\tau) \tag{14.3}$$

where $\frac{1}{p_\theta} = \frac{1-\theta}{p_0} + \frac{\theta}{p_1}$. Moreover, if we set by convention

$$L_\infty(\tau) = M$$

equipped with the operator norm (denoted $\| \ \|_\infty$ by convention), then (14.3) remains valid even if either $p_0 = \infty$ or $p_1 = \infty$.

As the reader may have guessed, the role of the 'realization' of M as operators on H is of limited influence. If we use a different realization $M \subset B(K)$ on another Hilbert space K but so that the underlying isomorphism on M preserves the trace τ, then obviously the resulting spaces $L_p(M, \tau)$ will be essentially identical (i.e. will correspond to each other via the underlying isomorphism). There is however a more convenient choice of realization of M, called the 'standard' one, as follows. For $a \in M$, we denote by $L(a) \in B(L_2(\tau))$ (resp. $R(a) \in B(L_2(\tau))$) the operator of left (resp. right) multiplication by a on $L_2(\tau)$, i.e. $L(a)x = ax$ (resp. $R(a)x = xa$). Then (assuming τ itself standard) $L\colon\ M \to B(L_2(\tau))$ is a σ-bicontinuous isometric isomorphism (of $*$-algebras) that allows us to identify M to $L(M)$. Similarly $R\colon\ M \to B(L_2(\tau))$ establishes an anti-isomorphism between M and $R(M)$ ('anti' because $R(xy) = R(y)R(x)$). These left and right realizations of M and M with reverse multiplication are sort of mirrors of each other, and actually the commutant of $L(M)$ (resp. $R(M)$) coincides with $R(M)$ (resp. $L(M)$). Choosing the left (here an arbitrary choice!) we are led to identify $a \in M$ with the operation of left multiplication by a.

Let $\dot{1}$ be the unit of M viewed as sitting in $L_2(\tau)$. We have obviously

$$\forall a \in M \qquad \tau(a) = \langle L(a)\dot{1}, \dot{1}\rangle \tag{14.4}$$

where the scalar product is in $L_2(\tau)$. Note that the unit vector $\dot{1}$ is cyclic, i.e. $L(M)\dot{1}$ is dense in $L_2(\tau)$.

In the commutative case, all von Neumann algebras equipped with a standard τ are isomorphic to some $L_\infty(\Omega, \mathcal{A}, \mathbb{P})$ in such a way that $\tau(x) = \int x(\omega)\mathbb{P}(d\omega)$. In that case $L_2(\tau) \simeq L_2(\Omega, \mathcal{A}, \mathbb{P})$ and the realization

$$L\colon\ L_\infty(\Omega, \mathcal{A}, \mathbb{P}) \to B(L_2(\Omega, \mathcal{A}, \mathcal{P}))$$

identifies a function $f \in L_\infty(\Omega, \mathcal{A}, \mathbb{P})$ with the operator of multiplication by f on $L_2(\Omega, \mathcal{A}, \mathbb{P})$ (and now $\dot{1}$ is the constant function identically equal to 1).

14.3 Conditional expectations: non-commutative martingales

Let $N \subset M$ be a von Neumann subalgebra of M. The restriction $\tau_{|N}$ is obviously a standard trace on N. By construction, we have an isometric 'inclusion'

$$L_2(N, \tau_{|N}) \subset L_2(M, \tau)$$

and more generally for any $1 \le p < \infty$

$$L_p(N, \tau_{|N}) \subset L_p(M, \tau)$$

so that we may (and do) identify $L_p(N, \tau_{|N})$ with a subspace of $L_p(M, \tau)$, namely the closure of N in $L_p(M, \tau)$. We can then introduce the conditional expectation $\mathbb{E}^N$ as the orthogonal projection from $L_2(M, \tau)$ onto $L_2(N, \tau_{|N})$. One then checks rather easily that, for any $1 \le p < \infty$, $\mathbb{E}^N$ extends (by density) to an operator of norm 1 on $L_p(\tau)$, which is again a contractive projection onto $L_p(N, \tau_{|N}) \subset L_p(M, \tau)$. Let $L_p(\tau)_+$ denote the closure of M_+ in $L_p(\tau)$. Then $\mathbb{E}^N$ is 'positivity preserving' (one often simply says 'positive'), i.e.

$$\forall x \in L_p(\tau)_+ \qquad \mathbb{E}^N(x) \ge 0.$$

Since any self-adjoint element of M is the difference of two positive ones, a fortiori, $\mathbb{E}^N(x)$ remains self-adjoint if $x = x^*$, or, equivalently, we have $\mathbb{E}^N(x)^* = \mathbb{E}^N(x^*)$ for any x in M. Of course, the self-adjointness of $\mathbb{E}^N$ on $L_2(M, \tau)$ extends to $L_p(\tau)$ and we have

$$\forall x \in L_p(\tau)\ \forall y \in L_{p'}(\tau) \qquad \tau(\mathbb{E}^N(x)y) = \tau(x\mathbb{E}^N(y)).$$

Moreover, the classical property of conditional expectations becomes in this setting:

$$\forall x \in L_p(\tau)\ \forall a, b \in N \qquad \mathbb{E}^N(axb) = a\mathbb{E}^N(x)b.$$

At this stage, the generalized notion of 'filtration' is immediate: a filtration is a family $(\mathcal{M}_n)_{n \ge 0}$ of von Neumann subalgebras of M such that $\mathcal{M}_0 \subset \mathcal{M}_1 \subset \mathcal{M}_2 \subset \cdots$. As before, a martingale $(f_n)_{n \ge 0}$ is a sequence in $L_1(M, \tau)$ such that

$$\forall n \ge 0 \qquad f_n = \mathbb{E}^{\mathcal{M}_n}(f_{n+1}).$$

Since we have a positive cone in $L_1(M, \tau)$, submartingales and supermartingales in $L_1(M, \tau)$ also have straightforward extensions. Let $\mathcal{M}_\infty$ be the von Neumann algebra generated by $\{\mathcal{M}_n \mid n \ge 0\}$, i.e. the σ-closure of the union of $\{\mathcal{M}_n \mid n \ge 0\}$. The basic martingale convergence theorems in L_p have immediate extensions:

Theorem 14.2. *Let $1 \le p < \infty$. For any f in $L_p(M, \tau)$, let $f_n = \mathbb{E}^{\mathcal{M}_n} f$. Then (f_n) is a martingale adapted to $(\mathcal{M}_n)$ and $f_n \to \mathbb{E}^{\mathcal{M}_\infty}(f)$ in $L_p(\tau)$. Moreover, if*

$1 < p < \infty$, any martingale (f_n) adapted to $(\mathcal{M}_n)$ that is bounded in $L_p(M, \tau)$ is of this form, i.e. f_n converges in $L_p(\tau)$ to a limit f_∞ and $f_n = \mathbb{E}^{\mathcal{M}_n}(f_\infty)$ for any $n \geq 0$.

We end this section with a few words about L_p-spaces associated to infinite traces. By a semi-finite trace on a von Neumann algebra M, we mean a positively homogeneous σ-additive functional $\tau\colon\ M_+ \to [0, \infty]$ that is tracial, i.e. such that $\tau(x^*x) = \tau(xx^*)$ for any x in M and moreover such that the (self-adjoint) subalgebra $\{x \in M \mid \tau(|x|) < \infty\}$ generates M as a von Neumann algebra (i.e. is σ-dense in M). We also require τ to be 'normal' in the sense that $\tau(P_\alpha) \to \tau(P)$ if P_α is a directed net in $\mathcal{P}$ admitting $P \in \mathcal{P}$ as its least upper bound (or its σ-limit). If $\tau(1) < \infty$, then τ can be uniquely extended to a σ-continuous linear form on M and after normalization we recover the preceding notion of generalized probability. The latter allows us to view $L_p(M, \tau)$ as already defined for any semi-finite trace such that $\tau(1) < \infty$. Then in the general semi-finite case, one can show that there is an increasing net of projections (P_α) in $\mathcal{P}$ with $\tau(P_\alpha) < \infty$ for all α and such that $P_\alpha \to 1$ for the σ-topology (or the strong operator topology).

Let $M_\alpha = P_\alpha M P_\alpha$ and $\tau_\alpha = \tau_{|M_\alpha}$. Then M_α is a (non-unital) von Neumann subalgebra of M with finite trace τ_α, and for any $\alpha \leq \beta$ we have a natural isometric inclusion

$$L_p(M_\alpha, \tau_\alpha) \subset L_p(M_\beta, \tau_\beta).$$

We can then define $L_p(M, \tau)$ as the completion of the union (or direct limit) of the spaces $L_p(M_\alpha, \tau_\alpha)$. Moreover, one can show that it does not depend on the particular choice of the net (P_α). This gives us a satisfactory definition of $L_p(M, \tau)$ in the semi-finite case. The typical example of this situation is the Schatten p-class $S_p(H)$ associated to $M = B(H)$ equipped with the semi-finite trace $T \in M_+ \to \mathrm{tr}(T)$ that is the usual trace when T is a trace class operator and $\mathrm{tr}(T) = \infty$ otherwise. In that case, $L_2(B(H), \mathrm{tr})$ (resp. $L_1(B(H), \mathrm{tr})$) coincides with the Hilbert-Schmidt (resp. trace) class. The space $L_p(B(H), \mathrm{tr})$ is usually denoted by $S_p(H)$.

When $\dim(H) < \infty$, then the usual trace $\tau(x) = \mathrm{tr}(x)$ is finite on $B(H)$. Thus, the simplest example of a generalized probability space is the space M_n of $n \times n$ complex matrices equipped with the normalized trace $\tau_n(x) = \frac{1}{n}\mathrm{tr}(x)$. Given a matrix $x = [x_{ij}]$ with eigenvalue $(\lambda_j(x))$, we have

$$\tau_n(x) = \frac{1}{n}\sum\nolimits_1^n x_{jj} = \frac{1}{n}\sum\nolimits_1^n \lambda_j(x).$$

14.4 Examples

If $H = \ell_2$ (resp. $H = \ell_2^n$), the space $S_p(H)$ appears as a non-commutative generalization of ℓ_p (resp. ℓ_p^n).

Returning to probability spaces, (M_n, τ_n) is the natural analogue of the space $[1, \ldots, n]$ equipped with the uniform probability $\mathbb{P} = \frac{1}{n}\sum_1^n \delta_j$.

Just like in classical probability where they form the foundation for stochastic independence, product spaces are very fruitful. Given two generalized probability spaces (M, τ) and (N, φ), with $M \subset B(H)$, $N \subset B(K)$ the product von Neumann algebra $M\overline{\otimes}N$ is defined as the σ-closure of the algebraic tensor product $M \otimes N$ viewed as sitting in $B(H \otimes_2 K)$. Then one can prove that $\tau \otimes \varphi$ extends to a standard trace on $M\overline{\otimes}N$ (assuming τ and φ were themselves standard).

This construction obviously extends to finite families (N_j, φ_j) of generalized probability spaces, yielding their product $\overline{\bigotimes}_{1\le j\le n} N_j$ equipped with $\bigotimes_{1\le j\le n}\varphi_j$. One can then generalize this to infinite families: say for an infinite sequence (N_j, φ_j) we will consider the inclusion

$$\overline{\bigotimes_{1\le j\le n}} N_j \to \overline{\bigotimes_{1\le j\le n+1}} N_j$$

defined by $x \to x \otimes 1$. One can then define the infinite product $\overline{\bigotimes}_{j\ge 1} N_j$ as well as $\bigotimes_{j\ge 1}\varphi_j$ as an inductive limit.

More precisely, we need to assume given for each j a unit vector $\xi_j \in H_j$ such that $\varphi_j(a) = \langle a\xi_j, \xi_j\rangle$ and ('cyclicity') $\overline{N_j\xi_j} = H_j$ (just as in (14.4)). We can then define isometric inclusions

$$\bigotimes\nolimits_{1\le j\le n} H_j \to \bigotimes\nolimits_{1\le j\le n+1} H_j$$

by $x \mapsto x \otimes \xi_{n+1}$. Completing the union of the resulting tower of spaces, this leads us to

$$H = \bigotimes\nolimits_{j\ge 1} H_j$$

and we can proceed as for finite products: firstly we define an embedding

$$N_1 \otimes \cdots \otimes N_n \subset B(H)$$

by viewing $N_1 \otimes \cdots \otimes N_n$ as acting on $H = (H_1 \otimes \cdots \otimes H_n) \otimes H_n'$ (here $H_n' = \bigotimes_{j>n} H_j$) by $T \to T \otimes id$; then we define $\bigotimes_{j\ge 1} N_j$ as the σ-closure of the union of the finite products $N_1 \otimes \cdots \otimes N_n$ viewed as sitting in $B(H)$.

For example we may consider the infinite product

$$M = \bigotimes_{j \geq 1} (N_j, \varphi_j)$$

when $(N_j, \varphi_j) = (M_2, \tau_2)$ for all $j \geq 1$. The resulting generalized probability space

$$(\mathcal{R}, \tau^{\infty}) = \bigotimes_{\mathbb{N}} (M_2, \tau_2)$$

is the non-commutative analogue of our familiar dyadic example

$$(\Delta, \nu) = \bigotimes_{\mathbb{N}} \left(\{-1, 1\}, \frac{\delta_1 + \delta_{-1}}{2} \right).$$

Just as in the classical setting, we have a natural filtration $\cdots \subset \mathcal{M}_n \subset \mathcal{M}_{n+1} \subset \cdots \subset \mathcal{R}$ obtained by identifying $\mathcal{M}_n$ with the product $(M_2, \tau_2)^{\otimes n} \otimes 1 \subset (\mathcal{R}, \tau^{\infty})$ (and setting, say, $\mathcal{M}_0 = \mathbb{C}1$). Note that now $\dim \mathcal{M}_n = 4^n$. The generalized probability space $(\mathcal{R}, \tau^{\infty})$ can be viewed in many ways as the analogue of the Lebesgue interval $([0, 1], dt)$. Indeed, since, as measure spaces, $([0, 1], dt)$ and (Δ, ν) are the 'same' object (i.e. are isomorphic), it is only a matter of taste (or context) to choose one or the other for measure theoretic purposes. Actually, the fundamental nature of $([0, 1], dt)$ is perhaps best expressed by the following classical fact.

Theorem 14.3. *The Lebesgue measure space* $([0, 1], dt)$ *is (up to isomorphism) the only countably generated probability space without any atom.*

Let (M, τ) be a generalized probability space. We will say that it is non-atomic if it contains no (non-zero) minimal projections (explicitly, $\forall P \neq 0\ P \in \mathcal{P}\ \exists Q \in \mathcal{P}$ such that $0 < Q < P$). We will say that it is countably generated if there is a countable subset of M generating M as a von Neumann algebra. Equivalently, this means that $L_2(M, \tau)$ (or $L_p(M, \tau)$ for $1 \leq p < \infty$) is norm-separable.

Unfortunately, Theorem 14.3 does not extend as stated to the non-commutative setting. First we need to restrict to 'factors,' i.e. algebras M with trivial centre, i.e. centre reduced to $\mathbb{C}1$. But even then, it is known that there are uncountably many mutually non-isomorphic examples of (M, τ) that are countably generated, non-atomic factors. However, if we restrict ourselves to 'amenable' (or equivalently 'injective') ones then we find a unique object.

We say that $M \subset B(H)$ is injective if there is a projection $P\colon\ B(H) \to M$ with $\|P\| = 1$. It is known that all commutative von Neumann algebras are injective.

We can now state Connes's far reaching generalization of Theorem 14.3:

Theorem 14.4 ([189]). *The space $(\mathcal{R}, \tau^\infty)$ is the unique countably generated, non-atomic generalized probability space that is also an injective factor.*

To illustrate the power of this statement, we now introduce the space (M, τ) associated to a discrete group G. Let $\lambda_G\colon\ G \to B(\ell_2(G))$ be the left regular representation of G, i.e. $\lambda_G(t)$ is the (unitary) operator of left translation by t on $\ell_2(G)$. Let $\{\delta_t \mid t \in G\}$ denote the canonical basis of $\ell_2(G)$. Let

$$\mathcal{L}(G) = \overline{\text{span}[\lambda_G(t) \mid t \in G]}^{\sigma}$$

and for any T in $\mathcal{L}(G)$ let

$$\tau_G(T) = \langle T\delta_e, \delta_e\rangle.$$

Then $(\mathcal{L}(G), \tau_G)$ is a generalized probability space.

It is known that $\mathcal{L}(G)$ is a factor iff G has infinite classes of conjugation (i.c.c. in short). The latter means that for any $t \neq e$, the set $\{xtx^{-1} \mid x \in G\}$ is infinite. Moreover, $\mathcal{L}(G)$ is injective iff G is amenable. Thus, Theorem 14.4 immediately implies

Corollary 14.5. *For any i.c.c. countable discrete amenable group G, the associated generalized probability space $(\mathcal{L}(G), \tau_G)$ is isomorphic to $(\mathcal{R}, \tau^\infty)$, i.e. there is a trace preserving isomorphism $\Phi\colon\ \mathcal{L}(G) \to \mathcal{R}$.*

For any subgroup $G_0 \subset G$, it is not hard to see that $\mathcal{L}(G_0)$ (resp. τ_{G_0}) can be identified with the von Neumann subalgebra $N_0 \subset \mathcal{L}(G)$ generated by $\lambda_G(G_0)$ (resp. with the restriction of τ_G to N_0). Thus we may associate to any subgroup $G_0 \subset G$, a conditional expectation, namely $\mathbb{E}^{N_0}$ on $\mathcal{L}(G)$. More generally, let $G_0 \subset G_1 \subset \cdots \subset G_n \subset \cdots$ be an increasing sequence of subgroups of G. Let $\mathcal{M}_n$ denote the von Neumann algebra generated by $\lambda_G(G_n)$. Then $(\mathcal{M}_n)_{n\geq 0}$ is a fruitful example of generalized filtration with a rich associated martingale theory. For instance, we may let G be a free group with a sequence of free generators (g_n) and let G_n be the subgroup generated by $g_1, \ldots, g_n$. Then G_n is a free group with n generators. More examples of this kind can be found in Voiculescu's 'Free Probability Theory' (see [96]).

14.5 Non-commutative Khintchin inequalities

Recall the notation $\Delta = \{-1, 1\}^{\mathbb{N}}$ with usual probability ν and coordinate functions (ε_n). Consider first a classical (commutative) L_p-space $B = L_p(\Omega, \mathcal{A}, m)$ on a general measure space $(\Omega, \mathcal{A}, m)$. For any $0 < p < \infty$ there are positive constants A_p, B_p such that for any finite set $x_1, \ldots, x_n$ in $B = L_p(m)$

$$A_p \left\| \left(\sum |x_j|^2 \right)^{1/2} \right\|_B \le \left(\int_\Delta \left\| \sum \varepsilon_j x_j \right\|_B^p d\nu \right)^{1/p} \le B_p \left\| \left(\sum |x_j|^2 \right)^{1/2} \right\|_B . \tag{14.5}$$

Indeed, this is an immediate consequence of the Khintchin inequality (5.7) raised at the p-th power and integrated with respect to μ.

As a corollary, the series $\sum_1^\infty \varepsilon_j x_j$ converges in $L_p(\Delta, \nu; B)$ (recall that this is equivalent to its a.s. convergence in B) iff $\sum_1^\infty |x_j|^2$ converges in $L_{p/2}$.

Assume now that $B = L_p(M, \tau)$ where (M, τ) is a semi-finite generalized measure space. The non-commutative analogue of (14.5) is as follows:

Theorem 14.6 ([338]). *Let $2 \le p < \infty$. There are positive constants A'_p, B'_p such that for any finite set $(x_1, \ldots, x_n)$ in $B = L_p(M, \tau)$ we have*

$$A'_p \left\| \left(\sum |x_j|_\bullet^2 \right)^{1/2} \right\|_B \le \left(\int \left\| \sum \varepsilon_j x_j \right\|_B^p d\nu \right)^{1/p} \le B'_p \left\| \left(\sum |x_j|_\bullet^2 \right)^{1/2} \right\|_B \tag{14.6}$$

where we use the notation

$$|x|_\bullet = \left(\frac{x^* x + x x^*}{2} \right)^{1/2} . \tag{14.7}$$

Note that if $x = a + ib$ with $a = a^*, b = b^*$ we have $|x|_\bullet = (a^2 + b^2)^{1/2}$. These inequalities were first proved by F. Lust-Piquard in [338]. The fact that $B'_p \in O(\sqrt{p})$ was later observed independently by Junge and the author (see [395, p. 106]). The restriction to $p \ge 2$ is surprising at first glance, but the striking discovery of [338] is that Theorem 14.6 fails for $p < 2$. In that case, the 'correct' form of the non-commutative Khintchin inequality is as follows:

Theorem 14.7 ([338, 339]). *For any $1 \le p \le 2$ there are positive constants A'_p, B'_p such that, for any finite set $(x_1, \ldots, x_n)$ in $B = L_p(M, \tau)$, we have*

$$A'_p |||(x_j)|||_p \le \left(\int \left\| \sum \varepsilon_j x_j \right\|_B^p d\nu \right)^{1/p} \le B'_p |||(x_j)|||_p , \tag{14.8}$$

with the notation

$$|||(x_j)|||_p = \inf \left\{ \left\| \left(\sum y_j^* y_j \right)^{1/2} \right\|_B + \left\| \left(\sum z_j z_j^* \right)^{1/2}] \right\|_B \right\} \qquad (1 \le p \le 2)$$

where the infimum runs over all possible decompositions $x_j = y_j + z_j$, $(y_j, z_j \in B = L_p(\tau))$. Moreover, we have $\inf_{1 \le p \le 2} A'_p > 0$.

Remark 14.8. A convenient way to 'unify' (14.5) and (14.6) is to define $|||\ |||_p$ differently for $p \le 2$ and $p \ge 2$. More explicitly, we set

$$|||(x_j)|||_p = \max\left\{\left\|\left(\sum x_j^* x_j\right)^{1/2}\right\|_p, \left\|\left(\sum x_j x_j^*\right)^{1/2}\right\|_p\right\} \qquad (2 \le p < \infty).$$

With this specific notation, (14.8) remains valid for any $1 \le p < \infty$.

The inequalities (14.6) and (14.8) are fundamental tools to investigate unconditional or almost unconditional convergence in $L_p(M, \tau)$. To give a more concrete example, consider an infinite matrix $\{a(i, j) \mid (i, j) \in \mathbb{N}\}$ with scalar entries. Let $\varepsilon(i, j) = \pm 1$ be random choices of signs chosen independently with equal probability $1/2$ (so this is our familiar object but here indexed by $\mathbb{N} \times \mathbb{N}$ instead of $\mathbb{N}$).

Corollary 14.9. *If $2 \le p < \infty$ (resp. $1 \le p \le 2$) the matrix $[\varepsilon(i, j)a(i, j)]$ is in S_p for almost all choices of signs $\varepsilon(i, j)$ iff* (i) *holds (resp. iff* (ii) *holds):*

(i) $$\sum_i \left(\sum_j |a(i, j)|^2\right)^{p/2} + \sum_j \left(\sum_i |a(i, j)|^2\right)^{p/2} < \infty$$

(ii) There is a decomposition $a(i, j) = b(i, j) + c(i, j)$ $(\forall i, j \in \mathbb{N})$ such that

$$\sum_i \left(\sum_j |b(i, j)|^2\right)^{p/2} + \sum_j \left(\sum_i |c(i, j)|^2\right)^{p/2} < \infty.$$

We note in passing that the case $0 < p < 1$, which remained open until recently, was settled by Éric Ricard (see [398] which completes the previous attempt in [396]). Thus (14.8) still holds when $0 < p < 1$.

14.6 Non-commutative Burkholder inequalities

Consider a generalized probability space (M, τ) equipped with a filtration $\mathcal{M}_0 \subset \mathcal{M}_1 \subset \mathcal{M}_2 \subset \cdots \subset M$. Let (f_n) be a martingale in $L_p(\tau)$ adapted to $(\mathcal{M}_n)$, and let $\xi_n = \pm 1$ be a sequence of signs. Let

$$\tilde{f}_n = f_0 + \sum_1^n \xi_k (f_k - f_{k-1}). \tag{14.9}$$

The non-commutative version of Burkholder's inequality (see §5.1) is as follows:

Theorem 14.10 ([401]). *Let $1 < p < \infty$. If (f_n) converges in $L_p(\tau)$ to a limit $f = f_0 + \sum_1^\infty f_n - f_{n-1}$ then for any choice of signs $\xi_n = \pm 1$ the transformed martingale $(\tilde{f}_n)$ converges in $L_p(\tau)$ to $\tilde{f} = f_0 + \sum_1^\infty \xi_n (f_n - f_{n-1})$ and there* *is a constant $C(p)$ depending only on p, such that*

$$\|\tilde{f}\|_p \le C(p)\|f\|_p. \tag{14.10}$$

Observing that $\tilde{\tilde{f}} = f$, we see that (14.10) (applied to $\tilde{f}$ instead of f) yields

$$\|f\|_p \le C(p)\|\tilde{f}\|_p,$$

so that we have actually

$$\frac{1}{C(p)}\|f\|_p \le \|\tilde{f}\|_p \le C(p)\|f\|_p. \tag{14.11}$$

One can then average (14.11) over all choices of signs $\xi_n = \pm 1$ and obtain

$$\frac{1}{C(p)}\|f\|_p \le \left(\int_\Delta \left\|\sum \varepsilon_n df_n\right\|_p^p d\nu\right)^{1/p} \le C(p)\|f\|_p, \tag{14.12}$$

with the usual convention $df_0 = f_0$. This leads to

Corollary 14.11 ([401]). *Assume $2 \le p < \infty$. A martingale (f_n) converges in $L_p(\tau)$ iff the series $\sum |df_n|_\bullet^2$ converges in $L_{p/2}(\tau)$. Moreover, if we let*

$$S_\bullet(f) = \left(\sum |df_n|_\bullet^2\right)^{1/2}$$

and $f = \lim f_n$, we have

$$A'_p C(p)^{-1}\|S_\bullet(f)\|_p \le \|f\|_p \le C(p)B'_p\|S_\bullet(f)\|_p. \tag{14.13}$$

Proof. Indeed, this is a rather easy combination of (14.12) and (14.6). □

In the case $1 < p \le 2$, the corresponding result is a bit more complicated: (14.13) fails when $p < 2$ and the 'correct' statement is the following consequence of (14.12) and (14.8) put together.

Corollary 14.12 ([401]). *Assume $1 < p \le 2$. A martingale (f_n) converges in $L_p(\tau)$ iff it can be decomposed as a sum $f_n = g_n + h_n$ adapted to the same filtration $(\mathcal{M}_n)$ such that $\sum dg_n^* dg_n$ and $\sum dh_n dh_n^*$ both converge in $L_{p/2}(\tau)$ (equivalently the square roots both converge in $L_p(\tau)$) and there is a constant $C'(p)$ such that*

$$C'(p)^{-1}[f]_p \le \|f\|_p \le C'(p)[f]_p \tag{14.14}$$

with the notation

$$[f]_p = \inf\left\{\left\|\left(\sum dg_n^* dg_n\right)^{1/2}\right\|_p + \left\|\left(\sum dh_n dh_n^*\right)^{1/2}\right\|_p\right\}$$

where the infimum runs over all possible decompositions of (f_n) as a sum of two $(\mathcal{M}_n)$-adapted martingales $f_n = g_n + h_n$.

Corollary 14.13 ([119, 137]). *The space $L_p(M,\tau)$ is UMD_p and hence UMD for any $1 < p < \infty$.*

Proof. Let $B = L_p(M,\tau)$ and let (f_n) be a B-valued martingale in $L_p(\Omega,\mathcal{A},\mathbb{P},B)$ adapted to a (classical sense) filtration $(\mathcal{A}_n)_{n\ge 0}$. Let then $\mathcal{M}_n = L_\infty(\Omega,\mathcal{A}_n,\mathbb{P};B)\overline{\otimes}(M,\tau)$ for $n\in\mathbb{N}$ and also for $n=\infty$. Note that $\mathcal{M}_\infty$ is a generalized probability space when equipped with the 'product trace' $\mathbb{P}\times\tau$ defined for all f in $\mathcal{M}_\infty = L_\infty(\Omega,\mathcal{A}_\infty,\mathbb{P})\otimes M$ by

$$(\mathbb{P}\times\tau)(f) = \int \tau(f(\omega))d\mathbb{P}(\omega).$$

One can identify $L_p(\mathcal{M}_\infty,\mathbb{P}\times\tau)$ with $L_p(\Omega,\mathcal{A}_\infty,\mathbb{P};B)$ for any $1\le p<\infty$. It then becomes clear that applied to the filtration $(\mathcal{M}_n)$ (14.11) implies that B is UMD_p. □

The non-commutative Burkholder-Rosenthal inequality (see § 5.9) requires to modify the definition of $\sigma(f)$: if (f_n) is a martingale adapted to $(\mathcal{M}_n)$ on a generalized probability space (M,τ), with $f_n = \mathbb{E}^{\mathcal{M}_n}(f)$ we define

$$\sigma(f) = \left(|f_0|_\bullet^2 + \sum\nolimits_1^\infty \mathbb{E}^{\mathcal{M}_{n-1}}(|f_n - f_{n-1}|_\bullet^2)\right)^{1/2}$$

where $|\ \ |_\bullet$ is as in (14.7). We also set

$$\sigma_p(f) = \left(\sum\nolimits_0^\infty |df_n|^p\right)^{1/p}$$

so that

$$\|\sigma_p(f)\|_p = \left(\sum\nolimits_0^\infty \|df_n\|_p^p\right)^{1/p}.$$

The non-commutative version of Theorem 5.52 can then be phrased identically: if $2\le p<\infty$ for any f in $L_p(M,\tau)$, the associated martingale $f_n = \mathbb{E}^{\mathcal{M}_n}(f)$ is such that $\sigma(f)$ and $\sigma_p(f)$ are well-defined and there is a positive constant ρ_p such that for any f in $L_p(M,\tau)$ we have

$$\rho_p^{-1}(\|\sigma(f)\|_p + \|\sigma_p(f)\|_p) \le \|f\|_p \le \rho_p(\|\sigma(f)\|_p + \|\sigma_p(f)\|_p). \quad (14.15)$$

14.7 Non-commutative martingale transforms

We can consider more general transforms than (14.9): let $(\varphi_n)_{n\ge 0}$ be a sequence in the unit ball of $M = L_\infty(M,\tau)$, assumed adapted to $(\mathcal{M}_n)_{n\ge 0}$, i.e. such that φ_n is $\mathcal{M}_n$-measurable for each $n \geq 0$. Then

$$\tilde{f}_n = \varphi_0 f_0 + \sum\nolimits_{k=1}^{n} \varphi_{k-1} df_k \tag{14.16}$$

is obviously a martingale such that

$$|d\tilde{f}_n|^2 = d\tilde{f}_n^* d\tilde{f}_n \leq df_n^* df_n = |df_n|^2,$$

and hence

$$\left(\sum |d\tilde{f}_n|^2\right)^{1/2} \leq \left(\sum |df_n|^2\right)^{1/2}.$$

But the reader should be warned that this requires that (φ_k) acts by left multiplication in (14.16). This problem disappears if we assume in addition

$$\forall k \qquad \varphi_{k-1} df_k = df_k \varphi_{k-1}. \tag{14.17}$$

In that case, we also have

$$\left(\sum |d\tilde{f}_n^*|^2\right)^{1/2} \leq \left(\sum |df_n^*|^2\right)^{1/2}$$

and hence, if $f_n \to f$ and $\tilde{f}_n \to \tilde{f}$ in $L_p(\tau)$, we have

$$\|S_\bullet(\tilde{f})\|_p \leq \|S_\bullet(f)\|_p,$$

from which we immediately deduce from (14.13) and (14.14):

Corollary 14.14 ([401]). *Assuming* (14.17), *the convergence of* (f_n) *in* $L_p(\tau)$ *implies that of* $(\tilde{f}_n)$ *and for any* $1 < p < \infty$ *the limits satisfy*

$$\|\tilde{f}\|_p \leq C''(p)\|f\|_p$$

where $C''(p)$ *depends only on* p.

Remark. By [410] the constant $C''(p)$ is $O(p)$ when $p \to \infty$ just like in the commutative case.

Remark. One may also consider more general transforms: assume given for each $n \geq 1$ an element T_n in $\mathcal{M}_{n-1} \otimes \mathcal{M}_{n-1}$ of the form $T_n = \sum_{j=1}^{N(n)} a_j(n) \otimes b_j(n)$. Let us denote for any x in $\mathcal{M}_n$

$$T_n[x] = \sum\nolimits_1^{N(n)} a_j(n) x b_j(n).$$

We define the transformed martingale $(\tilde{f}_n)$ by setting

$$\tilde{f}_n = f_0 + \sum\nolimits_1^n T_k[df_k].$$

Assume that for all $n \geq 1$ and all x in $\mathcal{M}_n$

$$\mathbb{E}_{n-1}|T_n[x]|_\bullet^2 \leq \mathbb{E}_{n-1}|x|_\bullet^2$$

and also

$$\|T_n[x]\|_p \le \|x\|_p.$$

Then we may invoke (14.15) to show that for any $2 \le p < \infty$ the convergence of (f_n) in $L_p(\tau)$ implies that of $(\tilde{f}_n)$. We refer to [122] for an interesting inequality, valid *even for* $p = \infty$, on this kind of multipliers for the filtration associated to free groups.

14.8 Non-commutative maximal inequalities

Recall that the maximal function f^* of a scalar martingale is defined as $\sup |f_n|$. In the non-commutative case, it seems hopeless to generalize this. Indeed, while there is a positive cone in (M, τ), it does not correspond to a lattice ordering so that there is no well defined notion of $\sup(x, y)$ for $x, y \in M_+$ (unless of course x and y commute). Nevertheless, it turns out that the 'dual Doob' inequality makes perfectly good sense, so that, arguing by duality, one can write down a satisfactory non-commutative maximal inequality, even though there is no maximal function!

To explain this we start with the non-commutative dual Doob inequality:

Theorem 14.15 ([292]). *For any $1 \le p < \infty$ there is a constant $D'(p)$ such that for any sequence $x_n \ge 0$ in $L_p(M, \tau)$ we have*

$$\left\|\sum\nolimits_0^N \mathbb{E}^{M_n}(x_n)\right\|_p \le D'(p) \left\|\sum\nolimits_0^N x_n\right\|_p, \qquad \forall N \ge 0$$

and the convergence in $L_p(\tau)$ of the right-hand side implies that of the left one.

As mentioned already in Theorem 1.26 the dual Doob inequality can be rephrased as the boundedness on $L_p(\Omega, \mathcal{A}, \mathbb{P}; \ell_1)$ of the operator $(x_n) \mapsto (\mathbb{E}^{\mathcal{A}_n} x_n)$.

In the non-commutative case, there is a priori no analogue of vector valued L_p-spaces such as $L_p(M, \tau; B)$ or $L_p(M, \tau; \ell_1)$. However, as explained in the next section, if B is given with an 'operator space structure', then there is a satisfactory definition of such spaces.

In the particular case when $B = \ell_1$ we can give an explicit definition for the space $L_p(M, \tau; \ell_1)$. Note that $L_p(M, \tau) = L_{2p}(M, \tau)L_{2p}(M, \tau)$. Indeed, if $x \in M$ the polar decomposition gives us $x = ab$ with $a = u|x|^{1/2}$, $b = |x|^{1/2}$, and moreover we find $\|x\|_p = \inf\{\|a\|_{2p}\|b\|_{2p}\}$ where the infimum runs over all possible decompositions $x = ab$. This can be extended to any x in $L_p(M, \tau)$. Then we can define the space $L_p(\tau; \ell_1)$ as the space of sequences $x = (x_n)$ in $L_p(\tau)$ that can be written as $x_n = a_n b_n$ with sequences (a_n), (b_n) in $L_{2p}(\tau)$ such that $\sum a_n a_n^*$ and $\sum b_n^* b_n$ converge in L_p, and we set

$$\|(x_n)\|_{L_p(\tau;\ell_1)} = \inf\left\{\left\|\sum a_n a_n^*\right\|_p \left\|\sum b_n^* b_n\right\|_p\right\}$$

where the infimum runs over all possible factorizations $x_n = a_n b_n$.

With this definition, the non-commutative version of Theorem 1.26 is:

Corollary 14.16 ([292]). *The mapping $(x_n) \mapsto \sum \mathbb{E}^{\mathcal{M}_n}(x_n)$ extends to a bounded linear map on $L_p(\tau;\ell_1)$ for any $1 \leq p < \infty$.*

At this stage, it is tempting to try to make sense of maximal inequalities by duality: We can define the space $L_{p'}(\tau;\ell_\infty)$ as formed of the sequences $y = (y_n)$ with $y_n \in L_{p'}(M,\tau)$ such that

$$\sup\left\{\sum |\langle x_n, y_n\rangle| \;\middle|\; (x_n) \in L_p(\tau;\ell_1),\ \|(x_n)\|_{L_p(\tau;\ell_1)} \leq 1\right\} \tag{14.18}$$

is finite, and we set $\|(y_n)\|_{L_{p'}(\tau;\ell_\infty)}$ equal to (14.18). Note that if we replace in (14.18) $\sum |\langle x_n, y_n\rangle|$ by $|\sum \langle x_n, y_n\rangle|$ we obtain the same norm. Indeed, this follows simply because for any z_n in the unit circle we have $\|(z_n x_n)\|_{L_p(\tau;\ell_1)} = \|(x_n)\|_{L_p(\tau;\ell_1)}$.

At this point, it is rather easy to check that $L_{p'}(\tau;\ell_\infty) = L_p(\tau;\ell_1)^*$ isometrically.

We can then state a 'non-commutative Doob inequality':

Corollary 14.17 ([292]). *For any $1 < p' < \infty$, the mapping $f \in L_{p'}(M,\tau) \mapsto (\mathbb{E}^{\mathcal{M}_n} f)_{n\geq 0}$ is bounded from $L_{p'}(M,\tau)$ to $L_{p'}(\tau;\ell_\infty)$.*

This would be of little use if (14.18) was the only available definition. Fortunately, there is another equivalent definition: The space $L_{p'}(\tau;\ell_\infty)$ is identical to the space of sequences (y_n) in $L_{p'}(M,\tau)$ that admit a factorization of the form

$$y_n = a\hat{y}_n b \tag{14.19}$$

with $a, b \in L_{2p'}(\tau)$ and $(\hat{y}_n) \in \ell_\infty(M)$. Moreover, we have

$$\|(y_n)\|_{L_{p'}(\tau;\ell_\infty)} = \inf\{\|a\|_{2p'}\|b\|_{2p'} \sup_n \|\hat{y}_n\|_\infty\} \tag{14.20}$$

where the infimum runs over all possible decompositions of the form (14.19).

Remark 14.18. There is an interesting divergence from the commutative case in the growth of the constants involved: in the commutative case the constant $D'(p)$ appearing in Theorem 14.15 grows like p (it is actually equal to p, see (1.33)), but in the non-commutative case the optimal growth is $O(p^2)$. See [294].

14.9 Martingales in operator spaces

An 'operator space' is, by definition, a closed subspace $E \subset B(\mathcal{H})$ of the space of bounded operators on a Hilbert space $\mathcal{H}$. Operator space theory is a rather recent development for which we refer the reader to [31] or to [80]. Since any Banach space embeds isometrically in $B(H)$ for some H, operator spaces do not seem at first glance like a true generalization of Banach spaces, but in operator space theory the morphisms are different: the bounded linear maps are replaced by the 'completely bounded' ones. A mapping $u : E \to F$ between operator spaces is called completely bounded (in short c.b.) if the maps $u_n : M_n(E) \to M_n(F)$ defined on the space $M_n(E)$ of $n \times n$-matrices with entries in E (equipped with the norm induced by $M_n(B(H))$) are uniformly bounded, and we define

$$\|u\|_{cb} = \sup_n \|u_n : M_n(E) \to M_n(F)\|.$$

The notions of isomorphism between two operator spaces E and F is modified accordingly by requiring that the isomorphism $u : E \to F$ as well as its inverse be completely bounded, and similarly for the notions of isomorphic or isometric embedding.

In that theory, it is possible to give a meaning to vector valued non-commutative L_p-spaces (see [395]) and hence to consider vector valued martingales also in the non-commutative setting. Since the martingale aspects of that theory are still only beginning to be explored, we will just give the flavor of what is going on.

Let $E \subset B(\mathcal{H})$ be an operator space and let H be another Hilbert space. We will equip the (algebraic) tensor product $B(H) \otimes E$ with the norm induced by $B(H \otimes_2 \mathcal{H})$. We denote the latter norm by $\|\cdot\|_{\min}$. Now if (M, τ) is a generalized probability space with $M \subset B(H)$ and if $1 \le p < \infty$, we introduce on $L_p(\tau) \otimes E$ the following norm:

$$\forall t \in L_p(\tau) \otimes E \qquad \|t\|_{L_p(\tau;E)} = \inf\left\{ \|a\|_{2p}\|b\|_{2p} \left\| \sum\nolimits_1^n x_k \otimes y_k \right\|_{M \otimes_{\min} E} \right\}$$

where the infimum runs over all possible decompositions of t of the form

$$t = \sum\nolimits_1^n a x_k b \otimes y_k$$

with $a, b \in L_{2p}(\tau)$ and $\sum_1^n x_k \otimes y_k \in M \otimes E$. We denote by $L_p(\tau; E)$ the completion of $L_p(\tau) \otimes E$ with respect to the corresponding norm.

It turns out that this definition enjoys many of the properties of the classical vector valued L_p-spaces, but there are two restrictions: firstly we need to assume that M is injective (i.e. there is a norm 1 projection from $B(H)$ onto M) and secondly we must always work with 'operator space structures' and not only

Banach spaces to define the usual notions such as quotient, dual or complex interpolation. In other words, even though our definition does produce a Banach space $L_p(\tau; E)$, its nice properties can only be understood in the framework of operator spaces where bounded maps are replaced by *completely bounded* ones. We refer to [395] for full details.

The restriction that M is injective is a bit puzzling at first glance, but it is required if we want, as in the classical case, two important properties preserved: Firstly when $E_2 \subset E_1$ is a subspace, we want to be able to identify (completely isometrically) $L_p(\tau; E_2)$ with a subspace of $L_p(\tau; E_1)$, and secondly we also want to identify $L_p(\tau; E_1/E_2)$ with $L_p(\tau; E_1)/L_p(\tau; E_2)$. With the preceding definition, the first one ('injectivity') always holds, but the second one ('projectivity') holds only when M injective.

When M is injective, the space $L_1(\tau; E)$ as defined previously can be identified with the *operator space version* of the projective tensor product of $L_1(\tau)$ (with a specific operator space structure, see [80, p. 139]) and E in the sense of [31, Chapter 7]. If we use this as an alternate definition, the second property ('projectivity') always holds. However, now the first one ('injectivity') only holds if the underlying M is injective. Thus, there are difficulties to prove a nice duality in the general non-injective case. However, as observed by Junge [292], in the particular case of the pair (ℓ_1, ℓ_∞) considered in the preceding section, these difficulties are not present. See also [293] for Junge's more general views on the spaces $L_p(\tau; E)$ in the non-injective case, using an embedding of M into an ultraproduct of injective (or matricial) algebras. Henceforth we will restrict to injective von Neumann algebras. The preceding construction leads naturally to the notion of E valued martingale: indeed, one can show that conditional expectations define contractive linear maps on $L_p(\tau; E)$ so we can define martingales exactly as in the commutative case. Then, this naturally leads to:

Definition. An operator space $E \subset B(\mathcal{H})$ is called $OUMD_p$ $(1 < p < \infty)$ with respect to a filtration $(\mathcal{M}_n)$ inside (M, τ) if there is a constant C such that for any $\xi_n = \pm 1$ and any martingale (f_n) in $L_p(\tau; E)$ adapted to $(\mathcal{M}_n)$ we have for any $N \geq 1$

$$\left\| f_0 + \sum\nolimits_1^N \xi_n (f_n - f_{n-1}) \right\|_{L_p(\tau;E)} \leq C \|f_N\|_{L_p(\tau;E)}.$$

If this holds for any filtration $(\mathcal{M}_n)$, we say that E is $OUMD_p$.

This notion was introduced rather recently in [395]. It was thoroughly investigated in Magdalena Musat's thesis [356]. However, many open questions remain, in particular it is unclear whether the property $OUMD_p$ really depends on $1 < p < \infty$. See Yanqi Qiu's paper [406] for a recent advance and an important update on the status of this problem.

14.10 Notes and remarks

Several references for this chapter are already in the text.

Early work on maximal inequalities for martingales in non-commutative L_1 was done by Cuculescu, see [17, 192].

The non-commutative Khintchin inequalities of Lust-Piquard [338] were the starting point of this whole development. Later on Q. Xu and the author proved the non-commutative Burkholder inequalities in [401], Q. Xu and Junge proved the non-commutative Burkholder-Rosenthal ones in [295], and Junge proved the non-commutative Doob inequality in [292]. The exact order of growth of the various relevant non-commutative constants was obtained in [294], and in [410, 412]. It should be emphasized that in several instances this order is quite different from the commutative case, in particular for the dual Doob inequality, see Remark 14.18. N. Randrianantoanina made many important additional contributions, [410–413], including a version of Doob's inequality for $p = 1$ in [414]. In particular N. Randrianantoanina and J. Parcet [373] obtained non-commutative versions of Gundy's decomposition. Additional, better weak-type 1-1 inequalities are proved in [412]. Recently, Hong, Junge and Parcet in [266] used certain asymmetric maximal functions to prove a non-commutative version of Burgess Davis's inequality (namely the left-hand side of (5.67)).

See [402] for a survey of non-commutative L_p-spaces. General references for further reading in non-commutative probability are [17, 65, 74, 96].

Bibliography

Books

[1] C. Ané, S. Blachère, D. Chafai, P. Fougères, I. Gentil, F. Malrieu, C. Roberto and G. Scheffer, *Sur les inégalités de Sobolev logarithmiques*, Panoramas and Synthèses 10, Société Mathématique de France, Paris, 2000.

[2] W. Arveson, *A short course on spectral theory*, Graduate Texts in Mathematics 209, Springer, New York, 2002.

[3] R. Bañuelos and C. Moore, *Probabilistic behavior of harmonic functions*, Birkhäuser, Basel, 1999.

[4] B. Beauzamy, *Introduction to Banach spaces and their geometry*, North-Holland, Amsterdam, 1985.

[5] Y. Benyamini and J. Lindenstrauss, *Geometric nonlinear functional analysis*, Vol. 1, American Mathematical Society, Providence, RI, 2000.

[6] J. Bergh and J. Löfström, *Interpolation spaces: an introduction*, Springer, Berlin, 1976.

[7] C. Bennett and R. Sharpley, *Interpolation of operators*, Academic Press, Boston, 1988.

[8] J. Bourgain, La propriété de Radon-Nikodym, *Publications mathématiques de l'Université Pierre et Marie Curie*, 36 (1979).

[9] J. Bourgain, *New classes of $\mathcal{L}^p$-spaces*, Lecture Notes in Mathematics 889, Springer, Berlin, 1981.

[10] R. Bourgin, *Geometric aspects of convex sets with the Radon-Nikodým property*, Lecture Notes in Mathematics 993, Springer, Berlin, 1983.

[11] L. Breiman, *Probability*, corrected reprint of the 1968 original *Classics in applied mathematics*, vol. 7, Society for Industrial and Applied Mathematics (SIAM), Philadelphia, 1992.

[12] Yu. A. Brudnyi and N. Ya. Krugljak, *Interpolation functors and interpolation spaces*, Vol. I, North-Holland, Amsterdam, 1991.

[13] M. Bruneau, *Variation totale d'une fonction*, Lecture Notes in Mathematics 413, Springer, New York, 1974.

[14] D. L. Burkholder, *Selected works of Donald L. Burkholder*, edited by Burgess Davis and Renming Song, Springer, New York, 2011.

[15] C. Carathéodory, *Conformal representation*, 2nd ed., Cambridge University Press, New York, 1952.

[16] G. Choquet, *Lectures on analysis, Vol. II: Representation Theory*, Benjamin, New York, 1969.

[17] I. Cuculescu and A. G. Oprea, *Non-commutative probability*, Kluwer, New York, 1994.

[18] C. Dellacherie and P. A. Meyer, *Probabilities and potential. B. Theory of martingales*, North-Holland, Amsterdam, 1982.

[19] R. Deville, G. Godefroy and V. Zizler, *Smoothness and renormings in Banach spaces*, Pitman Monographs and Surveys in Pure and Applied Mathematics 64, John Wiley, New York, 1993.

[20] J. Diestel, *Geometry of Banach spaces – Selected topics*, Springer Lecture Notes 485, Springer, New York, 1975.

[21] J. Diestel and J. J. Uhl Jr., *Vector measures*, Mathematical Surveys 15, American Mathematical Society, Providence, 1977.

[22] J. L. Doob, *Stochastic processes*, reprint of the 1953 original, Wiley Classics Library, Wiley-Interscience, New York, 1990.

[23] R. Dudley and R. Norvaiša, *Differentiability of six operators on nonsmooth functions and p-variation*, with the collaboration of Jinghua Qian, Lecture Notes in Mathematics 1703, Springer, Berlin, 1999.

[24] R. Dudley and R. Norvaiša, *Concrete functional calculus*, Springer Monographs in Mathematics, Springer, New York, 2011.

[25] P. Duren, *Theory of Hp spaces*, Academic Press, New York, 1970.

[26] R. Durrett, *Brownian motion and martingales in analysis*, Wadsworth Mathematics Series, Wadsworth, Belmont, CA, 1984.

[27] R. Durrett, *Probability: theory and examples*, 4th ed., Cambridge Series in Statistical and Probabilistic Mathematics, Cambridge University Press, Cambridge, 2010.

[28] H. Dym and H. P. McKean, *Fourier series and integrals*, Academic Press, New York, 1972.

[29] G. A. Edgar and L. Sucheston, *Stopping times and directed processes*, Cambridge University Press, Cambridge, 1992.

[30] R. E. Edwards and G. I. Gaudry, *Littlewood–Paley and multiplier theory*, Springer, New York, 1977.

[31] E. Effros and Z. J. Ruan, *Operator spaces*, Oxford University Press, Oxford, 2000.

[32] J. García-Cuerva and J. L. Rubio de Francia, *Weighted norm inequalities and related topics*, North-Holland, Amsterdam, 1985.

[33] J. Garnett, *Bounded analytic functions*, Academic Press, New York, 1981.

[34] A. M. Garsia, *Martingale inequalities: seminar notes on recent progress*, Mathematics Lecture Notes Series, W. A. Benjamin, Reading, MA, 1973.

[35] I. Gikhman and A. Skorokhod, *The theory of stochastic processes, III*, reprint of the 1974 edition, Springer, Berlin, 2007.

[36] L. Grafakos, *Classical Fourier analysis*, Springer, New York, 2008.

[37] L. Grafakos, *Modern Fourier analysis*, Springer, New York, 2009.

[38] P. Hàjek, S. Montesinos Santalucša, J. Vanderwerff and V. Zizler, *Biorthogonal systems in Banach spaces*, CMS Books in Mathematics/Ouvrages de Mathématiques de la SMC 26, Springer, New York, 2008.

[39] H. Helson, *Lectures on invariant subspaces*, Academic Press, New York, 1964.

[40] K. Hoffman, *Banach spaces of analytic functions*, Prentice Hall, Englewood Cliffs, NJ, 1962.

[41] L. Hörmander, *An introduction to complex analysis in several variables*, Van Nostrand, Princeton, NJ, 1966.

[42] L. Hörmander, *Notions of convexity*, Birkhäuser, Boston, 1994.

[43] R. Kadison and J. Ringrose, *Fundamentals of the theory of operator algebras, Vol. II, Advanced theory*, Academic Press, New York, 1986.

[44] J. P. Kahane, *Some random series of functions*, 2nd ed., Cambridge Studies in Advanced Mathematics 5, Cambridge University Press, Cambridge, 1985.

[45] N. Kalton, N. Peck and J. Roberts, *An F-space sampler*, London Mathematical Society Lecture Note Series 89, Cambridge University Press, Cambridge, 1984.

[46] B. Kashin and A. Saakyan, *Orthogonal series,* Translations of Mathematical Monographs 75, American Mathematical Society, Providence, RI, 1989.
[47] T. Kato, *Perturbation theory for linear operators*, reprint of the 1980 edition, Classics in Mathematics, Springer, Berlin, 1995.
[48] Y. Katznelson, *An introduction to harmonic analysis*, 3rd ed., Cambridge University Press, Cambridge, 2004.
[49] P. Koosis, *Introduction to H_p-spaces*, 2nd ed., Cambridge University Press, Cambridge, 1998.
[50] O. Kouba and A. Pallarès, Groupe de travail sur les espaces de Banach (B. Maurey et G. Pisier, 1986-1987), handwritten lecture notes.
[51] S. G. Krein, Yu. Petunin and E. M. Semenov, *Interpolation of linear operators*, Translations of Mathematical Monographs 54, American Mathematical Society, Providence, RI, 1982.
[52] S. Kwapień and W. Woyczyński, *Random series and stochastic integrals: single and multiple*, Birkhäuser, Boston, 1992.
[53] M. Ledoux, *The concentration of measure phenomenon*, Mathematical Surveys and Monographs 89, American Mathematical Society, Providence, RI, 2001.
[54] M. Ledoux and M. Talagrand, *Probability in Banach Spaces: isoperimetry and processes*, Springer, Berlin, 1991.
[55] Le Gall, *Mouvement brownien, martingales et calcul stochastique*, Springer, Heidelberg, 2013.
[56] D. Li and H. Queffélec, *Introduction . l'étude des espaces de Banach*, Analyse et probabilités, Société Mathématique de France, Paris, 2004.
[57] J. Lindenstrauss and L. Tzafriri. *Classical Banach spaces II*, Springer, New York, 1979.
[58] R. Long, *Martingale spaces and inequalities*, Peking University Press, Beijing, 1993.
[59] V. Mandrekar and B. Rüdiger, *Stochastic integration in Banach spaces*, Springer, New York, 2015.
[60] M. Marcus and G. Pisier, *Random Fourier series with applications to harmonic analysis*, Princeton University Press, Princeton, NJ, 1981.
[61] M. Métivier, *Semimartingales: a course on stochastic processes*, Walter de Gruyter, Berlin, 1982.
[62] M. Métivier and J. Pellaumail, *Stochastic integration*, Probability and Mathematical Statistics, Academic Press, New York, 1980.
[63] P. A. Meyer, *Probabilités et potentiel*, Hermann, Paris, 1966.
[64] P. A. Meyer, *Un cours sur les intégrales stochastiques*, Séminaire de Probabilités 10, Lecture Notes in Mathematics 511, Springer, Berlin, 1976.
[65] P. A. Meyer, *Quantum probability for probabilists*, Lecture Notes in Mathematics 1538, Springer, Berlin, 1993.
[66] P. Meyer-Nieberg, *Banach lattices*, Springer, Berlin, 1991.
[67] H. P. McKean Jr., *Stochastic integrals*, Academic Press, New York, 1969.
[68] V. Milman and G. Schechtman, *Asymptotic theory of finite-dimensional normed spaces*, with an appendix by M. Gromov, Lecture Notes in Mathematics 1200, Springer, Berlin, 1986.
[69] P. Mörters and Y. Peres, *Brownian motion*, with an appendix by Oded Schramm and Wendelin Werner, Cambridge University Press, Cambridge, 2010.
[70] P. F. X. Müller, *Isomorphisms between H^1 spaces*, Monografie Matematyczne (New Series) 66, Birkhäuser, Basel, 2005.
[71] J. Neveu, *Discrete-parameter Martingales*, translated from the French by T. P. Speed, rev. ed., North-Holland Mathematical Library 10, North-Holland, Amsterdam, 1975.
[72] A. Osękowski, *Sharp martingale and semimartingale inequalities*, Birkhäuser/Springer, Basel, 2012.
[73] M. I. Ostrovskii, *Metric embeddings, Bilipschitz and coarse embeddings into Banach spaces*, De Gruyter, Berlin, 2013.

[74] K. R. Parthasarathy, *An introduction to quantum stochastic calculus*, Monographs in Mathematics, Birkhäuser, Basel, 1992.
[75] M. C. Pereyra and L. Ward, *Harmonic analysis: from Fourier to wavelets*, Institute for Advanced Study (IAS), Princeton, NJ, 2012.
[76] K. Petersen, *Brownian motion, Hardy spaces, and bounded mean oscillation*, Cambridge University Press, Cambridge, 1977.
[77] R. R. Phelps, *Convex functions, monotone operators and differentiability*, 2nd ed., Lecture Notes in Mathematics 1364, Springer, Berlin, 1993.
[78] A. Pietsch, *Operator ideals*, North-Holland, Amsterdam, 1980.
[79] G. Pisier, *The volume of convex bodies and Banach space geometry*, Cambridge University Press, Cambridge, 1989.
[80] G. Pisier, *Introduction to operator space theory*, London Mathematical Society Lecture Note Series 294, Cambridge University Press, Cambridge, 2003.
[81] D. Revuz and M. Yor, *Continuous martingales and Brownian motion*, 3rd ed., Springer, Berlin, 1999.
[82] A. W. Roberts and D. E. Varberg, *Convex functions*, Pure and Applied Mathematics 57, Academic Press, New York, 1973.
[83] L. C. G. Rogers and D. Williams, *Diffusions, Markov processes, and martingales, Vol. 2, Itô calculus*, Cambridge University Press, Cambridge, 2000.
[84] W. Rudin, *Fourier analysis on groups*, Interscience, New York, 1962.
[85] E. M. Stein, *Topics in harmonic analysis related to the Littlewood–Paley theory*, Annals of Mathematics Studies 63, Princeton University Press, Princeton, NJ, 1970.
[86] E. Stein, *Harmonic analysis: real-variable methods, orthogonality, and oscillatory integrals*, Princeton University Press, Princeton, NJ, 1993.
[87] E. Stein and R. Shakarchi, *Fourier analysis: an introduction*, Princeton University Press, Princeton, NJ, 2003.
[88] E. Stein and R. Shakarchi, *Complex analysis*, Princeton University Press, Princeton, NJ, 2003.
[89] E. Stein and G. Weiss, *Introduction to Fourier analysis on Euclidean spaces*, Princeton University Press, Princeton, NJ, 1971.
[90] D. W. Stroock, *Probability theory: an analytic view*, Cambridge University Press, Cambridge, 1993.
[91] M. Takesaki, *Theory of operator algebras I*, Springer, New York, 1979.
[92] M. Talagrand, *The generic chaining: upper and lower bounds of stochastic processes*, Springer Monographs in Mathematics, Springer, Berlin, 2005.
[93] A. E. Taylor, *Introduction to functional analysis*, John Wiley, New York, 1958.
[94] A. E. Taylor and D. C. Lay, *Introduction to functional analysis*, 2nd ed., John Wiley, New York, 1980.
[95] H. Triebel, *Interpolation theory, function spaces, differential operators*, 2nd ed., Johann Ambrosius Barth, Heidelberg, 1995.
[96] D. Voiculescu, K. Dykema and A. Nica, *Free random variables*, CRM Monograph Series 1, American Mathematical Society, Providence, RI, 1992.
[97] F. Weisz, *Martingale Hardy spaces and their applications in Fourier analysis*, Lecture Notes in Mathematics 1568, Springer, Berlin, 1994.
[98] P. Wojtaszczyk, *A mathematical introduction to wavelets*, London Mathematical Society Student Texts 37, Cambridge University Press, Cambridge, 1997.
[99] A. Zygmund, *Trigonometric series*, Vols. I and II, 3rd ed., Cambridge University Press, Cambridge, 2002.

Articles and chapters

[100] D. Aldous, Unconditional bases and martingales in $L_p(F)$, *Math. Proc. Cambridge Philos. Soc.* **85** (1979), 117–123.

[101] M. E. Andersson, On the vector valued Hausdorff-Young inequality, *Ark. Mat.* **36** (1998), 1–30.

[102] T. Ando, Contractive projections in L_p-spaces, *Pacific J. Math.* **17** (1966), 391–405.

[103] N. H. Asmar, B. P. Kelly and S. Montgomery-Smith, A note on UMD spaces and transference in vector-valued function spaces, *Proc. Edinburgh Math. Soc.* (2) **39** (1996), 485–490.

[104] V. Aurich, Bounded holomorphic embeddings of the unit disk into Banach spaces, *Manuscripta Math.* **45** (1983), 61–67.

[105] V. Aurich, Bounded analytic sets in Banach spaces, *Ann. Inst. Fourier* (Grenoble) **36** (1986), 229–243.

[106] K. Azuma, Weighted sums of certain dependent random variables, *Tôhoku Math. J.* **19** (1967), 357–367.

[107] D. Bakry, L'hypercontractivité et son utilisation en théorie des semigroupes, in *Lectures on probability theory (Saint-Flour, 1992)*, Lecture Notes in Mathematics 1581, Springer, Berlin, 1994, 1–114.

[108] K. Ball, Markov chains, Riesz transforms and Lipschitz maps, *Geom. Funct. Anal.* **2** (1992), 137–172.

[109] K. Ball, E. Carlen and E. H. Lieb, Sharp uniform convexity and smoothness inequalities for trace norms, *Invent. Math.* **115** (1994), 463–482.

[110] J. Bastero and M. Romance, Random vectors satisfying Khinchine-Kahane type inequalities for linear and quadratic forms, *Math. Nach.* **278** (2005), 1015–1024.

[111] F. Baudier, Metrical characterization of super-reflexivity and linear type of Banach spaces, *Arch. Math. (Basel)* **89** (2007), 419–429.

[112] F. Baudier and G. Lancien, Embeddings of locally finite metric spaces into Banach spaces, *Proc. Am. Math. Soc.* **136** (2008), 1029–1033.

[113] F. Baudier and S. Zhang, (β)-distortion of some infinite graphs, *J. London Math. Soc.*, forthcoming.

[114] A. Beck, A convexity condition in Banach spaces and the strong law of large numbers, *Proc. Am. Math. Soc.* **13** (1962), 329–334.

[115] W. Beckner, Inequalities in Fourier analysis, *Ann. Math.* **102** (1975), 159–182.

[116] A. Benedek, A. Calderón and R. Panzone, Convolution operators on Banach space valued functions, *Proc. Natl. Acad. Sci. USA* **48** (1962), 356–365.

[117] J. Bergh, On the relation between the two complex methods of interpolation, *Indiana Univ. Math. J.* **28** (1979), 775–778.

[118] J. Bergh and J. Peetre, On the spaces $V_p(0 < p \leq \infty)$, *Boll. Un. Mat. Ital.* **10** (1974), 632–648.

[119] E. Berkson, T. A. Gillespie and P. S. Muhly, Abstract spectral decompositions guaranteed by the Hilbert transform, *Proc. London Math. Soc.* **53** (1986), 489–517.

[120] A. Bernal and J. Cerdà, Complex interpolation of quasi-Banach spaces with an A-convex containing space, *Ark. Mat.* **29** (1991), 183–201.

[121] A. Bernard and B. Maisonneuve, Décomposition atomique de martingales de la classe H1, in *Séminaire de Probabilités*, Vol. XI, Lecture Notes in Mathematics 581, Springer, Berlin, 1977, 303–326.

[122] P. Biane and R. Speicher, Stochastic calculus with respect to free Brownian motion and analysis on Wigner space, *Probab. Theory Related Fields* **112** (1998), 373–409.

[123] O. Blasco, Hardy spaces of vector-valued functions: duality, *Trans. Am. Math. Soc.* **308** (1988), 495–507.

[124] O. Blasco and A. Pełczyński, Theorems of Hardy and Paley for vector-valued analytic functions and related classes of Banach spaces, *Trans. Am. Math. Soc.* **323** (1991), 335–367.

[125] O. Blasco and S. Pott, Operator-valued dyadic BMO spaces, *J. Operator Theory* **63** (2010), 333–347.

[126] G. Blower, A multiplier characterization of analytic UMD spaces, *Studia Math.* **96** (1990), 117–124.

[127] G. Blower and T. Ransford, Complex uniform convexity and Riesz measures, *Can. J. Math.* **56** (2004), 225–245.

[128] A. Bonami, Étude des coefficients de Fourier des fonctions de $L^p(G)$, *Ann. Inst. Fourier (Grenoble)* **20** (1970), 335–402.

[129] A. Bonami and D. Lépingle, Fonction maximale et variation quadratique des martingales en présence d'un poids, *Séminaire de Probabilités*, Vol. XIII, Lecture Notes in Mathematics 721, Springer, Berlin, 1979, 294–306.

[130] J. Bourgain, A nondentable set without the tree property, *Studia Math.* **68** (1980), 131–139.

[131] J. Bourgain, New Banach space properties of the disc algebra and H^∞, *Acta Math.* **152** (1984), 1–48.

[132] J. Bourgain, On trigonometric series in super reflexive spaces, *J. London Math. Soc.* **24** (1981), 165–174.

[133] J. Bourgain, A Hausdorff-Young inequality for B-convex Banach spaces, *Pacific J. Math.* **101** (1982), 255–262.

[134] J. Bourgain, Some remarks on Banach spaces in which martingale difference sequences are unconditional, *Ark. Mat.* **21** (1983), 163–168.

[135] J. Bourgain, On martingales transforms in finite-dimensional lattices with an appendix on the K-convexity constant, *Math. Nachr.* **119** (1984), 41–53.

[136] J. Bourgain, Extension of a result of Benedek, Calderón and Panzone, *Ark. Mat.* **22** (1984), 91–95.

[137] J. Bourgain, Vector valued singular integrals and the H^1-BMO duality, in *Probability theory and harmonic analysis*, edited by Chao–Woyczynski, Dekker, New York, 1986, 1–19.

[138] J. Bourgain, The metrical interpretation of superreflexivity in Banach spaces, *Israel J. Math.* **56** (1986), 222–230.

[139] J. Bourgain, Pointwise ergodic theorems for arithmetic sets (Appendix: The return time theorem), *Publ. Math. Inst. Hautes Étud. Sci.* **69** (1989), 5–45.

[140] J. Bourgain, On the radial variation of bounded analytic functions on the disc, *Duke Math. J.* **69** (1993), 671–682.

[141] J. Bourgain and W. J. Davis, Martingale transforms and complex uniform convexity, *Trans. Am. Math. Soc.* **294** (1986), 501–515.

[142] J. Bourgain and M. Lewko, Sidonicity and variants of Kaczmarz's problem, preprint, Arxiv, April 2015.

[143] J. Bourgain and H. P. Rosenthal, Martingales valued in certain subspaces of L_1, *Israel J. Math.* **37** (1980), 54–75.

[144] J. Bourgain and H. P. Rosenthal, Applications of the theory of semi-embeddings to Banach space theory, *J. Funct. Anal.* **52** (1983), 149–188.

[145] J. Bourgain, H. P. Rosenthal and G. Schechtman, An ordinal L^p-index for Banach spaces, with application to complemented subspaces of L^p, *Ann. Math.* **114** (1981), 193–228.

[146] J. Bretagnolle, p-variation de fonctions aléatoires, I and II, in *Séminaire de Probabilités*, Vol. VI, Lecture Notes in Mathematics 258, Springer, Berlin, 1972, 51–71.

[147] J. Briët, A. Naor and O. Regev, Locally decodable codes and the failure of cotype for projective tensor products, *Elect. Res. Announ. Math. Sci.* **19** (2012), 120–130.

[148] J. Brossard, Comportement non-tangentiel et comportement brownien des fonctions harmoniques dans un demi-espace, Démonstration probabiliste d'un théorème de Calderón et Stein, *Sém. Probab. (Strasbourg)* **12** (1978), 378–397.

[149] A. Brunel and L. Sucheston, On J-convexity and some ergodic super-properties of Banach spaces, *Trans. Am. Math. Soc.* **204** (1975), 79–90.

[150] Q. Bu and P. Dowling, Observations about the projective tensor product of Banach spaces, III, $L_p[0,1]\widehat{\otimes}X$, $1 < p < \infty$, *Quaest. Math.* **25** (2002), 303–310.

[151] S. Bu, Deux remarques sur la propriété de Radon-Nikodym analytique, *Ann. Fac. Sci. Toulouse* **60** (1990), 79–89.

[152] S. Bu, Existence of radial limits of harmonic functions in Banach spaces, *Chin. Ann. Math. Ser. B* **13** (1992), 110–117. (Chinese summary in *Chin. Ann. Math. Ser. A* **13** (1992), 133.)

[153] S. Bu, On the analytic Radon-Nikodym property for bounded subsets in Banach spaces, *J. London Math. Soc.* **47** (1993), 484–496.

[154] S. Bu, A J-convex subset which is not PSH-convex, *Acta Math. Sci. (English Ed.)* **14** (1994), 446–450.

[155] S. Bu, A new characterization of the analytic Radon-Nikodym property for bounded subsets (English summary), *Northeast. Math. J.* **12** (1996), 227–229.

[156] S. Bu, The analytic Krein-Milman property in Banach spaces, *Acta Math. Sci. (English Ed.)* **18** (1998), 17–24.

[157] S. Bu, The existence of Jensen boundary points in complex Banach spaces, *Syst. Sci. Math. Sci.* **12** (1999), 8–12.

[158] S. Bu, A new characterisation of the analytic Radon-Nikodym property, *Proc. Am. Math. Soc.* **128** (2000), 1017–1022.

[159] S. Bu, The existence of radial limits of analytic functions with values in Banach spaces, *Chin. Ann. Math. Ser. B* **22** (2001), 513–518.

[160] S. Bu and B. Khaoulani, Une caractérization de la propriété de Radon–Nikodym analytique pour les espaces isomorphes à leur carrés, *Math. Ann.* **288** (1990), 345–360.

[161] S. Bu and C. Le Merdy, H_p-maximal regularity and operator valued multipliers on Hardy spaces, *Can. J. Math.* **59** (2007), 1207–1222.

[162] S. Bu and W. Schachermayer, Approximation of Jensen measures by image measures under holomorphic functions and applications, *Trans. Am. Math. Soc.* **331** (1992), 585–608.

[163] A. V. Bukhvalov and A. A. Danilevich, Boundary properties of analytic and harmonic functions with values in a Banach space (Russian), *Mat. Zametki* **31** (1982), 203–214, 317. (English translation in *Math. Notes* **31** (1982), 104–110 (1983).)

[164] D. L. Burkholder, Maximal inequalities as necessary conditions for almost everywhere convergence, *Z. Wahrscheinlichkeitstheorie Verw. Gebiete* **3** (1964), 75–88.

[165] D. L. Burkholder, Distribution function inequalities for martingales, *Ann. Probab.* **1** (1973), 19–42.

[166] D. L. Burkholder, Brownian motion and the Hardy spaces H^p, in *Aspects of contemporary complex analysis*, Academic Press, London, 1980, 97–118.

[167] D. L. Burkholder, A geometrical characterization of Banach spaces in which martingale difference sequences are unconditional, *Ann. Probab.* **9** (1981), 997–1011.

[168] D. L. Burkholder, Martingale transforms and the geometry of Banach spaces, in *Probability in Banach spaces, III*, Lecture Notes in Mathematics 860, Springer, Berlin, 1981, 35–50.

[169] D. L. Burkholder, A geometric condition that implies the existence of certain singular integrals of Banach space-valued functions, in *Conference on Harmonic Analysis in honor of Antoni Zygmund, Vols. I and II (Chicago, Ill., 1981)*, Wadsworth, Belmont, CA, 1983, 270–286.

[170] D. L. Burkholder, Boundary value problems and sharp inequalities for martingale transforms, *Ann. Probab.* **12** (1984), 647–702.

[171] D. L. Burkholder, Martingales and Fourier analysis in Banach spaces, in *Probability and analysis (Varenna, 1985), 61–108*, Lecture Notes in Mathematics 1206, Springer, Berlin, 1986, 61–108.

[172] D. L. Burkholder, Sharp inequalities for martingales and stochastic integrals, Colloque Paul Lévy sur les Processus Stochastiques (Palaiseau, 1987), *Astérisque* **157–158** (1988), 75–94.

[173] D. L. Burkholder, Explorations in martingale theory and its applications, in *École d'Été de Probabilités de Saint-Flour XIX – 1989*, Lecture Notes in Mathematics 1464, Springer, Berlin, 1991, 1–66.

[174] D. L. Burkholder, Martingales and singular integrals in Banach spaces, in *Handbook of the geometry of Banach spaces*, Vol. I, North-Holland, Amsterdam, 2001, 233–269.

[175] D. L. Burkholder and R. F. Gundy, Extrapolation and interpolation of quasi-linear operators on martingales, *Acta Math.* **124** (1970), 249–304.
[176] D. L. Burkholder, R. F. Gundy and M. L. Silverstein, A maximal function characterization of the class H^p, *Trans. Am. Math. Soc.* **157** (1971), 137–153.
[177] A. P. Calderón, Intermediate spaces and interpolation, the complex method, *Studia Math.* **24** (1964), 113–190.
[178] S. D. Chatterji, Martingale convergence and the Radon-Nikodym theorem in Banach spaces, *Math. Scand.* **22** (1968), 21–41.
[179] J. Cheeger and B. Kleiner, Characterization of the Radon-Nikodým property in terms of inverse limits, *Astérisque* **321** (2008), 129–138.
[180] J. Cheeger and B. Kleiner, Differentiability of Lipschitz maps from metric measure spaces to Banach spaces with the Radon-Nikodým property, *Geom. Funct. Anal.* **19** (2009), 1017–1028.
[181] C. H. Chu, A note on scattered C^*-algebras and the Radon-Nikodym property, *J. London Math. Soc.* **24** (1981), 533–536.
[182] F. Cobos and J. Peetre, Interpolation of compactness using Aronszajn-Gagliardo functors, *Israel J. Math.* **68** (1989), 220–240.
[183] R. R. Coifman, A real variable characterization of H_p, *Studia Math.* **51** (1974), 269–274.
[184] R. R. Coifman, M. Cwikel, R. Rochberg, Y. Sagher and G. Weiss, Complex interpolation for families of Banach spaces, in *Harmonic analysis in Euclidean spaces (Proc. Sympos. Pure Math., Williams Coll., Williamstown, Mass., 1978), Part 2*, Proc. Sympos. Pure Math. XXXV, American Mathematical Society, Providence, RI, 1979, 269–282.
[185] R. R. Coifman, R. Rochberg, G. Weiss, M. Cwikel and Y. Sagher, The complex method for interpolation of operators acting on families of Banach spaces, in *Euclidean harmonic analysis (Proc. Sem., Univ. Maryland, College Park, Md., 1979)*, Lecture Notes in Mathematics 779, Springer, Berlin, 1980, 123–153.
[186] R. R. Coifman, M. Cwikel, R. Rochberg, Y. Sagher and G. Weiss, A theory of complex interpolation for families of Banach spaces, *Adv. Math.* **43** (1982), 203–229.
[187] R. R. Coifman and S. Semmes, Interpolation of Banach spaces, Perron processes, and Yang-Mills, *Am. J. Math.* **115** (1993), 243–278.
[188] J. Conde, A note on dyadic coverings and nondoubling Calderón-Zygmund theory, *J. Math. Anal. Appl.* **397** (2013), 785–790.
[189] A. Connes, Classification of injective factors: cases II_1, II_∞, III_λ, $\lambda \neq 1$, *Ann. Math.* **104** (1976), 73–115.
[190] M. G. Cowling, G. I. Gaudry and T. Qian, A note on martingales with respect to complex measures, Miniconference on Operators in Analysis (Sydney, 1989), *Proc. Centre Math. Anal. Austral. Nat. Univ.* **24** (1990), 10–27.
[191] D. Cox, The best constant in Burkholder's weak-L1 inequality for the martingale square function, *Proc. Am. Math. Soc.* **85** (1982), 427–433.
[192] I. Cuculescu, Martingales on von Neumann algebras, *J. Multivariate Anal.* **1** (1971), 17–27.
[193] M. Cwikel, On $(L^{p_0}(A_0), L^{p_1}(A_1))_{\theta,q}$, *Proc. Am. Math. Soc.* **44** (1974), 286–292.
[194] M. Cwikel, Complex interpolation spaces, a discrete definition and reiteration, *Indiana Univ. Math. J.* **27** (1978), 1005–1009.
[195] M. Cwikel, Lecture notes on duality and interpolation spaces, arxiv 2008.
[196] M. Cwikel and S. Janson, Interpolation of analytic families of operators, *Studia Math.* **79** (1984), 61–71.
[197] M. Cwikel and S. Janson, Real and complex interpolation methods for finite and infinite families of Banach spaces, *Adv. Math.* **66** (1987), 234–290.
[198] M. Cwikel, M. Milman and Y. Sagher, Complex interpolation of some quasi-Banach spaces, *J. Funct. Anal.* **65** (1986), 339–347.
[199] M. Daher, Une remarque sur la propriété de Radon-Nikodým, *C. R. Acad. Sci. Paris Sér. I Math.* **313** (1991), 269–271.

[200] A. Daniluk, The maximum principle for holomorphic operator functions, *Integral Equations Operator Theory* **69** (2011), 365–372.

[201] B. Davis. On the integrability of the martingale square function, *Israel J. Math.* **8** (1970), 187–190.

[202] W. J. Davis, Moduli of complex convexity, in *Geometry of Banach spaces (Strobl, 1989)*, Cambridge University Press, Cambridge, 1990, 65–69.

[203] W. J. Davis, T. Figiel, W. B. Johnson and A. Pełczyński, Factoring weakly compact operators, *J. Funct. Anal.* **17** (1974), 311–327.

[204] W. J. Davis, D. J. H. Garling and N. Tomczak-Jaegermann, The complex convexity of quasi-normed linear spaces, *J. Funct. Anal.* **55** (1984), 110–150.

[205] W. J. Davis, W. B. Johnson and J. Lindenstrauss, The ℓ_1^n problem and degrees of non-reflexivity, *Studia Math.* **55** (1976), 123–139.

[206] W. J. Davis and J. Lindenstrauss, The ℓ_1^n problem and degrees of non-reflexivity, II, *Studia Math.* **58** (1976), 179–196.

[207] M. M. Day, Uniform convexity in factor and conjugate spaces, *Ann. Math.* **45** (1944), 375–385.

[208] M. M. Day, Some more uniformly convex spaces, *Bull. Am. Math. Soc.* **47** (1941), 504–507.

[209] C. Demeter, M. Lacey, T. Tao and C. Thiele, Breaking the duality in the return times theorem, *Duke Math. J.* **143** (2008), 281–355.

[210] S. Dilworth, Complex convexity and the geometry of Banach spaces, *Math. Proc. Cambridge Philos. Soc.* **99** (1986), 495–506.

[211] S. Dineen and R. Timoney, Complex geodesics on convex domains, in *Progress in functional analysis (Peniscola, 1990)*, North-Holland, Amsterdam, 1992, 333–365.

[212] J. Ding, J. Lee and Y. Peres, Markov type and threshold embeddings, *Geom. Funct. Anal.* **23** (2013), 1207–1229.

[213] R. Douglas, Contractive projections on an L_1-space, *Pacific J. Math.* **15** (1965), 443–462.

[214] P. Dowling and G. Edgar, Some characterizations of the analytic Radon-Nikodým property in Banach spaces, *J. Funct. Anal.* **80** (1988), 349–357.

[215] P. Dowling, Stability of Banach space properties in the projective tensor product, *Quaest. Math.* **27** (2004), 1–7.

[216] G. Edgar, A non-compact Choquet theorem, *Proc. Am. Math. Soc.* **49** (1975), 354–358.

[217] G. Edgar, Analytic martingale convergence, *J. Funct. Anal.* **69** (1986), 268–280.

[218] G. Edgar, Complex martingale convergence, in *Banach spaces. Proceedings, 1984*, Lecture Notes in Mathematics 1166, Springer, New York, 1985, 38–59.

[219] G. Edgar and R. F. Wheeler, Topological properties of Banach spaces, *Pacific J. Math.* **115** (1984), 317–350.

[220] P. Enflo, Banach spaces which can be given an equivalent uniformly convex norm, *Israel J. Math.* **13** (1972), 281–288.

[221] T. Fack and H. Kosaki, Generalized s-numbers of τ-measurable operators, *Pacific J. Math.* **123** (1986), 269–300.

[222] C. Fefferman and E. Stein, H_p spaces of several variables, *Acta Math.* **129** (1972), 137–193.

[223] T. Figiel, On the moduli of convexity and smoothness, *Studia Math.* **56** (1976), 121–155.

[224] T. Figiel, Uniformly convex norms on Banach lattices, *Studia Math.* **68** (1980), 215–247.

[225] T. Figiel, Singular integral operators: a martingale approach, in *Geometry of Banach spaces (Strobl, 1989)*, London Math. Soc. Lecture Note Ser. 158, Cambridge University Press, Cambridge, 1990, 95–110.

[226] T. Figiel, On equivalence of some bases to the Haar system in spaces of vector-valued functions, *Bull. Polish Acad. Sci. Math.* **36** (1988), 119–131.

[227] T. Figiel, J. Lindenstrauss and V. Milman, The dimension of almost spherical sections of convex bodies, *Acta Math.* **139** (1977), 53–94.

[228] T. Figiel and G. Pisier, Séries aléatoires dans les espaces uniformément convexes ou uniformément lisses, *C.R. Acad. Sci. Paris Sér. A* **279** (1974), 611–614.

[229] R. Fortet and E. Mourier, Les fonctions aléatoires comme éléments aléatoires dans les espaces de Banach, *Studia Math.* **15** (1955), 62–79.
[230] J. García-Cuerva, K. S. Kazaryan, V. I. Kolyada and J. L. Torrea, The Hausdorff-Young inequality with vector-valued coefficients and applications, *Russian Math. Surveys* **53** (1998), 435–513.
[231] D. J. H. Garling, Convexity, smoothness and martingale inequalities, *Israel J. Math.* **29** (1978), 189–198.
[232] D. J. H. Garling, Brownian motion and UMD-spaces, in *Probability and Banach spaces (Zaragoza, 1985)*, Lecture Notes in Mathematics 1221, Springer, Berlin, 1986, 36–49.
[233] D. J. H. Garling, On martingales with values in a complex Banach space, *Math. Proc. Camb. Philos. Soc.* **104** (1988), 399–406.
[234] D. J. H. Garling, Random martingale transform inequalities, in *Probability in Banach spaces 6 (Sandbjerg, 1986)*, Progr. Probab. 20, Birkhäuser Boston, Boston, 1990, 101–119.
[235] D. J. H. Garling and S. Montgomery-Smith, Complemented subspaces of spaces obtained by interpolation, *J. London Math. Soc.* **44** (1991), 503–513.
[236] J. Garnett and P. Jones, The distance in BMO to L^∞, *Ann. Math.* **108** (1978), 373–393.
[237] J. Garnett and P. Jones, BMO from dyadic BMO, *Pacific J. Math.* **99** (1982), 351–371.
[238] S. Geiss, A counterexample concerning the relation between decoupling constants and UMD-constants, *Trans. Am. Math. Soc.* **351** (1999), 1355–1375.
[239] S. Geiss, Contraction principles for vector valued martingales with respect to random variables having exponential tail with exponent $2 < \alpha < \infty$, *J. Theoret. Probab.* **14** (2001), 39–59.
[240] S. Geiss, S. Montgomery-Smith and E. Saksman, On singular integral and martingale transforms, *Trans. Am. Math. Soc.* **362** (2010), 553–575.
[241] N. Ghoussoub, G. Godefroy, B. Maurey and W. Schachermayer, Some topological and geometrical structures in Banach spaces, *Mem. Am. Math. Soc.* **70** (1987), 120 pp.
[242] N. Ghoussoub and W. B. Johnson, Counterexamples to several problems on the factorization of bounded linear operators, *Proc. Am. Math. Soc.* **92** (1984), 233–238.
[243] N. Ghoussoub, J. Lindenstrauss and B. Maurey, Analytic martingales and plurisubharmonic barriers in complex Banach spaces, in *Banach space theory (Iowa City, IA, 1987)*, Contemp. Math. 85, American Mathematical Society, Providence, RI, 1989, 111–130.
[244] N. Ghoussoub and B. Maurey, Counterexamples to several problems concerning G_δ-embeddings, *Proc. Am. Math. Soc.* **92** (1984), 409–412.
[245] N. Ghoussoub and B. Maurey, G_δ-embeddings in Hilbert space, *J. Funct. Anal.* **61** (1985), 72–97.
[246] N. Ghoussoub and B. Maurey, The asymptotic-norming and the Radon–Nikodým properties are equivalent in separable Banach spaces, *Proc. Am. Math. Soc.* **94** (1985), 665–671.
[247] N. Ghoussoub and B. Maurey, H_δ-embedding in Hilbert space and optimization on G_δ-sets, *Mem. Am. Math. Soc.* **62** (1986), 101.
[248] N. Ghoussoub and B. Maurey, G_δ-embeddings in Hilbert space. II, *J. Funct. Anal.* **78** (1988), 271–305.
[249] N. Ghoussoub and B. Maurey, Plurisubharmonic martingales and barriers in complex quasi-Banach spaces, *Ann. Inst. Fourier (Grenoble)* **39** (1989), 1007–1060.
[250] N. Ghoussoub, B. Maurey and W. Schachermayer, A counterexample to a problem on points of continuity in Banach spaces, *Proc. Am. Math. Soc.* **99** (1987), 278–282.
[251] N. Ghoussoub, B. Maurey and W. Schachermayer, Pluriharmonically dentable complex Banach spaces, *J. Reine Angew. Math.* **402** (1989), 76–127.
[252] N. Ghoussoub, B. Maurey and W. Schachermayer, Geometrical implications of certain infinite-dimensional decompositions, *Trans. Am. Math. Soc.* **317** (1990), 541–584.
[253] T. A. Gillespie, S. Pott, S. Treil and A. Volberg, Logarithmic growth for matrix martingale transforms, *J. London Math. Soc.* **64** (2001), 624–636.

[254] J. Globevnik, On complex strict and uniform convexity, *Proc. Am. Math. Soc.* **47** (1975), 175–178.

[255] Y. Gordon and D. R. Lewis, Absolutely summing operators and local unconditional structures, *Acta Math.* **133** (1974), 27–48.

[256] O. Guédon, Kahane-Khinchine type inequalities for negative exponent, *Mathematika* **46** (1999), 165–173.

[257] R. Gundy, A decomposition for L^1-bounded martingales, *Ann. Math. Stat.* **39** (1968), 134–138.

[258] R. Gundy and N. Varopoulos, A martingale that occurs in harmonic analysis, *Ark. Mat.* **14** (1976), 179–187.

[259] V. I. Gurarii and N. I. Gurarii, Bases in uniformly convex and uniformly smooth Banach spaces (Russian), *Izv. Akad. Nauk SSSR Ser. Mat.* **35** (1971), 210–215.

[260] U. Haagerup, The best constants in the Khintchine inequality, *Studia Math.* **70** (1981), 231–283.

[261] U. Haagerup and G. Pisier, Factorization of analytic functions with values in non-commutative L_1-spaces and applications, *Can. Math. J.* **41** (1989), 882–906.

[262] O. Hanner, On the uniform convexity of L^p and ℓ^p, *Ark. Mat.* **3** (1956), 239–244.

[263] C. S. Herz, Bounded mean oscillation and regulated martingales, *Trans. Am. Math. Soc.* **193** (1974), 199–215.

[264] C. S. Herz, H_p-spaces of martingales, $0 < p \leq 1$, *Z. Wahrscheinlichkeitstheorie Verw. Gebiete* **28** (1973/74), 189–205.

[265] J. Hoffmann-Jørgensen, Sums of independent Banach space valued random variables, *Studia Math.* **52** (1974), 159–186.

[266] G. Hong, M. Junge and J. Parcet, Algebraic Davis decomposition and asymmetric Doob inequalities, *Comm. Math. Physics*, forthcoming.

[267] R. E. Huff and P. D. Morris, Geometric characterizations of the Radon-Nikodym property in Banach spaces, *Studia Math.* **56** (1976), 157–164.

[268] T. Hytönen, The real-variable Hardy space and BMO, lecture notes of a course at the University of Helsinki, winter 2010, http://www.helsinki.fi/~tpehyton/Hardy/hardy.pdf.

[269] T. Hytönen and M. Lacey, Pointwise convergence of vector-valued Fourier series, *Math. Ann.* **357** (2013), 1329–1361.

[270] T. Hytonen, M. Lacey and I. Parissis, A variation norm Carleson theorem for vector valued Walsh-Fourier series, *Rev. Mat. Iberoam.* **30** (2014), 979–1014.

[271] A. Ionescu Tulcea and C. Ionescu Tulcea, Abstract ergodic theorems, *Trans. Am. Math. Soc.* **107** (1963), 107–124.

[272] K. Itô and M. Nisio, On the convergence of sums of independent Banach space valued random variables, *Osaka J. Math.* **5** (1968), 35–48.

[273] N. Jain and D. Monrad, Gaussian measures in B_p, *Ann. Probab.* **11** (1983), 46–57.

[274] R. C. James, Bases and reflexivity of Banach spaces, *Ann. Math.* **52** (1950), 518–527.

[275] R. C. James, Uniformly non-square Banach spaces, *Ann. Math.* **80** (1964), 542–550.

[276] R. C. James, Weak compactness and reflexivity, *Israel J. Math.* **2** (1964), 101–119.

[277] R. C. James, Reflexivity and the sup of linear functionals, *Israel J. Math.* **13** (1972), 289–300.

[278] R. C. James, Some self-dual properties of normed linear spaces, in *Symposium on Infinite-Dimensional Topology (Louisiana State Univ., Baton Rouge, LA, 1967)*, Annals of Mathematical Studies 69, Princeton University Press, Princeton, NJ, 1972, 159–175.

[279] R. C. James, Super-reflexive Banach spaces, *Can. J. Math.* **24** (1972), 896–904.

[280] R. C. James, Super-reflexive spaces with bases, *Pacific J. Math.* **41** (1972), 409–419.

[281] R. C. James, A nonreflexive Banach space that is uniformly nonoctahedral, *Israel J. Math.* **18** (1974), 145–155.

[282] R. C. James, Nonreflexive spaces of type 2, *Israel J. Math.* **30** (1978), 1–13.

[283] R. C. James and J. Lindenstrauss, The octahedral problem for Banach spaces, in *Proceedings of the Seminar on Random Series, Convex Sets and Geometry of Banach Spaces*, Various Publ. Ser. 24, Mat. Inst., Aarhus University, Aarhus, 1975, 100–120.

[284] R. C. James and J. J. Schäffer, Super-reflexivity and the girth of spheres, *Israel J. Math.* **11** (1972), 398–404.

[285] S. Janson, Minimal and maximal methods of interpolation, *J. Funct. Anal.* **44** (1981), 50–73.

[286] S. Janson, On hypercontractivity for multipliers on orthogonal polynomials, *Ark. Mat.* **21** (1983), 97–110.

[287] S. Janson, On complex hypercontractivity, *J. Funct. Anal.* **151** (1997), 270–280.

[288] S. Janson and P. Jones, Interpolation between Hp spaces: the complex method, *J. Funct. Anal.* **48** (1982), 58–80.

[289] W. B. Johnson and G. Schechtman, Diamond graphs and super-reflexivity, *J. Topol. Anal.* **1** (2009), 177–189.

[290] W. B. Johnson and L. Tzafriri, Some more Banach spaces which do not have local unconditional structure, *Houston J. Math.* **3** (1977), 55–60.

[291] R. L. Jones, J. M. Rosenblatt and M. Wierdl, Oscillation in ergodic theory: higher dimensional results, *Israel J. Math.* **135** (2003), 1–27.

[292] M. Junge, Doob's inequality for non-commutative martingales, *J. Reine Angew. Math.* **549** (2002), 149–190.

[293] M. Junge, Fubini's theorem for ultraproducts of noncommmutative L_p-spaces, *Can. J. Math.* **56** (2004), 983–1021.

[294] M. Junge and Q. Xu, On the best constants in some non-commutative martingale inequalities, *Bull. London Math. Soc.* **37** (2005), 243–253.

[295] M. Junge and Q. Xu, Noncommutative Burkholder/Rosenthal inequalities, *Ann. Probab.* **31** (2003), 948–995.

[296] M. Junge and Q. Xu, Noncommutative Burkholder/Rosenthal inequalities, II, Applications, *Israel J. Math.* **167** (2008), 227–282.

[297] M. I. Kadec and A. Pełczyński, Bases, lacunary sequences and complemented subspaces in the spaces L_p, *Studia Math.* **21** (1961/1962), 161–176.

[298] S. Kakutani, Markoff process and the Dirichlet problem, *Proc. Jpn. Acad.* **21** (1945), 227–233.

[299] N. Kalton, Differentials of complex interpolation processes for Köthe function spaces, *Trans. Am. Math. Soc.* **333** (1992), 479–529.

[300] N. Kalton, Lattice structures on Banach spaces, *Mem. Am. Math. Soc.* **103** (1993), 99 pp.

[301] N. Kalton, Complex interpolation of Hardy-type subspaces, *Math. Nachr.* **171** (1995), 227–258.

[302] N. Kalton, S. V. Konyagin and L. Vesely, Delta-semidefinite and delta-convex quadratic forms in Banach spaces, *Positivity* **12** (2008), 221–240.

[303] N. Kalton and S. Montgomery-Smith, Interpolation of Banach spaces, in *Handbook of the geometry of Banach spaces*, Vol. 2, North-Holland, Amsterdam, 2003, 1131–1175.

[304] N. H. Katz, Matrix valued paraproducts, *J. Fourier Anal. Appl.* **300** (1997), 913–921.

[305] S. V. Kisliakov, A remark on the space of functions of bounded p-variation, *Math. Nachr.* **119** (1984), 137–140.

[306] S. V. Kisliakov, Interpolation of H^p-spaces: some recent developments, in *Function spaces, interpolation spaces, and related topics (Haifa, 1995)*, Israel Math. Conf. Proc. 13, Bar-Ilan University, Ramat Gan, 1999, 102–140.

[307] B. Kloeckner, Yet another short proof of the Bourgain's distortion estimate for embedding of trees into uniformly convex Banach spaces, *Israel J. Math.* **200** (2014), 419–422.

[308] H. Koch, Adapted function spaces for dispersive equations, in *Singular phenomena and scaling in mathematical models*, Springer, Cham, 2014, 49–67.

[309] H. Kosaki, Applications of the complex interpolation method to a von Neumann algebra: noncommutative L_p-spaces, *J. Funct. Anal.* **56** (1984), 29–78.

[310] H. König, On the Fourier-coefficients of vector-valued functions, *Math. Nachr.* **152** (1991), 215–227.

[311] O. Kouba, H^1-projective spaces, *Q. J. Math. Oxford Ser.* **41** (1990), 295–312.

[312] J. L. Krivine, Sous-espaces de dimension finie des espaces de Banach réticulés, *Ann. Math.* **104** (1976), 1–29.

[313] J. L. Krivine and B. Maurey, Espaces de Banach stables, *Israel J. Math.* **39** (1981), 273–295.

[314] K. Kunen and H. P. Rosenthal, Martingale proofs of some geometrical results in Banach space theory, *Pacific J. Math.* **100** (1982), 153–175.

[315] S. Kwapień, Isomorphic characterizations of inner product spaces by orthogonal series with vector valued coefficients, *Studia Math.* **44** (1972), 583–595.

[316] S. Kwapień, On operators factorizable through Lp space, *Bull. Soc. Math. France (Mémoire)* **31–32** (1972), 215–225.

[317] G. Lancien, On uniformly convex and uniformly Kadec-Klee renormings, *Serdica Math. J.* **21** (1995), 1–18.

[318] R. Latała and K. Oleszkiewicz, On the best constant in the Khinchin-Kahane inequality, *Studia Math.* **109** (1994), 101–104.

[319] J. M. Lee, Biconcave-function characterisations of UMD and Hilbert spaces, *Bull. Austral. Math. Soc.* **47** (1993), 297–306.

[320] J. M. Lee, On Burkholder's biconvex-function characterization of Hilbert spaces, *Proc. Am. Math. Soc.* **118** (1993), 555–559.

[321] J. Lee and A. Naor, Embedding the diamond graph in L_p and dimension reduction in L_1, *Geom. Funct. Anal.* **14** (2004), 745–747.

[322] J. R. Lee, A. Naor, and Y. Peres, Trees and Markov convexity, *Geom. Funct. Anal.* **18** (2009), 1609–1659. (Conference version in *Proceedings of the Seventeenth Annual ACM-SIAM Symposium on Discrete Algorithms*, ACM, New York, 2006, 1028–1037.)

[323] K. de Leeuw, On L_p multipliers, *Ann. Math.* **81** (1965), 364–379.

[324] C. Le Merdy and Q. Xu, Strong q-variation inequalities for analytic semigroups, *Ann. Inst. Fourier (Grenoble)* **62** (2012), 2069–2097.

[325] E. Lenglart, D. Lépingle and M. Pratelli, Présentation unifiée de certaines inégalités de la théorie des martingales, with an appendix by Lenglart, in Seminar on Probability, XIV (Paris, 1978/1979), Lecture Notes in Mathematics 784, Springer-Verlag, Berlin, 1980, 26–52.

[326] D. Lépingle, Quelques inégalités concernant les martingales, *Studia Math.* **59** (1976), 63–83.

[327] D. Lépingle, La variation d'ordre p des semi-martingales, *Z. Wahrscheinlichkeitstheor. Verw.* **36** (1976), 295–316.

[328] M. Lévy, L'espace d'interpolation réel $(A_0, A_1)_{\theta,p}$ contient ℓ_p, *C. R. Acad. Sci. Paris Sér. A-B* **289** (1979), A675–A677.

[329] D. R. Lewis and C. Stegall, Banach spaces whose duals are isomorphic to $l_1(\Gamma)$, *J. Funct. Anal.* **12** (1973), 177–187.

[330] J. Lindenstrauss, On the modulus of smoothness and divergent series in Banach spaces, *Mich. Math. J.* **10** (1963), 241–252.

[331] J. Lindenstrauss and A. Pełczyński, Absolutely summing operators in L_p-spaces and their applications, *Studia Math.* **29** (1968), 275–326.

[332] J. Lindenstrauss and H. P. Rosenthal, The $\mathcal{L}_p$ spaces, *Israel J. Math.* **7** (1969), 325–349.

[333] V. I. Liokumovich, Existence of B-spaces with a non-convex modulus of convexity, *Izv. Vyssh. Uchebn. Zaved. Mat.* **12** (1973), 43–49.

[334] J. L. Lions, Une construction d'espaces d'interpolation, *C. R. Acad. Sci. Paris* **251** (1960), 1853–1855.

[335] J. L. Lions and J. Peetre, Sur une classe d'espaces d'interpolation, *Inst. Hautes Études Sci. Publ. Math.* **19** (1964), 5–68.

[336] A. E. Litvak, Kahane-Khinchin's inequality for quasinorms, *Can. Math. Bull.* **43** (2000), 368–379.

[337] A. Lubin, Extensions of measures and the von Neumann selection theorem, *Proc. Am. Math. Soc.* **43** (1974), 118–122.

[338] F. Lust-Piquard, Inégalités de Khintchine daus $C_p(1 < p < \infty)$, *C.R. Acad. Sci. Paris* **303** (1986), 289–292.

[339] F. Lust-Piquard and G. Pisier, Noncommutative Khintchine and Paley inequalities, *Arkiv Mat.* **29** (1991), 241–260.

[340] M. Manstavicius, p-variation of strong Markov processes, *Ann. Probab.* **32** (2004), 2053–2066.

[341] T. Martínez and J. L. Torrea, Operator-valued martingale transforms, *Tohoku Math. J.* **52** (2000), 449–474.

[342] J. Matoušek, On embedding trees into uniformly convex Banach spaces, *Israel J. Math.* **114** (1999), 221–237.

[343] B. Maurey, Systèmes de Haar, in *Séminaire Maurey–Schwartz*, Ecole Polytechnique, Paris, 74–75.

[344] B. Maurey, Théorèmes de factorisation pour les opérateurs linéaires à valeurs dans les espaces L^p, with an *Astérisque*, no. 11, Société Mathématique de France, Paris, 1974.

[345] B. Maurey, Type et cotype dans les espaces munis de structures locales inconditionnelles, in *Séminaire Maurey-Schwartz 1973–1974: Espaces L^p, applications radonifiantes et géométrie des espaces de Banach*, Exp. Nos. 24 and 25, Centre de Math., École Polytech., Paris, 1974.

[346] B. Maurey, Construction de suites symétriques, *C. R. Acad. Sci. Paris Sér. A-B* **288** (1979), A679–A681.

[347] B. Maurey, Type, cotype and K-convexity, in *Handbook of the geometry of Banach spaces*, Vol. II, North-Holland, Amsterdam, 2003, 1299–1332.

[348] B. Maurey and G. Pisier, Séries de variables aléatoires vectorielles indépendantes et propriétés géométriques des espaces de Banach, *Studia Math.* **58** (1976), 45–90.

[349] T. McConnell, On Fourier multiplier transformations of Banach-valued functions, *Trans. Am. Math. Soc.* **285** (1984), 739–757.

[350] T. Mei, BMO is the intersection of two translates of dyadic BMO, *C. R. Math. Acad. Sci. Paris* **336** (2003), 1003–1006.

[351] M. Mendel and A. Naor, Markov convexity and local rigidity of distorted metrics, *J. Eur. Math. Soc.* **15** (2013), 287–337.

[352] P. W. Millar, Path behavior of processes with stationary independent increments, *Z. Wahrscheinlichkeitstheorie Verw. Gebiete* **17** (1971), 53–73.

[353] M. Milman, Fourier type and complex interpolation. *Proc. Am. Math. Soc.* **89** (1983), 246–248.

[354] M. Milman, Complex interpolation and geometry of Banach spaces, *Ann. Mat. Pura Appl.* **136** (1984), 317–328.

[355] I. Monroe, On the γ-variation of processes with stationary independent increments, *Ann. Math. Statist.* **43** (1972), 1213–1220.

[356] M. Musat, On the operator space UMD property and non-commutative martingale inequalities, PhD Thesis, University of Illinois at Urbana-Champaign, 2002.

[357] A. Naor, Y. Peres, O. Schramm and S. Sheffield, Markov chains in smooth Banach spaces and Gromov-hyperbolic metric spaces, *Duke Math. J.* **134** (2006), 165–197.

[358] A. Naor and T. Tao, Random martingales and localization of maximal inequalities, *J. Funct. Anal.* **259** (2010), 731–779.

[359] F. Nazarov, G. Pisier, S. Treil and A. Volberg, Sharp estimates in vector Carleson imbedding theorem and for vector paraproducts, *J. Reine Angew. Math.* **542** (2002), 147–171.

[360] F. Nazarov and S. Treil, The weighted norm inequalities for Hilbert transform are now trivial, *C. R. Acad. Sci. Paris Sér. I Math.* **323** (1996), 717–722.

[361] F. Nazarov and S. Treil, The hunt for a Bellman function: applications to estimates for singular integral operators and to other classical problems of harmonic analysis, *St. Petersburg Math. J.* **8** (1997), 721–824.

[362] F. Nazarov, S. Treil and A. Volberg, The Bellman functions and two-weight inequalities for Haar multipliers, *J. Am. Math. Soc.* **12** (1999), 909–928.

[363] F. Nazarov, S. Treil and A. Volberg, Counterexample to infinite dimensional Carleson embedding theorem, *C. R. Acad. Sci. Paris Sér. I Math.* **325** (1997), 383–388.

[364] E. Nelson, The free Markoff field, *J. Funct. Anal.* **12** (1973), 211–227.

[365] E. Nelson, Notes on non-commutative integration, *J. Funct. Anal.* **15** (1974), 103–116.

[366] J. Neveu, Sur l'espérance conditionnelle par rapport à un mouvement brownien, *Ann. Inst. H. Poincaré Sect. B (N.S.)* **12** (1976), 105–109.

[367] G. Nordlander, The modulus of convexity in normed linear spaces, *Ark. Mat.* **4** (1960), 15–17.

[368] M. I. Ostrovskii, On metric characterizations of some classes of Banach spaces, *C. R. Acad. Bulgare Sci.* **64** (2011), 775–784.

[369] M. I. Ostrovskii, Embeddability of locally finite metric spaces into Banach spaces is finitely determined, *Proc. Am. Math. Soc.* **140** (2012), 2721–2730.

[370] M. I. Ostrovskii, Metric characterizations of superreflexivity in terms of word hyperbolic groups and finite graphs, *Anal. Geom. Metr. Spaces* **2** (2014), 154–168.

[371] M. I. Ostrovskii, Radon-Nikodym property and thick families of geodesics, *J. Math. Anal. Appl.* **409** (2014), 906–910.

[372] V. I. Ovchinnikov, The method of orbits in interpolation theory, *Math. Rep.* **1** (1984), 349–515.

[373] J. Parcet and N. Randrianantoanina, Gundy's decomposition for non-commutative martingales and applications,*Proc. London Math. Soc.* **93** (2006), 227–252.

[374] M. Pavlović, Uniform c-convexity of L_p, $0 < p1$, *Publ. Inst. Math. (Beograd) (N.S.)* **43** (1988), 117–124.

[375] M. Pavlović, On the complex uniform convexity of quasi-normed spaces, in *Math. Balkanica (N.S.)* **5** (1991), 92–98.

[376] J. Peetre, Sur la transformation de Fourier des fonctions à valeurs vectorielles, *Rend. Sem. Mat. Univ. Padova* **42** (1969), 15–26.

[377] M. C. Pereyra, Lecture notes on dyadic harmonic analysis: Second Summer School in Analysis and Mathematical Physics (Cuernavaca, 2000), *Contemp. Math.* **289**, 1–60.

[378] S. Petermichl, Dyadic shifts and a logarithmic estimate for Hankel operators with matrix symbol, *C. R. Acad. Sci. Paris Sér. I Math.* **330** (2000), 455–460.

[379] S. Petermichl, S. Treil and A. Volberg, Why the Riesz transforms are averages of the dyadic shifts? in *Proceedings of the 6th International Conference on Harmonic Analysis and Partial Differential Equations (El Escorial, 2000), Publ. Mat.*, extra volume (2002), 209–228.

[380] B. J. Pettis, A proof that every uniformly convex space is reflexive, *Duke Math. J.* **5** (1939), 249–253.

[381] R. R. Phelps, Dentability and extreme points in Banach spaces, *J. Funct. Anal.* **17** (1974), 78–90.

[382] M. Piasecki, A geometrical characterization of AUMD Banach spaces via subharmonic functions, *Demonstratio Math.* **30** (1997), 641–654.

[383] M. Piasecki, A characterization of complex AUMD Banach spaces via tangent martingales, *Demonstratio Math.* **30** (1997), 715–728.

[384] J. Picard, A tree approach to p-variation and to integration, *Ann. Probab.* **36** (2008), 2235–2279.

[385] S. Pichorides, On the best values of the constants in the theorems of M. Riesz, Zygmund and Kolmogorov, *Studia Math.* **44** (1972), 165–179.

[386] G. Pisier, Un exemple concernant la super-réflexivité, *Séminaire Maurey-Schwartz 1974-1975*, Annexe 2, http://www.numdam.org/.
[387] G. Pisier, Martingales with values in uniformly convex spaces, *Israel J. Math.* **20** (1975), 326–350.
[388] G. Pisier, Some applications of the complex interpolation method to Banach lattices, *J. Anal. Math. Jerusalem* **35** (1979), 264–281.
[389] G. Pisier, Holomorphic semigroups and the geometry of Banach spaces, *Ann. Math.* **115** (1982), 375–392.
[390] G. Pisier, On the duality between type and cotype, in *Martingale theory in harmonic analysis and Banach spaces (Cleveland, Ohio, 1981)*, Lecture Notes in Mathematics 939, Springer, Berlin, 1982, 131–144.
[391] G. Pisier, Probabilistic methods in the geometry of Banach spaces, in *Probability and analysis (Varenna, 1985)*, Lecture Notes in Mathematics 1206, Springer, Berlin, 1986, 167–241.
[392] G. Pisier, The dual J^* of the James space has cotype 2 and the Gordon-Lewis property, *Math. Proc. Cambridge Philos. Soc.* **103** (1988), 323–331.
[393] G. Pisier, The K_t-functional for the interpolation couple $L_1(A_0), L_\infty(A_1)$, *J. Approx. Theory* **73** (1993), 106–117.
[394] G. Pisier, Complex interpolation and regular operators between Banach lattices, *Arch. Math. (Basel)* **62** (1994), 261–269.
[395] G. Pisier, Noncommutative vector valued L_p-spaces and completely p-summing maps, *Soc. Math. France Astérisque* **237** (1998), 131 pp.
[396] G. Pisier, Remarks on the non-commutative Khintchine inequalities for $0 < p < 2$, *J. Funct. Anal.* **256** (2009), 4128–4161.
[397] G. Pisier, Complex interpolation between Hilbert, Banach and operator spaces, *Mem. Am. Math. Soc.* **208** (2010), 78 pp.
[398] G. Pisier and É. Ricard, The non-commutative Khintchine inequalities for $0 < p < 1$, *J. Inst. Math. Jussieu*, forthcoming.
[399] G. Pisier and Q. Xu, Random series in the real interpolation spaces between the spaces v_p, in *Geometrical aspects of functional analysis (1985/86)*, Lecture Notes in Mathematics 1267, Springer, Berlin, 1987, 185–209.
[400] G. Pisier and Q. Xu, The strong p-variation of martingales and orthogonal series, *Probab. Theory Related Fields* **77** (1988), 497–514.
[401] G. Pisier and Q. Xu, Non-commutative martingale inequalities, *Comm. Math. Phys.* **189** (1997), 667–698.
[402] G. Pisier and Q. Xu, Non-commutative L_p-spaces, in *Handbook of the geometry of Banach spaces*, Vol. II, North-Holland, Amsterdam, 2003, 1459–1517.
[403] D. Potapov, F. Sukochev and Q. Xu, On the vector-valued Littlewood-Paley-Rubio de Francia inequality, *Rev. Mat. Iberoam.* **28** (2012), 839–856.
[404] V. Pták, Biorthogonal systems and reflexivity of Banach spaces, *Czechoslovak Math. J.* **9** (1959), 319–326.
[405] Y. Qiu, On the UMD constants for a class of iterated $Lp(Lq)$ spaces, *J. Funct. Anal.* **262** (2012), 2409–2429.
[406] Y. Qiu, On the OUMD property for the column Hilbert space C, *Indiana Univ. Math. J.* **61** (2012), 2143–2156.
[407] Y. Qiu, A remark on the complex interpolation for families of Banach spaces, *Rev. Mat. Iber.* **31** (2015), 439–460.
[408] Y. Qiu, A non-commutative version of Lépingle-Yor martingale inequality, *Statist. Probab. Lett.* **91** (2014), 52–54.
[409] M. Raja, Finite slicing in superreflexive Banach spaces, *J. Funct. Anal.* **268** (2015), 2672–2694.

[410] N. Randrianantoanina, Non-commutative martingale transforms, *J. Funct. Anal.* **194** (2002), 181–212.

[411] N. Randrianantoanina, Square function inequalities for non-commutative martingales, *Israel J. Math.* **140** (2004), 333–365.

[412] N. Randrianantoanina, A weak-type inequality for non-commutative martingales and applications, *Proc. London Math. Soc.* **91** (2005), 509–544.

[413] N. Randrianantoanina, Conditioned square functions for noncommutative martingales, *Ann. Probab.* **35** (2007), 1039–1070.

[414] N. Randrianantoanina, A remark on maximal functions for noncommutative martingales, *Arch. Math. (Basel)* **101** (2013), 541–548.

[415] H. P. Rosenthal, On the subspaces of $L^p(p > 2)$ spanned by sequences of independent random variables, *Israel J. Math.* **8** (1970), 273–303.

[416] H. P. Rosenthal, Martingale proofs of a general integral representation theorem, in *Analysis at Urbana*, Vol. II, Cambridge University Press, Cambridge, 1989, 294–356.

[417] J. L. Rubio de Francia, Fourier series and Hilbert transforms with values in UMD Banach spaces, *Studia Math.* **81** (1985), 95–105.

[418] J. L. Rubio de Francia, Martingale and integral transforms of Banach space valued functions, in *Probability and Banach spaces (Zaragoza, 1985)*, Lecture Notes in Mathematics 1221, Springer, Berlin, 1986, 195–222.

[419] W. Schachermayer, For a Banach space isomorphic to its square the Radon-Nikodým property and the Krein-Milman property are equivalent, *Studia Math.* **81** (1985), 329–339.

[420] W. Schachermayer, The sum of two Radon-Nikodým-sets need not be a Radon-Nikodým-set, *Proc. Am. Math. Soc.* **95** (1985), 51–57.

[421] W. Schachermayer, Some remarks concerning the Krein-Milman and the Radon-Nikodým property of Banach spaces, in *Banach spaces (Columbia, Mo., 1984)*, Lecture Notes in Mathematics 1166, Springer, Berlin, 1985, 169–176.

[422] W. Schachermayer, The Radon-Nikodým property and the Krein-Milman property are equivalent for strongly regular sets, *Trans. Am. Math. Soc.* **303** (1987), 673–687.

[423] W. Schachermayer, A Sersouri and E. Werner, Moduli of nondentability and the Radon-Nikodým property in Banach spaces, *Israel J. Math.* **65** (1989), 225–257.

[424] J. J. Schäffer and K. Sundaresan, Reflexivity and the girth of spheres, *Math. Ann.* **184** (1969/1970), 163–168.

[425] J. Schwartz, A remark on inequalities of Calderón-Zygmund type for vector-valued functions, *Comm. Pure Appl. Math.* **14** (1961), 785–799.

[426] T. Simon, Small ball estimates in p-variation for stable processes, *J. Theoret. Probab.* **17** (2004), 979–1002.

[427] J. Stafney, The spectrum of an operator on an interpolation space, *Trans. Am. Math. Soc.* **144** (1969), 333–349.

[428] C. Stegall, The Radon-Nikodym property in conjugate Banach spaces, *Trans. Am. Math. Soc.* **206** (1975), 213–223.

[429] C. Stegall, The duality between Asplund spaces and spaces with the Radon-Nikodym property, *Israel J. Math.* **29** (1978), 408–412.

[430] S. Szarek, On the best constants in the Khinchin inequality, *Studia Math.* **58** (1976), 197–208.

[431] M. Talagrand, Concentration of measure and isoperimetric inequalities in product spaces, *Inst. Hautes Études Sci. Publ. Math.* **81** (1995), 73–205.

[432] N. Tomczak-Jaegermann, The moduli of smoothness and convexity and the Rademacher averages of trace classes $S_p(1 \leq p < \infty)$, *Studia Math.* **50** (1974), 163–182.

[433] D. A. Trautman, A note on MT operators, *Proc. Am. Math. Soc.* **97** (1986), 445–448.

[434] F. Watbled, Complex interpolation of a Banach space with its dual, *Math. Scand.* **87** (2000), 200–210.

[435] E. Werner, Nondentable solid subsets in Banach lattices failing RNP: applications to renormings, *Proc. Am. Math. Soc.* **107** (1989), 611–620.

[436] T. Wolff, A note on interpolation spaces, in *Harmonic analysis (Minneapolis, Minn., 1981)*, Lecture Notes in Mathematics 908, Springer, Berlin, 1982, 199–204.

[437] W. Woyczyński, On Marcinkiewicz-Zygmund laws of large numbers in Banach spaces and related rates of convergence, *Probab. Math. Statist.* **1** (1980), 117–131.

[438] Q. Xu, Espaces d'interpolation réels entre les espaces V_p: propriétés géométriques et applications probabilistes, *Publ. Math. Univ. Paris VII* **28** (1988), 77–123.

[439] Q. Xu, Real interpolation of some Banach lattices valued Hardy spaces, *Bull. Sci. Math.* **116** (1992), 227–246.

[440] Q. Xu, H^∞ functional calculus and maximal inequalities for semigroups of contractions on vector-valued L_p-spaces, preprint, Arxiv, 2014.

Index